The Principles of C (Volume II

Dmitry Ivanovich Mendeleyev
(Editor: T. A. Lawson), (Translator: George Kamensky)

Alpha Editions

This edition published in 2024

ISBN 9789362516251

Design and Setting By
Alpha Editions
www.alphaedis.com
Email - info@alphaedis.com

As per information held with us this book is in Public Domain.
This book is a reproduction of an important historical work.
Alpha Editions uses the best technology to reproduce historical work
in the same manner it was first published to preserve its original nature.
Any marks or number seen are left intentionally to preserve.

Contents

CHAPTER XV THE GROUPING OF THE ELEMENTS AND THE PERIODIC LAW	- 1 -
CHAPTER XVI ZINC, CADMIUM, AND MERCURY	- 48 -
CHAPTER XVII BORON, ALUMINIUM, AND THE ANALOGOUS METALS OF THE THIRD GROUP	- 75 -
CHAPTER XVIII SILICON AND THE OTHER ELEMENTS OF THE FOURTH GROUP	- 122 -
CHAPTER XIX PHOSPHORUS AND THE OTHER ELEMENTS OF THE FIFTH GROUP	- 182 -
CHAPTER XX SULPHUR, SELENIUM, AND TELLURIUM	- 245 -
CHAPTER XXI CHROMIUM, MOLYBDENUM, TUNGSTEN, URANIUM, AND MANGANESE	- 339 -
CHAPTER XXII IRON, COBALT, AND NICKEL	- 392 -
CHAPTER XXIII THE PLATINUM METALS	- 459 -
CHAPTER XXIV COPPER, SILVER, AND GOLD	- 498 -
APPENDIX I	- 564 -
APPENDIX II	- 585 -
APPENDIX III	- 609 -

CHAPTER XV

THE GROUPING OF THE ELEMENTS AND THE PERIODIC LAW

It is seen from the examples given in the preceding chapters that the sum of the data concerning the chemical transformations proper to the elements (for instance, with respect to the formation of acids, salts, and other compounds having definite properties) is insufficient for accurately determining the relationship of the elements, inasmuch as this may be many-sided. Thus, lithium and barium are in some respects analogous to sodium and potassium, and in others to magnesium and calcium. It is evident, therefore, that for a complete judgment it is necessary to have, not only qualitative, but also quantitative, exact and measurable, data. When a property can be measured it ceases to be vague, and becomes quantitative instead of merely qualitative.

Among these measurable properties of the elements, or of their corresponding compounds, are: (*a*) isomorphism, or the analogy of crystalline forms; and, connected with it, the power to form crystalline mixtures which are isomorphous; (*b*) the relation of the volumes of analogous compounds of the elements; (*c*) the composition of their saline compounds; and (*d*) the relation of the atomic weights of the elements. In this chapter we shall briefly consider these four aspects of the matter, which are exceedingly important for a natural and fruitful grouping of the elements, facilitating, not only a general acquaintance with them, but also their detailed study.

Historically the first, and an important and convincing, method for finding a relationship between the compounds of two different elements is by *isomorphism*. This conception was introduced into chemistry by Mitscherlich (in 1820), who demonstrated that the corresponding salts of arsenic acid, H_3AsO_4, and phosphoric acid, H_3PO_4, crystallise with an equal quantity of water, show an exceedingly close resemblance in crystalline form (as regards the angles of their faces and axes), and are able to crystallise together from solutions, forming crystals containing a mixture of the isomorphous compounds. Isomorphous substances are those which, with an equal number of atoms in their molecules, present an analogy in their chemical reactions, a close resemblance in their properties, and a similar or very nearly similar crystalline form: they often contain certain elements in common, from which it is to be concluded that the remaining elements (as in the preceding example of As and P) are analogous to each other. And inasmuch as crystalline forms are capable of exact measurement, the external form, or the relation of the molecules which causes their grouping into a crystalline form, is evidently as great a help in judging of the internal forces acting between the atoms as a comparison of reactions, vapour densities, and other like relations. We have already seen examples of this in the preceding pages. It will be sufficient to call to mind that the compounds of the alkali metals with the halogens RX, in a crystalline form, all belong to the cubic system and crystallise in octahedra or cubes—for example, sodium chloride, potassium chloride, potassium iodide, rubidium chloride, &c. The nitrates of rubidium and cæsium appear in anhydrous crystals of the same form as potassium nitrate. The carbonates of the metals of the alkaline earths are isomorphous with calcium carbonate—that is, they either appear in forms like calc spar or in the rhombic system in crystals analogous to aragonite.[1 bis] Furthermore, sodium nitrate crystallises in rhombohedra, closely resembling the rhombohedra of calc spar (calcium carbonate), $CaCO_3$, whilst potassium nitrate appears in the same form as aragonite, $CaCO_3$, and the number of atoms in both kinds of salts is the same: they all contain one atom of a metal (K, Na, Ca), one atom of a non-metal (C, N), and three atoms of oxygen. The analogy of form evidently coincides with an analogy of atomic composition. But, as we have learnt from the previous description of these salts, there is not any close resemblance in their properties. It is evident that calcium carbonate approaches more nearly to magnesium carbonate than to sodium nitrate, although their crystalline forms are all equally alike. Isomorphous substances which are perfectly analogous to each other are not only characterised by a close resemblance of form (homeomorphism), but also by the faculty of entering into analogous reactions, which is not the case with RNO_3 and RCO_3. The most important and direct method of recognising perfect isomorphism—that is, the absolute analogy of two compounds—is given by that property of analogous compounds of separating from solutions *in homogeneous crystals, containing the most varied proportions* of the analogous substances which enter into their composition. These quantities do not seem to be in dependence on the molecular or atomic weights, and if they are governed by any laws they must be analogous to those which apply to indefinite chemical compounds. This will be clear from the following examples. Potassium chloride and potassium nitrate are not isomorphous with each other, and are in an atomic sense composed in a different manner. If these salts be mixed in a solution and the solution be evaporated, independent crystals of the two salts will separate, each in that crystalline form which is proper to it. The crystals will not contain a mixture of the two salts. But if we mix the solutions of two isomorphous salts together, then, under certain circumstances, crystals will be obtained which contain both these substances. However, this cannot be taken as an absolute rule, for if we take a solution saturated at a high temperature with a mixture of potassium and sodium chlorides, then on evaporation sodium chloride only will separate, and on cooling only potassium chloride. The first will contain very little potassium chloride, and the latter very little sodium chloride. But if we take, for example, a mixture of solutions of magnesium sulphate and zinc sulphate, they cannot be separated from each other by evaporating the mixture, notwithstanding the rather considerable difference in the solubility of these salts. Again, the isomorphous salts, magnesium carbonate, and calcium carbonate are found together—that is, in one crystal—in nature. The angle of the rhombohedron of these magnesia-lime spars is intermediate between the angles proper to the two spars individually (for calcium carbonate, the angle of the rhombohedron is 105° 8'; magnesium carbonate, 107° 30'; $CaMg(CO_3)_2$, 106° 10'). Certain of these *isomorphous mixtures* of calc and magnesia spars appear in well-formed crystals, and in this case there not unfrequently exists a simple molecular proportion of strictly definite chemical combination between the component salts—for instance, $CaCO_3,MgCO_3$—whilst in other cases, especially in the absence of distinct crystallisation (in dolomites), no such simple molecular proportion is observable: this is also the case in many artificially prepared isomorphous mixtures. The microscopical and crystallo-optical researches of Professor Inostrantzoff and others show that in many cases there is really a mechanical, although microscopically minute, juxtaposition in one whole of the heterogeneous crystals of calcium carbonate (double refracting) and of the compound $CaMgC_2O_6$. If we suppose the adjacent parts to be microscopically small (on the basis of the researches of Mallard, Weruboff, and others), we obtain an idea of isomorphous mixtures. A formula of the following kind is given to isomorphous mixtures: for instance, for spars, RCO_3, where R = Mg, Ca, and where it may be Fe,Mn ..., &c. This means that the Ca is partially replaced by Mg or another metal. Alums form a common example of the separation of isomorphous mixtures from solutions. They are double sulphates (or seleniates) of alumina (or oxides isomorphous with it) and the alkalis, which crystallise in well-formed crystals. If aluminium sulphate be mixed with potassium sulphate, an alum separates, having the composition $KAlS_2O_8,12H_2O$. If sodium sulphate or ammonium sulphate, or rubidium (or thallium) sulphate be used, we obtain alums having the composition $RAlS_2O_8,12H_2O$. Not only do they all crystallise in the cubic system, but they also contain an equal atomic quantity of water of crystallisation ($12H_2O$). Besides which, if we mix solutions of the potassium and ammonium ($NH_4AlS_2O_8,12H_2O$) alums together, then the crystals which separate will contain various proportions of the alkalis taken, and separate crystals of the alums of one or the other kind will not be obtained, but each separate crystal will contain both potassium and ammonium. Nor is this all; if we take a crystal of a potassium alum and immerse it in a solution capable of yielding ammonia alum, the crystal of the potash alum will continue to grow and increase in size in this solution—that is, a layer of the ammonia or other alum will deposit itself upon the planes bounding the crystal of the potash alum. This is very distinctly seen if a colourless crystal of a common alum be immersed in a saturated violet solution of chrome alum, $KCrS_2O_8,12H_2O$, which then deposits itself in a violet layer over the colourless crystal of the alumina alum, as was observed even before Mitscherlich noticed it. If this crystal be then immersed in a solution of an alumina alum, a layer of this salt will form over the layer of chrome alum, so that one alum is able to incite the growth of the other. If the deposition proceed simultaneously, the resultant intermixture may be minute and inseparable, but its nature is understood from the preceding experiments; the attractive force of crystallisation of isomorphous substances is so nearly equal that the attractive power of an isomorphous substance induces a crystalline superstructure exactly the same as would be produced by the attractive force of like crystalline particles. From this it is evident that one isomorphous substance may *induce the crystallisation* of another. Such a phenomenon explains, on the one hand, the aggregation of different isomorphous substances in one crystal, whilst, on the other hand, it serves as a most exact indication of the nearness both of the molecular composition of isomorphous substances and of those forces which are proper to the elements which distinguish the isomorphous substances. Thus, for example, ferrous sulphate or green vitriol crystallises in the monoclinic system and contains seven molecules of water, $FeSO_4,7H_2O$, whilst copper vitriol crystallises with five molecules of water in the triclinic system, $CuSO_4,5H_2O$; nevertheless, it may be easily proved that both salts are perfectly isomorphous; that they are able to appear in identically the same forms and with an equal molecular amount of water. For instance, Marignac, by evaporating a mixture of sulphuric acid and ferrous sulphate under the receiver of an air-pump, first obtained crystals of the hepta-hydrated salt, and then of the penta-hydrated salt $FeSO_4,5H_2O$, which were perfectly similar to the crystals of copper sulphate. Furthermore, Lecoq de Boisbaudran, by immersing crystals of $FeSO_4,7H_2O$ in a supersaturated solution of copper sulphate, caused the latter to deposit in the same form as ferrous sulphate, in crystals of the monoclinic system, $CuSO_4,7H_2O$.

Hence it is evident that isomorphism—that is, the analogy of forms and the property of inducing crystallisation—may serve as a means for the discovery of analogies in molecular composition. We will take an example in order to render this clear. If, instead of aluminium sulphate, we add magnesium sulphate to potassium sulphate, then, on evaporating the solution, the double salt $K_2MgS_2O_8,6H_2O$ (Chapter XIV., Note [28]) separates instead of an alum, and the ratio of the component parts (in alums one atom of potassium per 2SO$_4$, and here two atoms) and the amount of water of crystallisation (in alums 12, and here 6 equivalents per 2SO$_4$) are quite different; nor is this double salt in any way isomorphous with the alums, nor capable of forming an isomorphous crystalline mixture with them, nor does the one salt provoke the crystallisation of the other. From this we must conclude that although alumina and magnesia, or aluminium and magnesium, resemble each other, they are not isomorphous, and that although they give partially similar double salts, these salts are not analogous to each other. And this is expressed in their chemical formulæ by the fact that the number of atoms in alumina or aluminium oxide, Al$_2$O$_3$, is different from the number in magnesia, MgO. Aluminium is trivalent and magnesium bivalent. Thus, having obtained a double salt from a given metal, it is possible to judge of the analogy of the given metal with aluminium or with magnesium, or of the absence of such an analogy, from the composition and form of this salt. Thus zinc, for example, does not form alums, but forms a double salt with potassium sulphate, which has a composition exactly like that of the corresponding salt of magnesium. It is often possible to distinguish the bivalent metals analogous to magnesium or calcium from the trivalent metals, like aluminium, by such a method. Furthermore, the specific heat and vapour density serve as guides. There are also indirect proofs. Thus iron gives ferrous compounds, FeX$_2$, which are isomorphous with the compounds of magnesium, and ferric compounds, FeX$_3$, which are isomorphous with the compounds of aluminium; in this instance the relative composition is directly determined by analysis, because, for a given amount of iron, FeCl$_2$ only contains two-thirds of the amount of chlorine which occurs in FeCl$_3$, and the composition of the corresponding oxygen compounds, *i.e.* of ferrous oxide, FeO, and ferric oxide, Fe$_2$O$_3$, clearly indicates the analogy of the ferrous oxide with MgO and of the ferric oxide with Al$_2$O$_3$.

Thus in the building up of similar molecules in crystalline forms we see one of the numerous means for judging of the internal world of molecules and atoms, and one of the weapons for conquests in the invisible world of molecular mechanics which forms the main object of physico-chemical knowledge. This method has more than once been employed for discovering the analogy of elements and of their compounds; and as crystals are measurable, and the capacity to form crystalline mixtures can be experimentally verified, this method is a numerical and measurable one, and in no sense arbitrary.

The regularity and simplicity expressed by the exact laws of crystalline form repeat themselves in the aggregation of the atoms to form molecules. Here, as there, there are but few forms which are essentially different, and their apparent diversity reduces itself to a few fundamental differences of type. There the molecules aggregate themselves into crystalline forms; here, the atoms aggregate themselves into molecular forms or into *the types of compounds*. In both cases the fundamental crystalline or molecular forms are liable to variations, conjunctions, and combinations. If we know that potassium gives compounds of the fundamental type KX, where X is a univalent element (which combines with one atom of hydrogen, and is, according to the law of substitution, able to replace it), then we know the composition of its compounds: K_2O, KHO, KCl, NH_2K, KNO_3, K_2SO_4, $KHSO_4$, $K_2Mg(SO_4)_2,6H_2O$, &c. All the possible derivative crystalline forms are not known. So also all the atomic combinations are not known for every element. Thus in the case of potassium, KCH_3, K_3P, K_2Pt, and other like compounds which exist for hydrogen or chlorine, are unknown.

Only a few fundamental types exist for the building up of atoms into molecules, and the majority of them are already known to us. If X stand for a univalent element, and R for an element combined with it, then eight atomic types may be observed:—

$$RX, RX_2, RX_3, RX_4, RX_5, RX_6, RX_7, RX_8.$$

Let X be chlorine or hydrogen. Then as examples of the first type we have: H_2, Cl_2, HCl, KCl, NaCl, &c. The compounds of oxygen or calcium may serve as examples of the type RX_2: OH_2, OCl_2, OHCl, CaO, $Ca(OH)_2$, $CaCl_2$, &c. For the third type RX_3 we know the representative NH_3 and the corresponding compounds N_2O_3, NO(OH), NO(OK), PCl_3, P_2O_3, PH_3, SbH_3, Sb_2O_3, B_2O_3, BCl_3, Al_2O_3, &c. The type RX_4 is known among the hydrogen compounds. Marsh gas, CH_4, and its corresponding saturated hydrocarbons, C_nH_{2n+2}, are the best representatives. Also CH_3Cl, CCl_4, $SiCl_4$, $SnCl_4$, SnO_2, CO_2, SiO_2, and a whole series of other compounds come under this class. The type RX_5 is also already familiar to us, but there are no purely hydrogen compounds among its representatives. Sal-ammoniac, NH_4Cl, and the corresponding $NH_4(OH)$, $NO_2(OH)$, $ClO_2(OK)$, as well as PCl_5, $POCl_3$, &c., are representatives of this type. In the higher types also there are no hydrogen compounds, but in the type RX_6 there is the chlorine compound WCl_6. However, there are many oxygen compounds, and among them SO_3 is the best known representative. To this class also belong $SO_2(OH)_2$, SO_2Cl_2, $SO_2(OH)Cl$, CrO_3, &c., all of an acid character. Of the higher types there are in general only oxygen and acid representatives. The type RX_7 we know in perchloric acid, $ClO_3(OH)$, and potassium permanganate, $MnO_3(OK)$, is also a member. The type RX_8 in a free state is very rare; osmic anhydride, OsO_4, is the best known representative of it.

The four lower types RX, RX$_2$, RX$_3$, and RX$_4$ are met with in compounds of the elements R with chlorine and oxygen, and also in their compounds with hydrogen, whilst the four higher types only appear for such acid compounds as are formed by chlorine, oxygen, and similar elements.

Among the oxygen compounds the *saline oxides* which are capable of forming salts either through the function of a base or through the function of an acid anhydride attract the greatest interest in every respect. Certain elements, like calcium and magnesium, only give one saline oxide—for example, MgO, corresponding with the type MgX$_2$. But the majority of the elements appear in several such forms. Thus copper gives CuX and CuX$_2$, or Cu$_2$O and CuO. If an element R gives a higher type RX$_n$, then there often also exist, as if by symmetry, lower types, RX$_{n-2}$, RX$_{n-4}$, and in general such types as differ from RX$_n$ by an even number of X. Thus in the case of sulphur the types SX$_2$, SX$_4$, and SX$_6$ are known—for example SH$_2$, SO$_2$, and SO$_3$. The last type is the highest, SX$_6$. The types SX$_5$ and SX$_3$ do not exist. But even and uneven types sometimes appear for one and the same element. Thus the types RX and RX$_2$ are known for copper and mercury.

Among the *saline* oxides only the *eight types* enumerated below are known to exist. They determine the possible formulæ of the compounds of the elements, if it be taken into consideration that an element which gives a certain type of combination may also give lower types. For this reason the rare type of the *suboxides* or quaternary oxides R$_4$O (for instance, Ag$_4$O, Ag$_2$Cl) is not characteristic; it is always accompanied by one of the higher grades of oxidation, and the compounds of this type are distinguished by their great chemical instability, and split up into an element and the higher compound (for instance, Ag$_4$O = 2Ag + Ag$_2$O). Many elements, moreover, form transition oxides whose composition is intermediate, which are able, like N$_2$O$_4$, to split up into the lower and higher oxides. Thus iron gives magnetic oxide, Fe$_3$O$_4$, which is in all respects (by its reactions) a compound of the suboxide FeO with the oxide Fe$_2$O$_3$. The independent and more or less stable saline compounds correspond with the following eight types :—

R$_2$O; salts RX, hydroxides ROH. Generally basic like K$_2$O, Na$_2$O, Hg$_2$O, Ag$_2$O, Cu$_2$O; if there are acid oxides of this composition they are very rare, are only formed by distinctly acid elements, and even then have only feeble acid properties; for example, Cl$_2$O and N$_2$O.

R$_2$O$_2$ or RO; salts RX$_2$, hydroxides R(OH)$_2$. The most simple basic salts R$_2$OX$_2$ or R(OH)X; for instance, the chloride Zn$_2$OCl$_2$; also an almost exclusively basic type; but the basic properties are more feebly developed than in the preceding type. For example, CaO, MgO, BaO, PbO, FeO, MnO, &c.

R_2O_3; salts RX_3, hydroxides $R(OH)_3$, $RO(OH)$, the most simple basic salts ROX, $R(OH)X_3$. The bases are feeble, like Al_2O_3, Fe_2O_3, Tl_2O_3, Sb_2O_3. The acid properties are also feebly developed; for instance, in B_2O_3; but with the non-metals the properties of acids are already clear; for instance, P_2O_3, $P(OH)_3$.

R_2O_4 or RO_2; salts RX_4 or ROX_2, hydroxides $R(OH)_4$, $RO(OH)_2$. Rarely bases (feeble), like ZrO_2, PtO_2; more often acid oxides; but the acid properties are in general feeble, as in CO_2, SO_2, SnO_2. Many intermediate oxides appear in this and the preceding and following types.

R_2O_5; salts principally of the types ROX_3, RO_2X, $RO(OH)_3$, $RO_2(OH)$, rarely RX_5. The basic character (X, a halogen, simple or complex; for instance, NO_3, Cl, &c.) is feeble; the acid character predominates, as is seen in N_2O_5, P_2O_5, Cl_2O_5; then X = OH, OK, &c., for example $NO_2(OK)$.

R_2O_6 or RO_3; salts and hydroxides generally of the type RO_2X_2, $RO_2(OH)_2$. Oxides of an acid character, as SO_3, CrO_3, MnO_3. Basic properties rare and feebly developed as in UO_3.

R_2O_7; salts of the form RO_3X, $RO_3(OH)$, acid oxides; for instance, Cl_2O_7, Mn_2O_7. Basic properties as feebly developed as the acid properties in the oxides R_2O.

R_2O_8 or RO_4. A very rare type, and only known in OsO_4 and RuO_4.

It is evident from the circumstance that in all the higher types the *acid hydroxides* (for example, $HClO_4$, H_2SO_4, H_3PO_4) and salts with a single atom of one element contain, like the higher saline type RO_4, *not more than four atoms of oxygen*; that the formation of the saline oxides is governed by a certain common principle which is best looked for in the fundamental properties of oxygen, and in general of the most simple compounds. The hydrate of the oxide RO_2 is of the higher type $RO_2 2H_2O = RH_4O_4 = R(HO)_4$. Such, for example, is the hydrate of silica and the salts (orthosilicates) corresponding with it, $Si(MO)_4$. The oxide R_2O_5, corresponds with the hydrate $R_2O_5 3H_2O = 2RH_3O_4 = 2RO(OH)_3$. Such is orthophosphoric acid, PH_3O_3. The hydrate of the oxide RO_3 is $RO_3 H_2O = RH_2O_4 = RO_2(OH)_2$—for instance, sulphuric acid. The hydrate corresponding to R_2O_7 is evidently $RHO = RO_3(OH)$—for example, perchloric acid. Here, besides containing O_4, it must further be remarked that *the amount of hydrogen in the hydrate is equal to the amount of hydrogen in the hydrogen compound*. Thus silicon gives SiH_4 and SiH_4O_4, phosphorus PH_3 and

PH_3O_4, sulphur SH_2 and SH_2O_4, chlorine ClH and $ClHO_4$. This, if it does not explain, at least connects in a harmonious and general system the fact that *the elements are capable of combining with a greater amount of oxygen, the less the amount of hydrogen which they are able to retain.* In this the key to the comprehension of all further deductions must be looked for, and we will therefore formulate this rule in general terms. An element R gives a hydrogen compound RH_n, the hydrate of its higher oxide will be RH_nO_4, and therefore the higher oxide will contain $2RH_nO_4 - nH_2O = R_2O_{8-n}$. For example, chlorine gives ClH, hydrate $ClHO_4$, and the higher oxide Cl_2O_7. Carbon gives CH_4 and CO_2. So also, SiO_2 and SiH_4 are the higher compounds of silicon with hydrogen and oxygen, like CO_2 and CH_4. Here the amounts of oxygen and hydrogen are equivalent. Nitrogen combines with a large amount of oxygen, forming N_2O_5, but, on the other hand, with a small quantity of hydrogen in NH_3. *The sum of the equivalents of hydrogen and oxygen*, occurring in combination with an atom of nitrogen, is, as always in the higher types, equal to *eight*. It is the same with the other elements which combine with hydrogen and oxygen. Thus sulphur gives SO_3; consequently, six equivalents of oxygen fall to an atom of sulphur, and in SH_2 two equivalents of hydrogen. The sum is again equal to eight. The relation between Cl_2O_7 and ClH is the same. This shows that the property of elements of combining with such different elements as oxygen and hydrogen is subject to one common law, which is also formulated in the system of the elements presently to be described.

In the preceding we see not only the regularity and simplicity which govern the formation and properties of the oxides and of all the compounds of the elements, but also a fresh and exact means for recognising the analogy of elements. Analogous elements give compounds of analogous types, both higher and lower. If CO_2 and SO_2 are two gases which closely resemble each other both in their physical and chemical properties, the reason of this must be looked for not in an analogy of sulphur and carbon, but in that identity of the type of combination, RX_4, which both oxides assume, and in that influence which a large mass of oxygen always exerts on the properties of its compounds. In fact, there is little resemblance between carbon and sulphur, as is seen not only from the fact that CO_2 is the *higher form* of oxidation, whilst SO_2 is able to further oxidise into SO_3, but also from the fact that all the other compounds—for example, SH_2 and CH_4, SCl_2 and CCl_4, &c.—are entirely unlike both in type and in chemical properties. This absence of analogy in carbon and sulphur is especially clearly seen in the fact that the highest saline oxides are of different composition, CO_2 for carbon, and SO_3 for sulphur. In Chapter VIII. we considered the limit to which carbon tends in its compounds, and in a similar manner there is for every element in its compounds a tendency to attain a certain highest limit RX_n. This view was particularly developed in the middle of the present century by Frankland in studying the metallo-organic compounds, *i.e.* those in which X is wholly or partially a hydrocarbon radicle; for instance, X = CH_3 or C_2H_5 &c. Thus, for example, antimony, Sb (Chapter XIX.) gives, with chlorine, compounds $SbCl_3$ and $SbCl_5$ and corresponding oxygen compounds Sb_2O_3 and Sb_2O_5, whilst under the action of CH_3I, C_2H_5I, or in general EI (where E is a hydrocarbon radicle of the paraffin series), upon antimony or its alloy with sodium there are formed SbE_3 (for example, $Sb(CH_3)_3$, boiling at about 81°), which, corresponding to the lower form of combination SbX_3, are able to combine further with EI, or Cl_2, or O, and to form compounds of the limiting type SbX_5; for example, SbE_4Cl corresponding to NH_4Cl with the substitution of nitrogen by antimony, and of hydrogen by the hydrocarbon radicle. The elements which are most chemically analogous are characterised by the fact of their giving compounds of similar form RX_n. The halogens which are analogous give both higher and lower compounds. So also do the metals of the alkalis and of the alkaline earths. And we saw that this analogy extends to the composition and properties of the nitrogen and hydrogen compounds of these metals, which is best seen in the salts. Many such groups of analogous elements have long been known. Thus there are analogues of oxygen, nitrogen, and carbon, and we shall meet with many such groups. But an acquaintance with them inevitably leads to the questions, what is the cause of analogy and what is the relation of one group to another? If these questions remain unanswered, it is easy to fall into error in the formation of the groups, because the notions of the degree of analogy will always be relative, and will not present any accuracy or distinctness Thus lithium is analogous in some respects to potassium and in others to magnesium; beryllium is analogous to both aluminium and magnesium. Thallium, as we shall afterwards see and as was observed on its discovery, has much kinship with lead and mercury, but some of its properties appertain to lithium and potassium. Naturally, where it is impossible to make measurements one is reluctantly obliged to limit oneself to approximate comparisons, founded on apparent signs which are not distinct and are wanting in exactitude. But in the elements there is one accurately measurable property, which is subject to no doubt—namely, that property which is expressed in their atomic weights. Its magnitude indicates the relative mass of the atom, or, if we avoid the conception of the atom, its magnitude shows the relation between the masses forming the chemical and independent individuals or elements. And according to the teaching of all exact data about the phenomena of nature, the mass of a substance is that property on which all its remaining properties must be dependent, because they are all determined by similar conditions or by those forces which act in the weight of a substance, and this is directly proportional to its mass. Therefore it is most natural to seek for a dependence between the properties and analogies of the elements on the one hand and their atomic weights on the other.

This is the fundamental idea which leads *to arranging all the elements according to their atomic weights*. A periodic repetition of properties is then immediately observed in the elements. We are already familiar with examples of this:—

F = 19, Cl = 35·5, Br = 80, I = 127,

Na = 23, K = 39, Rb = 85, Cs = 133,

Mg = 24, Ca = 340, Sr = 87, Ba = 137.

The essence of the matter is seen in these groups. The halogens have smaller atomic weights than the alkali metals, and the latter than the metals of the alkaline earths. Therefore, *if all the elements be arranged in the order of their atomic weights, a periodic repetition of properties is obtained*. This is expressed by the *law of periodicity, the properties of the elements, as well as the forms and properties of their compounds, are in periodic dependence or (expressing ourselves algebraically) form a periodic function of the atomic weights of the elements*. Table I. of *the periodic system of the elements*, which is placed at the very beginning of this book, is designed to illustrate this law. It is arranged in conformity with the eight types of oxides described in the preceding pages, and those elements which give the oxides, R_2O and consequently salts RX, form the 1st group; the elements giving R_2O_2 or RO as their highest grade of oxidation belong to the 2nd group; those giving R_2O_3 as their highest oxides form the 3rd group, and so on; whilst the elements of all the groups which are nearest in their atomic weights are arranged in series from 1 to 12. The even and uneven series of the same groups present the same forms and limits, but differ in their properties, and therefore two contiguous series, one even and the other uneven—for instance, the 4th and 5th—form a period. Hence the elements of the 4th, 6th, 8th, 10th, and 12th, or of the 3rd, 5th, 7th, 9th, and 11th, series form analogues, like the halogens, the alkali metals, &c. The conjunction of two series, one even and one contiguous uneven series, thus forms one large *period*. These periods, beginning with the alkali metals, end with the halogens. The elements of the first two series have the lowest atomic weights, and in consequence of this very circumstance, although they bear the general properties of a group, they still show many peculiar and independent properties. Thus fluorine, as we know, differs in many points from the other halogens, and lithium from the other alkali metals, and so on. These lightest elements may be termed *typical elements*. They include—

H.

Li, Be, B, C, N, O, F.

Na, Mg....

In the annexed table all the remaining elements are arranged, not in groups and series, but *according to periods*. In order to understand the essence of the matter, it must be remembered that here the atomic weight gradually increases along a given line; for instance, in the line commencing with K = 39 and ending with Br = 80, the intermediate elements have intermediate atomic weights, as is clearly seen in Table III., where the elements stand in the order of their atomic weights.

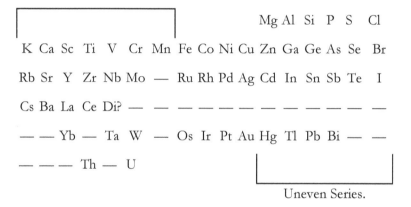

I. II. III. IV. V. VI. VII.   I. II. III. IV. V. VI. VII.

    Even Series.

                                         Mg Al Si P S Cl

K Ca Sc Ti V Cr Mn Fe Co Ni Cu Zn Ga Ge As Se Br

Rb Sr Y Zr Nb Mo — Ru Rh Pd Ag Cd In Sn Sb Te I

Cs Ba La Ce Di? — — — — — — — — — — — —

— — Yb — Ta W — Os Ir Pt Au Hg Tl Pb Bi — —

— — — Th — U

                                            Uneven Series.

The same degree of analogy that we know to exist between potassium, rubidium, and cæsium; or chlorine, bromine, and iodine; or calcium, strontium, and barium, also exists between the elements of the other vertical columns. Thus, for example, zinc, cadmium, and mercury, which are described in the following chapter, present a very close analogy with magnesium. For a true comprehension of the matter it is very important to see that all the aspects of the distribution of the elements according to their atomic weights essentially express one and the same fundamental *dependence—periodic properties*. The following points then must be remarked in it.

1. The composition of the higher oxygen compounds is determined by the groups: the first group gives R_2O, the second R_2O_2 or RO, the third R_2O_3, &c. There are eight types of oxides and therefore eight groups. Two groups give a period, and the same type of oxide is met with twice in a period. For example, in the period beginning with potassium, oxides of the

composition RO are formed by calcium and zinc, and of the composition RO_3 by molybdenum and tellurium. The oxides of the even series, of the same type, have stronger basic properties than the oxides of the uneven series, and the latter as a rule are endowed with an acid character. Therefore the elements which exclusively give bases, like the alkali metals, will be found at the commencement of the period, whilst such purely acid elements as the halogens will be at the end of the period. The interval will be occupied by intermediate elements, whose character and properties we shall afterwards describe. It must be observed that the acid character is chiefly proper to the elements with small atomic weights in the uneven series, whilst the basic character is exhibited by the heavier elements in the even series. Hence elements which give acids chiefly predominate among the lightest (typical) elements, especially in the last groups; whilst the heaviest elements, even in the last groups (for instance, thallium, uranium) have a basic character. Thus the basic and acid characters of the higher oxides are determined (*a*) by the type of oxide, (*b*) by the even or uneven series, and (*c*) by the atomic weight.[11 bis] The groups are indicated by Roman numerals from I. to VIII.

2. *The hydrogen compounds* being volatile or gaseous substances which are prone to reaction—such as HCl, H_2O, H_3N, and H_4C—are only formed by the elements of the uneven series and higher groups giving oxides of the forms R_2O_n, RO_3, R_2O_5, and RO_2.

3. If an element gives a hydrogen compound, RX_m, it forms an *organo-metallic compound* of the same composition, where $X = C_nH_{2n+1}$; that is, X is the radicle of a saturated hydrocarbon. The elements of the uneven series, which are incapable of giving hydrogen compounds, and give oxides of the forms RX, RX_2, R_X3, also give organo-metallic compounds of this form proper to the higher oxides. Thus zinc forms the oxide ZnO, salts ZnX_2 and zinc ethyl $Zn(C_2H_5)_2$. The elements of the even series do not seem to form organo-metallic compounds at all; at least all efforts for their preparation have as yet been fruitless—for instance, in the case of titanium, zirconium, or iron.

4. The atomic weights of elements belonging to contiguous periods differ approximately by 45; for example, K<Rb, Cr<Mo, Br<I. But the elements of the typical series show much smaller differences. Thus the difference between the atomic weights of Li, Na, and K, between Ca, Mg, and Be, between Si and C, between S and O, and between Cl and F, is 16. As a rule, there is a greater difference between the atomic weights of two elements of one group and belonging to two neighbouring series (Ti-Si = V-P = Cr-S = Mn-Cl = Nb-As, &c. = 20); and this difference attains a maximum with the heaviest elements (for example, Th-Pb = 26, Bi-Ta = 26, Ba-Cd = 25, &c.). Furthermore, the difference between the atomic

weights of the elements of even and uneven series also increases. In fact, the differences between Na and K, Mg and Ca, Si and Ti, are less abrupt than those between Pb and Th, Ta and Bi, Cd and Ba, &c. Thus even in the magnitude of the differences of the atomic weights of analogous elements there is observable a certain connection with the gradation of their properties.[12 bis]

5. According to the periodic system every element occupies a certain position, determined by the group (indicated in Roman numerals) and series (Arabic numerals) in which it occurs. These indicate the atomic weight, the analogues, properties, and type of the higher oxide, and of the hydrogen and other compounds—in a word, all the chief quantitative and qualitative features of an element, although there yet remain a whole series of further details and peculiarities whose cause should perhaps be looked for in small differences of the atomic weights. If in a certain group there occur elements, R_1, R_2, R_3, and if in that series which contains one of these elements, for instance R_2, an element Q_2 precedes it and an element T_2 succeeds it, then the properties of R_2 are determined by the properties of R_1, R_3, Q_2, and T_2. Thus, for instance, the atomic weight of $R_2 = ¼(R_1 + R_3 + Q_2 + T_2)$. For example, selenium occurs in the same group as sulphur, S = 32, and tellurium, Te = 125, and, in the 7th series As = 75 stands before it and Br = 80 after it. Hence the atomic weight of selenium should be ¼(32 + 125 + 75 + 80) = 78, which is near to the truth. Other properties of selenium may also be determined in this manner. For example, arsenic forms H_3As, bromine gives HBr, and it is evident that selenium, which stands between them, should form H_2Se, with properties intermediate between those of H_3As and HBr. Even the physical properties of selenium and its compounds, not to speak of their composition, being determined by the group in which it occurs, may be foreseen with a close approach to reality from the properties of sulphur, tellurium, arsenic, and bromine. *In this manner it is possible to foretell the properties of still unknown elements.* For instance in the position IV, 5—that is, in the IVth group and 5th series—an element is still wanting. These unknown elements may be named after the preceding known element of the same group by adding to the first syllable the prefix *eka-*, which means *one* in Sanskrit. The element IV, 5, follows after IV, 3, and this latter position being occupied by silicon, we call the unknown element ekasilicon and its symbol Es. The following are the properties which this element should have on the basis of the known properties of silicon, tin, zinc, and arsenic. Its atomic weight is nearly 72, higher oxide EsO_2, lower oxide EsO, compounds of the general form EsX_4, and chemically unstable lower compounds of the form EsX_2. Es gives volatile organo-metallic compounds—for instance, $Es(CH_3)_4$, $Es(CH_3)_3Cl$, and $Es(C_2H_5)_4$, which boil at about 160°, &c.; also a volatile and liquid chloride, $EsCl_4$, boiling at about 90° and of specific gravity about

1·9. EsO_2 will be the anhydride of a feeble colloidal acid, metallic Es will be rather easily obtainable from the oxides and from K_2EsF_6 by reduction, EsS_2 will resemble SnS_2 and SiS_2, and will probably be soluble in ammonium sulphide; the specific gravity of Es will be about 5·5, EsO_2 will have a density of about 4·7, &c. Such a prediction of the properties of ekasilicon was made by me in 1871, on the basis of the properties of the elements analogous to it: IV, 3, = Si, IV, 7 = Sn, and also II, 5 = Zn and V, 5 = As. And now that this element has been discovered by C. Winkler, of Freiberg, it has been found that its actual properties entirely correspond with those which were foretold. In this we see a most important confirmation of the truth of the periodic law. This element is now called germanium, Ge (*see* Chapter XVIII.). It is not the only one that has been predicted by the periodic law. We shall see in describing the elements of the third group that properties were foretold of an element ekaaluminium, III, 5, El = 68, and were afterwards verified when the metal termed 'gallium' was discovered by De Boisbaudran. So also the properties of scandium corresponded with those predicted for ekaboron, according to Nilson.

6. As a true law of nature is one to which there are no exceptions, the periodic dependence of the properties on the atomic weights of the elements gives a *new means for determining by the equivalent the atomic weight* or atomicity of imperfectly investigated but known elements, for which no other means could as yet be applied for determining the true atomic weight. At the time (1869) when the periodic law was first proposed there were several such elements. It thus became possible to learn their true atomic weights, and these were verified by later researches. Among the elements thus concerned were indium, uranium, cerium, yttrium, and others.

7. The periodic variability of the properties of the elements in dependence on their masses presents a distinction from other kinds of periodic dependence (as, for example, the sines of angles vary periodically and successively with the growth of the angles, or the temperature of the atmosphere with the course of time), in that the weights of the atoms do not increase gradually, but by leaps; that is, according to Dalton's law of multiple proportions, there not only are not, but there cannot be, any transitive or intermediate elements between two neighbouring ones (for example, between K = 39 and Ca = 40, or Al = 27 and Si = 28, or C = 12 and N = 14, &c.) As in a molecule of a hydrogen compound there may be either one, as in HF, or two, as in H_2O, or three, as in NH_3, &c., atoms of hydrogen; but as there cannot be molecules containing 2½ atoms of hydrogen to one atom of another element, so there cannot be any element intermediate between N and O, with an atomic weight greater than 14 or less than 16, or between K and Ca. Hence the periodic dependence of the

elements cannot be expressed by any algebraical continuous function in the same way that it is possible, for instance, to express the variation of the temperature during the course of a day or year.

8. The essence of the notions giving rise to the periodic law consists in a general physico-mechanical principle which recognises the correlation, transmutability, and equivalence of the forces of nature. Gravitation, attraction at small distances, and many other phenomena are in direct dependence on the mass of matter. It might therefore have been expected that chemical forces would also depend on mass. A dependence is in fact shown, the properties of elements and compounds being determined by the masses of the atoms of which they are formed. The weight of a molecule, or its mass, determines, as we have seen, (Chapter VII. and elsewhere) many of its properties independently of its composition. Thus carbonic oxide, CO, and nitrogen, N_2, are two gases having the same molecular weight, and many of their properties (density, liquefaction, specific heat, &c.) are similar or nearly similar. The differences dependent on the nature of a substance play another part, and form magnitudes of another order. But the properties of atoms are mainly determined by their mass or weight, and are in dependence upon it. Only in this case there is a peculiarity in the dependence of the properties on the mass, for this *dependence is determined by a periodic law*. As the mass increases the properties vary, at first successively and regularly, and then return to their original magnitude and recommence a fresh period of variation like the first. Nevertheless here as in other cases a small variation of the mass of the atom generally leads to a small variation of properties, and determines differences of a second order. The atomic weights of cobalt and nickel, of rhodium, ruthenium, and palladium, and of osmium, iridium, and platinum, are very close to each other, and their properties are also very much alike—the differences are not very perceptible. And if the properties of atoms are a function of their weight, many ideas which have more or less rooted themselves in chemistry must suffer change and be developed and worked out in the sense of this deduction. Although at first sight it appears that the chemical elements are perfectly independent and individual, instead of this idea of the nature of the elements, the notion of the dependence of their properties upon *their mass* must now be established; that is to say, the subjection of the individuality of the elements to a common higher principle which evinces itself in gravity and in all physico-chemical phenomena. Many chemical deductions then acquire a new sense and significance, and a regularity is observed where it would otherwise escape attention. This is more particularly apparent in the physical properties, to the consideration of which we shall afterwards turn, and we will now point out that Gustavson first (Chapter X., Note 28) and subsequently Potilitzin (Chapter XI., Note 66) demonstrated the direct dependence of the reactive power on the

atomic weight and that fundamental property which is expressed in the forms of their compounds, whilst in a number of other cases the purely chemical relations of elements proved to be in connection with their periodic properties. As a case in point, it may be mentioned that Carnelley remarked a dependence of the decomposability of the hydrates on the position of the elements in the periodic system; whilst L. Meyer, Willgerodt, and others established a connection between the atomic weight or the position of the elements in the periodic system and their property of serving as media in the transference of the halogens to the hydrocarbons. Bailey pointed out a periodicity in the stability (under the action of heat) of the oxides, namely: (*a*) in the even series (for instance, CrO_3, MoO_3, WO_3, and UO_3) the higher oxides of a given group decompose with greater ease the smaller the atomic weight, while in the uneven series (for example, CO_2, GeO_2, SnO_2, and PbO_2) the contrary is the case; and (*b*) the stability of the higher saline oxides in the even series (as in the fourth series from K_2O to Mn_2O_7) decreases in passing from the lower to the higher groups, while in the uneven series it increases from the Ist to the IVth group, and then falls from the IVth to the VIIth; for instance, in the series Ag_2O, CdO, In_2O_3, SnO_2, and then SnO_2, Sb_2O_5, TeO_3, I_2O_7. K. Winkler looked for and actually found (1890) a dependence between the reducibility of the metals by magnesium and their position in the periodic system of the elements. The greater the attention paid to this field the more often is a distinct connection found between the variation of purely chemical properties of analogous substances and the variation of the atomic weights of the constituent elements and their position in the periodic system. Besides, since the periodic system has become more firmly established, many facts have been gathered, showing that there are many similarities between Sn and Pb, B and Al, Cd and Hg, &c., which had not been previously observed, although foreseen in some cases, and a consequence of the periodic law. Keeping our attention in the same direction, we see that the most widely distributed elements in nature are those with small atomic weights, whilst in organisms the lightest elements exclusively predominate (hydrogen, carbon, nitrogen, oxygen), whose small mass facilitates those transformations which are proper to organisms. Poluta (of Kharkoff), C. C. Botkin, Blake, Brenton, and others even discovered a correlation between the physiological action of salts and other reagents on organisms and the positions occupied in the periodic system by the metals contained in them.

As, from the necessity of the case, the physical properties must be in dependence on the composition of a substance, *i.e.* on the quality and quantity of the elements forming it, so for them also a dependence on the atomic weight of the component elements must be expected, and consequently also on their periodic distribution. We shall meet with repeated proofs of this in the further exposition of our treatise, and for the present will content ourselves with citing the discovery by Carnelley in 1879 of the dependence of the magnetic properties of the elements on the position occupied by them in the periodic system. Carnelley showed that all the elements of the *even series* (beginning with lithium, potassium, rubidium, cæsium) belong to the number of magnetic (paramagnetic) substances; for example, according to Faraday and others,[17 bis] C, N, O, K, Ti, Cr, Mn, Fe, Co, Ni, Ce, are magnetic; and the elements of the *uneven series are diamagnetic*, H, Na, Si, P, S, Cl, Cu, Zn, As, Se, Br, Ag, Cd, Sn, Sb, I, Au, Hg, Tl, Pb, Bi.

Carnelley also showed that the *melting-point* of elements varies periodically, as is seen by the figures in Table III. (nineteenth column), where all the most trustworthy data are collected, and predominance is given to those having maximum and minimum values.

There is no doubt that many other physical properties will, when further studied, also prove to be in periodic dependence on the atomic weights,[19 bis] but at present only a few are known with any completeness, and we will only refer to the one which is the most easily and frequently determined—namely, the *specific gravity* in a solid and liquid state, the more especially as its connection with the chemical properties and relations of substances is shown at every step. Thus, for instance, of all the metals those of the alkalis, and of all the non-metals the halogens, are the most energetic in their reactions, and they have the lowest specific gravity among the adjacent elements, as is seen in Table III., column 17. Such are sodium, potassium, rubidium, cæsium among the metals, and chlorine, bromine, and iodine among the non-metals; and as such less energetic metals as iridium, platinum, and gold (and even charcoal or the diamond) have the highest specific gravity among the elements near to them in atomic weight; therefore the degree of the condensation of matter evidently influences the course of the transformations proper to a substance, and furthermore this dependence on the atomic weight, although very complex, is of a clearly periodic character. In order to account for this to some extent, it may be imagined that the lightest elements are porous, and, like a sponge, are easily penetrated by other substances, whilst the heavier elements are more compressed, and give way with difficulty to the insertion of other elements. These relations are best understood when, instead of the specific gravities referring to a unit of volume, the *atomic volumes of the elements*—that is, the quotient A/d of the atomic weight A by the specific gravity d—are taken for comparison. As, according to the entire sense of the atomic theory, the actual matter of a substance does not fill up its whole cubical contents, but is surrounded by a medium (ethereal, as is generally imagined), like the stars and planets which travel in the space of the heavens and fill it, with greater or less intervals, so the quotient A/d only expresses the *mean* volume corresponding to the sphere of the atoms, and therefore *[3root]A/d is the* *mean distance between the centres of the atoms.* For compounds whose molecules weigh M, the mean magnitude of the atomic volume is obtained by dividing the mean molecular volume M/d by the number of atoms n in the molecule. The above relations may easily be expressed from this point of view by comparing the atomic volumes. Those comparatively light elements which easily and frequently enter into reaction have the greatest atomic volumes: sodium 23, potassium 45, rubidium 57, cæsium 71, and the halogens about 27; whilst with those elements which enter into reaction with difficulty, the mean atomic volume is small; for carbon in the form of a diamond it is less than 4, as charcoal about 6, for nickel and cobalt less than 7, for iridium and platinum about 9. The remaining elements having atomic weights and properties intermediate between those elements mentioned above have also intermediate atomic volumes. Therefore *the specific gravities and specific volumes of solids and liquids stand in periodic dependence on the atomic weights*, as is seen in Table III., where both A (the atomic weight) and d (the specific gravity), and A/d (specific volumes of the atoms) are given (column 18).

Thus we find that in the large periods beginning with lithium, sodium, potassium, rubidium, cæsium, and ending with fluorine, chlorine, bromine, iodine, the extreme members (energetic elements) have a small density and large volume, whilst the intermediate substances gradually increase in density and decrease in volume—that is, as the atomic weight increases the density rises and falls, again rises and falls, and so on. Furthermore, the energy decreases as the density rises, and the greatest density is proper to the atomically heaviest and least energetic elements; for example, Os, Ir, Pt, Au, U.

In order to explain the relation between the volumes of the elements and of their compounds, the densities (column S) and volumes (column M/s) of some of the higher saline oxides arranged in the same order as in the case of the elements are given on p. 36. For convenience of comparison the volumes of the oxides are all calculated per two atoms of an element combined with oxygen. For example, the density of $Al_2O_3 = 4·0$, weight $Al_2O_3 = 102$, volume $Al_2O_3 = 25·5$. Whence, knowing the volume of aluminium to be 11, it is at once seen that in the formation of aluminium oxide, 22 volumes of it give 25·5 volumes of oxide. A distinct periodicity may also be observed with respect to the specific gravities and volumes of the higher saline oxides. Thus in each period, beginning with the alkali metals, the specific gravity of the oxides first rises, reaches a maximum, and then falls on passing to the acid oxides, and again becomes a minimum about the halogens. But it is especially important to call attention to the fact that the volume of the alkali oxides is less than that of the metal contained in them, which is also expressed in the last column, giving this difference for each atom of oxygen. Thus 2 atoms of sodium, or 46 volumes, give 24 volumes of Na₂O, and about 37 volumes of 2NaHO—that is, the oxygen and hydrogen in distributing themselves in the medium of sodium have not only not increased the distance between its atoms, but have brought them nearer together, have drawn them together by the force of their great affinity, by reason, it may be presumed, of the small mutual attraction of the atoms of sodium. Such metals as aluminium and zinc, in combining with oxygen and forming oxides of feeble salt-forming capacity, hardly vary in volume, but the common metals and non-metals, and especially those forming acid oxides, always give an increased volume when oxidised—that is, the atoms are set further apart in order to make room for the oxygen. The oxygen in them does not compress the molecule as in the alkalis; it is therefore comparatively easily disengaged.

	s	M/s	Volume of Oxygen
H₂O	1·0	18	?-22
Li₂O	2·0	15	-9
Be₂O₂	3·06	16	+2·6
B₂O₃	1·8	39	+10·0
C₂O₄	1·6	55	+10·6
N₂O₅	1·64	66	?+4
Na₂O	2·6	24	-22
Mg₂O₂	3·5	23	-4·5
Al₂O₃	4·0	26	+1·3
Si₂O₄	2·65	45	+5·2
P₂O₅	2·39	59	+6·2
S₂O₆	1·96	82	+8·7
Cl₂O₇	?1·92	95	+6

K$_2$O	2·7	35	−36
Ca$_2$O$_2$	3·25	34	−8
Sc$_2$O$_3$	3·86	35	?0
Ti$_2$O$_4$	4·2	38	+3
V$_2$O$_5$	3·49	52	+6·7
Cr$_2$O$_6$	2·74	73	+9·5
Cu$_2$O	5·9	24	+9·6
Zn$_2$O$_2$	5·7	23	+4·8
Ga$_2$O$_3$	?5·1	36	+4
Ge$_2$O$_4$	4·7	44	+4·5
As$_2$O$_5$	4·1	56	+6·0
Sr$_2$O$_2$	4·7	44	−13
Y$_2$O$_3$	5·0	45	?−2
Zr$_2$O$_4$	5·5	44	0
Nb$_2$O$_5$	4·7	57	+6
MoO$_6$	4·4	65	+6·8
Ag$_2$O	7·5	31	+11
Cd$_2$O$_2$	8·0	32	+3
In$_2$O$_3$	7·18	38	+2·7
Sn$_2$O$_4$	7·0	43	+2·7
Sb$_2$O$_5$	6·5	49	+2·6

TeO$_6$	5·1	68	+4·7
Ba$_2$O$_2$	5·7	52	-10
La$_2$O$_3$	6·5	50	+1
Ce$_2$O$_4$	6·74	50	+2
Ta$_2$O$_5$	7·5	59	+4·6
W$_2$O$_6$	6·8	68	+8·2
Hg$_2$O$_2$	11·1	39	+4·5
Pb$_2$O$_4$	8·9	53	+4·2
Th$_2$O$_4$	9·86	54	+2

As the volumes of the chlorides, organo-metallic and all other corresponding compounds, also vary in a like periodic succession with a change of elements, it is evidently possible to indicate the properties of substances yet uninvestigated by experimental means, and even those of yet undiscovered elements. It was possible by following this method to foretell, on the basis of the periodic law, many of the properties of scandium, gallium, and germanium, which were verified with great accuracy after these metals had been discovered. The periodic law, therefore, has not only embraced the mutual relations of the elements and expressed their analogy, but has also to a certain extent subjected to law the doctrine of the types of the compounds formed by the elements: it has enabled us to see a regularity in the variation of all chemical and physical properties of elements and compounds, and has rendered it possible to foretell the properties of elements and compounds yet uninvestigated by experimental means; thus it has prepared the ground for the building up of atomic and molecular mechanics.

Footnotes:

For instance the analogy of the sulphates of K, Rb, and Cs (Chapter XIII., Note 1).

[1 bis] The crystalline forms of aragonite, strontianite, and witherite belong to the rhombic system; the angle of the prism of CaCO$_3$ is 116° 10', of SrCO$_3$ 117° 19', and of BaCO$_3$ 118° 30'. On the other hand the crystalline forms of calc spar, magnesite, and calamine, which resemble

each other quite as closely, belong to the rhombohedral system, with the angle of the rhombohedra for $CaCO_3$ 105° 8', $MgCO_3$ 107° 10', and $ZnCO_3$ 107° 40'. From this comparison it is at once evident that zinc is more closely allied to magnesium than magnesium to calcium.

Solutions furnish the commonest examples of indefinite chemical compounds. But the isomorphous mixtures which are so common among the crystalline compounds of silica forming the crust of the earth, as well as alloys, which are so important in the application of metals to the arts, are also instances of indefinite compounds. And if in Chapter I., and in many other portions of this work, it has been necessary to admit the presence of definite compounds (in a state of dissociation) in solutions, the same applies with even greater force to isomorphous mixtures and alloys. For this reason in many places in this work I refer to facts which compel us to recognise the existence of definite chemical compounds in all isomorphous mixtures and alloys. This view of mine (which dates from the sixties) upon isomorphous mixtures finds a particularly clear confirmation in B. Roozeboom's researches (1892) upon the solubility and crystallising capacity of mixtures of the chlorates of potassium and thallium, $KClO_3$ and $TlClO_3$. He showed that when a solution contains different amounts of these salts, it deposits crystals containing either an excess of the first salt, from 98 p.c. to 100 p.c., or an excess of the second salt, from 63·7 to 100 p.c.; that is, in the crystalline form, either the first salt saturates the second or the second the first, just as in the solution of ether in water (Chapter I.); moreover, the solubility of the mixtures containing 36·3 and 98 p.c. $KClO_3$ is similar, just as the vapour tension of a saturated solution of water in ether is equal to that of a saturated solution of ether in water (Chapter I., Note 47). But just as there are solutions miscible in all proportions, so also certain isomorphous bodies can be present in crystals in all possible proportions of their component parts. Van 't Hoff calls such systems 'solid solutions.' These views were subsequently elaborated by Nernst (1892), and Witt (1891) applied them in explaining the phenomena observed in the coloration of tissues.

The cause of the difference which is observed in different compounds of the same type, with respect to their property of forming isomorphous mixtures, must not be looked for in the difference of their volumetric composition, as many investigators, including Kopp, affirm. The molecular volumes (found by dividing the molecular weight by the density) of those isomorphous substances which do give intermixtures are not nearer to each other than the volumes of those which do not give mixtures; for example, for magnesium carbonate the combining weight is 84, density 3·06, and volume therefore 27; for calcium carbonate in the form of calc spar the volume is 37, and in the form of aragonite 33; for strontium carbonate 41, for barium carbonate 46; that is, the volume of these closely allied isomorphous substances increases with the combining weight. The same is observed if we compare sodium chloride (molecular volume = 27) with potassium chloride (volume = 37), or sodium sulphate (volume = 55) with potassium sulphate (volume = 66), or sodium nitrate 39 with potassium nitrate 48, although the latter are less capable of giving isomorphous mixtures than the former. It is evident that the cause of isomorphism cannot be explained by an approximation in molecular volumes. It is more likely that, given a similarity in form and composition, the faculty to give isomorphous mixtures is connected with the laws and degree of solubility.

A phenomenon of a similar kind is shown for magnesium sulphate in Note 27 of the last chapter. In the same example we see what a complication the phenomena of dimorphism may introduce when the forms of analogous compounds are compared.

The property of solids of occurring in regular crystalline forms—the occurrence of many substances in the earth's crust in these forms—and those geometrical and simple laws which govern the formation of crystals long ago attracted the attention of the naturalist to crystals. The crystalline form is, without doubt, the expression of the relation in which the atoms occur in the molecules, and in which the molecules occur in the mass, of a substance. Crystallisation is determined by the distribution of the molecules along the direction of greatest cohesion, and therefore those forces must take part in the crystalline distribution of matter which act between the molecules; and, as they depend on the forces binding the atoms together in the molecules, a very close connection must exist between the atomic composition and the distribution of the atoms in the molecule on the one hand, and the crystalline form of a substance on the other hand; and hence an insight into the composition may be arrived at from the crystalline form. Such is the elementary and *a priori* idea which lies at the base of all researches into *the connection between composition and crystalline form*. Haüy in 1811 established the following fundamental law, which has been worked out by later investigators: That the fundamental crystalline form for a given chemical compound is constant (only the combinations vary), and that with a change of composition the crystalline form also changes, naturally with the exception of such limiting forms as the cube, regular octahedron, &c., which may belong to various substances of the regular system. The fundamental form is determined by the angles of certain fundamental geometric forms (prisms, pyramids, rhombohedra), or the ratio of the crystalline axes, and is connected with the optical and many other properties of crystals. Since the establishment of this law the description of definite compounds in a solid state is accompanied by a description (measurement) of its crystals, which forms an invariable, definite, and measurable character. The most important epochs in the further history of this question were made by the following discoveries:—Klaproth, Vauquelin, and others showed that aragonite has the same composition as calc spar, whilst the former belongs to the rhombic and the latter to the hexagonal system. Haüy at first considered that the composition, and after that the arrangement, of the atoms in the molecules was different. This is dimorphism (*see* Chapter XIV., Note 46). Beudant, Frankenheim, Laurent, and others found that the forms of the two nitres, KNO_3 and $NaNO_3$, exactly correspond with the forms of aragonite and calc spar; that they are able, moreover, to pass from one form into another; and that the difference of the forms is accompanied by a small alteration of the angles, for the angle of the prisms of potassium nitrate and aragonite is 119°, and of sodium nitrate and calc spar, 120°; and therefore dimorphism, or the crystallisation of one substance in different forms, does not necessarily imply a great difference in the distribution of the molecules, although some difference clearly exists. The researches of Mitscherlich (1822) on the dimorphism of sulphur confirmed this conclusion, although it cannot yet be affirmed that in dimorphism the arrangement of the atoms remains unaltered, and that only the molecules are distributed differently. Leblanc, Berthier, Wollaston, and others already knew that many substances of different composition appear in the same forms, and crystallise together in one crystal. Gay-Lussac (1816) showed that crystals of potash alum continue to grow in a solution of ammonia alum. Beudant (1817) explained this phenomenon as the *assimilation* of a foreign substance by a substance having a great force of crystallisation, which he illustrated by many natural and artificial examples. But Mitscherlich, and afterwards Berzelius and Henry Rose and others, showed that such an assimilation only exists with a similarity or approximate similarity of the forms of the individual substances and with a certain degree of chemical analogy. Thus was established the idea of *isomorphism* as an analogy of forms by reason of a resemblance of atomic composition, and by it was explained the variability of the composition of a number of minerals as isomorphous mixtures. Thus all the garnets are expressed by the general formula: $(RO)_3M_2O_3(SiO_2)_3$, where R = Ca, Mg, Fe, Mn, and M = Fe, Al, and where we may have either R and M separately, or their equivalent compounds, or their mixtures in all possible proportions.

But other facts, which render the correlation of form and composition still more complex, have accumulated side by side with a mass of data which may be accounted for by admitting the conceptions of isomorphism and dimorphism. Foremost among the former stand the phenomena of *homeomorphism*—that is, a nearness of forms with a difference of composition—and then the cases of polymorphism and hemimorphism— that is, a nearness of the fundamental forms or only of certain angles for substances which are near or analogous in their composition. Instances of homeomorphism are very numerous. Many of these, however, may be reduced to a resemblance of atomic composition, although they do not correspond to an isomorphism of the component elements; for example, CdS (greenockite) and AgI, $CaCO_3$ (aragonite) and KNO_3, $CaCO_3$ (calc spar) and $NaNO_3$, $BaSO_4$ (heavy spar), $KMnO_4$ (potassium permanganate), and $KClO_4$ (potassium perchlorate), Al_2O_3 (corundum) and $FeTiO_3$ (titanic iron ore), FeS_2 (marcasite, rhombic system) and FeSAs (arsenical pyrites), NiS and NiAs, &c. But besides these instances there are homeomorphous substances with an absolute dissimilarity of composition. Many such instances were pointed out by Dana. Cinnabar, HgS, and susannite, $PbSO_4 3PbCO_3$ appear in very analogous crystalline forms; the acid potassium sulphate crystallises in the monoclinic system in crystals analogous to felspar, $KAlSi_3O_8$; glauberite, $Na_2Ca(SO_4)_2$, augite, $RSiO_3$ (R = Ca, Mg), sodium carbonate, $Na_2CO_3, 10H_2O$, Glauber's salt, $Na_2SO_4, 10H_2O$, and borax, $Na_2BrO_7, 10H_2O$, not only belong to the same system (monoclinic), but exhibit an analogy of combinations and a nearness of corresponding angles. These and many other similar cases might appear to be perfectly arbitrary (especially as a *nearness* of angles and fundamental forms is a relative idea) were there not other cases where a resemblance of properties and a distinct relation in the variation of composition is connected with a resemblance of form. Thus, for example, alumina, Al_2O_3, and water, H_2O, are frequently found in many pyroxenes and amphiboles which only contain silica and magnesia (MgO, CaO, FeO, MnO). Scheerer and Hermann, and many others, endeavoured to explain such instances by *polymetric isomorphism*, stating that MgO may be replaced by $3H_2O$ (for example, olivine and serpentine), SiO_2 by Al_2O_3 (in the amphiboles, talcs), and so on. A certain number of the instances of this order are subject to doubt, because many of the natural minerals which served as the basis for the establishment of polymeric isomorphism in all probability no longer present their original composition, but one which has been altered under the influence of solutions which have come into contact with them; they therefore belong to the class of *pseudomorphs*, or false crystals. There is, however, no doubt of the existence of a whole series of natural and artificial homeomorphs, which differ from each other by atomic amounts of water, silica, and some other component parts. Thus, Thomsen (1874) showed a very striking instance. The metallic chlorides, RCl_2, often crystallise with water, and they do not then contain less than one molecule of water per atom of chlorine. The most familiar representative of the order $RCl_2, 2H_2O$ is $BaCl_2, 2H_2O$, which crystallises in the rhombic system. Barium bromide, $BaBr_2, 2H_2O$, and copper chloride, $CuCl_2, 2H_2O$, have nearly the same forms: potassium iodate, KIO_4; potassium chlorate, $KClO_4$; potassium permanganate, $KMnO_4$; barium sulphate, $BaSO_4$; calcium sulphate, $CaSO_4$; sodium sulphate, Na_2SO_4; barium formate, $BaC_2H_2O_4$, and others have almost the same crystalline form (of the rhombic system). Parallel with this series is that of the metallic chlorides containing $RCl_2,4H_2O$, of the sulphates of the composition $RSO_4,2H_2O$, and the formates $RC_2H_2O_4,2H_2O$. These compounds belong to the monoclinic system, have a close resemblance of form, and differ from the first series by containing two more molecules of water. The addition of two more molecules of water in all the above series also gives forms of the monoclinic system closely resembling each other; for example, $NiCl_2,6H_2O$ and $MnSO_4,4H_2O$. Hence we see that not only is $RCl_2,2H_2O$ analogous in form to RSO_4 and $RC_2H_2O_4$, but that their compounds with $2H_2O$ and with $4H_2O$ also exhibit closely analogous forms. From these examples it is evident that the conditions which determine a given form may be repeated not only in the presence of an isomorphous exchange—that is, with an equal number of atoms in the molecule—but also in the presence of an unequal number when there are peculiar and as yet ungeneralised relations in composition. Thus ZnO and Al_2O_3 exhibit a close analogy of form. Both oxides belong to the rhombohedral system, and the angle between the pyramid and the terminal plane of the first is 118° 7′, and of the second 118° 49′. Alumina, Al_2O_3, is also analogous in form to SiO_2, and we shall see that these analogies of form are conjoined with a certain analogy in properties. It is not surprising, therefore, that in the complex molecule of a siliceous compound it is sometimes possible to replace SiO_2 by means of Al_2O_3, as Scheerer admits. The oxides Cu_2O, MgO, NiO, Fe_3O_4, CeO_2, crystallise in the regular system, although they are of very different atomic structure. Marignac demonstrated the perfect analogy of the forms of K_2ZrF_6 and $CaCO_3$, and the former is even dimorphous, like the calcium carbonate. The same salt is isomorphous with R_2NbOF_5 and $R_2WO_2F_4$, where R is an alkali metal. There is an equivalency between $CaCO_3$ and K_2ZrF_6, because K_2 is equivalent to Ca, C to Zr, and F_6 to O_3, and with the isomorphism of the other two salts we find besides an equal contents of the alkali metal—an equal number of atoms on the one hand and an analogy to the properties of K_2ZrF_6 on the other. The long-known isomorphism of the corresponding compounds of potassium and ammonium, KX and NH_4X, may be taken as the simplest example of the fact that an analogy of form shows itself with an analogy of chemical reaction even without an equality in atomic composition. Therefore the ultimate progress of the entire doctrine of the correlation of composition and crystalline forms will only be arrived at with the accumulation of a sufficient number of facts collected on a plan corresponding with the problems which here present themselves. The first steps have already been made. The researches of the Geneva *savant*, Marignac, on the crystalline form and composition of many of the double fluorides, and the work of Wyruboff on the ferricyanides and other compounds, are particularly important in this respect. It is already evident that, with a definite change of composition, certain angles remain constant, notwithstanding that others are subject to alteration. Such an instance of the relation of forms was observed by Laurent, and named by him *hemimorphism* (an anomalous term) when the analogy is limited to certain angles, and *paramorphism* when the forms in general approach each other, but belong to different systems. So, for example, the angle of the planes of a rhombohedron may be greater or less than 90°, and therefore such acute and obtuse rhombohedra may closely approximate to the cube. Hausmannite, Mn_3O_4, belongs to the tetragonal system, and the planes of its pyramid are inclined at an angle of about 118°, whilst magnetic iron ore, Fe_3O_4, which resembles hausmannite in many respects, appears in regular octahedra—that is, the pyramidal planes are inclined at an angle of 109° 28'. This is an example of paramorphism; the systems are different, the compositions are analogous, and there is a certain resemblance in form. Hemimorphism has been found in many instances of saline and other substitutions. Thus, Laurent demonstrated, and Hintze confirmed (1873), that naphthalene derivatives of analogous composition are hemimorphous. Nicklès (1849) showed that in ethylene sulphate the angle of the prism is 125° 26', and in the nitrate of the same radicle 126° 95'. The angle of the prism of methylamine oxalate is 131° 20', and of fluoride, which is very different in composition from the former, the angle is 132°. Groth (1870) endeavoured to indicate in general what kinds of change of form proceed with the substitution of hydrogen by various other elements and groups, and he observed a regularity which he termed *morphotropy*. The following examples show that morphotropy recalls the hemimorphism of Laurent. Benzene, C_6H_6, rhombic system, ratio of the axes 0·891 : 1 : 0·799. Phenol, $C_6H_5(OH)$, and resorcinol, $C_6H_4(OH)_2$, also rhombic system, but the ratio of one axis is changed—thus, in resorcinol, 0·910 : 1 : 0·540; that is, a portion of the crystalline structure in one direction is the same, but in the other direction it is changed, whilst in the rhombic system dinitrophenol, $C_6H_3(NO_2)_2(OH)$ = O·833 : 1 : 0·753; trinitrophenol (picric acid), $C_6H_2(NO)_3(OH)$ = 0·937 : 1 : 0·974; and the potassium salt = 0·942 : 1 : 1·354. Here the ratio of the first axis is preserved—that is, certain angles remain constant, and the chemical proximity of the composition of these bodies is undoubted. Laurent compares hemimorphism with architectural style. Thus, Gothic cathedrals differ in many respects, but there is an analogy expressed both in the sum total of their common relations and in certain details—for example, in the windows. It is evident that we may expect many fruitful results for molecular mechanics (which forms a problem common to many provinces of natural science) from the further elaboration of the data concerning those variations which take place in crystalline form when the composition of a substance is subjected to a known change, and therefore I consider it useful to point out to the student of science seeking for matter for independent scientific research this vast field for work which is presented by the correlation of form and composition. The geometrical regularity and varied beauty of crystalline forms offer no small attraction to research of this kind.

The still more complex combinations—which are so clearly expressed in the crystallo-hydrates, double salts, and similar compounds—although they may be regarded as independent, are, however, most easily understood with our present knowledge as aggregations of whole molecules to which there are no corresponding double compounds, containing one atom of an element R and many atoms of other elements RX_n. The above types embrace all cases of direct combinations of atoms, and the formula $MgSO_4,7H_2O$ cannot, without violating known facts, be directly deduced from the types MgX_n or SX_n, whilst the formula $MgSO_4$ corresponds both with the type of the magnesium compounds MgX_2 and with the type of the sulphur compounds SO_2X_2, or in general SX_6, where X_2 is replaced by $(OH)_2$, with the substitution in this case of H_2 by the atom Mg, which always replaces H_2. However, it must be remarked that the sodium crystallo-hydrates often contain $10H_2O$, the magnesium crystallo-hydrates 6 and $7H_2O$, and that the type PtM_2X_6 is proper to the double salts of platinum, &c. With the further development of our knowledge concerning crystallo-hydrates, double salts, alloys, solutions, &c., in the *chemical sense* of feeble compounds (that is, such as are easily destroyed by feeble chemical influences) it will probably be possible to arrive at a perfect generalisation for them. For a long time these subjects were only studied by the way or by chance; our knowledge of them is accidental and destitute of system, and therefore it is impossible to expect as yet any generalisation as to their nature. The days of Gerhardt are not long past when only three types were recognised: RX, RX_2, and RX_3; the type RX_4 was afterwards added (by Cooper, Kekulé, Butleroff, and others), mainly for the purpose of generalising the data respecting the carbon compounds. And indeed many are still satisfied with these types, and derive the higher types from them; for instance, RX_5 from RX_3—as, for example, $POCl_3$ from PCl_3, considering the oxygen to be bound both to the chlorine (as in HClO) and to the phosphorus. But the time has now arrived when it is clearly seen that the forms RX, RX_2, RX_3, and RX_4 do not exhaust the whole variety of phenomena. The revolution became evident when Würtz showed that PCl_5 is not a compound of $PCl_3 + Cl_2$ (although it may decompose into them), but a whole molecule capable of passing into vapour, PCl_5 like PF_5 and SiF_4. The time for the recognition of types even higher than RX_8 is in my opinion in the future; that it will come, we can already see in the fact that oxalic acid, $C_2H_2O_4$, gives a crystal-hydrate with $2H_2O$; but it may be referred to the type CH_4, or rather to the type of ethane, C_2H_6, in which all the atoms of hydrogen are replaced by hydroxyl, $C_2H_2O_4 2H_2O = C_2(OH)_6$ (*see* Chapter XXII., Note 35).

The hydrogen compounds, R_2H, in equivalency correspond with the type of the suboxides, R_4O. Palladium, sodium, and potassium give such hydrogen compounds, and it is worthy of remark that according to the periodic system these elements stand near to each other, and that in those groups where the hydrogen compounds R_2H appear, the quaternary oxides R_4O are also present.

Not wishing to complicate the explanation, I here only touch on the general features of the relation between the hydrates and oxides and of the oxides among themselves. Thus, for instance, the conception of the ortho-acids and of the normal acids will be considered in speaking of phosphoric and phosphorous acids.

As in the further explanation of the periodic law only those oxides which give salts will be considered, I think it will not be superfluous to mention here the following facts relative to the peroxides. Of the *peroxides* corresponding with hydrogen peroxide, the following are at present known: H_2O_2, Na_2O_2, S_2O_7 (as HSO_4?), K_2O_4, K_2O_2, CaO_2, TiO_3, Cr_2O_7, $CuO_2(?)$, ZnO_2, Rb_2O_2, SrO_2, Ag_2O_2, CdO_2, CsO_2, Cs_2O_2, BaO_2, Mo_2O_7, SnO_3, W_2O_7, UO_4. It is probable that the number of peroxides will increase with further investigation. A periodicity is seen in those now known, for the elements (excepting Li) of the first group, which give R_2O, form peroxides, and then the elements of the sixth group seem also to be particularly inclined to form peroxides, R_2O_7; but at present it is too early, in my opinion, to enter upon a generalisation of this subject, not only because it is a new and but little studied matter (not investigated for all the elements), but also, and more especially, because in many instances only the hydrates are known—for instance, $Mo_2H_2O_8$—and they perhaps are only compounds of peroxide of hydrogen—for example, $Mo_2H_2O_8 = 2MoO_3 + H_2O_2$—since Prof. Schöne has shown that H_2O_2 and BaO_2 possess the property of combining together and with other oxides. Nevertheless, I have, in the general table expressing the periodic properties of the elements, endeavoured to sum up the data respecting all the known peroxide compounds whose characteristic property is seen in their capability to form peroxide of hydrogen under many circumstances.

The periodic law and the periodic system of the elements appeared in the same form as here given in the first edition of this work, begun in 1868 and finished in 1871. In laying out the accumulated information respecting the elements, I had occasion to reflect on their mutual relations. At the beginning of 1869 I distributed among many chemists a pamphlet entitled

'An Attempted System of the Elements, based on their Atomic Weights and Chemical Analogies,' and at the March meeting of the Russian Chemical Society, 1869, I communicated a paper 'On the Correlation of the Properties and Atomic Weights of the Elements.' The substance of this paper is embraced in the following conclusions: (1) The elements, if arranged according to their atomic weights, exhibit an evident *periodicity* of properties. (2) Elements which are similar as regards their chemical properties have atomic weights which are either of nearly the same value (platinum, iridium, osmium) or which increase regularly (*e.g.* potassium, rubidium, cæsium). (3) The arrangement of the elements or of groups of elements in the order of their atomic weights corresponds with their so-called *valencies*. (4) The elements, which are the most widely distributed in nature, have *small* atomic weights, and all the elements of small atomic weight are characterised by sharply defined properties. They are therefore typical elements. (5) The *magnitude* of the atomic weight determines the character of an element. (6) The discovery of many yet unknown elements may be expected. For instance, elements analogous to aluminium and silicon, whose atomic weights would be between 65 and 75. (7) The atomic weight of an element may sometimes be corrected by aid of a knowledge of those of the adjacent elements. Thus the combining weight of tellurium must lie between 123 and 126, and cannot be 128. (8) Certain characteristic properties of the elements can be foretold from their atomic weights.

The entire periodic law is included in these lines. In the series of subsequent papers (1870–72, for example, in the *Transactions* of the Russian Chemical Society, of the Moscow Meeting of Naturalists, of the St. Petersburg Academy, and Liebig's *Annalen*) on the same subject we only find applications of the same principles, which were afterwards confirmed by the labours of Roscoe, Carnelley, Thorpe, and others in England; of Rammelsberg (cerium and uranium), L. Meyer (the specific volumes of the elements), Zimmermann (uranium), and more especially of C. Winkler (who discovered germanium, and showed its identity with ekasilicon), and others in Germany; of Lecoq de Boisbaudran in France (the discoverer of gallium = ekaaluminium); of Clève (the atomic weights of the cerium metals), Nilson (discoverer of scandium = ekaboron), and Nilson and Pettersson (determination of the vapour density of beryllium chloride) in Sweden; and of Brauner (who investigated cerium, and determined the combining weight of tellurium = 125) in Austria, and Piccini in Italy.

I consider it necessary to state that, in arranging the periodic system of the elements, I made use of the previous researches of Dumas, Gladstone, Pettenkofer, Kremers, and Lenssen on the atomic weights of related elements, but I was not acquainted with the works preceding mine of De Chancourtois (*vis tellurique*, or the spiral of the elements according to their properties and equivalents) in France, and of J. Newlands (Law of Octaves—for instance, H, F, Cl, Co, Br, Pd, I, Pt form the first octave, and O, S, Fe, Se, Rh, Te, Au, Th the last) in England, although certain germs of the periodic law are to be seen in these works. With regard to the work of Prof. Lothar Meyer respecting the periodic law (Notes 12 and 13), it is evident, judging from the method of investigation, and from his statement (Liebig's *Annalen, Supt. Band 7*, 1870, 354), at the very commencement of which he cites my paper of 1869 above mentioned, that he accepted the periodic law in the form which I proposed.

In concluding this historical statement I consider it well to observe that no law of nature, however general, has been established all at once; its recognition is always preceded by many hints; the establishment of a law, however, does not take place when its significance is recognised, but only when it has been confirmed by experiment, which the man of science must consider as the only proof of the correctness of his conjectures and opinions. I therefore, for my part, look upon Roscoe, De Boisbaudran, Nilson, Winkler, Brauner, Carnelley, Thorpe, and others who verified the adaptability of the periodic law to chemical facts, as the true founders of the periodic law, the further development of which still awaits fresh workers.

This resembles the fact, well known to those having an acquaintance with organic chemistry, that in a series of homologues (Chapter VIII.) the first members, in which there is the least carbon, although showing the general properties of the homologous series, still present certain distinct peculiarities.

Besides arranging the elements (*a*) in a successive order according to their atomic weights, with indication of their analogies by showing some of the properties—for instance, their power of giving one or another form of combination—both of the *elements* and of their compounds (as is done in Table III. and in the table on p. 36), (*b*) according to periods (as in Table I. at the commencement of volume I. after the preface), and (*c*) according to groups and series or small periods (as in the same tables), I am acquainted with the following methods of expressing the periodic relations of the elements: (1) By a curve drawn through points obtained in the following manner: The elements are arranged along the horizontal axis as abscissæ at distances from zero proportional to their atomic weights, whilst the values for all the elements of some property—for example, the specific volumes or the melting points, are expressed by the ordinates. This method, although graphic, has the theoretical disadvantage that it does not in any way indicate the existence of a limited and definite number of elements in each period. There is nothing, for instance, in this method of expressing the law of periodicity to show that between magnesium and aluminium there can be no other element with an atomic weight of, say, 25, atomic volume 13, and in general having properties intermediate between those of these two elements. The actual periodic law does not correspond with a continuous change of properties, with a continuous variation of atomic weight—in a word, it does not express an uninterrupted function—and as the law is purely chemical, starting from the conception of atoms and molecules which combine in multiple proportions, with intervals (not continuously), it *above all* depends on there being but few types of compounds, which are arithmetically simple, *repeat themselves*, and offer no uninterrupted transitions, so that each period can only contain a definite number of members. For this reason there can be no other elements between magnesium, which gives the chloride $MgCl_2$, and aluminium, which forms AlX_3; there is a break in the continuity, according to the law of multiple proportions. The periodic law ought not, therefore, to be expressed by geometrical figures in which continuity is always understood. Owing to these considerations I never have and never will express the periodic relations of the elements by any geometrical figures. (2) *By a plane spiral.* Radii are traced from a centre, proportional to the atomic weights; analogous elements lie along one radius, and the points of intersection are arranged in a spiral. This method, adopted by De Chancourtois, Baumgauer, E. Huth, and others, has many of the imperfections of the preceding, although it removes the indefiniteness as to the number of elements in a period. It is merely an attempt to reduce the complex relations to a simple graphic representation, since the equation to the spiral and the number of radii are not dependent upon anything. (3) *By the lines of atomicity*, either parallel, as in Reynolds's and the Rev. S. Haughton's method, or as in Crookes's method, arranged to the right and left of an axis, along which the magnitudes of the atomic weights are counted, and the position of the elements marked off, on the one side the members of the even series (paramagnetic, like oxygen, potassium, iron), and on the other side the members of the uneven series (diamagnetic, like sulphur, chlorine, zinc, and mercury). On joining up these points a periodic curve is obtained, compared by Crookes to the oscillations of a pendulum, and, according to Haughton, representing a cubical curve. This method would be very graphic did it not require, for instance, that sulphur should be considered as bivalent and manganese as univalent, although neither of these elements gives stable derivatives of these natures, and although the one is taken on the basis of the lowest possible compound SX_2, and the other of the highest, because manganese can be referred to the univalent elements only by the analogy of $KMnO_4$ to $KClO_4$. Furthermore, Reynolds and Crookes place hydrogen, iron, nickel, cobalt, and others outside the axis of atomicity, and consider uranium as bivalent without the least foundation. (4) Rantsheff endeavoured to classify the elements in their periodic relations by a system dependent on solid geometry. He communicated this mode of expression to the Russian Chemical Society, but his communication, which is apparently not void of interest, has not yet appeared in print. (5) *By algebraic formulæ*: for example, E. J. Mills (1886) endeavours to express all the atomic weights by the logarithmic function A = 15(n - 0·9375t), in which the variables n and t are whole numbers. For instance, for oxygen n=2, t=1; hence A = 15·94; for antimony n=9, t=0; whence A=120, and so on. n varies from 1 to 16 and t from 0 to 59. The analogues are hardly distinguishable by this method: thus for chlorine the magnitudes of n and t are 3 and 7; for bromine 6 and 6; for iodine 9 and 9; for potassium 3 and 14; for rubidium 6 and 18; for cæsium 9 and 20; but a certain regularity seems to be shown. (6) A more natural method of expressing the dependence of the properties of elements on their atomic weights is obtained by *trigonometrical functions*, because this dependence is periodic like the functions of trigonometrical lines, and therefore Ridberg in Sweden (Lund, 1885) and F. Flavitzky in Russia (Kazan, 1887) have adopted a similar method of expression, which must be considered as worthy of being worked out, although it does not express the absence of intermediate elements—for instance, between magnesium and aluminium, which is essentially the most important part of the matter. (7) The investigations of B. N. Tchitchérin (1888, *Journal of the Russian Physical and Chemical Society*) form the first effort in the latter direction. He carefully studied the alkali metals, and discovered the following simple relation between their atomic volumes: they can all be expressed by A(2 - 0·0428An), where A is the atomic weight and n = 1 for lithium and sodium, ⅛ for potassium, ⅜ for rubidium, and ⅔ for cæsium. If n always = 1, then the volume of the atom would become zero at A = 46⅔, and would reach its maximum when A = 23⅓, and the density increases with the growth of A. In order to explain the variation of n, and the relation of the atomic weights of the alkali metals to those of the other elements, as also the atomicity itself, Tchitchérin supposes all atoms to be built up of a primary matter; he considers the relation of the central to the peripheric mass, and, guided by mechanical principles, deduces many of the properties of the atoms from the reaction of the internal and peripheric parts of each atom. This endeavour offers many interesting points, but it admits the hypothesis of the building up of all the elements from one primary matter, and at the present time such an hypothesis has not the least support either in theory or in fact. Besides which the starting-point of the theory is the specific gravity of the metals at a definite temperature (it is not known how the above relation would appear at other temperatures), and the specific gravity varies even under mechanical influences. L. Hugo (1884) endeavoured to represent the atomic weights of Li, Na, K, Rb, and Cs by geometrical figures—for instance, Li = 7 represents a central atom = 1 and six atoms on the six terminals of an octahedron; Na, is obtained by applying two such atoms on each edge of an octahedron, and so on. It is evident that such methods can add nothing new to our data respecting the atomic weights of analogous elements.

Many natural phenomena exhibit a dependence of a periodic character. Thus the phenomena of day and night and of the seasons of the year, and vibrations of all kinds, exhibit variations of a periodic character in dependence on time and space. But in ordinary periodic functions one variable varies continuously, whilst the other increases to a limit, then a period of decrease begins, and having in turn reached its limit a period of increase again begins. It is otherwise in the periodic function of the elements. Here the mass of the elements does not increase continuously, but abruptly, by steps, as from magnesium to aluminium. So also the valency or atomicity leaps directly from 1 to 2 to 3, &c., without intermediate quantities, and in my opinion it is these properties which are the most important, and it is their periodicity which forms the substance of the periodic law. It expresses *the properties of the real elements*, and not of what may be termed their manifestations visually known to us. The external properties of elements and compounds are in periodic dependence on the atomic weight of the elements only because these external properties are themselves the result of the properties of the real elements which unite to form the 'free' elements and the compounds. To explain and express the periodic law is to explain and express the cause of the law of multiple proportions, of the difference of the elements, and the variation of their atomicity, and at the same time to understand what mass and gravitation are. In my opinion this is still premature. But just as without knowing the cause of gravitation it is possible to make use of the law of gravity, so for the aims of chemistry it is possible to take advantage of the laws discovered by chemistry without being able to explain their causes. The above-mentioned peculiarity of the laws of chemistry respecting definite compounds and the atomic weights leads one to think that the time has not yet come for their full explanation, and I do not think that it will come before the explanation of such a primary law of nature as the law of gravity.

It will not be out of place here to turn our attention to the many-sided correlation existing between the undecomposable *elements and the compound carbon radicles*, which has long been remarked (Pettenkofer, Dumas, and others), and reconsidered in recent times by Carnelley (1886), and most originally in Pelopidas's work (1883) on the principles of the periodic system. Pelopidas compares the series containing eight hydrocarbon radicles, C_nH_{2n+1}, C_nH_{2n} &c., for instance, C_6H_{13}, C_6H_{12}, C_6H_{11}, C_6H_{10}, C_6H_9, C_6H_8, C_6H_7, and C_6H_6—with the series of the elements arranged in eight groups. The analogy is particularly clear owing to the property of

C_nH_{2n+1} to combine with X, thus reaching saturation, and of the following members with X_2, X_3 … X_8, and especially because these are followed by an aromatic radicle—for example, C_6H_5—in which, as is well known, many of the properties of the saturated radicle C_6H_{13} are repeated, and in particular the power of forming a univalent radicle again appears. Pelopidas shows a confirmation of the parallel in the property of the above radicles of giving oxygen compounds corresponding with the groups in the periodic system. Thus the hydrocarbon radicles of the first group—for instance, C_6H_{13} or C_6H_5—give oxides of the form R_2O and hydroxides RHO, like the metals of the alkalis; and in the third group they form oxides R_2O_3 and hydrates RO_2H. For example, in the series CH_3 the corresponding compounds of the third group will be the oxide $(CH)_2O_3$ or $C_2H_2O_3$—that is, formic anhydride and hydrate, CHO_2H, or formic acid. In the sixth group, with a composition of C_2, the oxide RO_3 will be C_2O_3, and hydrate $C_2H_2O_4$—that is, also a bibasic acid (oxalic) resembling sulphuric, among the inorganic acids. After applying his views to a number of organic compounds, Pelopidas dwells more particularly on the radicles corresponding with ammonium.

With respect to this remarkable parallelism, it must above all be observed that in the elements the atomic weight increases in passing to contiguous members of a higher valency, whilst here it decreases, which should indicate that the periodic variability of elements and compounds is subject to some higher law whose nature, and still more whose cause, cannot at present be determined. It is probably based on the fundamental principles of the internal mechanics of the atoms and molecules, and as the periodic law has only been generally recognised for a few years it is not surprising that any further progress towards its explanation can only be looked for in the development of facts touching on this subject.

[11 bis] True peroxides (*see* Note 7), like H_2O_2, BaO_2, S_2O_7 (Chapter XX.), must not be confused with true saline oxides even if the latter contain much oxygen (for instance, N_2O_5, CrO_3, &c.) although one and the other easily oxidise. The difference between them is seen in their fundamental properties: the saline oxides correspond to water, while the peroxides correspond in their reactions and origin to peroxide of hydrogen. This is clearly seen in the difference between Na_2O and Na_2O_2 (Chapter XII.). Therefore the peroxides should also have their periodicity. An element R, giving a highest degree of oxidation, R_2O_n, may give both a lower degree of oxidation, R_2O_{n-m} (where m is evidently less than n), and peroxides, R_2O_{n+1}, R_2O_{n+2}, or even more oxygen. This class of oxides, to which attention has only recently been turned (Berthelot, Piccini, &c.), may perhaps on further study give the possibility of generalising the capability of the elements to give unstable complex higher forms of combination,

such as double salts, and in my opinion should in the near future be the field of new and important discoveries. And in contemporary chemistry, salts, saline oxides, hydrogen compounds, and other combinations of the elements corresponding to them constitute an important and very complex problem for generalisation, which is satisfied by the periodic law in its present form, to which it has risen from its first state, in which it gave the means of foreseeing (*see* later on) the existence of unknown elements (Ga, Sc, and Ge), their properties, and many details respecting their compounds. Until those improvements in the periodic system which have been proposed by Prof. Flavitzky (of Kazan) and Prof. Harperath (of Cordoba, in the Argentine Republic), Ugo Alvisi (Italy), and others give similar practical results, I think it unnecessary to discuss them further.

The hydrides generalised by the periodic law are those to which metallo-organic compounds correspond, and they are themselves either volatile or gaseous. The hydrogen compounds like Na_2H, BaH, &c. are distinguished by other signs. They resemble alloys. They show (*see* end of last chapter) a systematic harmony, but they evidently should not be confused with true hydrides, any more than peroxides with saline oxides. Moreover, such hydrides have, like the peroxides, only recently been subjected to research, and have been but little studied. The best known of these compounds are given in the 16th column of Table III., and it will be seen that they already exhibit a periodicity of properties and composition.

[12 bis] The relation between the atomic weights, and especially the difference = 16, was observed in the sixth and seventh decades of this century by Dumas, Pettenkofer, L. Meyer, and others. Thus Lothar Meyer in 1864, following Dumas and others, grouped together the tetravalent elements carbon and silicon; the trivalent elements nitrogen, phosphorus, arsenic, antimony, and bismuth; the bivalent oxygen, sulphur, selenium, and tellurium; the univalent fluorine, chlorine, bromine, and iodine; the univalent metals lithium, sodium, potassium, rubidium, cæsium, and thallium, and the bivalent metals beryllium, magnesium, strontium and barium—observing that in the first the difference is, in general = 16, in the second about = 46, and the last about = 87–90. The first germs of the periodic law are visible in such observations as these. Since its establishment this subject has been most fully worked out by Ridberg (Note 10), who observed a periodicity in the variation of the differences between the atomic weights of two contiguous elements, and its relation to their atomicity. A. Bazaroff (1887) investigated the same subject, taking, not the arithmetical differences of contiguous and analogous elements, but the ratio of their atomic weights; and he also observed that this ratio alternately rises and falls with the rise of the atomic weights. I will here

remark that the relation of the eighth group to the others will be considered at the end of this work in Chapter XXII.

The laws of nature admit of no exceptions, and in this they clearly differ from such rules and maxims as are found in grammar, and other inventions, methods, and relations of man's creation. The confirmation of a law is only possible by deducing consequences from it, such as could not possibly be foreseen without it, and by verifying those consequences by experiment and further proofs. Therefore, when I conceived the periodic law, I (1869–1871, Note 9) deduced such logical consequences from it as could serve to show whether it were true or not. Among them was the prediction of the properties of undiscovered elements and the correction of the atomic weights of many, and at that time little known, elements. Thus uranium was considered as trivalent, $U = 120$; but as such it did not correspond with the periodic law. I therefore proposed to double its atomic weight—$U = 240$, and the researches of Roscoe, Zimmermann, and others justified this alteration (Chapter XXI.). It was the same with cerium (Chapter XVIII.) whose atomic weight it was necessary to change according to the periodic law. I therefore determined its specific heat, and the result I obtained was verified by the new determinations of Hillebrand. I then corrected certain formulæ of the cerium compounds, and the researches of Rammelsberg, Brauner, Clève, and others verified the proposed alteration. It was necessary to do one or the other—either to consider the periodic law as completely true, and as forming a new instrument in chemical research, or to refute it. Acknowledging the method of experiment to be the only true one, I myself verified what I could, and gave every one the possibility of proving or confirming the law, and did not think, like L. Meyer (Liebig's *Annalen, Supt. Band 7*, 1870, 364), when writing about the periodic law that 'it would be rash to change the accepted atomic weights on the basis of so uncertain a starting-point.' ('Es würde voreilig sein, auf so unsichere Anhaltspunkte hin eine Aenderung der bisher angenommenen Atomgewichte vorzunehmen.') In my opinion, the basis offered by the periodic law had to be verified or refuted, and experiment in every case verified it. The starting-point then became general. No law of nature can be established without such a method of testing it. Neither De Chancourtois, to whom the French ascribe the discovery of the periodic law, nor Newlands, who is put forward by the English, nor L. Meyer, who is now cited by many as its founder, ventured to foretell the *properties* of undiscovered elements, or to alter the 'accepted atomic weights,' or, in general, to regard the periodic law as a new, strictly established law of nature, as I did from the very beginning (1869).

When in 1871 I wrote a paper on the application of the periodic law to the determination of the properties of hitherto undiscovered elements, I did not think I should live to see the verification of this consequence of the law, but such was to be the case. Three elements were described—ekaboron, ekaaluminium, and ekasilicon—and now, after the lapse of twenty years, I have had the great pleasure of seeing them discovered and named Gallium, Scandium, and Germanium, after those three countries where the rare minerals containing them are found, and where they were discovered. For my part I regard L. de Boisbaudran, Nilson, and Winkler, who discovered these elements, as the true corroborators of the periodic law. Without them it would not have been accepted to the extent it now is.

Taking indium, which occurs together with zinc, as our example, we will show the principle of the method employed. The equivalent of indium to hydrogen in its oxide is 37·7—that is, if we suppose its composition to be like that of water; then In = 37·7, and the oxide of indium is In_2O. The atomic weight of indium was taken as double the equivalent—that is, indium was considered to be a bivalent element—and In = 2 × 37·7 = 75·4. If indium only formed an oxide, RO, it should be placed in group II. But in this case it appears that there would be no place for indium in the system of the elements, because the positions II., 5 = Zn = 65 and II., 6 = Sr = 87 were already occupied by known elements, and according to the periodic law an element with an atomic weight 75 could not be bivalent. As neither the vapour density nor the specific heat, nor even the isomorphism (the salts of indium crystallise with great difficulty) of the compounds of indium were known, there was no reason for considering it to be a bivalent metal, and therefore it might be regarded as trivalent, quadrivalent, &c. If it be trivalent, then In = 3 × 37·7 = 113, and the composition of the oxide is In_2O_3, and of its salts InX_3. In this case it at once falls into its place in the system, namely, in group III. and 7th series, between Cd = 112 and Sn = 118, as an analogue of aluminium or dvialuminium (dvi = 2 in Sanskrit). All the properties observed in indium correspond with this position; for example, the density, cadmium = 8·6, indium = 7·4, tin = 7·2; the basic properties of the oxides CdO, In_2O_3, SnO_2, successively vary, so that the properties of In_2O_3 are intermediate between those of CdO and SnO_2 or Cd_2O_2 and Sn_2O_4. That indium belongs to group III. has been confirmed by the determination of its specific heat, (0·057 according to Bunsen, and 0·055 according to me) and also by the fact that indium forms alums like aluminium, and therefore belongs to the same group.

The same kind of considerations necessitated taking the atomic weight of titanium as nearly 48, and not as 52, the figure derived from many analyses. And both these corrections, made on the basis of the law, have now been confirmed, for Thorpe found, by a series of careful experiments, the atomic weight of titanium to be that foreseen by the periodic law. Notwithstanding that previous analyses gave Os = 199·7, Ir = 198, and Pt

= 187, the periodic law shows, as I remarked in 1871, that the atomic weights should rise from osmium to platinum and gold, and not fall. Many recent researches, and especially those of Seubert, have fully verified this statement, based on the law. Thus a true law of nature anticipates facts, foretells magnitudes, gives a hold on nature, and leads to improvements in the methods of research, &c.

Meyer, Willgerodt, and others, guided by the fact that Gustavson and Friedel had remarked that metalepsis rapidly proceeds in the presence of aluminium, investigated the action of nearly all the elements in this respect. For example, they took benzene, added the metals to be experimented on to it, and passed chlorine through the liquid in diffused light. When, for instance, sodium, potassium, barium, &c. are taken, there is no action on the benzene; that is, hydrochloric acid is not disengaged; but if aluminium, gold, or, in general, any metal having this power of aiding chlorination (Halogen-überträger) is employed, then the action is clearly seen from the volumes of hydrochloric acid evolved (especially if the metallic chloride formed is soluble in benzene). Thus, in group I., and in general among the even and light elements, there are none capable of serving as agents of metalepsis; but aluminium, gallium, indium, antimony, tellurium, and iodine, which are contiguous members in the periodic system, are excellent transmitters (carriers) of the halogens.

The periodic relations enumerated above appertain to the real elements, and not to the elements in the free state as we know them; and it is very important to note this, because the periodic law refers to the real elements, inasmuch as the atomic weight is proper to the real element, and not to the 'free' element, to which, as to a compound, a molecular weight is proper. Physical properties are chiefly determined by the properties of molecules, and only indirectly depend on the properties of the atoms forming the molecules. For this reason the periods, which are clearly and quite distinctly expressed—for instance, in the forms of combination—become to some extent involved (complicated) in the physical properties of their members. Thus, for instance, besides the *maxima* and *minima* corresponding with the periods and groups, new molecules appear; thus, as regards the melting-point of germanium, a local maximum appears, which was, however, foreseen by the periodic law when the properties of germanium (ekasilicon) were forecast.

[17 bis] The relation of certain elements (for instance, the analogues of Pt) among diamagnetic and paramagnetic bodies is sometimes doubtful (probably partly owing to the imperfect purity of the reagents under investigation). This subject has been studied in some detail by Bachmetieff in 1889.

It is evident that many of the figures, especially those exceeding 1000°, have been determined with but little exactitude, and some, placed in Table III. with the sign (?), I have only given on the basis of rough and comparative determinations, calculated from the melting-points of silver and platinum, now established by many observers. In Table III., besides the large periods whose maxima correspond with carbon, silicon, titanium, ruthenium (?), and osmium (?), there are also small periods in the melting-points, and their maxima correspond with sulphur, arsenic, antimony. The minima correspond with the halogens and metals of the alkalis. A distinct periodicity is also seen in taking the coefficients of linear expansion (chiefly according to Fizeau); for instance, in the vertical series (according to the magnitude of the atomic weight), Fe, Co, Ni, Cu, the linear expansion in millionths of an inch = 12, 13, 17, and 29; for Rh, Pd, Ag, Cd, In, Sn, and Sb the coefficients are 8, 12, 19, 31, 46, 26, and 12, so that a maximum is reached at In. In the series Ir (7), Pt (5), Au (14), Hg (60), Tl (31), Pb (29), and Bi (14), the maximum is at Hg and the minimum at Pt. Raoul Pictet expressed this connection by the fact that he found the product $\alpha(t + 273)\sqrt[3]{(A/d)}$ to be nearly constant for all elements in the free state, and nearly equal to 0·045, and being the coefficient of linear expansion, $t + 273$, the melting-point calculated from the absolute zero (-273°), and $\sqrt[3]{(A/d)}$, the mean distance between the atoms, if A is the atomic weight and d the sp. gr. of an element. Although the above product is not strictly constant, nevertheless Pictet's rule gives an idea of the bond between magnitudes which ought to have a certain connection with each other. De Heen, Nadeschdin, and others also studied this dependence, but their deductions do not give a general and exact law.

Carnelley found a similar dependence in comparing the melting-points of the metallic chlorides, many of which he re-determined for this purpose. The melting-points (and boiling-points, in brackets) of the following chlorides are known, and a certain regularity is seen to exist in them, although the number (and degree of accuracy) of the data is insufficient for a generalisation:—

LiCl 598° BeCl$_2$ 600° BCl$_3$ -20°

NaCl 772° MgCl$_2$ 708° AlCl$_3$ 187°

KCl 734° CaCl$_2$ 719° ScCl$_2$?

⎡ CuCl 434° ZnCl$_2$ 262° GaCl$_3$ 76°

⎣ (993°)   (680°)   (217°)

AgCl 451°   CdCl$_2$ 541°   InCl$_3$?

TlCl 427°   PbCl$_2$ 498°   BiCl$_3$ 227°

(713°)   (908°)

We will also enumerate the following data given by Carnelley, which are interesting for comparison: HCl -112° (-102°); RbCl 710°, SrCl$_2$ 825°, CsCl 631°, BaCl$_2$ 860°, SbCl$_3$ 73° (223°), TeCl$_2$ 209° (327°), ICl 27°, HgCl$_2$ 276° (303°), FeCl$_3$ 306°, NbCl$_5$ 194° (240°), TaCl$_3$ 211° (242°), WCl$_6$ 190°. The melting-points of the bromides and iodides are higher or lower than those of the corresponding chlorides, according to the atomic weight of the element and number of atoms of the halogen, as is seen from the following examples:—1. KCl 734°, KBr 699°, KI 634°; 2. AgCl 454°, AgBr 427° AgI 527°; 3. PbCl$_2$ 498° (900°), PbBr$_2$ 499° (861°), PbI$_2$ 383° (906°); 4. SnCl$_4$ below -20° (114°), SnBr$_2$ 30° (201°), SnI$_4$ 146° (295°) (*see* Chapter II. Note 27, and Chapter XI. Note 47bis, &c.)

Laurie (1882) also observed a periodicity in the *quantity of heat* developed in the formation of the chlorides, bromides, and iodides (fig. 79), as is seen from the following figures, where the heat developed is expressed in thousands of calories, and referred to a molecule of chlorine, Cl$_2$, so that the heat of formation of KCl is doubled, and that of SnCl$_4$ halved, &c.: Na 195 (Ag 59, Au 12), Mg 151 (Zn 97, Cd 93, Hg 63), Al 117, Si 79 (Sn 64), K 211 (Li 187), Ca 170 (Sr 185, Ba 194), whence it is seen that the greatest amount of heat is evolved by the metals of the alkalis, and that in each period it falls from them to the halogens, which evolve very little heat in combining together. Richardson, by comparing the heats of formation of the fluorides also came to the conclusion that they are in periodic dependence upon the atomic weights of the combined elements.

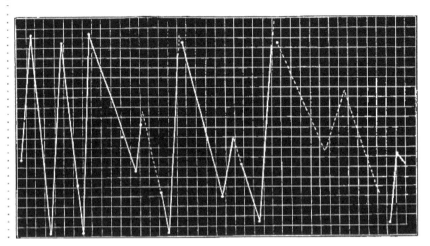

FIG. 79.—Laurie's diagram for expressing the periodic variation of the heat of formation of the chlorides. The abscissæ give the atomic weights from 0 to 210, and the ordinates the amounts of heat from 0 to 220 thousand calories evolved in the combination with Cl_2, (*i.e.* with 71 parts of chlorine). The apices of the curve correspond to Li, Na, K, Rb, Cs, and the lower extremities to F, Cl, Br, and I.

In this respect it may not be superfluous to remark (1) that Thomsen, whose results I have employed above, observed a correlation in the calorific equivalents of analogous elements, although he did not remark their periodic variation; (2) that the uniformity of many thermochemical deductions must gain considerably by the application of the periodic law, which evidently repeats itself in calorimetric data; and if these data frequently lead to true forecasts, this is due to the periodicity of the thermal as well as of many other properties, as Laurie remarked; and (3) that the heat of formation of the oxides is also subject to a periodic dependence which differs from that of the heat of formation of the chlorides, in that the greatest quantity corresponds with the bivalent metals of the alkaline earths (magnesium, calcium, strontium, barium), and not with the univalent metals of the alkalis, as is the case with chlorine, bromine, and iodine. This circumstance is probably connected with the fact that chlorine, bromine, and iodine are univalent elements, and oxygen bivalent (compare, for instance, Chapter XI., Note 13, Chapter XXII., Note 40, Chapter XXIV., Note 28[bis], &c.)

Keyser (1892), in investigating the spectra of the alkali metals and metals of the alkaline earths, came to the conclusion that in this respect also there is a regularity of a periodic character in dependence upon the atomic weights. Probably a closer and systematic study of many of the properties of the elements and of complex and simple bodies formed by them will more and more frequently lead to similar conclusions, and to extending the range of application of the periodic law.

[19 bis] Probably, besides thermo-chemical data (Note 19), the refractive index, cohesion, ductility, and similar properties of corresponding compounds or of the elements themselves will be found to exhibit a dependence of the magnitude of the atomic weight upon the periodic law.

Having occupied myself since the fifties (my dissertation for the degree of M.A. concerned the specific volumes, and is printed in part in the *Russian Mining Journal* for 1856) with the problems concerning the relations between the specific gravities and volumes, and the chemical compositions of substances, I am inclined to think that the direct investigation of specific gravities gives essentially the same results as the investigation of specific volumes, only that the latter are more graphic. Table III. of the periodic properties of the elements clearly illustrates this. Thus, for those members whose volume is the greatest among the contiguous elements, the specific gravity is least—that is, the periodic variation of both properties is equally evident. In passing, for instance, from silver to iodine we have a successive decrease of specific gravity and successive increase of specific volume. The periodic alternation of the rise and fall of the specific gravity and specific volume of the free elements was communicated by me in August 1869 to the Moscow Meeting of Russian Naturalists. In the following year (1870) L. Meyer's paper appeared, which also dealt with the specific volume of the elements.

In my opinion the mean volume of the atoms of compounds deserves more attention than has yet been paid to it. I may point out, for instance, that for feebly energetic oxides the mean volume of the atom is generally nearly 7; for example, the oxides SiO_2, Sc_2O_3, TiO_2, V_2O_5, as well as ZnO, Ga_2O_3, GeO_2, ZrO_2, In_2O_3, SnO_2, Sb_2O_5, &c., whilst the mean volume of the atom of the alkali and acid oxides is greater than 7. Thus we find in the magnitudes of the mean volumes of the atom in oxides and salts both a periodic variation and a connection with their energy of essentially the same character as occurs in the case of the free elements.

The volume of oxygen (judging by the table on p. 36) is evidently a variable quantity, forming a distinctly periodic function of the atomic weight and type of the oxide, and therefore the efforts which were formerly made to find the volume of the atom of oxygen in the volumes of its compounds may be considered to be futile. But since a distinct contraction takes place in the formation of oxides, and the volume of an oxide is frequently less than the volume in the free state of the element contained in it, it might be surmised that the volume of oxygen in a free state is about

15, and therefore the specific gravity of solid oxygen in a free state would be about O·9.

As an example we will take indium oxide, In_2O_3. Its sp. gr. and sp. vol. should be the mean of those of cadmium oxide, Cd_2O_2, and stannic oxide, Sn_2O_4, as indium stands between cadmium and tin. Thus in the seventies it was already evident that the volume of indium oxide should be about 38, and its sp. gr. about 7·2, which was confirmed by the determinations of Nilson and Pettersson (7·179) made in 1880.

As the distance between, and the volumes of, the molecules and atoms of solids and liquids certainly enter into the data for the solution of the problems of molecular mechanics, which as yet have only been worked out to any extent for the gaseous state, the study of the specific gravity of solids, and especially of liquids, has long had an extensive literature. With respect to solids, however, a great difficulty is met with, owing to the specific gravity varying not only with a change of isomeric state (for example, for silica in the form of quartz = 2·65, and in tridymite = 2·2) but also directly under mechanical pressure (for example, in a crystalline, cast, and forged metal), and even with the extent to which they are powdered, &c., which influences are imperceptible in liquids. Compare Chapter XIV., Note 55[bis].

Without going into further details, we may add to what has been said above that the conception of specific volumes and atomic distances has formed the subject of a large number of researches, but as yet it is only possible to lay down a few generalisations given by Dumas, Kopp, and others, which are mentioned and amplified by me in my work cited in Note 20, and in my memoirs on this subject.

1. Analogous compounds and their isomorphs have frequently approximately the same molecular volumes.

2. Other compounds, analogous in their properties, exhibit molecular volumes which increase with the molecular weight.

3. When a contraction takes place in combination in a gaseous state, then contraction is in the majority of instances also to be observed in the solid or liquid state—that is, the sum of the volumes of the reacting substances is greater than the volume of the resultant substance or substances.

4. In decomposition the reverse takes place to that which occurs in combination.

5. In substitution (when the volumes in a state of vapour do not vary) a very small change of volume generally takes place—that is, the sum of the

volumes of the reacting substances is almost equal to the sum of the resultant substances.

6. Hence it is impossible to judge the volume of the component substances from the volume of a compound, although it is possible to do so from the product of substitution.

7. The replacement of H_2 by sodium, Na_2, and by barium, Ba, as well as the replacement of SO_4 by Cl_2, scarcely changes the volume, but the volume increases with the replacement of Na by K, and decreases with the replacement of H_2, by Li_2 Cu, and Mg.

8. There is no need for comparing volumes in a solid and liquid state at the so-called corresponding temperatures—that is at temperatures at which vapour tension is equal in each case. The comparison of volumes at the ordinary temperature is sufficient for finding a regularity in the relations of volumes (this deduction was developed with particular detail by me in 1856).

9. Many investigators (Perseau, Schröder, Löwig, Playfair and Joule, Baudrimont, Einhardt) have sought in vain for a multiple proportion in the specific volumes of solids and liquids.

10. The truth of the above is seen very clearly in comparing the volumes of polymeric substances. The volumes of their molecules are equal in a state of vapour, but are very different in a solid and liquid state, as is seen from the close resemblance of the specific gravities of polymeric substances. But as a rule the more complex polymerides are denser than the simpler.

11. We know that the hydroxides of light metals have generally a smaller volume than the metals, whilst that of magnesium hydroxide is considerably greater, which is explained by the stability of the former and instability of the latter. In proof of this we may cite, besides the volumes of the true alkali metals, the volume of barium (36) which is greater than that of its stable hydroxide (sp. gr. 4·5, sp. vol. 30). The volumes of the salts of magnesium and calcium are greater than the volume of the metal, with the single exception of the fluoride of calcium. With the heavy metals the volume of the compound is always greater than the volume of the metal, and, moreover, for such compounds as silver iodide, AgI ($d = 5·7$), and mercuric iodide, HgI_2 ($d = 6·2$, and the volumes of the compounds 41 and 73), the volume of the compound is greater than the sum of the volumes of the component elements. Thus the sum of the volumes Ag + I = 36, and the volume of AgI = 41. This stands out with particular clearness on comparing the volumes K + I = 71 with the volume of KI, which is equal to 54, because its density = 3·06.

12. In such combinations, between solids and liquids, as solutions, alloys, isomorphous mixtures, and similar feeble chemical compounds, the sum of the reacting substances is always very nearly that of the resulting substance, but here the volume is either slightly larger or smaller than the original; speaking generally, the amount of contraction depends on the force of affinity acting between the combining substances. I may here observe that the present data respecting the specific volumes of solid and liquid bodies deserve a fresh and full elaboration to explain many contradictory statements which have accumulated on this subject.

CHAPTER XVI
ZINC, CADMIUM, AND MERCURY

These three metals give, like magnesium, oxides RO, which form feebly energetic bases, and like magnesium they are volatile. The volatility increases with the atomic weight. Magnesium can be distilled at a white heat, zinc at a temperature of about 930°, cadmium about 770°, and mercury about 351°. Their oxides, RO, are more easily reducible than magnesia, and mercuric oxide is the most easily reducible. The properties of their salts RX_2 are very similar to the properties of MgX_2. Their solubility, power of forming double and basic salts, and many other qualities are in many respects identical with those of MgX_2. The greater or less ease with which they are oxidised, the instability of their compounds, the density of the metals and their compounds, their scarcity in nature, and many other properties gradually change with the increase of atomic weight, as might be expected from the periodicity of the elements. Their principal characteristics, as contrasted with magnesium, find a general expression in the fact that zinc, cadmium, and mercury are heavy metals.

Zinc stands nearest to magnesium in atomic weight and in properties. Thus zinc sulphate, or white vitriol, easily crystallises with seven molecules of water, $ZnSO_4,7H_2O$. It is isomorphous with Epsom salts, and parts with difficulty with the last molecule of water; it forms double salts—for instance, $ZnK_2(SO_4)_2,6H_2O$—exactly as magnesium sulphate does. *Zinc oxide*, ZnO, is a white powder, almost insoluble in water, like magnesia, from which, however, it is distinguished by its solubility in solutions of sodium and potassium hydroxides. Zinc chloride is decomposed by water, combines with ammonium chloride, potassium chloride, &c., just like magnesium chloride, forms an oxychloride, and also combines with zinc oxide.[4 bis]

Zinc, like many heavy metals, is often *found in nature in combination with sulphur*, forming the so-called *zinc blende*, ZnS. It sometimes occurs in large masses, often crystallised in cubes; it is frequently translucent, and has a metallic lustre, although this is not so clearly developed as in many other metallic sulphides with which we shall hereafter become acquainted. The ores of zinc also comprise the carbonate, calamine, and silicate, *siliceous calamine*.

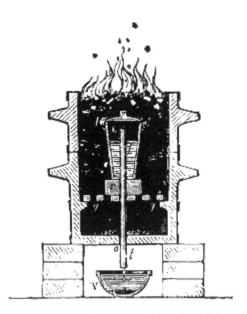

FIG. 80.—Distillation of zinc in a crucible placed in a furnace. *o c*, tube along which the vapour passes and condenses.

Metallic zinc (spelter) is most frequently obtained from the ores containing the carbonate—that is, from calamine, which is sometimes found in thick veins: for instance, in Poland, Galicia, in some places on the banks of the Rhine, and in considerable masses in Belgium and England. In Russia beds of zinc ore are met with in Poland and the Caucasus, but the output is small. In Sweden, as early as the fifteenth century, calamine was worked up into an alloy of zinc and copper (brass), and Paracelsus produced zinc from calamine; but the technical production of the metal itself, long ago practised in China, only commenced in Europe in 1807—in Belgium, when the Abbé Donnet discovered that zinc was volatile. From that time the production increased until it is now about 150 million kilograms in Germany alone.

The reduction of metallic zinc from its ores is based on the fact that zinc oxide is easily reduced by charcoal at a red heat: $ZnO + C = Zn + CO$. The zinc thus obtained is in a finely divided state and impure, being mixed with other metals reduced with it, but the greater portion is *converted into vapour*, from which it easily passes into a liquid or solid state. The reduction and distillation are carried on in earthenware retorts, filled with a mixture of the divided ore and charcoal. The vapours of zinc and gases formed during the reaction escape by means of a pipe leading downwards, and are led to a chamber where the vapours are cooled. By this means they do not come into contact with the air, because the neck of the retort is filled with gaseous carbonic oxide, and therefore the zinc does not oxidise; otherwise its vapour would burn in the air.[7 bis] The vapours of zinc, entering into the cooling chamber, condense into white zinc powder or zinc dust. When the neck of the retort is heated the zinc is obtained in a liquid state, and is cast into plates, in which form it is generally sold.

Commercial zinc is generally impure, containing a mixture of lead, particles of carbon, iron, and other metals carried over with the vapours, although they are not volatile at a temperature approaching 1000°. If it be required to obtain pure zinc from the commercial article, it is subjected to a further distillation in a crucible with a pipe passing through the bottom, the vapours formed by the heated zinc only having exit through the pipe cemented into the bottom of the crucible. Passing through this pipe, the vapours condense to a liquid, which is collected in a receiver. Zinc thus purified is generally re-melted and cast into rods, and in this form is often used for physical and chemical researches where a pure article is required.

Metallic zinc has a bluish-white colour; its lustre, compared with many other metals, is insignificant. When cast it exhibits a crystalline structure. Its specific gravity is about 7—that is, varies from 6·8 to 7·2, according to the degree of compression (by forging, rolling, &c.) to which it has been subjected. It is very ductile, considering its hardness. For this reason it chokes up files when being worked. Its malleability is considerable when pure, but in the ordinary impure condition in which it is sold, it is impossible to roll it at the ordinary temperature, as it easily breaks. At a temperature of 100°, however, it easily undergoes such operations, and can then be drawn into wire or rolled into sheets. If heated further it again becomes brittle, and at 200° may be even crushed into powder, so completely does it lose its molecular cohesion. It melts at 418°, and distils at 930°.

Zinc does not undergo any change in the atmosphere. Even in very damp air it only becomes slowly coated with a very thin white coating of oxide. For this reason it is available for all objects which are only in contact with air. Therefore sheet zinc may be used for roofing and many other purposes. This great unchangeability of zinc in the air shows its slight energy with regard to oxygen compared with the metals already mentioned, which are capable of reducing zinc from solutions. But zinc plays this part with regard to the remaining metals—for example, it reduces salts of lead, copper, mercury, &c. Although zinc is an almost unoxidisable metal at the ordinary temperature, it burns in the air on being heated, particularly when in the form of shavings or in the condition of vapour. At the ordinary temperature zinc does not decompose water—at any rate, if the metal be in a dense mass. But even at a temperature of 100° zinc begins little by little to decompose water; it easily displaces the hydrogen of acids at the ordinary temperature, and of alkalis on being heated.

In this respect the action of zinc varies a great deal with the degree of its purity. Weak sulphuric acid (corresponding with the composition $H_2SO_4,8H_2O$) at the ordinary temperature does not act at all on chemically pure zinc, and even a stronger solution acts very slowly. If the temperature be raised, and particularly if the zinc be previously slightly heated, so as to cover the surface with a film of oxide, chemically pure zinc acts on sulphuric acid. Thus, for example, one cubic centimetre of zinc in sulphuric acid having a composition $H_2SO_4,6H_2O$ at the ordinary temperature in two hours only dissolves to the extent of 0·018 gram, and at a temperature of 100° about 3·5 grams. If we compare this slow action with that rapid evolution of hydrogen which occurs in the case of commercial zinc, we see that the influence of those impurities in the zinc is very great. Every particle of charcoal or iron introduced into the mass of the zinc, and likewise the connection of the zinc with a piece of another electro-negative metal, assists such a dissolution. The slowness of the action of sulphuric acid on pure zinc (and likewise on amalgamated zinc) may also be explained by the fact that a layer of hydrogen collects on the surface of the metal, preventing contact between the acid and the metal.[10 bis]

The action of zinc on acids, and the consequent formation of zinc salts, interferes with its application in many cases, particularly for the preservation of liquids either containing or capable of developing acid. For this reason zinc vessels ought not to be used for the preparation or preservation of food, as this often contains acids which form poisonous salts with the zinc. Even ordinary water, containing carbonic acid, slowly attacks zinc.

Finely divided zinc, or *zinc dust*, obtained in the distillation of the metal when the receiver is not heated up to the melting point, on account of its presenting a large surface of contact and containing foreign matter (particularly zinc oxide), has in the highest degree the property of decomposing acids, and even water, which it easily decomposes, particularly if slightly heated. On this account zinc dust is often used in laboratories and factories as a reducing agent. A similar influence of the finely divided state is also noticed in other metals—for instance, copper and silver—which again shows the close connection between chemical and physico-mechanical phenomena. We must first of all turn to this close connection for an explanation of the widely spread application of zinc in galvanic batteries, where the chemical (latent, potential) energy of the acting substances is transformed into (evident, kinetic) galvanic energy, and through this latter into heat, light, or mechanical work.

Hermann and Stromeyer, in 1819, showed that *cadmium* is almost always found with zinc, and in many respects resembles it. When distilled the cadmium volatilises sooner, because it has a lower boiling point. Sometimes the zinc dust obtained by the first distillation of zinc contains as much as 5 per cent. of cadmium. When zinc blende, containing cadmium, is roasted, the zinc passes into the state of oxide, and the cadmium sulphide in the ore oxidises into cadmium sulphate, $CdSO_4$, which resists tolerably well the action of heat; therefore if roasted zinc blende be washed with water, a solution of cadmium sulphate will be obtained, from which it is very easy to prepare metallic cadmium. Hydrogen sulphide may be used for separating cadmium from its solutions; it gives *a yellow precipitate of cadmium sulphide*, CdS (according to the equation $CdSO_4 + H_2S = H_2SO_4 + CdS$), which, on account of its characteristic colour, is used as a pigment.[11 bis] Cadmium sulphide, when strongly heated in air, leaves cadmium oxide, from which the metal may be obtained in precisely the same way as in the case of zinc.

Cadmium is a white metal, and when freshly cut is almost as white and lustrous as tin. It is so soft that it may be easily cut with a knife, and so malleable that it can be easily drawn into wire, rolled into sheets, &c. Its specific gravity is 8·67, melting point 320°, boiling point 770°; its vapours burn, forming a brown powder of the oxide. Next to mercury it is the most volatile metal; hence Deville determined the density of its vapours compared with hydrogen, and found it to be equal to 57·1; therefore the molecule contains *one atom* whose weight = 112. V. Meyer found the like for zinc; the molecule of mercury also contains one atom.

Mercury resembles zinc and cadmium in many respects, but presents that distinction from them which is always noticed in all the heaviest metals (with regard to atomic weight and density) compared with the lighter ones—namely, that it oxidises with more difficulty, and its compounds are more easily decomposed.[12 bis] Besides compounds of the usual type RX_2, it also gives those of the lower type, RX, which are unknown for zinc and cadmium. Mercury therefore gives salts of the composition HgX (mercurous salts) and HgX_2 (mercuric salts), the oxides having the formulæ Hg_2O and HgO respectively.

Mercury is found *in nature* almost exclusively in combination with sulphur (like zinc and cadmium, but is still rarer than them) in the form known as cinnabar, HgS (Chapter XX., Note 29). It is far more rarely met with in the native or metallic condition, and this in all probability has been derived from cinnabar. Mercury ore is found only in a few places—namely, in Spain (in Almaden), in Idria, Japan, Peru, and California. About the year 1880 Minenkoff discovered a rich bed of cinnabar in the Bahmout district (near the station of Nikitovka), in the Government of Ekaterinoslav, so that now Russia even exports mercury to other countries. Cinnabar is now being worked in Daghestan in the Caucasus. Mercury ores are easily reduced to metallic mercury, because the combination between the metal and the sulphur is one of but little stability. Oxygen, iron, lime, and many other substances, when heated, easily destroy the combination. If iron is heated with cinnabar, iron sulphide is formed; if cinnabar is heated with lime, mercury and calcium sulphide and sulphate are formed, $4HgS + 4CaO = 4Hg + 3CaS + CaSO_4$. On being heated in the air, or roasted, the sulphur burns, oxidises, forming sulphurous anhydride, and vapours of metallic mercury are formed. Mercury is more easily distilled than all other metals, its boiling point being about 351°, and therefore its separation from natural admixtures, decomposed by one of the above-mentioned methods, is effected at the expense of a comparatively small amount of heat. The mixture of mercury vapour, air, and products of combustion obtained is cooled in tubes (by water or air), and the mercury condenses as liquid metal.

Mercury, as everybody knows, is a liquid metal at the ordinary temperature. In its lustre and whiteness it resembles silver. At -39° mercury is transformed into a malleable crystalline metal; at 0° its specific gravity is 13·596, and in the solid state at -40° it is 14·39. Mercury does not change in the air—that is to say, it does not oxidise at the ordinary temperature—but at a temperature approaching the boiling-point, as was stated in the Introduction, it oxidises, forming mercuric oxide. Both metallic mercury and its compounds in general produce salivation, trembling of the hands, and other unhealthy symptoms which are found in the workmen exposed to the influence of mercurial vapours or the dust of its compounds.

As many of the compounds of mercury decompose on being heated—for instance, the oxide or carbonate—and as zinc, cadmium, copper, iron, and other metals separate mercury from its salts, it is evident that mercury has less chemical energy than the metals already described, even than zinc and cadmium. Nitric acid, when acting on *an excess* of mercury at the ordinary temperature, gives mercurous nitrate, $HgNO_3$. The same acid, under the influence of heat and when in excess (nitric oxide being liberated), forms mercuric nitrate, $Hg(NO_3)_2$. This, both in its composition and properties, resembles the salts of zinc and cadmium. Dilute sulphuric acid does not act on mercury, but strong sulphuric acid dissolves it, with evolution of *sulphurous anhydride* (not hydrogen), and on being slightly heated with an excess of mercury it forms the sparingly soluble mercurous sulphate, Hg_2SO_4; but if mercury be strongly heated with an excess of the acid, the mercuric salt, $HgSO_4$, is formed. Alkalis do not act on mercury, but the non-metals chlorine, bromine, sulphur, and phosphorus easily combine with it. They form, like the acids, two series of compounds, HgX and HgX$_2$. The oxygen compound of the first series is the suboxide of mercury, or mercurous oxide, Hg$_2$O, and of the second order the oxide HgO, mercuric oxide. The chlorine compound corresponding with the suboxide is HgCl (calomel), and with the oxide HgCl$_2$ (corrosive sublimate or mercuric chloride). In the compounds HgX, mercury resembles the metals of the first group, and more especially silver. In the mercuric compounds there is an evident resemblance to those of magnesium, cadmium, &c. Here the atom of mercury is bivalent, as in the type RX$_2$. Every soluble mercurous compound (corresponding with the type of the suboxide of mercury), HgX, forms a white precipitate of calomel, HgCl, with hydrochloric acid or a metallic chloride, because HgCl is very slightly soluble in water, HgX + MCl = HgCl + MX. In soluble mercuric compounds, HgX$_2$, hydrochloric acid and metallic chlorides do not form a precipitate, because corrosive sublimate, HgCl$_2$, is soluble in water. Alkali hydroxides precipitate the yellow mercuric oxide from a solution of HgX$_2$, and the black mercurous oxide from HgX. Potassium iodide forms a dirty greenish precipitate, HgI, with mercurous salts, HgX, and a red precipitate, HgI$_2$, with the mercuric salts, HgX$_2$. These reactions distinguish the mercuric from the mercurous salts, which latter represent the transition from the mercuric salts to mercury itself, 2HgX = Hg + HgX$_2$. The salts, HgX, as well as HgX$_2$, are reduced by nascent hydrogen (*e.g.* from Zn + H$_2$SO$_4$), by such metals as zinc and copper, and also by many reducing agents—for example, hypophosphorous acid, the lowest grade of oxidation of phosphorus, by sulphurous anhydride, stannous chloride, &c. Under the action of these reagents the mercuric salts are first transformed into the mercurous salts, and the latter are then reduced to metallic mercury. This reaction is so delicate that it serves to detect the smallest quantity of mercury; for instance, in cases of poisoning, the mercury is detected by immersing a copper plate in the solution to be tested, the mercury being then deposited upon it (more readily on passing a galvanic current). The copper plate, on being rubbed, shows a silvery white colour; on being heated, it yields vapours of mercury, and then again assumes its original red colour (if it does not oxidise). The mercurous compounds, HgX, under the action of oxidising agents, even air, pass into mercuric compounds, especially in the presence of acids (otherwise a basic salt is produced), 2HgX + 2HX + O = 2HgX$_2$ + H$_2$O; but the mercuric compounds, when in contact with mercury, change more or less readily, and turn into mercurous compounds, HgX$_2$ + Hg = 2HgX. For this reason, in order to preserve solutions of mercurous salts, a little mercury is generally added to them.

The lowest oxygen compound of mercury—that is, *mercurous oxide*, Hg$_2$O—does not seem to exist, for the substance precipitated in the form of a black mass by the action of alkalis on a solution of mercurous salts gradually separates on keeping into the yellow mercuric oxide and metallic mercury, as does also a simple mechanical mixture of oxide, HgO, with mercury (Guibourt, Barfoed). The other compound of mercury with oxygen is already known to us as *mercuric oxide*, HgO, obtained in the form of a red crystalline substance by the oxidation of mercury in the air, and precipitated as a yellow powder by the action of sodium hydroxide on solutions of salts of the type HgX_2. In this case it is amorphous and more amenable to the action of various reagents (Chap. XI., Note 32) than when it is in the crystalline state. Indeed, on trituration, the red oxide is changed into a powder of a yellow colour. It is very sparingly soluble in water, and forms an alkaline solution which precipitates magnesia from the solution of its salts.

Mercury combines directly with chlorine, and the first product of combination is *calomel* or *mercurous chloride*, Hg_2Cl_2. This is obtained, as above stated, in the form of a white precipitate by mixing solutions of mercurous salts with hydrochloric acid or with metallic chlorides. A precipitate of calomel is also obtained by reducing a boiling aqueous solution of corrosive sublimate, $HgCl_2$, with sulphurous anhydride. It is likewise produced by heating corrosive sublimate with mercury.[22 bis] Calomel may be distilled (although in so doing it decomposes and recombines on cooling from a state of vapour); its vapour density equals 118 compared with hydrogen (= 1) (*see* Note 23). In the solid state its specific gravity is 7·0; it crystallises in rhombic prisms, is colourless, but has a yellowish tint, turns brown from the action of light, and when boiled with hydrochloric acid decomposes into mercury and corrosive sublimate. It is used as a strong purgative. *Corrosive sublimate or mercuric chloride*, $HgCl_2$, can be obtained from or converted into calomel by many methods. An excess of chlorine (for instance, *aqua regia*) converts calomel and also mercury into corrosive sublimate. It owes its name corrosive sublimate to its volatility, and, in medicine up to the present day, it is termed *Mercurius sublimatus seu corrosivus*. The vapour density, compared with hydrogen (= 1) is 135; therefore its molecule contains $HgCl_2$. It forms colourless prismatic crystals of the rhombic system, boils at 307°, and is soluble in alcohol. It is usually prepared by subliming a mixture of mercuric sulphate with common salt, $HgSO_4 + 2NaCl = Na_2SO_4 + HgCl_2$. Corrosive sublimate combines with mercuric oxide, forming an oxychloride or basic salt,[23 bis] of the composition $HgCl_2,2HgO$ (magnesium and zinc form similar compounds). This compound is obtained by mixing a solution of corrosive sublimate with mercuric oxide or with a solution of sodium bicarbonate. In general, with both mercurous and mercuric salts, there is a marked tendency to form basic salts.

Mercury has a remarkable power of forming very unstable compounds with ammonia, in which the mercury replaces the hydrogen, and, if a mercuric compound be taken, its atom occupies the place of two atoms of the hydrogen in the ammonia. Thus Plantamour and Hirtzel showed that precipitated mercuric oxide dried at a gentle heat, when continuously heated (up to 100°-150°) in a stream of dry ammonia, leaves a brown powder of *mercuric nitride*, N_2Hg_3, according to the equation $3HgO + 2NH_3 = N_2Hg_3 + 3H_2O$.[24 bis] This substance, which is attacked by water, acids, and alkalis (giving a white powder), is very explosive when struck or rubbed, evolving nitrogen, proving that the bond between the mercury and the nitrogen is very feeble. By the action of liquefied ammonia on yellow mercuric oxide Weitz also obtained an explosive compound, dimercurammonium hydroxide, N_2Hg_4O, which corresponds with an ammonium oxide, $(NH_4)_2O$, in which the whole of the hydrogen is replaced by mercury. A solution of ammonia reacts with mercuric oxide, forming the hydroxide, $NHg_2.OH$, to which a whole series of salts, NHg_2X, correspond; these are generally insoluble in water and capable of decomposing with an explosion. But salts of the same type, but with one atom of mercury, NH_2HgX, are more frequently and more easily formed; they were principally studied by Kane, although known much earlier. Thus, if ammonia be added to a solution of corrosive sublimate (or, still better, in reverse order), a precipitate is obtained known as white precipitate (*Mercurius præcipitatus albus*) or *mercurammonium chloride*, NH_2HgCl, which may also be regarded not only as sal-ammoniac with the substitution of H_2 by mercury, but also as HgX_2, where one X represents Cl and the other X represents the ammonia radicle, $HgCl_2 + 2NH_3 = NH_2.HgCl + NH_4Cl$. When heated, mercurammonium chloride decomposes, yielding mercurous chloride; when heated with dry hydrochloric acid it forms ammonium chloride and mercuric chloride. Other simple and double salts of mercurammonium, NH_2HgX, are also known. Pici (1890) showed that all the compounds HgH_2NX may be regarded as compounds of the above-named Hg_2NX with NH_4X because their sum equals $2HgH_2X$.[25 bis]

Mercury as a liquid metal is capable of dissolving other metals and forming metallic solutions. These are generally called 'amalgams.' The formation of these solutions is often accompanied by the development of a large amount of heat—for instance, when potassium and sodium are dissolved (Chapter XII., Note 39); but sometimes heat is absorbed, as, for instance, when lead is dissolved. It is evident that phenomena of this kind are exceedingly similar to the phenomena accompanying the dissolution of salts and other substances in water, but here it is easy to demonstrate that which is far more difficult to observe in the case of salts: the solution of metals in mercury is accompanied by the formation of definite chemical compounds of the mercury with the metals dissolved. This is shown by the fact that when pressed (best of all in chamois leather) such solutions leave solid, definite compounds of mercury with metals. It is, however, very difficult to obtain them in a pure state, on account of the difficulty of separating the last traces of mercury, which is mechanically distributed between the crystals of the compounds. Nevertheless, in many cases such compounds have undoubtedly been obtained, and their existence is clearly shown by the evident crystalline structure and characteristic appearance of many amalgams. Thus, for instance, if about 2½ p.c. of sodium be dissolved in mercury, a hard, crystalline amalgam is obtained, very friable and little changeable in air. It contains the compound $NaHg_5$ (Chapter XII., Note 39). Water decomposes it, with the evolution of hydrogen, but more slowly than other sodium amalgams, and this action of water only shows that the bond between the sodium and the mercury is weak, just like the connection between mercury and many other elements—for instance, nitrogen. Mercury directly and easily dissolves potassium, sodium, zinc, cadmium, tin, gold, bismuth, lead, &c., and from such solutions or alloys it is in most cases easy to extract definite compounds—thus, for instance, the compounds of mercury and silver have the compositions $HgAg$ and Ag_2Hg_3. Objects made of copper when rubbed with mercury become covered with a white coating of that metal, which slowly forms an amalgam; silver acts in the same way, but more slowly, and platinum combines with mercury with still greater difficulty. This metal only readily forms an amalgam when in the form of a fine powder. If salts of platinum in solution are poured on to an amalgam of sodium, the latter element reduces the platinum, and the platinum separated is dissolved by the mercury. Almost all metals readily form amalgams if their solutions are decomposed by a galvanic current, where mercury forms the negative pole. In this way an amalgam may even be made with iron, although iron in a mass does not dissolve in mercury. Some amalgams are found in nature— for instance, silver amalgams. Amalgams are used in considerable quantities in the arts. Thus the solubility of silver in mercury is taken advantage of for extracting that metal from the ore by means of amalgamation, and for silvering by fire. The same is the case with gold. Tin amalgam, which is incapable of crystallising and is obtained by dissolving tin in mercury, composes the brilliant coating of ordinary looking-glasses, which is made to adhere to the surface of the polished glass by simply pressing by mechanical means sheets of tin foil bathed in mercury on to the cleansed surface of the glass. (*See* 'The Nature of Amalgams,' by W. L. Dudley; Toronto, 1889.)

Footnotes:

Zinc sulphate is often obtained as a by-product—for instance, in the action of galvanic batteries containing zinc and sulphuric acid. When the anhydrous salt is heated it forms zinc oxide, sulphurous anhydride, and oxygen. The solubility in 100 parts of water at 0° = 43, 20° = 53, 40° = 63½, 60° = 74, 80° = 84½, 100° = 95 parts of anhydrous zinc sulphate—that is to say, it is closely expressed by the formula $43 + 0·52t$.

An admixture of iron is often found in ordinary sulphate of zinc in the form of ferrous sulphate, $FeSO_4$, isomorphous with the zinc sulphate. In order to separate it, chlorine is passed through the solution of the impure salt (when the ferrous salt is converted into ferric), the solution is then boiled, and zinc oxide is afterwards added, which, after some time has elapsed, precipitates all the ferric oxide. Ferric oxide of the form R_2O_3 is displaced by zinc oxide of the form RO.

Zinc oxide is obtained both by the combustion and oxidation of zinc, and by the ignition of some of its salts—for instance, those of carbonic and nitric acids; it is likewise precipitated by alkalis from a solution of ZnX_2 in the form of a gelatinous hydroxide. The oxide produced by roasting zinc blende (by burning in the air, when the sulphur is converted into sulphurous anhydride) contains various impurities. For purification, the oxide is mixed with water, and the sulphurous anhydride formed by roasting the blende is passed through it. Zinc bisulphite, $ZnSO_3,H_2SO_3$, then passes into solution. If a solution of this salt be evaporated, and the residue ignited, zinc oxide, free from many of its impurities, will remain. Zinc oxide is a light white powder, used as a paint instead of *white lead*; the basic salt, corresponding with magnesia alba, is used for the same purpose. V. Kouriloff (1890) by boiling the hydrate of the oxide with a 3 p.c. solution of peroxide of hydrogen obtained $Zn_2H_2O_4$ or the hydrate of the peroxide (= $ZnO_2ZnH_2O_2$ or a compound of $2ZnO$ with H_2O_2), which did not part with its oxygen at 100°, but only above 120°. Cadmium gives a similar compound of a yellow colour. Magnesium, although it does form such a compound, does so with great difficulty.

For the solution of one part of the oxide 55,400 parts of water are required. Nevertheless, even in such a weak solution, zinc oxide (hydroxide, ZnH_2O_2) changes the colour of red litmus paper. Zinc oxide is obtained in the wet way by adding an alkali hydroxide to a solution of a zinc salt—for instance: $ZnSO_4 + 2HKO = K_2SO_4 + ZnH_2O_2$. The gelatinous precipitate of zinc hydroxide is *soluble* in an excess of alkali, which clearly distinguishes it from magnesia. This solubility of zinc hydroxide in alkalis is due to the power of zinc oxide to form a compound, although an unstable one, with alkalis—that is to say, points to the fact that zinc oxide already partly belongs to the intermediate oxides. The oxides of the metals above mentioned (except BeO) do not show this property. The property which metallic zinc itself has of dissolving in caustic alkali with the disengagement of hydrogen (the solution is facilitated by contact with platinum or iron) depends on the formation of such a compound of the oxides of zinc and the alkali metals. The solution of zinc hydroxide, ZnH_2O_2, in potash (in a strong solution), proceeds when these hydrates are taken in proportion to $ZnH_2O_2 + KHO$. If such a solution be evaporated to dryness, water extracts only caustic potash from the fused residue. When a solution of zinc hydroxide in strong alkali is mixed with a large mass of water, nearly all the oxide of zinc is precipitated; and, therefore, in weak solutions, a large quantity of the alkali is required to effect solution, which points to the decomposition of the zinc-alkali compounds by water. If strong alcohol be added to a solution of zinc oxide in sodium hydroxide, the crystallo-hydrate, $2Zn(OH)(ONa),7H_2O$, separates.

Zinc chloride, $ZnCl_2$, is generally employed in the arts in the form of a solution obtained by dissolving zinc in hydrochloric acid. This solution is used for soldering metals, impregnating wood, &c. The reason why it is thus employed may be understood from its properties. When evaporated it first parts with its water of crystallisation; on being further heated, however, it loses all traces of water, and forms an oily mass of anhydrous salt which solidifies on cooling. This substance melts at 250°, commences to volatilise at about 400°, and boils at 730°. The soldering of metals—that is, the introduction of an easily fusible metal between two contiguous metallic objects—is hindered by any film of oxide upon them; and, as heated metals easily oxidise, they are naturally difficult to solder. Zinc chloride is used to prevent the oxidation. It fuses on being heated, and, covering the metal with an oily coating, prevents contact with the air; but even if any oxide has formed, the free hydrochloric acid generally existing in the zinc chloride solution dissolves it, and in this way the metallic surface of the metals to be soldered is preserved fit for the adhesion of the liquid solder, which, on cooling, binds the objects together. Much zinc chloride is used also for steeping wood (telegraph-posts and railway-sleepers) in order to preserve it from decaying quickly; this preservative action is in all probability mainly due to the poisonous character of zinc salts (corrosive sublimate is still more poisonous, and a still better agent to preserve wood from decay), since decay is due to the action of lower organisms.

The specific gravity of solutions containing p per cent. of zinc chloride, $ZnCl_2$, is as follows:

$p =$	10	20	30	40	50
15°/4° =	1·093	1·184	1·293	1·411	1·554

| $ds/dt =$ | -3 | -5 | -7 | -8 | -9 |

The last line shows the change of specific gravity for 1° in ten-thousandth parts for temperatures near 15°. More accurate determinations of Cheltzoff, personally communicated by him, led him to conclude that solutions of zinc chloride follow the same laws as the solutions of sulphuric acid, which will be considered in Chapter XX.: (1) from H_2O to $ZnCl_2,120H_2O$ $s = S_0 + 92·85p + 0·1748p^2$; (2) from thence to $ZnCl_2,40H_2O$ $s = S_0 + 93·96p - 0·0126p^2$; (3) thence to $ZnCl_2,25H_2O$ $s = 11481·5 + 96·45(p - 15·89) + 0·4567(p - 15·89)^2$; (4) thence to $ZnCl_2,10H_2O$ $s = 12212·1 + 104·82(p - 23·21) + 0·7992(p - 23·21)^2$; (5) thence to $p = 65$ p.c. $s = 14606·3 + 140·96(p - 43·05) + 1·4905(p - 43·05)^2$, where s is the specific gravity of the solution at 15°, containing p p.c. of $ZnCl_2$ by weight, taking water at 4° = 10000, and where $S_0 = 9991·6$ (specific gravity of water at 15°). The compound of zinc chloride with hydrochloric acid has been mentioned in Vol. I. Chapter X.

Zinc chloride has a great affinity for water; it is not only soluble in it, but in alcohol, and on being dissolved in water becomes considerably heated, like magnesium and calcium chlorides. Zinc chloride is capable of taking up water, not only in a free state, but also in chemical combination with many substances. Thus, for instance, it is used in organic researches for removing the elements of water from many of the organic compounds.

[4 bis] When mixed with zinc oxide it forms, with remarkable ease, a very hard mass of zinc oxychloride, which is applied in the arts; for instance, in painting, to resist the action of water, or for cementing such objects as are destined to remain in water. Zinc oxychloride, $ZnCl_2,3ZnO,2H_2O$ (= $Zn_2OCl_2,2ZnH_2O_2$), is also formed from a solution of zinc chloride by the action of a small quantity of ammonia on it after heating the precipitate obtained with the liquid for a considerable time; the admixture of ammonium salts with a mixture of a strong solution of zinc chloride with its oxide makes a similar mass, which does not solidify so rapidly, and is therefore more useful for some purposes. Moisture and cold do not change the hardened mass of oxychloride, and it also resists the action of many acids, and a temperature of 300°, which makes it a useful cement for many purposes. A solution of magnesium chloride with magnesium oxide forms a similar oxychloride. The mass solidifies best when there are equal quantities by weight of zinc in the chloride and oxide, and therefore when it has the composition Zn_2OCl_2 In preparing such a cement, naturally zinc oxide alone may be taken, and the requisite quantity of hydrochloric acid added to it. The capacity of $ZnCl_2$ to combine with water, ZnO, and HCl (and also with other metallic chlorides) indicates its property to combine with molecules of other substances, and therefore its

compounds with NH_3, and especially a compound, $ZnCl_2 2NH_3$, similar to sal-ammoniac, might be expected (*i.e.* $2NH_4Cl$, in which H_2 is replaced by Zn). And indeed it has long been known that $ZnCl_2$ absorbs ammonia and gives solid substances capable of dissociating with the disengagement of NH_3. Among these compounds Isambert and V. Kouriloff (1894) obtained $ZnCl_2 6NH_3$, $ZnCl_2 4NH_3$, $ZnCl_2 2NH_3$, and $ZnCl_2 NH_3$. The dissociation tension of the two last-mentioned compounds at 218° is equal to 43·6 mm. and 6·7 mm. $CdCl_2$ also forms similar compounds with NH_3 (Kouriloff, 1894).

This mineral has been given the name of 'mock-ore,' on account of its having the appearance (considerable density, 4·06, &c.) of ordinary metallic ores; it deceived the first miners, because it did not, like other ores, give metal when simply roasted in air and fused with charcoal. The white zinc oxide, formed by burning the vapours of zinc, was also called 'nihil album,' or 'white nothing,' on account of its lightness.

It may be here mentioned that by the word *ore* is meant a hard, heavy substance dug out of the earth, which is used in metallurgical works for obtaining the usual heavy metals long known and used. The natural compounds of sodium, or magnesium, are not called ores, because magnesium and sodium have not been long obtainable in quantity. The heavy metals, those which are easily reduced and do not easily oxidise, are exclusively those which are directly applied in manufactures. Ores either contain the metals themselves (for instance, ores of silver or bismuth), and the metals are then said to be in a native state, or else their sulphur compounds (blende, mock-ore, pyrites—as, for example, galena, PbS; zinc blende, ZnS; copper pyrites, $CuFeS$) or oxides (as the ores of iron), or salts (calamine, for instance). Zinc is incomparably rarer than magnesium, and is only well known because it is transformed from its ores into a metal which finds direct use in many branches of industry.

Ores, when extracted from the earth by the miners, are often enriched by sorting, washing, and other mechanical operations. The sulphurous ores (and likewise others) are then generally roasted. Roasting an ore means heating it to redness in air. The sulphur then burns, and passes off in the form of sulphurous anhydride, SO_2, and the metal oxidises. The roasting is carried on in order to obtain an oxide instead of a sulphur compound, the oxide being reducible by charcoal. These methods, introduced ages ago, are met with in nearly all metallurgical works for practically all ores. For this reason the preparatory treatment of zinc blende furnishes zinc oxide: this is already contained in calamine.

[7 bis] with very impure ores, especially such as contain lead (PbS often accompanies zinc), the vapour of the reduced zinc is allowed to pass directly into the air. It burns and gives ZnO, which is used as a pigment.

This zinc, although homogeneous, still contains certain impurities, to remove which it is necessary to prepare some salt of zinc in a pure state and transform it into carbonate, which latter is then distilled with charcoal; and, as thin sheets of zinc can only be obtained from very pure metal, they are frequently made use of in cases where pure zinc is required. In order to remove the arsenic from zinc, it was proposed to melt it and mix it with anhydrous magnesium chloride, by which means vapours of zinc chloride and arsenic chloride are formed. Perfectly pure zinc is made (V. Meyer and others) by decomposing, by means of the galvanic current, a solution of zinc sulphate to which an excess of ammonia has been added. The zinc used for Marsh's arsenic test (Chapter XIX.) is purified from As by fusing it with KNO_3 and then with $ZnCl_2$.

Cornices and other architectural ornaments, remarkable for their lightness and beauty, are stamped out of sheet zinc. Zinc-roofing does not require painting, but it melts during a conflagration, and even burns at a strong heat. Many iron vessels, &c., are covered with zinc ('galvanised') in order to prevent them from rusting.

Veeren (1891) proved this by simple experiments, finding that in vacuo the solution proceeds far more rapidly for both pure and commercial zinc, and still more rapidly in the presence of oxidising agents (which absorb the hydrogen) like CrO_3 and H_2O_2.

[10 bis] The addition of cupric sulphate, or, better still, a few drops of platinic chloride (the metals become reduced), to the sulphuric acid greatly accelerates the evolution of the hydrogen, because in this case, as with commercial zinc, galvanic couples are formed locally by the copper or platinum and the zinc, under the influence of which the zinc rapidly dissolves. The action of acids on metallic zinc of various degrees of purity has been the subject of many investigations, particularly important with reference to the application of zinc in galvanic batteries, whilst some investigations have direct significance for chemical mechanics, although from many points of view the matter is not clear. I consider it useful to mention certain of these investigations.

Calvert and Johnson made the following series of observations on the action of sulphuric acid of various degrees of concentration on 2 grams of pure zinc during two hours. In the cold the concentrated acid, H_2SO_4, does not act, $H_2SO_4,2H_2O$ dissolves about 0·002 gram, but principally forms hydrogen sulphide, which is obtained also when the dilution reaches $H_2SO_4,7H_2O$, when 0·035 gram of zinc is dissolved. When largely diluted with water, pure hydrogen begins to be disengaged. $H_2SO_4, 2H_2O$ at 130° gives a mixture of hydrogen sulphide and sulphurous anhydride dissolving 0·156 gram of zinc.

Bouchardat showed that if in a vessel made of glass or sulphur dilute sulphuric acid acting on a piece of zinc liberates one part of hydrogen, then the same acid with the same piece of zinc in the same time will liberate 4 parts of hydrogen if the vessel be made of tin—that is, zinc forms a galvanic couple with tin; in a leaden vessel 9 parts of hydrogen are set free, with a vessel of antimony or bismuth 13 parts, silver or platinum 38 parts, copper 50 parts, iron 43 parts. If a salt of platinum be added to the dilute sulphuric acid (1 part of acid and 12 parts of water), Millon determined that the rapidity of the action on the zinc is increased 149 times, and by the addition of copper sulphate is rendered 45 times greater than the action of pure sulphuric acid. The salts which are added are reduced to metals by the zinc, their contact serving to promote the reaction because they form local galvanic currents.

According to the observations of Cailletet, if, at the ordinary pressure, sulphuric acid with zinc liberates 100 parts of hydrogen, then with a pressure of 60 atmospheres 47 parts will be liberated and 1 part at a pressure of 120 atmospheres. With a reduced pressure under the receiver of an air-pump 168 parts are liberated. Helmholtz showed that a reduced pressure also exercises its influence on galvanic elements.

Debray, Löwel, and others showed that zinc liberates hydrogen and forms basic salts and zinc oxide with solutions of many salts—for instance, MCl_n, aluminium sulphate, and alum, Sodium and potassium carbonates scarcely act, because they form carbonates. The salts of ammonia act more strongly than the salts of potassium and sodium; the zinc remains bright. It is evident that this action is founded on the formation of double salts and basic salts.

The variation with concentration in the rate of the action of sulphuric acid on zinc (containing impurities) under otherwise uniform conditions is in evident connection with the electrical conductivity of the solution and its viscosity, although, when largely diluted, the action is almost proportional to the amount of acid in a known volume of the solution. Forging, casting the molten metal, and similar mechanical influences change the density and hardness of zinc, and also strongly influence its power of liberating hydrogen from acids. Kayander showed (1881) that when magnesium is submitted to the action of acids: (*a*) the action depends, not on the nature of the acid, but on its basicity; (*b*) the increase of the action is more rapid than the growth of the concentration; and (*c*) there is a decrease of action with the increase of the coefficient of internal friction and electrical conductivity.

Spring and Aubel (1887) measured the volume of hydrogen disengaged by an alloy of zinc and a small quantity of lead (0·6 p.c.), because the action of acids is then uniform. In order to deal with a known surface, spheres were taken (9·5 millimetres diameter) and cylinders (17 mm. dia.), the sides of which were covered with wax in order to limit the action to the end surfaces. During the commencement of the action of a definite quantity of acid the rapidity increases, attains a maximum, and then declines as the acid becomes exhausted. The results for 5, 10, and 15 per cent. of hydrochloric acid are given below. H denotes the number of cubic centimetres of hydrogen, D the time in seconds elapsing after the zinc spheres have been plunged into the acid. At 15° were obtained:

H =	50	100	200	400	600	800	1000
5 p.c. D =	5714	1152	1755	2731	3908	6234	15462
10 p.c. D =	301	455	649	995	1573	2746	6748
15 p.c. D =	106	151	233	440	826	1604	4289

At 35°:

5 p.c. D =	426	705	1058	1700	2525	4132	8499
10 p.c. D =	96	148	239	460	835	1594	3735
15 p.c. D =	44	64	112	255	505	1011	2457

At 55°:

5 p.c. D =	178	276	408	699	1164	2105	5093
10 p.c. D =	34	60	113	258	491	970	2457
15 p.c. D =	24	35	58	136	239	610	1593

In consequence of the complex character of the phenomenon, the authors themselves do not consider their determinations as being conclusive, and only give them a relative significance; and in this connection it is remarkable that hydrobromic acid under similar conditions (with an equivalent strength) gives a greater (from 2 to 5 times) rapidity of action than hydrochloric acid, but sulphuric acid a far smaller velocity (nearly 25 times smaller). It is also remarkable that during the reaction the metal becomes much more heated than the acid.

It may be mentioned that zinc dust and zinc itself, when heated with hydrated lime and similar hydrates, disengages hydrogen: this method has even been proposed for obtaining hydrogen for filling war balloons.

It may be here remarked that sulphate of zinc (especially in the presence of mineral acids) does not give a precipitate of sulphide of zinc, or is only slightly precipitated by sulphuretted hydrogen.

[11 bis] Sulphide of cadmium appears in two varieties of a similar chemical but different physical character: one is of a lemon colour, and the other bright red. Kloboukoff (1890) studied the physical properties of these varieties more closely. The sp. gr. of the former is 3·906, and of the latter 4·513. They belong to different crystallographic systems. The first variety may be converted into the second by friction or pressure, but the second cannot be converted into the first variety by these means.

Amongst the compounds of cadmium very closely allied to the compounds of zinc, we must mention *cadmium iodide*, CdI_2, which is used in medicine and photography. This salt crystallises very well: it is prepared by the direct action of iodine, mixed with water, on metallic cadmium. One part of cadmium iodide at 20° requires for its solution 1·08 part of water. It may be remarked that cadmium chloride at the same temperature requires 0·71 part of water to dissolve it, so that the iodine compound of this metal is less soluble than the chloride, whilst the reverse relation holds in the case of the corresponding compounds of the alkali or alkaline earthy metals. Cadmium sulphate crystallises well, and has the composition $3CdSO_4,8H_2O$, thus differing from zinc sulphate.

Cadmium oxide is soluble, although sparingly, in alkalis, but in the presence of tartaric and certain other acids the alkaline solution of cadmium oxide does not change when boiled, whilst a *diluted* solution in that case deposits cadmium oxide: this may also serve for separating zinc compounds from those of cadmium. Cadmium is precipitated from its salts by zinc, which fact may also be taken advantage of for separating cadmium; for this reason, in an alloy of zinc and cadmium, acids first of all extract the zinc. Cadmium is in all respects less energetic than zinc. Thus, for instance, it decomposes water with difficulty, and this only when strongly heated. It even acts but slowly on acids, but then displaces hydrogen from them. It is necessary here to call attention to the fact that for alkali and alkaline earthy metals (of the even series) the highest atomic weight determines the greatest energy; but cadmium (of the uneven series), whilst having a larger atomic weight than zinc, is less energetic. The salts of cadmium are colourless, like those of zinc. De Schulten obtained a crystalline oxychloride, Cd(OH)Cl by heating marble with a solution of cadmium chloride in a sealed tube at 200°.

[12 bis] According to its atomic weight, mercury follows gold in the periodic system, just as cadmium follows silver and zinc follows copper:—

Ni =   59   Cu =   63   Zn =   65

Pd = 106   Ag = 108   Cd = 112

Pt = 196   Au = 198   Hg = 200

Eventually we shall see the near relation of platinum, palladium, and nickel, and also of gold, silver, and copper, but we will now point out the parallelism between these three groups. The relation between the physical and also chemical properties is here strikingly similar. Nickel, palladium, and platinum are very difficult to fuse (far more so than iron, ruthenium, and osmium, which stand before them). Copper, silver, and gold melt far more easily in a strong heat than the three preceding metals, and zinc, cadmium, and mercury melt still more easily. Nickel, palladium, and platinum are very slightly volatile; copper, silver, and gold are more volatile; and zinc, cadmium, and mercury are among the most volatile metals. Zinc oxidises more easily than copper, and is reduced with more difficulty, and the same is true for mercury as compared with gold. These properties for cadmium and silver are intermediate in the respective groups. Relations of this kind clearly show the nature of the periodic law.

Thus thallium, lead, and bismuth, following mercury according to their atomic weights, form, besides compounds of the highest types, TlX_3, PbX_4, and BiX_5, also the lower ones TlX, PbX_2, and BiX_3.

During the condensation of the vapours of mercury in works, a part forms a black mass of finely-divided particles, which gives metallic mercury when worked up in centrifugal machines, or on pressure, or on re-distillation. In mercury we observe a tendency to easily split up into the finest drops, which are difficult to unite into a dense mass. It is sufficient to shake up mercury with nitric and sulphuric acids in order to produce such a mercury *powder*. The mercury separated (for instance, reduced by substances like sulphurous anhydride) from solutions, forms such a powder. According to the experiments of Nernst, this disintegrated mercury when entering into reactions develops more heat than the dense liquid metal—that is to say, the work of disintegration reappears in the form of heat. This example is instructive in considering thermochemical deductions.

Mercury may sometimes be obtained in a perfectly pure state from works (in iron bottles holding about 35 kilos), but after being used in laboratories (for baths, calibration, &c.) it contains impurities. It may be purified mechanically in the following way: a paper filter with a fine hole (pricked with a needle) is placed in a glass funnel and mercury is poured into it, which slowly trickles through the hole, leaving the impurities upon the filter. Sometimes it is squeezed through chamois leather or through a block of wood (as in the well-known experiment with the air-pump). It may be purified from many metals by contact with dilute nitric acid, if small drops of mercury are allowed to pass through a long column of it (from the fine end of a funnel); or by shaking it up with sulphuric acid in air. Mercury may be purified by the action of an electric current, if it be covered with a solution of $HgNO_3$. But the complete purification of mercury for barometers and thermometers can only be attained by distillation, best in a vacuum (the vapour-tension of mercury is given in Chapter II., Note 27). For this purpose Weinhold's apparatus is most often used. The principle of this apparatus is very ingenious, the distillation being effected in a Torricellian vacuum continuously supplied with fresh mercury, whilst the condensed mercury is continuously removed. This process of distillation requires very little attention, and gives about one kilo of pure mercury per hour.

If the volume of *liquid* mercury at 0° be taken as 1000000, then, according to the determinations of Regnault (recalculated by me in 1875), at t it will be $1000000 + 180·1t + 0·02t^2$.

All salts of mercury, when mixed with sodium carbonate and heated, give mercurous or mercuric carbonates; these decompose on being heated, forming carbonic anhydride, oxygen, and vapours of mercury.

Spring (1888) showed that, solid dry HgCl is gradually decomposed in contact with metallic copper. According to the determinations of Thomsen, the formation of a gram of mercurial compounds from their elements develops the following amounts of heat (in thousands of units): $Hg_2 + O$, 42; $Hg + O$, 31; $Hg + S$, 17; $Hg + Cl$, 41; $Hg + Br$, 34; $Hg + I$, 24; $Hg + Cl_2$, 63; $Hg + Br_2$, 51; $Hg + I_2$, 34; $Hg + C_2N_2$, 19. These numbers are less than the corresponding ones for potassium, sodium, calcium, barium, and for zinc and cadmium—for instance, $Zn + O$, 85; $Zn + Cl_2$, 97; $Zn + Br_2$, 76; $Zn + I_2$, 49; $Cd + Cl_2$, 93; $Cd + Br_2$, 75; $Cd + I_2$, 49.

This salt easily forms the crystallo-hydrate $HgNO_3,H_2O$, corresponding with ortho-nitric acid, H_3NO_4 (the terms ortho-, pyro-, and meta-acids are explained in the chapter on Phosphorus), with the substitution of Hg for H. In an aqueous solution this salt can only be preserved in the presence of free mercury, otherwise it forms basic salts, which will be mentioned hereafter (Chapter VI., Note 59).

Mercuric nitrate, $Hg(NO_3)_2, 8H_2O$, crystallises from a concentrated solution of mercury in an excess of boiling nitric acid. Water decomposes this salt; at the ordinary temperature crystals of a basic salt of the composition $Hg(NO_3)_2, HgO, 2H_2O$ are formed, and with an excess of water the insoluble yellow basic salt $Hg(NO_3)_2, H_2O, 2HgO$. These three salts correspond with the type of ortho-nitric acid, $(H_3NO_4)_2$, in which mercury is substituted for 1, 2 and 3 times H_2. As all these salts still contain water, it is possible that they correspond with the tetrahydrate $= N_2O_5 + 4H_2O = N_2O(OH)_8$ if ortho-nitric acid $= N_2O_5 + 3H_2O = 2NO(OH)_3$.

To obtain the mercuric salt a large excess of strong sulphuric acid must be taken and strongly heated. With a small quantity of water colourless crystals of $HgSO_4, H_2O$ may be obtained. An excess of water, especially when heated, forms the basic salt (as in Note [20]), $HgSO_4, 2HgO$, which corresponds with trihydrated sulphuric acid, $SO_3 + 3H_2O = S(OH)_6$, with the substitution of H_6 by $3Hg$, which in mercuric salts is equivalent to H_6. Le Chatelier (1888) gives the following ratio between the amounts of equivalents per litre:

$HgSO_4$ 0·318 0·890 1·80 2·02

SO_3 0·752 1·42   2·10 2·40

—that is, the relative amount of free acid decreases as the strength of the solution increases.

The question of the molecular weight of calomel—that is, whether the mercury in the salts of the suboxide is monatomic or diatomic—long occupied the minds of chemists, although it is not of very great importance. It is only recently (1894) that this question can be considered as answered, thanks to the researches of V. Meyer and Harris, in favour of diatomicity—that is, that calomel is analogous to peroxide of hydrogen and contains Hg_2Cl_2 (like O_2H_2) in its molecule if corrosive sublimate contains $HgCl_2$ (like water OH_2). As a matter of fact, direct experiment gives the vapour density of calomel as about 118—that is, indicates that its molecule contains $HgCl$, whilst the molecule of the sublimate, judging also by the vapour density (nearly 136), contains $HgCl_2$; it might therefore be concluded that the mercury in the suboxide is not only monovalent (corresponding to H) but also monatomic, whilst in the oxide it is divalent and diatomic. Instances of a variable atomicity, as shown by the vapour density, are known in N_2O, NO, and NH_3, CO and CO_2, PCl_3 and PCl_5, and it might therefore be supposed that the present was a similar instance. But there are also instances of a variable equivalency which do not correspond to a variation of atomicity—for example, OH_2 (water) and OH (peroxide of hydrogen), CH_4 (methane), C_2H_5 (ethyl), and CH_2 (ethylene),

&c. Here, according to the law of substitution, the residues of OH_2 and CH_4 combine together and give molecules; $OHOH = O_2H_2$ (peroxide of hydrogen) and $CH_3CH_3 = C_2H_6$ (ethane), &c. The same may be assumed also to be the relation of calomel to sublimate; the residue HgCl, which is combined with Cl in sublimate, corresponds to $HgCl_2$, and in calomel it may be supposed that this residue is combined with itself, forming the molecule Hg_2Cl_2. On this view of the composition of the molecule of calomel it would follow that in the state of vapour it breaks up into two molecules, $HgCl_2$ and Hg, when the vapour density would be about 118 (because that of sublimate is about 136 and that of mercury about 100), and that in cooling this mixture (like a mixture of HCl and NH_3) again gives Hg_2Cl_2. It was therefore necessary to prove that calomel is decomposed in the state of vapour. This was not effected for a long time, although Odling, as far back as the thirties, showed that gold becomes amalgamated (*i.e.* absorbs metallic mercury) in the vapour of calomel, but not in the vapour of sublimate. Recently, however, V. Meyer and Harris (1894) have shown that a greater amount of the vapour of mercury than of calomel passes (at about 465°) through a porous clay cell, containing calomel. This proves that the vapour of calomel contains a mixture of the vapours of Hg and $HgCl_2$, as would follow from the second hypothesis. Moreover, on introducing a heated piece of KHO into the vapour of calomel, Meyer observed the formation, not of suboxide (black), but of oxide of mercury (yellow). Therefore the molecular formula of calomel must be taken as Hg_2Cl_2 (and not HgCl).

[22 bis] Calomel (in Japanese 'Keyfun') has been prepared in Japan (and China) for many centuries, by heating mercury in clay crucibles with sea salt, which contains $MgCl_2$ and gives HCl. The vapour of the mercury reacts with this HCl and the oxygen of the air and forms calomel: $2Hg + 2HCl + O = Hg_2Cl_2 + H_2O$. The calomel collects on the lid of the crucible in the form of a sublimate (Divers, 1894).

$HgCl_2$ is partially converted into calomel even in the act of dissolving in ordinary water, especially under the action of light.

[23 bis] As feebly energetic bases (for instance, the oxides MgO, ZnO, PbO, CuO, Al_2O_3, Bi_2O_3, &c.), mercuric oxide (*see* Notes 20, 21) and mercurous oxide easily give basic salts, which are usually directly formed by the action of water on the normal salt, according to the general equation (for mercuric compounds, RX_2):

$nRX_2 \quad + mH_2O = 2mHX + (n-m)RX_2mRO$

neutral salt   water   acid   basic salt

or else are produced directly from the normal salt and the oxide or its hydroxide. Thus mercurous nitrate, when treated with water, forms basic salts of the composition $6(HgNO_3),Hg_2O,H_2O$, $2(HgNO_3),Hg_2O,H_2O$, and $3(HgNO_3),Hg_2O,H_2O$, the first two of which crystallise well. Naturally it is possible either to refer similar salts to the type of hydrates—for instance, the second salt to the hydrate $N_2O_5,4H_2O$—or to view it as a compound, $HgNO_3,HgHO$, but our present knowledge of basic salts is not sufficiently complete to admit of generalisations. However, it is already possible to view the subject in the following aspects: (1) basic salts are principally formed from feeble bases; (2) certain metals (mentioned above) form them with particular ease, so that one of the causes of the formation of many basic salts must depend on the property of the metal itself; (3) those bases which readily form basic salts as a rule also readily form double salts; (4) in the formation of basic salts, as also everywhere in chemistry, where sufficient facts have accumulated, we clearly see the conditions of equally balanced heterogeneous systems, such as we saw, for instance, in the formation of double salts, crystallo-hydrates, &c.

The mercuric salts often form double salts (confirming the third thesis), and mercuric chloride easily combines with ammonia, forming $Hg(NH_4)_2Cl_4$, or in general $HgCl_2nMCl$. If a mixture of mercurous and potassium sulphates be dissolved in dilute sulphuric acid, the solution easily yields large colourless crystals of a double salt of the composition $K_2SO_4,3HgSO_4,2H_2O$. Boullay obtained crystalline compounds of mercuric chloride with hydrochloric acid, and mercuric iodide with hydriodic acid; and Thomsen describes the compound $HgBr_2,HBr,4H_2O$ as a well-crystallised salt, melting at 13°, and having, in a molten state, a specific gravity 3·17 and a high index of refraction. Moreover, the capacity of salts for forming basic compounds has been considerably cleared up since the investigation (by Würtz, Lorenz, and others) of glycol, $C_2H_4(OH)_2$ (and of polyatomic alcohols resembling it), because the ethers $C_2H_4X_2$, corresponding with it, are capable of forming compounds containing $C_2H_4X_2nC_2H_4O$.

On the other hand, there is reason to think that the property of forming basic salts is connected with the polymerisation of bases, especially colloidal ones (*see* the chapter on Silica, Lead Salts, and Tungstic Acid).

Mercuric iodide, HgI_2, is obtained first as a yellow, and then as a red, precipitate on mixing solutions of mercuric salts and potassium iodide, and is soluble in an excess of the latter (in consequence of the formation of the double salt, $HgKI_3$); of ammonium chloride (for a similar reason), &c. It crystallises at the ordinary temperature in square prisms of a red colour. On being heated, these change into yellow rhombic crystals, isomorphous with mercuric chloride. This yellow form of mercuric iodide is very unstable, and when cooled and triturated easily again assumes the more stable red form. When fused, a yellow liquid is obtained. *Mercuric cyanide*, $Hg(CN)_2$, forms one of the most stable metallic cyanides. It is obtained by dissolving mercuric oxide in prussic acid, and by boiling Prussian blue with water and mercuric oxide, ferric oxide being then obtained in the precipitate. Mercuric cyanide is a colourless crystalline substance, soluble in water, and distinguished by its great stability; sulphuric acid does not liberate prussic acid from it, and even caustic potash does not remove the cyanogen (a complex salt is probably produced), but the halogen acids disengage HCN. Like the chloride, it combines with mercuric oxide, forming the oxycyanide, $Hg_2O(CN)_2$, and it shows a very marked tendency to form double compounds—for example, $K_2Hg(CN)_4$. The alkali chlorides and iodides form similar compounds—for instance, the salt $HgKI(CN)_2$ crystallises very well, and is produced by directly mixing solutions of potassium iodide and mercuric cyanide.

Wells (1889) and Vare obtained and investigated many such double salts, and showed the possibility of the formation, not only of $HgCl_2MCl$ and $HgCl_2 2MCl$ where M is a metal of the alkalis—for example, Cs—but also of $HgCl_2 3MCl, 2(HgCl_2)MCl$, and in general $nHgX_2 mMX$, where X stands for various haloids.

[24 bis] *See* Chapter XIX., Note 6 bis: Hg_3P_2. In studying the metallic nitrides it is necessary to keep the corresponding phosphides in mind.

Hg_3N_2 is similar in composition to Mg_3N_2, &c. (Chapter XIV.) The readiness with which mercuric nitride explodes shows that the connection between the nitrogen and the mercury is very unstable, and explains the circumstance that the so-called *mercury fulminate*, or *fulminating mercury*, is an exceedingly explosive substance. This substance is prepared in large quantities for explosive mixtures; it enters into the composition of percussion caps, which explode when struck, and ignite gunpowder. Mercury fulminate was discovered by Howard, and from that time has been prepared in the following way: one part of mercury is dissolved in twelve parts of nitric acid, of sp. gr. 1·36, and when the whole of the mercury is dissolved, 5·5 parts of 90 p.c. alcohol are added, and the mass is shaken. A reaction then commences, accompanied by a rise in temperature due to the oxidation of the alcohol. As a matter of fact, many oxidation products are produced during the action of the nitric acid on the alcohol (glycolic acid, ethers, &c.) When the reaction becomes tolerably vigorous, the same quantity of alcohol is added as at the commencement, when a grey precipitate of the fulminate separates. This salt has the composition $C_2Hg(NO_2)N$. It explodes when struck or heated. The mercury in it may be replaced by other metals—for instance, copper or zinc, and also silver. The silver salt, $C_2Ag_2(NO_2)N$, is obtained in a precisely analogous manner, and is even more explosive. Under the action of alkali chlorides, only half the silver is replaced by the alkali metal, but if the whole of the silver be replaced by an alkali metal, then the salt decomposes. This is evidently because combinations of this kind proceed in virtue of the formation of substances in which mercury, and metals akin to it, are connected in an unstable way with nitrogen. Potassium and other light metals are incapable of entering into such connection and therefore, the substitution of potassium for mercury entails the splitting-up of the combination. Investigations of the fulminates were carried on by Gay-Lussac and Liebig, but only the investigations of L. N. Shishkoff fully cleared up the composition and relation of these substances to the other carbon compounds. Shishkoff showed that fulminates correspond with the nitro-acid, $C_2H_2(NO_2)N$. The explosiveness of the group depends partly on its containing at the same time NO_2 and carbon; we already know that all such nitrogen compounds are explosive. If we imagine that the NO_2 is replaced by hydrogen, we shall have a substance of the composition C_2H_3N. This is acetonitrile—that is, acetic acid + NH_3 - $2H_2O$, or ethenyl nitrile, as shown in Chapter VI. The formation of an acetic compound by the action of nitric acid on alcohol is easily understood, because acetic acid is produced by the oxidation of alcohol, and the production of the elements of ammonia, indispensable for the formation of a nitrile, is accounted for by the fact that nitric acid under the action of reducing substances in many cases forms ammonia. Moreover a certain analogy has been found between fulminating acid and hydroxylamine, but details upon this subject must be looked for in works on organic chemistry. The explosiveness of fulminating mercury, the rapidity of its decomposition (gunpowder, and even guncotton, burn more slowly and explode less violently), and the force of its explosion, are such that a small quantity (loosely covered) will shatter massive objects.

The investigations of Abel on the communication of explosion from one substance to another are remarkable. If guncotton be ignited in an open space, it burns quietly; but if fulminating mercury be exploded by the side of it, the decomposition of the guncotton is effected instantaneously, and it then shatters the objects upon which it lies, so rapid is the decomposition. Abel explains this by supposing that the explosion of the fulminating salt brings the molecules of guncotton into a uniform or as it were harmonious state of vibration, which causes the rapid decomposition of the whole mass. This rapid decomposition of explosive substances defines the distinction between explosion and combustion. Besides this, Berthelot showed that from that form of powerful molecular concussion which takes place during the explosion of fulminating mercury, the state of strain and stability of equilibrium of substances which are endothermal, or capable of decomposing with the disengagement of heat—for instance, cyanogen, nitro compounds, nitrous oxide, &c.—is generally destroyed.

Thorpe showed that carbon bisulphide, CS_2, also an endothermal substance, decomposes into sulphur and charcoal, when fulminating mercury is exploded in contact with it.

[25 bis] The capacity for replacing hydrogen in chloride of ammonium by metals also belongs to Zn and Cd. Kvasnik (1892), by the action of ammonia upon alcoholic solutions of $CdCl_2$ and $ZnCl_2$, obtained substances of the general formula $M(NH_3Cl)_2$, formed as it were from two molecules of sal-ammoniac by the substitution of two atoms of hydrogen by a diatomic metal. These substances appear as white, finely crystalline powders. Under the action of heat half the ammonia passes off, and a compound of the composition $MClNH_3Cl$ is formed. The compounds of cadmium and zinc are distinguished from each other by the former being more volatile than the latter.

We may further remark that in the series Mg, Zn, Cd, and Hg the capacity to form double salts of diverse composition increases with the atomic weight. Thus, according to Wells and Walden's observations (1893), the ratio $n : m$ for the type nMClmRCl$_2$ (M = K, Li, Na … R = Mg, Zn …) is for Mg 1 : 1, for Zn 3 : 1, 2 : 1, and 1 : 1; for Cd, besides this, salts are known with the ratio 4 : 1, and for Hg 3 : 1, 2 : 1, 1 : 1, 2 : 3, 1 : 2, and 1 : 5.

I consider it appropriate here to call attention to the want of an element (ekacadmium) between cadmium and mercury in the periodic system (Chapter XV.) But as in the ninth series there is not a single known element, it may be that this series is entirely composed of elements incapable of existing under present conditions. However, until this is proved in one way or another, it may be concluded that the properties of ekacadmium will be between those of cadmium and mercury. It ought to have an atomic weight of about 155, to form an oxide EcO, a slightly stable oxide Ec_2O. Both ought to be feeble bases, easily forming double and basic salts. The volume of the oxide will be nearly 17·5, because the volume of cadmium oxide is about 16, and that of mercuric oxide 19. Therefore the density of the oxide will approach $171 \div 17·5 = 9·7$. The metal ought to be easily fusible, oxidising when heated, of a grey colour, with a specific volume, about 14 (cadmium = 13, mercury = 15), and, therefore, its specific gravity $(155 \div 14)$ will nearly = 11. Such a metal is unknown. But in 1879 Dahl, in Norway, discovered in the island of Oterö, not far from Kragerö, in a vein of Iceland spar in a nickel mine, traces of a new metal which he called norwegium, and which presented a certain resemblance to ekacadmium. Perfect purity of the metal was not attained, and therefore the properties ascribed to norwegium must be regarded as approximate, and likely to undergo considerable alteration on further study. A solution of the roasted mineral in acid was twice precipitated by sulphuretted hydrogen, and again ignited; the oxide obtained was easily reduced. When the metal was dissolved in hydrochloric acid largely diluted with water, and the solution boiled, the basic salt was precipitated, and thus freed from the copper which remained in the solution. The reduced metal had a density 9·44, and easily oxidised. If the composition NgO be assigned to the oxide, then Ng = 145·9. It fused at 254°; the hydroxide was soluble in alkalis and potassium carbonate. In any case, if norwegium is not a mixture of other metals, it belongs to the uneven series, because the heavy metals of the even series are not easily reducible. Brauner thinks that norwegium oxide is Ng_2O_3, the atom Ng = 219, and places it in Group VI., series 11, but then the feebly acid higher oxide, NgO_3, ought to be formed.

Amongst the metals accompanying zinc which have been named, but not authentically separated, must be included the *actinium* of Phipson (1881). He remarked that certain sorts of zinc give a white precipitate of zinc sulphide which blackens on exposure to light and then becomes white in the dark again. Its oxide, closely resembling in many ways cadmium oxide, is insoluble in alkalis, and it forms a white metallic sulphide, blackening on exposure to light. As no further mention has been made of it since 1882, its existence must be regarded as doubtful.

CHAPTER XVII
BORON, ALUMINIUM, AND THE ANALOGOUS METALS OF THE THIRD GROUP

If the elements of small atomic weight which we have hitherto discussed be placed in order, it will be clearly seen that, judging by the formulæ of their higher compounds, one element is wanting between beryllium and carbon. For lithium gives LiX, beryllium forms BeX_2, and then comes carbon giving CX_4. Evidently to complete the series we must look for an element forming RX_3, and having an atomic weight greater than 9 and less than 12. And *boron* is such a one; its atomic weight is 11, and its compounds are expressed by BX_3. Lithium and beryllium are metals; carbon has no metallic properties; boron appears in a free state in several forms which are intermediate between the metals and non-metals. Lithium gives an energetic caustic oxide, beryllium forms a very feeble base; hence one would expect to find that the oxide of boron, B_2O_3, has still more feeble basic properties and some acid properties, all the more as CO_2 and N_2O_5, which follow after B_2O_3 in their composition and in the periodic system, are acid oxides. And, indeed, the only known *oxide of boron* exhibits a feeble basic character, together with the properties of a feeble acid oxide. This is even seen from the fact that a solution of boron oxide reddens blue litmus and acts on turmeric paper as an alkali, and these reactions may be used for determining the presence of B_2O_3 in solutions. By themselves the alkali borates have an alkaline reaction, which clearly indicates the feeble acid character of boric acid. If they are mixed in solution with hydrochloric acid, boric acid is liberated, and if a piece of turmeric paper be immersed in this solution and then dried, the excess of hydrochloric acid volatilises, while the boric acid remains on the paper and communicates a *brown coloration* to it, just like alkalis.

Boron trioxide or boric anhydride enters into the composition of many minerals, in the majority of cases in small quantities as an isomorphous admixture, not replacing acids but bases, and most frequently alumina (Al_2O_3), for as a rule the amount of alumina decreases as that of the boric anhydride increases in them. This substitution is explained by the similarity between the atomic composition of the oxides of aluminium (alumina) and boron. The subdivision of oxides into basic and acid can in no way be sharply defined, and here we meet with the most conclusive proof of the fact, for the oxides of boron and aluminium belong to the number of intermediate oxides, closely approaching the limit separating the basic from the acid oxides. Their type (Chapter XV.) R_2O_3 is intermediate between those of the basic oxides R_2O and RO and those of the acid oxides R_2O_5

and RO$_3$. If we turn our attention to the chlorides, we remark that lithium chloride is soluble in water, is not volatile, and is not decomposed by water; the chlorides of beryllium and magnesium are more volatile, and although not entirely, still are decomposed by water; whilst the chlorides of boron and aluminium are still more volatile and are decomposed by water. Thus the position of boron and aluminium in the series of the other elements is clearly defined by their atomic weights, and shows us that we must not expect any new and distinct functions in these elements.

Boron was originally known in the form of sodium borate, Na$_2$B$_4$O$_7$,10H$_2$O, or *borax*, or *tincal*, which was exported from Asia, where it is met with in solution in certain lakes of Thibet; it has also been discovered in California and Nevada, U.S.A. Boric acid was afterwards found in sea-water and in certain mineral springs. Its presence may be discovered by means of the green coloration which it communicates to the flame of alcohol, which is capable of dissolving free boric acid. Many of the boron compounds employed in the arts are obtained from the impure boric acid which is extracted in Tuscany from the so-called *suffioni*. In these localities, which present the remains of volcanic action, steam mixed with nitrogen, hydrogen sulphide, small quantities of boric acid, ammonia, and other substances, issue from the earth.[3 bis] The boric acid partially volatilises with the steam, for if a solution of boric acid be boiled, the distillate will always contain a certain amount of this substance.

If boric acid be introduced into an excess of a strong hot solution of sodium hydroxide, then, on slowly cooling, the salt NaBO$_2$,4H$_2$O crystallises out. This salt contains an equivalent of Na$_2$O to one equivalent B$_2$O$_3$. It might be termed a neutral salt did it not possess strongly alkaline reactions and easily split up into the alkali and the more stable borax or biborate of sodium mentioned above, which contains 2B$_2$O$_3$ to Na$_2$O. This salt is prepared by the action of boric acid on a solution of sodium carbonate. Borax may be perfectly purified by crystallisation. If a saturated and hot solution of borax be mixed with strong hydrochloric acid, common salt and a normal crystalline hydrate of *boric acid* are formed. The composition of this hydrate is B(HO)$_3$, according to the form BX$_3$—that is, of the composition B$_2$O$_3$,3H$_2$O. This is the easiest method of obtaining pure boric acid. The water is easily expelled from this hydrate; it loses half at 100° and the remainder on further heating, and the remaining B$_2$O$_3$ or boric anhydride fuses at 580° (according to Carnelley), forming at first a ductile (easily drawn out into threads), tenacious mass and then a colourless liquid solidifying to a transparent glass, which absorbs moisture from the atmosphere and then becomes cloudy. Only the alkaline salts of boric acid are soluble in water, but all borates are soluble in acids, owing to their easy decomposability and the solubility of boric acid itself. Although boric anhydride, B_2O_3, absorbs $3H_2O$ from damp air, still in the presence of water it always combines with a less quantity of bases (borax only contains $\frac{1}{6}$). However, fused boric anhydride forms a crystalline compound with magnesium of the same type as the hydrate $(MgO)_3B_2O_3$ (Ebelmann), and even with sodium it forms $(Na_2O)_3B_2O_3$ or Na_3BO_3 (Benedict). As a rule, the salts of boric acid contain less base, although they are all able to form saline compounds with bases when fused. Generally, vitreous fluxes are formed by this means, which when fused recall ordinary aqueous solutions in many respects. Some of them crystallise on solidifying, and then they have, like salts, a definite composition. The property of boric anhydride of forming higher grades of combination with basic oxides when fused explains the power of fused borax to dissolve metallic oxides, and the experiments of Ebelmann on the preparation of artificial crystals of the precious stones by means of boric anhydride. Boric anhydride is, although with difficulty, volatile at a high temperature, and therefore if it dissolves an oxide, it may be partially driven off from such a solution by prolonged and powerful ignition; in which case the oxides previously in solution separate out in a crystalline form, and frequently in the same forms as those in which they occur in nature—for example, crystals of alumina, which by itself fuses with difficulty, have been obtained in this manner. It dissolves in molten boric anhydride, and separates out in natural rhombohedric crystals. In this way Ebelmann also obtained *spinel*—that is, a compound of magnesium and aluminium oxides which occurs in nature.

Free *boron* was obtained (1809) by Davy, Gay-Lussac, and Thénard when they obtained the metals of the alkalis, for boric anhydride when fused with sodium gives up its oxygen to the sodium, and free boron is liberated as an *amorphous* powder like charcoal. It is of a brown colour, specific gravity 2·45 (Moissan), and when dry does not alter in the air at the ordinary temperature; but it burns when ignited to 700°, and in so doing combines not only with the oxygen of the air, but also with the nitrogen. However, the combustion is never complete, because the boric anhydride formed on the surface covers the remaining mass of the boron, and so preserves it from the action of the oxygen. Acids, even sulphuric (forming SO_2) and phosphoric (forming phosphorus), easily oxidise amorphous boron, especially when heated, converting it into boric acid. Alkalis have the same action on it, only in this case hydrogen is evolved. Boron decomposes steam at a red heat, also with evolution of hydrogen.

Amorphous boron, like charcoal, dissolves in certain molten metals. The property of fused *aluminium of dissolving boron* in considerable quantity is very striking; on cooling such a solution, the boron partially combined with the aluminium separates out in a crystalline form, and its properties are then exceedingly remarkable. The crystalline boron may be obtained by heating (to 1,300°) the pulverulent boron with aluminium in a well-closed crucible, the access of air being prevented as far as possible. After cooling, crystals are observed on the surface of the aluminium, and may easily be separated by dissolving the latter in hydrochloric acid, which does not act on the crystals. The specific gravity of the crystals is 2·68; they are partially transparent, but are for the most part coloured dark brown; they contain about 4 p.c. of carbon and up to 7 p.c. of aluminium, so that they cannot be considered as pure boron. Nevertheless, the properties of this *crystalline* substance, which was obtained by Wöhler and Deville, are very remarkable. It most closely resembles *the diamond in its properties*—in fact, these crystals have the lustre and high refracting power proper to the diamond only, whilst their hardness competes with that of the diamond. Their powder polishes even the diamond, and like the diamond scratches the sapphire and corundum. Crystalline boron is much more stable with respect to chemical reagents than the amorphous variety, and as it resembles the diamond, so amorphous boron, on the other hand, distinctly recalls certain of the properties of charcoal; thus a certain resemblance exists between boron and carbon in a free state, which is further justified by the proximity of their positions in the periodic system.

Among the other compounds of boron, those with nitrogen and the halogens are the most remarkable. As already mentioned above, amorphous boron combines directly with *nitrogen* at a red heat. If it be heated in a glass tube in a stream of nitric oxide, perfect combustion takes place, $5B + 3NO = B_2O_3 + 3BN$. If the residue be treated with nitric acid, the boric anhydride dissolves, whilst the *boron nitride* remains as an extremely light white powder, which is sometimes partially crystalline and greasy to the touch, like talc. It is infusible and unchanged, even at the melting-point of nickel. In general, it is remarkable for its great stability with respect to chemical reagents. Nitric and hydrochloric acids, as well as alkaline solutions, and hydrogen and chlorine at a red heat, have no action on it. When fused with potash, it evolves ammonia, and when ignited in steam it also yields ammonia: $2BN + 3H_2O = B_2O_3 + 2NH_3$.

No less remarkable is the compound of boron with fluorine—*boron fluoride*, BF_3. It is produced in many instances when compounds of boron and of fluorine are brought together. The most convenient method of preparing it is by heating a mixture of calcium fluoride with boric anhydride and sulphuric acid, $3CaF_2 + B_2O_3 + 3H_2SO_4 = 3CaSO_4 + 3H_2O + 2BF_3$. It is a colourless liquefiable *gas* (the liquid boils at -100°), which on coming into contact with damp air forms white fumes, owing to its combining with water. One volume of water dissolves as much as 1,050 volumes of this gas (Bazaroff), forming a liquid which disengages boron fluoride when heated, and distils over unaltered. Boron fluoride chars organic matter, owing to its taking up the water from it, and in this respect it acts like sulphuric acid. The behaviour of boron fluoride with water must be understood as a reversible reaction, since with water it yields hydrofluoric and boric acids, whilst they, acting on one another, re-form boron fluoride and water. A state of equilibrium is set up between these four substances (and between two reversible reactions) which is distinctly dependent on the mass of the water.[14 bis] When boron fluoride is in great excess, the equilibrated system, which is capable of distilling over (sp. gr. of the liquid 1·77), has a composition $BF_3,2H_2O$ (or $B_2O_3,H_2O,6HF$). It has also its corresponding salts. It is a caustic liquid, having the properties of a powerful acid; but it does not act on glass, which shows that there is no free hydrofluoric acid present. Under the action of water this system changes, with the formation of boric acid and hydroborofluoric acid (HBF_4) according to the equation $4BF_3H_4O_2 = 3HBF_4 + BH_3O_3 + 5H_2O$. This hydroborofluoric acid has its corresponding salts—for instance, KBF_4. On evaporating the aqueous solution this free acid decomposes, with the evolution of hydrofluoric acid, and a stable system is again obtained: $2HBF_4 + 5H_2O = B_2F_6H_{10}O_5 + 2HF$. The resultant solution (containing $2BF_3,5H_2O$, sp. gr. 1·58), which is identical with that formed by the evaporation of a solution of boric acid with hydrofluoric acid, again only contains a compound of boron fluoride with water. Probably there are various other possible and more or less stable states of equilibrium and definite compounds of boron fluoride, hydrofluoric acid, and water.

Nothing of this kind occurs with boron chloride, because hydrochloric acid does not act on boric acid. However, amorphous boron at 400° burns in chlorine, and at 410° forms *boron chloride*, BCl_3. The boron burns in the chlorine, forming a gas which, in a freezing mixture, condenses into a liquid boiling at 17°, and gives up its excess of chlorine, if there be any, to mercury. The specific gravity of this liquid is 1·42 at 6°. Boron chloride may also be directly obtained from boric anhydride by the simultaneous action of charcoal and chlorine at a high temperature: $B_2O_3 + 3C + 3Cl_2 = 2BCl_3 + 3CO$. It is also obtained by the action of phosphoric chloride on boric anhydride in a closed tube at 200° It is completely decomposed by water, like the chloranhydride of an acid, boric acid being formed; hence it fumes in the air: $2BCl_3 + 6H_2O = 2BH_3O_3 + 6HCl$. Boron forms with bromine a similar compound, BBr_3, specific gravity at 6° = 2·64, boiling at 90°. The vapour densities of the fluoride, chloride, and bromide of boron show that they contain three atoms of the halogen in the molecule—that is, that boron is a trivalent element forming BX_3.[16 bis]

As in the first group lithium is followed by sodium, giving a more basic oxide, so in the second group beryllium is followed by magnesium, and so also in the third group there is, besides the lightest element, boron, whose basic character is scarcely defined, *aluminium*, Al = 27, whose oxide, alumina, has somewhat distinct basic properties, which, although not so powerful as in magnesium oxide, are more distinct than in boric anhydride. Among the elements of the third group, aluminium is the most widely distributed in nature; it will be sufficient to mention that it enters into the composition of clay to demonstrate the universal distribution of aluminium in the earth's crust.

Alumina is so named from its being the metal of alums (*alumen*).

Clay, which is so widely distributed and familiar to everybody, is the insoluble residue obtained after the action of water containing carbonic acid on many rocks, and especially on the felspars contained in some of them. Felspar is a compound containing potash or soda, alumina, and silica. The primary rocks, like granite, contain many similar compounds (*see* Chapter XVIII.: Felspars). Felspar is acted on by water containing carbonic acid, all the alkalis (potash and soda), and a portion of the silica passing into the water as substances which are soluble and carried away by it, whilst the alumina and silica left from the felspar remain on the spot where the solution has taken place. This is the original method of the formation of clay in its primary deposits among rocks along whose crevices the atmospheric water has permeated. Such primary deposits often contain a white pure clay, termed *kaolin* or *porcelain clay*. But such clay is a rarity, because the conditions for its formation are rarely met with. The water, whilst acting chemically on rocks, at the same time destroys them *mechanically*, and carries off the finely divided residues of disintegration with it. Clay is most easily subjected to this mechanical action of water, because it is composed of grains of exceedingly small size and void of any visible crystalline structure, which easily remain suspended in water. The cloudy water of running mountain streams generally contains particles of clay in suspension, owing to the above-described chemical and mechanical action of the water on the minerals contained in the mountain rocks. Together with these minute particles of clay the water carries away the coarser components on which it is not able to act—for example, splinters of rock, grains of mica, quartz, &c. They were originally held together by those minerals which form clay. When the water acts on these binding minerals, a sandy mass is formed which water bears away. The cloudy water in which the particles of clay and sand are held in suspension carries them to, and deposits them at, the estuaries of rivers, lakes, seas, and oceans. The coarser particles are first deposited and form sand and similar disintegrated rocky matter, whilst the clay, owing to its finely divided state, is carried on further, and is only deposited in the still parts of the rivers, lakes, &c. Such disintegrations of rocks and separations of clay from sand have been gradually going on during the millions of years of the earth's existence, and are now proceeding, and have been the cause of the formation of the immense deposits of sandstone and clay now forming a part of the earth's strata. Such beds of clay may have been transferred by currents and streams from one locality to another, so that we must distinguish between primary and secondary deposits of clay. In places these beds of clay have, owing to long exposure under water, and perhaps partially owing to the action of heat, undergone compression, and have formed the rocky masses known as clay slates and schists, which sometimes form entire mountains. Roofing slates belong to this class of rocks.

From what has been said above it will be evident that these deposits can never consist of a chemically pure and homogeneous substance, but will contain all kinds of extraneous insoluble finely divided matter, and especially sand—that is, fragments of rock, chiefly quartz (SiO_2). It is, however, possible to considerably purify clay from these impurities, owing to the fact that they are the result of mechanical disintegration, whilst the clay has been formed as a residue of the chemical alteration of rocky matter, and therefore its particles are incomparably more minute than the particles of sand and other rock fragments mixed with it. This difference in the size of the grains causes the clay to remain longer in suspension when shaken up in water than the coarser grains of sand. If clay be shaken up in water, and especially if it be previously boiled in it, and if after the first portion has settled the cloudy water be decanted, it will give a deposit of a very much purer clay than the original. This method is employed for purifying kaolin designed for the manufacture of the best kinds of china, earthenware, &c. A similar method is also employed in the investigation of earths for determining the *composition of soils* chiefly composed of a mixture of sand, clay, limestone, and mould. The limestone is soluble in dilute acids, but neither the clay nor sand passes into solution by this means, and therefore the limestone is easily separated in the investigation of soils. The clay is separated from the sand by a mechanical method similar to that described above, and termed *levigation*.

By treating clay with strong sulphuric acid, which dissolves the alumina in it, and then (by means of an alkaline carbonate) dissolving the silica which was combined with the alumina in the clay (but not that occurring in the form of sand, &c., which is hardly dissolved by carbonate of soda solution at all even on boiling), we may form an idea of the proportion between the component parts of a clay; and by igniting it at a high temperature, we may determine the amount of water held in it. In the purer sorts of clay dried at 100° (sp. gr. of pure kaolin is about 2·5) this proportion is about $2SiO_2 : 2H_2O : Al_2O_3$. In this case the conversion of felspar into kaolin is expressed by the equation:—

$K_2O,Al_2O_3,6SiO_2 = Al_2O_3,2SiO_2 + K_2O,4SiO_2;$

    Felspar   Kaolin

the compound $K_2O,4SiO_2$ passes into solution.

But as a rule clays contain from 45 to 60 p.c. of silica, from 20 to 30 p.c. of alumina, and about 12 p.c. of water; and it cannot be supposed that clays are always homogeneous, because they are an aggregation of residues (of silico-aluminous compounds) which are unacted on by water. Nevertheless, clays always contain a hydrous compound of alumina and silica, which is able to give up the alumina contained by it as a base to strong sulphuric acid, forming aluminium sulphate, which is soluble in water. After this treatment the silica remains, and is soluble in a solution of an alkaline carbonate.

Clay is the source from which alumina, Al_2O_3, and the majority of the compounds of aluminium are prepared. Among these compounds the most important are the alums—that is, the double sulphates of potassium (and allied metals) and aluminium, $AlK(SO_4)_2,12H_2O$. When clay is treated with sulphuric acid diluted with a certain amount of water, aluminium sulphate, $Al_2(SO_4)_3$, is formed; and if potassium carbonate or sulphate be added to this solution, a double salt or alum is obtained in solution. The alums crystallise easily, and are prepared on a very large manufacturing scale owing to their being employed in the process of dyeing. Alums are soluble in water, and, on the addition of ammonia to their solutions, they give *hydrous alumina*, or *aluminium hydroxide*, as a white gelatinous precipitate, which is insoluble in water but easily soluble in acids, even when dilute, and in aqueous soda or potash. The solubility of alumina in acids indicates the basic character of the oxide, and its solubility in alkalis and its power of forming compounds with them shows the weakness of this basic character. However, the feeblest acids, even carbonic acid, take up the alkali from such a solution, and the alumina then separates out in a precipitate as the hydroxide. It must also be remembered as characteristic of the salt-forming properties of alumina that it does not combine with such feeble acids as carbonic, sulphurous, or hypochlorous, &c.—that is, its compounds with these acids are decomposed by water. It is also important to observe that the hydroxide is not soluble in aqueous ammonia.

Alumina, Al_2O_3—that is, the anhydrous aluminium oxide—is met with in nature, sometimes in a somewhat pure state, having crystallised in transparent crystals, which are often coloured by impurities (chromic, cobaltic, and ferric compounds). Such are the ruby and sapphire, the former red and the latter blue. They have a specific gravity 4·0, are distinguished by their very great hardness, which is second only to that of the diamond, and they represent the purest form of alumina. They are found in Ceylon and other islands of the Indian Archipelago, embedded in a rock matrix.[18 bis] *Corundum* is the same crystallised anhydrous alumina coloured brown by a trace of oxide of iron. A very much larger portion of this impurity occurs in *emery*, which is found in crystalline masses in Asia Minor and in Massachusetts, and owing to its extreme hardness is employed for polishing stones and metals. In this anhydrous and crystalline state the aluminium oxide is a substance which very powerfully resists the action of reagents, and is insoluble both in solutions of the alkalis and in strong acids. It is only capable of passing into solution after being fused with alkalis. Alumina may be obtained in this form by artificial means if the hydroxide be ignited and then fused in the oxyhydrogen flame. Alumina also occurs in nature in combination with water—as, for instance, in the rather rare minerals *hydrargillite* (sp. gr. 2·3), $Al_2O_3,3H_2O = 2Al(HO)_3$, and *diaspore*, $Al_2O\_3,H_2O = 2AlO(HO)$ (sp. gr. 3·4). A less pure hydrate, mixed with ferric oxide, sometimes occurs in masses (at Baux in the south of France) and is termed *bauxite*; it contains $Al_2O_3,2H_2O = Al_2O(HO)_4$ (sp. gr. 2·6). When bauxite is ignited with sodium carbonate, carbonic anhydride is liberated and the alumina then combines with the sodium oxide, forming a saline aluminate of the oxides of aluminium and sodium. This is taken advantage of in practice for the preparation of pure alumina compounds on a large scale, for bauxite is found in large masses (in the South of France, in Austria, and in Carolina in South America), and the resultant compound of alumina and sodium is soluble in water and does not contain ferric oxide. This solution when subjected to the action of carbonic anhydride gives a precipitate of aluminium hydroxide, which with acids forms aluminium salts. If aqueous ammonia be added to a solution of aluminium sulphate a gelatinous precipitate is formed, which at first remains suspended in the liquid and then on settling forms a gelatinous mass, which itself indicates the *colloidal property of aluminium hydroxide*. The following points are characteristic of this colloidal state: (1) in an anhydrous state such a colloidal substance is insoluble in water, as alumina is; (2) in the hydrated state, it is gelatinous and insoluble in water; and (3) it is also capable of existing in solutions, from which it separates out in a non-crystalline state, forming a substance resembling glue. These different states of colloids were distinguished by Graham, who gave them the following very characteristic names. He called the gelatinous form of the hydrate *hydrogel*, *i.e.* a gelatinous hydrate, and the soluble form of the aqueous compound, *hydrosol*, from the Latin for a soluble hydrate. Alumina readily and frequently assumes these states. The gelatinous hydrate of alumina is its hydrogel. It is, as has been already mentioned, insoluble in water, and, like all similar hydrogels, shows not the faintest sign of crystallisation; it is apt to vary in many of its properties with the amount of water it contains, and loses its water on ignition, leaving a white powder of the anhydrous oxide. The hydrogel of alumina is soluble both in acids and alkalis. It may also be obtained by the evaporation of its solutions in such feebly energetic acids as volatile acetic acid. These properties are very frequently made use of in the arts, and especially in *the processes of dyeing*, because the hydrogel of alumina in precipitating attracts a number of colouring matters from their solutions, the precipitate being thus coloured by the dyes attracted. The preparation of fixed dyes and the employment of aluminous compounds (mordants) in the processes of dyeing are founded on this fact. When precipitated upon the fibres of tissues (calicoes, linens, &c.) the aluminium hydroxide renders them impermeable to water; this may be taken advantage of for the preparation of waterproof tissues.

The hydrosol of alumina—*i.e.* the soluble aluminium hydroxide—is more difficult to obtain. In order to obtain this soluble variety of alumina, Graham took a solution of its hydrogel in hydrochloric acid—that is, a solution of aluminium chloride, which is able to dissolve a still further quantity of the hydrogel of alumina, forming a basic salt having probably one of the compositions $Al(HO)Cl_2$ or $Al(HO)_2Cl$. When such a solution, considerably diluted with water, is subjected to dialysis—that is, to diffusion through a membrane—the hydrochloric acid diffuses through the membrane and leaves the alumina in the form of hydrosol. The resultant solution, even when only containing two or three per cent. of alumina, passes into the hydrogel state with such facility that it is sufficient to transfer it from one vessel to another which has not been previously washed with water, for the entire mass to solidify into a jelly. But a solution containing not more than one-half per cent. of alumina may even be boiled without coagulating; however, after the lapse of several days this solution will of its own accord yield the hydrogel of alumina.[25 bis]

With respect to alumina as a base, it is very important to observe that it is not only capable of combining with other bases but that it does not give salts with feeble volatile acids (like carbonic and hypochlorous); it forms salts which are easily decomposed by water, especially when heated, as well as double and basic salts, *so that it forms a clear example of a feeble base*. To these characteristics of alumina we must add that it not only gives compounds of the type AlX_3, but also the polymeric type Al_2X_6, even when X is a simple univalent haloid like chlorine. Deville and Troost showed (1857) that the vapour density of aluminium chloride (at about 400°) is 9·37 with respect to air—that is, nearly 135 with respect to hydrogen, and therefore the formula of its molecule is expressed by Al_2Cl_6, and not $AlCl_3$, although in the case of boron, arsenic, and antimony, which give oxides R_2O_3 of the same composition as Al_2O_3, the chlorine compounds form non-polymeric molecules, BCl_3, $AsCl_3$, $SbCl_3$. This duplication (polymerisation) of the form AlX_3 is connected with the facility with which the salts of aluminium combine with other salts to form double salts and with aluminium hydroxide itself to form basic salts.

Aluminium sulphate, $Al_2(SO_4)_3$, which is obtained by treating clay or the hydrates of alumina with sulphuric acid, crystallises in the cold with $27H_2O$, or at the ordinary temperature in pearly crystals, which are greasy to the touch and contain $16H_2O$. Its solutions act like sulphuric acid—for instance, they evolve hydrogen with zinc, forming basic salts, which are sometimes met with in nature (*aluminite*, $Al_2O_3,SO_3,9H_2O$, *alumiane*, $Al_2O_3,2SO_3$, and others), and may be obtained by the decomposition of normal salts and by the direct solution of the hydroxide in normal salts: these exhibit a varying composition, $(Al_2O_3)_n(SO_3)_m(H_2O)_q$, where m/n is less than 3. Aluminium sulphate is now prepared (from the pure hydrate obtained from bauxite, Note 21) in large quantities for dyeing purposes (instead of alums) as a mordant. With solutions of the alkali sulphates (potassium, sodium, ammonium, rubidium, and cæsium sulphates), the normal salt easily forms double salts, termed *alums*—for example, the ordinary crystalline alum contains $KAl(SO_4)_2,12H_2O$, or $K_2SO_4,Al_2(SO_4)_3,24H_2O$. In the ammonium alums (which leave a residue of alumina when ignited) the potassium is replaced by ammonium (NH_4). Alums are used in large quantities, because there is scarcely any other salt which crystallises so easily. In this respect the alums formed by potassium and ammonium are equally convenient to purify, because they present a considerable difference in their solubility at the ordinary and higher temperatures. If the crystallisation be conducted rapidly, the salt separates in minute crystals, but if it be slowly deposited, especially in large masses, as in factories, then crystals several centimetres long are sometimes obtained. At a higher temperature alums are very much more soluble, and crystallise with greater difficulty, and are therefore less easily freed from impurities; at 0° 100 parts of water dissolve 3 parts, at 30° 22 parts, at 70° 90 parts, and at 100° 357 parts of potassium alum. The solubility of ammonium alum is slightly less. The specific gravity of potassium alum is 1·74, of ammonium alum 1·63, and of sodium alum 1·60. Alums easily part with their water of crystallisation; thus potash alum partially effloresces when exposed to the air, and loses 9 mol. H_2O under the receiver of an air-pump. At 100°, dry air passed over alums takes up nearly all their water. As we have already mentioned (Chapter XV.), the law of isomorphous substitutions exhibits itself more clearly in the alums than in any other salts, and all alums not only contain the same amount of water of crystallisation, $MR(SO_4)_2,12H_2O$ (where M = K, NH_4, Na; R = Al, Fe, Cr), and appear in crystals whose planes are inclined at equal angles, but they also give every possible kind of isomorphous mixture. The aluminium in them is easily replaced by iron, chromium, indium and sometimes by other metals, whilst the potassium may be substituted by sodium, rubidium, ammonium, and thallium, and the sulphuric acid may be replaced by selenic and chromic acids.

Aluminium chloride, Al_2Cl_6, is obtained, like other similar chlorides, (for instance $MgCl_2$) either directly from chlorine and the metal, or by heating to redness an intimate mixture of the amorphous anhydrous oxide and charcoal in a stream of dry chlorine.[33 bis] The resultant sublimate is very volatile, and forms a crystalline, easily fusible mass, which deliquesces in the air and easily dissolves in water, with the evolution of a large amount of heat.[34 bis] On evaporating this solution, hydrochloric acid and aluminium hydroxide are liberated. But if the solution be heated in a closed tube, with an excess of hydrochloric acid, then, on cooling, crystals of $AlCl_3,6H_2O$ are obtained—that is, aluminium chloride both combines with water and is decomposed by it. And the faculty of the type AlX_3 for combining with other molecules is seen in the compounds of $AlCl_3$ with many other chlorine compounds. Thus, for example, a mixture of aluminium chloride with sulphur tetrachloride gives Al_2Cl_6,SCl_4, under the action of chlorine, whilst with phosphorus pentachloride it forms $AlCl_3,PCl_5$; it also combines with NOCl. Thus, the compounds $AlCl_3,NOCl$, $AlCl_3,POCl_3$, $AlCl_3,3NH_3$, $AlCl_3,KCl$, $AlCl_3,NaCl$ are known. The compound of aluminium and sodium chlorides, $AlNaCl_4$, is very fusible and much more stable in the air than aluminium chloride itself. It seems to be of the same type as the alums. This compound, $AlNaCl_4$, is employed in the extraction of metallic aluminium, as we shall presently proceed to describe. Aluminium bromide, which is obtained by the direct combination of metallic aluminium with bromine, closely resembles the chloride; it melts at 90°, volatilises at 270°, and its vapour density indicates the formula Al_2Br_6. Aluminium iodide is obtained by heating iodine with finely divided aluminium in a closed tube; it is so easily decomposed by oxygen that its vapour even explodes when mixed with it.

Metallic Aluminium was first prepared by Wöhler in 1822 as a grey powder by the action of potassium on aluminium chloride. He afterwards (in 1845) obtained it as a white compact metal, unoxidisable in the air, and only slowly attacked by acids. Owing to the vast and wide occurrence of clay, many efforts have been made in investigating in detail the methods for the extraction of this metal. These efforts were brought to a successful issue (1854) by Sainte-Claire Deville, who is also renowned for his doctrine of dissociation. Experiments on a large scale have proved that metallic aluminium, although possessed of great lightness, strength, and durability, is not so generally suitable for technical purposes as was at first thought. Nitric and many other acids, indeed, do not act on it, but the alkalis, alkaline substances, and even salts—for instance, moist table salt—

humidity, &c.,[36 bis] tarnish it, and hence objects made of aluminium suffer at the surfaces, alter, and cannot, as was hoped, replace the precious metals, from which it differs in its extreme lightness. But the alloys made with aluminium (especially with copper, for example aluminium bronze) are very valuable in their properties and applications.

The Deville method for the preparation of metallic aluminium is based on the decomposition of the above-mentioned compound of sodium and aluminium chlorides by metallic sodium. The compound is obtained by passing the vapour of aluminium chloride (evolved from a mixture of alumina, extracted from bauxite or cryolite, with charcoal ignited in a stream of chlorine) over red-hot salt, when the compound $AlNaCl_4$, is itself volatilised, and may in this manner be obtained pure. A mixture of this compound with salt and fluor spar, or with cryolite, is heated with a certain excess of sodium, cut into small lumps. On a large scale this operation is carried on in special furnaces with a small access of air and at a high temperature. The decomposition takes place chiefly according to the equation $NaAlCl_4 + 3Na = 4NaCl + Al$. Neither charcoal nor zinc will reduce the oxygen compounds of aluminium; even sodium and potassium do not act on alumina. Moreover, metallic aluminium, like magnesium, is able to reduce even the metals of the alkalis from their oxygen compounds. This is connected with the fact that the atom of oxygen evolves more heat in combining with Al (and Mg) than it does in combining with other metals; whilst on the other hand, chlorine (and the other halogens) evolve more heat in combining with the metals of the alkalis.[36 tri]

Since the close of the eighties the metallurgy of aluminium has taken a new direction, based upon the action of an electric current upon cryolite at a high temperature, and the solution of oxide of aluminium (obtained from bauxite or in the form of corundum) in it; under these conditions metallic aluminium is reduced at the negative pole (cathode) in a sufficiently pure state, and if the cathode be copper, forms alloys with it. Such are Hall's and Cowle's (both in the United States) and the Neuhausen process (where the current is obtained from a dynamo worked by the Falls of the Rhine at Schaffhausen). As an example, we will describe (in the words of Prof. D. P. Konovaloff, who became acquainted with this process at the Chicago Exhibition), Hall's process as applied near Pittsburg, where it gives about 1,500 kilos of Al a day. An iron box (about 1 metre long and ½ metre wide), provided with a well rammed down charcoal lining, is charged with a mixture of cryolite and Al_2O_3 (from bauxite), over which salt is strewn, and a current of 5,000 ampères at 20 volts is passed through the mixture. The anode is composed of a carbon cylinder (about 9 cm. in diameter), while the charcoal lining forms the cathode. When the temperature inside the box is raised to a red heat by the current, the mixture fuses and the Al_2O_3

begins to decompose. The Al liberated collects at the bottom of the box, whilst the oxygen evolved burns the charcoal anode. When the decomposition is at an end, and the resistance of the mass increases, a fresh quantity of Al_2O_3 is added, and this is continued until the amount of impurities accumulated in the furnace and passing into the metal becomes too great.[37 bis]

Aluminium has a white colour resembling that of tin—that is, it is greyer than silver and has the feebly dull lustre of tin, but compared to tin and pure silver, aluminium is very hard. Its density is 2·67—that is, it is nearly four times lighter than silver and three times lighter than copper. It melts at an incipient red heat (600°), and in so doing is but slightly oxidised. At the ordinary temperature it does not alter in the air, and in a compact mass it burns with great difficulty at a white heat, but in thin sheets, into which it may be rolled, or as a very fine wire, it burns with a brilliant white light, since it forms an infusible and non-volatile oxide. Aluminium itself is non-volatile at a furnace heat. These properties render Al a very good reducing agent, and N. N. Beketoff showed that it reduces the oxides of the alkali metals (Chapter XIII., Note 42 bis). Dilute sulphuric acid has scarcely any action on it, but the strong acid dissolves it, especially with the aid of heat. Nitric acid, dilute or strong, has no action whatever on it. On the other hand, hydrochloric acid dissolves aluminium with great ease, as do also solutions of caustic soda and potash. In the latter cases hydrogen is evolved.

Aluminium forms alloys with different metals with great ease. Among them the copper alloy is of practical use. It is called *aluminium bronze*. This alloy is prepared by dissolving 11 p.c. by weight of metallic aluminium in molten copper at a white heat. The formation of the alloy is accompanied by the development of a considerable quantity of heat, so that it glows to a bright white heat. This alloy, which corresponds with the formula $AlCu_3$, presents an exceedingly homogeneous mass, especially if perfectly pure copper be taken. It is distinguished for its capacity to fill up the most minute impressions of the mould into which it may be cast, and by its extraordinary elasticity and toughness, so that objects cast from it may be hammered, drawn, &c., and at the same time it is fine-grained and exceedingly hard, takes an excellent polish, and, what is most important, its surface then remains almost unchangeable in the air, and has a colour and lustre which may be compared to that of gold alloys. Hence aluminium bronze is much used in the arts for making spoons, watches, vessels, forks, knives, and for ornaments, &c. No less important is the fact that the admixture of one-thousandth part of aluminium with steel renders its castings homogeneous (free from cavities) to an extent that could not be arrived at by other means, nor does the quality of the steel in any respect deteriorate by this admixture, but rather is it improved. In a pure state, aluminium is only employed for such objects as require the hardness of metals with comparative lightness, such as telescopes and various physical apparatus and small articles.

According to the periodic system of the elements, the analogues of magnesium are zinc, cadmium, and mercury in the second group. So also in the third group, to which aluminium belongs, we find its corresponding analogues *gallium*, *indium*, and *thallium*. They are all three so rarely and sparingly met with in nature that they could only be discovered by means of the spectroscope. This fact shows that they are partially volatile, as should be the case according to the property of their nearest neighbours, the very volatile zinc, cadmium and mercury. As with them, in gallium, indium, and thallium the density of the metal, decomposability of compounds, &c., rises with the atomic weight. But here we find a peculiarity which does not exist in the second group. In the latter, the fusibility increases with the atomic weight of magnesium, zinc, cadmium, and mercury; indeed, the heaviest metal—mercury—is a liquid. In the third group it is not so. In order to understand this it is sufficient to turn our attention to the elements of the further groups of the uneven series—for instance, to group V., containing phosphorus, arsenic, and antimony, or to group VI., with sulphur, selenium, and tellurium, and also to group VII., where chlorine, bromine and iodine are situated. In all these instances the fusibility decreases with a rise of atomic weight; the members of the higher series, the elements of a high atomic weight, fuse with greater difficulty than the lighter elements. The representatives of the uneven series of group III., aluminium, gallium, indium, thallium, forming, as they do, a transition, all show an intermediate behaviour. Here the most fusible of all is the medium metal gallium,[38 bis] which fuses at the heat of the hand; whilst indium, thallium, and aluminium fuse at much higher temperatures.

Zinc (group II.), which has an atomic weight 65, should be followed in group III. by an element with an atomic weight of about 69. It will be in the same group as Al and should consequently give R_2O_3, RCl_3, $R_2(SO_4)_3$, alums and similar compounds analogous to those of aluminium. Its oxide should be more easily reducible to metal than alumina, just as zinc oxide is more easily reduced than magnesia. The oxide R_2O_3 should, like alumina, have feeble but clearly expressed basic properties. The metal reduced from its compounds should have a greater atomic volume than zinc, because in the fifth series, proceeding from zinc to bromine, the volume increases. And as the volume of zinc = 9·2, and of arsenic = 18, that of our metal should be near to 12. This is also evident from the fact that the volume of aluminium = 11, and of indium = 14, and our metal is situated in group III., between aluminium and indium. If its volume = 11·5 and its atomic weight be about 69, then its density will be nearly 5·9. The fact that zinc is more volatile than magnesium gives reason for thinking that the metal in question will be more volatile than aluminium, and therefore for expecting its discovery by the aid of the spectroscope, &c.

These properties were indicated by me for the analogue of aluminium in 1871, and I named it (*see* Chapter XV.) eka-aluminium. In 1875, Lecoq de Boisbaudran, who had done much work in spectrum analysis, discovered a new metal in a zinc blende from the Pyrenees (Pierrefitte). He recognised its individuality and difference from zinc, cadmium, indium, and the other companions of zinc by means of the spectroscope; but he only obtained some fractions of a centigram of it in a free state. Consequently only a few of its reactions were determined, as, for instance, that barium carbonate precipitates the new oxide from its salts (alumina, as is known, is also precipitated). Lecoq de Boisbaudran named the newly discovered metal *gallium*. As one would expect the same properties for eka-aluminium as were observed in gallium, I pointed out this fact at the time in the Memoirs of the Paris Academy of Sciences. All the subsequent observations of Lecoq de Boisbaudran confirmed the identity between the properties of gallium and those indicated for eka-aluminium. Immediately after this the ammonium alum of gallium was obtained, but the most convincing proof of all was found in the fact that the density of gallium although first apparently different (4·7) from that indicated above, afterwards, when the metal was carefully purified from sodium (which was first used as a reducing agent), proved to be just that (5·9) which would have been looked for in the analogue of aluminium; and, what was very important, the equivalent (23·3) and atomic weight (69·8) determined by the specific heat (0·08) were shown by experiment to be such as would be expected. These facts confirmed the universality and applicability of the periodic system of the elements. It must be remarked that previous to it there was no means of either foretelling the properties or even the existence of undiscovered elements.

Much more light has been thrown on that element of the aluminium group which follows after cadmium (its position in the periodic system is III., 7, that is, it is in group III. in the 7th series). This is *indium*, In, which also occurs in small quantities in certain zinc ores. It was discovered (1863) by Reich and Richter (and more fully investigated by Winkler) in the Freiberg zinc ores, and was named indium from the fact that it gives to the flame of a gas-burner a blue coloration, owing to the indigo blue spectral lines proper to it. The equivalent (*see* Chapter XV., Note 15), specific heat, and other properties of the metal confirm the atomic weight In = 113.

Inasmuch as we found among the analogues of magnesium in group II. a metal, mercury, heavier and more easily reduced than the rest, and giving two grades of oxidation, so we should expect to find a metal among the analogues of aluminium in group III. which would be heavy, easily reduced, and give two grades of oxidation, and would have an atomic weight greater than 200. Such is *thallium*. It forms compounds of a lower type, TlX, besides the higher unstable type TlX$_3$, just as mercury gives HgX$_2$ and HgX. In the form of the thallic oxide, Tl$_2$O$_3$, the base is but feebly energetic, as would be expected by analogy with the oxides Al$_2$O$_3$, Ga$_2$O$_3$, and In$_2$O$_3$, whilst in thallous oxide, Tl$_2$O, the basic properties are sharply defined, as might be expected according to the properties of the type R$_2$O (Chapter XV.). *Thallium* was discovered in 1861 by Crookes and by Lamy in certain pyrites. When pyrites are employed in the manufacture of sulphuric acid, they are burned, and give besides sulphurous anhydride the vapours of various substances which accompany the sulphur, and are volatile. Among these substances arsenic and selenium are found, and together with them, thallium. These substances accumulate in a more or less considerable quantity in the tubes through which the vapours formed in the combustion of the pyrites have to pass. When the methods of spectrum analysis were discovered (1860), a great number of substances were subjected to spectroscopic research, and it was observed that those sublimations which are obtained in the combustion of certain pyrites contained an element having a very sharply-defined and characteristic spectrum—namely, in the green portion of the spectra it gave a well-defined band (wave-length 535 millionth millimetres) which did not correspond with any then known element.

Under the action of a galvanic current solutions of thallium salts deposit the metal in the form of a heavy powder. It is of a grey colour like tin, is soft like sodium, and has a metallic lustre. Its specific gravity is 11·8, it melts at 290°, and volatilises at a high temperature. When heated slightly above its melting point it forms an insoluble (in water) higher oxide, Tl$_2$O$_3$, as a dark-coloured powder, generally however accompanied by the lower oxide Tl$_2$O, which is also black but soluble in water and alcohol. This solution has a distinctly alkaline reaction. This *thallous oxide*, melts at 300°, and is easily obtained from the hydroxide TlHO by igniting it without access of air (in the presence of air the incandescent thallous oxide partly passes into thallic oxide). *Thallous hydroxide*, TlOH, crystallises with one molecule H$_2$O in yellow prisms which are very easily soluble in water. Metallic thallium may be used for its preparation, as the metal in the presence of water attracts oxygen from the air and forms the hydroxide. But metallic thallium does not decompose water, although it gives a hydroxide which is soluble in water.[41 bis] All the other data for the chemical and physical properties of thallium, of its two grades of oxidation and of their corresponding salts, are expressed by the position occupied by this metal in virtue of its atomic weight Tl = 204, between mercury Hg = 200, and lead Pb = 206.

Gallium, indium, and thallium belong to the uneven series, and there should be elements of the even series in group III. corresponding with calcium, strontium, and barium in group II. These elements should in their oxides R_2O_3 present basic characters of a more energetic kind than those shown by alumina, just as calcium, strontium, and barium give more energetic bases than magnesium, zinc, and cadmium. Such are *yttrium* and *ytterbium*, which occur in a rare Swedish mineral called *gadolinite*, and are therefore termed the gadolinite metals. To these belong also the metal *lanthanum*, which accompanies the two other metals *cerium* and *didymium* in the mineral *cerite*, and it therefore belongs to the cerite metals. All these metals and certain others accompanying them, give basic oxides R_2O_3. At first their formula was supposed to be RO, but the application of the periodic system required their being counted as elements of groups III. and IV., which was also confirmed by the determination of the specific heats of these metals, and better still by the fact that Nilson and Clève, in their researches on the gadolinite metals (1879), discovered that they contain a peculiar and very rare element, *scandium*, which by the magnitude of its atomic weight, Sc = 44, and in all its properties, exactly corresponds with the metal (previously foretold on the basis of the periodic system) *ekaboron*, whose properties were determined by taking the cerite and gadolinite metals as forming oxides R_2O_3.

The brevity of this work and the great rarity of the above-mentioned elements will give me the right to exclude their description, all the more as the principles of the periodic system enable many of their properties to be foreseen, and as their practical uses (cerium oxalate is used in medicine, and didymium oxide in the manufacture of glass, a mixture of the oxides of lanthanum and similar metals is employed for giving a bright light, as this mixture emits a brilliant white light when brought to incandescence) are very limited, by reason of their great rarity in nature, and the difficulty of separating them from one another.

Footnotes:

Borax is either directly obtained from lakes (the American lakes give about 2,000 tons and the lakes of Thibet about 1,000 tons per annum), or by heating native calcium borate (*see* Note 2) with sodium carbonate (about 4,000 tons per annum), or it is obtained (up to 2,000 tons) from the Tuscan impure boric acid and sodium carbonate (carbonic anhydride is evolved). Borax gives supersaturated solutions with comparative ease (Gernez), from which it crystallises, both at the ordinary and higher temperatures, in octahedra, containing $Na_2B_4O_7,5H_2O$. Its sp. gr. is 1·81. But if the crystallisation proceeds in open vessels, then at temperatures below 56°, the ordinary prismatic crystallo-hydrate $B_4Na_2O_7,10H_2O$ is obtained. Its sp. gr. is 1·71, it effloresces in dry air at the ordinary temperature, and at 0° 100 parts of water dissolve about 8 parts of this crystallo-hydrate, at 50° 27 parts, and at 100° 201 parts. Borax fuses when heated, loses its water and gives an anhydrous salt which at a red heat fuses into a mobile liquid and solidifies into a transparent amorphous *glass* (sp. gr. 2·37), which before hardening acquires the pasty condition peculiar to common molten glass. Molten borax dissolves many oxides and on solidifying acquires characteristic tints with the different oxides; thus oxide of cobalt gives a dark blue glass, nickel a yellow, chromium a green, manganese an amethyst, uranium a bright yellow, &c. Owing to its fusibility and property of dissolving oxides, borax is employed in soldering and brazing metals. Borax frequently enters into the composition of strass and fusible glasses.

We may mention the following among the minerals which contain boron: calcium borate, $(CaO)_3(B_2O_3)(H_2O)_6$, found and extracted in Asia Minor, near Brusa; *boracite* (stassfurtite), $(MgO)_6(B_2O_3)_8,MgCl_2$, at Stassfurt, in the regular system, large crystals and amorphous masses (specific gravity 2·95), used in the arts; *ereméeffite* (Damour), $AlBO_3$ or $Al_2O_3B_2O_3$, found in the Adulchalonsk mountains in colourless, transparent prisms (specific gravity 3·28) resembling apatite; *datholite*, $(CaO)_2(SiO_2)_2B_2O_3,H_2O$; and ulksite, or the boron-sodium carbonate from which a large quantity of borax is now extracted in America (Note 1). As much as 10 p.c. of boric anhydride sometimes enters into the composition of tourmalin and axinite.

This green coloration is best seen by taking an alcoholic solution of volatile ethyl borate, which is easily obtained by the action of boron chloride on alcohol.

[3 bis] P. Chigeffsky showed in 1884 (at Geneva) that in the evaporation of saline solutions many salts are carried off by the vapour—for instance, if a solution of potash containing about 17–20 grams of K_2CO_3 per litre be boiled, about 5 milligrams of salt are carried off for every litre of water evaporated. With Li_2CO_3 the amount of salt carried over is infinitesimal, and with Na_2CO_3 it is half that given by K_2CO_3. The volatilisation of B_2O_3 under these circumstances is incomparably greater—for instance, when a solution containing 14 grams of B_2O_3 per litre is boiled, every litre of water evaporated carries over about 350 milligrams of B_2O_3. When Chigeffsky passed steam through a tube containing B_2O_3 at 400°, it carried over so much of this substance that the flame of a Bunsen's burner into which the steam was led gave a distinct green coloration; but when, instead of steam, air was passed through the tube there was no coloration whatever. By

placing a tube with a cold surface in steam containing B_2O_3, Chigeffsky obtained a crystalline deposit of the hydrate $B(OH)_3$ on the surface of the tube. Besides this, he found that the amount of B_2O_3 carried over by steam increases with the temperature, and that crystals of $B(OH)_3$ placed in an atmosphere of steam (although perfectly still) volatilise, which shows that this is not a matter of mechanical transfer, but is based on the capacity of B_2O_3 and $B(OH)_3$ to pass into a state of vapour in an atmosphere of steam.

How it is that these vapours containing boric acid are formed in the interior of the earth is at present unknown. Dumas supposes that it depends on the presence of *boron sulphide*, B_2S_3 (others think boron nitride), at a certain depth in the earth. This substance may be artificially prepared by heating a mixture of boric acid and charcoal in a stream of carbon bisulphide vapour, and by the direct combination of boron and the vapour of sulphur at a white heat. The almost non-crystalline compound B_2S_3, sp. gr. 1·55, thus obtained is somewhat volatile, has an unpleasant smell, and is very easily decomposed by water, forming boric acid and hydrogen sulphide, $B_2S_3 + 3H_2O = B_2O_3 + 3H_2S$. It is supposed that a bed of boron sulphide lying at a certain depth below the surface of the earth comes into contact with sea water which has percolated through the upper strata, becomes very hot, and gives steam, hydrogen sulphide, and boric acid. This also explains the presence of ammonia in the vapours, because the sea water certainly passes through crevices containing a certain amount of animal matter, which is decomposed by the action of heat and evolves ammonia. There are several other hypotheses for explaining the presence of the vapours of boric acid, but owing to the want of other known localities the comparison of these hypotheses is at present hardly possible. The amount of boric anhydride in the vapours which escape from the Tuscan fumerolles and suffioni is very inconsiderable, less than one-tenth per cent., and therefore the direct extraction of the acid would be very uneconomical, hence the heat contained in the discharged vapours is made use of for evaporating the water. This is done in the following manner. Reservoirs are constructed over the crevices evolving the vapours, and the water of some neighbouring spring is passed into them. The vapours are caused to pass through these reservoirs, and in so doing they give up all their boric acid to the water and heat it, so that after about twenty-four hours it even boils; still this water only forms a very weak solution of boric acid. This solution is then passed into lower basins and again saturated by the vapours discharged from the earth, by which means a certain amount of the water is evaporated and a fresh quantity of boric acid absorbed; the same process is repeated in another reservoir, and so on until the water has collected a somewhat considerable amount of boric acid. The solution is drawn from the last reservoir A into settling vessels B D, and then into a series of vessels *a*, *b*, *c*. In these vessels, which are made of lead, the solution is also evaporated by the vapours escaping from the earth, and attains a density of 10° to 11° Baumé. It is allowed to settle in the vessel C, in which it cools and crystallises, yielding (not quite pure) crystalline boric acid. At temperatures above 100°, for instance, with superheated steam, boric acid volatilises with steam very easily.

FIG. 81.—Extraction of boric acid in Tuscany.

Metals, like Na, K, Li, give salts of the type of borax, MBO_2 or MH_2BO_3. A solution of borax, $Na_2B_4O_7$, has an alkaline reaction, decomposes ammonia salts with the liberation of ammonia (Bolley), absorbs carbonic anhydride like an alkali, dissolves iodine like an alkali (Georgiewics), and seems to be decomposed by water. Thus Rose showed that strong solutions of borax give a precipitate of silver borate with silver nitrate, whilst dilute solutions precipitate silver oxide, like an alkali. Georgiewics even supposes (1888) boric anhydride to be entirely void of acid properties; for all acids, on acting on a mixture of solutions of potassium iodide and iodate, evolve iodine, but boric acid does not do this. With dilute solutions of sodium hydroxide Berthelot obtained a development of heat equal to 11½ thousand calories per equivalent of alkali (40 grams sodium hydroxide) when the ratio $Na_2O : 2B_2O_3$ (as in borax) was taken, and only 4 thousand calories when the ratio was $Na_2O :$

B_2O_3, whence he concludes that water powerfully decomposes those sodium borates in which there is more alkali than in borax. Laurent (1849) obtained a sodium compound, $Na_2O,4B_2O_3,10H_2O$, containing twice as much boric anhydride as borax, by boiling a mixture of borax with an equivalent quantity of sal-ammoniac until the evolution of ammonia entirely ceased.

Hence it is evident that feeble acids are as prone to, and as easily, form acid salts (that is, salts containing much acid oxide) as feeble bases are to give basic salts. These relations become still clearer on an acquaintance with such feeble acids as silicic, molybdic, &c. This variety of the proportions in which bases are able to form salts recalls exactly the variety of the proportions in which water combines with crystallo-hydrates. But the want of sufficient data in the study of these relations does not yet permit of their being generalised under any common laws.

With respect to the feeble acid energy of boric anhydride I think it useful to add the following remarks. Carbonic anhydride is absorbed by a solution of borax, and displaces boric anhydride; but it is also displaced by it, not only on fusion, but also on solution, as the preparation of borax itself shows. Sulphuric anhydride is absorbed by boric acid, forming a compound $B(HSO_4)_3$, where HSO_4 is the radicle of sulphuric acid (D'Ally). With phosphoric acid, boric acid forms a stable compound, BPO_4, or $B_2O_3P_2O_5$, undecomposable by water, as Gustavson and others have shown. With respect to tartaric acid, boric anhydride is able to play the same part as antimonious oxide. Mannitol, glycerol, and similar polyhydric alcohols also seem able to form particularly characteristic compounds with boric anhydride. All these aspects of the subject require still further explanation by a method of fresh and detailed research.

Ditte determined the sp. gr.:—

	0°	12°	80°
B_2O_3	1·8766	1·8470	1·6988
$B(OH)_3$	1·5463	1·5172	1·3828
Solubility	1·95	2·92	16·82

The last line gives the solubility, in grams, of boric acid, $B(OH)_3$, per 100 c.c. of water, also according to the determinations of Ditte.

It is evident that, in the presence of basic oxides, water competes with them, which fact in all probability determines both the amount of water in the salts of boric acid as well as their decomposition by an excess of water.

In confirmation of the above-mentioned competing action between water and bases, I think it useful to point out that the crystallo-hydrate of borax containing 5H$_2$O may be represented as B(HO)$_3$, or rather as B$_2$(OH)$_6$, with the substitution of one atom of hydrogen by sodium, since Na$_2$B$_4$O$_7$,5H$_2$O = 2B$_2$(OH)$_5$(ONa). The composition of the acid boric salts is very varied, as is seen from the fact that Reychler (1893) obtained (Cs$_2$O)3B$_2$O$_3$, (Rb$_2$O)2B$_2$O$_3$ (corresponding to borax) and (Li$_2$O)B$_2$O$_3$, and that Le Chatelier and Ditte obtained, for CaO, MgO, &c., (RO)B$_2$O$_3$, (RO)$_2$3B$_2$O$_3$, (RO)2B$_2$O$_3$, (RO)$_2$B$_2$O$_3$, and even (RO)$_3$B$_2$O$_3$.

A glass can only be formed by those slightly volatile oxides which correspond with feeble acids, like silica, phosphoric and boric anhydrides, &c., which themselves give glassy masses, like quartz, glacial phosphoric acid, and boric anhydride. They are able, like aqueous solutions and like metallic alloys, to solidify either in an amorphous form or to yield (or even be wholly converted into) definite crystalline compounds. This view illustrates the position of solutions amongst the other chemical compounds, and allows all alloys to be regarded from the aspect of the common laws of chemical reactions. I have therefore frequently recurred to it in this work, and have since the year 1850 introduced it into various provinces of chemistry.

If boric acid in its aqueous solutions proves to be exceedingly feeble, unenergetic, and easily displaced from its salts by other acids, yet in an anhydrous state, as anhydride, it exhibits the properties of an energetic acid oxide, and it *displaces* the anhydrides of other acids. This of course does not mean that the acid then acquires new chemical properties, but only depends on the fact that the anhydrides of the majority of acids are much more volatile than boric anhydride, and therefore the salts of many acids— even of sulphuric acid—are decomposed when fused with boric anhydride.

By itself boric acid is used in the arts in small quantity, chiefly for the preservation of meat and fish (which must be afterwards well washed in water) and of milk, and for soaking the wicks of stearin candles; the latter application is based on the fact that the wicks, which are made of cotton twist, contain an ash which is infusible by itself but which fuses when mixed with boric acid.

Amorphous boron is prepared by mixing 100 parts of powdered boric anhydride with 50 parts of sodium in small lumps; this mixture is thrown into a powerfully heated cast-iron crucible, covered with a layer of ignited salt, and the crucible covered. Reaction proceeds rapidly; the mass is stirred with an iron rod, and poured directly into water containing hydrochloric acid. The action is naturally accompanied by the formation of sodium borate, which is dissolved, together with the salt, by the water, whilst the boron settles at the bottom of the vessel as an insoluble powder. It is washed in water, and dried at the ordinary temperature. Magnesium, and even charcoal and phosphorus, are also able to reduce boron from its oxide. Boron, in the form of an amorphous powder, very easily passes through filter-paper, remains suspended in water, and colours it brown, so that it appears to be soluble in water. Sulphur precipitated from solutions shows the same (colloidal) property. When borax is fused with magnesium powder, it gives a brown powder of a compound of boron and magnesium, Mg_2B (Winkler, 1890), but when a mixture of 1 part of magnesium and 3 parts of B_2O_3 is heated to redness (Moissan, 1892), it forms amorphous boron in the form of a chestnut-coloured powder, which, after being washed with water, hydrochloric and hydrofluoric acids, is fused again with B_2O_3 in an atmosphere of hydrogen in order to prevent the access of the nitrogen of the air, which is easily absorbed by incandescent amorphous boron.

Sabatier (1891) considers that a certain amount of gaseous hydride of boron is evolved in the action of hydrochloric acid upon the alloys of magnesium and boron, because the gas disengaged burns with a green flame. Still, the existence of hydride of boron cannot be regarded as certain.

Under the action of the heat of the electric furnace boron forms with carbon a *carbide*, BC, as Mühlhäuser and Moissan showed in 1893.

At first boron nitride was obtained by heating boric acid with potassium cyanide or other cyanogen compounds. It may be more simply prepared by heating anhydrous borax with potassium ferrocyanide, or by heating borax with ammonium chloride. For this purpose one part of borax is intimately mixed with two parts of dry ammonium chloride, and the mixture heated in a platinum crucible. A porous mass is formed, which after crushing and treating with water and hydrochloric acid, leaves boron nitride. *Boron fluoride*, BF, is known, corresponding to BN; this body was obtained by Besson and Moissan (1891). The action of phosphorus upon iodide of boron, BI_3, forms PBI_2, and when heated to 500° in hydrogen it forms BP, which gives PH_3 with fused KHO.

When fused with potassium carbonate it forms potassium cyanate, BN + K_2CO_3 = KBO_2 + KCNO. All this shows that boron nitride is a nitrile of boric acid, BO(OH) + NH_3 - $2H_2O$ = BN. The same is expressed by saying that boron nitride is a compound of the type of the boron compounds BX_3, with the substitution of X_3 by nitrogen, as the trivalent radicle of ammonia, NH_3.

Boron fluoride is frequently evolved on heating certain compounds occurring in nature containing both boron and fluorine. If calcium fluoride is heated with boric anhydride, calcium borate and boron fluoride are formed, and the latter, as a gas, is volatilised: $2B_2O_3 + 3CaF_2 = 2BF_3 + Ca_3B_2O_6$. The calcium borate, however, retains a certain amount of calcium fluoride.

In order to avoid the formation of silicon fluoride the decomposition should not be carried on in glass vessels, which contain silica, but in lead or platinum vessels. Boron fluoride by itself does not corrode glass, but the hydrofluoric acid liberated in the reaction may bring a part of the silica into reaction. Boron fluoride should be collected over mercury, as water acts on it, as we shall see afterwards.

[14 bis] It appears to me that from this point of view it is possible to understand the apparently contradictory results of different investigators, especially those of Gay-Lussac (and Thénard), Davy, Berzelius, and Bazaroff. In the form in which the reaction of BF_3 on water is given here, it is evident that the act of solution in water is accompanied by complex but direct chemical transformations, and I think that this example should prove the justness of those observations upon the nature of solutions which are given in Chapter I.

They are called fluoborates. They may be prepared directly from fluorides and borates. Such compounds of halogens with oxygen salts are known in nature (for instance, apatite and boracite), and may be artificially prepared. The composition of the fluoborates—for example, $K_4BF_3O_2$—may be expressed as that of a double salt, $BO(OK),3KF$. If an excess of water decomposes them (Bazaroff), this does not prove that they do not exist as such, for many double salts are decomposed by water.

Fluoboric acid contains boron fluoride and water, hydrofluoboric acid, boron fluoride, and hydrofluoric acid. It is evident that on the one side the competition between water and hydrofluoric acid, and, on the other hand, their power to combine, are among the forces which act here. From the fact that hydroborofluoric acid, HBF_4, can only exist in an aqueous solution, it must be assumed that it forms a somewhat stable system only in the presence of $3H_2O$.

[16 bis] Iodide of boron, BI_3, was obtained by Moissan (1891), by heating a mixture of the vapours of HI and BCl_3 in a tube, or by the action of iodine vapour (at 750°) or HI upon amorphous boron. BI_3 is a solid substance which dissolves in benzol and CS_2, reacts with water, melts at 43°, boils at 210°, has a density 3·3 at 50°, and partially decomposes in the light. Besson (1891) obtained $BIBr_2$ (boiling at 125°), and BI_2Br (boiling at 180°) by heating (300–400°) a mixture of the vapours of HI and BBr_3, and showed that NH_3 combines with BBr_3 and BI_3 in various proportions.

The process of *levigation* is based on the difference in the diameters of the particles of clay and sand. In density these particles differ but little from each other, and therefore a stream of water of a certain velocity can only carry away the particles of a certain diameter, whilst the particles of a larger diameter cannot be borne away by it. This is due to the resistance to falling offered by the water. This resistance to substances moving in it increases with the velocity, and therefore a substance falling into water will only move with an increasing velocity until its weight equals the resistance offered by the water, and then the velocity will be uniform. And as the weight of the minute particles of clay is small, the maximum velocity attained by them in falling is also small. A detailed account of the theory of falling bodies in liquid, and of the experiments bearing on this subject, may be found in my work, *Concerning the Resistance of Liquids and Aeronautics*, 1880. The minute particles of clay remain suspended longer in water, and take longer to fall to the bottom. Heavy particles, although of small dimensions, fall more quickly, and are borne away by water with greater difficulty than the lighter. In this way gold and other heavy ores are washed free from sand and clay, and the coarser portions and heavier particles are left behind. A current of water of a certain velocity cannot carry away with it particles of more than a definite diameter and density, but by increasing the velocity of the current a point may be arrived at when it will bear away larger particles. A description of apparatus for the observation of phenomena of this kind is given by Schöne in his memoir in the Transactions of the Moscow Society of Natural Sciences for 1867. In order to be able accurately to vary the velocity of the current of water, a cylinder is employed in which the earth to be experimented on is placed, and water is introduced through the conical bottom of the cylinder. The rate at which the water rises in the cylinder will vary according to the quantity of water flowing per unit of time into the vessel, and consequently particles of various sizes will be carried away by the water flowing over the upper edges of the vessel. Schöne showed by direct experiment that a current of water having a velocity of 0·1 mm. per second will carry away particles having a diameter of not more than 0·0075 mm., that is, only the most minute; with a velocity $v = 0.2$ mm. per second, particles having a diameter $d = 0.011$ mm. are carried away; with $v = 0.3$ mm., $d = 0.0146$ mm.; with $v = 0.4$ mm., $d = 0.017$ mm.; with $v = 0.5$ mm., $d = 0.02$ mm.; with $v = 1$ mm., $d = 0.03$ mm.; with $v = 4$ mm., $d = 0.07$ mm.; with $v = 10$ mm., $d = 0.137$ mm.; with $v = 12$ mm., $d = 0.15$ mm.; and therefore if the current does not exceed one of these velocities, it will only carry away or wash away particles having a diameter less than that indicated. The sand and other particles mixed with the clay will then remain in the vessel. The very minute particles obtained after levigation are all considered as clay, although not only clay but other rock residue may also exist in it as very fine particles.

However, this is very seldom the case, and the fine mud separated from all clays has practically the same composition as the purest kinds of kaolin.

The relation between the amounts of clay and sand in soils used for the cultivation of plants is very important, because a soil rich in clay is denser, heavier, shrinks up under the action of heat, and does not readily yield to the plough in dry or wet weather, whilst a soil rich in sand is friable, crumbling, easily parts with its moisture and dries rapidly, but is comparatively easily worked. Neither crumbling sand nor pure clay can be regarded as a good *cultivating soil*. The difference in the amounts of clay and sand in a soil has also a purely chemical signification. Sand is easily permeated by the air, because its particles are not closely packed together. Hence the chemical change of manures proceeds very easily in sandy soils. But on the other hand such soils do not retain the nutritious principles contained in the manure, nor the water necessary for the nourishment of plants by means of their roots. Solutions of nutritious substances, containing salts of potassium, phosphoric acid, &c., when passed through sand only leave a portion moistening the surface of its particles. The sand has only to be washed with pure water and all the adhering films of solution are washed away. It is not so with clay. If the above solutions be passed through a layer of clay the retention of the nutritive substances of these solutions will be very marked; this is partly because of the very large surface which the minute particles of clay expose. The nutritive elements dissolved in water are retained by the particles of clay in a peculiar manner—that is, the absorptive power of clay is very great compared to that of sand—and this has a great significance in the economy of nature (Chapter XIII., p. 547). It is evident that for cultivation the most convenient soils in every respect will be those containing a definite mixture of clay and sand, and indeed the most fertile soils have this composition. The study of fertile soils, which is so important for a knowledge of the natural conditions for the application of fertilisers, belongs, strictly speaking, to the province of agriculture. In Russia the first foundation of a scientific fertilisation has been laid by Dokuchaeff. As an example only, we will give the composition of four soils; (1) The black earth of the Simbirsk Government; (2) a clay soil from the Smolensk Government; (3) a more sandy soil from the Moscow Government; and (4) a peaty soil from near St. Petersburg. These analyses were made in the laboratory of the St. Petersburg University about 1860, in connection with experiments on fertilisation (conducted by me) by the Imperial Free Economical Society. 10,000 grams of air-dried soil contain the following quantities (in grams) of substances capable of dissolving in acids, and of serving for the nourishment of plants.

	(1)	(2)	(3)	(4)
Na_2O	11	5	4	4
K_2O	58	10	7	5
MgO	92	33	19	7
CaO	134	17	14	11
P_2O_5	7	1	7	3
N	44	11	13	16
S	13	7	7	6
Fe_2O_3	341	155	111	46

By chemical and mechanical analysis, the chief component parts per 100 parts of air-dried soil are

Clay	46	29	12	10
Sand	40	67	86	84
Organic matter	3·7	1·7	0·6	4·1
Hygroscopic water	6·3	1·3	0·8	1·9
Weight of a litre in grams	1150	1270	1350	960

The black earth excels the other soils in many respects, but naturally its stores are also exhausted by cultivation if nothing be returned to it in the form of fertilisers; and the improvement of a soil (for instance, by the addition of marl or peat, and by drainage and watering), and its fertilisation, if carried on in conformity with its composition and with the properties of the plants to be cultivated, are capable of rendering not only every soil fit for cultivation, but also of improving its value, so that in the course of time whole countries (like Holland) may clearly improve their agricultural position, whilst under the ordinary *régime* of continued exhaustion of the soil, entire regions (as, for instance, many parts of Central Asia) may be rendered unfit for any agriculture.

Everyone knows that a mixture of clay and water is endowed with the property of taking a given form when subjected to a moderate pressure. This plasticity of clay renders it an invaluable material for practical purposes. From clay are moulded and manufactured a variety of objects, beginning with the common brick and ending with the most delicate china works of art. This *plasticity of clay* increases with its purity. When articles made of clay are dried, the well-known hard mass is obtained; but water washes it away, and furthermore, the cohesion of its particles is not sufficiently great for it to resist the impression of blows, shocks, &c. If such an article be subjected to the action of heat, its volume first decreases, then it begins to lose water, and it shrinks still further (in the case of a compact mass approximately by $\frac{1}{5}$ of its linear measurement). On the other hand, a great coherence of particles is obtained, and thus burnt clay has the hardness of stone. Pure clay, however, shrinks so considerably when burnt that the form given to it is destroyed and cracks easily form; such vessels are also porous, so that they will not hold water. The addition of sand—that is, silica in fine particles—or of *chamotte*—that is, already burnt and crushed clay—renders the mass much more dense and incapable of cracking in the furnace. Nevertheless, such clay articles (bricks, earthenware vessels, &c.) are still porous to liquids after being burnt, because the clay in the furnace is only baked and does not fuse. In order to obtain articles impervious to water the clay must either be mixed with substances which form a glassy mass in the furnace, permeating the clay and filling up its pores, or else only the surface of the article is covered with such a glassy fusible substance. In the first case the purest kinds of clay give what is known as china, in the second case porcelain or 'faïence.' So, for instance, by covering the surface of clay articles with a layer of the oxides of lead and tin, the well-known white glaze is obtained, because the oxides of these metals give a white gloss when fused with silica and clay. In the preparation of china, fluor spar and finely ground silica is mixed up into the clay; these ingredients give a mass which is infusible but softens in the furnace, so that all the particles of the clay cohere in this softened mass, which hardens on cooling. A glaze composed of glassy substances, which only fuse at a high temperature, is also applied to the surface of china articles.

[18 bis] Frémy (1890) obtained transparent rubies, which crystallised in rhombohedra, and resembled natural rubies in their hardness, colour, size, and other properties. He heated together a mixture of anhydrous alumina containing more or less caustic potash, with barium fluoride and bichromate of potassium. The latter is added to give the ruby its colour, and is taken in small quantity (not more than 4 parts by weight to 100 parts of alumina). The mixture is put into a clay crucible, and heated (for from 100 hours to 8 days) in a reverberatory furnace at a temperature approaching 1,500°. At the end of the experiment the crucible was found to contain a crystalline mass, and the walls were covered with crystals of the ruby of a beautiful rose colour. It was found that the access of moist air

was indispensable for the reaction. According to Frémy, the formation of the ruby may be here explained by the formation of fluoride of aluminium which under the action of the moist air at the high temperature of the furnace gives the ruby and hydrofluoric acid gas.

The effects of purely mechanical subdivision on the solubility of alumina are evident from the fact that native anhydrous alumina, when converted into an exceedingly fine powder by means of levigation, dissolves in a mixture of strong sulphuric acid and a small quantity of water, especially when heated in a closed tube at 200°, or when fused with acid sulphate of potassium (*see* Chapter XIII., Note 9).

The preparation of crystallised alumina is given on p. 65, and in Note 18 bis. When alumina, moistened with a solution of cobalt salt, is ignited, it forms a blue mass called Thénard's salt. This coloration is taken advantage of not only in the arts, but also for distinguishing alumina from other earthy substances resembling it.

The treatment of bauxite is carried on on a large scale, chiefly in order to obtain alumina from alkaline solutions, free from ferric oxide, because in dyeing it is necessary to have salts of aluminium which do not contain iron. But this end, it would seem, may also be obtained by igniting alumina containing ferric oxide in a stream of chlorine mixed with hydrocarbon vapours, as ferric chloride then volatilises. K. Bayer observed that in the treatment of bauxite with soda, about 4 molecules of sodium hydroxide pass into solution to 1 molecule of alumina, and that on agitating this solution (especially in the presence of some already precipitated aluminium hydroxide), about two-thirds of the alumina is precipitated, so that only 1 molecule of alumina to 12 molecules of sodium hydroxide remains in solution. This solution is evaporated directly, and used again. He therefore treats bauxite directly with a solution of NaHO at 170° in a closed boiler, and on cooling adds hydrated alumina to the resultant solution. The greater part of the dissolved alumina then precipitates on this hydrated alumina, and the solution is used over again. The hydroxide which separates from the alkaline solution contains $Al(OH)_3$. All these properties bear a great resemblance to those of boric acid. It may be taken for granted that the relation between sodium hydroxide and alumina in solution varies with the mass of water.

If lime be added to a solution of alumina in alkali (sodium aluminate) calcium aluminate is precipitated, from which acids first extract the lime, leaving aluminium hydroxide, which is easily soluble in acids (Loewig). When sodium aluminate is mixed with a solution of sodium bicarbonate, a double carbonate of the alkali and aluminium is precipitated, which is easily soluble in acids.

These coloured precipitates of alumina are termed *lakes*, and are employed in dyeing tissues and in the formation of various pigments—such as pastels, oil colours, &c. Thus, if organic colouring matters, such as logwood, madder, &c., are added to a solution of any aluminium salt, and then an alkali is added, so that alumina may be precipitated, these pigments, which are by themselves soluble in water, will come down with the precipitate. This shows that alumina is able to combine with the colouring matter, and that this compound is not decomposed by water. The dyes then become insoluble in water. If a dye be mixed with starch paste and aluminium acetate, and then, by means of engraved blocks having a design in relief, we transfer this mixture to a fabric which is then heated, the aluminium acetate will leave the hydrogel of alumina which binds the colouring matter, and water will no longer be able to wash the pigment from the material—that is, a so-called 'fixed' dye is obtained. In the case of dyeing a fabric a uniform tint, it is first soaked in a solution of aluminium acetate and then dried, by which means the acetic acid is driven off, while the hydrogel of alumina adheres to the fibres of the material. If the latter be then passed through a solution of a dye in water, the former will be attracted to the portions covered with alumina, and closely adhere to them. If certain parts of the material be protected by the application of an acid, such as tartaric, $C_4H_6O_6$, oxalic, citric, &c. (these acids being non-volatile), the alumina will be dissolved in those parts, and the pigment will not adhere, so that after washing, a white design will be obtained on those parts which have been so protected.

In dye-works the aluminium acetate is generally obtained in solution by taking a solution of alum, and mixing it with a solution of lead acetate. In this case lead sulphate is precipitated and aluminium acetate remains in solution, together with either acetate or sulphate of potassium, according to the amount of acetate of lead first taken. The complete decomposition will be as follows: $KAl(SO_4)_2 + 2Pb(C_2H_3O_2)_2 = KC_2H_3O_2 + Al(C_2H_3O_2)_3 + 2PbSO_4$, or the less complete decomposition, $2KAl(SO_4)_2 + 3Pb(C_2H_3O_2)_2 = 2Al(C_2H_3O_2)_3 + K_2SO_4 + 3PbSO_4$. If the resultant solution of aluminium acetate be evaporated or further boiled, the acetic acid passes off and the hydrogel of alumina remains.

As the salt of potassium obtained in the solution passes away with the water used for washing, and the salt of lead precipitated has no practical use, this method for the preparation of aluminium acetate cannot be considered economical; it is retained in the process of dyeing mainly because both the salts employed, alum and sugar of lead, easily crystallise, and it is easy to judge of their degree of purity in this form. Indeed, it is very important to employ pure reagents in dyeing, because if impurity is present—such as a small quantity of an iron compound—the tint of the dye changes; thus madders give a red colour with alumina, but if oxide of iron be present the red changes into a violet tint. The aluminium hydroxide is soluble in alkalis, whilst ferric oxide is not. Therefore sodium aluminate—that is, the dissolved compound of alumina and caustic soda—obtained, as already described, from bauxite, is sometimes employed in dyeing. Every aluminium salt gives a solution containing sodium aluminate free from iron, when it is mixed with excess of caustic soda. This solution, when mixed with a solution of ammonium chloride, gives a precipitate of the hydrogel of alumina: $Al(OH)_3 + 3NaHO + 3NH_4Cl = Al(OH)_3 + 3NaCl + 3NH_4OH$. There was originally free soda, and on the addition of sal-ammoniac there is free ammonia, and this does not dissolve alumina, therefore the hydrogel of the latter is precipitated.

Another direct method for the preparation of pure aluminium compounds consists in the treatment of *cryolite* containing aluminium fluoride together with sodium fluoride, $AlNa_3F_6$. This mineral is exported from Greenland, and is also found in the Urals. It is crushed and heated in reverberatory furnaces with lime, and the resultant mass is treated with water; sodium aluminate is then obtained in solution, and calcium fluoride in the precipitate $AlNa_3F_6 + 3CaO = 3CaF_2 + AlNa_3O_3$.

Crum first prepared a solution of basic acetate of alumina—that is, a salt containing as large as possible an excess of aluminium hydroxide with as small as possible a quantity of acetic acid. The solution must be dilute—that is, not contain more than one part of alumina per 200 of water—and if this solution be heated in a closed vessel (so that the acetic acid cannot evaporate) to the boiling point of water, for one and a half to two days, then the solution, which apparently remains unaltered, loses its original astringent taste, proper to solutions of all the salts of alumina, and has instead the purely acid taste of vinegar. The solution then no longer contains the salt, but acetic acid and the hydrosol of alumina in an uncombined state; they may be isolated from each other by evaporating the acetic acid in shallow vessels at the ordinary temperature, and with a thin layer of liquid the alumina does not separate as a precipitate. When the acid vapours cease to come off there remains a solution of the hydrosol of alumina, which is tasteless and has no action on litmus paper. When concentrated, this solution acquires a more and more gluey consistency, and when completely evaporated over a water-bath it leaves a non-crystalline glue-like hydrate, whose composition is $Al_2H_4O_5 = Al_2O_3, 2H_2O$. The smallest quantity of alkalis, and of many acids and salts, will convert the hydrosol into the hydrogel of alumina—that is, convert the aluminium hydroxide from a soluble into an insoluble form, or, as it is said, cause the hydrate to coagulate or gelatinise. The smallest amount of sulphuric acid and its salts will cause the alumina to gelatinise—that is, cause the hydrogel to separate. Many such colloidal solutions are known (Vol. I. p. 98, Note 57).

In a dialyser, Vol. I. p. 63, Note 18.

[25 bis] The different states in which the hydrates of alumina occur and are prepared resemble similar varieties of the hydrates of the oxides of iron and chromium, of molybdic and tungstic acids, as well as of phosphoric and silicic acids, of many sulphides, proteid substances, &c. We shall therefore have occasion to recur to this subject in the further course of this work.

The most remarkable peculiarity of Graham's solution is that it solidifies on litmus paper, and leaves a blue ring on it, which shows the alkaline— that is, basic—character of the alumina in such a solution. If in the dialysis the basic hydrochloric acid salt be replaced by a similar acetic acid salt, a hydrosol of alumina is obtained which does not act upon litmus.

Compounds of alumina with bases (aluminates, *see* Note 21) are sometimes met with in nature. Such are spinel (*see* p. 65), $MgO,Al_2O_3 = MgAl_2O_4$, chrysoberyl, $BeAl_2O_4$, and others. Magnetic oxide of iron, $FeO,Fe_2O_3 = Fe_3O_4$, and compounds like it, belong to the same class. Here we evidently have a case of combination 'by analogy,' as in solutions and alloys, accompanied by the formation of strictly definite saline compounds, and such instances form a clear transition from so-called solutions and certain mixtures to the type of true salts.

Not only aluminium acetate (Note 24), but also every other aluminium salt with a volatile acid, parts with its acid on heating an aqueous solution—that is, is decomposed by water, and forms either basic salts or a hydrate of alumina. By dissolving aluminium hydroxide in nitric acid we may easily obtain a well-crystallising *aluminium nitrate*, $Al(NO_3)_3,9H_2O$, which fuses at 73° without decomposing (Ordway), gives a basic salt, $2Al_2O_3,6HNO_3$, at 100°, and at 140° leaves the aluminium hydroxide perfectly free from the elements of nitric acid. But the solutions of this salt, like those of the acetate, are also able to yield aluminium hydroxide. From all this it is evident that we must suppose that the solutions of this and similar salts contain an equilibrated dissociated system, containing the salt, the acid, and the base, and their compounds with water, as well as partly the molecules of water itself. Such examples much more clearly confirm those conceptions of solutions which are given in the first chapter than a general preliminary acquaintance with the subject can do.

As an example of native basic salts we may cite *alunite*, or alum-stone (sp. gr. 2·6), which sometimes occurs in crystals, but more frequently in fibrous masses. It has been found in masses in the Caucasus (at Zaglik, forty versts distance from Elizabetpol), and at Tolfa, near Rome. Its composition is $K_2O,3Al_2O_3,4SO_3,6H_2O$ (alunite contains $9H_2O$). It is soluble in water but not decomposed by it, but after being slightly ignited it gives up alum to it. It may be artificially prepared by heating a mixture of alum with aluminium sulphate in a closed tube at 230°.

As the colloidal properties are particularly sharply developed in those oxides (Al_2O_3, SiO_2, MoO_3, SnO_2, &c.) which show (like water also) the properties of feeble bases and feeble acids, there is probably some causal reason for this coincidence, all the more so since among organic substances—gelatins, albumins, &c.—the representatives of the colloids also have the property of feebly combining with bases and acids.

Since Deville's experiments the question of the density of aluminium chloride has been frequently re-investigated. The subject has more especially occupied the attention of Nilson, Pettersson, Friedel and Crafts, and V. Meyer and his collaborators. In general, it has been found that at low temperatures (up to 440°) the density is constant, and indicates a molecule Al_2Cl_6; whilst depolymerisation probably (although it is not yet certain) takes place at higher temperatures, and the molecule $AlCl_3$ is obtained. Along with this there has been, and still is, a difference of opinion as to the vapour density of aluminium ethyl and methyl—whether for instance, $Al(CH_3)_3$ or $Al_2(CH_3)_6$ expresses the molecule of the latter. The interest of these researches is intimately connected with the question of the valency of aluminium, if we hold to the opinion that elements in their various compounds have a constant and strictly definite valency. In this case the formula $AlCl_3$ or $Al(CH_3)_3$ would show that Al is trivalent, and that consequently the compounds of aluminium are $Al(OH)_3$, AlO_3Al, and, in general, AlX_3. But if the molecule be Al_2Cl_6, it is—for the followers of the doctrine of the invariable valency of the elements—incompatible with the idea of the trivalency of aluminium, and they assume it to be quadrivalent like carbon, likening Al_2Cl_6 to ethane $C_2H_6 = CH_3CH_3$, although this does not explain why Al does not form $AlCl_4$, or, in general, AlX_4. In this work another supposition is introduced; according to this, although aluminium, as an element of group III., gives compounds of the type AlX_3, this does not exclude the possibility of these molecules combining with others, and consequently with *each other*—that is, forming Al_2X_6; just as the molecules of univalent elements exist either as H_2, Cl_2, &c., or as Na, and the molecules of bivalent elements either as Zn, or as S_2, or even S_6. In the first place it must be recognised that the limiting form does not exhaust all power of combination, it only exhausts the capacity of the element for combining with X's, but the saturated substance may afterwards combine with *whole molecules*, which fact is best proved by the capacity of substances to form crystalline compounds with water, ammonia, &c. But in some substances this faculty for further combinations is less developed (for instance, in carbon tetrachloride, CCl_4), whilst in others it is more so. AlX_3 combines with many other molecules. Now if a limiting form, which does not combine with new X's, nevertheless combines with other whole molecules, it will naturally in some instances combine with itself, will polymerise. In this manner the mind clearly grasps the idea that the same forces which cause S_2 to unite itself to Cl_2, or C_2H_4 to Cl_2, &c., also unite molecules of a similar kind together; thus *polymerisation* ceases to be an isolated fragmentary phenomenon, and chemical combinations 'by analogy' acquire a particular and important interest. In conformity with these views the following proposition may be made concerning the compounds of aluminium. They are of the type AlX_3 in the limit, like BX_3, but those limiting forms are still able to combine to form AlX_3,RZ, and the aluminium chloride is a compound of this kind— *i.e.* $(AlX_3)_2$. In boron, for example, in BCl_3, this tendency to form further compounds is less developed. Hence boron chloride appears as BCl_3, and not $(BCl_3)_2$. Polymerisation is not only possible when a substance has not attained the limit (although it is more probable then), but also when the limiting form has been reached, if only the latter has the faculty of combining with other whole molecules. We may therefore conclude that aluminium, like boron, is trivalent in the same sense that lithium and sodium are univalent, magnesium bivalent, and carbon tetravalent. In a word, there is no reason to consider that aluminium is capable of forming compounds AlX_4, and in that way to explain the existence of the molecule Al_2Cl_6. Furthermore, there are many reasons for thinking that AlF_3, Al_2O_3, and other empirical formulæ do not express the molecular weights of these compounds, but that they are much higher: Al_nF_{3n}, $Al_{2n}O_{3n}$. In recent years convincing proofs of the truth of the above statements have been obtained, and of the independent existence of AlX_3 in a state of vapour; for Comb has determined the vapour density of the volatile acetyl of aluminium acetate $Al(C_3H_7O_2)_3$ (which melts at 193°, boils at 315°, and distils without a trace of decomposition), and has found that it exactly corresponds to the above molecular composition. On the other hand, Louise and Roux (1889) by employing the method of 'freezing point depression' of solutions (Chapter I., Note 49) found that the molecules $Al_2(C_2H_5)_6$ and $Al_2(C_5H_{11})_6$, &c., correspond to the type Al_2X_6. Thus it may now be accepted that the molecular composition of the compounds of aluminium in their simplest form is AlX_3, but that they may polymerise and give Al_2X_6 or, in general, Al_2X_{3n}.

In the case of gallium, as a close analogue of aluminium, Lecoq de Boisbaudran (1880) showed that probably the molecule gallium chloride contains Ga_2Cl_6 at low temperatures and high pressures, and that it dissociates into $GaCl_3$ at high temperatures and low pressures. The molecule of indium chloride seems to exist only in the simplest form, $InCl_3$.

The pure salt ($16H_2O$) is not hygroscopic. In the presence of impurities the amount of water increases to $18H_2O$, and the salt becomes hygroscopic.

The common form of crystals of alums is octahedral, but if this solution contains a certain small excess of alumina above the ratio $2Al(OH)_3$ to K_2SO_4, and not more sulphuric acid than $3H_2SO_4$ to $2Al(OH)_3$, then it easily forms combinations of the cube and octahedron, and these alums are called 'cubic' alums. They are valued by the dyer because they can contain no iron in solution, for oxide of iron is precipitated before alumina, and if the latter be in excess there can be no oxide of iron present. These alums were long exported from Italy, where they were prepared from alunite (Note 28).

[33 bis] It is also formed by the action of hydrochloric acid upon metallic aluminium (Nilson and Pettersson), by heating alumina in a mixture of the vapours of naphthaline and HCl (Faure, 1889), and by the action of dry HCl upon an alloy of 14 p.c., or more of Al and copper (Mobery).

Aluminium chloride fuses at 178°, boils at 183° (pressure 755 mm., at 168° under a pressure of 250 mm., and at 213° under 2,278 mm.), according to Friedel and Crafts, so that it boils immediately after fusion. According to Seubert and Pallard (1892), Al_2Cl_6 fuses at 193°. Aluminium bromide fuses at about 92°, and the iodide at 185° according to Weber, at 125° according to Deville and Troost.

All these halogen compounds of aluminium are soluble in water. *Aluminium fluoride*, AlF_3 (Al_nF_{3n}), is insoluble in water. It is obtained by dissolving alumina in hydrofluoric acid; a solution is then formed, but it contains an excess of hydrofluoric acid. When this solution is evaporated, crystals containing Al_2F_6, HF, H_2O are obtained. They are also insoluble in water. By saturating the above solution with a large quantity of alumina, and then evaporating, we obtain crystals having the composition $Al_2F_6, 7H_2O$. All these compounds, when ignited, leave insoluble anhydrous aluminium fluoride. It forms colourless rhombohedra, which are non-volatile, of sp. gr. 3·1, and are decomposed by steam into alumina and hydrofluoric acid. The acid solution apparently contains a compound which has its corresponding salts; by the addition of a solution of potassium fluoride, a gelatinous precipitate of AlK_3F_6 is obtained. A similar compound occurs in nature—namely, $AlNa_3F_6$, or *cryolite*, sp. gr. 3·0.

[34 bis] In this respect aluminium chloride resembles the chloranhydrides of the acids, and probably in the aqueous solution the elements of the hydrochloric acid are already separated, at least partially, from the aluminium hydroxide. The solution may also be obtained by the action of aluminium hydroxide on hydrochloric acid.

Here we see an instance in confirmation of what has been said in Note 30—*i.e.* the action of the molecule $AlCl_3$. We will cite still another instance confirming the power of alumina to enter into complex combinations. Alumina, moistened with a solution of calcium chloride, gives, when ignited, an anhydrous crystalline substance (tetrahedral), which is soluble in acids, and contains $(Al_2O_3)_6(CaO)_{10}CaCl_2$. Even clay forms a similar stony substance, which might be of practical use.

Among the most complex compounds of aluminium, *ultramarine*, or *lapis lazuli*, must be mentioned. It occurs in nature near Lake Baikal, in crystals, some colourless and others of various tints—green, blue, and violet. When heated it becomes dull and acquires a very brilliant blue colour. In this form it is used for ornaments (like malachite), and as a brilliant blue pigment. At the present time ultramarine is prepared artificially in large quantities, and this process is one of the most important conquests of science; for the blue tint of ultramarine has been the object of many scientific researches, which have culminated in the manufacture of this native substance. The most characteristic property of ultramarine is that when placed in sulphuric acid it evolves hydrogen sulphide and becomes colourless. This shows that the blue colour of ultramarine is due to the presence of sulphides. If clay be heated in a furnace with sodium sulphate and charcoal (forming sodium sulphide) without access of air, a white mass is obtained, which becomes green when heated in the air, and when treated with water leaves a colourless substance known as 'white ultramarine.' When ignited in the air it absorbs oxygen and turns blue. The coloration is ascribed to the presence of metallic sulphides or polysulphides, but it is most probable that silicon sulphide, or its oxysulphide, SiOS, is present. At all events the sulphides play an important part, but the problem is not yet quite settled. The formula $Na_8Al_6Si_6O_{24}S$ is ascribed to white ultramarine. The green probably contains more sulphur, and the blue a still larger quantity. The last is supposed to contain $Na_8Al_6Si_6O_{24}S_3$. It is more probable (according to Guckelberger, 1882) that the composition of the blue varies between $Si_{18}Al_{18}Na_{20}S_6O_{71}$ and $Si_{18}Al_{12}Na_{20}S_6O_{69}$. The latter may be expressed as $(Al_2O_3)_6(SiO_2)_{18}(Na_2O)_{10}S_6O_5$, which would indicate the presence of insufficiently-oxidised sulphur in ultramarine.

At the ordinary temperature aluminium does not decompose water, but if a small quantity of iodine, or of hydriodic acid and iodine, or of aluminium iodide and iodine, is added to the water, then hydrogen is abundantly evolved. It is evident that here the reaction proceeds at the expense of the formation of Al_2I_6, and that this substance, with water, gives aluminium hydroxide and hydriodic acid, which, with aluminium, evolves hydrogen. Aluminium probably belongs to those metals having a greater affinity for oxygen than for the halogens (Note 36 tri).

[36 bis] As an example we may mention that if mercury comes in contact with metallic aluminium and especially if it be rubbed upon the surface of aluminium moistened with a dilute acid, the Al becomes rapidly oxidised (Al_2O_3 being formed). The oxidation is accompanied by a very curious appearance, as it were of wool (or fur) formed by threads of oxide of aluminium growing upon the metal. This was first pointed out by Cass in 1870, and subsequently by A. Sokoleff in 1892. This interesting and curious phenomenon has not to my knowledge been further studied.

I think it necessary, however, to add that according to Lubbert and Rascher's researches (1891), wine, coffee, milk, oil, urine, earth, &c., have no more action upon aluminium vessels than upon copper, tin, and other similar articles. In the course of four months ordinary vinegar dissolved 0·35 grm. of Al per sq. centimetre, whilst a 5 per cent. solution of common salt dissolved about 0·05 grm. of aluminium. Ditte (1890) showed that Al is acted upon by nitric and sulphuric acids, although only slowly (owing to the formation of a layer of gas, as in Chapter XVI., Note 10) and that the reaction proceeds much more rapidly in vacuo or in the presence of oxidising agents. Al is even oxidised by water on the surface, but the thin coating of alumina formed prevents further action. In the course of twelve hours nitric acid sp. gr. 1·383 dissolved at 17° about 20 grms. of aluminium (containing only a small amount of Si, 1¼ p.c.) from a sq. metre of surface (Le Rouart, 1891).

[36 tri] In addition to the data given in Chapters XI., XIII., and in Chapter XV., Note 19, the following are the amounts of heat in thousands of units, evolved in the formation of the oxides and chlorides from the metals taken in gram-atomic quantities:

Na_2O 100; MgO 140*; $⅓Al_2O_3$ 120*; $⅓Fe_2O_3$ 63*;

Na_2Cl_2 195; $MgCl_2$ 151; $⅓Al_2Cl_6$ 107; $⅓Fe_2Cl_6$ 64.

The asterisks following the oxides of Mg, Al and Fe call attention to the fact that the existing data refer to the formation of the hydrates of these metals, from which the heat of formation of the anhydrous oxides may easily be assumed, because the heat of hydration (for example, $MgO + H_2O$) has not yet been determined.

Cryolite under the action of the current at about 1,000° gives off the vapour of Na which reduces the Al, but it recombines with the liberated fluorine and again passes into the fused mass. It is important to obtain aluminium at as low a temperature as possible, but the action proceeds far more easily with the solution (alloy) of oxide of aluminium in cryolite.

[37 bis] The cost of working this process can be brought as low as 20 cents per lb. or about 2½ fcs. per kilo. In England, Castner, prior to the introduction of the electric method, obtained Al by taking a mixture of 1,200 parts of the double salt $NaAlCl_4$, 600 parts of cryolite, and 350 parts of Na, and obtained about 120 parts of Al, so that the cost of this process is about 1½ time that of the electric method.

Buchner found that sulphide of aluminium, Al_2S_3, is more suitable for the preparation of Al by the electrolytic method than Al_2O_3, but since the formation of Al_2S_3 by heating a mixture of Al_2O_3, and charcoal in sulphur vapour proceeds with difficulty, Gray (1894) proposed to prepare Al_2S_3 by heating a mixture of charcoal, sulphate of aluminium, and sodium fluoride. The resultant molten mixture of NaF and Al_2S_3 gives aluminium directly under the action of an electric current.

Aluminium, when heated to the high temperature of the electric furnace, dissolves carbon and forms an alloy which, according to Moissan, when rapidly treated with *cold* hydrochloric acid leaves a compound C_3Al_4 in the form of a yellow crystalline transparent powder, sp. gr. 2·36 (*see* Chapter VIII. Note 12 bis). This *carbide of aluminium* C_3Al_4 corresponds to methane CH_4, for Al replaces H_3 and carbon O_2 or H_4, that is, it is equal to three molecules of CH_4 with the substitution of twelve atoms of H in it by four of Al, or, what is the same thing, it is the duplicated molecule of Al_2O_3 with the substitution of O_6 by C_3. And indeed C_3Al_4 under the action of water forms marsh gas and hydrate of alumina: $C_3Al_4 + 12H_2O - 3CH_4 + 4Al(OH)_3$. This decomposition gives a new aspect of the synthesis of hydrocarbons, and quite agrees with what should follow from the action of water upon the metallic carbides as applied by me for explaining the origin of naphtha (Chapter VIII., Notes 57, 58, and 59). Frank (1894) by heating Al with carbon obtained a similar although not quite pure compound, which (like CaC_2) evolves acetylene with hydrochloric acid *i.e.* probably has the composition AlC_3.

[38 bis] The same is the case in group IV. of the uneven series, where tin is the most fusible. Thus the temperature of fusion rises on both sides of tin (silicon is very infusible; germanium, 900°; tin, 230°; lead, 326°); as it also does in group III., starting from gallium, for indium fuses at 176°, less easily than gallium but more easily than thallium (294°). Aluminium also fuses with greater difficulty than gallium.

The spectrum of gallium is characterised by a brilliant violet line of wave-length = 417 millionths of a millimetre. The metal can be separated from the solution, containing a mixture of the many metals occurring in the zinc blende, by making use of the following reactions: it is precipitated by sodium carbonate in the first portions; it gives a sulphate which, on boiling, easily decomposes into a basic salt, very slightly soluble in water; and it is deposited in a metallic state from its solutions by the action of a galvanic current. It fuses at +30°, and, when once fused, remains liquid for some time. It oxidises with difficulty, evolves hydrogen from hydrochloric acid and from potassium hydroxide, and, like all feeble bases (for instance, alumina and indium oxide), it easily forms basic salts. The hydroxide is soluble in a solution of caustic potash, and slightly so in caustic ammonia. Gallium forms volatile $GaCl_3$ and $GaCl_2$ (Nilson and Pettersson).

The vapour density of indium chloride, $InCl_3$ (Note 31), determined by Nilson and Pettersson, confirms this atomic weight. Indium is separated from zinc and cadmium, with which it occurs, by taking advantage of the fact that its hydroxide is insoluble in ammonia, that the solutions of its salts give indium when treated with zinc (hence indium is dissolved after zinc by acids) and that they give a precipitate with hydrogen sulphide even in acid solutions. Metallic indium is grey, has a sp. gr. of 7·42, fuses at 176°, and does not oxidise in the air; when ignited, it first gives a black suboxide, In_4O_3, then volatilises and gives a brown oxide, In_2O_3, whose salts, InX_3, are also formed by the direct action of acids on the metal, hydrogen being evolved. Caustic alkalis do not act on indium, from which it is evident that it is less capable of forming alkaline compounds than aluminium is; however, with potassium and sodium hydroxides, solutions of indium salts give a colourless precipitate of the hydroxide, which is soluble in an excess of the alkali, like the hydroxides of aluminium and zinc. Its salts do not crystallise. Nilson and Pettersson (1889), by the action of HCl upon In, obtained volatile crystalline, $InCl_2$, and by treating this compound with In, InCl also.

Thallium was afterwards found in certain micas and in the rare mineral crookesite, containing lead, silver, thallium, and selenium. Its isolation depends on the fact that in the presence of acids thallium forms thallous compounds, TlX. Among these compounds the chloride and sulphate are only slightly soluble, and give with hydrogen sulphide a black precipitate of the sulphide Tl_2S, which is soluble in an excess of acid, but insoluble in ammonium sulphide.

[41 bis] The best method of preparing thallous hydroxide, TlOH, is by the decomposition of the requisite quantity of baryta by thallous sulphate, which is slightly soluble in water; barium sulphate is then obtained in the precipitate and thallous hydroxide in solution. This solubility of the

hydroxide is exceedingly characteristic, and forms one of the most important properties of thallium. These lower (thallous) compounds are of the type TlX, and recall the salts of the alkalis. The salts TlX are colourless, do not give a precipitate with the alkalis or ammonia, but are precipitated by ammonium carbonate, because thallous carbonate, Tl_2CO_3, is sparingly soluble in water. Platinic chloride gives the same kind of precipitate as it does with the salts of potassium—that is, thallous platinochloride, $PtTl_2Cl_6$. All these facts, together with the isomorphism of the salts TlX with those of potassium, again point out what an important significance the types of compounds have in the determination of the character of a given series of substances. Although thallium has a greater atomic weight and greater density than potassium, and although it has a less atomic volume, nevertheless thallous oxide is analogous to potassium oxide in many respects, for they both give compounds of the same type, RX. We may further remark that thallous fluoride, TlF, is easily soluble in water as well as thallous silicofluoride, $SiTl_2F_6$, but that thallous cyanide, TlCN, is sparingly soluble in water. This, together with the slight solubility of thallous chloride, TlCl, and sulphate, Tl_2SO_4, indicates an analogy between TlX and the salts of silver, AgX.

As regards the higher oxide or the *thallic oxide*, Tl_2O_3, the thallium is trivalent in it—that is, it forms compounds of the type TlX_3. The hydroxide, TlO(OH), is formed by the action of hydrogen peroxide on thallous oxide, or by the action of ammonia on a solution of thallic chloride, $TlCl_3$. It is obtained as a brown precipitate, insoluble in water but easily soluble in acids, with which it gives thallic salts, TlX_3. Thallic chloride, which is obtained by cautiously heating the metal in a stream of chlorine, forms an easily fusible white mass, which is soluble in water and able to part with two-thirds of its chlorine when heated. An aqueous solution of this salt yields colourless crystals containing one equivalent of water. It is evident from the above that all the thallic salts can easily be reduced to thallous salts by reducing agents such as sulphurous anhydride, zinc, &c. Besides these salts, thallic sulphate, $Tl_2(SO_4)_3,7H_2O$, thallic nitrate $Tl(NO_3)_3,4H_2O$, &c., are known. These salts are decomposed by water, like the salts of many feeble basic metals—for example, aluminium.

The specific heat of cerium determined (1870) by me, and afterwards confirmed by Hillebrand, corresponds with that atomic weight of cerium according to which the composition of two oxides should be Ce_2O_3 and CeO_2. Hillebrand also obtained metallic lanthanum and didymium by decomposing their salts by a galvanic current, and he found their specific heats to be near that of cerium and about 0·04, and it is therefore justifiable to give them an atomic weight near that of cerium, as was done on the basis of the periodic law. Up to 1870 yttrium oxide was also given the formula RO. Having re-determined the equivalent of yttrium oxide (with respect to water), and found it to be 74·6, I considered it necessary to also ascribe to it the composition Y_2O_3, because then it falls into its proper place in the periodic system. If the equivalent of the oxide to water be 74·6, it contains 58·6 of metal per 16 of oxygen, and consequently one part by weight of hydrogen replaces 29·3 of yttrium, and if it be regarded as bivalent (oxide RO), it would not, by its atomic weight 58·6, find a place in the second group. But if it be taken as trivalent—that is, if the formula of its oxide be R_2O_3 and salts RX_3—then Y = 88, and a position is open for it in the third group in the sixth series after rubidium and strontium. These alterations in the atomic weights of the cerite and gadolinite metals were afterwards accepted by Clève and other investigators, who now ascribe a formula R_2O_3 to all the newly discovered oxides of these metals. But still the position in the periodic system of certain elements—for example of holmium, thulium, samarium, and others—has not yet been determined for want of a sufficient knowledge of their properties in a state of purity.

So, for example, in 1871, in the *Journal of the Russian Physico-Chemical Society* (p. 45) and in Liebig's *Annalen*, Supt. Band viii. 198, I deduced, on the basis of the periodic law, an atomic weight 44 for ekaboron, and Nilson in 1888 found that of scandium, which is ekaboron, to be Sc = 44·03, The periodic law showed that the specific gravity of the ekaboron oxide would be about 8·5, that it would have decided but feeble basic properties and that it would give colourless salts. And this proved to be the case with scandium oxide. In describing scandium, Clève and Nilson acknowledge that the particular interest attached to this element is due to its complete identity with the expected element ekaboron. And this accurate foretelling of properties could only be arrived at by admitting that alteration of the atomic weights of the cerite and gadolinite metals which was one of the first results of the application of the periodic system of the elements to the interpretation of chemical facts. In my first memoirs, namely, in the *Bulletin of the St. Petersburg Academy of Sciences*, vol. viii. (1870), and in Liebig's *Annalen* (*l. c.* p. 168) and others, I particularly insisted on the necessity of altering the then accepted atomic weights of cerium, lanthanum, and didymium. Clève, Höglund, Hillebrand and Norton, and more especially Brauner, and others accepted the proposed alteration, and gave fresh proofs in favour of the proposed alterations of these atomic weights. The study of the fluorides was particularly important. Placing cerium in the fourth group, the composition of its highest oxide would then be CeO_2, and its compounds CeX_4 and the lower oxide, Ce_2O_3 or CeX_3. Brauner obtained the fluoride CeF_4,H_2O corresponding with the first, and a double crystalline salt, $3KF,2CeF_4,2H_2O$, without any admixture of compound of the lower grade CeX_3, which generally occur together with the majority of salts corresponding with CeX_4. It will be seen from these formulæ and from the tables of the elements, that cerium and didymium do not belong to the third group, which is now being described, but we mention them here for convenience, as all the cerite and gadolinite metals have much in common. These metals, which are rare in nature, resemble each other in many respects, always accompany each other, are with difficulty isolated from each other, and stand together in the periodic system of the elements; they have acquired a peculiar interest owing to their having been in 1870 the objects of the study of Marignac, Delafontaine, Soret, Lecoq de Boisbaudran, Brauner, Clève, Nilson, the professors of Upsala, and others.

The cerite and gadolinite metals occur in rare siliceous minerals from Sweden, America, the Urals, and Baikal, such as cerite (in Sweden), gadolinite, and orthite; and in still rarer minerals formed by titanic, niobic, and tantalic acids, such as euxenite in Norway and America, and samarskite in Norway, the Urals and America, and in a few rare fluorides and phosphates. Among the latter, monazite is found in somewhat considerable quantities in Brazil and North Carolina; this contains the phosphate of cerium, $CePO_4$ (= $Ce_2O_3P_2O_5$), together with didymium, thorium and lanthanum (according to W. Edron and Shapleigh's analyses), and is now used for preparing that mixture of the oxides of the rare metals (especially ThO_2, Ce_2O_3, La_2O_3, &c.), which is employed for incandescent burners (Auer von Welsbach), as it has been found by experiment that these oxides when raised to incandescence in a non-luminous gas flame, give a far more brilliant flame with a smaller consumption of gas, besides being suitable for such non-luminous gases as water gas. The insufficiency of material to work upon, and the difficulty of separating the oxides from each other, are the chief reasons why the composition of the compounds of these rare metals is so imperfectly known. Cerite is the most accessible of these minerals. Besides silica it contains more than 50 p.c. of the oxides of cerium, lanthanum (from 4 p.c.), and didymium. The decomposition of its powder by sulphuric acid gives sulphates, all of which are soluble in water. The other minerals mentioned above are also decomposed in the same manner. The solution of sulphates is precipitated with free oxalic acid, which forms salts insoluble in water and dilute acids with all the cerite and gadolinite oxides. The oxides themselves are obtained by igniting the oxalates. When ignited in the air the cerium passes from its ordinary oxide Ce_2O_3 into the higher oxide CeO_2, which is so feeble a base that its salts are decomposed by water, and it is insoluble in dilute nitric acid. Therefore it is always possible to remove all the cerium oxide by repeated ignitions and solutions in sulphuric acid. The further separation of the metals is mainly based on four methods employed by many investigators.

(*a*) A solution of the mixed salts is treated with an excess of solid potassium sulphate. Double salts, such as $Ce_2(SO_4)_3,3K_2SO_4$, are thus formed. The gadolinite metals, namely yttrium, ytterbium, and erbium, then remain in solution—that is, their double salts are soluble in a solution of potassium sulphate, whilst the cerite metals—namely, cerium, lanthanum, and didymium—are precipitated, that is, their double salts are insoluble in a saturated solution of potassium sulphate. This ordinary method of separation, however, appears from the researches of Marignac to be so untrustworthy that a considerable amount of didymium and the other metals remain in the soluble portion, owing to the fact that, although individually insoluble, they are dissolved when mixed together. Thus erbium and terbium occur both in the solution and precipitate. Nevertheless, beryllium, yttrium, erbium, and ytterbium belong to the soluble, and scandium, cerium, lanthanum, didymium, and thorium to the insoluble portion. The insoluble salt of scandium, for example (*i.e.* insoluble in a solution of potassium sulphate), has a composition $Sc_2(SO_4)_3, 3K_2SO_4$.

(*b*) The oxides obtained by the ignition of the oxalates are dissolved in nitric acid (the nitrates of the cerite metals easily form double salts with those of the alkali metals, and as some—for example, the ammonio-lanthanum salt—crystallise very well, they should be studied and applied to the analytical separation of these metals), the solution is then evaporated to dryness, and the residue fused. All nitrates are destroyed by heat; those of aluminium and iron, &c., very easily, those of the cerite and gadolinite metals also easily (although not so easily as the above) but in different degrees and sequence; so that by carrying on the decomposition carefully from the beginning it is possible to destroy the nitrate of only one metal without touching the others, or leaving them as insoluble basic salts. This method, like the preceding and the two following, must be repeated as many as seventy times to attain a really constant product of fixed properties, that is, one in which the decomposed and undecomposed portions contain one and the same oxide. This method, due to Berlin and worked out by Bunsen, has given in the hands of Marignac and Nilson the best results, especially for the separation of the gadolinite metals, ytterbium and scandium.

(*c*) A solution of the salts is partially precipitated by ammonia; that is, the solution is mixed with a small quantity of ammonia insufficient for the precipitation of the entire quantity of the bases (fractional precipitation). Thus, the didymium hydroxide is first precipitated from a mixture of the salts of didymium and lanthanum. A partial separation may be effected by repeating the solution of the precipitate and fractional precipitation, but a perfectly pure product is scarcely attainable.

(*d*) The formates having different degrees of solubility (lanthanum formate 420 parts of water per one of salt, didymium formate 221, cerium formate 360, yttrium and erbium formates easily soluble) give a possible means of separating certain of the gadolinite metals from each other by a method of fractional solution and precipitation, as Bunsen, Bahr, Clève, and others have pointed out.

(*e*) Crookes (1893) took advantage of the fractional precipitation of alcoholic solutions of the chlorides by amylene, and by this means separated, for example, erbium, terbium, and others.

(*f*) Lastly, oxide of thorium ThO_2 (Chapter VIII., Note [59]) is separated by means of its solubility in a solution of sodium carbonate.

A good method of separating these metals is not known, for they are so like each other. There are also only a few *methods of distinguishing* them from each other, and we can only add the following four to the above.

ᵃ The faculty of oxidising into a higher oxide. This is very characteristic for cerium, which gives the oxides Ce_2O_3 and CeO_2 or Ce_2O_4. Didymium also gives one colourless oxide, Di_2O_3, which is capable of forming salts (of a lilac colour), and another, according to Brauner, Di_2O_5 which is dark brown and does not form salts, so far as is known, and (like ceric oxide) acts as an oxidising agent, like the higher oxides of tellurium, manganese, lead, and others. Lanthanum, yttrium, and many others are not capable of such oxidation. The presence of the higher oxides may be recognised by ignition in a stream of hydrogen, by which means the higher oxides are reduced to the lower, which then remain unaltered.

ᵇ The majority of the salts of the gadolinite and cerite metals are colourless, but those of didymium and erbium are rose-coloured, the salts of the higher oxide of cerium, CeX_4, yellow, of the higher oxide of terbium, yellow, &c. Thus, the first metals obtained from gadolinite were yttrium, giving colourless, and erbium, giving rose-coloured, salts. Afterwards it was found that the salts of erbium of former investigators contained numerous colourless salts of scandium, ytterbium, &c., so that a coloration sometimes indicates the presence of a small impurity, as was long known to be the case in minerals, and therefore this point of distinction cannot be considered trustworthy.

ᶜ In a solid state and in solutions, the salts of didymium, samarium, holmium, &c., give characteristic absorption spectra, as we pointed out in Chapter [XIII.], and this naturally is connected with the colour of these salts. The most important point is, that those metals which do not give an absorption spectrum—for example, lanthanum, yttrium, scandium, and ytterbium—may be obtained free from didymium, samarium, and the other metals giving absorption spectra, because the presence of the latter may be easily recognised by means of the spectroscope, whilst the presence of the former in the latter cannot be distinguished, and therefore the purification of the former can be carried further than that of the latter. We may further remark that the sensitiveness of the spectrum reaction for didymium is so great that it is possible with a layer of solution half a metre thick to recognise the presence of 1 part of didymium oxide (as salt) in 40,000 parts of water. Cossa determined the presence of didymium (together with cerium and lanthanum) in apatites, limestones, bones, and the ashes of plants by this method. The main group of dark lines of didymium correspond with wave-lengths of from 580 to 570 millionths mm.; and the secondary to about 520, 730, 480, &c. The chief absorption bands of samarium are 472–486, 417, 500, and 559. Besides which, Crookes applied the investigation of the spectra of the phosphorescent light which is emitted by certain earths in an almost perfect vacuum, when an electric discharge is passed through it, to the discovery and characterisation of these rare metals. But it would seem that the smallest admixture of other oxides (for example, bismuth, uranium) so powerfully influences these spectra that the fundamental distinctions of the oxides cannot be determined by this method. Besides which, the spectra obtained by the passage of sparks through solutions or powders of the salts are determined and applied to distinguishing the elements, but as spectra vary with the temperature and elasticity (concentration) this method cannot be considered as trustworthy.

[d] The most important point of distinction of individual metallic oxides is given by the direct *determination of their equivalent with respect to water*—that is, the amount of the oxide by weight which combines (like water) with 80 parts by weight of sulphuric anhydride, SO_3, for the formation of a normal salt. For this purpose the oxide is weighed and dissolved in nitric acid, sulphuric acid is then added, and the whole is evaporated to dryness over a water-bath and then heated over a naked flame sufficiently strongly to drive off the excess of sulphuric acid, but so as not to decompose the salt (the product would in that case not be perfectly soluble in water); then, knowing the weight of the oxide and of the anhydrous sulphate, we can find the equivalent of the oxide. The following are the most trustworthy figures in this connection: scandium oxide 45·35 (Nilson), yttrium oxide 75·7 (Clève; according to my determination, 1871—74·6), cerous oxide— that is, the lower form of oxidation of cerium, according to various investigators (Bunsen, Brauner, and others) from 108 to 111, the higher oxide of cerium from 85 to 87, lanthanum oxide, according to Brauner, 108, didymium oxide (in salts of the ordinary lower form of oxidation) about 112 (Marignac, Brauner, Clève), samarium oxide about 116 (Clève), ytterbium oxide 131·3 (Nilson). It may not be superfluous here to draw attention to the fact that the equivalent of the oxides of all the gadolinite and cerite metals for water distribute themselves into four groups with a

somewhat constant difference of nearly 30. In the first group is scandium oxide with equivalent 45, in the second, yttrium oxide 76, in the third, lanthanum, cerium, didymium, and samarium oxides with equivalent about 110, and, in the fourth, erbium, ytterbium, and thorium oxides with equivalent about 131. The common difference of period is nearly 45. And if we ascribe the type R_2O_3 to all the oxides—that is, if we triple the weight of the equivalent of the oxide—we shall obtain a difference of the groups nearly equal to 90, which, for two atoms of the metal, forms the ordinary periodic difference of 45. If one and the same type of oxide R_2O_3 be ascribed to all these elements (as now generally accepted, in many cases there being insufficiently trustworthy data), then the atomic weights should be Sc = 44, Y = 89, La = 138, Ce = 140, Di = 144, (neodymium 140, praseodymium 144), Sm = 150, Yb = 173, also terbium 147, holmium 162, alphayttrium 157, erbium 166, thulium 170, decipium 171. It should be observed that there may be instances of basic salts. If, for example, an element with an atomic weight 90 gave an oxide RO_2, but salts ROX_2, then by counting its oxide as R_2O_3 its atomic weight would be 159.

All the points distinguishing many gadolinite and cerite elements have not been sufficiently well established in certain cases (for example, with decipium, thulium, holmium, and others). At present the most certain are yttrium, scandium, cerium, and lanthanum. In the case of didymium, for example, there is still much that is doubtful. Didymium, discovered in 1842 by Mosander after lanthanum, differs from the latter in its absorption spectrum and the lilac-rose colour of its salts. Delafontaine (1878) separated samarium from it. Welsbach showed that it contains two particular elements, neodymium (salts bluish-red) and praseodymium (salts apple-green), and Becquerel (1887) by investigating the spectra of crystals, recognised the presence of six individual elements. Probably, therefore, many of the now recognised elements contain a mixture of various others, and as yet there is not enough confirmation of their individuality. As regards yttrium, scandium, cerium, and lanthanum, which have been established without doubt, I think that, owing to their great rarity in nature and chemical art, it would be superfluous to describe them further in so elementary a work as the present. We may add that Winkler (1891) obtained a hydrogen compound of lanthanum, whose composition (according to Brauner) is La_2H_3, as would be expected from the composition of Na_2H, Mg_2H_2, &c. C. Winkler (1891), on reducing CeO_2 with magnesium, also remarked a rapid absorption of hydrogen, and showed that a *hydride of cerium*, CeH_2, corresponding to CaH, and the other similar hydrides of metals of the alkaline earths, is formed (Chapter XIV., Note 63).

CHAPTER XVIII
SILICON AND THE OTHER ELEMENTS OF THE FOURTH GROUP

Carbon, which gives the compounds CH_4, and CO_2, belongs to the fourth group of elements. The nearest element to carbon is silicon, which forms the compounds SiH_4 and SiO_2; its relation to carbon is like that of aluminium to boron or phosphorus to nitrogen. As carbon composes the principal and most essential part of animal and vegetable substances, so is silicon almost an invariable component part of the rocky formations of the earth's crust. Silicon hydride, SiH_4, like CH_4, has no acid properties, but silica, SiO_2, shows feeble acid properties like carbonic anhydride. In a free state silicon is also a non-volatile, slightly energetic non-metal, like carbon. Therefore the form and nature of the compounds of carbon and silicon are very similar. In addition to this resemblance, silicon presents one exceedingly important distinction from carbon: namely, the nature of the higher degree of oxidation. That is, silica, silicon dioxide, or silicic anhydride, SiO_2 is a solid, non-volatile, and exceedingly infusible substance, very unlike carbonic anhydride, CO_2, which is a gas. This expresses the essential peculiarity of silicon. The cause of this distinction may be most probably sought for in the polymeric composition of silica compared with carbonic anhydride. The molecule of carbonic anhydride contains CO_2, as seen by the density of this gas. The molecular weight and vapour density of silica, were it volatile, would probably correspond with the formula SiO_2, but it might be imagined that it would correspond to a far higher atomic weight of Si_nO_{2n}, principally from the fact that SiH_4 is a gas like CH_4, and $SiCl_4$ is a liquid and volatile, boiling at 57°—that is, even lower than CCl_4, which boils at 76°. In general, analogous compounds of silicon and carbon have nearly the same boiling points if they are liquid and volatile. From this it might be expected that silicic anhydride, SiO_2, would be a gas like carbonic anhydride, whilst in reality silica is a hard non-volatile substance,[1 bis] and therefore it may with great certainty be considered that in this condition it is polymeric with SiO_2, as on polymerisation—for instance, when cyanogen passes into paracyanogen, or hydrocyanic acid into cyanuric acid (Chapter IX.)—very frequently gaseous or volatile substances change into solid, non-volatile, and physically denser and more complex substances. We will first make acquaintance with free silicon and its volatile compounds, as substances in which the analogy of silicon with carbon is shown, not only in a chemical but also in a physical sense.

Free silicon can be obtained in an amorphous or crystalline state. Amorphous silicon is produced, like aluminium, by decomposing the double fluoride of sodium and silicon (sodium silicofluoride) by means of sodium: $Na_2SiF_6 + 4Na = 6NaF + Si$. By treating the mass thus obtained with water the sodium fluoride may be extracted and the residue will consist of brown, powdery silicon. In order to free it from any silica which might be formed, it is treated with hydrofluoric acid. This silicon powder is not lustrous; when heated it easily ignites, but does not completely burn. It fuses when very strongly heated, and has then the appearance of carbon. Crystalline silicon is obtained in a similar way, but by substituting an excess of aluminium for the sodium: $3Na_2SiF_6 + 4Al = 6NaF + 4AlF_3 + 3Si$. The part of the aluminium remaining in the metallic state dissolves the silicon, and the latter separates from the solution on cooling in a crystalline form. The excess of aluminium after the fusion is removed by means of hydrochloric and hydrofluoric acid. The best silicon crystals are obtained from molten zinc; 15 parts of sodium silicofluoride are mixed with 20 parts of zinc and 4 parts of sodium, and the mixture is thrown into a strongly heated crucible, a layer of common salt being used to cover it; when the mass fuses it is stirred, cooled, treated with hydrochloric acid, and then washed with nitric acid. Silicon, especially when crystalline, like graphite and charcoal, does not in any way act on the above-mentioned acids. It forms black, very brilliant, regular octahedra having a specific gravity of 2·49; it is a bad conductor of electricity, and does not burn even in pure oxygen (but it burns in gaseous fluorine). The only acid which acts on it is a mixture of hydrofluoric and nitric acids; but caustic alkalis dissolve in it like aluminium, with evolution of hydrogen, thus showing its acid character. In general silicon strongly resists the action of reagents, as do also boron and carbon. Crystalline silicon was obtained in 1855 by Deville, and amorphous silicon in 1826 by Berzelius.[4 bis]

Silicon hydride, SiH_4, analogous to marsh gas was obtained first of all in an impure state, mixed with hydrogen, by two methods: by the action of an alloy of silicon and magnesium on hydrochloric acid, and by the action of the galvanic current on dilute sulphuric acid, using electrodes of aluminium, containing silicon. In these cases silicon hydride is set free, together with hydrogen, and the presence of the hydride is shown by the fact that the hydrogen separated ignites spontaneously on coming into contact with the air, forming water and silica. The formation of silicon hydride by the action of hydrochloric acid on magnesium silicide is perfectly akin to the formation of phosphuretted hydrogen by the action of hydrochloric acid on calcium phosphide, to the formation of hydrogen sulphide by the action of acids on many metallic sulphides, and to the formation of hydrocarbons by the action of hydrochloric acid on white cast iron. On heating silicon hydride—that is, on passing it through an incandescent tube, it is decomposed into silicon and hydrogen, just like the hydrocarbons, but the caustic alkalis, although without action on the latter, react with silicon hydride according to the equation: $SiH_4 + 2KHO + H_2O = SiK_2O_3 + 4H_2$.

Silicon chloride, SiCl$_4$, is obtained from amorphous anhydrous silica (made by igniting the hydrate) mixed with charcoal, heated to a white heat in a stream of dry chlorine—that is, by that general method by which many other chloranhydrides having acid properties are obtained. Silicon chloride is purified from free chlorine by distillation over metallic mercury. Free silicon forms the same substance when treated with dry chlorine. It is a volatile colourless liquid, which boils at 59° and has a specific gravity of 1·52. It fumes strongly in air, has a pungent smell, and in general has the characteristic properties of the acid chloranhydrides. It is completely decomposed by water, forming hydrochloric acid and silicic acid, according to the equation: SiCl$_4$ + 4H$_2$O = Si(OH)$_4$ + 4HCl.

The most remarkable of the haloid compounds of silicon is *silicon fluoride*, SiF$_4$. It is a gaseous substance only liquefied by intense cold, -100°, and is obtained (Chapter XI.) directly by the action of hydrofluoric acid on silica and its compounds (SiO$_2$ + 4HF = 2H$_2$O + SiF$_4$), and also by heating fluorspar with silica (2CaF$_2$ + 3SiO$_2$ = 2CaSiO$_3$ + SiF$_4$). In order to prepare silicon fluoride, sand or broken glass is mixed with an equal quantity by weight of fluorspar and 6 parts by weight of strong sulphuric acid, and the mixture is gently heated. It fumes strongly in air, reacting with the aqueous vapours, although it is produced from silica and hydrofluoric acid with the separation of water. It is evident that a reverse reaction occurs here; that is to say, the water reacts with the silicon fluoride, but the reaction is not complete. This phenomenon is similar to that which occurs when water decomposes aluminium chloride, but at the same time hydrochloric acid dissolves aluminium hydroxide and forms the same aluminium chloride. The relative amount of water present (together with the temperature) determines the limit and direction of the reaction. The faculty which silicon fluoride has of reacting with water is so great that it takes up the elements of water from many substances—for instance, like sulphuric acid, it chars paper. Water dissolves about 300 volumes of this gas, but in this case it is not a common dissolution which takes place, but a reaction. During the first absorption of silicon fluoride by water, silicic acid is separated in the form of a jelly, but a certain quantity of the silicon fluoride also remains in the liquid, because the hydrofluoric acid formed dissolves the other part of the silica and forms the so-called *hydrofluosilicic acid*: H$_2$SiF$_6$ = SiF$_4$ + 2HF = SiH$_2$O$_3$ + 6HF - 3H$_2$O. That is to say, a metasilicic acid, SiH$_2$O$_3$, in which O$_3$ is replaced by F$_6$. This view of the composition of hydrofluosilicic acid may be admitted, because it forms a whole series of crystallisable and well defined salts. In general, the whole reaction of water on silicon fluoride may be expressed by the equation: 3SiF$_4$ + 3H$_2$O = SiO(OH)$_2$ + 2SiH$_2$F$_6$. Hydrofluosilicic acid and silicic acid resemble each other as much, and differ as much, in their chemical character as water and hydrofluoric acid. For this reason silicic acid is a feebler acid than hydrofluosilicic acid, and in addition to this the former is insoluble, and the latter soluble, in water. Hydrofluosilicic acid is also formed if silicic acid be dissolved in a solution of hydrofluoric acid. It is incapable of volatilising without decomposition, and on heating the concentrated acid silicon fluoride is evolved, leaving an aqueous solution of hydrofluoric acid. This is the reason why solutions of hydrofluosilicic acid corrode glass. This decomposition may be further accelerated by the addition of sulphuric acid, or even of other acids. Hydrofluosilicic acid, when acting on potassium and barium salts, gives precipitates, because the salts of these metals are but sparingly soluble in water: thus $2KX + H_2SiF_6 = 2HX + K_2SiF_6$. The potassium salt is obtained in the form of very fine octahedra, but the precipitate does not form quickly, and at first appears as a jelly. Nevertheless, the decomposition is complete, and it is taken advantage of for obtaining their corresponding acids from salts of potassium.[10 bis]

Silicon, having so much in common with carbon, is also able to combine with it in the proportion given by the law of substitution, that is, it forms a carbide of silicon CSi, called *carborundum* and obtained by Mühlhäuser and Acheson in the United States, and by Moissan in France (1891), and others, by reducing silica with carbon in the electrical furnace at a temperature of about 2500°, *i.e.* by the action of an electrical current upon a mixture of carbon and SiO_2 with NaCl. After treating the resultant mass with acids and washing with water, carborundum is obtained in transparent, lustrous grains of a greenish color, possessing great hardness (greater than corundum) and therefore used for polishing the hardest kinds of steel and stones. The specific gravity is about 3·1. Carborundum does not alter at a red heat, does not burn, and apparently approaches the diamond in its properties. (Moissan obtained, 1894, a similar very hard compound for boron, B_6C, sp. gr. 2·5.)

According to the principle of substitution, if silicon forms SiH_4, a series of hydrates, or hydroxyl derivatives, ought to exist corresponding to it. The first hydrate of an alcoholic character ought to have the composition $SiH_3(OH)$; the second hydrate $SiH_2(OH)_2$; the third, $SiH(OH)_3$;[11 bis] and the last, $Si(OH)_4$. The last is a hydrate of silica, because it is equal to $SiO_2 + 2H_2O$); and it is formed by the action of water on silicon chloride, when all four atoms of chlorine are replaced by four hydroxyl groups. It does not, however, remain in this state, but easily loses part of its water.

Silica or silicic anhydride, both in the free state and in combination with other oxides, enters into the composition of most of the rocky formations of the earth's crust. These silicious compounds are substances varying so much in their properties, crystalline forms, and relations to one another that they are comprised in a special branch of natural science (like the carbon compounds), and are treated of in works on mineralogy; so that, in dealing with them further, we shall only give a short description of these various compounds. It is first of all necessary to turn to the description of silica itself, especially as it is not unfrequently met with in nature in a separate state, and often forms whole masses of rocky formations, called 'quartz.' In an anhydrous condition silica appears in the greatest variety of natural forms—sometimes in well-formed crystals, hexagonal prisms, terminated by hexagonal pyramids. If the crystals are colourless and transparent, they are called *rock crystal*. This is the purest form of silica. Prismatic crystals of rock crystal sometimes attain considerable size, and as they are remarkable for their unchangeability, great hardness, and high index of refraction, they are used for ornaments, for seals, making necklaces, &c. Rock crystal coloured with organic matter in contact with which it has been produced has a brown or greyish colour, and then bears the name of *cairngorm* or *smoky quartz*. In this form it has the same uses as rock crystal, especially as it is often found in large masses. The same mineral, frequently occurs, coloured red or pink by manganese or iron oxides, especially in aqueous formations, and is then known as *amethyst*. When finely coloured the amethyst is used as a precious stone, but amethysts most frequently occur as small crystals in the cavities formed in other rocky formations, and especially in those formed in silica itself. A similar anhydrous silica is often found in transparent non-crystalline masses, having the same specific gravity as rock crystal itself (2·66). In this case it is called *quartz*. Sometimes it forms complete rocky formations, but more often penetrates or is interspersed through other rocky formations, together with other siliceous compounds. Thus, in granite, quartz is mixed with felspar and similar substances. Sometimes the colouring of quartz is so considerable that it is hardly transparent in thin sheets, but it is often found in transparent masses slightly coloured with various tints. The existence in nature of enormous masses of quartz proves that it resists the action of water. When water destroys rocky formations, the siliceous minerals which they contain are partly dissolved and partly transformed into clay, &c. But the quartz remains untouched, in the form of grains in which it existed in the rocky formation; sometimes, when crushed, it is carried away by the water and deposited. This is the nature of *sand*. Naturally, sometimes other rocky substances which are not changed by water, or only slightly acted on by it, are found in sand; but as these latter are more or less changed by the continuous action of water, it is not unusual to find sand which consists almost entirely of pure quartz. Common sand is generally coloured yellow or reddish-brown by foreign mineral matter, consisting principally of ferruginous minerals and clays. The purest or so-called quartz sand is, however, rarely found, and is recognised by the absence of colour, and also by the test that when shaken in water it does not form any turbidity: this shows the absence of clay; when fused with bases it forms a colourless glass, and on this account is a valuable material for the manufacture of glass. Sands were formed at all periods of the earth's existence; the ancient ones, compressed by strata of more recent formation and permeated with various substances (deposited from the infiltrating water), are sometimes solidified into rock, called *sandstone*, composing, in some places, whole mountain chains, and serviceable as a most excellent building material, on account of the slight change it undergoes under the influence of atmospheric agencies, and on account of the facility with which it may be wrought from rocky formations into immense regularly-shaped flags—the latter property is due to the primary laminar structure of the sand formations deposited, as above-mentioned, by water. Many grindstones and whetstones are made from such rocks.

Perfectly pure anhydrous silica is not only known in the condition of rock crystal and quartz having a specific gravity of 2·6, but also in another special form, having other chemical and physical properties. This variety of silica has a specific gravity of 2·2, and is formed by fusing rock crystal or heating silicic acid.[12 bis] Silicic acid, when heated to a dull red heat, parts entirely with the water it contains, and leaves an exceedingly fine amorphous mass of silica (easily levigated, but difficult to moisten); it is characterised by such excessive friability that, when lightly blown on, a large mass of it rises into the air like a cloud of dust. A mass of anhydrous silica maybe poured in this way from one vessel to another like a liquid, and like the latter it takes a horizontal position in the vessel containing it. Anhydrous silica, like quartz, does not fuse in the heat of a furnace, but it fuses in the oxyhydrogen flame to a colourless glassy mass exactly similar to that formed in the same way from rock crystal. In this condition silica has a specific gravity of 2·2.[13 bis] Both forms of silica are insoluble in ordinary acids, and even when they are in the state of powder, alkalis in solution act very slowly and feebly on them; rock crystal offers much greater resistance to the action of alkalis than the powder obtained by heating the hydrate. The latter is quite soluble, although but slowly, in hot alkaline solutions. This last property appertains in a greater degree to anhydrous silica having a specific gravity of 2·2 than to that which has a specific gravity of 2·6. Hydrofluoric acid more easily transforms the former into silicon fluoride than it does the latter. Both varieties of silica, when taken in the form of powder, easily combine with bases, forming, on being fused with an alkali, a vitreous slag, which is a salt corresponding with silica. Glass is such a salt, formed of alkalis and alkaline earthy bases; if the glass does not contain any of the latter—that is, if only alkaline glass be taken—a mass soluble in water is obtained. In order to obtain such *soluble glass*, potassium or sodium carbonates, or better a mixture of the two (fusion mixture), is fused with fine sand. A still better and further saturation of the alkalis with silica is effected by the action of alkaline solutions on the silicon hydrate met with in nature; for instance, an alkaline solution is often made use of to act on the so-called *tripoli*, or collection of siliceous skeletons of the lowest microscopical infusoria, which is sometimes found in considerable layers in the form of a sandy mass. Tripoli is used for polishing, not only on account of the considerable hardness of the silica, but also because the microscopic bodies of the infusoria have a pointed shape, which, however, is not angular, so that they do not scratch metals like sand. The alkaline solutions of silica obtained by boiling tripoli with caustic soda under pressure contain various proportions of silica and alkali.[14 bis] In order that it may contain the greatest amount of silica, silicic acid should be added to the heated solution. Silicic acid is formed by taking any solution containing silica and alkali, and adding to it, by degrees, some acid—for instance, sulphuric or hydrochloric; if the experiment be carried on carefully and the solution be concentrated, the whole mass thickens to a jelly, due to the gelatinous form of the *silicic acid* separated from the salt by the action of the acid. The decomposition may be expressed by the following equation: $Si(ONa)_4 + 4HCl = 4NaCl + Si(OH)_4$. The hydrate separated, $Si(OH)_4$, easily loses part of the water and forms a jelly, the whole mass gelatinising if the solution be strong enough.

Neither of the two varieties of anhydrous silica, nor the various natural gelatinous hydrates, are directly soluble in water. There is, however, a condition of silica known which is soluble in water, *soluble silica*, and silica is found in this state in nature. Small quantities of soluble silica are met with in all waters. Certain mineral springs, and especially hot springs—of which the best known are the Geysers of Iceland and those in the North American National Park (Yellowstone Valley)—contain a considerable amount of silica in solution. Such water, permeating the objects it meets with—for instance, wood—penetrates into them and deposits silica inside them, that is, transforms them into a petrified condition. Siliceous stalactites, and also many (if not all) forms of silica are formed by such water. The absorption of silica by plants by means of their roots, and also by the lower organisms having siliceous bodies, is due also to their nourishing themselves with the solutions containing silica continually formed in nature. Thus, in plants, in the straws of the grasses, in hard shave-grass, and especially in the knots of bamboo and other straw-like plants, a considerable quantity of silica is deposited, which must previously have been absorbed by the plants.

Silicic acid is a colloid. The gelatinous silicon hydrate is its hydrogel, the soluble hydrate is the hydrosol (Chapter XII.) Both varieties may be easily obtained from the alkaline silicates and from water-glass. The very same substances—that is, aqueous solutions of soluble glass and acid—taken in the same proportion, may produce either the gelatinous or the soluble silica, according to the way these solutions are mixed together. If the acid be added little by little to the *alkaline silicate*, with continuous stirring, a moment arrives when the whole mass thickens to a jelly, hydrogel; in this case the silicic acid is formed in the midst of the alkaline solution and becomes insoluble. But if the mixing be done in the reverse order—that is, if the soluble glass be added to the acid, or if a quantity of acid be rapidly poured into the solution of the salt—then the separation of the silica takes place in the midst of the acid liquid, and it is obtained in the form of the soluble hydrate, the hydrosol.

The hydrosol of silica prepared by mixing an excess of hydrochloric acid with a solution of sodium silicate, may be freed from the admixtures both of hydrochloric acid and salt, sodium chloride, *by means of dialysis*, as Graham showed (in 1861) in enquiring into the nature of colloids (Chapter [L](#)), and making many other important chemical investigations. The solution, containing the acid, salt, and silica, all dissolved in water, is poured into a dialyser—that is, a vessel with a porous diaphragm surrounded by water. Certain substances pass more easily through the diaphragm than others. This may be represented thus: the passage through the diaphragm proceeds in both directions, and if the solutions on each side of the diaphragm be equally strong, there will be equal numbers of molecules of the soluble substance passing into either side in a given time, some passing quickly and others slowly. The metallic chlorides and hydrochloric acid belong to the series of crystalloids which easily pass through a diaphragm, and therefore the hydrochloric acid and sodium chloride contained in the above-mentioned dialyser pass from the solution through the diaphragm into the water of the external vessel with considerable rapidity. The aqueous solution of colloidal silica also penetrates through the diaphragm, but very much more slowly. But if the amount of the substance dissolved is not equal on either side of the diaphragm, the whole system strives to attain a state of equilibrium; that is, the given substance penetrates through the diaphragm from the side where it is in excess to the part where there is a smaller quantity of it. All substances which are soluble in water have the faculty of penetrating through a membrane swollen in water, but the velocity of penetration is not equal, and in this respect the dialyser separates substances like a sieve. The silica passes less rapidly through the diaphragm than the sodium chloride and hydrochloric acid, so that by repeatedly changing the external water it is easy to effect the extraction of the chlorine compounds from the dialyser, which will finally only contain a solution of silica. This extraction (of HCl and NaCl) may be so complete that the liquid taken from the dialyser will not give any precipitate with a solution of silver nitrate. Graham obtained in this way soluble silica having a distinctly acid reaction, which, however, disappeared on the addition of a very minute quantity of alkali; for ten parts of silica in the solution it was sufficient to take one part of alkali in order to give the liquid an alkaline reaction, so slightly energetic are the acid properties of silicic acid. The solution of silica obtained by this method becomes gelatinous on standing, on being heated, or on evaporation under the receiver of an air-pump, &c. The hydrosol is transformed into the hydrogel, the soluble hydrate into the gelatinous.

Thus in addition to the gelatinous form of the silicic acid, there exists also a variety of this substance, soluble in water, as is the case with alumina. Such variation in properties and exactly the same relations with regard to water characterise an immense series of other substances having a great significance in nature. The number of such substances is especially great among organic compounds, and particularly in those classes of them which compose the principal material of the bodies of animals and plants. It is sufficient to mention, for instance, the gelatin which is familiar to all as carpenter's and other glues, and in the form of size and jelly. The same substance is also known in the solution which is used to join objects together. In a peculiar insoluble condition it enters into the composition of hides and bones. These various forms of gelatin differ in the same way as the different varieties of silica. The property of forming a jelly is exactly the same as in silica, and the adhesiveness of the solutions of both substances is identical; soluble silica adheres like a solution of gelatin. The same properties are again shown by starch, rosin, and albumin, and by a series of similar substances. The diaphragms used in dialysis are also insoluble, gelatinous, forms of colloids. The bodies of animals and plants consist largely of similar matter, insoluble in water, corresponding with the gelatinous or insoluble silicon hydrate, or with glue. The albumin which coagulates when eggs are boiled is a typical form of the gelatinous condition of such substances in the body. These slight indications are sufficient in order to show how great is the significance of those transformations which are so well marked in silica. The facts discovered by *Graham* in 1861–1864 comprise the most essential acquisitions in the general association of these phenomena of nature in the history of organic forms. The facility of transit from hydrogel to hydrosol is the first condition of the possibility of the development of organisms. The blood contains hydrosols, and the hydrogels of the same substances are contained in the muscles and tissues, and especially on the surface, of the body. All tissues are formed from the blood, and in that case the hydrosols are converted into hydrogels. The absence of crystallisation, the property, apparently under the influence of feeble agencies, of passing from the soluble condition to the insoluble, to the gelatinous condition of the hydrogel, constitute the fundamental properties of all colloids.

Silica, as regards its *salt forming properties*, stands in the series of oxides on the boundary line on the side of the acids in just such a place as alumina occupies on the side of the bases—that is, aluminium hydroxide is the representative of the feeblest bases and silicic acid is the least energetic of acids (at least in the presence of water—that is, in aqueous solutions); in alumina, however, the basic properties are distinctly expressed, while in silica the acid properties preponderate. Like all feeble acid oxides it is capable of forming, with other acids, saline compounds which are but slightly stable and are very easily decomposed in the presence of water. The chief peculiarity of the silicates consists in the number of their types. The salts formed with nitric or sulphuric acid exist in one, two, and three tolerably stable forms, but for acids like silicic acid the number of forms is very great, almost unlimited. The natural silicates in particular furnish proof of this fact; they contain various bases in combination with silica, and for one and the same base there often exist various degrees of combination. As feeble bases are capable of forming basic salts in addition to normal salts— that is, a compound of a normal salt with a feeble base (either the hydroxide or the oxide)—so the feeble acid oxides (although not all) form, in addition to normal salts, highly acid salts—that is, normal salts *plus* acid (hydrate or anhydride). Such acids are boric, phosphoric, molybdic, chromic, and especially silicic, acid.

In order to explain these relations it is necessary first to recollect the existence of the various hydrates of silica, or silicic acids, and then to turn our attention to the similarity between silicon compounds and metallic alloys. Silica is an oxide having the appearance of, and in many respects the same properties as, those oxides which combine with it, and if two metals are capable of forming homogeneous alloys in which there exist definite or indefinite compounds, it is permissible to assume a similar power of forming alloys in the case of analogous oxides. Such alloys are found in indefinite, amorphous masses in the form of glass, lava, slags, and a number of similar siliceous compounds which do not contain any definite types of combination, but nevertheless are homogeneous throughout their mass. By slow cooling, or under other circumstances, definite crystalline compounds may—and sometimes do—separate from this homogeneous mass, as also sometimes definite crystalline alloys separate from metallic alloys.

The formation of crystalline rocks in nature is partly of such a nature. By aqueous or igneous agency, but in any case in a liquid condition, those oxides which form the earth's crust and her crystalline minerals came into mutual contact. First of all they formed a shapeless mass, of which lava, glass, slags and solutions are examples, but little by little, or else suddenly, some definite compounds of certain oxides existing in this alloy or in the shapeless mass were formed. This is entirely similar to two metals forming a homogeneous alloy, and under known circumstances (for instance, on cooling the alloy, or in the case of aqueous solution when the two metals are simultaneously liberated from the solution), definite crystalline compounds are separated. In any case there is no doubt that there is less distinction between silica and bases, than between bases and such anhydrides as, for instance, sulphuric or nitric, or even carbonic, as is seen on comparing the physical and chemical properties of silica and various kinds of oxides. Alumina, especially, is exceedingly near akin to silica; not only in the hydrated state, but also in the anhydrous condition, there exists a certain similarity between the crystalline forms of alumina and silica, in the uncombined state. Both are very hard, transparent, inactive, non-volatile, infusible, and crystallise in the hexagonal system—in a word, they are remarkably similar, and for this reason they are capable, like two kindred metals, of entering into many different degrees of combination. Isomorphous mixtures—that is, differing by the substitution of oxides akin both in their physical and chemical characters—are very frequently met with among minerals, and the study of the latter gave the principal impetus to the study of isomorphism. Thus, in a whole series of minerals, lime and magnesia are found in variable and interchangeable proportions. Exactly the same may be said of potassium and sodium, of alumina and ferric oxide, of manganous, ferrous, magnesium oxides, &c. Such isomorphism does not, however, extend without change of form and properties beyond certain rather narrow limits. What I mean by this is that lime is not always replaced totally, but often only in small quantities, by magnesia, or by the manganous and ferrous oxides, without changing the crystalline form. The same may be observed with regard to potassium and lithium, which may be in part, but not completely, replaced by sodium. On the total substitution of one metal for another, often (although not invariably) the entire nature of the substance is changed; for instance, *enstatite* (or bronzite) is a magnesium bisilicate with a small isomorphous substitution of calcium for magnesium; its composition is expressed by the formula $MgSiO_3$, it belongs to the rhombic system. On the entire substitution of calcium, *wollastonite*, $CaSiO_3$, of the monoclinic system, is obtained; when manganese is substituted, *rhodonite*, of the triclinic system, is produced; but in all of them the angles of the prism are 86° to 88°.

The most remarkable complex siliceous compounds are the *felspars*, which enter into nearly all the primary rocks like porphyry, granite, gneiss, &c. These felspars always contain, in addition to silica and alumina, oxides presenting more marked basic properties, such as potash, soda, and lime. Thus the *orthoclase* (adularia), or ordinary felspar (monoclinic) of the granites, contains $K_2O,Al_2O_3,6SiO_2$; *albite* contains the same substances, only with Na_2O instead of K_2O (it already appertains to the triclinic system); *anorthite* contains lime, and its composition is $CaO,Al_2O_3,2SiO_2$. On expressing the two last as containing equal quantities of oxygen, we have:—

Albite $Na_2\ Al_2\ Si_6\ O_{16}$

Anorthite $Ca_2\ Al_4\ Si_4\ O_{16}$

It is then evident that on the conversion of albite into anorthite, Na_2Si_2 is replaced by Ca_2Al_2, and this sum, both in chemical energy and in the form of oxide, may be considered as corresponding with the first, because sodium and silicon are extreme elements in chemical character (from groups I. and IV.), and calcium and aluminium are means between them (from groups II. and III.), and actually both these felspar minerals are not only of one (triclinic) system, but form (Tchermak, Schuster) all possible kinds of definite compounds (isomorphous mixtures) between themselves, as indicated by their composition and all their properties. Thus oligoclase, andesine, labradorite, &c. (plagioclases), are nothing more than mutual combinations of albite and anorthite. Labradorite consists of albite, in combination with 1 to 2 molecules of anorthite. The class of *zeolites* corresponds to the felspars; they are hydrated compounds of a similar composition to the felspars. Thus *natrolite* contains $Na_2O,Al_2O_3,3SiO_2,2H_2O$, and *analcime* presents the same composition, but contains $4SiO_2$ instead of $3SiO_2$. In general, the felspars and zeolites contain $RO,Al_2O_3,nSiO_2$, where n varies considerably.

Such complex silicates are generally insoluble in water, and if they undergo change in it, it is but very slow, and more often only in the presence of carbonic acid. Some of the silicates which are insoluble in water are easily and directly decomposed by acids; for instance, the zeolites and those fused silicates which contain a large quantity of energetic bases—such as lime. Many of the silicates, like glass, are hardly changed by acids, particularly if they contain much silica, whilst fusion with alkalis leads to the formation of compounds rich in bases, after which acids decompose the alloys formed.

According to the periodic law, the nearest analogues of silicon ought to be elements of the uneven series, because silicon, like sodium, magnesium, and aluminium, belongs to the uneven series. Immediately after silicon follows ekasilicon or *germanium*, Ge = 72, whose properties were predicted (1871) before Winkler (1886) in Freiberg, Saxony (Chapter XV. § 5), discovered this element in a peculiar silver ore called *argyrodite*, Ag_6GeS_5. Easily reduced from the oxide by heating with hydrogen and charcoal, and separated from its solutions by zinc, metallic germanium proved to be greyish white, easily crystallisable (in octahedra), brittle, fusible (under a coating of fused borax) at about 900°, and easily oxidisable; the specific gravity = 5·469, the atomic weight = 72·3, and the specific heat = 0·076, as might be expected for this element according to the periodic law. The corresponding *germanium dioxide*, GeO$_2$, is a white powder having a specific gravity of 4·703; water, especially when boiling, dissolves this dioxide (1 part of GeO$_2$ requires for solution 247 parts of water at 20°, 95 parts at 100°). It forms soluble salts with alkalis and is but sparingly soluble in acids. In a stream of chlorine the metal forms *germanium chloride*, GeCl$_4$, which boils at 86°, and has a specific gravity of 1·887 at 18°; water decomposes it, forming the oxide. All these properties of germanium, showing its analogy to silicon and tin, form a most beautiful demonstration of the truth of the periodic law.

The increase of atomic weight from silicon 28 to germanium 72 is 44—that is, about the same difference as there is in the atomic weights of chlorine and bromine; between germanium and its next analogue, *tin* (Sn = 118), the difference is 46—that is, almost as much as the amount by which the atomic weight of iodine exceeds that of bromine.

Metallic tin is rarely met with in *nature*; it occurs in the veins of ancient formations, almost exclusively in the form of oxide, SnO$_2$, called *tin-stone*. The best known tin deposits are in Cornwall and in Malacca. In Russia, tin ores have been found in small quantities on the shores of Lake Ladoga, in Pitkarand. The crushed ore may easily be separated from the earthy matter accompanying it by washing on inclined tables, as the tin-stone has a specific gravity of 6·9, whilst the impurities are much lighter. *Tin oxide is very easily reduced* to metallic tin by heating with charcoal. For this reason tin was known in ancient times, and the Phœnicians brought it from England. Metallic tin is cast into ingots of considerable weight or into thin sticks or rods. Tin has a white colour, rather duller than that of silver. It fuses easily at 232°, and crystallises on cooling. Its specific gravity is 7·29. The crystalline structure of ordinary tin is noticed in bending tin rods, when a peculiar sound is heard, produced by the fracture of the particles of tin along the surfaces of crystalline structure.

When pure tin is cooled to a low temperature it splits up into separate crystals, the bond between the particles is lost, the tin assumes a grey colour, becomes less brilliant—in a word, its properties become changed, as Fritzsche showed. This depends on the peculiar structure which the tin then acquires, and is particularly remarkable because it is effected by cold in a solid.[33 bis] If such tin be fused, or even simply heated, it becomes like ordinary tin, but is again changed when cooled. When in this condition tin has a specific gravity of 7·19. Similarly, tin is obtained by the action of the galvanic current on a solution of tin chloride; it then appears in crystals of the cubic system, and has a specific gravity of 7·18—that is, the same as when cooled.

Tin is softer than silver and gold, and is only surpassed by lead in this respect. In addition to this it is very ductile, but its tenacity is very slight, so that wire made from it will bear but little strain. In consequence of its ductility it is easily worked, by forging and rolling into very thin sheets (tin foil), which are used for wrapping many articles to preserve them from moisture, &c. In this case, however, and in many others, lead is mixed with the tin, which, within certain limits, does not alter the ductility. Whilst so soft at the ordinary temperatures tin becomes brittle at 200°, before fusing. Tin powder may be easily obtained if the metal be fused and then stirred whilst cooling. At a white heat tin may be distilled, but with more difficulty than zinc. If molten tin comes into contact with oxygen, it oxidises, forming stannic oxide, SnO_2, *and its vapour burns* with a white flame. *At ordinary temperatures tin does not oxidise*, and this very important property of tin allows it to be applied in many cases for covering other metals to prevent their oxidising. This is termed *tinning*. Iron and copper are frequently tinned. Iron and steel sheets, coated with tin, bear the name of tin plate (for the most part made in England), and are used for numerous purposes. Tin plate is prepared by immersing iron sheets, previously thoroughly cleansed by acid and mechanical means, into molten tin.[34 bis]

Tin with copper forms *bronze*, an alloy which is most extensively used in the arts. Bronze has various colours and a variety of physical properties, according to the relative amount of copper and tin which it contains. With an excess of copper the alloy has a yellow colour; the admixture of tin imparts considerable hardness and elasticity to the copper. An alloy containing 78 parts of copper and about 22 per cent. of tin is so elastic that it is used for casting bells, which naturally require a very elastic and hard alloy. For casting statues and various large or small ornamental articles alloys containing 2 to 5 p.c. of tin, 10 to 30 p.c. of zinc, and 65 to 85 p.c. of copper are used. Tin is also often used alloyed with lead, for making various objects—for instance, drinking vessels.

Tin decomposes the vapour of water when heated with it, liberating the hydrogen and forming stannic oxide. Sulphuric acid, diluted with a considerable quantity of water, does not act, or at all events acts very slightly, on tin, but tin reduces hot strong sulphuric acid, when not only sulphurous anhydride but also sulphuretted hydrogen is evolved. Hydrochloric acid acts very easily on tin, with evolution of hydrogen and formation of stannous chloride, $SnCl_2$, in solution, which, with an excess of hydrochloric acid and access of air, is converted into stannic chloride: $SnCl_2 + 2HCl + O = SnCl_4 + H_2O$.[36 bis] Nitric acid diluted with a considerable quantity of water dissolves tin at the ordinary temperature, whilst the nitric acid itself is reduced, forming, amongst other products, ammonia and hydroxylamine. Here the tin passes into solution in the form of stannous nitrate. Stronger nitric acid (also more dilute, when heated) transforms the tin into its highest grade of oxidation, SnO_2, but the latter then appears as the so-called metastannic acid, which does not dissolve in nitric acid, and therefore the tin does not pass into solution. Feeble acids—for instance, carbonic and organic acids—do not act on tin even in the presence of oxygen, because tin does not form any powerful bases.

It is important to remark as a characteristic of tin that it is reduced from its solutions by many metals which are more easily oxidised, as, for instance, by zinc.

In combination, tin appears in the two types, SnX_4 and SnX_2, compounds of the intermediate type, Sn_2X_6, being also known, but these latter pass with remarkable facility in most cases into compounds of the higher and lower types, and therefore the form SnX_3 cannot be considered as independent.

Stannous oxide, SnO, in an anhydrous condition is obtained by boiling solutions of stannous salts with alkalis, the first action of the alkali being to precipitate a white hydrate of stannous oxide, $Sn(OH)_2SnO$. The latter when heated parts with water as easily as the hydrate of copper oxide. In this form stannous oxide is a black crystalline powder (specific gravity 6·7) capable of further oxidation when heated. The hydrate is freely soluble in acids, and also in potassium and sodium hydroxides, but not in aqueous ammonia. This property indicates the feeble basic properties of this lower oxide, which acts in many cases as a reducing agent. Among the compounds corresponding with stannous oxide the most remarkable and the one most frequently used is stannous chloride or *chloride of tin*, $SnCl_2$, also called proto-chloride of tin (because it is the lowest chloride, containing half as much Cl as $SnCl_4$). It is a transparent, colourless, crystalline substance, melting at 250° and boiling at 606°. Water dissolves it, without visible change (in reality partial decomposition occurs, as we shall see presently). It is also soluble in alcohol. It is obtained by heating tin in dry hydrochloric acid gas, the hydrogen being then liberated, or by dissolving metallic tin in hot strong hydrochloric acid and then evaporating quickly. On cooling, crystals of the monoclinic system are obtained having the composition $SnCl_2,2H_2O$. An aqueous solution of this substance absorbs oxygen from the atmosphere, and gives a precipitate containing stannic oxide. From this it follows that a solution of stannous chloride will act as a reducing agent, a fact frequently made use of in chemical investigations—for example, for reducing metals from their solutions—since even mercury may be reduced to a metallic state from its salts by means of stannous chloride. This reducing property is also employed in the arts, especially in the dyeing industry, where this substance in the form of a crystalline salt finds an extensive application, and is known as *tin salt* or tin crystals.

Stannic oxide, SnO_2, occurring in nature as *tin-stone*, or *cassiterite*, is formed during the oxidation or combustion of heated tin in air as a white or yellowish powder which fuses with difficulty. It is prepared in large quantities, being used as a white vitreous mixture for coating ordinary tiles and similar earthenware objects with a layer of easily fusible glass or enamel. Acid solutions of stannic oxide treated with alkalis, and alkaline solutions treated with acids, give a precipitate of stannic hydroxide, $Sn(OH)_4$, also known as stannic acid, which, when heated, gives up water and leaves the anhydride, SnO_2, which is insoluble in acids, clearly showing the feebleness of its basic character. When fused with alkali hydroxides (not with their carbonates or acid sulphates), an alkaline compound is obtained which is soluble in water. Stannic hydroxide, like the hydrates of silica, is a colloidal substance, and presents several different modifications, depending on the method of preparation, but having an identical composition; the various hydroxides have also a different appearance, and act differently with reagents. For instance, a distinction is made between ordinary stannic acid and metastannic acid. *Stannic acid* is produced by precipitation by soda or ammonia from a freshly-prepared solution of stannic chloride, $SnCl_4$, in water; on drying the precipitate thus obtained, a non-crystalline mass is formed, which is freely soluble in strong hydrochloric or nitric acids, and also in potassium and sodium hydroxides. This ordinary stannic acid may be still better obtained from sodium stannate by the action of acids. *Metastannic acid* is insoluble in sulphuric and nitric acids. It is obtained in the form of a heavy white powder by treating tin with nitric acid; hydrochloric acid does not dissolve it immediately, but changes it to such an extent that, after pouring off the acid, water extracts the stannic chloride, $SnCl_4$, already formed. Dilute alkalis not only dissolve metastannic acid, but also transform it into salts, which, slowly, yet completely, dissolve in *pure water*, but are insoluble even in dilute alkali hydroxides. Dilute hydrochloric acid, especially when boiling, changes the ordinary hydrate into metastannic acid. On this depends, by the way, the formation of a white precipitate, stannic hydroxide, from solutions of stannous and stannic chlorides diluted with water. The stannic oxide first dissolved changes under the influence of hydrochloric acid into metastannic acid, which is insoluble in water in the presence of hydrochloric acid. Solutions of metastannic acid differ from solutions of ordinary stannic acid, and in the presence of alkali they change into solutions of ordinary acid, so that metastannic acid corresponds principally with the acid compounds of stannic oxide, and ordinary stannic acid with the alkaline compounds. Graham obtained a soluble colloidal hydroxide; it is subject to the same transformations that are in general peculiar to colloids.

Stannic oxide shows the properties of a slightly energetic and intermediate oxide (like water, silica, &c.); that is to say, it forms saline compounds both with bases and with acids, but both are easily decomposed, and are but slightly stable. But still the acid character is more clearly developed than the basic, as in silica, germanic oxide, and lead dioxide. This determines the character of the compounds SnX_4, corresponding to stannic chloride, $SnCl_4$ (also called tetrachloride of tin). It is obtained in an anhydrous condition by the direct action of chlorine on tin, and is then easily purified, because it is a liquid boiling at 114°, and therefore can be easily distilled. Its specific gravity is 2·28 (at 0°), and it fumes in the open air (spiritus fumans libavii), reacting on the moisture of the air, thus showing the properties of a chloranhydride. Water however does not at first decompose it, but dissolves it, and on evaporation gives the crystallo-hydrate $SnCl_4,5H_2O$. If but little water be taken, crystals containing $SnCl_4,3H_2O$ are formed, which part with one-third of the water when placed under the receiver of the air-pump. A large quantity of water however, especially on heating, causes a precipitate of metastannic acid and formation of HCl.

The alkali compounds of stannic oxide—that is, the compounds in which it plays the part of an acid, corresponding in this respect to the compounds of silica and other anhydrides of the composition RO_2—are very easily formed and are used in the arts. Their composition in most cases corresponds with the formula SnM_2O_3—that is, $SnO(MO)_2$, similar to $CO(MO)_2$, where M = K, Na. Acids, even feeble acids like carbonic, decompose the salts, like the corresponding compounds of alumina or silica. In order to obtain *potassium stannate*, which crystallises in rhombohedra, and has the composition $SnK_2O_3,3H_2O$, potassium hydroxide (8 parts) is fused, and metastannic acid (3 parts) gradually added. *Sodium stannate* is prepared in practice in large quantities by heating a solution of caustic soda with lead oxide and metallic tin. In this last case an alkaline solution of lead oxide is formed, and the tin acts on the solution in such a way as to reduce the lead to the metallic state, and itself passes into solution. It is very remarkable that lead displaces tin when in combination with acids, whilst tin, on the contrary, displaces lead from its alkali compounds. By dissolving the mass obtained in water, and adding alcohol, sodium stannate is precipitated, which may then be dissolved in water and purified by re-crystallisation. In this case it has the composition $SnNa_2O_3,3H_2O$ if separated from strong solutions, and $SnNa_2O_3,10H_2O$ when crystallised at a low temperature from dilute solutions. In the arts this salt is used as a mordant in dyeing operations. With a cold solution of sodium hydroxide metastannic acid forms a salt of the composition $(NaHO)_2,5SnO_2,3H_2O$, from which Frémy drew his conclusions concerning the polymerism of metastannic acid. Tin, like other metals and many metalloids, gives a peroxide form of combination or *perstannic oxide*. This substance was obtained by Spring (1889) in the form of a hydrate,

$H_2Sn_2O_7 = 2(SnO_3)H_2O$, by mixing a solution of $SnCl_2$, containing an excess of HCl, with freshly prepared peroxide of barium. A cloudy liquid is then obtained, and this after being subjected to dialysis leaves a gelatinous mass which on drying is found to have the composition $Sn_2H_2O_7$. Above 100° this substance gives off oxygen and leaves SnO_2. It is evident that SnO_3 bears the same relation to SnO_2 as H_2O_2 to H_2O or ZnO_2 to ZnO, &c.

Tin occupies the same position amongst the analogues of silicon as cadmium and indium amongst the analogues of magnesium and aluminium respectively, and as in each of these cases the heavier analogues with a high atomic weight and a special combination of properties—namely, mercury and thallium—are known, so also for silicon we have *lead* as the heaviest analogue (Pb = 206), with a series of both kindred and special properties. The higher type, PbX_4—for instance, PbO_2—is in a chemical sense far less stable than the lower type, PbX. The ordinary compounds of lead correspond with the latter, and in addition to this, PbO, although not particularly energetic, is still a decided base easily forming basic salts, $PbX_2(PbO)_n$. Although the compounds PbX_4, are unstable they offer many points of analogy with the corresponding compounds of tin SnO_2; this is seen, for instance, in the fact that PbO_2 is a feeble acid, giving the salt PbK_2O_3, that $PbCl_4$ is a liquid like $SnCl_4$ which is not affected by sulphuric acid, and that PbF_4 gives double salts, like SnF_4 or SiF_4 (Brauner 1894. See Chapter II., Note 49 bis); $Pb(C_2H_5)_4$ also resembles $Sn(C_2H_5)_4$ &c. All this shows that lead is a true analogue of tin, as Hg is of cadmium.[41 bis]

Lead is found in nature in considerable masses, in the form of galena, *lead sulphide*, PbS. The specific gravity of galena is 7·58, colour grey; it crystallises in the regular system, and has a fine metallic lustre. Both the native and artificial sulphides are insoluble in acids (hydrogen sulphide gives a black precipitate with the salts PbX_2).[42 bis] When heated, lead melts, and in the open air is either totally or partially transformed into white lead sulphate, $PbSO_4$, as it also is by many oxidising agents (hydrogen peroxide, potassium nitrate). Lead sulphate is also insoluble in water, and lead is but rarely met with in this form in nature. The chromates, vanadates, phosphates, and similar salts of lead are also somewhat rare. The carbonate, $PbCO_2$, is sometimes found in large masses, especially in the Altai region. Lead sulphide is often worked for extracting the silver which it contains; and as the lead itself also finds manifold industrial applications, this work is carried out on an exceedingly large scale. Many methods are employed. Sometimes the lead sulphide is decomposed by heating it with cast iron. The iron takes up the sulphur from the lead and forms easily-fusible iron sulphide, which does not mix with the heavier reduced lead. But another process is more frequently used: the lead ore (it must be clean; that is, free from earthy matter, which may be easily removed by washing) is heated in a reverberatory furnace to a moderate temperature with a free access of air. During this operation part of the lead sulphide oxidises and forms lead sulphate, $PbSO_4$, and lead oxide. When the oxidation of part of the lead has been attained, it is necessary to shut off the air supply and increase the temperature, then the oxidised compounds of the lead enter into reaction with the remaining lead sulphide, with formation of sulphurous anhydride and metallic lead. At first from $PbS + O_3$, $PbO + SO_2$ are formed, and also from $PbS + O_4$ lead sulphate $PbSO_4$, and then PbO and $PbSO_4$ react with the remaining PbS, according to the equations $2PbO + PbS = 3Pb + SO_2$ and also $PbSO_4 + PbS = 2Pb + 2SO_2$.

The appearance of lead is well known; its specific gravity is 11·3; the bluish colour and well-marked metallic lustre of freshly-cut lead quickly disappear when exposed to the air, because it becomes coated with a layer—although a very thin layer—of oxide and salts formed by the moisture and acids in the atmosphere. It melts at 320°, and crystallises in octahedra on cooling. Its softness is apparent from the flexibility of lead pipes and sheets, and also from the fact that it may be cut with a knife, and also that it leaves a grey streak when rubbed on paper. On account of its being so soft, lead naturally cannot be applied in many cases where most metals may be used; but on the other hand it is a metal which is not easily changed by chemical reagents, and as it is capable of being soldered and drawn into sheets, &c., lead is most valuable for many technical uses. Lead pipes are used for conveying water and many other liquids, and sheet lead is used for lining all kinds of vessels containing liquids—(acids, for instance) which act on other metals. This particularly refers to sulphuric and hydrochloric acids, because at a low temperature they do not act on lead, and if they form lead sulphate, $PbSO_4$, and chloride, $PbCl_2$, these salts being insoluble in water and in acids, cover the lead and protect it from further corrosion. All soluble preparations of lead are poisonous. At a white heat lead may be partially distilled; the vapours oxidise and burn. Lead may also be easily oxidised at low temperatures. Lead only decomposes water at a white heat, and does not liberate hydrogen from acids, with the exception only of very strong hydrochloric acid and then only when boiling. Sulphuric acid diluted with water does not act on it, or only acts very feebly at the surface; but strong sulphuric acid, when heated, is decomposed by it, with the evolution of sulphurous anhydride. The best solvent for lead is nitric acid, which transforms it into a soluble salt, $Pb(NO_3)_2$.

Although acids thus have directly but little effect on lead, and this is one of its most important practical properties, *yet when air has free access, lead (like copper) very easily reacts with many acids*, even with those which are comparatively feeble. The action of acetic acid on lead is particularly striking and often applied in practice. If lead be plunged into acetic acid it does not change at all and does not pass into solution, but if part of the lead be immersed in the acid, and the other part remain in contact with the air, or if lead be merely covered with a thin layer of acetic acid in such a way that the air is practically in contact with the metal, then it unites with the oxygen of the air to form oxide, which combines with the acetic acid and forms lead acetate, soluble in water. The formation of lead oxide is especially marked from the fact that with a sufficient quantity of air not only is the normal lead acetate formed but also the basic salts.

When oxidising in the presence of air, when heated or in the presence of an acid at the ordinary temperature, lead forms compounds of the type PbX_2. *Lead oxide*, PbO, known in industry as *litharge*, silberglätte (this name is due to the fact that silver is extracted from the lead ores of this kind) and massicot. If the lead is oxidised in air at a high temperature, the oxide which is formed fuses, and on cooling is easily obtained in fused masses which split up into scales of a yellowish colour, having a specific gravity of 9·3; in this form it bears the name of litharge. Litharge is principally used for making lead salts, for the extraction of metallic lead, and also for the preparation of drying oils—for instance, from linseed oil. When oxidised carefully and slightly heated, lead forms a powdery (not fused) oxide known under the name of *massicot*. It is best prepared in the laboratory by heating lead nitrate, or lead hydroxide. It has a yellow colour, and differs from litharge in the greater difficulty with which it forms lead salts with acids. Thus, for instance, when massicot is moistened with water it does not attract the carbonic acid of the air so easily as litharge does. It may, however, be imagined that the cause of the difference depends only on the formation of dioxide on the surface of the lead oxide, on which the acids do not act. In any case lead oxide is comparatively easily soluble in nitric and acetic acids. It is but slightly soluble in water, but communicates an alkaline reaction to it, since it forms the hydroxide. This hydroxide is obtained in the shape of a white precipitate by the action of a small quantity of an alkali hydroxide on a solution of a lead salt. An excess of alkali dissolves the hydroxide separated, which fact demonstrates the comparatively indistinct basic properties of lead oxide. The normal lead hydroxide, which should have the composition $Pb(OH)_2$, is unknown in a separate state, but it is known in combination with lead oxide as $Pb(OH)_2, 2PbO$ or $Pb_3O_2(OH)_2$. The latter is obtained in the form of brilliant, white, octahedral crystals when basic lead acetate is mixed with ammonia and gently heated. The basic qualities of this hydroxide are shown distinctly by its absorbing the carbonic anhydride of the air. When an alkaline solution of the hydroxide is boiled, it deposits lead oxide in the form of a crystalline powder.

Lead oxide forms but few soluble salts—for instance, the nitrate and the acetate. The majority of its salts (sulphate, $PbSO_4$; carbonate, $PbCO_3$; iodide, PbI_2, &c.) are insoluble in water. These salts are colourless or light yellow if the acid be colourless. In lead oxide *the faculty of forming basic salts*, $PbX_2 n PbO$ or $PbX_2 n PbH_2O_2$, is strongly developed. A similar property was observed in magnesium and also in the salts of mercury, but lead oxide forms basic salts with still greater facility, although double salts are in this case more rarely formed.

Amongst the soluble lead salts, that best known and most often applied in practical chemistry is *lead nitrate*, obtained directly by dissolving lead or its oxide in nitric acid. The normal salt, $Pb(NO_3)_2$, crystallises in octahedra, dissolves in water, and has a specific gravity of 4·5. When a solution of this salt acts on white lead or is boiled with litharge, the basic salt, having a composition $Pb(OH)(NO_3)$, is formed in crystalline needles, sparingly soluble in cold water but easily dissolved in hot water, and therefore in many respects resembling lead chloride. When the nitrate is heated, either lead oxide is obtained or else the oxide in combination with peroxide.

Lead chloride, $PbCl_2$, is precipitated from the soluble salts of lead when a strong solution is treated with hydrochloric acid or a metallic chloride. It is soluble in considerable quantities in hot water, and therefore if the solutions be dilute or hot, the precipitation of lead chloride does not occur, and if a hot solution be cooled, the salt separates in brilliant prismatic crystals. It fuses when heated (like silver chloride), but is insoluble in ammonia. This salt is sometimes met with in nature, and when heated in air is capable of exchanging half its chlorine for oxygen, forming the basic salt or lead oxychloride, $PbCl_2 PbO$, which may also be obtained by fusing $PbCl_2$ and PbO together. The reaction of lead chloride with water vapour leads to the same conclusion, showing the feeble basic character of lead $2PbCl_2 + H_2O = PbCl_2,PbO + 2HCl$. When ammonia is added to an aqueous solution of lead chloride a white precipitate is formed, which parts with water on being heated, and has the composition $Pb(OH)Cl,PbO$. This compound is also formed by the action of metallic chlorides on other soluble basic salts of lead.

Lead carbonate, or *white lead*, is the most extensively used basic lead salt. It has the valuable property of 'covering,' which only to a certain extent appertains to lead sulphate and other white powdery substances used as pigments. This faculty of 'covering' consists in the fact that a small quantity of white lead mixed with oil spreads uniformly, and if such a mixture be spread over a surface (for instance, of wood or metal) the surface is quickly covered—that is, light does not penetrate through even a very thin layer of superposed white lead; thus, for example, the grain of the wood remains invisible. White lead, or *basic lead carbonate*, after being dried at 120°, has a composition $Pb(OH)_2,2PbCO_3$. It may be obtained by adding a solution of sodium carbonate to a solution of one of the basic salts of lead—for instance, the basic acetate—and likewise by treating this latter with carbonic acid. For this purpose the solution of basic acetate is poured into the vessel f; it is prepared in the vat A, containing litharge, into which the pump P delivers the solution of the acetate, which remains after the action of carbonic anhydride on the basic salt. In A a basic salt is formed having a composition approaching to $Pb_4(OH)_6(C_2H_3O_2)_2$; carbonic anhydride, $2CO_2$, is passed through this solution and precipitates white lead, $Pb_2(OH)_2(CO_3)_2$, and normal lead acetate, $Pb(C_2H_3O_2)_2$, remains in the solution, and is pumped back into the vat A containing lead oxide, where the normal salt is again (on being agitated) converted into the basic salt. This is run into the vessel E, and thence into f. Into the latter carbonic anhydride is delivered from the generator D, and forms a precipitate of white lead.[53 bis]

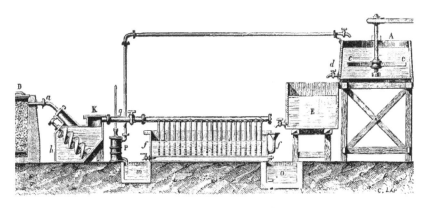

FIG. 82.—Manufacture of white lead.

In order to mark the transition from lead oxide, PbO, into lead dioxide PbO_2 (plumbic anhydride), it is necessary to direct our attention to the intermediate oxide, or *red lead*, Pb_3O_4. In the arts it is used in considerable quantities, because it forms a very durable yellowish-red paint used for colouring the resins (shellac, colophony, &c.) composing sealing wax. It also forms a very good cheap oil paint, used especially for painting metals, more particularly because drying oils—for instance, hemp seed, linseed oils—very quickly dry with red lead and with lead salts. Red lead is prepared by slightly heating massicot, for which purpose two-storied stoves are used. In the lower story the lead is turned into massicot, and in the higher one, having the lower temperature (about 300°), the massicot is transformed into red lead. Frémy and others showed the instability of red lead prepared by various methods, and its decomposition by acids, with formation of lead dioxide, which is insoluble in acids, and a solution of the salts of lead oxide.

The artificial production (synthesis) of red lead by double decomposition was most important. For this purpose Frémy mixed an alkaline solution of potassium plumbate, K_2PbO_3 (prepared by dissolving the dioxide in fused potash),[54 bis] with an alkaline solution of lead oxide. In this way a yellow precipitate of minium hydrate is formed, which, when slightly heated, loses water and turns into bright red anhydrous minium Pb_3O_4.

Minium is the first and most ordinary means of producing *lead dioxide*, or plumbic anhydride, PbO_2, because when red lead is treated with dilute nitric acid it gives up lead oxide, and PbO_2 remains, on which dilute nitric acid does not act. The composition of minium is Pb_3O_4, and therefore the action of nitric acid on it is expressed by the equation: $Pb_3O_4 + 4HNO_3 = PbO_2 + 2Pb(NO_3)_2 + 2H_2O$. The dioxide may also be obtained by treating lead hydroxide suspended in water with a stream of chlorine. Under these conditions the chlorine takes up the hydrogen from the water, and the oxygen passes over to the lead oxide. When a strong solution of lead nitrate is decomposed by the electric current, the appearance of crystalline lead dioxide is also observed upon the positive pole; it is also found in nature in the form of a black crystalline substance having a specific gravity of 9·4. When artificially produced it is a fine dark powder, resisting the action of acids, but nevertheless when treated with strong sulphuric acid it evolves oxygen and forms lead sulphate, and with hydrochloric acid it evolves chlorine. The oxidising property of lead dioxide depends of course on the facility of its transition into the more stable lead oxide, which is easily understood from the whole history of lead compounds. In the presence of alkalis it transforms chromium oxide into chromic acid, whilst lead chromate, $PbCrO_4$, is formed, remaining, however, in solution, on account of its being soluble in caustic alkalis. The oxidising action of lead dioxide on sulphurous anhydride is most striking, as it immediately absorbs it, with formation of lead sulphate. This is accompanied by a change of colour and development of heat, $PbO_2 + SO_2 = PbSO_4$. When triturated with sulphur the mixture explodes, the sulphur burning at the expense of the oxygen of the lead dioxide. *Tetrachloride of lead*, $PbCl_4$, belongs to the same class of lead compounds as PbO_2. This chloride is formed by the action of strong hydrochloric acid upon PbO_2, or, in the cold, by passing a stream of chlorine through water containing $PbCl_2$ in suspension. The resultant yellow solution gives off chlorine when heated. With a solution of sal ammoniac (Nicolukin, 1885) it gives a precipitate of a double salt, $(NH_4)_2PbCl_6$ (very slightly soluble in a solution of sal ammoniac), which when treated with strong sulphuric acid (Friedrich, 1890) gives $PbCl_4$ as a yellow liquid sp. gr. 3·18, which solidifies at -18°, and when heated gives $PbCl_2 + Cl_2$. It is not acted upon by H_2SO_4 like $SnCl_4$. Tetrafluoride of lead (Brauner) belongs to the same class of compounds, it easily forms double salts and decomposes with the evolution of fluorine (Chapter II., Note 49 bis).[56 bis]

Amongst the elements of the second and third groups it was observed that the elements were more basic in the even than in the uneven series. It is sufficient to remember calcium, strontium, and barium in the even, and magnesium, zinc, and cadmium in the uneven series. In addition to this, in the even series, as the atomic weight increases, in the same type of oxidation the basic properties increase (the acid properties decrease); for example, in the second group, calcium, strontium, barium. The same also appears in the fourth and all the following groups. In the even series of the fourth group titanium, zirconium, cerium, and thorium are found. All their highest oxides, RO_2, even the lightest, titanic oxide, TiO_2, have more highly developed basic properties than silica, SiO_2, and in addition to this the basic properties are more distinctly seen in zirconium dioxide, ZrO_2, than in titanic oxide, TiO_2, although the acid property of combining with bases still remains. In the heaviest oxides, cerium dioxide, CeO_2, and thorium dioxide, ThO_2, no acid properties are observed, these being both purely basic oxides. In Chapter XVII. (Note 43) we already pointed out this higher oxide of cerium. As the above-mentioned elements are rather rare in nature, have but little practical application, and do not present any new forms of combination, it is unadvisable to dwell on them in this treatise.

Titanium is found in nature in the form of its anhydride or oxide, TiO_2, mixed with silicon in many minerals, but the oxide is also found separately in the form of semi-metallic *rutile* (sp. gr. 4·2). Another titanic mineral is found as a mixture in other ores, known as *titanic iron ore* (in the Thuensky mountains of the southern Ural; it is known as *thuenite*), $FeTiO_3$. This is a salt of ferrous oxide and titanic anhydride. It crystallises in the rhombohedric system, has a metallic lustre, grey colour, sp. gr. 4·5. The third mineral in which titanium is found in considerable quantities in nature is *sphene* or *titanite*, $CaTiSiO_5 = CaO,SiO_2,TiO_2$, sp. gr. 3·5, colour yellow, green, or the like, crystallises in tablets. The fourth, but rare, titanic mineral is *peroffskite*, calcium titanate, $CaTiO_3$; it forms blackish-grey or brown cubic crystals, sp. gr. 4·02, and occurs in the Ural and other localities. It may be prepared artificially by fusing sphene in an atmosphere of water vapour and carbonic anhydride. At the end of the last century Klaproth showed the distinction between titanic compounds and all others then known.

The comparatively rare element *zirconium*, $Zr = 90$, is very similar to titanium, but has a more basic character. It is rarer in nature than titanium, and is found principally in a mineral called *zircon*, $ZrSiO_4 = ZrO_2.SiO_2$, crystallising in square prisms, sp. gr. 4·5. It has considerable hardness and a characteristic brownish-yellow colour, and is occasionally found in the form of transparent crystals, as a precious stone called hyacinth. Metallic zirconium was obtained, by Berzelius and Troost, by the action of aluminium on potassium zirconofluoride in the same way that silicon is prepared; it forms a crystalline powder, similar in appearance to graphite and antimony, but having a very considerable hardness, not much lustre, sp. gr. 4·15. In many respects it resembles silicon; it does not fuse when heated, and even oxidises with difficulty, but liberates hydrogen when fused with potash. When fused with silica it liberates silicon. With carbon in the electrical furnace it forms ZrC_2, with hydrogen it gives ZrH_2 (like CaH_2, Winkler, Vol. I., p. 621); hydrochloric and nitric acids act feebly on it, but aqua regia easily dissolves it. It is distinguished from silicon by the fact that hydrofluoric acid acts on it with great facility, even in the cold and when diluted, whilst this acid does not act on silicon at all.

The very similar element *thorium* (Th = 232) was distinguished by Berzelius from zirconium. It is very rarely met with, in *thorite* and *orangeite*, $ThSiO_4,2H_2O$. The latter is isomorphous with zircon (sp. gr. 4·8).

Footnotes:

Chloroform, $CHCl_3$, boils at 60°, and silicon chloroform, $SiHCl_3$, at 34°; silicon ethyl, $Si(C_2H_5)_4$, boils at about 150°, and its corresponding carbon compound, $C(C_2H_5)_4$, at about 120°; ethyl orthosilicate, $Si(OC_2H_5)_4$, boils at 160°, and ethyl orthocarbonate, $C(OC_2H_5)_4$, at 158°. The specific volumes in a liquid state—that is, those of the silicon compounds—generally are slightly greater than those of the carbon compounds; for example, the volumes of CCl_4 = 94, $SiCl_4$ = 112, $CHCl_3$ = 81, $SiHCl_3$ = 82, of $C(OC_2H_5)_4$ = 186, and $Si(OC_2H_5)_4$ = 201. The corresponding salts have also nearly equal specific volumes; for example, $CaCO_3$ = 37, $CaSiO_3$ = 41. It is impossible to compare SiO_2 and CO_2, because their physical states are so widely different.

[1 bis] But silica fuses and volatilises (Moissan) in the heat of the electric furnace, about 3000°, SiO_2 is also partially volatile at the temperature attained in the flame of detonating gas (Cremer, 1892).

A property of intercombination is observable in the atoms of carbon, and a faculty for intercombination, or polymerisation, is also seen in the unsaturated hydrocarbons and carbon compounds in general. In silicon a property of the same nature is found to be particularly developed in silica, SiO_2, which is not the case with carbonic anhydride. The faculty of the molecules of silica for combining both with other molecules and among themselves is exhibited in the formation of most varied compounds with bases, in the formation of hydrates with a gradually decreasing proportion of water down to anhydrous silica, in the colloid nature of the hydrate (the molecules of colloids are always complex), in the formation of polymeric ethereal salts, and in many other properties which will be considered in the sequel. Having come to this conclusion as to the polymeric state of silica since the years 1850–1860, I have found it to be confirmed by all subsequent researches on the compounds of silica, and, if I mistake not, this view has now been very generally accepted.

It was only after Gerhardt, and in general subsequently to the establishment of the true atomic weights of the elements (Chapter [VII.](#)), that a true idea of the atomic weight of silicon and of the composition of silica was arrived at from the fact that the molecules of $SiCl_4$, SiF_4, $Si(OC_2H_5)_4$, &c., never contain less than 28 parts of silicon.

The question *of the composition of silica* was long the subject of the most contradictory statements in the history of science. In the last century Pott, Bergmann, and Scheele distinguished silica from alumina and lime. In the beginning of the present century Smithson for the first time expressed the opinion that silica was an acid, and the minerals of rocks salts of this acid. Berzelius determined the presence of oxygen in silica—namely, that 8 parts of oxygen were united with 7 of silicon. The composition of silica was first expressed as SiO (and for the sake of shortness S only was sometimes written instead). An investigation in the amount of silica present in crystalline minerals showed that the amount of oxygen in the bases bears a very varied proportion to the amount of oxygen in the silica, and that this ratio varies from 2 : 1 to 1 : 3. The ratio 1 : 1 is also met with, but the majority of these minerals are rare. Other more common minerals contain a larger proportion of silica, the ratio between the oxygen of the bases and the oxygen of the silica being equal to 1 : 2, or thereabouts; such are the augites, labradorites, oligoclase, talc, &c. The higher ratio 1 : 3 is known for a widely distributed series of natural silicates—for example, the felspars. Those silicates in which the amount of oxygen in the bases is equal to that in the silica are termed *monosilicates*; their general formula will be $(RO)_2SiO_2$ or $(R_2O_3)_2(SiO_2)_3$. Those in which the ratio of the oxygen is equal to 1 : 2 are termed *bisilicates*, and their general formula will be $ROSiO_2$ or $R_2O_3(SiO_2)_3$. Those in which the ratio is 1 : 3 will be *trisilicates*, and their general formula $(RO)_2(SiO_2)_3$ or $(R_2O_3)_2(SiO_2)_9$.

In these formulæ the now established composition of SiO_2—that is, that in which the atom of Si = 28—is employed. Berzelius, who made an accurate analysis of the composition of felspar, and recognised it as a trisilicate formed by the union of potassium oxide and alumina with silica, in just the same manner as the alums are formed by sulphuric acid, gave silica the same formula as sulphuric anhydride—that is, SiO_3. In this case the formula of felspar would be exactly similar to that of the alums—that is, $KAl(SiO_4)_2$, like the alums, $KAl(SO_4)_2$. If the composition of silica be represented as SiO_3, the atom of silicon must be recognised as equal to 42 (if $O = 16$; or if $O = 8$, as it was before taken to be, $Si = 21$).

The former formulæ of silica, SiO ($Si = 14$) and SiO_3 ($Si = 42$), were first changed into the present one, SiO_2 ($Si = 28$), on the basis of the following arguments:—An excess of silica occurs in nature, and in siliceous rocks free silica is generally found side by side with the silicates, and one is therefore led to the conclusion that it has formed acid salts. It would therefore be incorrect to consider the trisilicates as normal salts of silica, for they contain the largest proportion of silica; it is much better to admit another formula with a smaller proportion of oxygen for silica, and it then appears that the majority of minerals are normal or slightly basic salts, whilst some of the minerals predominating in nature contain an excess of silica—that is, belong to the order of acid salts.

At the present time, when there is a general method (Chapter VII.) for the determination of atomic weights, the volumes of the volatile compounds of silica show that its atomic weight $Si = 28$, and therefore silica is SiO_2. Thus, for example, the vapour density of silicon chloride with respect to air is, as Dumas showed (1862), 5·94, and hence with respect to hydrogen it is 85·5, and consequently its molecular weight will be 171 (instead of 170 as indicated by theory). This weight contains 28 parts of silicon and 142 parts of chlorine, and as an atom of the latter is equal to 35·5, the molecule of silicon chloride contains $SiCl_4$. As two atoms of chlorine are equivalent to one of oxygen, the composition of silica will be SiO_2—that is, the same as stannic oxide, SnO_2, or titanic oxide, TiO_2, and the like, and also as carbonic and sulphurous anhydrides, CO_2 and SO_2. But silica bears but little physical resemblance to the latter compounds, whilst stannic and titanic oxides resemble silica both physically and chemically. They are non-volatile, crystalline insoluble, are colloids, also form feeble acids like silica, &c., and they might therefore be expected to form analogous compounds, and be isomorphous with silica, as Marignac (1859) found actually to be the case. He obtained stannofluorides, for example an easily soluble strontium salt, $SrSnF_6,2H_2O$, corresponding with the already long known silicofluorides, such as $SrSiF_6,2H_2O$. These two salts are almost identical in crystalline form (monoclinic; angle of the prism, 83° for the former and 84° for the latter; inclination of the axes, 103° 46' for the latter and 103° 30' for the former), that is, they are isomorphous. We may here add that the specific volume of silica in a solid form is 22·6, and of stannic oxide 21·5.

A similar form of silicon is obtained by fusing SiO_2 with magnesium, when an alloy of Si and Mg is also formed (Gattermann). Warren (1888) by heating magnesium in a stream of SiF_4 obtained silicon and its alloy with magnesium. Winkler (1890) found that Mg_5Si_3 and Mg_2Si are formed when

SiO_2 and Mg are heated together at lower temperatures, whilst at a high temperature Si only is formed.

[4 bis] It is very remarkable that silicon decomposes carbonic anhydride at a white heat, forming a white mass which, after being treated with potassium hydroxide and hydrofluoric acid, leaves a very stable yellow substance of the formula SiCO, which is formed according to the equation, $3Si + 2CO_2 = SiO_2 + 2SiCO$. It is also slowly formed when silicon is heated with carbonic oxide. It is not oxidised when heated in oxygen. A mixture of silicon and carbon when heated in nitrogen gives the compound Si_2C_2N, which is also very stable. On this basis Schützenberger recognises a group, C_2Si_2, as capable of combining with O_2 and N, like C.

We may add that Troost and Hautefeuille, by heating amorphous silicon in the vapour of $SiCl_4$, obtained crystalline silicon, and probably at the same time lower compounds of Si and Cl were temporarily formed. In the vapour of $TiCl_4$ under the same conditions crystalline titanium is formed (Levy, 1892).

This alloy, as Beketoff and Cherikoff showed, is easily obtained by directly heating finely divided silica (the experiment may be conducted in a test tube) with magnesium powder (Chapter XIV., Notes 17, 18). The substance formed, when thrown into a solution of hydrochloric acid, evolves spontaneously inflammable and impure silicon hydride, so that the self-inflammability of the gas is easily demonstrated by this means.

In 1850–60 Wöhler and Buff obtained an alloy of silicon and magnesium by the action of sodium on a molten mixture of magnesium chloride, sodium silicofluoride, and sodium chloride. The sodium then simultaneously reduces the silicon and magnesium.

Friedel and Ladenburg subsequently prepared silicon hydride in a pure state, and showed that it is not spontaneously inflammable in air, at the ordinary pressure, but that, like PH_3, and like the mixture prepared by the above methods, it easily takes fire in air under a lower pressure or when mixed with hydrogen. They prepared the pure compound in the following manner: Wöhler showed that when dry hydrochloric acid gas is passed through a slightly heated tube containing silicon it forms a very volatile colourless liquid, which fumes strongly in air; this is a mixture of silicon chloride, $SiCl_4$, and *silicon chloroform*, $SiHCl_3$, which corresponds with ordinary chloroform, $CHCl_3$. This mixture is easily separated by distillation, because silicon chloride boils at 57°, and silicon chloroform at 36°. The formation of the latter will be understood from the equation $Si + 3HCl = H_2 + SiHCl_3$. It is an anhydrous inflammable liquid of specific gravity 1·6. It forms a transition product between SiH_4 and $SiCl_4$, and may be obtained from silicon hydride by the action of chlorine and $SbCl_5$, and is itself also transformed into silicon chloride by the action of chlorine. Gattermann obtained $SiHCl_3$ by heating the mass obtained after the action (Note 4) of Mg upon SiO_2, in a stream of chlorine (with HCl) at about 470°. Friedel and Ladenburg, by acting on anhydrous alcohol with silicon chloroform, obtained an ethereal compound having the composition $SiH(OC_2H_5)_3$. This ether boils at 136°, and when acted on by sodium disengages silicon hydride, and is converted into ethyl orthosilicate, $Si(OC_2H_5)_4$, according to the equation $4SiH(OC_2H_5)_3 = SiH_4 + 3Si(OC_2H_5)_4$ (the sodium seems to be unchanged), which is exactly similar to the decomposition of the lower oxides of phosphorus, with the evolution of phosphuretted hydrogen. If we designate the group C_2H_5, contained in the silicon ethers by Et, the parallel is found to be exact:

$4PHO(OH)_2 = PH_3 + 3PO(OH)_3$; $4SiH(OEt)_3 = SiH_4 + 3Si(OEt)_4$.

The amorphous silica is mixed with starch, dried, and then charred by heating the mixture in a closed crucible. A very intimate mixture of silica and charcoal is thus formed. In Chapter XI., Note 13, we saw that elements like silicon disengage more heat with oxygen than with chlorine, and therefore their oxygen compounds cannot be directly decomposed by chlorine, but that this can be effected when the affinity of carbon for oxygen is utilised to aid the action. When the mass obtained by the action of Mg upon SiO_2 is heated to 300° in a current of chlorine, it easily forms $SiCl_4$ (Gattermann): besides which two other compounds, corresponding to $SiCl_4$, are formed, namely: Si_2Cl_6, which boils at 145° and solidifies at -1°, and Si_3Cl_8, which boils at about 212°. These substances, which answer to corresponding carbon compounds (C_2H_6 and C_3H_8), act upon water and form corresponding oxygen compounds; for instance, $Si_2Cl_6 + 4H_2O = (SiO_2H)_2 + 6HCl$ gives the analogue of oxalic acid $(CO_2H)_2$. This substance is insoluble in water, decomposes under the action of friction and heat with an explosion, and should be called *silico-oxalic acid*, $Si_2H_2O_4$ (*see* later, Note 11 bis).

Silicon chloride shows a similar behaviour with alcohol. This is accompanied by a very characteristic phenomenon; on pouring silicon chloride into anhydrous alcohol a momentary evolution of heat is observed, owing to a reaction of double decomposition, but this is immediately followed by a powerful cooling effect, due to the disengagement of a large amount of hydrochloric acid—that is, there is an absorption of heat from the formation of gaseous hydrochloric acid. This is a very instructive example in this respect; here two processes occurring simultaneously—one chemical and the other physical—are divided from each other by time, the latter process showing itself by a distinct fall in temperature. In the majority of cases the two processes proceed simultaneously, and we only observe the difference between the heat developed and absorbed. In acting on alcohol, silicon chloride forms ethyl orthosilicate, $SiCl_4 + 4HOC_2H_5 = 4HCl + Si(OC_2H_5)_4$. This substance boils at 160°, and has a specific gravity 0·94. Another salt, ethyl metasilicate, $SiO(OC_2H_5)_2$, is also formed by the action of silicon chloride on anhydrous alcohol; it volatilises above 300°, having a sp. gr. 1·08. It is exceedingly interesting that these two ethereal salts are both volatile, and both correspond with silica, SiO_2: the first ether corresponds to the hydrate $Si(OH)_4$, orthosilic acid, and the second to the hydrate $SiO(OH)_2$, metasilicic acid. As the nature of hydrates may be judged from the composition of salts, so also, with equal right, can ethereal salts serve the same purpose. The composition of an ethereal salt corresponds with that of an acid in which the hydrogen is replaced by a hydrocarbon radicle—for instance, by C_2H_5. And, therefore, it may be truly said that there exist at least the two silicic acids above mentioned. We shall afterwards see that there are really several such hydrates; that these ethereal salts actually correspond with hydrates of silica is clearly shown from the fact that they are decomposed by water, and that in moist air they give alcohol and the corresponding hydrate, although the hydrate which is obtained in the residue always corresponds with the second ethereal salt only—that is, it has the composition $SiO(OH)_2$; this form corresponds also to carbonic acid in its ordinary salts. This hydrate is formed as a vitreous mass when the ethyl silicates are exposed to air, owing to the action of the atmospheric moisture on them. Its specific gravity is 1·77.

Silicon bromide, $SiBr_4$, as well as silicon bromoform, $SiHBr_3$, are substances closely resembling the chlorine compounds in their reactions, and they are obtained in the same manner. Silicon iodoform, $SiHI_3$, boils at about 220°, has a specific gravity of 3·4, reacts in the same manner as silicon chloroform, and is formed, together with silicon iodide, SiI_4, by the action of a mixture of hydrogen and hydriodic acid on heated silicon. Silicon iodide is a solid at the ordinary temperature, fusing at about 120°; it may be distilled in a stream of carbonic anhydride, but easily takes fire in air, and behaves with water and other reagents just like silicon chloride. It may be obtained by the direct action of the vapour of iodine on heated silicon. Besson (1891) also obtained $SiCl_3I$ (boils at 113°), $SiCl_2I_2$ (172°), and $SiClI_3$ (220°), and the corresponding bromine compounds. All the halogen compounds of Si are capable of absorbing $6NH_3$ and more. Besides which Besson obtained $SiSCl_2$ by heating Si in the vapour of chloride of sulphur; this compound melts at 74°, boils at 185°, and gives with water the hydrate of SiO_2, HCl, and H_2S.

This property of calcium fluoride of converting silica into a gas and a vitreous fusible slag of calcium silicate is frequently taken advantage of in the laboratory and in practice in order to remove silica. The same reaction is employed for preparing silicon fluoride on a large scale in the manufacture of hydrofluosilicic acid (see sequel).

The amount of heat developed by the solution of silicic acid, $SiO_2 n H_2O$, in aqueous hydrofluoric acid, $xHF n H_2O$, increases with the magnitude of x and normally equals $x5,600$ heat units, where x varies between 1 and 8. However, when $x = 10$ the maximum amount of heat is developed (= 49,500 units), and beyond that the amount decreases (Thomsen).

In reality, however, it would seem that the reaction is still more complex, because the aqueous solution of silicon fluoride does not yield a hydrate of silica, but a fluo-hydrate (Schiff), $Si_2O_3(OH)F$, corresponding to the (pyro) hydrate $Si_2O_3(OH)_2$, equal to $SiO(OH)_2SiO_2$, so that the reaction of silicon fluoride on water is expressed by the equation: $5SiF_4 + 4H_2O = 3SiH_2F_6 + Si_2O_3(OH)F + HF$. However, Berzelius states that the hydrate, when well washed with water, contains no fluorine, which is probably due to the fact that an excess of water decomposes $Si_2O_3(OH)F$, forming hydrofluoric acid and the compound $Si_2O_3(OH)_2$. Water saturated with silicon fluoride disengages silicon fluoride and hydrofluoric acid when treated with hydrochloric acid, the gelatinous precipitate being simultaneously dissolved. It may be further remarked that hydrofluosilicic acid has been frequently regarded as $SiO_2,6HF$, because it is formed by the solution of silica in hydrofluoric acid, but only two of these six hydrogens are replaced by metals. On concentration, solutions of the acid begin to decompose when they reach a strength of $6H_2O$ per H_2SiF_6, and therefore the acid may be regarded as $Si(OH)_4,2H_2O,6HF$, but the corresponding salts contain less water, and there are even anhydrous salts, R_2SiF_6, so that the acid itself is most simply represented as H_2SiF_6.

If gaseous silicon fluoride be passed directly into water, the gas-conducting tube becomes clogged with the precipitated silicic acid. This is best prevented by immersing the end of the tube under mercury, and then pouring water over the mercury; the silicon fluoride then passes through the mercury, and only comes into contact with the water at its surface, and consequently the gas-conducting tube remains unobstructed. The silicic acid thus obtained soon settles, and a colourless solution with a pleasant but distinctly acid taste is procured.

Mackintosh, by taking 9 p.c. of hydrofluoric acid, observed that in the course of an hour its action on opal attained 77 p.c. of the possible, and did not exceed 1½ p.c. of its possible action on quartz during the same time. This shows the difference of the structure of these two modifications of silica, which will be more fully described in the sequel.

[10 bis] The sodium salt is far more soluble in water, and crystallises in the hexagonal system. The magnesium salt, $MgSiF_6$, and calcium salt are soluble in water. The salts of hydrofluosilicic acid may be obtained not only by the action of the acid on bases or by double decompositions, but also by the action of hydrofluoric acid on metallic silicates. Sulphuric acid decomposes them, with evolution of hydrofluoric acid and silicon fluoride, and the salts when heated evolve silicon fluoride, leaving a residue of metallic fluoride, R_2F_2.

See Note 4 bis. Probably Schützenberger had already obtained CSi in his researches together with other silicon compounds. An amorphous, less hard compound of the same alloy is also obtained together with the hard crystalline CSi.

[11 bis] The following consideration is very important in explaining the nature of the lower hydrates which are known for silicon. If we suppose water to be taken up from the first hydrates (just as formic acid is $CH(OH)_3$, *minus* water), we shall obtain the various lower hydrates corresponding with silicon hydride. When ignited they should, like phosphorous and hypophosphorous acids, disengage silicon hydride, and leave a residue of silica behind—*i.e.* of the oxide corresponding to the highest hydrate—just as organic hydrates (for example, formic acid with an alkali) form carbonic anhydride as the highest oxygen compound. Such imperfect hydrates of silicon, or, more correctly speaking, of silicon hydride, were first obtained by Wöhler (1863) and studied by Geuther (1865), and were named after their characteristic colours. (*See* Note 66).

Leucone is a white hydrate of the composition $SiH(OH)_3$. It is obtained by slowly passing the vapour of silicon chloroform into cold water: $SiHCl_3$ + $3H_2O$ = $SiH(OH)_3$ + $3HCl$. But this hydrate, like the corresponding hydrate of phosphorus or carbon, does not remain in this state of hydration, but loses a portion of its water. The carbon hydrate of this nature, $CH(OH)_3$, loses water and forms formic acid, $CHO(OH)$; but the silicon hydrate loses a still greater proportion of water, $2SiH(OH)_3$, parting with $3H_2O$, and consequently leaving $Si_2H_2O_3$. This substance must be an anhydride; all the hydrogen previously in the form of hydroxyl has been disengaged, two remaining hydrogens being left from SiH_4. The other similar hydrate is also white, and has the composition Si_3H_2O (nearly). It may be regarded as the above white hydrate + SiO_2. A yellow hydrate, known as *chryseone* (silicone), is obtained by the action of hydrochloric acid on an alloy of silicon and calcium; its composition is about $Si_6H_4O_3$. Most probably, however, chryseone has a more complex composition, and stands in the same relation to the hydrate $SiH_2(OH)_3$ as leucone does to the hydrate $SiH(OH)_3$, because this very simply expresses the transition of the first compound into the second with the loss of water, $SiH_2(OH)_3 - H_2 +$

$H_2O = SiH(OH)_3$. When these lower hydrates are ignited without access of air, they are decomposed into hydrogen, silicon, and silica—that is, it may be supposed that they form silicon hydride (which decomposes into silicon and hydrogen) and silica (just as phosphorous and hypophosphorous acids give phosphoric acid and phosphuretted hydrogen). When ignited in air, they burn, forming silica. They are none of them acted on by acids, but when treated with alkalis they evolve hydrogen and give silicates; for example, leucone: $SiH_2O_3 + 4KHO = 2SiK_2O_3 + H_2O + 2H_2$. They have no acid properties.

Two modifications of rock crystal are known. They are very easily distinguished from each other by their relation to polarised light; one rotates the plane of polarisation to the right and the other to the left—in the one the hemihedral faces are right and in the other they are left; this opposite rotatory power is taken advantage of in the construction of polarisers. But, with this physical difference—which is naturally dependent on a certain difference in the distribution of the molecules—there is not only no observable difference in the chemical properties, but not even in the density of the mass. Perfectly pure rock crystal is a substance which is most invariable with respect to its specific gravity. The numerous and accurate determinations made by Steinheil on the specific gravity of rock crystal show that (if the crystal be free from flaws) it is very constant and is equal to 2·66.

[12 bis] Several other modifications are known as minute crystals. For example, there is a particular mineral first found in Styria and known as *tridymite*. Its specific gravity 2·3 and form of crystals clearly distinguish it from rock crystal; its hardness is the same as that of quartz—that is, slightly below that of the ruby and diamond.

There is a distinct rise of temperature (about 4°) when amorphous silica is moistened with water. Benzene and amyl alcohol also give an observable rise of temperature. Charcoal and sand give the same result, although to a less extent.

[13 bis] Silica also occurs in nature in two modifications. The opal and tripoli (infusorial earth) have a specific gravity of about 2·2, and are comparatively easily soluble in alkalis and hydrofluoric acid. Chalcedony and flint (tinted quartzose concretions of aqueous origin), agate and similar forms of silica of undoubted aqueous origin, although still containing a certain amount of water, have a specific gravity of 2·6, and correspond with quartz in the difficulty with which they dissolve. This form of silica sometimes permeates the cellulose of wood, forming one of the ordinary kinds of petrified wood. The silica may be extracted from it by the action of hydrofluoric acid, and the cellulose remains behind, which clearly shows

that silica in a soluble form (see sequel) has permeated into the cells, where it has deposited the hydrate, which has lost water, and given a silica of sp. gr. 2·6. The quartzose stalactites found in certain caves are also evidently of a similar aqueous origin; their sp. gr. is also 2·6. As crystals of amethyst are frequently found among chalcedonies, and as Friedau and Sarrau (1879) obtained crystals of rock crystal by heating soluble glass with an excess of hydrate of silica in a closed vessel, there is no doubt but that rock crystal itself is formed in the wet way from the gelatinous hydrate. Chroustchoff obtained it directly from soluble silica. Thus this hydrate is able to form not only the variety having the specific gravity 2·2 but also the more stable variety of sp. gr. 2·6; and both exist with a small proportion of water and in a perfectly anhydrous state in an amorphous and crystalline form. All these facts are expressed by recognising silica as dimorphous, and their cause must be looked for in a difference in the degree of polymerisation.

Deposits of perfectly white tripoli have been discovered near Batoum, and might prove of some commercial importance.

[14 bis] Alkaline solutions, saturated with silica and known as *soluble glass*, are prepared on a large scale for technical purposes by the action of potassium (or sodium) hydroxide in a steam boiler on tripoli or infusorial earth, which contains a large proportion of amorphous silica. All solutions of the alkaline silicates have an alkaline reaction, and are even decomposed by carbonic acid. They are chiefly used by the dyer, for the same purposes as sodium aluminate, and also for giving a hardness and polish to stucco and other cements, and in general to substances which contain lime. A lump of chalk when immersed in soluble glass, or better still when moistened with a solution and afterwards washed in water (or better in hydrofluosilicic acid, in order to bind together the free alkali and make it insoluble), becomes exceedingly hard, loses its friability, is rendered cohesive, and cannot be levigated in water. This transformation is due to the fact that the hydrate of silica present in the solution acts upon the lime, forming a stony mass of calcium silicate, whilst the carbonic acid previously in combination with the lime enters into combination with the alkali and is washed away by the water.

The equation given above does not express the actual reaction, for in the first place silica has the faculty of forming compounds with bases, and therefore the formula $SiNa_4O_4$ is not rightly deduced, if one may so express oneself. And, in the second place, silica gives several hydrates. In consequence of this, the hydrate precipitated does not actually contain so high a proportion of water as $Si(OH)_4$, but always less. The insoluble gelatinous hydrate which separates out is able (before, but not after, having been dried) to dissolve in a solution of sodium carbonate. When dried in air its composition corresponds with the ordinary salts of carbonic acid—that is, SiH_2O_3, or $SiO(OH)_2$. If gradually heated it loses water by degrees, and, in so doing, gives various degrees of combination with it. The existence of these degrees of hydration, having the composition $SiH_2O_3 nSiO_2$, or, in general, $nSiO_2 mH_2O$, where $m < n$, must be recognised, because most varied degrees of combination of silica with bases are known. The hydrate of silica, when not dried above 30°, has a composition of nearly $H_4Si_3O_8 =$ $(H_2SiO_3)2SiO_2$, but at 60° contains a greater proportion of silica—that is, it loses still more water; and at 100° a hydrate of the composition $SiH_2O_3 2SiO_2$, and at 250° a hydrate having approximately a composition $SiH_2O_3 7SiO_2$ is obtained.

These data show the complexity of the molecules of anhydrous silica. The hydrates of silica easily lose water and give the hydrates $(SiO_2)_n(H_2O)_m$, where m becomes smaller and smaller than n. In the natural hydrates, this decrement of water proceeds quite consecutively, and, so to say, imperceptibly, until n becomes incomparably greater than m, and when the ratio becomes very large, anhydrous silica of the two modifications 2·6 and 2·2 is obtained. The composition $(SiO_2)_{10}, H_2O$ still corresponds with 2·9 p.c. of water, and natural hydrates often contain still less water than this. Thus some opals are known which contain only 1 p.c. of water, whilst others contain 7 and even 10 p.c. As the artificially prepared gelatinous hydrate of silica when dried has many of the properties of native opals, and as this hydrate always loses water easily and continually, there can be no doubt that the transition of $(SiO_2)_n(H_2O)_m$ into anhydrous silica, both amorphous and crystalline (in nature, chalcedony), is accomplished gradually. This can only be the case if the magnitude of n be considerable, and therefore the molecule of silica in the hydrate is undoubtedly complex, and hence the anhydrous silica of sp. gr. 2·2 and 2·6 does not contain SiO_2, but a complex molecule, Si_nO_{2n}—that is, the structure of silica is polymeric and complex, and not simple as represented above by the formula SiO_2.

The presence of an excess of acid aids the retention of the silica in the solution, because the gelatinous silica obtained in the above manner, but not heated to 60°—that is, containing more water than the hydrate H_2SiO_3—is more soluble in water containing acid than in pure water. This would seem to indicate a feeble tendency of silica to combine with acids, and it might even have been imagined that in such a solution the hydrate of silica is held in combination by an excess of acid, had Graham not obtained soluble silica perfectly free from acid, and if there were not solutions of silica free from any acid in nature. At all events a tolerably strong solution of free silica or silicic acid may be obtained from soluble glass diluted with water. The solution, besides silica, will contain sodium chloride and an excess of the acid taken. If this solution remains for some time exposed to the air, or in a closed vessel, and under various other conditions, it is found that, after a time, insoluble gelatinous silica separates out—that is, the soluble form of silica is unstable, like the soluble form of alumina. The analogous forms of molybdic or tungstic acids may be heated, evaporated, and kept for a long period of time without the soluble form being converted into the insoluble.

See Chapter I., Note 18. A solution of water-glass mixed with an excess of hydrochloric acid is poured into the dialyser, and the outer vessel is filled with water, which is continually renewed. The water carries off the sodium chloride and hydrochloric acid, and the hydrosol remains in the dialyser.

A similar process occurs in plants—for example, when they secrete a store of material for the following year in their bulbs, roots, &c. (for instance, the potato in its tubers), the solutions from the leaves and stems penetrate into the roots and other parts in the form of hydrosols, where they are converted into hydrogels—that is, into an insoluble form, which is acted on with difficulty and is easily kept unaltered until the period of growth—for example, until the following spring—when they are reconverted into hydrosols, and the insoluble substance re-enters into the sap, and serves as a source of the hydrogels in the leaves and other portions of plants.

As regards their chemical composition the colloids are very complex—that is, they have a high molecular weight and a large molecular volume—in consequence of which they do not penetrate through membranes, and are easily subject to variation in their physical and chemical properties (owing to their complex structure and polymerism?) They have but little chemical energy, and are generally feeble acids, if belonging to the order of oxides or hydrates, such as the hydrates of molybdic and tungstic acids (Chapter XXI.). But now the number of substances capable, like colloids, of passing into aqueous solutions and of easily separating out from them, as well as of appearing in an insoluble form, must be supplemented by various other substances, among which soluble gold and silver (Chapter XXIV.) are of particular interest. So that now it may be said that the capacity of forming colloid solutions is not limited to a definite class of compounds, but is, if not a general, at all events, an exceedingly widely distributed phenomenon.

This is in accordance with the generally-accepted representation of the relations between salts and the hydrates of acids, but it is of little help in the study of siliceous compounds. Generally speaking, it becomes necessary to explain the property of $(SiO_2)_n$ to combine with $(RO)_m$, where n may be greater than m, and where R may be H_2, Ca, &c. Here we are aided by those facts which have been attained by the investigation of carbon compounds, especially with respect to glycol. Glycol is a compound having the composition $C_2H_6O_2$, only differing from alcohol, C_2H_6O, by an extra atom of oxygen. This hydrate contains two hydroxyl groups, which may be successively replaced by chlorine, &c. Hence the composition of glycol should be represented as $C_2H_4(OH)_2$. It has been found that glycol forms so-called polyglycols. Their origin will be understood from the fact that glycol as a hydrate has a corresponding anhydride of the composition C_2H_4O, known as ethylene oxide. This substance is ethane, C_2H_6, in which two hydrogens are replaced by one atom of oxygen. Ethylene oxide is not the only anhydride of glycol, although it is the simplest one, because $C_2H_4O = C_2H_4(OH)_2 - H_2O$. Various other anhydrides of glycol are possible, and have actually been obtained, of the composition $nC_2H_4(OH)_2 - (n - 1)H_2O = (C_2H_4)_nO_{n-1}(OH)_2$. These imperfect anhydrides of glycol, or *polyglycols*, still contain hydroxyls like glycol itself, and therefore are of an alcoholic character in the same sense as glycol itself. They are obtained by various methods, and, amongst others, by the direct combination of ethylene oxide with glycol, because $C_2H_4(OH)_2 + (n - 1)C_2H_4O = (C_2H_4)_nO_{n-1}(OH)_2$. The most important circumstance, from a theoretical point of view, is that these polyglycols may be distilled without undergoing decomposition, and that the general formula given above expresses their actual molecular composition. Hence we have here a direct combination of the anhydride with the hydrate, and, moreover, a repeated one. The formula A_nH_2O may be used to express the composition of glycol and polyglycols with respect to ethylene oxide in the most simple manner, if A stand for ethylene oxide. When $n = 1$ we have glycol, when n is greater than 1 a polyglycol. Such also is the relationship of the salts of hydrate of silica, if A stand for silica, and if we imagine that H_2O may also be taken m times. Such a representation of the *polysilicic acids* corresponds with the representation of the polymerism of silica. Laurent supposed the existence of several polymeric forms, Si_2O_4, Si_3O_6, &c., besides silica, SiO_2.

For us the latter have not a saline character, only because they are not regarded from this point of view, but an alloy of sodium and zinc is, in a broad sense, a salt in many of its reactions, for it is subject to the same double decompositions as sodium phosphide or sulphide, which clearly have saline properties. The latter (sodium phosphide), when heated with ethyl iodide, forms ethyl phosphide, and the former—*i.e.* the alloy of zinc and sodium—gives zinc ethyl; that is, the element (P, S, Zn) which was united with the sodium passes into combination with the ethyl: $RNa + EtI = REt + NaI$. By combining sodium successively with chlorine, sulphur, phosphorus, arsenic, antimony, tin, and zinc, we obtain substances having less and less the ordinary appearance of salts, but if the alloy of sodium and zinc cannot be termed a salt, then perhaps this name cannot be given to sodium sulphide, and the compounds of sodium with phosphorus. The following circumstance may also be observed: with chlorine, sodium gives only one compound (with oxygen, at the most three), with sulphur five, with phosphorus probably still more, with antimony naturally still more, and the more analogous an element is to sodium, the more varied are the proportions in which it is able to combine with it, the less are the alterations in the properties which take place by this combination, and the nearer does the compound formed approach to the class of compounds known as indefinite chemical compounds. In this sense a siliceous alloy, containing silica and other acids, is a salt. The oxide to a certain extent plays the same part as the sodium, whilst the silica plays the part of the acid element which was taken up successively by zinc, phosphorus, sulphur, &c., in the above examples. Such a comparison of the silica compounds with alloys presents the great advantage of including under one category the definite and indefinite silica compounds which are so analogous in composition—that is, brings under one head such crystalline substances as certain minerals, and such amorphous substances as are frequently met with in nature, and are artificially prepared, as glass, slags, enamels, &c.

If the compounds of silica are substances like the metallic alloys, then (1) the chemical union between the oxides of which they are composed must be a feeble one, as it is in all compounds formed between analogous substances. In reality such feeble agencies as water and carbonic acid are able, although slowly, to act on and destroy the majority of the complex silica compounds in rocks, as we saw in the preceding chapter; (2) their formation, like that of alloys, should not be accompanied by a considerable alteration of volume; and this is actually the case. For example, felspar has a specific gravity of about 2·6, and therefore, taking its composition to be $K_2O,Al_2O_3,6SiO_2$, we find its volume, corresponding with this formula, to be 556·8. 2·6 = 214, the volume of K_2O = 35, of Al_2O_3 = 26, and of SiO_2 = 22·6. Hence the sum of the volumes of the component oxides, 35 + 26 + 6 × 22·6 = 196, which is very nearly equal to that of the felspar; that is, its formation is attended by a slight expansion, and not by contraction, as is the case in the majority of other cases when combinations determined by strong affinities are accomplished. In the case in question the same phenomenon is observed as in solutions and alloys—that is, as in cases of feeble affinities. So also the specific gravity of glass is directly dependent on the amount of those oxides which enter into its composition. If in the preceding example we take the sp. gr. of silica to be, not 2·65, but 2·2, its volume = 27·3, and the sum of the volumes will be = 224—that is, greater than that of orthoclase.

It is, however, easy to imagine, and experience confirms the supposition, that in a complex siliceous compound containing for instance sodium and calcium, the whole of the sodium may be replaced by potassium, and *at the same time* the whole of the calcium by magnesium, because then the substitution of potassium for the sodium will produce a change in the nature of the substance contrary to that which will occur from the calcium being replaced by magnesium. That increase in weight, decrease in density, increase of chemical energy, which accompanies the exchange of sodium for potassium will, so to speak, be compensated by the exchange of calcium for magnesium, because both in weight and in properties the sum of Na + Ca is very near to the sum of K + Mg. *Pyroxene* or *augite* can be taken as an example; its composition may be expressed by the formula $CaMgSi_2O_6$; that is, it corresponds with the acid H_2SiO_3; it is a bisilicate. In many respects it closely resembles another mineral called '*spodumene*' (they are both monoclinic). This latter has the composition $Li_6Al_8Si_{15}O_{45}$. On reducing both formulæ to an equal contents of silica the following distinction will be observed between them: spodumene $(Li_2O)_6(Al_2O_3)_8 30SiO_2$; augite $(CaO)_{15}(MgO)_{15} 30SiO_2$. That is, the difference between them consists in the sum of the magnesia and lime $(MgO)_{15} + (CaO)_{15}$ replacing the sum of the lithium oxide and alumina $(Li_2O)_6 + (Al_2O_3)_8$; and in the chemical relation these sums are near to one another, because magnesium and calcium, both in forms of oxidation and in energy (as bases), in all respects occupy a position intermediate between lithium and aluminium, and therefore the sum of the first may be replaced by the sum of the second.

If we take the composition of spodumene, as it is often represented to be, $Li_2O,Al_2O_3,4SiO_2$, the corresponding formula of augite will be $(CaO)_2,(MgO)_2,4SiO_2$, and also the amount of oxygen in the sum of $Li_2OAl_2O_3$ will be the same as in $(CaO)_2(MgO)_2$. I may remark, for the sake of clearness, that lithium belongs to the first, aluminium to the third group, and calcium and magnesium to the intermediate second group; lithium, like calcium, belongs to the even series, and magnesium and aluminium to the uneven.

The representation of the substitutions of analogous compounds here introduced was first deduced by me in 1856. It finds much confirmation in facts which have been subsequently discovered—for example, with respect to tourmalin. Wülfing (1888), on the basis of a number of analyses (especially of those by Röggs), states that all varieties contain an isomorphous mixture of alkali and magnesia tourmalin; into the composition of the former there enters $12SiO_2,3B_2O_3,8Al_2O_3,2Na_2O,4H_2O$, and of the latter $12SiO_2,3B_2O_3,5Al_2O_3,12MgO,3H_2O$. Hence it is seen that the former contains in addition the sum of $3Al_2O_3,2Na_2O,H_2O$, whilst in the latter this sum of oxides is replaced by $12MgO$, in which there is as much oxygen as in the sum of the more clearly-defined base $2Na_2O$ and less basic

$3Al_2O_3H_2O$—that is, the relation is just the same here as between augite and spodumene.

With respect to the silica compounds of the various oxides, it must be observed that only the *alkali salts* are known in a soluble form; all the others only exist in an insoluble form, so that a solution of the alkali compounds of silica, or soluble glass, gives a precipitate with a solution of the salts of the majority of other metals, and this precipitate will contain the silica compounds of the other bases. The maximum amount of the gelatinous hydrate of silica, which dissolves in caustic potash, corresponds with the formation of a compound, $2K_2O,9SiO_2$. But this compound is partially decomposed, with the precipitation of hydrate of silica, on cooling the solution. Solutions containing a smaller amount of silica may be kept for an indefinite time without decomposing, and silica does not separate out from the solution; but such compounds crystallise from the solutions with difficulty. However, a crystalline bisilicate (with water) has been obtained for sodium having the composition Na_2O,SiO_2—*i.e.* corresponding to sodium carbonate. The whole of the carbonic acid is evolved, and a similar soluble sodium metasilicate is obtained on fusing 3·5 parts of sodium carbonate with 2 parts of silica. If less silica be taken a portion of the sodium carbonate remains undecomposed; however, a substance may then be obtained of the composition $Si(ONa)_4$, corresponding with orthosilicic acid. It contains the maximum amount of sodium oxide capable of combining with silica under fusion. It is a sodium orthosilicate, $(Na_2O)_2,SiO_2$.

Calcium carbonate, and the carbonates of the alkaline earths in general, also evolve all their carbonic acid when heated with silica, and in some instances even form somewhat fusible compounds. Lime forms a fusible slag of *calcium silicate*, of the composition CaO,SiO_2 and $2CaO,3SiO_2$. With a larger proportion of silica the slags are infusible in a furnace. The magnesium *slags* are less fusible than those with lime, and are often formed in smelting metals. Many compounds of the metals of the alkaline earths with silica are also met with in nature. For instance, among the magnesium compounds there is *olivine*, $(MgO)_2,SiO_2$, sp. gr. 3·4, which occurs in meteorites, and sometimes forms a precious stone (peridote), and occurs in slags and basalts. It is decomposed by acids, is infusible before the blow-pipe, and crystallises in the rhombic system. *Serpentine* has the composition $3MgO,2SiO_2,2H_2O$; it sometimes forms whole mountains, and is distinguished for its great cohesiveness, and is therefore used in the arts. It is generally tinted green; its specific gravity is 2·5; it is exceedingly infusible, even before the blowpipe. It is acted on by acids. Among the magnesium compounds of silica, *talc* is very widely used. It is frequently met with in rocks which are widely distributed in nature, and sometimes in compact masses; it can be used for writing like a slate pencil or chalk, and being greasy to the touch, is also known as *steatite*. It crystallises in the rhombic system, and resembles mica in many respects; like it, it is divisible into laminæ, greasy to the touch, and having a sp. gr. 2·7. These laminæ are very soft, lustrous, and transparent, and are infusible and insoluble in acids. The composition of talc approaches nearly to $6MgO,5SiO_2,2H_2O$.

Among the crystalline silicates the following minerals are known:— *Wollastonite* (tabular-spar), crystallises in the monoclinic system; sp. gr. 2·8; it is semi-transparent, difficultly fusible, decomposed by acids, and has the composition of a metasilicate, $CaOSiO_2$. But isomorphous mixtures of calcium and magnesium silicates occur with particular frequency in nature. The *augites* (sp. gr. 3·3), diallages, hyperstenes, hornblendes (sp. gr. 3·1), amphiboles, common asbestos, and many similar minerals, sometimes forming the essential parts of entire rock formations, contain various relative proportions of the bisilicates of calcium and magnesium partially mixed with other metallic silicates, and generally anhydrous, or only containing a small amount of water. In the pyroxenes, as a rule, lime predominates, and in the amphiboles (also of the monoclinic system) magnesia predominates. Details upon this subject must be looked for in works upon mineralogy.

The majority of the siliceous minerals have now been obtained artificially under various conditions. Thus N. N. Sokoloff showed that slags very frequently contain peridote. Hautefeuille, Chroustchoff, Friedel, and Sarasin obtained felspar identical in all respects with the natural minerals. The details of the methods here employed must be looked for in special works on mineralogy; but, as an example, we will describe the method of the preparation of felspar employed by Friedel and Sarasin (1881). From the fact that felspar gives up potassium silicate to water even at the ordinary temperature (Debray's experiments), they concluded that the felspar in granites had an aqueous origin (and this may be supposed to be the case from geological data); then, in the first place, its formation could not be accomplished unless in the presence of an excess of a solution of potassium silicate. In order to render this argument clear I may mention, as an example, that carnallite is decomposed by water into easily soluble magnesium chloride and potassium chloride, and therefore if it is of aqueous origin it could not be formed otherwise than from a solution containing an excess of magnesium chloride, and, in the second place, from a strongly-heated solution; again, felspar itself and its fellow-components in granites are anhydrous. On these facts were based experiments of heating hydrates of silica with alumina and a solution of potassium silicate in a closed vessel. The mixture was placed in a sealed platinum tube, which was enclosed in a steel tube and heated to dull redness. When the mixture contained an excess of silica the residue contained many crystals of rock crystal and tridymite, together with a powder of felspar, which formed the main product of the reaction when the proportion of hydrate of silica was decreased, and a mixture of a solution of potassium silicate with alumina precipitated together with the silica by mixing soluble glass with aluminium chloride was employed. The composition, properties, and forms of the resultant felspar proved it to be identical with that found in nature. The experiments approach very nearly to the natural conditions, all the more as felspar and quartz are obtained together in one mixture, as they so often occur in nature.

The application of *cements* is based on this principle; they are those sorts of 'hydraulic' lime which generally form a stony mass, which hardens even under water, when mixed with sand and water.

The hydraulic properties of cements are due to their containing calcareous and silico-aluminous compounds which are able to combine with water and form hydrates, which are then unacted on by water. This is best proved, in the first place, by the fact that certain slags containing lime and silica, and obtained by fusion (for example, in blast-furnaces), solidify like cements when finely ground and mixed with water; and, in the second place, by the method now employed for the manufacture of artificial cements (formerly only peculiar and comparatively rare natural products were used). For this purpose a mixture of lime and clay is taken, containing about 25 p.c. of the latter; this mixture is then heated, not to fusion, but until both the carbonic anhydride and water contained in the clay are expelled. This mass when finely ground forms Portland cement, which hardens under water. The process of hardening is based on the formation of chemical compounds between the lime, silica, alumina, and water. These substances are also found combined together in various natural minerals—for example, in the zeolites, as we saw above. In all cases cement which has set contains a considerable amount of water, and its hardening is naturally due to hydration—that is, to the formation of compounds with water. Well-prepared and very finely-ground cement hardens comparatively quickly (in several days, especially after being rammed down), with 3 parts (and even more) of coarse sand and with water, into a stony mass which is as hard and durable as many stones, and more so than bricks and limestone. Hence not only all maritime constructions (docks, ports, bridges, &c.), but also ordinary buildings, are made of Portland cement, and are distinguished for their great durability. A combination of ironwork (ties, girders) and cement is particularly suitable for the construction of aqueducts, arches, reservoirs, &c. Arches and walls made of such cements may be much less thick than those built up of ordinary stone. Hence the production and use of cement rapidly increases from year to year. The origin of accurate data respecting cements is chiefly due to Vicat. In Russia Professor Schuliachenko has greatly aided the extension of accurate data concerning Portland cement. Many works for the manufacture of cement have already been established in various parts of Russia, and this industry promises a great future in the arts of construction.

Glass presents a similar complex composition, like that of many minerals. The ordinary sorts of white glass contain about 75 p.c. of silica, 13 p.c. of sodium oxide, and 12 p.c of lime; but the inferior sorts of glass sometimes contain up to 10 p.c. of alumina. The mixtures which are used for the manufacture of glass are also most varied. For example, about 300 parts of pure sand, about 100 parts of sodium carbonate, and 50 of limestone are taken, and sometimes double the proportion of the latter. Ordinary *soda-glass* contains sodium oxide, lime, and silica as the chief component parts. It is generally prepared from sodium sulphate mixed with charcoal, silica, and lime (Chapter [XII.](#)), in which case the following reaction takes place at a high temperature: $Na_2SO_4 + C + SiO_2 = Na_2SiO_3 + SO_2 + CO$. Sometimes potassium carbonate is taken for the preparation of the better qualities of glass. In this case a glass, *potash-glass*, is obtained containing potassium oxide instead of sodium oxide. The best-known of these glasses is the so-called Bohemian glass or crystal, which is prepared by the fusion of 50 parts of potassium carbonate, 15 parts of lime, and 100 parts of quartz. The preceding kinds of glass contain lime, whilst crystal glass contains lead oxide instead. Flint glass—that is, the lead glass used for optical instruments—is prepared in this manner, naturally from the purest possible materials. *Crystal-glass*—*i.e.* glass containing lead oxide—is softer than ordinary glass, more fusible and has a higher index of refraction. However, although the materials for the preparation of glass be most carefully sorted, a certain amount of iron oxides falls into the glass and renders it greenish. This coloration may be destroyed by adding a number of substances to the vitreous mass, which are able to convert the ferrous oxide into ferric oxide; for example, manganese peroxide (because the peroxide is deoxidised to manganous oxide, which only gives a pale violet tint to the glass) and arsenious anhydride, which is deoxidised to arsenic, and this is volatilised. The manufacture of glass is carried on in furnaces giving a very high temperature (often in regenerative furnaces, Chapter [IX.](#)). Large clay crucibles are placed in these furnaces, and the mixture destined for the preparation of the glass, having been first roasted, is charged into the crucibles. The temperature of the furnace is then gradually raised. The process takes place in three separate stages. At first the mass intermixes and begins to react; then it fuses, evolves carbonic acid gas, and forms a molten mass; and, lastly, at the highest temperature, it becomes homogeneous and quite liquid, which is necessary for the ultimate elimination of the carbonic anhydride and solid impurities, which latter collect at the bottom of the crucible. The temperature is then somewhat lowered, and the glass is taken out on tubes and blown into objects of various shapes. In the manufacture of window-glass it is blown into large cylinders, which are then cut at the ends and across, and afterwards bent back in a furnace into the ordinary sheets. After being worked up, all glass objects have to be subjected to a slow cooling (*annealing*) in special furnaces, otherwise they are very brittle, as is seen in the so-called 'Rupert's drops,' formed by dropping molten glass into water; although these drops preserve their form, they are so brittle that they break up into a fine powder if a small piece be knocked off them. Glass objects have frequently to be polished and chased. In the manufacture of mirrors and many massive objects the glass is cast and then ground and polished. Coloured glasses are either made by directly introducing into the glass itself various oxides, which give their characteristic tints, or else a thin layer of a coloured glass is laid on the surface of ordinary glass. Green glasses are formed by the oxides of chromium and copper, blue by cobalt oxide, violet by manganese oxide, and red glass by cuprous oxide and by the so-called purple of Cassius—*i.e.* a compound of gold and tin—which will be described later. A yellow coloration is obtained by means of the oxides of iron, silver, or antimony, and also by means of carbon, especially for the brown tints for certain kinds of bottle-glass.

From what has been said about glass it will be understood that it is impossible to give a definite formula for it, because it is a non-crystalline or amorphous alloy of silicates; but such an alloy can only be formed within certain limits in the proportions between the component oxides. With a large proportion of silica the glass very easily becomes clouded when heated; with a considerable proportion of alkalis it is easily acted on by moisture, and becomes cloudy in time on exposure to the air; with a large proportion of lime it becomes infusible and opaque, owing to the formation of crystalline compounds in it; in a word, a certain proportion is practically attained among the component oxides in order that the glass formed may have suitable properties. Nevertheless, it may be well to remark that the composition of common glass approaches to the formula $Na_2O, CaO, 4SiO_2$.

The coefficient of cubical expansion of glass is nearly equal to that of platinum and iron, being approximately $0 \cdot 000027$. The specific heat of glass is nearly $0 \cdot 18$, and the specific gravity of common soda glass is nearly $2 \cdot 5$, of Bohemian glass $2 \cdot 4$, and of bottle glass $2 \cdot 7$. Flint glass is much heavier than common glass, because it contains the heavier oxide of lead, its specific gravity being $2 \cdot 9$ to $3 \cdot 2$.

It must be recollected that although acids seem to act only feebly on the majority of silicates, nevertheless a finely-levigated powder of siliceous compounds is acted on by strong acids, especially with the aid of heat, the basic oxides being taken up and gelatinous silica left behind. In this respect sulphuric acid heated to 200° with finely-divided siliceous compounds in a closed tube acts very energetically.

Such elements as silicon, tin, and lead were only brought together under one common group by means of the periodic law, although the quadrivalency of tin and lead was known much earlier. Generally silicon was placed among the non-metals, and tin and lead among the metals.

At first (February 1886) the want of material to work on, the absence of a spectrum in the Bunsen's flame, and the solubility of many of the compounds of germanium, presented difficulties in the researches of Professor Winkler, who, on analysing argyrodite by the usual method, obtained a constant loss of 7 p.c., and was thus led to search for a new element. The presence of arsenic and antimony in the accompanying minerals also impeded the separation of the new metal. After fusion with sulphur and sodium carbonate, argyrodite gives a solution of a sulphide which is precipitated by an *excess* of hydrochloric acid; germanium sulphide is soluble in ammonia and then precipitated by hydrochloric acid, as a *white* precipitate, which is dissolved (or decomposed) by water. After being oxidised by nitric acid, dried and ignited germanium sulphide leaves the oxide GeO_2, which is reduced to the metal when ignited in a stream of hydrogen.

G. Kobb determined the spectrum of germanium, when the metal was taken as one of the electrodes of a powerful Ruhmkorff's coil. The wave-lengths of the most distinct lines are 602, 583, 518, 513, 481, 474, millionths of a millimetre.

If germanium or germanium sulphide be heated in a stream of hydrochloric acid, it forms a volatile liquid, boiling at 72°, which Winkler regarded as germanium chloride, $GeCl_2$, or germanium chloroform, $GeHCl_3$. It is decomposed by water, forming a white substance, which may perhaps be the hydrate of germanious oxide, GeO, and acts as a powerfully reducing agent in a hydrochloric acid solution.

Under certain circumstances germanium gives a blue coloration like that of ultramarine, as Winkler showed, which might have been expected from the analogy of germanium with silicon.

Winkler expressed this in the following words (*Jour. f. pract. Chemie*, 1886, 34, 182–183): '… es kann keinem Zweifel mehr unterliegen, dass das neue Element nichts Anderes, als das vor fünfzehn Jahren von *Mendeléeff* prognosticirte *Ekasilicium* ist.'

'Denn einen schlagenderen Beweis für die Richtigkeit der Lehre von der Periodicität der Elemente, als den, welchen die Verkörperung des bisher hypothetischen "Ekasilicium" in sich schliesst, kann es kaum geben, und er bildet in Wahrheit mehr, als die blosse Bestätigung einer kühn aufgestellten Theorie, er bedeutet eine eminente Erweiterung des chemischen Gesichtfeldes, einen mächtigen Schritt in's Reich der Erkenntniss.'

[33 bis] Emilianoff (1890) states that in the cold of the Russian winter 30 out of 200 tin moulds for candles were spoilt through becoming quite brittle.

The tin deposited by an electric current from a neutral solution of $SnCl_2$ easily oxidises and becomes coated with SnO (Vignon, 1889).

[34 bis] If after this the coating of tin be rapidly cooled—for instance, by dashing water over it—it crystallises into diverse star-shaped figures, which become visible when the sheets are first immersed in dilute aqua regia and then in a solution of caustic soda.

The coating of iron by tin, guards it against the direct access of air, but it only preserves the iron from oxidation so long as it forms a perfectly continuous coating. If the iron is left bare in certain places, it will be powerfully oxidised at these spots, because the tin is electro-negative with respect to the iron, and thus the oxidation is confined entirely to the iron in the presence of tin. Hence a coating of tin over iron objects only partially preserves them from rusting. In this respect a coating of zinc is more effectual. However, a dense and invariable alloy is formed over the surface of contact of the iron and tin, which binds the coating of tin to the remaining mass of the iron. Tin may be fused with cast iron, and gives a greyish-white alloy, which is very easily cast, and is used for casting many objects for which iron by itself would be unsuitable owing to its ready oxidisability and porosity. The coating of copper objects by tin is generally done to preserve the copper from the action of acid liquids, which would attack the copper in the presence of air and convert it into soluble salts. Tin is not acted on in this manner, and therefore copper vessels for the preparation of food should be tinned.

The ancient Chinese alloys, containing about 20 p.c. of tin (specific gravity of alloys about 8·9), which have been rapidly cooled, are distinguished for their resonance and elasticity. These alloys were formerly manufactured in large quantities in China for the musical instruments known as *tom-toms*. Owing to their hardness, alloys of this nature are also employed for casting guns, bearings, &c., and an alloy containing about 11 p.c. of tin (corresponding with the ratio $Cu_{15}Sn$) is known as gun-metal. The addition of a small quantity of phosphorus, up to 2 p.c., renders bronze still harder and more elastic, and the alloy so formed is now used under the name of phosphor-bronze.

The alloy $SnCu_3$ is brittle, of a bluish colour, and has nothing in common with either copper or tin in its appearance or properties. It remains perfectly homogeneous on cooling, and acquires a crystalline structure (Riche). All these signs clearly indicate that the alloy $SnCu_3$ is a product of chemical combination, which is also seen to be the case from its density, 8·91. Had there been no contraction, the density of the alloy would be 8·21. It is the heaviest of all the alloys of tin and copper, because the density of tin is 7·29 and of copper 8·8. The alloy $SnCu_4$, specific gravity 8·77, has similar properties. All the alloys except $SnCu_3$ and $SnCu_4$ split up on cooling; a portion richer in copper solidifies first (this phenomenon is termed the *liquation* of an alloy), but the above two alloys do not split up on cooling. In these and many similar facts we can clearly distinguish a *chemical union between the metals* forming an alloy. The alloys of tin and copper were known in very remote ages, before iron was used. The alloys of zinc and tin are less used, but alloys composed of zinc, tin, and copper frequently replace the more costly bronze. Concerning the alloys of lead *see* Note 46.

An excellent proof of the fact that alloys and solutions are subject to law is given, amongst others, by the application of Raoult's method (Chapter I., Note 49) to solutions of different metals in tin. Thus Heycock and Neville (1889) showed that the temperature of solidification of molten tin (226°·4) is lowered by the presence of a small quantity of other metals in proportion to the concentration of the solution. The following were the reductions of the temperature of solidification of tin obtained by dissolving in it atomic proportions of different metals (for example, 65 parts of zinc in 11,800 parts of tin); Zn 2°·53, Cu 2°·47, Ag 2°·67, Cd 2°·16, Pb 2°·22, Hg 2°·3, Sb 2° [rise], Al 1°·34. As Raoult's method (Chapter VII.) enables the molecular weight to be determined, the almost perfect identity of the resultant figures (except for aluminium) shows that the molecules of copper, silver, lead, and antimony contain *one atom in the molecule*, like zinc, mercury, and cadmium. They obtained the same result (1890) for Mg, Na, Ni, Au, Pd, Bi and In. It should here be mentioned that Ramsay (1889) for the same purpose (the determination of the molecular weight of metals on the basis of their mutual solution) took advantage of the variation of the vapour tension of mercury (*see* Vol. I., p. 134), containing various metals in solution, and he also found that the above-mentioned metals contain but one atom in the molecule.

[36 bis] The action of a mixture of hydrochloric acid and tin forms an excellent means of reducing, wherein both the hydrogen liberated by the mixture (at the moment of separation) and the stannous chloride act as powerful reducing and deoxidising agents. Thus, for instance, by this

mixture nitro-compounds are transformed into amido-compounds—that is, the elements of the group NO_2 are reduced to NH_2.

Many volatile compounds of tin are known, whose molecular weights can therefore be established from their vapour densities. Among these may be mentioned stannic chloride, $SnCl_4$, and stannic ethide, $Sn(C_2H_5)_4$ (the latter boils at about 150°). But V. Meyer found the vapour density of stannous chloride, $SnCl_2$, to be variable between its boiling point (606°) and 1100°, owing, it would seem, to the fact that the molecule then varies from Sn_2Cl_4 to $SnCl_2$, but the vapour density proved to be less than that indicated by the first and greater than that shown by the second formula, although it approaches to the latter as the temperature rises—that is, it presents a similar phenomenon to that observed in the passage of N_2O_4 into NO_2.

When rapidly boiled, an alkaline solution of stannous oxide deposits tin and forms stannic oxide, $2SnO = Sn + SnO_4$, which remains in the alkaline solution.

Weber (1882) by precipitating a solution of stannous chloride with sodium sulphite (this salt as a reducing agent prevents the oxidation of the stannous compound) and dissolving the washed precipitate in nitric acid, obtained crystals of *stannous nitrate*, $Sn(NO_3)_2,20H_2O$, on refrigerating the solution. This crystallo-hydrate easily melts, and is deliquescent. Besides this, a more stable anhydrous basic salt, $Sn(NO_3)_2,SnO$, is easily formed. In general, stannous oxide as a feeble base easily forms basic salts, just as cupric and lead oxides do. For the same reason SnX_2 easily forms double salts. Thus a potassium salt, SnK_2Cl_4,H_2O, and especially an ammonium salt, $Sn(NH_4)_2Cl_4,H_2O$, called *pink salt*, are known. Some of these salts are used in the arts, owing to their being more stable than tin salts alone. Stannous bromide and iodide, $SnBr_2$ and SnI_2, resemble the chloride in many respects.

Among other stannous salts a sulphate, $SnSO_4$, is known. It is formed as a crystalline powder when a solution of stannous oxide in sulphuric acid is evaporated under the receiver of an air-pump. The feeble basic character of the stannous oxide is clearly seen in this salt. It decomposes with extreme facility, when heated, into stannic oxide and sulphurous anhydride, but it easily forms double salts with the salts of the alkali metals.

In gaseous hydrochloric acid, stannous chloride, $SnCl_2,2H_2O$, forms a liquid having the composition $SnCl_2,HCl,3H_2O$ (sp gr. 2·2, freezes at -27°), and a solid salt, $SnCl_2,H_2O$ (Engel).

Frémy supposes the cause of the difference to consist in a difference of polymerisation, and considers that the ordinary acid corresponds with the oxide SnO_2, and the meta-acid with the oxide Sn_5O_{10}, but it is more probable that both are polymeric but in a different degree. Stannic acid with sodium carbonate gives a salt of the composition Na_2SnO_3. The same salt is also obtained by fusing metastannic acid with sodium hydroxide, whilst metastannic acid gives a salt, $Na_2SnO_3,4SnO_2$ (Frémy), when treated with a dilute solution of alkali; moreover, stannic acid is also soluble in the ordinary stannate, Na_2SnO_3 (Weber), so that both stannic acids (like both forms of silica) are capable of polymerisation, and probably only differ in its degree. In general, there is here a great resemblance to silica, and Graham obtained a solution of stannic acid by the direct dialysis of its alkaline solution. The main difference between these acids is that the meta-acid is soluble in hydrochloric acid, and gives a precipitate with sulphuric acid and stannous chloride, which do not precipitate the ordinary acid. Vignon (1889) found that more heat is evolved in dissolving stannic acid in KHO than metastannic.

The formation of the compound $SnCl_4,3H_2O$ is accompanied by so great a contraction that these crystals, although they contain water, are heavier than the anhydrous chloride $SnCl_4$. The penta-hydrated crystallo-hydrate absorbs dry hydrochloric acid, and gives a liquid of specific gravity 1·971, which at 0° yields crystals of the compound $SnCl_4,2HCl,6H_2O$ (it corresponds with the similar platinum compound), which melt at 20° into a liquid of specific gravity 1·925 (Engel).

Stannic chloride combines with ammonia ($SnCl_4,4NH_3$), hydrocyanic acid, phosphoretted hydrogen, phosphorus pentachloride ($SnCl_4,PCl_5$), nitrous anhydride and its chloranhydride ($SnCl_4,N_2O_3$ and $SnCl_4,2NOCl$), and with metallic chlorides (for example, K_2SnCl_6, $(NH_4)_2SnCl_6$, &c.) In general, a highly-developed faculty for combination is observed in it.

Tin does not combine directly with iodine, but if its filings be heated in a closed tube with a solution of iodine in carbon bisulphide, it forms stannic iodide, SnI_4, in the form of red octahedra which fuse at 142° and volatilise at 295°. The fluorine compounds of tin have a special interest in the history of chemistry, because they give a series of double salts which are isomorphous with the salts of hydrofluosilicic acid, SiR_2F_6, and this fact served to confirm the formula SiO_2 for silica, as the formula SnO_2 was indubitable. Although *stannic fluoride*, SnF_4, is almost unknown in the free state, its corresponding double salts are very easily formed by the action of hydrofluoric acid on alkaline solutions of stannic oxide; thus, for example, a crystalline salt of the composition SnK_2F_6,H_2O is obtained by dissolving stannic oxide in potassium hydroxide and then adding hydrofluoric acid to the solution. The barium salt, $SnBaF_6,3H_2O$, is sparingly soluble like its corresponding silicofluoride. The more soluble salt of strontium, $SnSrF_6,2H_2O$, crystallises very well, and is therefore more important for the purposes of research; it is isomorphous with the corresponding salt of silicon (and titanium); the magnesium salt contains 6H$_2$O.

Stannic sulphide, SnS$_2$, is formed, as a yellow precipitate, by the action of sulphuretted hydrogen on acid solutions of stannic salts; it is easily soluble in ammonium and potassium sulphides, because it has an acid character, and then forms thiostannates (see Chapter XX.). In an anhydrous state it has the form of brilliant golden yellow plates, which may be obtained by heating a mixture of finely-divided tin, sulphur, and sal-ammoniac for a considerable time. It is sometimes used in this form under the name of mosaic gold, as a cheap substitute for gold-leaf in gilding wood articles. On ignition it parts with a portion of its sulphur, and is converted into stannous sulphide SnS. It is soluble in caustic alkalis. Hydrochloric acid does not dissolve the anhydrous crystalline compound, but the precipitated powdery sulphide is soluble in boiling strong hydrochloric acid, with the evolution of hydrogen sulphide.

[41 bis] Although this has long been generally recognised from the resemblance between the two metals, still from a chemical point of view it has only been demonstrated by means of the periodic law.

Mixed ores of copper compounds together with PbS and ZnS are frequently found in the most ancient primary rocks. As the separation of the metals themselves is difficult, the ores are separated by a method of selection or mechanical sorting. Such mixed ores occur in Russia, in many parts of the Caucasus, and in the Donetz district (at Nagolchik).

[42 bis] Lead sulphide in the presence of zinc and hydrochloric acid is completely reduced to metallic lead, all the sulphur being given off as hydrogen sulphide.

Lead sulphate, PbSO$_4$, occurs in nature (*anglesite*) in transparent brilliant crystals which are isomorphous with barium sulphate, and have a specific gravity of 6·3. The same salt is formed on mixing sulphuric acid or its soluble salts with solutions of lead salts, as a heavy white precipitate, which is insoluble in water and acids, but dissolves in a solution of ammonium tartrate in the presence of an excess of ammonia. This test serves to distinguish this salt from the similar salts of strontium and barium.

According to J. B. Hannay (1894) the last named decomposition (PbS + PbSO$_4$ = 2Pb + 2SO$_2$) is really much more complicated, and in fact a portion of the PbS is dissolved in the Pb, forming a slag containing PbO, PbS and PbSO$_4$, whilst a portion of the lead volatilises with the SO$_2$ in the form of a compound PbS$_2$O$_2$, which is also formed in other cases, but has not yet been thoroughly studied.

Besides these methods for extracting lead from PBS in its ores, roasting (the removal of the S in the form of SO_2) and smelting with charcoal with a blast in the same manner as in the manufacture of pig iron (Chapter XXII.) are also employed.

We may add that PbS in contact with Zn and hydrochloric acid (which has no action upon PbS alone) entirely decomposes, forming H_2S and metallic lead: $PbS + Zn + 2HCl = Pb + ZnCl_2 + H_2S$.

As lead is easily reduced from its ores, and the ore itself has a metallic appearance, it is not surprising that it was known to the ancients, and that its properties were familiar to the alchemists, who called it 'Saturn.' Hence metallic lead, reduced from its salts in solution by zinc, having the appearance of a tree-like mass of crystals, is called 'arbor saturni,' &c.

Freshly laid new lead pipes contaminate the water with a certain amount of lead salts, arising from the presence of oxygen, carbonic acid, &c., in the water. But the lead pipes under the action of running water soon become coated with a film of salts—lead sulphate, carbonate, chloride, &c.—which are insoluble in water, and the water pipes then become harmless.

Lead is used in the arts, and owing to its considerable density, it is cast, mixed with small quantities of other metals, into shot. A considerable amount is employed (together with mercury) in extracting gold and silver from poor ores, and in the manufacture of chemical reagents, and especially of lead chromate. *Lead chromate*, $PbCrO_4$, is distinguished for its brilliant yellow colour, owing to which it is employed in considerable quantities as a dye, mainly for dyeing cotton tissues yellow. It is formed on the tissue itself, by causing a soluble salt of lead to react on potassium chromate. Lead chromate is met with in nature as 'red lead ore.' It is insoluble in water and acetic acid, hut it dissolves in aqueous potash. The so-called pewter vessels often consist of an alloy of 5 parts of tin and 1 part of lead, and solder is composed of 1 to 2 parts of tin with ½ part of lead. Amongst the alloys of lead and tin, Rudberg states that the alloy $PbSn_3$ stands out from the rest, since, according to his observations, the temperature of solidification of the alloy is 187°.

The normal lead acetate, known in trade as *sugar of lead*, owing to its having a sweetish taste, has the formula $Pb(C_2H_3O_2)_2, 3H_2O$. This salt only crystallises from acid solutions. It is capable of dissolving a further quantity of lead oxide or of metallic lead in the presence of air. A basic salt of the composition $Pb(C_2H_3O_2)_2, PbH_2O_2$ is then formed which is soluble in water and alcohol. As in this salt the number of atoms is even and the same as in the hydrate of acetic acid, $C_2H_4O_2, H_2O = C_2H_3(OH)_3$, it may be represented as this hydrate in which two of hydrogen are replaced by lead—that is, as $C_2H_3(OH)(O_2Pb)$. This basic salt is used in medicine as a remedy for inflammation, for bandaging wounds, &c., and also in the manufacture of white lead. Other basic acetates of lead, containing a still greater amount of lead oxide, are known. According to the above representation of the composition of the preceding lead acetate, a basic salt of the composition $(C_2H_3)_2(O_2Pb)_3$ would be also possible, but what appear to be still more basic salts are known. As the character of a salt also depends on the property of the base from which it is formed, it would seem that lead forms a hydroxide of the composition HOPbOH, containing two water residues, one or both of which may be replaced by the acid residues. If both water residues are replaced, a normal salt, XPbX, is obtained, whilst if only one is replaced a basic salt, XPbOH, is formed. But lead does not only give this normal hydroxide, but also polyhydroxides, $Pb(OH), nPbO$, and if we may imagine that in these polyhydroxides there is a substitution of both the water residues by acid residues, then the power of lead for forming basic salts is explained by the properties of the base which enters into their composition.

Few compounds are known of the lower type PbX, and still fewer of the intermediate type PbX_3. To the first type belongs the so-called lead suboxide, Pb_2O, obtained by the ignition of lead oxalate, C_2PbO_4, without access of air. It is a black powder, which easily breaks up under the action of acids, and even by the simple action of heat, into metallic lead and lead oxide. This is the character of all suboxides. They cannot be regarded as independent salt-forming oxides, neither can those forms of oxidation of lead which contain more oxygen than the oxide of lead, PbO, and less than the dioxide, PbO_2. As we shall see, at least two such compounds are formed. Thus, for example, an oxide having the composition Pb_2O_3 is known, but it is decomposed by the action of acids into lead oxide, which passes into solution, and lead dioxide, which remains behind. Such is red lead. (See further on.)

In the boiling of drying oils, the lead oxide partially passes into solution, forming a saponified compound capable of attracting oxygen and solidifying into a tar-like mass, which forms the oil paint. Perhaps, however, glycerine partially acts in the process.

Ossovetsky by saturating drying oil with the salts of certain metals obtained oil colours of great durability.

A mixture of very finely-divided litharge with glycerine (50 parts of litharge to 5 c.c. of anhydrous glycerine) forms a very quick (two minutes) setting cement, which is insoluble in water and oils, and is very useful in setting up chemical apparatus. The hardening is based on the reaction of the lead oxide with glycerine (Moraffsky).

It is very instructive to observe that lead not only easily forms basic salts, but also salts containing several acid groups. Thus, for example, lead carbonate occurs in nature and forms compounds with lead chloride and sulphate. The first compound, known as *corneous lead, phosgenite*, has the composition $PbCO_3,PbCl_2$; it occurs in nature in bright cubical crystals, and is prepared artificially by simply boiling lead chloride with lead carbonate. A similar compound of normal salts, $PbSO_4,PbCO_3$, occurs in nature as *lanarkite* in monoclinic crystals. *Leadhillite* contains $PbSO_4,3PbCO_3$, and also occurs in yellowish, monoclinic, tabular crystals. We will turn our attention to these salts of lead, because it is very probable that their formation is allied to the formation of the basic salts, and the following considerations may lead to the explanation of the existence of both. In describing silica we carefully developed the conception of polymerisation, which it is *also indispensable to recognise in the composition of many other oxides*. Thus it may be supposed that PbO_2 is a similar polymerised compound to SiO_2—*i.e.* that the composition of lead peroxide will be Pb_nO_{2n}, because lead methyl, $PbMe_4$, and lead ethyl, $PbEt_4$, are volatile compounds, whilst PbO_2 is non-volatile, and is very like silica in this respect, and not in the least like carbonic anhydride. Still more should a polymeric structure, Pb_nO_n, be ascribed to lead oxide, since it differs as little from lead dioxide in its physical properties as carbonic oxide does from carbonic anhydride, and being an unsaturated compound is more likely to be capable of intercombination (polymerisation) than lead dioxide. These considerations respecting the complexity of lead oxide could have no real significance, and could not be accepted, were it not for the existence of the above-mentioned basic and mixed salts. The oxide apparently corresponds with the composition Pb_nX_{2n}, and since, according to this representation, the number of X's in the salts of lead is considerable, it is obvious that they may be diverse. When a part of these X's is replaced by the water residue (OH) or by oxygen, $X_2 = O$, and the other parts by an *acid residue*, X, then basic salts are obtained, but if a part of the X's is replaced by acid residues of one kind, and the other part by acid residues of another kind, then those mixed salts about which we are now speaking are formed. Thus, for example, we may suppose, for a comparison of the composition of the majority of the salts of lead, that $n = 12$, and then the above-mentioned compounds will present themselves in the following form:—Lead oxide, $Pb_{12}O_{12}$, its crystalline hydrate, $Pb_{12}O_8(OH)_8$, lead chloride, $Pb_{12}Cl_{24}$, lead oxychloride, $Pb_{12}Cl_{12}O_6$, the other oxychloride, $Pb_{12}(OH)_8Cl_6O_6$, mendipite (*see* Note 51), $Pb_{12}Cl_8O_8$, normal lead carbonate, $Pb_{12}(CO_3)_{12}$, crystalline basic salt, $Pb_{12}(OH)_6(CO_3)_6$, white lead, $Pb_{12}(CO_3)_8(HO)_8$, corneous lead, $Pb_{12}Cl_{12}(CO_3)_6$, lanarkite, $Pb_{12}(CO_3)_6(SO_4)_6$, leadhillite, $Pb_{12}(CO_3)_9(SO_4)_3$, &c. The number 12 is only taken to avoid fractional quantities. Possibly the polymerisation is much higher than this. The theory of the polymerisation of oxides introduced by me in the first edition of this work (1869) is now beginning to be generally accepted.

A similar basic salt having a white colour, and therefore used as a substitute for white lead, is also obtained by mixing a solution of basic lead acetate with a solution of lead chloride. Its formation is expressed by the equation: $2PbX(OH),PbO + PbCl_2 = 2Pb(OH)Cl,PbO + PbX_2$. Similar basic compounds of lead are met with in nature—for instance, *mendipite*, $PbCl,2PbO$, which appears in brilliant yellowish-white masses. The ignition of red lead with sal-ammoniac results in similar polybasic compounds of lead chloride, forming the *Cassel's*, or *mineral yellow* of the composition $PbCl_2nPbO$. *Lead iodide*, PbI_2, is still less soluble than the chloride, and is therefore obtained by mixing potassium iodide with a solution of a lead salt. It separates as a yellow powder, which may be dissolved in boiling water, and on cooling separates in very brilliant crystalline scales of a golden yellow colour. The salts $PbBr_2$, PbF_2, $Pb(CN)_2$, $Pb_2Fe(CN)_6$ are also insoluble in water, and form white precipitates.

It is remarkable that a peculiar kind of attraction exists between boiled linseed oil and white lead, as is seen from the following experiments. White lead is triturated in water. Although it is heavier than water, it remains in suspension in it for some time and is thoroughly moistened by it, so that the trituration may be made perfect; boiled linseed oil is then added, and shaken up with it. A mixture of the oil and white lead is then found to settle at the bottom of the vessel. Although the oil is much lighter than the water it does not float on the top, but is retained by the white lead and sinks under the water together with it. There is not, however, any more perfect combination nor even any solution. If the resultant mass be then treated with ether or any other liquid capable of dissolving the oil, the latter passes into solution and leaves the white lead unaltered.

It may be regarded as a salt corresponding with the normal hydrate of carbonic acid, $C(OH)_4$, in which three-quarters of the hydrogen is replaced by lead. A salt is also known in which all the hydrogen of this hydrate of carbonic acid is replaced by lead—namely, the salt containing CO_4Pb_2. This salt is obtained as a white crystalline substance by the action of water and carbonic acid on lead. The normal salt, $PbCO_3$, occurs in nature under the name of white lead ore (sp. gr. 6·47), in crystals, isomorphous with aragonite, and is formed by the double decomposition of lead nitrate with sodium carbonate, as a heavy white precipitate. Thus both these salts resemble white lead, but the first-named salt is exclusively used in practice, owing to its being very conveniently prepared, and being characterised by its great covering capacity, or 'body,' due to its fine state of division.

[53 bis] One of the many methods by which white lead is prepared consists in mixing massicot with acetic acid or sugar of lead, and leaving the mixture exposed to air (and re-mixing from time to time), containing carbonic acid, which is absorbed from the surface by the basic salt formed. After repeated mixings (with the addition of water), the entire mass is converted into white lead, which is thus obtained very finely divided.

If lead hydroxide be dissolved in potash and sodium hypochlorite be added to the solution, the oxygen of the latter acts on the dissolved lead oxide, and partially converts it into dioxide, so that the so-called lead sesquioxide is obtained; its empirical formula is Pb_2O_3. Probably it is nothing but a lead salt—*i.e.* is referable to the type of dioxide of lead, or its hydroxide, $PbO(OH)_2$, in which two atoms of hydrogen are replaced by lead, $PbO(O_2Pb)$. The brown compound precipitated by the action of dilute acids—for example, nitric—splits up, even at the ordinary temperature, into insoluble lead dioxide and a solution of a lead salt. This compound evolves oxygen when it is heated. It dissolves in hydrochloric acid, forming a yellow liquid, which probably contains compounds of the composition $PbCl_2$ and $PbCl_4$, but even at the ordinary temperature the latter soon loses the excess of chlorine, and then only lead chloride, $PbCl_2$, remains. In order to see the relation between red lead and lead sesquioxide, it must be observed that they only differ by an extra quantity of lead oxide—that is, red lead is a basic salt of the preceding compound, and if the compound Pb_2O_3 may be regarded as PbO_3Pb, then red lead should be looked on as PbO_3Pb,PbO—that is, as basic lead plumbate.

[54 bis] Frémy obtained potassium plumbate in the following manner. Pure lead dioxide is placed in a silver crucible, and a strong solution of pure caustic potash is poured over it. The mixture is heated and small quantities are removed from time to time for testing, which consists in dissolving in a small quantity of water and decomposing the resultant solution with nitric acid. There is a certain moment during the heating when a considerable amount of insoluble lead dioxide is precipitated on the addition of the nitric acid; the solution then contains the salt in question, and the heating must be stopped, and a small amount of water added to dissolve the potassium plumbate formed. On cooling the salt separates in somewhat large crystals, which have the same composition as the stannate—that is, $PbO(KO)_2,3H_2O$.

Lead dioxide is often called lead peroxide, but this name leads to error, because PbO_2 does not show the properties of true peroxides, like hydrogen or barium peroxides, but is endowed with acid properties—that is, it is able to form true salts with bases, which is not the case with true peroxides. Lead dioxide is a normal salt-forming compound of lead, as Bi_2O_5 is for bismuth, CeO_2 for cerium, and TeO_3 for tellurium, &c. They all evolve chlorine when treated with hydrochloric acid, whilst true peroxides form hydrogen peroxide. The true lead peroxide, if it were obtained, would probably have the composition Pb_2O_5, or, in combination with peroxide of hydrogen, $H_2Pb_2O_7 = H_2O_2 + Pb_2O_5$, judging from the peroxides corresponding with sulphuric, chromic, and other acids, which we shall afterwards consider.

As a proof of the fact, that the form PbO_2, or PbX_4, is the highest normal form of any combination of lead, it is most important to remark that it might be expected that the action of lead chloride, $PbCl_2$, on zinc-ethyl, $ZnEt_2$, would result in the formation of zinc chloride, $ZnCl_2$, and lead-ethyl, $PbEt_2$, but that in reality the reaction proceeds otherwise. Half of the lead is set free, and lead tetrethyl, $PbEt_4$, is formed as a colourless liquid, boiling at about 200° (Butleroff, Frankland, Buckton, Cahours, and others). The type PbX_4 is not only expressed in $PbEt_4$ and PbO_2, but also in PbF_4, obtained by Brauner.

According to Carnelley and Walker, the hydrate $(PbO_2)_3, H_2O$ is then formed; it loses water at 230°. The anhydrous dioxide remains unchanged up to 280°, and is then converted into the sesquioxide, Pb_2O_3, which again loses oxygen at about 400°, and forms red lead, Pb_3O_4. Red lead also loses oxygen at about 550°, forming lead oxide, PbO, which fuses without change at about 600°, and remains constant as far as the limit of the observations made (about 800°).

The best method for preparing pure lead dioxide consists in mixing a hot solution of lead chloride with a solution of bleaching powder (Fehrman).

[56 bis] The plumbates of Ca and other similar metals, mentioned in Chapter III., Note 7, also belong to the form PbX_4.

The compounds of titanium are generally obtained from rutile; the finely-ground ore is fused with a considerable amount of acid potassium sulphate, until the titanic anhydride, as a feeble base, passes into solution. After cooling, the resultant mass is ground up, dissolved in cold water, and treated with ammonium hydrosulphide; a black precipitate then separates out from the solution. This precipitate contains TiO_2 (as hydrate) and various metallic sulphides—for example, iron sulphide. It is first washed with water and then with a solution of sulphurous anhydride until it becomes colourless. This is due to the iron sulphide contained in the precipitate, and rendering it black, being converted into dithionate by the action of the sulphurous acid. The titanic acid left behind is nearly pure. The considerable volatility of titanium chloride may also be taken advantage of in preparing the compounds of titanium from rutile. It is formed by heating a mixture of rutile and charcoal in dry chlorine; the distillate then contains *titanium chloride*, $TiCl_4$. It may be easily purified, owing to its having a constant boiling point of 136°. Its specific gravity is 1·76; it is a colourless liquid, which fumes in the air, and is perfectly soluble in water if it be not heated. When hot water acts on titanic chloride, a large proportion of titanic acid separates out from the solution and passes into metatitanic acid. A similar decomposition of acid solutions of titanic acid is accomplished whenever they are heated, and especially in the presence of sulphuric acid, just as with metastannic acid, which titanic acid resembles in many respects. On igniting the titanic acid a colourless powder of the anhydride, TiO_2, is obtained. In this form it is no longer soluble in acids or alkalis, and only fuses in the oxyhydrogen flame; but, like silica, it dissolves when fused with alkalis and their carbonates; as already mentioned, it dissolves when fused with a considerable excess of acid potassium sulphate—that is, it then reacts as a feeble base. This shows the basic character of titanic anhydride; it has at once, although feebly developed, both basic and acid properties. The fused mass, obtained from titanic anhydride and alkali when treated with water, parts with its alkali, and a residue is obtained of a sparingly-soluble poly-titanate, $K_2TiO_3 n TiO_2$. The hydrate, which is precipitated by ammonia from the solutions obtained by the fusion of TiO_2 with acid potassium sulphate, when dried forms an amorphous mass of the composition $Ti(OH)_4$. But it loses water over sulphuric acid, gradually passing into a hydrate of the composition $TiO(OH)_2$, and when heated it parts with a still larger proportion of water; at 100° the hydrate $Ti_2O_3(OH)_2$ is obtained, and at 300° the anhydride itself. The higher hydrate, $Ti(OH)_4$, is soluble in dilute acid, and the solution may be diluted with water; but on boiling the sulphuric acid solution (though not the solution in hydrochloric acid), all the titanic acid separates in a modified form, which is, however, not only insoluble in dilute acids, but even in strong sulphuric acid. This hydrate has the composition $Ti_2O_3(OH)_2$, but shows different properties from those of the hydrate of the same composition described above, and therefore this modified hydrate is called *metatitanic acid*. It is most important to note the property of the ordinary gelatinous hydrate (that precipitated from acid solutions by ammonia) of dissolving in acids, the more so since silica does not show this property. In this property a transition apparently appears between the cases of common solution (based on a capacity for unstable combination) and the case of the formation of a hydrosol (the solubility of germanium oxide, GeO_2, perhaps presents another such instance). If titanium chloride be added drop by drop to a dilute solution of alcohol and hydrogen peroxide, and then ammonia be added to the resultant solution, a yellow precipitate of *titanium trioxide*, TiO_3H_2O, separates out, as Piccini, Weller, and Classen showed. This substance apparently belongs to the category of true peroxides.

Titanium chloride absorbs ammonia and forms a compound, $TiCl_4, 4NH_3$, as a red-brown powder which attracts moisture from the air and when ignited forms *titanium nitride*, Ti_3N_4. Phosphuretted hydrogen, hydrocyanic acid, and many similar compounds are also absorbed by titanium chloride, with the evolution of a considerable amount of heat. Thus, for example, a yellow crystalline powder of the composition $TiCl_4, 2HCN$ is obtained by passing dry hydrocyanic acid vapour into cold titanium chloride. Titanium chloride combines in a similar manner with cyanogen chloride, phosphorus pentachloride, and phosphorus oxychloride, forming molecular compounds, for example $TiCl_4, POCl_3$. This faculty for further combination probably stands in connection, on the one hand, with the capacity of titanium oxide to give polytitanates, $TiO(MO)_2, nTiO_2$; on the other hand, it corresponds with the kindred faculty of stannic chloride for the formation of poly-compounds (Note 41), and lastly it is probably related to the remarkable behaviour of titanium towards nitrogen. Metallic titanium, obtained as a grey powder by reducing potassium titanofluoride, K_2TiF_6, (sp. gr. 3·55 K. Hofman 1893), with iron in a charcoal crucible, combines directly with nitrogen at a red heat. If titanic anhydride be ignited in a stream of ammonia, all the oxygen of the titanic oxide is disengaged, and the compound TiN_2 is formed as a dark violet substance having a copper-red lustre. A compound Ti_5N_6 is also known; it is obtained by igniting the compound Ti_3N_4 in a stream of hydrogen, and is of a golden-yellow colour with a metallic lustre. To this order of compounds also belongs the well-known and chemically historical compound known as *titanium nitrocyanide*; its composition is Ti_5CN_4. This substance appears as infusible, sometimes well-formed, cubical crystals of sp. gr, 4·3, and having a red copper colour and metallic lustre; it is found in blast furnace slag. It is insoluble in acids but is acted on by chlorine at a red heat, forming titanium chloride. It was at first regarded as metallic titanium; it is formed in the blast furnace at the expense of those cyanogen compounds (potassium cyanide and others) which are always present, and at the expense of the titanium compounds which accompany the ores of iron. Wöhler, who investigated this compound, obtained it artificially by heating a mixture of titanic oxide with a small quantity of charcoal, in a stream of nitrogen, and thus proved the direct power for combination between nitrogen and titanium. When fused with caustic potash, all the nitrogen compounds of titanium evolve ammonia and form potassium titanate. Like metals they are able to reduce many oxides—for example, oxides of copper—at a red heat. Among the alloys of titanium, the crystalline compound Al_4Ti is remarkable. It is obtained by directly dissolving titanium in fused aluminium; its specific gravity is 3·11. The crystals are very stable, and are only soluble in aqua regia and alkalis.

The formula ZrO was first given to the oxide of zirconium as a base, in this case Zr = 45 whilst the present atomic weight is Zr = 90—that is, the formula of the oxide is now recognised as being ZrO_2. The reasons for ascribing this formula to the compounds of zirconium are as follows. In the first place, the investigation of the crystalline forms of the zirconofluorides—for example, K_2ZrF_6, $MgZrF_6,5H_2O$—which proved to be analogous in composition and crystalline form with the corresponding compounds of titanium, tin, and silicon. In the second place, the specific heat of Zr is 0·067, which corresponds with the combining weight 90. The third and most important reason for doubling the combining weight of zirconium was given by Deville's determination of the vapour density of *zirconium chloride*, $ZrCl_4$. This substance is obtained by igniting zirconium oxide mixed with charcoal in a stream of dry chlorine, and is a colourless, saline substance which is easily volatile at 440°. Its density referred to air was found to be 8·15, that is 117 in relation to hydrogen, as it should be according to the molecular formula of this substance above-cited. It exhibits, however, in many respects, a saline character and that of an acid chloranhydride, for zirconium oxide itself presents very feebly developed acid properties but clearly marked basic properties. Thus zirconium chloride dissolves in water, and on evaporation the solution only partially disengages hydrochloric acid—resembling magnesium chloride, for example. Zirconium was discovered and characterised as an individual element by Klaproth.

Pure compounds of zirconium are generally prepared from zircon, which is finely ground, but as it is very hard it is first heated and thrown into cold water, by which means it is disintegrated. Zircon is decomposed or dissolved when fused with acid potassium sulphate, or still more easily when fused with acid potassium fluoride (a double soluble salt, K_2ZrF_6, is then formed); however, zirconium compounds are generally prepared from powdered zircon by fusing it with sodium carbonate and then boiling in water. An insoluble white residue is obtained consisting of a compound of the oxides of sodium and zirconium, which is then treated with hydrochloric acid and the solution evaporated to dryness. The silica is thus converted into an insoluble form, and zirconium chloride obtained in solution. Ammonia precipitates *zirconium hydroxide* from this solution, as a white gelatinous precipitate, $ZrO(OH)_2$. When ignited this hydroxide loses water and in so doing undergoes a spontaneous recalescence and leaves a white infusible and exceedingly hard mass of *zirconium oxide*, ZrO_2, having a specific gravity of 5·4 (in the electrical furnace ZrO_2 fuses and volatilises like SiO_2, Moissan). Owing to its infusibility, zirconium oxide is used as a substitute for lime and magnesia in the Drummond light. This oxide, in contradistinction to titanium oxide, is soluble, even after prolonged ignition, in hot strong sulphuric acid. The hydroxide is easily soluble in acids. The composition of the salts is ZrX_4, or $ZrOX_2$, or $ZrOX_2,ZrO_2$, just as with those of its analogues. But although zirconium oxide forms salts in the same way with acids, it also gives salts with bases. Thus it liberates carbonic anhydride when fused with sodium carbonate, forming the salts $Zr(NaO)_4$, $ZrO(NaO)_2$, &c. Water, however, destroys these salts and extracts the soda.

Thorium has also been found in the form of oxide in certain pyrochlores, euxenites, monazites, and other rare minerals containing salts of niobium and phosphates. The compounds of thorium are prepared by decomposing thorite or orangeite with strong sulphuric acid at its boiling point; this renders the silica insoluble, and the thorium oxide passes into solution when the residue is treated with cold water, after having been previously boiled with water (boiling water does not dissolve the oxide of thorium). Lead and other impurities are separated by passing sulphuretted hydrogen through the solution, and the thorium hydroxide is then precipitated by ammonia. If this hydroxide be dissolved in the smallest possible amount of hydrochloric acid, and oxalic acid be then added, thorium oxalate is obtained as a white precipitate, which is insoluble in an excess of oxalic acid; this reaction is taken advantage of for separating this metal from many others. It, however, resembles the cerite metals (Chapter XVII., Note 43) in this and many other respects. The thorium hydroxide is gelatinous; on ignition it leaves an infusible oxide, ThO_2, which, when fused with borax, gives crystals of the same form as stannic oxide or titanic anhydride; sp. gr. 9·86. But the basic properties are much more developed in thorium oxide than in the preceding oxides, and it does not even disengage carbonic acid when fused with sodium carbonate—that is, it is a much more energetic base than zirconium oxide. The hydrate, ThO_2, however, is soluble in a solution of Na_2CO_3 (Chapter XVII., Note 43). Thorium chloride, $ThCl_4$ is obtained as a distinctly crystalline sublimate when thorium oxide, mixed with charcoal, is ignited in a stream of dry chlorine. When heated with potassium, thorium chloride gives a metallic powder of thorium having a sp. gr. 11·1. It burns in air, and is but slightly soluble in dilute acids. The atomic weight of thorium was established by Chydenius and Delafontaine on the basis of the ismorphism of the double fluorides.

CHAPTER XIX
PHOSPHORUS AND THE OTHER ELEMENTS OF THE FIFTH GROUP

Nitrogen is the lightest and most widely distributed representative of the elements of the fifth group, which form a higher saline oxide of the form R_2O_5, and a hydrogen compound of the form RH_3. Phosphorus, arsenic, bismuth, and antimony belong to the uneven series of this group. *Phosphorus* is the most widely distributed of these elements. There is hardly any mineral substance composing the mass of the earth's crust which does not contain some—it may be a small—amount of phosphorus compounds in the form of the salts of phosphoric acid. The soil and earthy substances in general usually contain from one to ten parts of phosphoric acid in 10,000 parts. This amount, which appears so small, has, however, a very important significance in nature. No plant can attain its natural growth if it be planted in an artificial soil completely free from phosphoric acid. Plants equally require the presence of potash, magnesia, lime, and ferric oxide, among basic, and of carbonic, sulphuric, nitric, and phosphoric anhydrides, among acid oxides. In order to increase the fertility of a more or less poor soil, the above-named nutritive elements are introduced into it by means of fertilisers. Direct experiment has proved that these substances are undoubtedly necessary to plants, but that they must be all present simultaneously and in small quantities, and that an excess, like an insufficiency, of one of these elements is necessarily followed by a bad harvest, or an imperfect growth, even if all the other conditions (light, heat, water, air) are normal. The phosphoric compounds of the soil accumulated by plants pass into the organism of animals, in which these substances are assimilated in many instances in large quantities. Thus the chief component part of bones is calcium phosphate, $Ca_3P_2O_8$, and it is on this that their hardness depends.

Phosphorus was first extracted by Brand in 1669, by the ignition of evaporated urine. After the lapse of a century Scheele, who knew of the existence of a more abundant source of phosphorus in bones, pointed out the method which is now employed for the extraction of this element. Calcium phosphate in bones permeates a nitrogenous organic substance, which is called ossein, and forms a gelatin. When bones are treated exclusively for the extraction of phosphorus, neglecting the gelatin, they are burnt, in which case all the ossein is burnt away. When, however, it is desired to preserve the gelatin, the bones are immersed in cold dilute hydrochloric acid, which dissolves the calcium phosphate and leaves the gelatin untouched; calcium chloride and acid calcium phosphate, $CaH_4(PO_4)_2$, are then obtained in the solution. When the bones are directly burnt in an open fire their mineral components only are left as an ash, containing about 90 per cent. of calcium phosphate, $Ca_3(PO_4)_2$, mixed with a small amount of calcium carbonate and other salts. This mass is treated with sulphuric acid, and then the same substance is obtained in the solution as was obtained from the unburnt bones immersed in hydrochloric acid— *i.e.* the acid calcium phosphate soluble in water, in which reaction naturally the chief part of the sulphuric acid is converted into calcium sulphate:

$$Ca_3(PO_4)_2 + 2H_2SO_4 = 2CaSO_4 + CaH_4(PO_4)_2.$$
$$Ca_3(PO_4)_2 + 4HCl = 2CaCl_2 + CaH_4(PO_4)_2.$$

On evaporating the solution, crystallisable acid calcium phosphate is obtained. The extraction of the phosphorus from this salt consists in *heating it with charcoal to a white heat*. When heated, the acid phosphate, $CaH_4(PO_4)_2$, first parts with water, and forms the metaphosphate, $Ca(PO_3)_2$, which for the sake of simplicity may be regarded, like the acid salt, as composed of pyrophosphate and phosphoric anhydride, $2Ca(PO_3)_2 = Ca_2P_2O_7 + P_2O_5$. The latter, with charcoal, gives phosphorus and carbonic oxide, $P_2O_5 + 5C = P_2 + 5CO$. So that in reality a somewhat complicated process takes place here, yielding ultimately products according to the following equation:

$$2CaH_4(PO_4)_2 + 5C = 4H_2O + Ca_2P_2O_7 + P_2 + 5CO.$$

After the steam has come over, phosphorus and carbonic oxide distil over from the retort and calcium pyrophosphate remains behind.[1 bis]

As phosphorus melts at about 40°, it condenses at the bottom of the receiver in a molten liquid mass, which is cast under water in tubes, and is sold in the form of sticks. This is common or *yellow phosphorus*. It is a transparent, yellowish, waxy substance, which is not brittle, almost insoluble in water, and easily undergoes change in its external appearance and properties under the action of light, heat, and of various substances. It crystallises (by sublimation or from its solution in carbon bisulphide) in the regular system, and (in contradistinction to the other varieties) is easily soluble in carbon bisulphide, and also partially in other oily liquids. In this it recalls common sulphur. Its specific gravity is 1·84. It fuses at 44°, and passes into vapour at 290°; it is easily inflammable, and must therefore be handled with great caution; careless rubbing is enough to cause phosphorus to ignite. Its application in the manufacture of matches is based on this.[2 bis] It emits light in the air owing to its slow oxidation, and is therefore kept under water (such water is phosphorescent in the dark, like phosphorus itself). It is also very easily oxidised by various oxidising agents and takes up the oxygen from many substances.[3 bis] Phosphorus enters into direct combination with many metals and with sulphur, chlorine, &c., with development of a considerable amount of heat. It is very poisonous although not soluble in water.

Besides this, there is a red variety of phosphorus, which differs considerably from the above. *Red phosphorus* (sometimes wrongly called *amorphous phosphorus*) is partially formed when ordinary phosphorus remains exposed to the action of light for a long time. It is also formed in many reactions; for example, when ordinary phosphorus combines with chlorine, bromine, iodine, or oxygen, a portion of it is converted into red phosphorus. Schrötter, in Vienna, investigated this variety of phosphorus, and pointed out by what methods it may be produced in considerable quantities. Red phosphorus is a powdery red-brown opaque substance of specific gravity 2·14. It does not combine so energetically with oxygen and other substances as yellow phosphorus, and evolves less heat in combining with them. Common phosphorus easily oxidises in the air; red phosphorus does not oxidise at all at the ordinary temperature; hence it does not phosphoresce in the air, and may be very conveniently kept in the form of powder. It does not, like yellow phosphorus, fuse at 44°. After being converted into vapour at 290° or 300°, it again passes into the ordinary variety when slowly cooled. Red phosphorus is not soluble in carbon bisulphide and other oily liquids, which permits of its being freed from any admixture of the ordinary phosphorus. It is not poisonous, and is used in many cases for which the ordinary phosphorus is unsuitable or dangerous; for example, in the manufacture of matches, which are then not poisonous or inflammable by accidental friction, and therefore the red variety has now replaced the ordinary phosphorus.[4 bis]

The heads of the 'safety' matches do not contain any phosphorus, but only substances capable of burning and of supporting combustion. Red phosphorus is spread over a surface on the box, and it is the friction against this phosphorus which ignites the matches. There is no danger of the matches taking fire accidentally, nor are they poisonous. This red phosphorus is prepared by heating the ordinary phosphorus at 230° to 270°; it is evident that this must be done in an atmosphere incapable of supporting combustion—for example, in nitrogen, carbonic anhydride, steam, &c. On a large scale, ordinary phosphorus is placed in closed iron vessels,[5 bis] and immersed in a bath of different proportions of tin and lead, by which means the temperature of 250° necessary for the conversion is easily attained. It is kept at this temperature for some time. The temperature is at first cautiously raised, and the air is thus partially expelled by the heat, and also by the evolution of steam (the phosphorus is damp when put in), whilst the remaining oxygen is also partially absorbed by the phosphorus, so that an atmosphere of nitrogen is produced in the iron vessel. Red phosphorus enters into all the reactions proper to yellow phosphorus, only with greater difficulty and more slowly; and, as its vapour tension (volatility) is less than that of the yellow variety, it may be supposed that a polymerisation takes place in the passage of the yellow into the red modification, just as in the passage of cyanogen into paracyanogen, or of cyanic acid into cyanuric acid (Chapter IX. Notes 39 bis and 48).

The vapour of phosphorus is colourless; its density remains constant between 300° and 1000° (Dumas, 1833; Mitscherlich, Deville, and Troost, 1859, and others). The density with respect to air has been determined as from 4·3 to 4·5. Hence, referred to hydrogen, it is $4·4 \times 14·4 = 63$, corresponding with a molecular weight 124, *i.e.* the molecule of phosphorus in a state of vapour contains P_4. The reader will remember that the molecule of nitrogen contains N_2, of sulphur S_6 or S_2, and of oxygen O_2 or O_3.

The chemical energy of phosphorus in a free state more nearly approaches that of sulphur than nitrogen. Phosphorus is combustible and inflames at 60°; but having in the act of combination parted with a portion of its energy in the form of heat it becomes analogous to nitrogen, so long as there is no question of its reduction back again into phosphorus. Nitric acid is easily reduced to nitrogen, whilst phosphoric acid is reduced with very much greater difficulty. All the compounds of phosphorus are less volatile than those of nitrogen. Nitric acid, HNO_3, is easily distilled; metaphosphoric acid, HPO_3, is generally said to be non-volatile; triethylamine, $N(C_2H_5)_3$, boils at 90°, and triethylphosphine, $P(C_2H_5)_3$, at 127°.

Phosphorus not only combines easily and directly with oxygen, but also with chlorine, bromine, iodine, sulphur, and with certain metals, and red phosphorus when heated combines with hydrogen also.[6 bis] So, for instance, when fused with sodium under naphtha, phosphorus gives the compound Na_3P_2. Zinc, absorbing the vapour of phosphorus, gives the phosphide Zn_3P_2 (sp. gr. 4·76); tin, SnP; copper, Cu_2P; even platinum combines with phosphorus (PtP_2, sp. gr. 8·77).[6 tri] Iron, when combined even with a small quantity of phosphorus, becomes brittle. Some of these compounds of phosphorus are obtained by the action of phosphorus on the solutions of metallic salts, and by the ignition of metallic oxides in the vapour of phosphorus, or by heating mixtures of phosphates with charcoal and metals. Phosphides do not exhibit the external properties of salts, which are so clearly seen in the chlorides and still distinctly observable in the sulphides. *The phosphides of the metals* of the alkalis and of the alkaline earths are even immediately and very easily decomposed by water, whereas this is found to be the case with only a very few sulphides, and still more rarely and indistinctly with the chlorides. We may take calcium phosphide as an example.[7 bis] Phosphorus is laid in a deep crucible, and covered with a clay plug, over which lime is strewn. At a red heat the vapours of phosphorus combine with the oxygen of the lime and form phosphoric anhydride, which forms a salt with another portion of the lime, whilst the liberated calcium combines with the phosphorus and forms calcium phosphide. Its composition is not quite certain; it may be CaP (corresponding with liquid phosphuretted hydrogen). This substance is remarkable for the following reaction: if we take water—or, better still, a dilute solution of hydrochloric acid—and throw calcium phosphide into it, bubbles of gas are evolved, which take fire spontaneously in the air and form white rings. This is owing to the fact that the liquid hydrogen phosphide, PH_2, is first formed, thus, $CaP + 2HCl = CaCl_2 + PH_2$, which, owing to its instability, very easily splits up into the solid phosphide, P_2H, and gaseous phosphide, PH_3; $5PH_2 = P_2H + 3PH_3$; the latter corresponds with ammonia. The mixture of the gaseous and liquid phosphides takes fire spontaneously in the air, forming phosphoric acid. The same hydrogen phosphides are formed when water acts on sodium phosphide (P_2Na_3). A similar mixture of gaseous liquid and solid phosphuretted hydrogen (Retgers 1894) is formed by heating (in a glass tube) red phosphorus in a stream of dry hydrogen. Hence we see that there are *three compounds of phosphorus with hydrogen*. (1) The first or solid yellow phosphide, P_2H (more probably P_4H_2), is obtained by the action of strong hydrochloric acid on sodium phosphide; it takes fire when struck or at 175°. (2) The liquid, PH_2, or more correctly expressed as the molecule, P_2H_4, is a colourless liquid which takes fire spontaneously in the air, boils at 30°, is very unstable, and is easily decomposed (by light or hydrochloric acid) into the two other phosphides of hydrogen. It is prepared by passing the gases evolved by the action of water on calcium phosphide through a freezing mixture. And, lastly, (3), gaseous hydrogen phosphide, *phosphine*, PH_3, which is distinguished as being the most stable. It is a colourless gas, which does not take fire in the air. It has an odour of garlic, and is very poisonous. It resembles ammonia in many of its properties.[8 bis] It is easily decomposed by heat, like ammonia, forming phosphorus and hydrogen; but it is very slightly soluble in water, and does not saturate acids, although it forms compounds with some of them which resemble ammonium salts in their form and properties. Among them the *compound with hydriodic acid*, PH_4I, analogous to ammonium iodide, is remarkable. This compound crystallises on sublimation in well-formed cubes, like sal-ammoniac, which it resembles in many respects. However, this compound does not enter into those reactions of double decomposition which are proper to sal-ammoniac, because its saline properties are very feebly developed. Phosphuretted hydrogen also combines, like ammonia, with certain chloranhydrides; but they are decomposed by water, with the evolution of phosphine. Ogier (1880) showed that hydrochloric acid also combines with phosphine under a pressure of 20 atmospheres at +18°, and under the ordinary pressure at -35°, forming the crystalline phosphonium chloride PH_4Cl, corresponding to sal-ammoniac. Hydrobromic acid does the same with greater ease, and hydriodic acid with still greater facility, forming phosphonium iodide, PH_4I.

Phosphuretted hydrogen, or phosphine, PH_3, is generally prepared by the action of caustic potash on phosphorus. Small pieces of phosphorus are dropped into a flask containing a strong solution of caustic potash and heated. Potassium hypophosphite, H_2KPO_2, is then obtained in solution; gaseous phosphuretted hydrogen is evolved:

$$P_4 + 3KHO + 3H_2O = 3(KH_2PO_2) + PH_3.$$

Liquid phosphuretted hydrogen (and free hydrogen) is also formed, together with the phosphine, so that the gaseous product, on escaping from the water into the air, takes fire spontaneously, forming beautiful white rings of phosphoric acid. In this experiment, as in that with calcium phosphide, it is the liquid, P_2H_4, that takes fire; but the phosphine set light to by it also burns, $PH_3 + O_4 = PH_3O_4$. The same phosphuretted hydrogen, PH_3, may be obtained pure, and not spontaneously combustible, by igniting the hydrates of phosphorous acid ($4PH_3O_3 = PH_3 + 3PH_3O_4$) and hypophosphorous acid ($2PH_3O_2 = PH_3 + PH_3O_4$); or, more simply, by the decomposition of calcium phosphide by hydrochloric acid, because then all the liquid phosphide, P_2H_4, is decomposed into non-volatile P_2H and gaseous PH_3. Pure phosphine liquefies when cooled to $-90°$, boils at $-85°$, and solidifies at $-135°$ (Olszewski). When phosphorus burns in an excess[10 bis] of *dry* oxygen, then only *phosphoric anhydride*, P_2O_5 is formed. It is prepared by dropping pieces of phosphorus through a wide tube, fixed into the upper neck of a large glass globe, on to a cup suspended in the centre of the globe. These lumps are set alight by touching them with a hot wire, and the phosphorus burns into P_2O_5. The dry air necessary for its combustion is forced into the globe through a lateral neck, and the white flakes of phosphoric anhydride formed are carried by the current of air through a second lateral neck into a series of Woulfe's bottles, where they settle as friable white flakes. Phosphoric anhydride may also be formed by passing dry air through a solution of phosphorus in carbon bisulphide. All the materials for the preparation of this substance must be carefully dried, because it *combines* with great eagerness *with water*, at the same time developing a large amount of heat and forming metaphosphoric acid, HPO_3, from which the water cannot be separated by heat. Phosphoric anhydride is a colourless snow-like substance, which attracts moisture from the air with the utmost avidity. It fuses at a red heat, and then *volatilises*. Its affinity for water is so great that it takes it up from many substances. Thus it converts sulphuric acid into sulphuric anhydride, and carbohydrates (wood, paper) are carbonised, and give up the elements of water when brought into contact with it.

When moist phosphorus slowly oxidises in the air, it not only forms phosphorous and phosphoric acids, but also *hypophosphoric acid*, $H_4P_2O_6$, which when in a dry state easily splits up at 60° into phosphorous and metaphosphoric acids ($H_4P_2O_6 = H_3PO_3 + HPO_3$), but differs from a mixture of these acids in that it forms well-characterised salts, of which the sodium salt, $H_2Na_2P_2O_6$, is but slightly soluble in water (the sodium salts of phosphoric and phosphorous acids are easily soluble), and that it does not act as a reducing agent, like mixtures containing phosphorous acid.

Judging by the general law of the formation of acids (Chapter XV.), the series of phosphorus compounds should include the following *ortho-acids* and their corresponding anhydrides, answering to phosphuretted hydrogen, H_3P:—

H_3PO_4, phosphoric acid, and $\quad$ P_2O_5, anhydride,

H_3PO_3, phosphorous acid, and $\quad$ P_2O_3, anhydride,

H_3PO_2, hypophosphorous acid, and P_2O, anhydride.

The last of these (the analogue of N_2O) is almost unknown. Phosphoric anhydride (P_2O_5) with a small quantity of water does not at first give orthophosphoric acid, PH_3O_4, but a compound P_2O_5,H_2O, or PHO_3, whose composition corresponds with that of nitric acid; this is *metaphosphoric acid*. Even with an excess of water, combining with phosphoric anhydride, this metaphosphoric acid, and not the ortho-, passes at first into solution. Metaphosphoric acid in solution only passes into orthophosphoric acid when the solution is heated or after a lapse of time.

Orthophosphoric acid is obtained by oxidising phosphorus with nitric acid until the phosphorus entirely passes into solution and the lower oxides of nitrogen cease to be evolved. The reaction takes place best with dilute nitric acid, and when aided by heat. The resultant solution is evaporated to a syrup. If a weighed quantity of phosphorus (dried in a current of dry carbonic anhydride) be taken, a crystalline mass of the acid can be obtained by evaporating the solution until it consists only of the quantity of phosphoric acid corresponding with the amount of phosphorus taken (from 31 parts of P, 98 parts of solution). The acid fuses at +39°; specific gravity of the liquid 1·88. Phosphorus pentachloride, PCl_5, and oxychloride, $POCl_3$ (see further on), give orthophosphoric acid and hydrochloric acid with water. The two other varieties of phosphoric acid, with which we shall presently become acquainted, give the same ortho-acid when under the influence of acids, with particular ease when boiled and more slowly in the cold. By itself orthophosphoric acid (either in solution or when dry) does not pass into the other varieties; it does not oxidise, and therefore presents the limiting and stable form. When heated to 300°, it loses water and passes into pyrophosphoric acid, $2H_3PO_4 = H_2O + H_4P_2O_7$, whilst at a red heat it loses twice as much water and is converted into metaphosphoric acid, $H_3PO_4 = H_2O + HPO_3$. In aqueous solution orthophosphoric acid differs clearly from pyro- or metaphosphoric acids, because the solutions of these latter acids give different reactions: thus orthophosphoric acid does not precipitate albumin, does not give a precipitate with barium chloride, and forms a yellow precipitate of silver orthophosphate, Ag_3PO_4, with silver nitrate (in the presence of alkalis, but not otherwise); whilst a solution of pyrophosphoric acid, $H_4P_2O_7$, although it does not precipitate albumin or barium chloride, gives a white precipitate of silver pyrophosphate, $Ag_4P_2O_7$, with silver nitrate; and a solution of metaphosphoric acid, HPO_3, precipitates both albumin and barium chloride, and gives a white precipitate of silver metaphosphate, $AgPO_3$, with silver nitrate. These points of distinction were studied by Graham, and are exceedingly instructive. They show that the solution of a substance does not determine the maximum of chemical combination with water, that solutions may contain various degrees of combination with water, and that there is a clear difference between the water serving for solution and that entering into chemical combination. Graham's experiments also showed that the water whose removal or combination determines the conversion of ortho- into meta- and pyrophosphoric acids differs distinctly from water of crystallisation, for he obtained the salts of ortho-, meta-, and pyrophosphoric acids with water of crystallisation, and they differed in their reactions, like the acids themselves. This water of crystallisation was expelled with greater ease than the water of constitution of the hydrates in question.[14 bis]

Orthophosphoric acid has a pleasant acid taste and a distinctly acid reaction; it is used as a medicine, and is not poisonous (phosphorous acid is poisonous). Alkalis, like sodium, potassium, and ammonium hydroxides, saturate the acid properties of phosphoric acid when taken in the ratio $2NaHO : H_3PO_4$—that is, when salts of the composition HNa_2PO_4 are formed. When taken in the ratio $NaHO : H_3PO_4$, a solution having an acid reaction is obtained, and when $3NaHO : H_3PO_4$—that is, when the salt Na_3PO_4 is formed—an alkaline reaction is obtained. Hence many chemists (Berzelius) even regarded the salts of composition R_2HPO_4 as normal, and considered phosphoric acid to be bibasic. But the salt Na_2HPO_4 also shows a feeble alkaline reaction, so that it is impossible to judge the characteristic peculiarities of acids by the reactions on litmus paper, as we already know from many examples. Orthophosphoric acid is tribasic, because it contains three equivalents of hydrogen replaceable by metals, forming salts, such as NaH_2PO_4, Na_2HPO_4, and Na_3PO_4. It is also tribasic, because with silver nitrate its soluble salts always give Ag_3PO_4, a salt with three equivalents of silver, and because by double decomposition with barium chloride it forms a salt of the composition $Ba_3(PO_4)_2$, and silver and barium hardly ever give basic salts. With the metals of the alkalis, phosphoric acid forms soluble salts, but the normal salts of the metals of the alkaline earths, $R_3(PO_4)_2$ and even $R_2H_2(PO_4)$, are insoluble in water, but dissolve in feeble acids, such as phosphoric and acetic, because they then form soluble acid salts, especially $RH_4(PO_4)_2$.

Phosphoric anhydride, or any of its hydrates, when ignited with an excess of sodium hydroxide, carbonate, &c., forms normal or *trisodium orthophosphate*, Na_3PO_4, but when a solution of sodium carbonate is decomposed by orthophosphoric acid, only the salt Na_2HPO_4 is formed; and when an excess of sodium chloride is ignited with orthophosphoric acid, hydrochloric acid is evolved, and the acid salt H_2NaPO_4 alone is formed. These facts clearly indicate the small energy of phosphoric acid with respect to the formation of the tri-metallic salt, which is seen further from the fact that the salt Na_3PO_4 has an alkaline reaction, decomposes in the presence of water and carbonic acid, forming Na_2HPO_4, corrodes glass vessels in which it is boiled or evaporated, just like solutions of the alkalis, disengages, like them, ammonia from ammonium chloride, and crystallises from solutions, as $Na_3PO_4,12H_2O$, only in the presence of an excess of alkali. At 15° the crystals of this salt require five parts of water for solution; they fuse at 77°.

Disodium orthophosphate, or common sodium phosphate, Na_2HPO_4, is more stable both in solution and in the solid state. As it is used in medicine and in dyeing, it is prepared in considerable quantities, most frequently from the impure phosphoric acid obtained by the action of sulphuric acid on bone ash. The solution thus formed—which contains, besides phosphoric and sulphuric acids, salts of sodium, calcium, and magnesium—is heated, and sodium carbonate added so long as carbonic anhydride is disengaged. A precipitate is formed containing the insoluble salts of magnesium and calcium, whilst the solution contains sodium phosphate, Na_2HPO_4, with a small quantity of other salts, from which it may be easily purified by crystallisation. At the ordinary temperature its solutions, especially in the presence of a small amount of sodium carbonate, give finely-formed inclined prismatic crystals, $Na_2HPO_4,12H_2O$; when the crystallisation takes place above 30° they only contain $7H_2O$. The former crystals even lose a portion of their water of crystallisation at the ordinary temperature (the salt effloresces), and form the second salt with $7H_2O$; whilst under the receiver of an air-pump and over sulphuric acid they also part with this water. When ignited they lose the last molecule of water of constitution, and give sodium pyrophosphate, $Na_4P_2O_7$.

Monosodium orthophosphate, NaH_2PO_4, crystallises with one equivalent of water; its solution has an acid reaction. At 100° the salt only loses this water of crystallisation, and at about 200° it parts with all its water, forming the metaphosphate $NaPO_3$. It is prepared from ordinary sodium phosphate by adding phosphoric acid until the solution does not give a precipitate with barium chloride, and then evaporating and crystallising the solution. The solution of this salt does not absorb carbonic anhydride, and does not give a precipitate with salts of calcium, barium, &c.

As a hydrate, orthophosphoric acid should be expressed, after the fashion of other hydrates, as containing three water residues (hydroxyl groups), *i.e.* as $PO(OH)_3$. This method of expression indicates that the type PX_5, seen in PH_4I, is here preserved, with the substitution of X_2 by oxygen and X_3 by three hydroxyl groups. The same type appears in $POCl_3$, PCl_5, PF_5, &c. And if we recognise phosphoric acid as $PO(OH)_3$, we should expect to find three anhydrides corresponding with it: (1) $[PO(OH)_2]_2O$, in which two of the three hydroxyls are preserved; this is pyrophosphoric acid, $H_4P_2O_7$. (2) $PO(OH)O$, where only one hydroxyl is preserved. This is metaphosphoric acid. (3) $(PO)_2O_3$ or P_2O_5, that is, perfect phosphoric anhydride. Therefore, *pyro- and metaphosphoric acids are imperfect anhydrides* (or anhydro-acids) *of orthophosphoric acid.*

Pyrophosphoric acid, $H_4P_2O_7$, is formed by heating orthophosphoric acid to 250° when it loses water.[19 bis] Its normal salts are formed by igniting the dimetallic salts of orthophosphoric acid of the types HM_2PO_4. Thus from the disodium salt we obtain sodium pyrophosphate, $Na_4P_2O_7$ (it crystallises from water with $10H_2O$, is very stable, fuses when heated, has an alkaline reaction, and does not form ortho-salts when its solution is boiled): and from the monosodium salt NaH_2PO_4 the acid salt $Na_2H_2P_2O_7$ (easily soluble in water) is formed; this has an acid reaction, and when ignited further gives the meta-salt.

Metaphosphoric acid, HPO_3 (the analogue of nitric acid), is formed by the ignition of the pyro- and ortho-acids (or, better, of their ammonium salts), as a vitreous, hygroscopic, fused mass (glacial phosphoric acid, *acidum phosphoricum glaciale*), soluble in water and volatilising without decomposition. It is also formed in the first slow action of cold water on the anhydride, but metaphosphoric acid gradually changes into the ortho-acid when its solution is boiled, or when it is kept for any length of time, especially in the presence of acids.

In order to see the relation between phosphoric acid and the lower acids of phosphorus, it is simplest to imagine the substitution of hydroxyl in H_3PO_4 or $PO(OH)_3$ by hydrogen. Then from orthophosphoric acid, $PO(OH)_3$, we shall obtain phosphorous acid, $POH(OH)_2$, and hypophosphorous acid, POH(OH); and, furthermore, phosphorous acid should be bibasic if orthophosphoric acid was tribasic, and hypophosphorous acid should be monobasic. This conclusion[21 bis] is, in fact, true, and hence all the acids of phosphorus may be referred to one common type, PX_5, whose representatives are PH_4I and PCl_5, $POCl_3$, PCl_2F_3, &c.

Phosphorous acid, PH_3O_3, is generally obtained from phosphorus trichloride, PCl_3, by the action of water: $PCl_3 + 3H_2O = 3HCl + PH_3O_3$. Both acids formed are soluble in water, but are easily separated, because hydrochloric acid is volatile whilst phosphorous acid volatilises with difficulty, and if a small amount of water be originally taken the hydrochloric acid nearly all passes off directly. Concentrated solutions of phosphorous acid give crystals of H_3PO_3, which fuse at 70°, attract moisture from the air, and deliquesce when ignited, giving phosphine and phosphoric acid, and are oxidised into orthophosphoric acid by many oxidising agents. In its salts only two hydrogen atoms are replaced by metals (Würtz); the salts of the alkaline metals are soluble, and give precipitates with salts of the majority of other metals.

The monobasic *hypophosphorous acid*, PH_3O_2, gives salts PH_2O_2Na, $(PH_2O_2)_2Ba$, &c.; the two remaining atoms of hydrogen (which exist in the same form as in phosphine, PH_3) are not replaceable by metals, and this determines the property of these salts of evolving phosphuretted hydrogen when heated (especially with alkalis). In acting on substances liable to reduction it is this hydrogen which acts, and, for example, *reduces* gold and mercury from the solutions of their salts, or converts cupric into cuprous salts. In all these instances the hypophosphorous acid is converted into phosphoric acid. Under the action of zinc and sulphuric acid it gives phosphine, PH_3. Nevertheless, neither hypophosphorous acid nor its dry salts absorb oxygen from the air. The salts of hypophosphorous acid are more soluble than those of the preceding acids of phosphorus. Thus the sodium salt $PNaH_2O_2$ does not give a precipitate with barium chloride, and the salts of calcium, barium, and many other metals are soluble. The hypophosphites are prepared by boiling an alkali with phosphorus so long as phosphuretted hydrogen is evolved. The acid itself is obtained from barium hypophosphite (prepared in the same manner by boiling phosphorus in baryta water), by decomposing its solution with sulphuric acid. By concentration of the solution of hypophosphorous acid (it must not be heated above 130°, at which temperature it decomposes) a syrup is formed which is able to crystallise. In the solid state hypophosphorous acid fuses at +17°, and has the properties of a clearly defined acid.

The types PX_3 and PX_5, which are evident for the hydrogen and oxygen compounds of phosphorus, are most clearly seen in its halogen compounds, to the consideration of which we will proceed, fixing our attention more especially on the chlorine compounds, as being the most important from the historical, theoretical, and practical point of view.

Phosphorus burns in chlorine, forming phosphorous chloride, PCl_3, and with an excess of chlorine, phosphoric chloride, PCl_5. The oxychloride, $POCl_3$, as the simplest chloranhydride according to the type PX_5, and also phosphoric chloride, correspond with orthophosphoric acid, $PO(OH)_3$, while phosphorous chloride, PCl_3, corresponds with phosphorous acid and the type PX_3. Phosphoric oxychloride, $POCl_3$, is a colourless liquid, boiling at 110°. Phosphorus trichloride is also a colourless liquid, boiling at 76°, whilst phosphoric chloride is a solid yellowish substance, which volatilises without melting at about 168°. They are all heavier than water, and form types of the *chloranhydrides* or chlorine compounds of the non-metallic elements whose hydrates are acids, just as NaCl or $BaCl_2$ are types of halogen metallic salts.

If a piece of phosphorus be dropped into a flask containing chlorine, it burns when touched with a red-hot wire, and combines with the chlorine. If the phosphorus be in excess, liquid *phosphorus trichloride*, PCl_3, is always formed, but if the chlorine be in excess the solid pentachloride is obtained. The trichloride is generally prepared in the following manner. Dry chlorine (passed through a series of Woulfe's bottles containing sulphuric acid) is led into a retort containing sand and phosphorus. The retort is heated, the phosphorus melts, spreads through the sand, and gradually forms the trichloride, which distils over into a receiver, where it condenses. *Phosphoric chloride* or *phosphorus pentachloride*, PCl_5, is prepared by passing dry chlorine into a vessel containing phosphorus trichloride (purified by distillation). Phosphorous chloride combines directly with oxygen, but more rapidly with ozone or with the oxygen of potassium chlorate ($3PCl_3 + KClO_3 = 3POCl_3 + KCl$), forming *phosphorus oxychloride*, $POCl_3$ (Brodie). This compound is also formed by the first action of water on phosphoric chloride; for example, if two vessels, one containing phosphoric chloride and the other water, are placed under a bell jar, after a certain time the crystals of the chloride disappear and hydrochloric acid passes into the water. The aqueous vapour acts on the pentachloride, and the following reaction occurs: $PCl_5 + H_2O = POCl_3 + 2HCl$, the result being that liquid phosphorus oxychloride is found in one vessel, and a solution of hydrochloric acid in the other. However, an excess of water directly transforms phosphoric chloride into orthophosphoric acid, $PCl_5 + 4H_2O = PH_3O_4 + 5HCl$, since $POCl_3$ reacts with water ($3H_2O$), forming $3HCl$ and phosphoric acid $PO(OH)_3$.

The above chlorine compounds serve not only as a type of the chloranhydrides, but also as a means for the preparation of other *acid chloranhydrides.* Thus the conversion of acids XHO into chloranhydrides, XCl, is generally accomplished by means of *phosphorus pentachloride.* This fact was discovered by Chancel, and adopted by Gerhardt as an important method for studying organic acids. By this means organic acids, containing, as we know, RCOOH (where R is a hydrocarbon group, and where carboxyl may repeat itself several times by replacing the hydrogen of hydrocarbon compounds), are converted into their chloranhydrides, RCOCl. With water they again form the acid, and resemble the chloranhydrides of mineral acids in their general properties.

Since carbonic acid, $CO(OH)_2$, contains two hydroxyl groups, its perfect chloranhydride, $COCl_2$, *carbonic oxychloride, carbonyl chloride* or *phosgene gas,* contains two atoms of chlorine, and differs from the chloranhydrides of organic acids in that in them one atom of chlorine is replaced by the hydrocarbon radicle RCOCl, if R be a monatomic radicle giving a hydrocarbon RH. It is evident, on the one hand, that in RCOCl the hydrogen is replaced by the radicle COCl, which is also able to replace several atoms of hydrogen (for example, $C_2H_4(COCl)_2$ corresponds with the bibasic succinic acid); and, on the other hand, that the reactions of the chloranhydrides of organic acids will answer to the reactions of carbonyl chloride, as the reactions of the acids themselves answer to those of carbonic acid. Carbonyl chloride is obtained directly from dry carbon monoxide and chlorine exposed to the action of light, and forms a colourless gas, which easily condenses into a liquid, boiling at +8°, specific gravity 1·43, and having the suffocating odour belonging to all chloranhydrides. Like all chloranhydrides, it is immediately decomposed by water, forming carbonic anhydride, according to the equation $COCl_2 + H_2O = CO_2 + 2HCl$, and thus expresses the type proper to all chloranhydrides of both mineral and organic acids.

In order to show the general method for the preparation of acid chloranhydrides, we will take that of acetic acid, $CH_3 \cdot COOH$, as an example. Phosphorus pentachloride is placed in a glass retort, and acetic acid poured over it; hydrochloric acid is then evolved, and the substance distilling over directly after is a very volatile liquid, boiling at 50°, and having all the properties of the chloranhydrides. With water it forms hydrochloric and acetic acids. The reaction here taking place may be explained thus: the substitution of the oxygen taken from the acetic acid (from its carboxyl) by two atoms of chlorine from the PCl_5 should be as follows: $CH_3 \cdot COOH + PCl_5 = CH_3 \cdot COHCl_2 + POCl_3$. But the compound $CH_3 \cdot COHCl_2$ does not exist in a free state (because it would indicate the possibility of the formation of compounds of the type CX_6, and carbon only gives those of the type CX_4); it therefore splits up into HCl and the chloranhydride $CH_3 \cdot COCl$. The general scheme for the reaction of phosphorus pentachloride with hydrates ROH is exactly the same as with water; namely, ROH with PCl$_5$, gives POCl$_3$ + HCl + RCl—that is a chloranhydride.[28 bis]

Containing, as they do, chlorine, which easily reacts with hydrogen, phosphorus pentachloride, trichloride, and oxychloride enter into reaction with ammonia, and give a series of amide and nitrile compounds of phosphorus. Thus, for example, when ammonia acts on the oxychloride we obtain sal-ammoniac (which is afterwards removed by water) and an orthophosphoric triamide, PO(NH$_2$)$_3$, as a white insoluble powder on which dilute acids and alkalis do not act, but which, when fused with potassium hydroxide, gives potassium phosphate and ammonia like other amides. When ignited, the triamide liberates ammonia and forms the nitrile PON, just as urea, CO(NH$_2$)$_2$, gives off ammonia and forms the nitrile CONH. This nitrile, called *monophosphamide*, PON, naturally corresponds with metaphosphoric acid, namely, with its ammonium salt. NH$_4$PO$_3$ - H$_2$O = PO$_2$·NH$_2$, an as yet unknown amide, and PO$_2$·NH$_2$ - H$_2$O gives the nitrile PON. This relation is confirmed by the fact that PON, moistened with water, gives metaphosphoric acid when ignited. It is the analogue of nitrous oxide, NON. It is a very stable compound, more so than the preceding.

The most important analogue of phosphorus is *arsenic*, the metallic aspect of which and the general character of its compounds of the types AsX$_3$ and AsX$_5$ at once recall the metals. The hydrate of its highest oxide, arsenic acid (ortho-arsenic acid), H$_3$AsO$_4$, is an oxidising agent, and gives up a portion of its oxygen to many other substances; but, nevertheless, it is very like phosphoric acid. Mitscherlich established the conception of isomorphism by comparing the salts of these acids.

Arsenic occurs *in nature*, not only combined with metals, but also, although rarely, native and also in combination with sulphur in two minerals—one red, *realgar*, As$_2$S$_2$, and the other yellow, *orpiment*, As$_2$S$_3$ (Chapter XX., Note 29 29). Arsenic occurs, but more rarely, in the form of salts of arsenic acid—for instance, the so-called cobalt and nickel blooms, two minerals which are found accompanying other cobalt ores, are the arsenates of these metals. Arsenic is also found in certain clays (ochres) and has been discovered in small quantities in some mineral springs, but it is in general of rarer occurrence in nature than phosphorus. Arsenic is most frequently extracted from arsenical pyrites, FeSAs, which, when roasted without access of air, evolves the vapour of arsenic, ferrous sulphide being left behind. It is also obtained by heating arsenious anhydride with charcoal, in which case carbonic oxide is evolved. In general, the oxides and other compounds are very easily reduced. Solid *arsenic* is a steel-grey brittle *metal*, having a bright lustre and scaly structure. Its specific gravity is 5·7. It is opaque and infusible, but volatilises as a yellow vapour which on cooling deposits rhombohedral crystals.[30 bis] The vapour density of arsenic is 150 times greater than that of hydrogen—that is, its molecule, like that of phosphorus, contains 4 atoms, As_4. When heated in the air, arsenic easily oxidises into white arsenious anhydride, As_2O_3, but even at the ordinary temperature it loses its lustre (becomes dull), owing to the formation of a coating of a lower oxide. The latter appears to be as volatile as arsenious anhydride, and it is probable that it is owing to the presence of this compound that the vapours of arsenious compounds, when heated with charcoal (for example, in the reducing flame of a blow-pipe), have the characteristic smell of garlic, because the vapour of arsenic itself has not this odour.

Arsenic easily combines with bromine and chlorine; nitric acid and aqua regia also oxidise it into the higher oxide, or rather its hydrate, arsenic acid. As far as is known, it does not decompose steam, and it acts exceedingly slowly on those acids, like hydrochloric, which are not capable of oxidising.

Arseniuretted hydrogen, *arsine*, AsH_3, resembles phosphuretted hydrogen in many respects. This colourless gas, which liquefies into a mobile liquid at -40°, has a disagreeable garlic-like odour, is only slightly soluble in water, and is exceedingly poisonous. Even in a small quantity it causes great suffering, and if present to any considerable amount in air it even causes death. The other compounds of arsenic are also poisonous, with the exception of the insoluble sulphur compound and some compounds of arsenic acid. Arseniuretted hydrogen, AsH_3, is obtained by the action of water on the alloy of arsenic and sodium, sodium hydroxide and arseniuretted hydrogen being formed. It is also formed by the action of sulphuric acid on the alloy of arsenic and zinc: $Zn_3As_2 + 3H_2SO_4 = 2AsH_3 + 3ZnSO_4$. The oxygen compounds of arsenic are very easily reduced by the action of hydrogen at the moment of its evolution from acids, and the reduced arsenic then combines with the hydrogen; hence, if a certain amount of an oxygen compound of arsenic be put into an apparatus containing zinc and sulphuric acid (and thus serving for the evolution of hydrogen), the hydrogen evolved will contain arseniuretted hydrogen. In this case it is diluted with a considerable amount of hydrogen. But its presence in the most minute quantities may be easily recognised from the fact that it is *easily decomposed* by heat (200° according to Brunn) into metallic arsenic and hydrogen, and therefore if such impure hydrogen he passed through a moderately-heated tube metallic arsenic will be deposited as a bright layer on the part of the tube which was heated (*see* Note 30 bis). This reaction is so sensitive that it enables the most minute traces of arsenic to be discovered; hence it is employed in medical jurisprudence, as a test in poisoning cases. It is easy to discover the presence of arsenic in common zinc, copper, sulphuric and hydrochloric acids, &c. by this method. It is obvious that in testing for poison by Marsh's apparatus it is necessary to take zinc and sulphuric acid quite free from arsenic. The arsenic deposited in the tube may be driven as a volatile metal from one place to another in the current of hydrogen evolved, owing to its volatility. This forms a distinction between arseniuretted and antimoniuretted hydrogen, which is decomposed by heat in just the same way as arseniuretted hydrogen, but the mirror given by Sb is not so volatile as that formed by As.

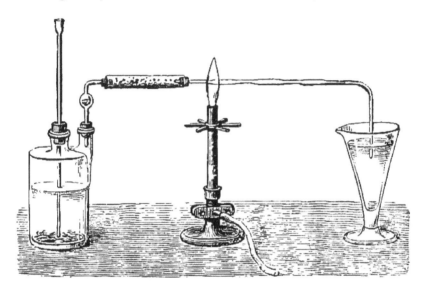

FIG. 84.—Formation and decomposition of arseniuretted hydrogen. Hydrogen is evolved in the Woulfe's bottle, and when the gas comes off, a solution containing arsenic is poured through the funnel. The presence of AsH_3 is recognised from the deposition of a mirror of arsenic when the gas-conducting tube is heated. If the escaping hydrogen be lighted, and a porcelain dish be held in the flame, a film of arsenic is deposited on it. The gas is dried by passing through the tube containing calcium chloride. This apparatus is used for the detection of arsenic by Marsh's test.

If hydrogen contains arseniuretted hydrogen, it also gives metallic arsenic when it burns, because in the reducing flame of hydrogen the oxygen attracted combines entirely with the hydrogen and not with the arsenic, so that if a cold object, such as a piece of china, be held in the hydrogen flame the arsenic will be deposited upon it as a metallic spot.

The most common compound of arsenic is the solid and volatile *arsenious anhydride*, As_2O_3, which corresponds with phosphorous and nitrous anhydrides. This very poisonous, colourless, and sweet-tasting substance is generally known under the name of arsenic, or *white arsenic*. The corresponding hydrate is as yet unknown; its solutions, when evaporated, yield crystals of arsenious anhydride. It is chiefly prepared for the dyer, and is also used as a vermin killer, and sometimes in medicine; it is a product from which all other compounds of arsenic can be prepared. It is obtained as a by-product in roasting cobalt and other ores containing arsenic. Arsenical pyrites are sometimes purposely roasted for the extraction of arsenious anhydride. When arsenical ores are burnt in the air, the sulphur and arsenic are converted into the oxides As_2O_3 and SO_2. The former is a solid at the ordinary temperature, and the latter gaseous, and therefore the arsenious anhydride is deposited as a sublimate in the cooler portion of the flues through which the vapours escape from the furnace. It collects in condensing chambers especially constructed in the flues. The deposit is collected, and after being distilled gives arsenious anhydride in the form of a vitreous non-crystalline mass. This is one of the varieties of arsenious anhydride, which is also known in two crystalline forms. When sublimed— *i.e.* when it rapidly passes from the state of vapour to the solid state—it appears in the regular system in the form of octahedra. It is obtained in the same form when it is crystallised from acid solutions. The specific gravity of the crystals is 3·7. The other crystalline form (in prisms) belongs to the rhombohedral system, and is also formed by sublimation when the crystals are deposited on a heated surface, or when it is crystallised from alkaline solutions.

Solutions of arsenious anhydride have a sweet metallic taste, and give *a feeble acid reaction*. Its solubility increases with the admixture of acids and alkalis. This shows the property of arsenious anhydride of forming salts with acids and alkalis. And in fact compounds of it with hydrochloric acid (Note 31), sulphuric anhydride (*see* further on), and with the alkali oxides are known. If silver nitrate be added to a solution of arsenious anhydride, it does not give any precipitate unless a certain amount of the arsenious anhydride is saturated with an alkali—for instance, ammonia. It then gives a precipitate of silver arsenite, Ag_3AsO_3. This is yellow, soluble in an excess of ammonia, and anhydrous; it distinctly shows that arsenious acid is tribasic, and that it differs in this respect from phosphorous acid, in which only two atoms of hydrogen can be replaced by metals. The feeble acid character of arsenious anhydride is confirmed by the formation of saline compounds with acids. In this respect the most remarkable example is the anhydrous compound with sulphuric acid, having the composition As_2O_3,SO_3. It is formed in the roasting of arsenical pyrites in those spaces where the arsenious anhydride condenses, a portion of the sulphurous anhydride being converted into sulphuric anhydride, SO_3, at the expense of the oxygen of the air. The compound in question forms colourless tabular crystals, which are decomposed by water with formation of sulphuric acid and arsenious anhydride.

Antimony (stibium), Sb = 120, is another analogue of phosphorus. In its external appearance and the properties of its compounds it resembles the metals still more closely than arsenic. In fact, antimony has the appearance, lustre, and many of the characteristic properties of the metals. Its oxide, Sb_2O_3, exhibits the earthy appearance of rust or of lime, and has distinctly basic properties, although it corresponds with nitrous and phosphorous anhydride, and is able, like them, to give saline compounds with bases. At the same time antimony presents, in the majority of its compounds, an entire analogy with phosphorus and arsenic. Its compounds belong to the type SbX_3 and SbX_5. It is found in nature chiefly in the form of sulphide, Sb_2S_3. This substance sometimes occurs in large masses in mineral veins and is known in mineralogy under the name of antimony glance or *stibnite*, and commercially as *antimony* (Chapter XX., Note 29). The most abundant deposits of antimony ore occur in Portugal (near Oporto on the Douro). Besides which antimony partially or totally replaces arsenic in some minerals; thus, for example, a compound of antimony sulphide and arsenic sulphide with silver sulphide is found in red silver ore. But in every case antimony is a rather rare metal found in few localities. In Russia it is known to occur in Daghestan in the Caucasus. It is extracted chiefly for the preparation of alloys with lead and tin, which are used for casting printing type. Some of its compounds are also used in medicine, the most important in this respect being antimony pentasulphide, Sb_2S_5 (*sulfur auratum antimonii*), and tartar emetic, which is a double salt derived from tartaric acid and has the composition $C_4H_4K(SbO)O_6$. Even the native antimony sulphide is used in large quantities as a purgative for horses and dogs. Metallic antimony is extracted from the glance, Sb_2S_2, by roasting, when the sulphur burns away and the antimony oxidises, forming the oxide Sb_2O_3, which is then heated with charcoal, and thus reduced to a *metallic state*. The reduction may be carried on in the laboratory on a small scale by fusing the sulphide with iron which takes up the sulphur.[40 bis]

Metallic antimony has a white colour and a brilliant lustre; it remains untarnished in the air, for the metal does not oxidise at the ordinary temperature. It crystallises in rhombohedra, and always shows a distinctly crystalline structure which gives it quite a different aspect from the majority of the metals yet known. It is most like tellurium in this respect. Antimony is brittle, so that it is very easily powdered; its specific gravity is 6·7, it melts at about 432°, but only volatilises at a bright red heat. When heated in the air—for instance, before the blow-pipe—it burns and gives white odourless fumes, consisting of the oxide. This oxide is termed antimonious oxide, although it might as well be termed antimonious anhydride. It is given the first name because in the majority of cases its compounds with acids are used, but it forms compounds with the alkalis just as easily.

Antimonious oxide, like arsenious anhydride, crystallises either in regular octahedra or in rhombic prisms; its specific gravity is 5·56; when heated it becomes yellow and then fuses, and when further heated in air it oxidises, forming an oxide of the composition Sb_2O_4. Antimonious oxide is insoluble in water and in nitric acid, but it easily dissolves in strong hydrochloric acid and in alkalis, as well as in tartaric acid or solutions of its acid salts. When dissolved in the latter it forms tartar emetic. It is precipitated from its solutions in alkalis and acids (by the action of acids on the former and alkalis on the latter). It occurs native but rarely. As a base it gives salts of the type SbOX (as if the basic salts = SbX_3, Sb_2O_3) and hardly ever forms salts, SbX_3. In the antimonyl salts, SbOX, the group SbO is univalent, like potassium or silver. The oxide itself is $(SbO)_2O$, the hydroxide, SbO(OH), &c.; tartar emetic is a salt in which one hydrogen of tartaric acid is replaced by potassium and the other by antimonyl, SbO. Antimonious oxide is very easily separated from its salts by any base, but it must be observed that this separation does not take place in the presence of tartaric acid, owing to the property of tartaric acid of forming a soluble double salt—*i.e.* tartar emetic.

If metallic antimony, or antimonious oxide, be oxidised by an excess of nitric acid and the resultant mass be carefully evaporated to dryness, *metantimonic acid*, $SbHO_3$, is formed. Its corresponding potassium salt, $2SbKO_3,5H_2O$, is prepared by fusing metallic antimony with one-fourth its weight of nitre and washing the resultant mass with cold water. This potassium salt is only slightly soluble in water (in 50 parts) and the sodium salt is still less so. An ortho-acid, SbH_3O_4, also appears to exist;[41 bis] it is obtained by the action of water on antimony pentachloride, but it is very unstable, like the pentachloride, $SbCl_5$, itself, which easily gives up Cl_2, leaving antimony trichloride, $SbCl_3$, and this is decomposed by water, forming an oxychloride—SbOCl, only slightly soluble in water. When antimonic acid is heated to an incipient red heat, it parts with water and forms the anhydride, Sb_2O_5, of a yellow colour and specific gravity 6·5.

The heaviest analogue of nitrogen and phosphorus is *bismuth*, Bi = 208. Here, as in the other groups, the basic, metallic, properties increase with the atomic weight. Bismuth does not give any hydrogen compound and the highest oxide, Bi_2O_5, is a very feeble acid oxide. Bismuthous oxide, Bi_2O_3, is a base, and bismuth itself a perfect metal. To explain the other properties of bismuth it must further be remarked that in the eleventh series it follows mercury, thallium and lead, whose atomic weights are near to that of bismuth, and that therefore it resembles them and more especially its nearest neighbour, lead. Although PbO and PbO_2, represent types different from Bi_2O_3 and Bi_2O_5, they resemble them in many respects, even in their external appearance, moreover the lower oxides both of Pb and Bi are basic and the higher acid, which easily evolve oxygen. But judging by the formula, Bi_2O_3 is a more feeble base than PbO. They both easily give basic salts.

Bismuth forms compounds of two types, BiX_3 and BiX_5, which entirely recall the two types we have already established for the compounds of lead. Just as in the case of lead, the type PbX_2, is basic, stable, easily formed, and passes with difficulty into the higher and lower types, which are unstable, so also in the case of bismuth the type of combination BiX_3 is the usual basic form. The higher type of combination, BiX_5, in fact behaves toward this stable type, BiX_3, in exactly the same manner as lead dioxide does to the monoxide; and bismuthic acid is obtained by the action of chlorine on bismuth oxide suspended in water, in exactly the same way as lead dioxide is obtained from lead oxide. It is an oxidising agent like lead dioxide, and even the acid character in bismuthic acid is only slightly more developed than in lead dioxide. Here, as in the case of lead (minium), intermediate compounds are easily formed in which the bismuth of the lower oxide plays the part of a base combined with the acid which is formed by the higher form of the oxidation of bismuth.

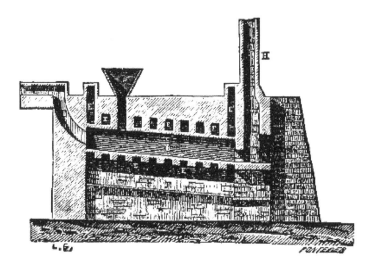

FIG. 85.—Furnace used for the extraction of bismuth from its ores.

In nature, bismuth occurs in only a few localities and in small quantities, most frequently in a native state, and more rarely as oxide and as a compound of bismuth sulphide with the sulphides of other metals, and sometimes in gold ores. It is extracted from its native ores by simple fusion in the furnace shown in fig. 85. This furnace contains an inclined iron retort, into the upper extremity of which the ore is charged, and the molten *metal* flows from the lower extremity. It is refined by re-melting, and the pure metal may be obtained by dissolving in nitric acid, decomposing the resultant salt with water, and reducing the precipitate by heating it with charcoal. Bismuth is a metal which crystallises very well from a molten state. Its specific gravity is 9·8; it melts at 269°, and if it be melted in a crucible, allowed to cool slowly, and the crust broken and the remaining molten liquid poured out, perfect rhombohedral crystals of bismuth are obtained on the sides of the crucible.[44 bis] It is brittle, has a grey-coloured fracture with a reddish lustre, is not hard, and is but very slightly ductile and malleable; it volatilises at a white heat and easily oxidises. It recalls antimony and lead in many of its properties. When oxidised in air, or when the nitrate is ignited, bismuth forms the *oxide*, Bi_2O_3, as a white powder which fuses when heated and resembles massicot. The addition of an excess of caustic potash to a solution of a bismuthous salt gives a white precipitate of the hydroxide, $BiO(OH)$, which loses its water and gives the anhydrous oxide when boiled with a solution of caustic potash. Both the hydroxide and oxide easily dissolve in acids and form bismuthous salts.

Bismuthous oxide, Bi_2O_3, is a feeble and unenergetic base. The normal hydroxide of the oxide Bi_2O_3 is $Bi(OH)_3$; it parts with water and forms a metahydroxide (bismuthyl hydroxide), $BiO(OH)$. Both of these hydroxides have their corresponding saline compounds of the composition BiX_3 and $BiOX$. And the form $BiOX$ is nothing else but the type of the basic salt, because $3ROX = RX + R_2O_3$. It is evident that in the type BiX_3 the bismuth replaces three atoms of hydrogen. And indeed with phosphoric acid solutions of the bismuthous salts give a precipitate of the composition $BiPO_4$. On the other hand, in the form of compounds $BiOX$ or $Bi(OH)_2X$, the univalent group (BiO) or (BiH_2O_2) is combined with X. Many bismuth salts are formed according to the type $BiOX$. For instance the carbonate, $(BiO)_2CO_3$, which corresponds with the other carbonates M_2CO_3. It is obtained as a white precipitate when a solution of sodium carbonate is added to a solution of a bismuth salt. The compound radicle BiO is not a special natural grouping, as it was formerly represented to be; it is simply a mode of expression for showing the relation between the compound in question and the compounds of other oxides.

Three *salts of nitric acid* are known containing bismuthous oxide. If metallic bismuth or its oxide be dissolved in nitric acid, it forms a colourless transparent solution containing a salt which separates in large transparent crystals containing $Bi(NO_3)_3,5H_2O$. When heated at 80° these crystals melt in their water of crystallisation, and in so doing lose a portion of their nitric acid together with water, forming a salt whose empirical formula is $Bi_2N_2H_2O_9$. If the preceding salt belongs to the type BiX_3, this one should belong to the form $BiOX$, because it = $BiO(NO_3)$ + $Bi(H_2O_2)(NO_3)$. This salt may be heated to 150° without change. When the first colourless crystalline salt dissolves in water *it is decomposed*. There is no decomposition if an excess of acid be added to the water—that is to say, the salt is able to exist in an acid solution without decomposing, without separation of the so-called basic salt—but by itself it cannot be kept in solution; water decomposes this salt, acting on it like an alkali. In other words the basic properties of bismuthic oxide are so feeble that even water acts by taking up a portion of the acid from it. Here we see one of the most striking facts, long since observed, confirming that action of water on salts about which we have spoken in Chapter X. and elsewhere. This action on water may be expressed thus:—$BiX_3 + 2H_2O = Bi(OH)_2X + 2XH$. A salt of the type $Bi(OH)_2X$ is obtained in the precipitate. But if the quantity of acid, HX, be increased, the salt BiX_3 is again formed and passes into solution. The quantity of the salt BiOX which passes into solution on the addition of a given quantity of acid depends indisputably on the amount (mass) of water (Muir). The solution, which is perfectly transparent with a small amount of water, becomes cloudy and deposits the salt of the type BiOX, when diluted. The white flaky precipitate of $Bi(OH)_2NO_3$ formed from the normal salt $Bi(NO_3)_3$ by mixing it with five parts of water, and in general with a small amount of water, is used in medicine under the name of magistery of bismuth.

Metallic bismuth is used in the preparation of fusible alloys. The addition of bismuth to many metals renders them very hard, and at the same time generally lowers their melting point to a considerable extent. Thus Wood's metal, which contains one part of cadmium, one part of tin, two parts of lead, and four parts of bismuth, fuses at about 60°, and in general many alloys composed of bismuth, tin, lead, and antimony melt below or about the boiling point of water.

Just as in group II., side by side with the elements zinc, cadmium, and mercury in the uneven series, we found calcium, strontium, and barium in the even series; and as in group IV., parallel to silicon, germanium, tin, and lead, we noticed thallium, zirconium, cerium, and thorium; so also in group V. we find, beside those elements of the uneven series just considered by us, a series of analogues in the even series, which, with a certain degree of similarity (mainly quantitative, or relative to the atomic weights), also present a series of particular (qualitative) independent points of distinction. In the even series are known *vanadium*, which stands between titanium and chromium, *niobium*, between zirconium and molybdenum, and *tantalum*, situated near tungsten (an element of group VI. like chromium and molybdenum). Just as bismuth is similar in many respects to its neighbour lead, so also do these neighbouring elements resemble each other, even in their external appearance, not to mention the quality of their compounds, naturally taking into account the differences of type corresponding with the different groups. The occurrence in group V. determines the type of the oxides, R_2O_3 and R_2O_5, and the development of an acid character in the higher oxides. The occurrence in the even series determines the absence of volatile compounds, RH_3, for these metals, and a more basic character of the oxides of a given composition than in the uneven series, &c. Vanadium, niobium, and tantalum belong to the category of rare metals, and are exceedingly difficult to obtain pure, more especially owing to their similarity to, and occurrence with, chromium, tungsten and other metals, and also in combination among themselves; therefore it is natural that they have been far from completely studied, although since 1860 chemists have devoted not a little time to their investigation. The researches carried out by Marignac, at Geneva, on niobium, and by Sir Henry Roscoe, at Manchester, on vanadium deserve special attention. The undoubted external resemblance of the compounds of chromium and vanadium, as well as the want of completeness in the knowledge of the compounds of vanadium, long caused its oxides to be considered analogous in atomic composition to those formed by chromium. The higher oxide of vanadium was therefore supposed to have the formula VO_3. But the fact of the matter is, that the chemical analogy of the elements does not hold in one direction only; vanadium is at one and the same time the analogue of chromium, and consequently of the elements like sulphur of group VI, and also the analogue of phosphorus, arsenic, and antimony; just as bismuth stands in respect to lead and antimony. Investigation has shown that the compounds of vanadium are always accompanied by those of phosphorus as well as of iron, and that it is even more difficult to separate it from the compounds of phosphorus than from those of iron and tungsten. We should have to extend our description considerably if we wished to give the complete history, even of vanadium alone, not to mention niobium and tantalum, all the more as questions would not unfrequently arise concerning the compounds of these elements which have not yet been fully elucidated. We shall therefore limit ourselves to pointing out the most important features in the history of these elements, the more so since the minerals themselves in which they occur are exceedingly rare and only accessible to a few investigators.

An important point in the history of the members of this group is the circumstance that they form volatile compounds with chlorine, similar to the compounds of the elements of the phosphorus group, namely, of the type RX_5. The vapour densities of the compounds of this kind were determined, and served as the most important basis for the explanation of the atomic composition of these molecules. In this we see the power of general and fundamental laws, like the law of Avogadro-Gerhardt. An oxychloride, VOCl$_3$, is known for vanadium, which is the perfect analogue of phosphorus oxychloride. It was formerly considered to be vanadium chloride, for just as in the case of uranium (Chapter XXI.), its lower oxide, VO, was considered to be the metal, because it is exceedingly difficultly reduced—even potassium does not remove all the oxygen, besides which it has a metallic appearance, and decomposes acids like a metal; in a word, it simulates a metal in every respect. *Vanadium oxychloride* is obtained by heating the trioxide, V$_2$O$_3$, mixed with charcoal, in a current of hydrogen; the lower oxide of vanadium is then formed, and this, when heated in a current of dry chlorine, gives the oxychloride VOCl$_3$ as a reddish liquid which does not act on sodium and may be purified by distillation over this metal. It fumes in the air, giving reddish vapours; it reacts on water, forming hydrochloric and vanadic acids; hence, on the one hand it is very similar to phosphorus oxychloride, and on the other hand to chromium oxychloride, CrO$_2$Cl$_2$ (Chapter XXI.). It is of a yellow colour, its specific gravity is 1·83, it boils at 120°, and its vapour density is 86 with respect to hydrogen; therefore the above formula expresses its molecular weight.

Vanadic anhydride, V$_2$O$_5$, is obtained either in small quantities from certain clays where it accompanies the oxides of iron (hence some sorts of iron contain vanadium) and phosphoric acid, or from the rare minerals: *volborthite*, CuHVO$_4$, or basic vanadate of copper; *vanadinite*, PbCl$_2$3Pb$_3$(VO$_4$)$_2$; lead vanadate, Pb$_3$(VO$_4$)$_2$, &c. The latter salts are carefully ignited for some time with one-third of their weight of nitre; the fused mass thus formed is powdered and boiled in water: the yellow solution obtained contains potassium vanadate. The solution is neutralised with acid, and barium chloride added; a meta-salt, Ba(VO$_3$)$_2$, is then precipitated as an almost insoluble white powder, which gives a solution of vanadic acid when boiled with sulphuric acid. (The precipitate is at first yellow, as long as it remains amorphous, but it afterwards becomes crystalline and white.) The solution thus obtained is neutralised with ammonia, which thus forms ammonium (meta) vanadate, NH$_4$VO$_3$, which, when evaporated, gives colourless crystals, insoluble in water containing sal-ammoniac; hence this salt is precipitated by adding solid sal-ammoniac to the solution. Ammonium vanadate, when ignited, leaves vanadic acid behind. In this it differs from the corresponding chromium salt, which is deoxidised into chromium oxide when ignited. In general, vanadic acid has but a small oxidising action. It is reduced with difficulty, like phosphoric or sulphuric acid, and in this differs from arsenic and chromic acids. Vanadic acid, like chromic acid, separates from its solution as the anhydride V$_2$O$_5$, and not in a hydrous state. Vanadic anhydride, V$_2$O$_5$, forms a reddish-brown mass, which easily fuses and re-solidifies into transparent crystals having a violet lustre (another point of resemblance to chromic acid); it dissolves in water, forming a yellow solution with a slightly acid reaction.

Niobium and tantalum occur as acids in rare minerals, and are mainly extracted from *tantalite* and *columbite*, which are found in Bavaria, Finland, North America, and in the Urals. These minerals are composed of the ferrous salts of niobic and tantalic acids; they contain about 15 per cent. of ferrous oxide in isomorphous mixture with manganous oxide, in combination with various proportions of tantalic and niobic anhydrides. These minerals are first fused with a considerable amount of potassium bisulphate, and the fused mass is boiled in water, which dissolves the ferrous and potassium salts and leaves an insoluble residue of impure niobic and tantalic acids. This raw product is then treated with ammonium sulphide, in order to extract the tin and tungsten, which pass into solution. The residue containing the acids (according to Marignac) is then treated with hydrofluoric acid, in which it entirely dissolves, and potassium fluoride is added to the resultant hot solution; on cooling, a sparingly soluble double fluoride of potassium and tantalum separates out in fine crystals, while the much more soluble niobium salt remains in solution. The difference in the solubility of these double salts in water acidified with hydrofluoric acid (in pure water the solution becomes cloudy after a certain time) is so great that the tantalum compound requires 150 parts of water for its solution, and the niobium compound only 13 parts. The Greenland columbite (specific gravity 5·36) only contains niobic acid, and that from Bodenmais, Bavaria (specific gravity 6·06) almost equal quantities of tantalic and niobic acids. Having isolated tantalic and niobic salts, Marignac found that the relation between the potassium and fluorine in them is very variable—that is, that there exist various double salts of fluoride of potassium, and of the fluorides of the metals of this group, but that with an excess of hydrofluoric acid both the tantalum and niobium compounds contain seven atoms of fluorine to two of potassium, whence it must be concluded that the simplest formula for these double salts will be $K_2RF_7 = RF_5,2KF$; that is, that the type of the higher compounds of niobium and tantalum is RX_5, and hence is similar to phosphoric acid. A chloride, $TaCl_5$, may be obtained from pure tantalic acid by heating it with charcoal in a current of chlorine. This is a yellow crystalline substance, which melts at 211°, and boils at 241°; its vapour density with respect to hydrogen is 180, as would follow from the formula $TaCl_5$. It is completely decomposed by water into tantalic and hydrochloric acids. *Niobium pentachloride* may be prepared in the same manner; it fuses at 194°, and boils at 240°. When treated with water this substance gives a solution containing niobic acid, which only separates out on boiling the solution. Delafontaine and Deville found its vapour density to be 9·3 (air = 1), as is shown by its formula $NbCl_5$.

Footnotes:

Dry bones contain about one-third of gelatinous matter and about two-thirds of ash, chiefly calcium phosphate. The salts of phosphoric acid are also found in the mass of the earth as separate minerals; for example, the *apatites* contain this salt in a crystalline form, combined with calcium chloride or fluoride, $CaR_2,3Ca_3(PO_4)_2$, where $R = F$ or Cl, sometimes in a state of isomorphous mixture. This mineral often crystallises in fine hexagonal prisms; sp. gr. 3·17 to 3·22. Vivianite is a hydrated ferrous phosphate, $Fe_3(PO_4)_2,8H_2O$. Phosphates of copper are frequently found in copper mines; for example, *tagilite*, $Cu_3(PO_4)_2,Cu(OH)_2,2H_2O$. Lead and aluminium form similar salts. They are nearly all insoluble in water. The turquoise, for instance, is hydrated phosphate of alumina, $(Al_2O_3)_2,P_2O_55H_2O$, coloured with a salt of copper. Sea and other waters always contain a small amount of phosphates. The ash of sea-plants, as well as of land-plants, always contains phosphates. Deposits of calcium phosphate are often met with; they are termed *phosphorites* and *osteolites*, and are composed of the fossil remains of the bones of animals; they are used for manure. Of the same nature are the so-called guano deposits from Baker's Island, and entire strata in Spain, France, and in the Governments of Orloff and Kursk in Russia. It is evident that if a soil destined for cultivation contain very little phosphoric acid, the fertilisation by means of these minerals will be beneficial, but, naturally, only if the other elements necessary to plants be present in the soil.

FIG. 83.—Preparation of phosphorus. The mixture is calcined in the retort c. The vapours of phosphorus pass through *a* into water without coming into contact with air. The phosphorus condenses in the water, and the gases accompanying it escape through *i*.

[1 bis] By subjecting the pyrophosphate to the action of sulphuric or hydrochloric acid it is possible to obtain a fresh quantity of the acid salt from the residue, and in this manner to extract all the phosphorus. It is usual to take burnt bones, but mineral phosphorites, osteolites, and apatites may also be employed as materials for the extraction of phosphorus. Its extraction for the manufacture of matches is everywhere extending, and in Russia, in the Urals, in the Government of Perm, it has attained such proportions that the district is able to supply other countries with phosphorus. A great many methods have been proposed for facilitating the extraction of phosphorus, but none of them differ essentially from the usual one, because the problem is dependent on the liberation of phosphoric acid by the action of acids, and on its ultimate reduction by charcoal. Thus the calcium phosphate may be mixed directly with charcoal and sand, and phosphorus will be liberated on heating the mixture, because the silica displaces the phosphoric anhydride, which gives carbonic oxide and phosphorus with the charcoal. It has also been proposed to pass hydrochloric acid over an incandescent mixture of calcium phosphate and charcoal; the acid then acts just as the silica does, liberating phosphoric anhydride, which is reduced by the charcoal. It is necessary to prevent the access of air in the condensation of the vapours of phosphorus, because they take fire very easily; hence they are condensed under water by causing the gaseous products to pass through a vessel full of water. For this purpose the condenser shown in fig. 83 is usually employed.

Vernon (1891) observed that ordinary (yellow) phosphorus is dimorphous. If it be melted and by careful cooling be brought in a liquid form to as low a temperature as possible, it gives a variety which melts at 45°·3 (the ordinary variety fuses at 44°·3), sp. gr. 1·827 (that of the ordinary variety is 1·818) at 13°, crystallises in rhombic prisms (instead of in forms belonging to the cubical system). This is similar to the relation between octahedral and prismatic sulphur (Chapter XX.).

[2 bis] According to Herr Irinyi (an Hungarian student), the first phosphorus matches were made in Austria at Roemer's works in 1835.

The absorption of the oxygen of the atmosphere at a constant ordinary temperature by a large surface of phosphorus proceeds so uniformly, regularly, and rapidly, that it may serve, as Ikeda (Tokio, 1893) has shown, for demonstrating the law of the velocity (rate) of reaction, which is considered in theoretical chemistry, and shows that the rate of reaction is proportional to the active mass of a substance—*i.e.* $dx/dt = k(A - x)$ where t is the time, A the initial mass of the reacting substance—in this case oxygen—x the amount of it which has entered into reaction, and k the coefficient of proportionality. Ikeda took a test-tube (diameter about 10 mm.), and covered its outer surface with a coating of phosphorus (by melting it in a test-tube of large diameter, inserting the smaller test-tube, and, when the phosphorus had solidified, breaking away the outer test-tube), and introduced it into a definite volume of air, contained in a Woulfe's bottle (immersed in a water bath to maintain a constant temperature), one of whose orifices was connected with a mercury manometer showing the fall of pressure, x. Knowing that the initial pressure of the oxygen (in air nearly $750 \times \cdot 0209$) was about 155 mm. = A, the coefficient of the rate of reaction k is given, by the law of the variation of the rate of reaction with the mass of the reacting substance, by the equation: $k = \frac{1}{t} \log \frac{A}{A-x}$, where t is the time, counting from the commencement, of the experiment in minutes. When the surface of the phosphorus was about 11 sq. cm., the following results were actually obtained.

$t=$	10	20	30	40	50	60 minutes
$x=$	10·5	21·5	31·1	40·7	49·1	57·3 mm
10,000 $k=$	32	32	32	33	33	33

The constancy of k is well shown in this case. The determination takes a comparatively short time, so that it may serve as a lecture experiment, and demonstrates one of the most important laws of chemical mechanics.

[3 bis] Not only do oxidising agents like nitric, chromic, and similar acids act upon phosphorus, but even the alkalis are attacked—that is, phosphorus acts as a reducing agent. In fact it reduces many substances, for instance, copper from its salts. When phosphorus is heated with sodium carbonate, the latter is partially reduced to carbon. If phosphorus be placed under water slightly warmed, and a stream of oxygen be passed over it, it will burn under the water.

The thermochemical determinations for phosphorus and its compounds date from the last century, when Lavoisier and Laplace burnt phosphorus in oxygen in an ice calorimeter. Andrews, Despretz, Favre, and others have studied the same subject. The most accurate and complete data are due to Thomsen. To determine the heat of combustion of yellow phosphorus, Thomsen oxidised it in a calorimeter with iodic acid in the presence of water, and a mixture of phosphorous and phosphoric acids was thus formed (was not any hypophosphoric acid formed?—Salzer), and the iodic acid converted into hydriodic acid. It was first necessary to introduce two corrections into the calorimetric result obtained, one for the oxidation of the phosphorous into phosphoric acid, knowing their relative amounts by analysis, and the other for the deoxidation of the iodic acid. The result then obtained expresses the conversion of phosphorous into hydrated phosphoric acid. This must be corrected for the heat of solution of the hydrate in water, and for the heat of combination of the anhydride with water, before we can obtain the heat evolved in the reaction of P_2 with O_5 in the proportion for the formation of P_2O_5. It is natural that with so complex a method there is a possibility of many small errors, and the resultant figures will only present a certain degree of accuracy after repeated corrections by various methods. Of such a kind are the following figures determined by Thomsen, which we express in thousands of calories:—$P_2 + O_5 = 370$; $P_2 + O_3 + 3H_2O = 400$; $P_2 + O_5 +$ a mass of water $= 405$. Hence we see that $P_2O_5 + 3H_2O = 30$; $2PH_3O_4 +$ an excess of water $= 5$. Experiment further showed that crystallised PH_3O_4, in dissolving in water, evolves 2·7 thousand calories, and that fused (39°) PH_3O_4 evolves 5·2 thousand calories, whence the heat of fusion of H_3PO_4 $= 2·5$ thousand calories. For phosphorous acid, H_3PO_3, Thomsen obtained $P_2 + O_3 + 3H_2O = 250$, and the solution of crystallised H_3PO_3 in water $= -0·13$, and of fused $H_3PO_3 = +2·9$. For hypophosphorous acid, H_3PO_2, the heats of solution are nearly the same (-0·17 and +2·1), and the heat of formation $P_2 + O + 3H_2O = 75$; hence its conversion into $2H_3PO_3$ evolves 175 thousand calories, and the conversion of $2H_3PO_3$ into $2H_3PO_4 = 150$ thousand calories. For the sake of comparison we will take the combination of chlorine with phosphorus, also according to Thomsen, per 2 atoms of phosphorus, $P_2 + 3Cl_2 = 151$, $P_2 + 5Cl_2 = 210$ thousand calories. In their reaction on an excess of water (with the formation of a solution), $2PCl_3 = 130$, $2PCl_5 = 247$, and $2POCl_3 = 142$ thousand calories.

Besides which we will cite the following data given by various observers: heat of fusion for P (that is, for 31 parts of phosphorus by weight) -0·15 thousand calories; the conversion of yellow into red phosphorus for P, from +19 to +27 thousand calories; $P + H_3 = 4·3$, $HI + PH_3 = 24$, $PH_3 + HBr = 22$ thousand calories.

At the ordinary temperature (20° C.) phosphorus is not oxidised by pure oxygen; oxidation only takes place with a slight rise of temperature, or the dilution of the oxygen with other gases (especially nitrogen or hydrogen), or a decrease of pressure.

[4 bis] Ordinary phosphorus takes fire at a temperature (60°) at which no other known substance will burn. Its application to the manufacture of matches is based on this property. In order to illustrate the easy inflammability of common (yellow) phosphorus, its solution in carbon bisulphide may be poured over paper; this solvent quickly evaporates, and the free phosphorus spread over a large surface takes fire spontaneously, notwithstanding the cooling effect produced by the evaporation of the bisulphide. The majority of *phosphorus matches* are composed of common phosphorus mixed with some oxidising substance which easily gives up

oxygen, such as lead dioxide, potassium chlorate, nitre, &c. For this purpose common phosphorus is carefully triturated under warm water containing a little gum; lead dioxide and potassium nitrate are then added to the resultant emulsion, and the match ends, previously coated with sulphur or paraffin, are dipped into this preparation. After this the matches are dipped into a solution of gum and shellac, in order to preserve the phosphorus from the action of the air. When such a match containing particles of yellow phosphorus is rubbed over a rough surface, it becomes (especially at the point of rupture of the brittle gummy coating) slightly heated, and this is sufficient to cause the phosphorus to take fire and burn at the expense of the oxygen of the other ingredients.

In the so-called 'safety' or Swedish matches (which are not poisonous, and do not take fire from accidental friction) a mixture of red phosphorus and glass forms the surface on which the matches are struck, and the matches themselves do not contain any phosphorus at all, but a mixture of antimonious sulphide, Sb_2S_3 (or similar combustible substances) and potassium chlorate (or other oxidising agents). The combustion, when once started by contact with the red phosphorus, proceeds by itself at the expense of the inflammatory and combustible elements contained in the tip of the match. The mixture applied on the match itself must not be liable to take fire from a blow or friction. The mixture forming the heads of the 'safety' matches has the following approximate composition: 55–60 parts of chlorate of potassium, 5–10 parts of peroxide of manganese (or of $K_2Cr_2O_7$), about 1 part of sulphur or charcoal, about 1 part of pentasulphide of antimony, Sb_2S_5, and 30–40 parts of rouge and powdered glass. This mixture is stirred up in gum or glue, and the matches are dipped into it. The paper on which the matches are struck is coated with a mixture of red phosphorus and trisulphide of antimony, Sb_2S_3, stirred up in dextrine.

[5 bis] Phosphorus only acts on iron at a red heat. The boiler is provided with a safety valve and gas-conducting tube, which is immersed in mercury or other liquid to prevent the admission of air into the boiler.

The specific heat of the yellow variety is 0·189—that is, greater than that of the red variety, which is 0·170. The sp. gr. of the yellow is 1·84, and of the red prepared at 260° 2·15, and of that prepared at 580° and above (*i.e.* 'metallic' phosphorus, *see* below) = 2·34. At 230° the pressure of the vapour of ordinary phosphorus = 514 millimetres of mercury, and of the red = 0—that is to say, the red phosphorus does not form any vapour at this temperature; at 447° the vapour tension of ordinary phosphorus is at first = 5500 mm., but it gradually diminishes, whilst that of red phosphorus is equal to 1636 mm.

Hittorf, by heating the lower portion of a closed tube containing red phosphorus to 530° and the upper portion to 447°, obtained crystals of the so-called 'metallic' phosphorus at the upper extremity. As the vapour tensions (according to Hittorf, at 530° the vapour tension of yellow phosphorus = 8040 mm., of red = 6139 mm., and of metallic = 4130 mm.) and reactions are different, *metallic phosphorus* may be regarded as a distinct variety. It is still less energetic in its chemical reaction than red phosphorus, and it is denser than the two preceding varieties: sp. gr. = 2·34. It does not oxidise in the air; is crystalline, and has a metallic lustre. It is obtained when ordinary phosphorus is heated with lead for several hours at 400° in a closed vessel, from which the air has been exhausted. The resultant mass is then treated with dilute nitric acid, which first dissolves the lead (phosphorus is electro-negative to lead, and does not, therefore, act on the nitric acid at first) and leaves brilliant rhombohedral crystals of phosphorus of a dark violet colour with a slight metallic lustre, which conduct an electric current incomparably better than the yellow variety; this also is characteristic of the metallic state of phosphorus.

The researches of Lemoine partially explain the passage of yellow (ordinary) phosphorus into its other varieties. He heated a closed glass globe containing either ordinary or red phosphorus, in the vapour of sulphur (440°), and then determined the amount of the red and yellow varieties after various periods of time, by treating the mixture with carbon bisulphide. It appeared that after the lapse of a certain time a mixture of definite and equal composition is obtained from both—that is, between the red and yellow varieties a state of equilibrium sets in like that of dissociation, or that observed in double decompositions. But at the same time, the progress of the transformation appeared to be dependent on the relative quantity of phosphorus taken per volume of the globe (*i.e.* upon the pressure). Neglecting the latter, we will cite as an example the amounts of the red phosphorus transformed into the ordinary, and of the ordinary not converted into red, per 30 grams of red or yellow taken per litre capacity of the globe, heated to 440°. When red phosphorus was taken, 4·75 grams of yellow phosphorus were formed after two hours, four grams after eight hours, three grams after twenty-four hours, and the last limit remained constant on further heating. When thirty grams of yellow phosphorus were taken, five grams remained unaltered after two hours, four grams after eight hours, and after twenty-four hours and more three grams as before. Troost and Hautefeuille showed that liquid phosphorus in general changes more easily into the red than does phosphorus vapour, which, however, is able, although slowly, to deposit red phosphorus.

The question presents itself as to whether phosphorus in a state of vapour is the ordinary or some other variety? Hittorf (1865) collected many data for the solution of this problem, which leave no doubt that (as experimental figures show) the density of the vapour of phosphorus is always the same, although the vapour tension of the different varieties and their mixtures is very variable. This shows that the different varieties of phosphorus only occur in a liquid and solid state, as indeed is implied in the idea of polymerisation. Strictly speaking, the vapour of phosphorus is a particular state of this substance, and the molecular formula P_4 refers only to it, and not to any other definite state of phosphorus. But Raoult's solution method showed that in a benzene solution the fall of the freezing point indicates for ordinary phosphorus a molecule P_4, judging by the determinations of Paterno and Nasini (1888), Hirtz (1890), and Beckmann (1891), who obtained for sulphur by the same method a molecular weight $= S_6$, in conformity with the vapour density. Further research in this direction will perhaps show the possibility of finding the molecular weight of red phosphorus, if a means be discovered for dissolving it without converting it into the yellow variety.

I think it will not be out of place here to draw the reader's attention to the fact that red phosphorus, which we must recognise as polymeric with the yellow, stands nearer to nitrogen, whose molecule is N_2, in its small inclination towards chemical reactions, although judging by its small vapour tension it must be more complex than ordinary (yellow and white) phosphorus.

[6 bis] Retgers (see further on) showed this in 1894, and observed that As when heated also combines with hydrogen.

[6 tri] The capacity of mercury (Chapter XVI., Note 25 bis) to give unstable compounds with nitrogen gives rise to the supposition that similar compounds exist with phosphorus also. Such a compound was obtained by Granger (1892) by heating mercury with iodide of phosphorus in a closed tube at 275°-300°. After removing the iodide of mercury formed, there remain fine rhombic crystals having a metallic lustre, and composition Hg_3P_2. This compound is stable, does not alter at the ordinary temperature and only decomposes at a red heat; when heated in air it burns with a flame. Nitric and hydrochloric acids do not act upon it, but it is easily decomposed by aqua regia. A phosphide of copper, Cu_2P_2, was obtained by Granger (1893) by heating a mixture of water, finely divided copper and red phosphorus in a sealed tube to 130°. The excess of copper was afterwards washed away by a solution of NH_3 in the presence of air.

The metallic compounds of phosphorus possess a great chemical interest, because they show a transition from metallic alloys (for instance, of Sb, As) to the sulphides, halogen salts, and oxides, and on the other hand to the nitrides. Although there are already many fragmentary data on the subject, the interesting province of the metallic phosphides cannot yet be regarded as in any way generalised. The varied applications (phosphor-iron, phosphor-bronze, &c.), which the phosphides have recently acquired should give a strong incentive to the complete and detailed study of this subject, which would, in my opinion, help to the explanation of chemical relations beginning with alloys (solutions) and ending with salts and the compounds of hydrogen (hydrides), because the phosphor-metals, as is proved by direct experiment, stand in the same relation to phosphuretted hydrogen as the sulphides do towards sulphuretted hydrogen, or as the metallic chlorides to hydrochloric acid.

[7 bis] Many other compounds of phosphorus are also capable of forming phosphuretted hydrogen. Thus BP also gives PH_3 (*see* Chapter XVII., Note 12). According to Lüpke (1890) phosphuretted hydrogen is formed by phosphide of tin. The latter is prepared by treating molten tin covered with a layer of carbonate of ammonium, with red phosphorus; 200–300 c.c. of water are then poured into a flask, 3–5 grams of this phosphide of tin dropped in, and after driving out the air by a stream of carbonic acid, hydrochloric acid (sp. gr. 1·104) is poured in. The disengagement of phosphuretted hydrogen takes place on heating the flask in a water bath. The following is another easy method for preparing PH_3. A mixture of 1 part of zinc dust (fume) and 2 parts of red phosphorus are heated in an atmosphere of hydrogen (the mixture burns in air). Combination takes place accompanied by a flash, and a grey mass of Zn_3P_2 is formed which gives PH_3 when treated with dilute H_2SO_4.

The spontaneous inflammability of the hydride PH_2 in air is very remarkable, and it is particularly interesting that its analogues in composition, $P(C_2H_5)_2$ (the formula must be doubled) and $Zn(C_2H_5)_2$, also take fire spontaneously in air.

[8 bis] The analogy between PH_3 and NH_3 is particularly clear in the hydrocarbon derivatives. Just as NH_2R, NHR_2, and NR_3, where R is CH_3, and other hydrocarbon radicles, correspond to NH_3, so there are actually similar compounds corresponding to PH_3. These compounds form a branch of organic chemistry.

The periodic law and direct experiment (the molecular weight) show that PH_3 is the normal compound of P and H and that it is more simple than PH_2 or P_2H_4, just as methane, CH_4, is more simple than ethane, C_2H_6, whose empirical composition is CH_3. The formation of liquid phosphuretted hydrogen may be understood from the law of substitution. The univalent radicle of PH_3 is PH_2, and if it is combined with H in PH_3 it replaces H in liquid phosphuretted hydrogen, which thus gives P_2H_4. This substance corresponds with free amidogen (hydrazine), N_2H_4 (Chapter VI.)

Probably P_2H_4 is able to combine with HI, and perhaps also with 2HI, or other molecules—that is, to give a substance corresponding to phosphonium iodide.

Phosphonium iodide, PH_4I, may be prepared, according to Baeyer, in large quantities in the following manner:—100 parts of phosphorus are dissolved in dry carbon bisulphide in a tubulated retort: when the mixture has cooled, 175 parts of iodide are added little by little, and the carbon bisulphide is then distilled off, this being done towards the end of the operation in a current of dry carbonic anhydride at a moderate temperature. The neck of the retort is then connected with a wide glass tube, and the tubulure with a funnel furnished with a stopcock, and containing 50 parts of water. This water is added drop by drop to the phosphorous iodide, and a violent reaction takes place, with the evolution of hydriodic acid and phosphonium iodide. The latter collects as crystals in the glass tube and the retort itself. It is purified by further distillations; more than 100 parts may be obtained. Baeyer expresses the reaction by the equation $P_2I + 2H_2O = PH_4I + PO_2$; and the compound PO_2 may be represented as phosphorous phosphoric anhydride: $P_2O_5 + P_2O_3 = 4PO_2$. As a better proportion we may take 400 grams of phosphorus, 680 grams of iodine, and 240 grams of water, and express the formation thus: $13P + 9I + 21H_2O = 3H_4P_2O_7 + 7PH_4I + 2HI$ (Chapter XI., Note [77]).

Phosphonium iodide and even phosphine act as reducing agents in solutions of many metallic salts. Cavazzi showed that with a solution of sulphurous anhydride phosphine gives sulphur and phosphoric acid.

The air must first be expelled from the flask by hydrogen, or some other gas which will not support combustion, as otherwise an explosion might take place owing to the spontaneous inflammability of the phosphuretted hydrogen.

The combustion of phosphuretted hydrogen in oxygen also takes place under water when the bubbles of both gases meet, and it is very brilliant. The phosphuretted hydrogen obtained by the action of phosphorus on caustic potash always contains free hydrogen, and often even the greater part of the gas evolved consists of hydrogen.

Pure phosphuretted hydrogen (not containing hydrogen or liquid or solid phosphides) is obtained by the action of a solution of potash on phosphonium iodide: $PH_4I + KHO = PH_3 + KI + H_2O$ (in just the same way as ammonia is liberated from ammonium chloride). The reaction proceeds easily, and the purity of the gas is seen from the fact that it is entirely absorbed by bleaching powder and is not spontaneously inflammable. Its mixture with oxygen explodes when the pressure is diminished (Chapter XVIII., Note [8]). The vapours of bromine, nitric acid,

&c., cause it to again acquire the property of inflaming in the air; that is, they partially decompose it, forming the liquid hydride, P_2H_4. Oppenheim showed that when red phosphorus is heated at 200° with hydrochloric acid in a closed tube it forms the compound $PCl_3(H_3PO_3)$, together with phosphine.

[10 bis] If there be a deficiency of oxygen, *phosphorous anhydride* P_2O_3 is formed. It was obtained by Thorpe and Tutton (1890) and is easily volatilised, melts at 22°·5, boils without change (in an atmosphere of N_2 or CO_2) at 173°, and is therefore easily separated from P_2O_3, which volatilises with difficulty. The vapour density shows that the molecular weight is double, *i.e.* P_4O_6 (like As_2O_3). Although colourless, phosphorous anhydride (its density in a state of fusion at 24° = 1·936) turns yellow and reddens in sun-light (possibly red phosphorus separates out ?), and decomposes at 400° forming hypophosphorous anhydride P_2O_4 (Note 11) and phosphorus. It passes into P_2O_5 in air and oxygen, and when slightly heated in oxygen becomes luminous, and ultimately takes fire. Cold water slowly transforms P_2O_3 into phosphoric acid, but hot water gives an explosion and leads to the formation of PH_3, ($P_4O_6 + 6H_2O = PH_3 + 3PH_3O_4$). Alkalis act in the same manner. It takes fire in chlorine and forms $POCl_3$ and PO_2Cl, and combines with sulphur at 160°, forming $P_2S_2O_3$ (the molecular formula is double this) a substance which volatilises in vacuo and is decomposed by water into H_2S and phosphoric acid, *i.e.* it may be regarded as P_2O_5, in which O_2 has been replaced by two atoms of sulphur. Judging from the above, the mixture of P_2O_3 and P_2O_5 formed in the combustion of phosphorus in air is transformed into P_2O_5 in an excess of oxygen.

Salzer proved the existence of hypophosphoric acid (it is also called subphosphoric acid), in which many chemists did not believe. Drawe (1888) and Rammelsberg (1892) investigated its salts. It may be obtained in a free state by the following method. The solution of acid produced by the slow oxidation of moist phosphorus is mixed with a solution (25 p.c.) of sodium acetate. A salt, $Na_2H_2P_2O_6,6H_2O$, crystallises out on cooling; it is soluble in 45 parts of water, and gives a precipitate of $Pb_2P_2O_6$ with lead salts ($Ag_4P_2O_6$ with salts of silver). The lead salt is decomposed by a current of hydrogen sulphide, when lead sulphide is precipitated, while the solution, evaporated under the receiver of an air-pump, gives crystals of $H_4P_2O_6,2H_2O$, which easily lose water and give $H_4P_2O_6$. The salts in which the H_4 is replaced by Ni_2, or $NiNa_2$, or $CdNa_2$, &c., are insoluble in water.

In order to see the relation between phosphoric acid and hypophosphoric acid which does not contain the elements of phosphorous acid (because it does not reduce either gold or mercury from their solutions), but which nevertheless is capable of being oxidised (for example, by potassium permanganate) into phosphoric acid, it is simplest to apply the law of substitution. This clearly indicates the relation between oxalic acid, $(COOH)_2$, and carbonic acid, $OH(COOH)$. The relation between the above acids is exactly the same if we express phosphoric acid as $OH(POO_2H_2)$, because in this case $P_2H_4O_6$, or $(POO_2H_2)_3$, will correspond with it just as oxalic does with carbonic acid. A similar relationship exists between hyposulphuric or dithionic acid, $(SO_2OH)_2$, and sulphuric acid, $OH(SO_2OH)$, as we shall find in the following chapter. Dithionic acid corresponds with the anhydride S_2O_5, intermediate between SO_2 and SO_3; oxalic acid with C_2O_3, intermediate between CO and CO_2; hypophosphoric acid corresponds with the anhydride P_2O_4, intermediate between P_2O_3 and P_2O_5, and the analogue of N_2O_4.

Besides the hydrates enumerated, a compound, PH_3O, should correspond with PH_3. This hydrate, which is analogous to hydroxylamine, is not known in a free state, but it is known as triethylphosphine oxide, $P(C_2H_5)_3O$, which is obtained by the oxidation of triethylphosphine, $P(C_2H_5)_3$. It must be observed that there may also be lower oxides of phosphorus corresponding with PH_3, like N_2O and NO, and there are even indications of the formation of such compounds, but the data concerning them cannot be considered as firmly established.

Phosphoric acid, being a soluble and almost non-volatile substance, cannot be prepared like hydrochloric and nitric acids by the action of sulphuric acid on the alkali phosphates, although it is partially liberated in the process. For this purpose the salts of barium or lead may be taken, because they give insoluble salts, thus $Ba_3(PO_4)_2 + 3H_2SO_4 = 3BaSO_4 + 2H_3PO_4$. Bone ash contains, besides calcium phosphate, sodium and magnesium phosphates, and fluorides and other salts, so that it cannot give directly a pure phosphoric acid.

If this is not done the orthophosphoric acid, PH_3O_4, loses a portion of its water, and then, as with an excess of water, it does not crystallise.

[14 bis] The difference between the reactions of ortho-, meta- and pyrophosphoric acids, established by Graham (see p. 163), is of such importance for the theory of hydrates and for explaining the nature of solutions, that in my opinion its influence upon chemical thought has been far from exhausted. At the present time many such instances are known both in organic (for instance, the difference between the reactions of the solutions of certain anhydrides and hydrates of acids), and inorganic chemistry (for example, the difference between the rose and purple cobalt compounds, Chapter XXII. &c.) They essentially recall the long known and generalised difference between C_2H_4 (ethylene), C_2H_6O (ethyl alcohol = ethylene + water), and $C_4H_{10}O$ (ethyl ether = 2 ethylene + water = 2

alcohol - water); but to the present day the numerous analogous phenomena existing among inorganic substances are only considered as a simple difference in degrees of affinity, distinguishing the water of constitution (hydration), crystallisation, and solution without penetrating into the difference of the structure or distribution of the elements, which exists here and gives rise to a distinct isomerism of solutions. In my opinion the progress of chemistry, especially with regard to solutions, should make rapid strides when the cause of the isomerism of solutions, for instance, of ortho- and pyrophosphoric acids, has become as clear to us as the cause of many well-studied instances of the isomerism, polymerism, and metamerism of organic compounds. Here it forms one of those many important problems which remain for the chemistry of the future in a state of only indistinct presentiments and in the form of facts empirically known but insufficiently comprehended.

Silver orthophosphate, Ag_3PO_4, is yellow, sp. gr. 7·32, and insoluble in water. When heated it fuses like silver chloride, and if kept fused for some length of time it gives a white pyrophosphate (the decomposition which causes this is not known). It is soluble in aqueous solutions of phosphoric, nitric, and even acetic acids, of ammonia, and many of its salts. If silver nitrate acts on a dimetallic orthophosphate—for instance, Na_2HPO_4—it still gives Ag_3PO_4, nitric acid being disengaged: $Na_2HPO_4 + 3AgNO_3 = Ag_3PO_4 + 2NaNO_3 + HNO_3$. When alcohol is added to silver orthophosphate, Ag_3PO_4, dissolved in syrupy phosphoric acid, it precipitates a white salt (the alcohol takes up the free phosphoric acid) having the composition Ag_2HPO_4, which is immediately decomposed by water into the normal salt and phosphoric acid.

The researches of Thomsen showed that in very dilute aqueous solutions the majority of monobasic acids—nitric, acetic, hydrochloric, &c. (but hydrofluoric acid more and hydrocyanic less)—HX evolve the following amounts of heat (in thousands of calories) with caustic soda: $NaHO + 2HX = 14$; $NaHO + HX = 14$; $2NaHO + HX = 14$; that is, if n be a whole number $n NaHO + HX = 14$ and $NaHO + n HX = 14$. Hence reaction here only takes place between one molecule of NaHO and one molecule of acid, and the remaining quantity of acid or alkali does not enter into the reaction. In the case of bibasic acids, H_2R'' (sulphuric, dithionic, oxalic, sulphuretted hydrogen, &c.), $NaHO + 2H_2R'' = 14$; $NaHO + H_2R'' = 14$; $2NaHO + H_2R'' = 28$; $n NaHO + H_2R'' = 28$; that is, with an excess of acid ($NaHO + 2H'_2R''$) 14 thousand units of heat are developed, and with an excess of alkali 28. When phosphoric acid is taken (but not all tribasic acids—for instance, not citric) the general character of the phenomenon is similar to the preceding, namely, $NaHO + 2H_3PO_4 = 14·7$; $NaHO + H_3PO_4 = 14·8$; $2NaHO + H_3PO_4 = 27·1$; $3NaHO + H_3PO_4 =$

34·0; 6NaHO + H_3PO_4 = 35·3; or, in general terms, NaHO + nH_3PO_4 = 14 (approximately) and nNaHO + H_3PO_4 = 35 and not 42, which shows a peculiarity of phosphoric acid. In the case of energetic acids, when one equivalent (23 grams) of sodium (in the form of hydroxide) replaces one equivalent (1 gram) of hydrogen (with the formation of water and in dilute solutions), 14,000 heat units are evolved; and this is true for phosphoric acid when in H_3PO_4, Na or Na_2 replaces H or H_2, but when Na_3 replaces H_3 less heat is developed. This will be seen from the following scheme based on the preceding figures: H_3PO_4 + NaHO = 14·8; NaH_2PO_4 + NaHO = 12·3; Na_2HPO_4 + NaHO = 5·9; with Na_3PO_4 + NaHO, a very small amount of heat is evolved, as may be judged from the fact that Na_3PO_4 + 3NaHO = 1·3, but still heat is evolved. It must be supposed that in acting on phosphoric acid in the presence of a large quantity of water, a certain portion of the sodium hydroxide remains as alkali uncombined with the acid. Thus, on increasing the mass of the alkali, heat is still evolved, and a fresh interchange between Na and H takes place. Hence water shows a decomposing action on the alkali phosphates. The same decomposing action of water is seen, but to a less extent, with Na_2HPO_4, as may be judged both from the reactions of this salt and from the amount of heat developed by NaH_2PO_4 with NaHO. Such an explanation is in accordance with many facts concerning the decomposition of salts by water already known to us. Recent researches made by Berthelot and Louguinine have confirmed the above deductions made by me in the first edition (1871) of this work. At the present time views of this nature are somewhat generally accepted, although they are not sufficiently strictly applied in other cases. As regards PH_3O_4 it may be said that: on the substitution of the first hydrogen this acid acts as a powerful acid (like HCl, HNO_3, H_2SO_4); on the substitution of the second hydrogen as a weaker acid (like an organic acid); and on the substitution of the third, as an alcohol, for instance phenol, having the properties of a feeble acid.

$Na_2HPO_4,12H_2O$ has a sp. gr. 1·53. Poggiale determined the solubility in 100 parts of water (1) of the anhydrous ortho-salt Na_2HPO_4, and (2) of the corresponding pyro-salt $Na_4P_2O_7$:—

	0°	20°	40°	80°	100°
I.	1·5	11·1	30·9	81	108
II.	3·2	6·2	13·5	30	40

At temperatures of 20° to 100° the ortho-salt is so very much less soluble that this difference alone already indicates the deeply-seated alteration in constitution which takes place in the passage from the ortho- to the pyro-salts.

The *ammonium orthophosphates* resemble the sodium salts in many respects, but the instability of the di- and tri-metallic salts is seen in them still more clearly than in the sodium salts; thus $(NH_4)_3PO_4$, and even $(NH_4)_2HPO_4$, lose ammonia in the air (especially when heated, even in solutions); $NH_4H_2PO_4$ alone does not disengage ammonia and has an acid reaction. The crystals of the first salt contain $3H_2O$, and are only formed in the presence of an excess of ammonia; both the others are anhydrous, and may be obtained like the sodium salts. When ignited these salts leave metaphosphoric acid behind; for example, $(NH_4)_2HPO_4 = 2NH_3 + H_2O + HPO_3$. Ammonia also enters into the composition of many double phosphates. Ammonium sodium orthophosphate, or simply phosphate, $NH_4NaHPO_4,4H_2O$, crystallises in large transparent crystals from a mixture of the solutions of disodium phosphate and ammonium chloride (in which case sodium chloride is obtained in the mother liquid), or, better still, from a solution of monosodium phosphate saturated with ammonia. It is also formed from the phosphates in urine when it ferments. This salt is frequently used in testing metallic compounds by the blow-pipe, because when ignited it leaves a vitreous metaphosphate, $NaPO_3$, which, like borax, dissolves metallic oxides, forming characteristic tinted glasses.

When a solution of trisodium phosphate is added to a solution of a magnesium salt it gives a white precipitate of the normal orthophosphate $Mg_2(PO_4)_2,7H_2O$. If the trisodium salt be replaced by the ordinary salt, Na_2HPO_4, a precipitate is also formed, and $MgHPO_4,7H_2O$ is obtained. It might be thought that the normal salt $Mg_3(PO_4)_2$ would be precipitated if disodium phosphate was added to ammonia and a salt of magnesium, but in reality *ammonium magnesium orthophosphate*, $MgNH_4PO_4,6H_2O$, is precipitated as a crystalline powder, which loses ammonia and water when ignited, and gives a pyrophosphate, $Mg_2P_2O_7$. This salt occurs in nature as the mineral struvite, and in various products of the changes of animal matter. If we consider that the above salt parts with ammonia with difficulty, and that the corresponding salt of sodium is not formed under the same conditions ($MgNaPO_4,9H_2O$ is obtained by the action of magnesia on disodium phosphate), if we turn our attention to the fact that the salts of calcium and barium do not form double salts as easily as magnesium, and remember that the salts of magnesium in general easily form double ammonium salts, we are led to think that this salt is not really a normal, but an acid salt, corresponding with Na_2HPO_4, in which Na_2 is replaced by the equivalent group NH_3Mg.

The common normal *calcium phosphate*, $Ca_3(PO_4)_2$, occurs in minerals, in animals, especially in bones, and also probably in plants, although the ash of many portions of plants, as a rule, contains less lime than the formation of the normal salt requires. Thus 100 parts of the ash (from 5,000 parts of grain) of rye grain contain 47·5 of phosphoric anhydride and only 2·7 of lime, and even the ash of the whole of the rye (including the straw) contains twice as much phosphoric anhydride as lime, and the normal salt contains almost equal weights of these substances. Only the ash of grasses, and especially of clover, and of trees, contains in the majority of cases more lime than is required for the formation of $Ca_3P_2O_8$. This salt, which is insoluble in water, dissolves even in such feeble acids as acetic and sulphurous, and even in water containing carbonic acid. The latter fact is of immense importance in nature, since by reason of it rain water is able to transfer the calcium phosphates in the soil into solutions which are absorbed by plants. The solubility of the normal salt in acids takes place by virtue of the formation of an acid salt, which is evident from the quantity of acid required for its solution, and more especially from the fact that the acid solutions when evaporated give crystalline scales of the acid calcium phosphate, $CaH_4(PO_4)_2$, soluble in water. This solubility of the acid salt forms the basis of the treatment by acids of bones, phosphorites, guano, and other natural products containing the normal salt and employed for fertilising the soil. The perfect decomposition requires at least $2H_2SO_4$ to $Ca_3(PO_4)_2$, but in reality less is taken, so that only a portion of the normal salt is converted into the acid salt. Hydrochloric acid is sometimes used. (In practice such mixtures are known as *superphosphates*). Certain experiments, however, show that a thorough grinding, the presence of organic, and especially of nitrogenous, substances, and the porous structure of some calcium phosphates (for example, in burnt bones), render the treatment of phosphoric manures by acids superfluous—that is, the crop is not improved by it.

In this sense the ortho-acid itself might be regarded as an anhydro-acid, counting $P(HO)_5$ as the perfect hydrate, if PH_5 existed; but as in general the normal hydrates correspond with the existing hydrogen compounds with the addition of up to 4 atoms of oxygen, therefore PH_3O_4 is the normal acid, just as SH_2O_4 and $ClHO_4$; while NHO_3, CH_2O_3 are meta-acids, or higher normal acids (NH_3O_4 and CH_4O_4) with the loss of a molecule of water.

In order to see the relation between the ortho-, pyro-, and metaphosphoric acids, the first thing to remark in them is that the anhydride P_2O_5 is combined with 3, 2, and 1 molecules of water. In the absence of data for the molecular weight of ortho- and pyrophosphoric acids it is necessary to mention that all existing data for metaphosphoric acid indicate (Note 21) that its molecule is much more complex and contains at least $H_3P_3O_9$ or $H_6P_6O_{18}$. The explanation of the problems which here present themselves can, it seems to me, be only looked for after a detailed study of the phenomena of the polymerisations of mineral substances, and of those complex acids, such as phosphomolybdic, which we shall hereafter describe (Chapter XXI.) A similar instance is exhibited in the solubility of hydrate of silica (produced by the action of silicon fluoride on water) in fused metaphosphoric acid, with the formation, on cooling, of an octahedral compound (sp. gr., 3·1) containing SiO_2,P_2O_5. A certain indication (but no proof) that ordinary orthophosphoric acid is polymerised is given by Staudenmaier (1893), who obtained a salt, $K_5H_4P_3O_{12}$, by the action of a solution of KH_2PO_4 upon K_2CO_3; and a compound, $KH_3P_2O_8$, corresponding to the doubled molecule of H_3PO_4, by the action of KH_2PO_4 upon H_3PO_4 itself.

[19 bis] According to Watson (1893) the ortho-acid is partially transformed into the pyro-acid at 230°, whilst at 260° the latter begins to volatilise. At 300° the meta-acid only is formed.

The method of preparation of the acid itself consists in converting the sodium salt, $Na_4P_2O_7$, by double decomposition with water and a salt of lead, into insoluble lead pyrophosphate, $Pb_2P_2O_7$, which is then suspended in water and decomposed by sulphuretted hydrogen; lead sulphide is thus precipitated, and pyrophosphoric acid remains in solution. This solution cannot be heated, or the pyro-acid will pass into the ortho-, but must be evaporated under the receiver of an air-pump. It concentrates to a syrup and crystallises, and when ignited in this form loses water, and forms metaphosphoric acid. It resembles orthophosphoric acid in many respects; its salts with the alkalis are also soluble, and the others insoluble in water but soluble in acids. When heated in solution with acid it gives orthophosphoric acid, as well as when fused with an excess of alkali.

Witt heated ammonium chloride with phosphoric acid (hydrochloric acid was evolved), ignited the residue to drive off ammonia, and obtained pyrophosphoric acid in the residue.

As when using phenolphthalein as an indicator in neutralising by an alkali metaphosphoric acid is monobasic, and orthophosphoric acid is bibasic, it is possible by means of this difference to follow the transition of meta- into orthophosphoric acid. Sabatier (1888) carried on an investigation of this nature, and found that the rate of transformation is dependent on the temperature, and is subject to the general laws of the rate of chemical transformations which belongs to physical chemistry.

Metaphosphoric acid has a particular interest in respect to the variations to which its salts are subject. The metaphosphates are formed by the ignition of the acid orthophosphates, MH_2PO_4, or MNH_4HPO_4, or of the acid pyrophosphates, $M_2H_2P_2O_7$, or $M_2(NH_4)_2P_2O_7$, water and ammonia being given off in the process. The properties of the metaphosphates, which have a similar composition to nitrates—for instance, $NaPO_3$, or

$Ba(PO_3)_2$—vary according to the duration of the ignition to which the ortho-, or pyrophosphates from which they are prepared have been subjected. When the salts NaH_2PO_4 or NH_4NaHPO_4 are strongly ignited, a salt $NaPO_3$ is formed, which deliquesces in the air, and gives a gelatinous precipitate with salts of the alkaline earths. But, as Graham (in 1830–40), and many others, especially Fleitmann and Henneberg (in 1840–50), and Tamman (in the nineties), observed, under other conditions the salts of the same composition acquire other properties. The above chemists recognise five polymeric forms of metaphosphates, $(HPO_3)_n$. We will follow the nomenclature and researches of Fleitmann.

Monometaphosphoric acid. The salts are distinguished for their insolubility in water; even the salts $NaPO_3$, KPO_3, are insoluble. They are obtained by igniting the monometallic orthophosphates—for example, RH_2PO_4—up to the temperature at which all water is evolved (316°), but not to fusion. No double salts are known.

Dimetaphosphoric acid, on the contrary, easily forms double salts—for example, $KNaP_2O_6$, and also the copper potassium salt, &c. The copper salt is obtained by evaporating a solution of copper oxide in orthophosphoric acid. A blue ortho-salt, $CuRHO_4$, first separates from the solution, then a light-blue pyro-salt, $Cu_2P_2O_7$; and above 350°, when metaphosphoric acid itself begins to volatilise, the dimetaphosphate, CuP_2O_6, is formed. The residue is washed with water, and decomposed with a hot solution of sodium sulphide, when the sodium salt, $Na_2P_2O_6$, is obtained in solution. This salt, when evaporated with alcohol, gives crystals containing 2 mol. H_2O, which, however, retain their solubility (in 7 parts of water) after the water is driven off at 100°. When fused, these crystals give a deliquescent salt (hexa-metaphosphate). The solution of the salt has a neutral reaction, which only after prolonged boiling becomes acid, owing to the formation of orthophosphate, NaH_2PO_4. The soluble salts of dimetaphosphoric acid give the insoluble silver salt, $Ag_2P_2O_6$, with silver nitrate, and a precipitate of $BaP_2O_6 2H_2O$ with barium chloride.

Trimetaphosphoric acid is obtained as the sodium salt $Na_3P_3O_9$ when any other metaphosphate of sodium is fused and *slowly* cooled, then dissolved in a slight excess of warm water, and the resultant solution evaporated. The crystals contain 6 mol. H_2O, and dissolve in four parts of water. An acid reaction is only obtained, as with the preceding salt, after prolonged boiling with water. The acid is a true analogue of nitric acid, because *all its metallic salts are soluble.*

Hexametaphosphoric acid. Fleitmann so named the ordinary metaphosphoric acid (glacial) which attracts moisture. The deliquescent sodium salt is obtained, like the trimetaphosphate, only by *rapid* cooling. It is also formed by fusing silver oxide with an excess of phosphoric acid. The sodium salt is soluble in water, and gives viscous, elastic precipitates with salts of Ba, Ca, and Mg. Lubert (1893) obtained salts of Ag, Pb, &c.

Jawein and Thillot (1889), who investigated the sodium salts of metaphosphoric acid by Raoult's method, came to the conclusion that the salts of di- and tri-metaphosphoric acid behave in such a manner that their molecule must be represented as non-polymerised $NaPO_3$, whilst those of hexametaphosphoric acid behave as $(NaPO_3)_4$. At all events, the series of salts which Fleitmann and Henneberg regard as monometaphosphates—*i.e.* as non-polymerised—are most probably the most polymerised, because they are insoluble.

According to Tamman's researches, vitreous metaphosphoric acid contains a mixture consisting chiefly of two varieties, differing in the solubility and degree of stability of their salts. The least stable corresponds to Fleitmann's hexa-acid, and gives three isomeric salts. Tamman came to the conclusion that there exist polymers also in the form of penta-, ortho-, and deca-metaphosphoric acids. Without going into details upon this subject, I do not think it superfluous to point out that the undoubted capability of metaphosphoric acid to polymerise should be connected with its faculty of combining with water, whilst the degree of polymerisation and the number of polymeric forms cannot yet be considered as sufficiently explained.

[21 bis] The bibasity of H_3PO_3, established by Würtz, has been proved by many direct experiments (see, for instance, Note 22), among which we may mention that Amat (1892) took a mixture of the aqueous solutions of Na_2HPO_3 and NaHO and added absolute alcohol to it. Two layers were formed; the upper, alcoholic, contained all the excess of NaHO, whilst the lower only contained the salt Na_2HPO_3, which was therefore unable to react with the excess of NaHO. Amat also obtained NaH_2PO_3 by saturating H_3PO_3 with soda until he obtained a neutral reaction with methyl-orange. The replacement of one atom of H by sodium here, as in phosphoric acid (Note 16), gives more heat than the replacement of the second atom. For the third atom there is no formation of a salt, and therefore no evolution of heat. The monometallic salts—for example, NaH_2PO_3—or the ammonia salts, when heated to 160°, give, as Amat had previously shown, a salt of bibasic pyrophosphorous acid, $Na_2H_2P_2O_5$.

Phosphorous acid, when subjected to the action of nascent hydrogen (zinc and sulphuric acid), evolves phosphine, and when boiled with an excess of alkali it evolves hydrogen ($PH_3O_3 + 3KHO = PK_3O_4 + 2H_2O + H_2$); owing to its liability to oxidation, it is a reducing agent—for instance, it reduces cupric chloride to cuprous chloride, and precipitates silver from the nitrate and mercury from its salts.

These reactions are perhaps connected with the fact that in this acid one atom of hydrogen should be considered as in the same condition as in phosphuretted hydrogen, which is expressed by the formula $PHO(OH)_2$, if we represent it as PH_4X, with the substitution of two of the hydrogen atoms by oxygen and of HX by two of hydroxyl. The direct passage of phosphorous chloride into phosphorous acid would, however, indicate that all the three atoms of hydrogen in it occur in the form of hydroxyl, because no difference is known between the three atoms of chlorine in PCl_3—they all react alike, as a rule. However, Menschutkin, by acting on alcohol, C_2H_5OH, with phosphorous chloride, obtained hydrochloric acid and a substance $P(C_2H_5O)Cl_2$, and from it by the action of bromine he obtained ethyl bromide, C_2H_5Br, and a compound $PBrOCl_2$, which proves, to a certain extent, the existence of a difference between the three atoms of chlorine in phosphorous chloride. If we turn our attention to the formation of phosphine by the ignition of phosphorous acid, we see that $4PH_3O_3$ only evolve 3H in the form of PH_3, and therefore the residue—that is, $3PH_3O_4$—will still contain one hydrogen of the same nature as in phosphine, because in $4PH_3O_3$ we should recognise four such hydrogens as in phosphine. We arrive at the same conclusion by examining the decomposition of hypophosphorous acid, $2PH_3O_2 = PH_3 + PH_3O_4$. In the two molecules of the monobasic hypophosphorous acid taken, there are only two atoms of hydrogen replaceable by metals, whilst in the molecule of the resultant phosphoric acid there are three. Perhaps relations of this nature determine the relative stability of the dimetallic salts of orthophosphoric acid.

Calcium hypophosphite is used in medicine. According to Cavazzi, a mixture of sodium hypophosphite, NaH_2PO_2, and sodium nitrate explodes violently.

Fluorine and bromine give PX_3 and PX_5, like chlorine. With respect to iodine PI_5 is, in a chemical sense, a very unstable substance, and generally *phosphorus tri-iodide* only is formed (from yellow or red phosphorus and iodine in the requisite proportions). It is a red crystalline substance, fuses at 55°, is easily decomposed by water, forming phosphorous and hydriodic acids, and when heated it evolves iodine vapours and forms phosphorus di-iodide, PI_2. This substance may be obtained in the same manner as the preceding by taking a smaller proportion of iodine (8 parts of iodine to 1 part of phosphorus, whilst the tri-iodide requires 12·3); it also forms red crystals, which melt at 110°. When decomposed by water it not only gives phosphorous and hydriodic acids, but also phosphine and a yellow substance (a lower oxide of phosphorus). In its composition di-iodide of phosphorus corresponds with liquid phosphuretted hydrogen, PH_2, and probably its molecular weight is much higher: P_2I_4 or P_3I_6, &c. As the iodine compounds of phosphorus give hydriodic and phosphorous acids with water, and as both these substances are reducing agents in the presence of water (and hydrates), iodide of phosphorus also acts as a reducing agent.

In a liquid state the density of phosphorous chloride at $10° = 1·597$, and therefore its molecular volume $= 137·5/1·597 = 86·0$, and that of phosphorus oxychloride is equal to $153·5/1·693 = 90·7$; hence the addition of oxygen has produced considerable increase in volume, just as in the conversion of sulphur dichloride, SCl_2, into sulphuryl chloride, $SOCl_2$, the volume changes from 64 to 71. It is the same with the boiling-points; phosphorus trichloride boils at 70°, the oxychloride at 100°, sulphur dichloride at 64°, and sulphuryl chloride at 78°—that is, the addition of oxygen raises the boiling points.

The vapour density of phosphorus trichloride and oxychloride corresponds with their formulæ (Cahours, Würtz)—namely, is equal to half the molecular weight referred to hydrogen. But it is not so with phosphorus pentachloride. Cahours showed that the vapour density of phosphorus pentachloride referred to air $= 3·65$, to hydrogen $= 52·6$, whilst according to the formula PCl_5 it should be $= 104·2$. Hence this formula corresponds with four, and not with two, molecules. This shows that the vapour of phosphoric chloride contains two and not one molecule, that in a state of vapour it splits up, like sal-ammoniac, sulphuric acid, &c. The products of disruption must here be phosphorous chloride, PCl_3, and chlorine, Cl_2, bodies which easily re-form phosphoric chloride, PCl_5, at a lower temperature. This decomposition of phosphoric chloride in its conversion into vapour is confirmed by the fact that the vapour of this almost colourless substance shows the greenish-yellow colour proper to chlorine. This dissociation of phosphoric chloride has been considered by some chemists as a sign that phosphorus, like nitrogen, does not give volatile compounds of the type PX_5, and that such substances are only obtained as unstable molecular compounds which break up when distilled; for example, PH_3,HI, PCl_3,Cl_2, NH_3,HCl, &c. To prove that the molecule PCl_5 actually exists, Würtz in 1870 observed that when mixed with the vapour of phosphorous chloride the vapour of phosphoric chloride distils over (from 160° to 190°) perfectly colourless, and has a density which is really near to the formula—namely, to 104—and the same density was determined for the pentachloride in an atmosphere of chlorine. Hence at low temperatures and in admixture with one of the products of dissociation, there is no longer that decomposition which occurs at higher temperatures—that is, we have here a case of dissociation proceeding at moderate temperatures.

An important proof in favour of the type PX_5 is exhibited by phosphorus pentafluoride PF_5, obtained by Thorpe as a colourless gas which only corrodes glass after the lapse of time; it may be kept over mercury, and has a normal density. It is formed when liquid arsenic trifluoride, AsF_3, is added to phosphoric chloride surrounded by a freezing mixture: $3PCl_5 + 5AsF_3 = 3PF_5 + 5AsCl_3$.

In general, fluorine and phosphorus give stable compounds: PF_3, POF_3, and PF_5, as would be expected from the fact that in passing from Cl to I (*i.e.* as the atomic weight of the halogen increases) the stability of the compounds with P and the tendency to give PX_5 (Note 24) decreases. *Phosphorus trifluoride* is obtained by heating a mixture of ZnF_2 and PBr_3, by the action of AsF_3 upon PCl_3, by heating phosphide of copper with PbF_2, &c. It is a strong-smelling gas, which liquefies at -10° under a pressure of 40 atmospheres, giving a colourless liquid. It dissolves easily in (is absorbed by, reacts with) water, and acts upon glass; when mixed with Cl_2 it combines with it (Poulenc, 1891), forming PCl_2F_3, a colourless gas of normal density, which is transformed into a liquid at 8°, decomposes into $PF_3 + Cl_2$ at 250°, and, with a small amount of water, gives *oxy-fluoride* of phosphorus, POF_3 (with a large amount of water it gives PH_3O_4), which Moissan (1891) obtained by the action of dry HF upon P_2O_5, and Thorpe and Tutton (1890) by heating a mixture of cryolite and P_2O_5. It is a gas of normal density, like PF_3, and was obtained by Moissan by the action of fluorine upon PF_3 (PSF_3, *see* Chapter XX., Note 20). Thus the forms PX_3 and PX_5 not only exist in many solid and non-volatile substances, but also as vapours.

Phosphorus oxychloride is obtained by the action of phosphoric chloride on hydrates of acids (because alkalis decompose phosphorus oxychloride), according to the equation $PCl_5 + RHO = POCl_3 + RCl + HCl$, where RHO is an acid. The reaction only proceeds according to this equation with monobasic acids, but then RCl is volatile, and therefore a mixture is obtained of two volatile substances, the acid chloride and phosphorus oxychloride, which are sometimes difficult to separate; whilst if the hydrate be polybasic the reaction frequently proceeds so that an anhydride is formed: $RH_2O_2 + PCl_5 = RO + POCl_3 + 2HCl$. If the anhydride be non-volatile (like boric), or easily decomposed (like oxalic), it is easy to obtain pure oxychloride. Thus phosphorus oxychloride is often prepared by acting on boric or oxalic acid with phosphoric chloride. It is also formed when the vapour of phosphoric chloride is passed over phosphoric anhydride, $P_2O_5 + 3PCl_5 = 5POCl_3$. This forms an excellent example in proof of the fact that the formation of one substance from two does not necessarily show that the resultant compound contains the molecules of these substances in its molecule. But other oxychlorides of phosphorus are also formed by the interaction of phosphoric anhydride and chloride; thus at 200° the chloranhydride, PO_2Cl, or chloranhydride of metaphosphoric acid, is formed (Gustavson). The chloranhydride of pyrophosphoric acid, $P_2O_3Cl_4$, was obtained (Hayter and Michaelis), together with NOCl, &c., by the action of NO upon cold PCl_3, as a fuming liquid boiling at 210°.

The direct action of the sun's rays, or of magnesium light, is necessary to start the reaction between carbonic oxide and chlorine, but when once started it will proceed rapidly in diffused light. An excess of chlorine (which gives its coloration to the colourless phosgene) aids the completion of the reaction, and may afterwards be removed by metallic antimony. Porous substances, like charcoal, aid the reaction. Phosgene may be prepared by passing a mixture of carbonic anhydride and chlorine over incandescent charcoal. Lead or silver chloride, when heated in a current of carbonic oxide, also partially form phosgene gas. Carbon tetrachloride, CCl_4, also forms it when heated with carbonic anhydride (at 400°), with phosphoric anhydride (200°), and most easily of all with sulphuric anhydride ($2SO_3 + CCl_4 = COCl_2 + S_2O_5Cl_2$, this is pyrosulphuryl chloride). Chloroform, $CHCl_3$, is converted into carbonyl chloride when heated with $SO_2(OH)Cl$ (the first chloranhydride of sulphuric acid); $CHCl_3 + SO_3HCl = COCl_2 + SO_2 + 2HCl$ (Dewar), and when oxidised by chromic acid.

Among the reactions of phosgene we may mention the formation of urea with ammonia, and of carbonic oxide when heated with metals.

We are already acquainted with some of the chloranhydrides of the inorganic acids—for instance, BCl_3, and $SiCl_4$—and here we shall describe those which correspond with sulphuric acid in the following chapter. It may be mentioned here that when hydrochloric acts on nitric acid (aqua regia, Vol. I. p. [467](#)) there is formed, besides chlorine, the oxychlorides NOCl and NO_2Cl, which may be regarded as chloranhydrides of nitric and nitrous acids (nitrogen chloride, Vol. I. p. [476](#)). The former boils at -5°, the latter at +5°, the specific gravity of the first at $-12° = 1·416$, and at $-18° = 1·433$ (Geuther), and of the second $= 1·3$; the first is obtained from nitric oxide and chlorine, the second from nitric peroxide and chlorine, and also by the action of phosphoric chloride on nitric acid. If the gases evolved by aqua regia be passed into cold and strong sulphuric acid, they form crystals of the composition $NHSO_3$ (like chamber crystals), which melt at 86°, and with sodium chloride form acid sodium sulphate and the oxychloride NOCl. This chloranhydride of nitric acid is termed *nitrosyl chloride*.

Cyanogen chloride, CNCl, is the gaseous chloranhydride of cyanic acid; it is formed by the action of chlorine on aqueous mercury cyanide, $Hg(CN)_2 + 2Cl_2 = HgCl_2 + 2CNCl$. When chlorine acts on cyanic acid, it forms not only this cyanogen chloride, but also polymerides of it—a liquid, boiling at 18°, and a solid, boiling at 190°. The latter corresponds with cyanuric acid, and consequently contains $C_3N_3Cl_3$. Details concerning these substances must be looked for in works on organic chemistry.

[28 bis] This reaction indeed proceeds very easily and completely with a number of hydroxides, if they do not react on hydrochloric acid and phosphorus oxychloride, which is the case when they have alkaline properties. When the hydroxide is bibasic and is present in excess, it not unfrequently happens that the elements of water are taken up: $R(OH)_2 + PCl_5 = RO + 2HCl + POCl_3$. The anhydride RO may then be converted into chloranhydride, $RO + PCl_5 = RCl_2 + POCl_3$—that is, phosphorus pentachloride brings about the substitution of O by Cl_2. Thus carbonyl chloride, $COCl_2$, boron chloride, $2BCl_3$, and succinic chloride, $C_4H_4O_2Cl_2$, &c., are respectively obtained by the action of phosphoric chloride on carbonic, boric, and succinic anhydrides. Phosphorus pentachloride reacts in a similar manner on the aldehydes, RCHO, forming $RCHCl_2$, and on the chloranhydrides themselves—for example, with acetic chloride, $CH_3.COCl$ (when heated in a closed tube), it forms a substance having the composition CH_3CCl_3.

Phosphorus trichloride and oxychloride act in a similar manner to phosphoric chloride. When phosphorus trichloride acts on an acid, $3RHO + PCl_3 = 3RCl + P(HO)_3$. If a salt is taken, then by the action of phosphorus oxychloride a corresponding chloranhydride and salt of orthophosphoric acid are easily formed: $3R(KO) + POCl_3 = 3RCl + PO(KO)_3$. The chloranhydride RCl is always more volatile than its corresponding acid, and distils over before the hydrate RHO. Thus acetic acid boils at 117°, and its chloranhydride at 50°. Phosphoric and phosphorous acids are very slightly volatile, whilst their chloranhydrides are comparatively easily converted into vapour. The faculty of the chloranhydrides to react at the expense of their own chlorine determines their great importance in chemistry. For instance, suppose we require to know the molecular formula of some hydrate which does not pass into a state of vapour and does not give a chloranhydride with hydrochloric acid—that is, which has not any basic or alkaline properties; we must then endeavour to obtain this chloranhydride by means of phosphoric chloride, and it frequently happens that the corresponding chloranhydride is volatile. The resultant chloranhydride is then converted into vapour, and its composition is determined; and if we know its composition we are able to decide that of its corresponding hydrate. Thus, for example, from the formula of silicon chloride, $SiCl_4$, or of boron chloride, BCl_3, we can judge the composition of their corresponding hydrates, $Si(HO)_4$, $B(HO)_3$. Having obtained the chloranhydride RCl or RCl_n, it is possible by its means to obtain many other compounds of the same radicle R according to the equation MX + RCl = MCl + RX. M may be = H, K, Ag, or other metal. The reaction proceeds thus if M forms a stable compound with chlorine—for example, silver chloride, hydrochloric acid, and R, an unstable substance. Hence, a chloranhydride is frequently employed for the formation of other compounds of a given radicle; for instance, with ammonia they form amides RNH_2, and with salts ROK, with anhydrides R_2O, &c.

The reaction of ammonia on phosphorus pentachloride is more complex than the preceding. This is readily understood: to the oxychloride, $POCl_3$, tere corresponds a hydrate $PO(OH)_3$, and a salt $PO(NH_4O)_3$, and consequently also an amide $PO(NH_2)_3$, whilst the pentachloride, PCl_5, has no corresponding hydrate $P(OH)_5$, and therefore there is no amide $P(NH_2)_5$. The reaction with ammonia will be of two kinds: either instead of 5 mol. NH_3, only 3 mol. NH_3 or still less will act; *i.e.* $PCl_2(NH_2)_3$, $PCl_3(NH_2)_2$, &c. are formed; or else the pentachloride will act like a mixture of chlorine with the trichloride, and then as the result there will be obtained the products of the action of chlorine on those amides which are formed from phosphorus trichloride and ammonia. It would appear that both kinds of reaction proceed simultaneously, but both kinds of products are unstable, at all events complex, and in the result there is obtained a mixture containing sal-ammoniac, &c. The products of the first kind should react with water, and we should obtain, for example, $PCl_3(NH_2)_2 + 2H_2O =$ 3HCl and $PO(HO)(NH_2)_2$. This substance has not actually been obtained, but the compound $PONH(NH_2)$ derived from it by elimination of the elements of water is known, and is termed *diphosphamide*; it is, however, more probable that it is a nitrile than an amide, because only amides contain the group NH_2. It is a colourless, stable, insoluble powder, which possibly corresponds with pyrophosphoric acid, more especially since when heated it evolves ammonia and gives and leaves phosphoryl nitride, PON—that is, the nitrile of metaphosphoric acid. The amide corresponding with the pyrophosphate $P_2O_3(NH_4O)_4$ should be $P_2O_3(NH_2)_4$, and the nitriles corresponding to the latter would be $P_2O_2N(NH_2)_3$, $P_2ON_2(NH_2)_2$, and $P_2N_3(NH_2)$. The composition of the first is the same as that of the above diphosphamide. The third pyrophosphoric nitrile has a formula $P_2N_4H_2$, and this is the composition of the body known as *phospham*, PHN_2 (in a certain sense this is the analogue of N_3H polymerised, Chapter VI.) Indeed, phospham has been obtained by heating the products of the action of ammonia on phosphoric chloride, as an insoluble and alkaline powder, which gives ammonia and phosphoric acid when subjected to the action of water. The same substance is obtained by the action of ammonium chloride on phosphoric chloride ($PNCl_2$ is first formed, and reacts further with ammonia, forming phospham), and by igniting the mass which is formed by the action of ammonia on phosphorus trichloride. Formerly the composition of phospham was supposed to be PHN_2, now there is reason to think that its molecular weight is $P_3H_3N_6$.

The above compounds correspond with normal salts, but nitriles and amides corresponding to acid salts are also possible, and they will be acids. For example, the amide $PO(HO)_2(NH_2)$, and its nitrile, will be either $PN(HO)_2$ or $PO(HO)(NH)$, but at all events of the composition PNH_2O_2, and having acid properties. The ammonium salt of this *phosphonitrilic acid* (it is called phosphamic acid), $PNH(NH_4)O_2$, is obtained by the action of ammonia on phosphoric anhydride, $P_2O_5 + 4NH\_3 = H_2O + 2PNH(NH_4)O_2$. A non-crystalline soluble mass is thus formed, which is dissolved in a dilute solution of ammonia and precipitated with barium chloride, and the resultant barium salt is then decomposed with sulphuric acid, and thus a solution of the acid of the above composition is obtained.

It is evident from the theory of the formation of amides and nitriles (Chapter IX.) that very many compounds of this kind can correspond with the acids of phosphorus; but as yet only a few are known. The easy transitions of the ortho-, meta-, and pyrophosphoric acids, by means of the hydrogen of ammonia, into the lower acids, and conversely, tend to complicate the study of this very large class of compounds, and it is rarely that the nature of a product thus obtained can be judged from its composition; and this all the more that instances of isomerism and polymerism, of mixture between water of crystallisation and of constitution, &c., are here possible. Many data are yet needed to enable us to form a true judgment as to the composition and structure of such compounds. As the best proof of this we will describe the very interesting and most fully investigated compound of this class, $PNCl_2$, called *chlorophosphamide*, or nitrogen chlorophosphorite. It is formed in small quantities when the vapour of phosphoric chloride is passed over ignited sal-ammoniac. Besson (1892) heated the compound PCl_58NH_3 (which is easily and directly formed from PCl_5 and NH_3) under a pressure of about 50 mm. (of mercury) to 200°, and obtained brilliant crystals of $PNCl_2$, which melted at 106° (in the residue after the distillation of sal-ammoniacal phospham). The chlorine in it is very stable—quite different from that in phosphoric chloride. Indeed, the resultant substance is not only insoluble in water (though soluble in alcohol and ether), but it is not even moistened by it, and distils over, together with steam, without being decomposed. In a free state it easily crystallises in colourless prisms, fuses at 114°, boils at 250° (Gladstone, Wichelhaus), and when fused with potash gives potassium chloride and the amidonitrile of phosphoric acid. Judging from its formula and the simplicity of its composition and reactions, it might be thought that the molecular weight of this substance would be expressed by the formula PCl_2N, that it corresponds with PON and with PCl_5 (like $POCl_3$), with the substitution of Cl_3 by N, just as in $POCl_3$ two atoms of chlorine are replaced by oxygen; but all these surmises are incorrect, because its vapour density (referred to hydrogen—Gladstone, Wichelhaus) = 182—that is, the molecular formula must be three times greater, $P_3N_3Cl_6$. The polymerisation (tripling) is here of exactly the same kind as with the nitriles.

It is necessary to remark that, although arsenic is so closely analogous to phosphorus (especially in the higher forms of combination, RX_3 and RX_5), at the same time it exhibits a certain resemblance and even isomorphism with the corresponding compounds of sulphur (especially the metallic compounds of the type MAs, corresponding with MS). Thus compounds containing metals, arsenic, and sulphur are very frequently met with in nature. Sometimes the relative amounts of arsenic and sulphur vary, so that an isomorphous substitution between the arsenides and sulphides must be recognised. Besides FeS_2 (ordinary pyrites), and $FeAs_2$, iron forms an arsenical pyrites containing both sulphur and arsenic, which from its composition, FeAsS or FeS_2FeAs_2, resembles the two preceding.

[30 bis] According to Retgers (1893) the arsenic mirror (see further on) is an unstable variety of metallic arsenic, whilst the brown product which is formed together with it in Marsh's apparatus is a lower hydride AsH. Schuller and McLeod (1894), however, recognise a peculiar yellow variety of arsenic.

Hydrochloric acid dissolves arsenious anhydride in considerable quantities, and this is probably owing to the formation of unstable compounds in which the arsenious anhydride plays the part of a base. A compound called *arsenious oxychloride*, having the composition AsOCl, is even known. It is formed when arsenious anhydride is added little by little to boiling arsenic trichloride, $As_2O_3 + AsCl_3 = 3AsOCl$. It is a transparent substance, which fumes in air, and combines with water to form a crystalline mass having the composition $As_2(OH)_4Cl_2$. When heated it decomposes into arsenious chloride and a fresh oxychloride of a more complex composition, $As_6O_8Cl_2$. Arsenic trichloride, when treated with a small quantity of water, forms the crystalline compound, $As_2(HO)_4Cl_2$, mentioned above. These compounds resemble the basic salts of bismuth and aluminium. The existence of these compounds shows that arsenic is of a more metallic or basic character than phosphorus. Nevertheless *arsenic trichloride*, $AsCl_3$, resembles phosphorus trichloride in many respects. It is obtained by the direct action of chlorine on arsenic, or by distilling a mixture of common salt, sulphuric acid, and arsenious anhydride. The latter mode of preparation already indicates the basic properties of the oxide.

Arsenious chloride is a colourless oily liquid, boiling at 130°, and having a sp. gr. of 2·20. It fumes in air like other chloranhydrides, but it is much more slowly and imperfectly decomposed by water than phosphorus trichloride. A considerable quantity of water is required for its complete decomposition into hydrochloric acid and arsenious anhydride. It forms an excellent example of the transition from true metallic chlorides to true chloranhydrides of the acids. It hardly combines with chlorine, *i.e.* if $AsCl_5$ is formed it is very unstable. *Arsenic tribromide*, $AsBr_3$, is formed as a crystalline substance, fusing at 20° and boiling at 220°, by the direct action of metallic arsenic on a solution of bromine in carbon bisulphide, the latter being then evaporated. The specific gravity of arsenic tribromide is 3·36. Crystalline arsenic tri-iodide, AsI_3, having a sp. gr. 4·39, may be obtained in a like manner; it may be dissolved in water, and on evaporation separates out from the solution in an anhydrous state—that is, it is not decomposed—and consequently behaves like metallic salts. *Arsenic trifluoride*, AsF_3, is obtained by heating fluor spar and arsenious anhydride with sulphuric acid. It is a fuming, colourless, and very poisonous liquid, which boils at 63° and has a sp. gr. of 2·73. It is decomposed by water. It is very remarkable that fluorine forms a pentafluoride of arsenic also, although this compound has not yet been obtained in a separate state, but only in combination with potassium fluoride. This compound, K_3AsF_8, is formed as prismatic crystals when potassium arsenate, K_3AsO_4, is dissolved in hydrofluoric acid.

Arsenic acid, H_3AsO_4, corresponding with orthophosphoric acid, is formed by oxidising arsenious anhydride with nitric acid, and evaporating the resultant solution until it attains a sp. gr. of 2·2; on cooling it separates in crystals having the above composition. This hydrate corresponds with the normal salts of arsenic acid; but on dissolving in water (without heating), and on cooling a strong solution, crystals containing a greater amount of water, namely, $(AsH_3O_4)_2,H_2O$, separate. This water, like water of crystallisation, is very easily expelled at 100°. At 120° crystals having a composition identical with that of pyrophosphoric acid, $As_2H_4O_7$, separate, but water, on dissolving this hydrate with the development of heat, forms a solution in no way differing from a solution of ordinary arsenic acid, so that it is not an independent pyroarsenic acid that is formed. Neither is there any true analogue of metaphosphoric acid, although the compound $AsHO_3$ is formed at 200°, and on solidifying forms a mass having a pearly lustre and sparingly soluble in cold water; but on coming into contact with warm water it becomes very hot, and gives ordinary orthoarsenic acid in solution. Arsenic acid forms three series of salts, which are perfectly analogous to the three series of orthophosphates. Thus the normal salt, K_3AsO_4, is formed by fusing the other potassium arsenates with potassium carbonate; it is soluble in water and crystallises in needles which do not contain water. Di-potassium arsenate, K_2HAsO_4, is formed in solution by mixing potassium carbonate and arsenic acid until carbonic anhydride ceases to be evolved; it does not crystallise, and has an alkaline reaction; hence it corresponds perfectly with the sodium phosphate. As was mentioned above, arsenic acid itself acts as an oxidising agent; for example, it is used in the manufacture of aniline dyes for oxidising the aniline, and it is prepared in large quantities for this purpose. When sulphuretted hydrogen is passed through its solution, sulphuric acid and arsenious anhydride are obtained in solution. Arsenic acid is very easily soluble in water, and its solution has an exceedingly acid reaction, and when boiled with hydrochloric acid evolves chlorine, like selenic, chromic, manganic, and certain other higher metallic acids.

Arsenic anhydride, As_2O_5, is produced when arsenic acid is heated to redness. It must be carefully heated, as at a bright red heat it decomposes into oxygen and arsenious anhydride. Arsenic anhydride is an amorphous substance almost entirely insoluble in water, but it attracts moisture from the air, deliquesces, and passes into the acid. Hot water produces this transformation with great ease.

The formation of arseniuretted hydrogen is accompanied by the absorption of 37,000 heat units, while phosphine evolves 18,000 (Ogier), and ammonia 27,000. Sodium (0·6 p.c.) amalgam, with a strong solution of As_2O_3, gives a gas containing 86 vols. of arsenic and 14 vols. of hydrogen (Cavazzi).

This spot, or the metallic ring which is deposited on the heated tube, may easily be tested as to whether it is really due to arsenic or proceeds from some other substance reduced in the hydrogen flame—for instance, carbon or antimony. The necessity for distinguishing arsenic from antimony is all the more frequently encountered in medical jurisprudence, from the fact that preparations of antimony are very frequently used as medicine, and antimony behaves in the hydrogen apparatus just like arsenic, and therefore in making an investigation for poisoning by arsenic it is easy to mistake it for antimony. The best method to distinguish between the metallic spots of arsenic and antimony is to test them with a solution of sodium hypochlorite, free from chlorine, because this will dissolve arsenic and not antimony. Such a solution is easily obtained by the double decomposition of solutions of sodium carbonate and bleaching powder. A solution of potassium chlorate acts in the same manner, only more slowly. Further particulars must be looked for in analytical works.

Arseniuretted hydrogen, like phosphuretted hydrogen, is only slightly soluble in water, has no alkaline properties—that is, it does not combine with acids—and acts as a reducing agent. When passed into a solution of silver nitrate it gives a blackish brown precipitate of metallic silver, the arsenic being oxidised. If acting on copper sulphate and similar salts, arseniuretted hydrogen sometimes forms arsenides—*i.e.* it reduces the metallic salt with its hydrogen, and is itself reduced to arsenic. Sulphuric, and even hydrochloric, acid reduces arseniuretted hydrogen to arsenic, and it is still more easily decomposed by arsenious chloride, and with phosphorous chloride it gives the compound PAs. Arseniuretted hydrogen gives metallic arsenic with an acid solution of arsenious anhydride (Tivoli).

According to Mitscherlich's determination, the vapour density of arsenious anhydride is 199 (H = 1)—that is, it answers to the molecular formula As_4O_6. Probably this is connected with the fact that the molecule of free arsenic contains As_4. V. Meyer and Biltz, however, showed (1889) that at a temperature of about 1,700° the vapour density of arsenic corresponds with the molecule As_2, and not As_4, as at lower temperatures.

Arsenious anhydride is obtained in an amorphous form after prolonged heating at a temperature near to that at which it volatilises, or, better still, by heating it in a closed vessel. It then fuses to a colourless liquid, which on cooling forms a transparent vitreous mass, whose specific gravity is only slightly less than that of the crystalline anhydride. On cooling, this vitreous mass undergoes an internal change, in which it crystallises and becomes opaque, and acquires the appearance of porcelain. The following difference between the vitreous and opaque varieties is very remarkable: when the vitreous variety is dissolved in strong and hot hydrochloric acid it gives crystals of the anhydride on cooling, and this crystallisation *is accompanied by the emission of light* (which is visible in the dark), and the entire liquid glows as the crystals begin to separate. The opaque variety does not emit light when the crystals separate from its hydrochloric acid solution. It is also remarkable that the vitreous variety passes into the opaque form when it is pounded—that is, under the action of a series of blows. Thus, several varieties of arsenious anhydride are known, but as yet they are not characterised by any special chemical distinctions, and even differ but little in their specific gravities, so that it cannot be said that the above differences are due to any isomeric transformation—that is, to an arrangement of the atoms in the molecule—but probably only depend on a difference in the distribution of the molecules, or, in other terms, are physical and not chemical variations. One part of the vitreous anhydride requires twelve parts of boiling water for its solution, or twenty-five parts at the ordinary temperature. The opaque variety is less soluble, and at the ordinary temperature requires about seventy parts of water for its solution.

Arsenious anhydride does not oxidise in air, either in a dry state or in solution, but in the presence of alkalis it absorbs oxygen from the air, and acts as an excellent reducing agent. This probably is connected with the fact that arsenic acid is much more energetic than arsenious acid, and that it is arsenic acid which is formed by the oxidation of the latter in the presence of alkalis. Arsenious anhydride is easily reduced to arsenic by many metals, even by copper.

The feebleness of the acid properties of arsenious anhydride is seen in the fact that if it be dissolved in ammonia water, and then a still stronger solution of ammonia be added, prismatic crystals separate having the composition of ammonium metarsenite, NH_4AsO_3. This ammonium salt deliquesces in air, and loses all its ammonia. The magnesium salt is tri-metallic, $Mg_3(AsO_3)_2$; it is insoluble in water, and is formed by mixing an ammoniacal solution of arsenious anhydride with an ammoniacal solution of a magnesium salt. It is insoluble even in ammonia, although it dissolves in an excess of acids. Magnesium hydroxide gives the same salt with arsenious solutions, and hence magnesia is one of the best antidotes for arsenic poisoning. *The arsenites of copper* are much used in the manufacture of colours, more especially of pigments. They are distinguished by their insolubility in water and by their remarkably vivid green colour, but at the same time by their poisonous character. Not only do such pigments applied to wall papers or other materials easily dust off from them, but they give exhalations containing AsH_3. The cupric salts, CuX_2, when mixed with an alkaline solution of arsenious acid, give a green precipitate of a copper salt called *Scheele's green*. Its composition is probably $CuHAsO_3$. Ammonia dissolves it, and gives a colourless solution, containing cuprous arsenate—that is, the cupric compound is reduced and the arsenic subjected to a further oxidation. The so-called *Schweinfurt green* was still more used, especially in former times; it is an insoluble green cupric salt, which resembles the preceding in many respects, but has a different tint. It is prepared by mixing boiling solutions of arsenious acid and cupric acetate. Arsenious acid forms an insoluble compound with ferric hydroxide, resembling the phosphate; and this is the reason why freshly precipitated oxide of iron is employed as an *antidote for arsenic*. The freshly precipitated oxide of iron, taken immediately after poisoning by arsenic, converts the arsenious acid into an insoluble state, by forming a compound on which the acids of the stomach have no action, so that the poisoning cannot proceed. It is remarkable that the inhabitants of certain mountainous countries accustom themselves to taking arsenic, as a means which, according to their experience, helps to overcome the fatigue of mountain ascents. Arsenious anhydride and certain of its salts are also used in medicine, naturally only in small quantities. When taken internally arsenic passes into the blood, and is mainly excreted by the urine.

Adie (1889) obtained compounds of As_2O_3 with 1, 2, 4, and 8 SO_3 by the direct action of ordinary and Nordhausen sulphuric acid upon As_2O_3.

Weber had previously obtained $As_2O_3SO_3$ (which disengages SO_3 at 225°), and also other $As_2O_3nSO_3$ (where n = 3, 6, and 8), by the action of the vapours of SO_3 upon As_2O_3 at a definite temperature. The compound $As_2O_3,8SO_3$ loses SO_3 at 100°. Oxide of antimony, Sb_2O_3, gives similar compounds. Adie (1891) also obtained (by the action of SO_3 upon H_3PO_4) a compound $H_3PO_43SO_3$ in the form of a viscous liquid decomposed by water.

Printers' type consists of an alloy known as 'type-metal,' containing usually about 15 parts of antimony to 85 parts of lead; sometimes (for example, for stereotypes) from 10 to 15 per cent. Bi or 8 per cent. Sn and even Cu is added. The hardness of the alloy, which is essential for printing, evidently depends upon the presence of antimony, but an excess must be avoided, since this renders the alloy brittle, and the type after a time loses its sharpness.

[40 bis] Antimony is prepared in a state of greater purity by heating with charcoal the oxide obtained by the action of nitric acid on the impure commercial metallic antimony. This is based on the fact that by the action of the acid, antimony forms the oxide Sb_2O_3, which is but slightly soluble in water. The arsenic, which is nearly always present, forms soluble arsenious and arsenic acids, and remains in solution. The purest antimony is easily obtained from tartar emetic, by heating it with a small quantity of nitre. Metallic antimony also occurs, although rarely, native; and as it is very easily obtained, it was known to the alchemists of the fifteenth century. Very pure metallic antimony may be deposited by the electric current from a solution of antimonious sulphide in sodium sulphide after the addition of sodium chloride to the solution.

As antimonious oxide answers to the type SbX_3, it is evident that compounds may exist in which antimony will replace three atoms of hydrogen; such compounds have been to some extent obtained, but they are easily converted by water into substances corresponding with the ordinary formulæ of the compounds of antimony. Thus tartar emetic, $C_4H_4(SbO)KO_6$, loses water when heated, and forms $C_4H_2SbKO_6$—that is, tartaric acid, $C_2H_6O_6$, in which one atom of hydrogen is replaced by potassium and three by antimony. But this substance is reconverted into tartar emetic by the action of water.

A similar compound is seen in that *intermediate oxide of antimony* which is formed when antimonious oxide is heated in air: its composition is SbO_2 or Sb_2O_4. This oxide may be regarded as orthantimonic acid, $SbO(HO)_3$, in which three atoms of hydrogen are replaced by antimony in that state in which it occurs in oxide of antimony—*i.e.* $SbO(SbO_3) = Sb_2O_4$. Oxide of antimony is also formed when antimonic acid is ignited; it then loses water and oxygen, and gives this intermediate oxide as a white infusible powder, of sp. gr. 6·7. It is somewhat soluble in water, and gives a solution which turns litmus paper red.

[41 bis] Beilstein and Blaese (1889), after preparing many salts of antimonic acid, came to the conclusion that it is monobasic, but all the salts still contain water, so that their general type is mostly: $MSbO_3 3H_2O$, for example, M = Li, Hg (salts of the suboxide), ½ Pb, &c. The type of the ortho-salts, M_2SbO_4, is quite unknown, although it is reproduced in the thio-compounds, for instance, Schlippe's salt, Na_2SbS_4, but this salt also contains water of crystallisation, $9H_2O$ (Chapter XX., Note 29).

Among the other compounds of antimony, *antimoniuretted hydrogen*, SbH_3, resembles arseniuretted hydrogen in its mode of formation and properties (it splits up at 150°, Brunn 1890; when liquified, it boils at -65° and solidifies at -92°), whilst the halogen compounds differ in many respects from those of arsenic. When chlorine is passed over an excess of antimony powder, it forms *antimony trichloride*, $SbCl_3$, but if the chlorine be in excess it forms the *pentachloride*, $SbCl_5$. The trichloride is a crystalline substance which melts at 72° and distils at 230°, whilst the pentachloride is a yellow liquid, which splits up into chlorine and the trichloride when heated; at 140° it begins to give off chlorine abundantly, carrying away the vapour of the trichloride with it; and at 200° the decomposition is complete, and pure antimonious chloride only passes over. This property of antimony pentachloride has caused it to be applied in many cases for the transference of chlorine; all the more that when it has given up its chlorine, it leaves the trichloride, which is able to absorb a fresh amount of chlorine; and therefore many substances which are unable to react directly with gaseous chlorine do so with antimony pentachloride, and in the presence of a small quantity of it chlorine will act on them, just as oxygen is able, in the presence of nitrogen oxides, to oxidise substances which could not be oxidised by means of free oxygen. Thus carbon bisulphide is not acted on by chlorine at low temperatures—this reaction requires a high temperature—but in the presence of antimony pentachloride its conversion into carbon tetrachloride takes place at low temperatures. Antimony tri- and pentachloride, having the character of chloranhydrides, fume in air, attract moisture, and are decomposed by water, forming antimonious and antimonic acids. But in the first action of water the trichloride does not evolve all its chlorine as hydrochloric acid, which is intelligible in view of the fact that antimonious anhydride is also a base, and is therefore able to react with acids; indeed antimony sulphide dissolved in an excess of hydrochloric acid (hydrogen sulphide is evolved) gives an aqueous solution of antimony trichloride, which, when carefully distilled, even gives the anhydrous compound. Antimony trichloride is only decomposed by an excess of water, and then not completely, for with a large quantity of water it forms *powder of algaroth*—*i.e.* antimony oxychloride. The first action of water consists in the formation of *oxychloride*, SbOCl—that is, a salt corresponding to oxide of antimony as a base. If antimony oxide or antimony chloride be dissolved in an excess of hydrochloric acid, and the solution diluted with a considerable amount of water, then this same powder of algaroth is precipitated. The composition varies with the relative amount of water; namely, between the limits SbOCl and $Sb_4O_5Cl_2$. The latter compound is, as it were, a basic salt of the former, because its composition = $2(SbOCl)Sb_2O_3$.

With bromine and iodine, antimony forms compounds similar to those with chlorine. Antimonious bromide, $SbBr_3$, crystallises in colourless prisms, melts at 94°, and boils at 270°; antimonious iodide, SbI_3, forms red crystals of sp. gr. 5·0; antimony trifluoride, SbF_3 separates from a solution of antimonious oxide in hydrofluoric acid, and SbF_5 is formed by a similar treatment of antimonic acid. The latter gives easily-soluble double salts with the fluorides of the metals of the alkalis.

De Haën (1887) obtained very stable double soluble salts, SbF_3,KCl (100 parts of water dissolve 57 parts of salt), SbF_3,K_2SO_4, &c., which he proposed to make use of in the arts as very easily crystallisable and soluble salts of antimony.

Engel, by passing hydrochloric acid gas into a saturated solution of antimonious chloride at 0°, obtained a compound $HCl,2SbCl_3,2H_2O$, and with the pentachloride a compound $SbCl_5,5HCl,10H_2O$. Bismuth trichloride, $BiCl_3$, gives a similar compound.

Saunders (1892) obtained $5RbCl,3SbCl_3$ and $RbCl,SbCl_3$. Ditte and Metzner (1892) showed that Sb and Bi dissolve in hydrochloric acid only owing to the participation of the oxygen of the air or of that dissolved in the acid.

Metallic bismuth is very easily obtained when the compounds of the oxide are reduced by powerful reducing agents, but when less powerful reducing agents—for example, stannous oxide—are taken, bismuth suboxide is formed as a black crystalline powder. It is a compound of the type BiX_2, its composition being BiO; it is decomposed by acids into the metal and oxide, which passes into solution.

The type BiX_5 is represented by the pentoxide, Bi_2O_5, its metahydrate, Bi_2O_5,H_2O, or $BiHO_3$, known as bismuthic acid, and the pyrohydrate, $Bi_2H_4O_7$. *Bismuth pentoxide* is obtained by the prolonged passage of chlorine through a boiling solution of potassium hydroxide (sp. gr. 1·38), containing bismuth oxide in suspension; the precipitate is washed with water, with boiling nitric acid (but not for long, as otherwise the bismuthic acid is decomposed), then again with water, and finally the resultant bright red powder of the hydrate BiHO$_3$ is dried at 125°. The prolonged action of nitric acid on bismuthic anhydride, Bi$_2$O$_5$, results in the formation of the compound Bi$_2$O$_4$,H$_2$O, which decomposes in moist air, forming Bi$_2$O$_3$. The density of bismuthic anhydride is 5·10, of the tetroxide, Bi$_2$O$_4$, 3·60, and of bismuthic acid, BiHO$_3$, 5·75. *Pyrobismuthic acid*, Bi$_2$H$_4$O$_7$, forms a brown powder, which loses a portion of its water at 150°, and decomposes on further heating, with the evolution of oxygen and water. It is obtained by the action of potassium cyanide on a solution of bismuth nitrate. The meta-salts of bismuthic acid are known, for example KBiO$_3$. They generally occur, however, in combinations with metabismuthic acid itself. Thus André (1891) took a solution of the double salt of BiBr$_3$ and KBr, treated it with bromine after adding ammonia, and obtained a red-brown precipitate, which after being washed (for several weeks) had the composition KBiO$_3$,HBiO$_3$ When washed with dilute nitric acid this salt gave bismuthic acid.

[44 bis] Hérard (1889) obtained a peculiar variety of bismuth by heating pure crystalline bismuth to a bright red heat in a stream of nitrogen. A greenish vapour was deposited in the cooler portions of the apparatus in the form of a grey powder, which under the microscope had the appearance of minute globules. An atmosphere of nitrogen is necessary for this transformation, other gases such as hydrogen and carbonic oxide do not favour the transition. The resultant amorphous bismuth fuses at 410° (the crystalline variety at 269°), sp. gr. 9·483. (Does it not contain a nitride?)

Basic bismuth carbonate is employed for whitening the skin (veloutine, &c.)

With an excess of water a further quantity of acid is separated and a still more basic salt formed. The ultimate product, on which an excess of water has apparently no action whatever, is a substance having the composition BiO(NO$_3$).BiO(OH). In the latter salt we see the limit of change, and this limit appears to show that the type of the saline compounds of bismuthic oxide is of the form Bi$_2$X$_6$, and not BiX$_3$; but it is very probable, on the basis of the examples which we considered in the case of lead, that this type should be still further polymerised in order to give a correct idea of the type of the bismuthous compounds. If we refer all the bismuthous compounds to this type, Bi$_2$X$_6$, we shall obtain the following expression for the composition of the nitrates: normal salt, Bi$_2$(NO$_3$)$_6$, first basic salt, Bi$_2$O(OH)$_2$(NO$_3$)$_2$, magistery of bismuth, Bi$_2$(OH)$_4$(NO$_3$)$_2$, and the limiting form Bi$_2$O$_2$(OH)(NO$_3$).

The general character of bismuthous oxide in its compounds is well-exemplified in the nitrate; bismuthous chloride, $BiCl_3$, which is obtained by heating bismuth in chlorine, or by dissolving it in aqua regia, and then distilling without access of air, is also decomposed by water in exactly the same manner, and forms basic salts—for instance, first, BiOCl, like the above salt of nitric acid. Bismuth chloride boils at 447° and probably its formula is $BiCl_3$. Polymerisation may take place in some compounds and not in others. A volatile compound of the composition $Bi(C_2H_5)_3$ is also known as a liquid which is insoluble in water and decomposes with explosion when heated at 130°. Double salts containing chloride of bismuth are: $2(KCl)BiCl_3 2H_2O$ (from a solution of Bi_2O_3 and KCl in hydrochloric acid) and $KClBiCl_3 H_2O$. Bigham (1892) also obtained $KBr(SO_4)_2$ in tabular crystals by treating the above-named double salt with strong sulphuric acid. The composition of this salt recalls that of alum.

As the metals contained in alloys like the above (bismuth, lead, tin, cadmium) are difficultly volatile and their alloys are fusible, they may be employed in the place of mercury in many physical experiments conducted at or above 70°, and they offer the advantage that they do not give any vapour having an appreciable tension (mercury at 100°, 0·75 mm.) Bismuth expands in passing into a molten state, but it has a temperature of maximum density. According to Luedeking the mean coefficient of expansion of liquid bismuth is 0·0000442 (between 270° and 303°), and of solid bismuth 0·0000411.

Although, guided by Brauner, who showed that didymium gives a higher oxide, Di_2O_5, I place this element in the fifth group, still I am not certain as to its position, because I consider that the questions relating to this metal are still far from being definitely answered.

When the vapours of vanadium oxychloride are heated with zinc in a closed tube at 400°, they lose a portion of their chlorine and form a green crystalline mass of sp. gr. 2·88, which is deliquescent in air and has the composition $VOCl_2$. Only its vapour density is unknown, and it would be extremely important to determine whether its molecular composition is that given above, or whether it corresponds with the formula $V_2O_2Cl_4$. Another less volatile oxychloride, VOCl, is formed with it as a brown insoluble substance, which is, however, soluble in nitric acid like the preceding. Roscoe obtained a still less chlorinated substance, namely, $(VO)_2Cl$; but it may only consist of a mixture of VO and VOCl. At all events, we here find a graduated series such as is met with in the compounds of very few other elements.

Strong acids and alkalis dissolve vanadic anhydride in considerable quantities, forming yellow solutions. When it is ignited, especially in a current of hydrogen, it evolves oxygen and forms the lower oxides; V_2O_4 (acid solutions of a green colour, like the salts of chromic oxide), V_2O_3, and the lowest oxide, VO. The latter is the metallic powder which is obtained when the vanadium oxychloride is heated in an excess of hydrogen, and was formerly mistaken for metallic vanadium. When a solution of vanadic acid is treated with metallic zinc it forms a blue solution, which seems to contain this oxide. It acts as a reducing agent (and forms a close analogue to chromous oxide, CrO). Metallic *vanadium* can only be obtained from vanadium chloride which is quite free from oxygen. Moissan (1893) obtained it by reducing the oxide with carbon in the electric furnace, and considered it to be most infusible of the metals in the series Pt, Cr, Mo, U, W, and V (he also obtained a compound of vanadium and carbon). The specific gravity of this metal is 5·5. It is of a grey-white colour, is not decomposed by water, and is not oxidised in air, but burns when strongly heated, and can be fused in a current of hydrogen (forming perhaps a compound with hydrogen). It is insoluble in hydrochloric acid, but easily dissolves in nitric acid, and when fused with caustic soda it forms sodium vanadate.

As regards the salts of vanadic acid, three different classes are known; the first correspond with metavanadic acid, $VMO_3 = M_2OV_2O_5$, the second correspond with the dichromates—that is, have the composition $V_4M_2O_{11}$, which is equal to $M_2O + 2V_2O_5$—and the third correspond with orthovanadic acid, VM_3O_4 or $3M_2O + V_2O_5$. The latter are formed when vanadic anhydride is fused with an excess of an alkaline carbonate.

Vanadic acid gives the so-called 'complex' acids (which are considered more fully in Chapter XXI. in speaking of Mo and W)—*i.e.* acids formed of two acids assimilated into one. Thus Friedheim (1890) obtained phosphor-vanadic acid, and Schmitz-Dumont (1890) a similar arseno-vanadic acid. The former is obtained by heating V_2O_5 with sirupy phosphoric acid. The resultant golden-yellow tabular crystals have the composition $H_2OV_2O_5P_2O_59H_2O$, and there are corresponding salts—for example, $(NH_4)_2V_2O_5P_2O_5$ with 3 and $7H_2O$, &c. These salts cannot be separated by crystallisation, so that there are 'complexes' of these acids in a whole series of salts (and also in nature). It may be supposed (Friedheim) that V_2O_5 here, as it were, plays the part of a base, or that those acids may be looked upon as double salts. Among the true double salts of vanadium (Nb and Ta) very many are known among the fluorides, such as VF_32NH_4F, VOF_22NH_4F, $VO_2F,3NH_4F$, &c. (Pettersson, Piccini, and Georgi, 1890–92).

Vanadium was discovered at the beginning of this century by Del-Rio, and afterwards investigated by Sefström, but it was only in 1868 that Roscoe established the above formulæ of the vanadic compounds.

The researches made by Roscoe were preceded by those of Marignac in 1865, on the *compounds* of *niobium* and *tantalum*, to which were also ascribed different formulæ from those now recognised. Tantalum was discovered simultaneously with vanadium by Hatchett and Ekeberg, and was afterwards studied by Rose, who in 1844 discovered niobium in it. Notwithstanding the numerous researches of Hermann (in Moscow), Kobell, Rose, and Marignac, still there is not yet any certainty as to the purity of, and the properties ascribed to, the compounds of these elements. They are difficult to separate from each other, and especially from the cerite metals and titanium, &c., which accompany them. Before the investigations of Rose the highest oxide of tantalum was supposed to belong to the type TaX_6—that is, its composition was taken as TaO_3, and to the lower oxide was ascribed a formula TaO_2. Rose gave the formula TaO_2 to the higher oxide, and discovered a new element called niobium in the substance previously supposed to be the lower oxide. He even admitted the existence of a third element occurring together with tantalum and niobium, which he named pelopium, but he afterwards found that pelopic acid was only another oxide of niobium, and he considered it probable that the higher oxide of this element is NbO_2, and the lower Nb_2O_3. Hermann found that niobic acid which was considered pure contained a considerable quantity of tantalic acid, and besides this he admitted the existence of another special metallic acid, which he called ilmenic acid, after the locality (the Ilmen mountains of the Urals) of the mineral from which he obtained it. V. Kobell recognised still another acid, which he called dianic acid, and these diverse statements were only brought into agreement in the sixties by Marignac. He first of all indicated an accurate method for the separation of tantalic and niobic compounds, which are always obtained in admixture.

If niobic acid be mixed with a small quantity of charcoal and ignited in a stream of chlorine, a difficultly-fusible and difficultly-volatile oxychloride, $NbOCl_3$ separates. The vapour density of this compound with respect to air is 7·5, and this vapour density perfectly confirms the accuracy of the formulæ given by Marignac, and indicates the quantitative analogy between the compounds of niobium and tantalum, and those of phosphorus and arsenic, and consequently also of vanadium. In their qualitative relations (as is evident also from the correspondence of the atomic weights), the compounds of tantalum and niobium exhibit a great analogy with the compounds of molybdenum and tungsten. Thus zinc, when acting on acid solutions of tantalic and niobic compounds, gives a blue coloration, exactly as it does with those of tungsten and molybdenum (also titanium). These acids form the same large number of salts as those of tungsten and molybdenum. The anhydrides of the acids are also insoluble in water, but as colloids are sometimes held in solution, just like those of titanic and molybdic acids. Furthermore, niobium is in every respect the nearest analogue of molybdenum, and tantalum of tungsten. *Niobium* is obtained by reducing the double fluoride of niobium and sodium, with sodium. It is difficult to obtain in a pure state. It is a metal on which hydrochloric acid acts with some energy, as also does hydrofluoric acid mixed with nitric acid, and also a boiling solution of caustic potash. *Tantalum*, which is obtained in exactly the same way, is a much heavier metal. It is infusible, and is only acted on by a mixture of hydrofluoric and nitric acids. Rose in 1868 showed that in the reduction of the double fluoride, $NbF_5,2KF$, by sodium, a greyish powder is obtained after treating with water. The specific gravity of this powder is 6·8, and he considers it to be niobium hydride, NbH. Neither did he obtain metallic niobium when he reduced with magnesium and aluminium, but an alloy, Al_3Nb, having a sp. gr. of 4·5.

Niobium, so far as is known, unites in three proportions with oxygen. NbO, which is formed when $NbOF_3,2KF$ is reduced by sodium; NbO_2, which is formed by igniting niobic acid in a stream of hydrogen, and niobic anhydride, Nb_2O_5, a white infusible substance, which is insoluble in acids, and has a specific gravity of 4·5. Tantalic anhydride closely resembles niobic anhydride, and has a specific gravity of 7·2. *The tantalates and niobates* present the type of ortho-salts—for example, $Na_2HNbO_4,6H_2O$, and also of pyro-salts, such as $K_3HNb_2O_7,6H_2O$, and of meta-salts—for example, $KNbO_3,2H_2O$. And, besides these, they give salts of a more complex type, containing a larger amount of the elements of the anhydride; thus, for instance, when niobic anhydride is fused with caustic potash it forms a salt which is soluble in water, and crystallises in monoclinic prisms, having the composition $K_8Nb_6O_{19},16H_2O$. There is a perfectly similar isomorphous salt of tantalic acid. Tantalite is a salt of the type of metatantalic acid, $Fe(TaO_3)_2$. The composition of Yttrotantalite appears to correspond with orthotantalic acid.

CHAPTER XX
SULPHUR, SELENIUM, AND TELLURIUM

The acid character of the higher oxides RO_3 of the elements of group VI. is still more clearly defined than that of the higher oxides of the preceding groups, whilst feeble basic properties only appear in the oxides RO_3 of the elements of the even series, and then only for those elements having a high atomic weight—that is, under those two conditions in which, as a rule, the basic characters increase. Even the lower types RO_2 and R_2O_3, &c., formed by the elements of group VI., are acid anhydrides in the uneven series, and only those of the elements of the even series have the properties of peroxides or even of bases.

Sulphur is the typical representative of group VI., both on account of the fact that the acid properties of the group are clearly defined in it, and also because it is more widely distributed in nature than any of the other elements belonging to this group. As an element of the uneven series of group VI., sulphur gives H_2S, sulphuretted hydrogen, SO_3, sulphuric anhydride, and SO_2, sulphurous anhydride. And in all of them we find acid properties—SO_3 and SO_2 are anhydrides of acids, and H_2S is an acid, although a feeble one. As an element sulphur has all the properties of a true non-metal; it has not a metallic lustre, does not conduct electricity, is a bad conductor of heat, is transparent, and combines directly with metals—in short it has all the properties of the non-metals, like oxygen and chlorine. Furthermore, sulphur exhibits a great qualitative and quantitative *resemblance to oxygen*, especially in the fact that, like oxygen, it combines *with two atoms of hydrogen*, and forms compounds resembling oxides with metals and non-metals. From this point of view sulphur is bivalent, if the halogens are univalent. The chemical character of sulphur is expressed by the fact that it forms a very slightly stable and feebly energetic acid with hydrogen. The salts corresponding with this acid are the sulphides, just as the oxides correspond to water and the chlorides to hydrochloric acid. However, as we shall afterwards see more fully, the sulphides are more analogous to the former than to the latter. But although combining with metals, like oxygen, sulphur also forms chemically stable compounds with oxygen, and this fact impresses a peculiar character on all the relations of this element.

Sulphur belongs to the number of those elements which *are very widely distributed in nature*, and occurs both free and combined in various forms. The atmosphere, however, is almost entirely free from compounds of sulphur, although a certain amount of them should be present, if only from the fact that sulphurous anhydride is emitted from the earth in volcanic eruptions, and in the air of cities, where much coal is burnt, since this always contains FeS_2. Sea and river water generally contain more or less sulphur in the form of sulphates. The beds of gypsum, sodium sulphate, magnesium sulphate, and the like are formations of undoubtedly aqueous origin. The sulphates contained in the soil are the source of the sulphur found in plants, and are indispensable to their growth. Among vegetable substances, the proteïds always contain from one to two per cent. of sulphur. From plants the albuminous substances, together with their sulphur, pass into the animal organism, and therefore the decomposition of animal matter is accompanied by the odour of sulphuretted hydrogen, as the product into which the sulphur passes in the decomposition of the albuminous substances. Thus a rotten egg emits sulphuretted hydrogen. Sulphur occurs largely in nature, as the various insoluble sulphides of the metals. Iron, copper, zinc, lead, antimony, arsenic, &c., occur in nature combined with sulphur. These *sulphides* frequently have a metallic lustre, and in the majority of cases occur crystallised, and also very often several sulphides occur combined or mixed together in these crystalline compounds. If they are yellow and have a metallic lustre they are called pyrites. Such are, for example, copper pyrites, $CuFeS_2$, and iron pyrites, FeS_2, which is the commonest of all. They are all also known as glances or blendes if they are greyish and have a metallic lustre—for example, zinc blende, lead glance, PbS, antimony glance, Sb_2S_3, &c. And, lastly, sulphur occurs *native*. It occurs in this form in the most recent geological formations in admixture with limestone and gypsum, and most frequently in the vicinity of active or extinct volcanoes. As the gases of volcanoes contain sulphur compounds—namely, sulphuretted hydrogen and sulphurous anhydride, which by reacting on one another may produce sulphur, which also frequently appears in the craters of volcanoes as a sublimate—it might be imagined that the sulphur was of volcanic origin. But on a nearer acquaintance with its mode of occurrence, and more especially considering its relation to gypsum, $CaSO_4$, and limestone, the present general opinion leads to the conclusion that the 'native' sulphur has been formed by the reduction of the gypsum by organic matter and that its occurrence is only indirectly connected with volcanic agencies. Near Tetush, on the Volga, there are beds containing gypsum, sulphur, and asphalt (mineral tar). In Europe the most important deposits of sulphur are in the south of Sicily from Catania to Girgenti. There are very rich deposits of sulphur in Daghestan near Cherkai and Cherkat in Khyut, near Mount Kanabour-bam, near Petrovsk, and in the Kira Koumski steppes in the Trans-Caspian provinces, which are able to supply the whole of Russia with this mineral. Abundant deposits of sulphur have also been found in Kamtchatka in the neighbourhood of the volcanoes. The method of separation of the sulphur from its earthy impurities is based on the fact that sulphur melts when it is heated. The fusion is carried on at the expense of a portion of the sulphur, which is burnt, so that the remainder may melt and run from the mass of the earth. This is carried on in special furnaces called calcaroni, built up of unhewn stone in the neighbourhood of the mines.

FIG. 86.—Refining sulphur by sublimation.

Sulphur is purified by distillation in special retorts (see fig. 86) by passing the vapour into a chamber G built of stone. The first portions of the vapour entering into the condensing chamber are condensed straightway from the vapour into a solid state, and form a fine powder known as *flowers of sulphur*. But when the temperature of the receiver attains the melting point of sulphur, it passes into a liquid state and is cast into moulds (like sealing wax), and is then known under the name of *roll sulphur*.

In an uncombined state sulphur exists in *several modifications*, and forms a good example of the facility with which an alteration of properties can take place without a change of composition—that is, as regards the material of a substance. Common sulphur has the well-known yellow colour. This colour fades as the temperature falls, and at -50° sulphur is almost colourless. It is very brittle, so that it may be easily converted into a powder, and it presents a crystalline structure, which, by the way, shows itself in the unequal expansion of lumps of sulphur by heat. Hence when a piece of sulphur is heated by the warmth of the hand, it emits sounds and sometimes cracks, which probably also depends on the bad heat-conducting power of this substance. It is easily obtained in a crystalline form by artificial means, because although insoluble in water it dissolves in carbon bisulphide, and in certain oils. Solutions of sulphur in carbon bisulphide when evaporated at the ordinary temperature yield well-formed transparent crystals of sulphur in the form of rhombic octahedra, in which form it occurs native. The specific gravity of these crystals is 2·045. Fused sulphur, cast into moulds and cooled, has, after being kept a long time, a specific gravity 2·066; almost the same as that of the crystalline sulphur of the above form, which shows that common sulphur is the same as that which crystallises in octahedra. The specific heat of octahedral sulphur is 0·17; it melts at 114°, and forms a bright yellow mobile liquid. On further heating, the fused sulphur undergoes an alteration, which we shall presently describe, first observing that the above octahedral state of sulphur is its most stable form. Sulphur may be kept at the ordinary temperature in this form for an indefinite length of time, and many other modifications of sulphur pass into this form after being left for a certain time at ordinary temperature.

If sulphur be melted and then slightly cooled, so that it forms a crust on the surface and over the sides of the crucible, while the internal mass remains liquid, then the sulphur takes another crystalline form as it solidifies. This may be seen by breaking the crust, and pouring out the remaining molten sulphur. It is then found that the sides of the crucible are covered with *prismatic crystals* of the monoclinic system; they have a totally different appearance from the above-described crystals of rhombic sulphur. The prismatic crystals are brown, transparent, and less dense than the crystals of rhombic sulphur, their specific gravity being only 1·93, and their melting point higher—about 120°. These crystals of sulphur cannot be kept at the ordinary temperature, which is indeed evident from the fact that in time they turn yellow; the specific gravity also changes, and they pass completely into the ordinary modification. This is accompanied by a considerable development of heat, so that the temperature of the mass may rise 12°. Thus sulphur is *dimorphous*—that is, it exists in two crystalline forms, and in both forms it has independent physical properties. However, no chemical reactions are known which distinguish the two modifications of sulphur, just as there are none distinguishing aragonite from calcspar.

If molten sulphur be heated to 158° it loses its mobility and becomes thick and very dark-coloured, so that the crucible in which it is heated may be inverted without the sulphur running out. When heated above this temperature the sulphur again becomes liquid, and at 250° it is very mobile, although it does not acquire its original colour, and at 440° it boils. These modifications in the properties of sulphur depend not only on the variations of temperature, but also on a change of structure. If sulphur, heated to about 350°, be poured in a thin stream into cold water, it does not solidify into a solid mass, but retains its brown colour and *remains soft*, may be stretched out into threads, and is elastic, like guttapercha. But in this soft and ductile state, also, it does not remain for a long time. After the lapse of a certain period this soft transparent sulphur hardens, becomes opaque, passes into the ordinary yellow modification of sulphur, and in so doing develops heat, just as in the conversion of the prismatic into the octahedral variety. The soft sulphur is characterised by the fact that a certain portion of it is insoluble in carbon bisulphide. When soft sulphur is immersed in this liquid, only a portion of common sulphur passes into solution, whilst a certain portion is quite insoluble and remains so for a long time. The maximum proportion of insoluble sulphur is obtained by heating slightly above 170°. It melts at 114°. An exactly similar *insoluble amorphous sulphur* is obtained in certain reactions in the wet way, when sulphur separates out from solutions. Thus sodium thiosulphate, $Na_2S_2O_3$, when treated with acids, gives a precipitate of sulphur, which is insoluble in carbon bisulphide. The action of water on sulphur chloride also gives a similar modification of sulphur. Certain sulphides, when treated with nitric acid, also yield sulphur in this form.

At temperatures of 440° to 700° the vapour density of sulphur is 6·6 referred to air—*i.e.* about 96 referred to hydrogen. Hence, at these temperatures *the molecule of sulphur contains six atoms*, it has the composition S_6. The agreement between the observations of Dumas, Mitscherlich, Bineau, and Deville confirms the accuracy of this result. But in this respect the properties of sulphur were found to be variable. When heated to higher temperatures, that is to say, *above* 800°, the vapour density of sulphur is found to be one-third of this quantity, *i.e.* about 32 referred to hydrogen. At this temperature *the molecule of sulphur*, like that of hydrogen, oxygen, nitrogen, and chlorine, *contains two atoms*; hence the molecular formula is then S_2. This variation in the vapour density of sulphur evidently corresponds with a polymeric modification, and may be likened to the transformation of ozone, O_3, into oxygen, O_2, or better still, of benzene, C_6H_6, into acetylene, C_2H_2.

In its faculty for combination, sulphur most closely resembles oxygen and chlorine; like them, it combines with nearly all elements, with the development of heat and light, forming sulphur compounds, but as a rule this only takes place at a high temperature. At the ordinary temperature it does not enter into reactions, owing, amongst other things, to the fact that it is a solid. In a molten state it acts on most metals and on the halogens. It burns in air at about 300°, and with carbon at a red heat, but it does not combine with nitrogen.

Fine wires, or the powders of the greater number of metals, burn in the vapour of sulphur. The direct combination of hydrogen with sulphur is restricted by a limit—that is, at a given temperature and under other given conditions it does not proceed unrestrictedly; there is no explosion or recalescence. Sulphuretted hydrogen, H_2S, decomposes at its temperature of combination—that is, it is easily dissociated. The same phenomenon is repeated here as with water, except that the temperatures at which the attraction of hydrogen for sulphur begins and ceases are much lower than in the case of oxygen and hydrogen. The temperature at which combination takes place is here, as in many other instances, nearly the same as that at which dissociation begins. Hence *sulphuretted hydrogen* is formed in a small quantity by the direct ignition of a mixture of the vapour of sulphur and hydrogen. However, the temperature must not be high, because otherwise the whole of the sulphuretted hydrogen is decomposed; but at lower temperatures a small amount of sulphuretted hydrogen is formed by direct combination. Sulphuretted hydrogen however, like all other hydrogen compounds, may be easily obtained by the double decomposition of its corresponding metallic compounds, the replacement of the metal by hydrogen being effected by the action of acids on the sulphides. The metallic sulphides are, as a rule, easily formed. A sulphide, when mixed with a non-volatile acid, may give, by double decomposition, a salt of the acid taken and sulphuretted hydrogen, $M_2S + H_2SO_4 = H_2S + M_2SO_4$. However, it is not all sulphides nor solutions of all acids that will evolve sulphuretted hydrogen, which fact is exceedingly characteristic, because, for example, all carbonates evolve carbonic anhydride when treated with any acid. Sulphuric acid will only evolve sulphuretted hydrogen from those sulphides which contain a metal capable of decomposing the acid with the evolution of hydrogen. Thus zinc, iron, calcium, magnesium, manganese, potassium, sodium, &c., form sulphides which evolve sulphuretted hydrogen when treated with sulphuric acid, and the metals themselves evolve hydrogen with acids. The sulphides of those metals which do not liberate hydrogen from acids do not generally act on acids—that is, do not form sulphuretted hydrogen with them; such are, for example, the sulphides of lead, silver, copper, mercury, tin, &c. Therefore, the *modus operandi* of the formation of sulphuretted hydrogen by the action of acids on metallic sulphides may be looked on as a phenomenon of the combination of hydrogen, at the moment of its evolution, with the sulphur, which is combined with the metal. Such a representation is all the more simple as all the circumstances under which sulphuretted hydrogen is formed are exactly similar to the conditions of the formation of hydrogen itself. Thus the usual mode of preparing sulphuretted hydrogen is by the action of *sulphuric acid on ferrous sulphide*, in which the same apparatus and method are employed as in the preparation of hydrogen, only replacing the metallic iron or zinc by ferrous sulphide or zinc sulphide. The reaction between sulphide of iron and sulphuric acid takes place at the ordinary temperature, and is accompanied by just as small a development of heat as in the liberation of hydrogen itself, $FeS + H_2SO_4 = FeSO_4 + H_2S$.

In nature sulphuretted hydrogen is formed in many ways. The most usual mode of its formation is by the decomposition of albuminous substances containing sulphur, as mentioned above. Another method is by the reducing action of organic matter on sulphates, and by the action of water and carbonic acid on the sulphides formed by this reduction. Volcanic eruptions are a third source of sulphuretted hydrogen in nature. Although sulphuretted hydrogen is formed in small quantities everywhere, it nevertheless soon disappears from the atmosphere, owing to its being easily decomposed by oxidising agencies. Many mineral waters contain sulphuretted hydrogen, and smell of it; they are called 'sulphur waters.'

Sulphuretted hydrogen, at the ordinary temperature, is a colourless gas, having a very unpleasant odour. It has, as its composition H_2S shows, a specific gravity seventeen times greater than hydrogen, and therefore it is somewhat heavier than air. Sulphuretted hydrogen *liquefies* at about -74°, and at the ordinary temperature when subjected to a pressure of 10 to 15 atmospheres; at -85° it is converted into a solid crystalline mass.[15 bis] The easy liquefaction of sulphuretted hydrogen is evidently allied to its solubility. One volume of water at 0° dissolves 4·37 volumes of sulphuretted hydrogen, at 10° 3·58 volumes, and at 20° 2·9 volumes. The solutions impart a very feeble red coloration to litmus paper. This gas is poisonous. One part in fifteen hundred parts of air will kill birds. Mammalia die in an atmosphere containing $1/200$ of this gas.

Sulphuretted hydrogen is very easily *decomposed* into its component parts by the action of heat or a series of electric sparks. Hence it is not surprising that sulphuretted hydrogen undergoes change under the action of many substances having a considerable affinity for hydrogen and oxygen. Very many metals evolve hydrogen with sulphuretted hydrogen, so that in this respect it presents the property of an acid; for instance, $2H_2S + Sn = 2H_2 + SnS_2$. This may be taken advantage of for determining the composition of sulphuretted hydrogen, because a given volume then leaves the same volume of hydrogen. On the other hand, oxygen, chlorine, and even iodine decompose sulphuretted hydrogen, removing the hydrogen from it and leaving free sulphur, so that in this reaction the sulphur is replaced by the above-named elements; for example, $H_2S + Br_2 = 2HBr + S$. In no other hydrogen compound is it so easy to show the *substitution*, both of hydrogen and of the element combined with it, as in hydrogen sulphide. This clearly proves the feeble union between the elements forming this gas. Compounds containing a considerable amount of oxygen, with which they easily part, can accomplish the separation of the sulphur very easily. Such are, for instance, nitrous acid, chromic acid, and even ferric oxide and the higher oxides like it. Thus, if sulphuretted hydrogen be passed into a solution of chromic acid or an acid solution of ferric oxide, water is formed, *and the sulphur is separated in a free state*. Thus, sulphuretted hydrogen acts as a *reducing agent*, in virtue of the hydrogen it contains. Salts of iodic, chlorous, chloric, and other acids are reduced by sulphuretted hydrogen, their oxygen acting mainly on its hydrogen; but in the presence of an excess of a powerful oxidising agent a portion of the sulphur may also be oxidised to sulphurous anhydride. The reducing action of sulphuretted hydrogen is frequently applied in chemical manipulations for the preparation of lower oxides, and for the conversion of certain oxygen compounds into hydrogen compounds: thus, the higher oxides of nitrogen are converted into ammonia by it, and in the presence of alkalis the nitro-compounds are converted into ammonia derivatives. The reaction of sulphuretted hydrogen on sulphurous anhydride belongs to this class of phenomena, the chief products of which are sulphur and water, $2H_2S + SO_2 = 2H_2O + S_3$.

The acid character of sulphuretted hydrogen is clearly seen in its action on alkalis and salts.[19 bis] Thus lead oxide and its salts in the presence of sulphuretted hydrogen form water or an acid, and sulphide of lead: $PbX_2 + H_2S = PbS + 2HX$. This reaction takes place even in the presence of powerful acids, because lead sulphide is one of those sulphides which are unacted on by acids, and in solutions the reaction is a complete one. This reaction is taken advantage of for the preparation of many acids, by first converting into a lead salt, and then submitting this salt to the action of sulphuretted hydrogen. For example, lead formate with sulphuretted hydrogen gives formic acid. Sulphuretted hydrogen in acting on a number of metallic acid substances in solution or in an anhydrous state also forms corresponding sulphates: (1) if it does not reduce the acid; (2) if the sulphur compound corresponding with the anhydride of the acid be insoluble in water, the reaction proceeds in solutions; (3) if the sulphuretted hydrogen and the acid taken do not come in contact with an alkali, on which they would be able to act first; and (4) if the sulphur compound be not decomposed by water. Thus solutions of arsenious acid give a precipitate of arsenious sulphide, As_2S_3, with sulphuretted hydrogen. This reaction proceeds not only in the presence of water, but also of acids, because the latter do not decompose the resultant sulphur compounds. The type of the decomposition is the same as with bases—that is, the sulphur and oxygen change places: $RO_n + nH_2S = RS_n + nH_2O$. Some sulphides corresponding with acid anhydrides are decomposed by water, and therefore are not formed in the presence of water. Such, for example, are the sulphides of phosphorus.

The metallic sulphides corresponding with the metallic oxides have either a feeble alkaline or a feeble acid character, according to the character of the corresponding oxide, and therefore by combining together they are able to form saline substances—that is, salts in which the oxygen is replaced by sulphur. Thus sulphuretted hydrogen having the properties of a feeble acid has, at the same time, the properties of water, and forms the type of the sulphur derivatives, which may also be formed by means of sulphuretted hydrogen, just as the oxides may be formed by the aid of water. But as sulphuretted hydrogen has acid properties, it combines more easily with the basic metallic sulphides. Hence, for instance, there exists a compound of sulphuretted hydrogen with potassium sulphide, potassium hydrosulphide, $2KHS = K_2S + H_2S$, just as there are potassium hydroxides; but there are scarcely any compounds of sulphuretted hydrogen with the sulphides corresponding with acids. Thus the sulphides of the metals may be regarded either as salts of sulphuretted hydrogen or as oxides of the metals in which the oxygen is replaced by sulphur. In general terms the sulphides exhibit the same degrees of difference with respect to their solubility in water as do the oxides. Thus the oxides of the alkali metals, and of some of the metals of the alkaline earths, are soluble in water, whilst those of nearly all the other metals are insoluble. The same may be said as to the sulphides; the sulphides of the metals of the alkalis and certain of the alkaline earths are soluble in water, whilst those of the other metals are insoluble. Those metals, like aluminium, whose oxides—for example, Al_2O_3—have intermediate properties and do not form compounds with feeble acids, at least in a wet way, also do not form sulphides by this method, although these may be obtained indirectly. And in general the sulphides of the metals are easily formed in a wet way, and with particular ease if they are insoluble in water. In this case their salts enter into double decomposition with sulphuretted hydrogen, or with soluble sulphides, and give an insoluble sulphide—for instance, a salt of lead gives lead sulphide with sulphuretted hydrogen. By the action of sulphuretted hydrogen on a salt of a metal, a free acid must be formed besides the metallic sulphide. Thus if a metal M be in a state of combination MX_2, then by the action of sulphuretted hydrogen there will be formed, besides MS, an acid 2HX. It is evident that sulphuretted hydrogen will not precipitate an insoluble sulphide from the salts of those metals whose sulphides react with free acid, such as zinc, iron, manganese, &c. The reaction $FeCl_2 + H_2S = FeS + 2HCl$, and the like, do not take place because the acid acts on the ferrous sulphide. Antimonious sulphide is not acted on by dilute hydrochloric acid, but it is decomposed by strong acid, and therefore in presence of an excess of hydrochloric acid antimonious chloride does not entirely react with hydrogen sulphide, whilst the reaction $2SbCl_3 + 3H_2S = Sb_2S_3 + 6HCl$ is a complete one in a dilute solution and with a small quantity of acid. Those metallic sulphides which are decomposed by acids may be obtained in a wet way by the double decomposition of the salts of the metals, not with hydrogen sulphide, but with soluble metallic sulphides, such as sulphide of ammonium or of potassium, because then no free acid is formed, but a salt of the metal (potassium or ammonium) which was taken as a soluble sulphide. So, for example, $FeCl_2 + K_2S = FeS + 2KCl$.

Metallic sulphides may be obtained by many other means besides the action of sulphuretted hydrogen on salts and oxides, or by the simple combination of metals with sulphur when heated or fused. Thus they may also be formed by the reduction of sulphates by heating them with charcoal or other means. Charcoal takes up the oxygen from many sulphates, leaving corresponding sulphides. Thus sodium sulphate, Na_2SO_4, when heated with charcoal, forms sodium sulphide, Na_2S. Besides which metallic sulphides are also obtained by heating metals or their oxides in the vapours of many sulphur compounds—for example, in the vapour of carbon bisulphide, CS_2, when the carbon takes up the oxygen and the sulphur combines with the metal. The sulphides formed in this manner are often crystalline, and often appear with those properties and in that crystalline form in which they occur in nature. Besides which we must mention that many of the sulphides of the metals are oxidised in air at the ordinary, and especially at a higher, temperature, forming either SO_2 and the oxide of the metal or sulphates. This oxidation proceeds with particular ease, even at the ordinary temperature, when a metallic sulphide is precipitated from its solutions, as a fine powder containing water. The sulphides of iron and manganese, &c., are very easily oxidised in this manner. But if these hydrates be ignited, they lose their water (the ignition must be carried on in a stream of hydrogen to prevent their oxidation during the process), become denser, and are no longer oxidised at the ordinary temperature. Those sulphides whose corresponding sulphates are decomposed by heat part with their sulphur in the form of sulphurous anhydride when they are ignited in air, and the metal, as a rule, remains behind as oxide. This is taken advantage of in the treatment of sulphurous ores. The process is called *roasting*.

Hydrogen not only forms sulphuretted hydrogen with sulphur, but it also combines with it in several other proportions, just as it combines with oxygen, forming not only water but also hydrogen peroxide. Moreover these *polysulphides of hydrogen* are also unstable, like hydrogen peroxide, and are also obtained from the corresponding polysulphides of the metals of the alkaline earths, just as hydrogen peroxide is obtained from barium peroxide. Thus calcium forms not only calcium sulphide, CaS, but also as bi-, tri-, and pentasulphide, CaS_5, and all these compounds are soluble in water. Sodium also combines with sulphur in the same proportions, forming sulphides from Na_2S to Na_2S_5. If an acid be added to a solution of a polysulphide, it gives sulphur, sulphuretted hydrogen, and a salt of the metal. For instance, $MS_5 + 2HCl = MCl_2 + H_2S + 4S$. If we reverse the operation, and pour a solution of a polysulphide into an acid, sulphur is not precipitated, but an oily liquid is formed which is heavier than water and insoluble in it. This is the polysulphide of hydrogen: $MS_5 + 2HCl = MCl_2 + H_2S_5$. As Rebs showed (1888), whatever polysulphide be taken—of sodium, for instance—it always gives one and the same *hydrogen pentasulphide*, of specific gravity 1·71 (15°). It can only be preserved in the absence of water and at low temperatures, and then not for long: for, especially in the presence of alkalis and when slightly warmed, it splits up very easily into sulphuretted hydrogen and sulphur.

The soluble sulphides and polysulphides of the metals of the alkalis and alkaline earths—for example, of ammonium, potassium, and calcium,—have the appearance and properties of salts, just as the hydrated oxides have, whilst the sulphides of the metals of the higher groups resemble their oxides and have not at all the appearance of salts, and this is more especially the case with regard to the crystalline forms in which they frequently occur in nature.

As the acids derived from chlorine, phosphorus, and carbon are the oxidised hydrogen compounds of these elements, so also we can form an idea of the acid hydrates of sulphur, or of *the normal acids of sulphur*, by representing them as the oxidised products of sulphuretted hydrogen—

HCl	H_2S	H_3P	H_4C
HClO	$H_2SO(?)$	$H_3PO(?)$	H_4CO
$HClO_2$	$H_2SO_2(?)$	H_3PO_2	H_4CO_2
$HClO_3$	H_2SO_3	H_3PO_3	H_4CO_3
$HClO_4$	H_2SO_4	H_3PO_4	H_4CO_4

In the case of chlorine, if not all the hydrates, at all events salts of all the normal hydrates are known, whilst in the case of sulphur only the acids H_2S, H_2SO_3 and H_2SO_4 are known. But, on the other hand, the latter are obtained not only as hydrates but also as stable anhydrides, SO_2 and SO_3, which are formed with the evolution of heat from sulphur and oxygen; 32 parts of sulphur in combining with 32 parts of oxygen—that is, in forming SO_2—evolve 71,000 heat units, and if the oxidation proceeds to the formation of SO_3, 103,000 heat units are evolved. These figures may be compared with those which correspond with the passage of carbon into CO and CO_2, when 29,000 and 97,000 units of heat are evolved. This determines the stability of the higher oxides of sulphur, and also expresses the peculiarity of sulphur as an element which, although an analogue of oxygen, forms stable compounds with it, and thus fundamentally differs from chlorine. The higher and lower oxides of chlorine are powerful oxidising agents, whilst the higher oxide of sulphur, SO_3, has but feeble oxidising powers, and the lower oxide, SO_2, frequently acts as a reducing agent, and is formed by the direct combustion of sulphur, just as carbonic anhydride, CO_2, proceeds from the combustion of carbon. In the combustion of sulphur, and also in the oxidation (roasting) of the sulphides and polysulphides by their ignition in air, *sulphurous oxide*, or *sulphurous anhydride*, or *sulphur dioxide*, SO_2,[31 bis] is exclusively formed. It is prepared on a large scale by burning sulphur or roasting iron pyrites or other sulphides for the manufacture of sulphuric acid (Chapter VI.), and for direct application in the manufacture of wine or for bleaching tissues and other purposes. In the latter instances its application is based on the fact that sulphurous anhydride acts on certain vegetable matters, and has the property of a reducing and feeble acid.[32 bis]

In the laboratory—that is, on a small scale—sulphurous anhydride is best prepared by deoxidising sulphuric acid by heating it with charcoal, or copper, sulphur, mercury, &c. Charcoal produces this decomposition of sulphuric acid at but moderately high temperatures; it is itself converted into carbonic anhydride,[32 tri] and therefore when sulphuric acid is heated with charcoal it evolves a mixture of sulphurous and carbonic anhydrides: $C + 2H_2SO_4 = CO_2 + 2SO_2 + 2H_2O$. The metals which are unable to decompose water, and which do not, therefore, expel hydrogen from sulphuric acid, are frequently capable of decomposing sulphuric acid, with the evolution of sulphurous anhydride, just as they decompose nitric acid, forming the lower oxides of nitrogen. These metals are silver, mercury, copper, lead, and others. Thus, for example, the action of copper on sulphuric acid may be expressed by the following equation: $Cu + 2H_2SO_4 = CuSO_4 + SO_2 + 2H_2O$. In the laboratory this reaction is carried on in a flask with a gas-conducting tube, and does not take place unless aided by heat.

In its physical and chemical properties sulphurous anhydride presents a great *resemblance to carbonic anhydride*. It is a heavy gas, somewhat considerably soluble in water, very easily condensed into a liquid; it forms normal and acid salts, does not evolve oxygen under the direct action of heat, although such metals as sodium and magnesium burn in it, just as in carbonic anhydride. It has a suffocating odour, which is well known owing to its being evolved when sulphur or sulphur matches are burnt. In characterising the properties of sulphurous anhydride, it is very important to remember

(Chapter II.) also that it is more easily liquefied (at -10°, or at 0° under two atmospheres pressure) than carbonic anhydride (thirty-six atmospheres at 0°), that it is more soluble than carbonic anhydride (Vol. I. p. 79); at 0°, 100 vols. of water dissolve 180 vols. of carbonic anhydride and 688 vols. of sulphuric anhydride), that the molecular weight of $SO_2 = 64$ and of $CO_2 = 44$, and that the density of liquid sulphurous anhydride at $0° = 1·43$ (molecular volume = 45) and of carbonic anhydride = 0·95 (molecular volume = 49). Although sulphur dioxide is the anhydride of an acid, nevertheless, like carbonic anhydride, it does not form any stable compounds with water, but gives a solution from which it may be entirely expelled by the action of heat. The acid character of sulphurous anhydride is clearly expressed by the fact that it is entirely absorbed by alkalis, with which it forms acid and normal salts easily soluble in water. With salts of barium, calcium, and the heavy metals, the normal salts of the alkalis, M_2SO_3, give precipitates exactly like those formed by the carbonates. In general, the salts of sulphurous acid are closely analogous to the corresponding carbonates.

Acid sodium sulphite, $NaHSO_3$, may be obtained by passing sulphurous anhydride into a solution of sodium hydroxide. It is also formed by saturating a solution of sodium carbonate with the gas (carbonic anhydride is then given off), and as the solubility of the acid sulphite is much greater than that of the carbonate, a further quantity of the latter may be dissolved after the passage of the sulphurous anhydride, so that ultimately a very strong solution of the sulphite may be formed in this manner, from which it may be obtained in a crystalline form, either by cooling and evaporating (without heating, for then the salt would give off sulphurous anhydride) or by adding alcohol to the solution. When exposed to the air this salt loses sulphurous anhydride and attracts oxygen, which converts it into sodium sulphate. The acid sulphites of the alkali metals are able to combine not only with oxygen, but also with many other substances—for example, a solution of the sodium salt dissolves sulphur, forming sodium thiosulphate, gives crystalline compounds with the aldehydes and ketones, and dissolves many bases, converting them into double sulphites. Having the faculty of attracting or absorbing oxygen, acid sodium sulphite is also able to absorb chlorine, and is therefore employed, like sodium thiosulphate, for the removal of chloride (as an antichlor), especially in the bleaching of fabrics, when it is necessary to remove the last traces of the chlorine held in the tissues, which might otherwise have an injurious effect on them. If a solution of an alkali hydroxide be divided into two parts, and one half is saturated with sulphurous anhydride, and then the other half added to it, a normal salt will be obtained in the solution, having an alkaline reaction, like a solution of sodium carbonate. The acid salt has a neutral reaction.[36 bis] Like sodium carbonate, *normal sodium sulphite* has the composition

$Na_2SO_3,10H_2O$, and its maximum solubility is at 33°—in a word, it very closely resembles sodium carbonate. Although this salt does not give off sulphurous anhydride from its solution, it is able, like the acid salt, to absorb oxygen from the air, and is then converted into sodium sulphate.

Besides the acid character we must also point out the reducing character of sulphurous anhydride. The reducing action of sulphurous acid, its anhydride and salts, is due to their faculty of passing into sulphuric acid and sulphates. The reducing action of the sulphites is particularly energetic, so that they even convert nitric oxide into nitrous oxide: $K_2SO_3 + 2NO = K_2SO_4 + N_2O$. The salts of many of the higher oxides are converted into those of the lower—for example, FeX_3 into FeX_2, CuX_2 into CuX, HgX_2 into HgX; thus $2FeX_3 + SO_2 + 2H_2O = 2FeX_2 + H_2SO_4 + 2HX$. In the presence of water, sulphurous anhydride is oxidised by chlorine ($SO_2 + 2H_2O + Cl_2 = H_2SO_4 + 2HCl$), iodine, nitrous acid, hydrogen peroxide, hypochlorous acid, chloric acid, and other oxygen compounds of the halogens, chromic, manganic, and many other metallic acids and higher oxides, as well as all peroxides. Free oxygen in the presence of spongy platinum is able to oxidise sulphurous anhydride even in the absence of water, in which case sulphuric anhydride SO_3 is formed, so that the latter may be prepared by passing a mixture of sulphurous anhydride and oxygen over incandescent spongy platinum, or, as it is now prepared on a large scale in chemical works, by passing this mixture over asbestos or pumice stone moistened with a solution of platinum salt and ignited. Sulphurous anhydride is completely absorbed by certain higher oxides—for instance, by barium peroxide and lead dioxide ($PbO_2 + SO_2 = PbSO_4$).

There are, however, cases where sulphurous anhydride acts as an oxidising agent—that is, it is *deoxidised* in the presence of substances which are capable of absorbing oxygen with still greater energy than the sulphurous anhydride itself. This oxidising action proceeds with the formation of sulphuretted hydrogen or of sulphides, while the reducing agent is oxidised at the expense of the oxygen of the sulphurous anhydride. In this respect, the action of stannous salts is particularly remarkable. Stannous chloride, $SnCl_2$, in an aqueous solution gives a precipitate of stannic sulphide, SnS_2, with sulphurous anhydride—that is, the latter is deoxidised to sulphuretted hydrogen, while SnX_2 is oxidised into SnX_4. A solution of sulphurous anhydride has also an oxidising action on zinc. The zinc passes into solution, but no hydrogen is evolved, because a salt of *hydrosulphurous acid*, ZnS_2O_4, is formed. The free acid is still less stable than the salt.

The faculty of sulphurous anhydride of combining with various substances is evident from the above-cited reactions, where it combines with hydrogen and with oxygen, and this faculty also appears in the fact that, like carbonic oxide, it combines with chlorine, forming a chloranhydride of sulphuric acid, SO_2Cl_2, to which we shall afterwards return. The same faculty for combination also appears in the salts of sulphurous acid, in their liability to oxidation and in the exceedingly characteristic formation of a peculiar series of salts obtained by Pelouze and Frémy. At a temperature of -10° or below, nitric oxide NO is absorbed by alkaline solutions of the alkali sulphites, forming a peculiar series of *nitrosulphates*. At a higher temperature these salts are not formed but the nitric oxide is reduced to nitrous oxide. But in the cold the liquid saturated with nitric oxide after a certain time gives prismatic crystals resembling those of nitre. The composition of the potassium salt is $K_2SN_2O_3$—that is, the salt contains the elements of potassium sulphite and of nitric oxide.

There are also several other substances, formed by the oxides of nitrogen and sulphur, which belong to this class of complex and, under some circumstances, unstable compounds. In the manufacture of sulphuric acid, both these classes of oxides come into contact with each other in the lead chambers, and if there be insufficient water for the formation of sulphuric acid they give crystalline compounds, termed *chamber crystals*. As a rule, the composition of the crystals is expressed by the formula $NHSO_3$. This is a compound of the radicles NO_2 of nitric acid, and HSO_3 of sulphuric acid, or nitro-sulphuric acid, $NO_2.SHO_3$, if sulphuric acid be expressed as $OH.SHO_3$ and nitric by $NO_2.OH$. The tabular crystals of this substance fuse at about 70°, are formed both by the direct action of nitrous anhydride or nitric peroxide (but not NO, which is not absorbed by sulphuric acid) on sulphuric acid (Weltzien and others), and especially on sulphuric acid containing an anhydride and the lower oxides of sulphur and nitric acid.

Thiosulphuric acid, $H_2S_2O_3$—that is, a compound of sulphurous acid and sulphur—also belongs to the products of combination of sulphurous acid. In the same way that sulphurous acid, H_2SO_3, gives H_2SO_4 with oxygen, so it gives $H_2S_2O_3$ with sulphur. In a free state it is very unstable, and it is only known in the form of its salts proceeding from the direct action of sulphur on the normal sulphites; if endeavours be made to separate it in a free state, it immediately splits up into those elements from which it might be formed—that is, into sulphur and sulphurous acid. The most important of its salts is the *sodium thiosulphate* (known as hyposulphite), $Na_2S_2O_3,5H_2O$, which occurs in colourless crystals, and is unacted on by atmospheric oxygen either when in a dry state or in solution. Many other salts of this acid are easily formed by means of this salt,[41 bis] although this cannot be done with all bases, for such bases as alumina, ferric oxide, chromium oxide, and others do not give compounds with thiosulphuric acid, just as they do not form stable compounds with carbonic acid. Whenever these salts might be formed, they (like the acid) split up into sulphurous acid and sulphur, and furthermore the elements of thiosulphuric acid in many cases act in a reducing manner, forming sulphuric acid and taking up the oxygen from reducible oxides. Thus when treated with a thiosulphate the soluble ferric salts give a precipitate of sulphur and form ferrous salts. The thiosulphates of the metals of the alkalis are obtained directly by boiling a solution of their sulphites with sulphur: $Na_2SO_3 + S = Na_2S_2O_3$. The same salts are formed by the action of sulphurous anhydride on solutions of the sulphides; thus sodium sulphide dissolved in water gives sulphur and sodium thiosulphate when a stream of sulphurous anhydride is passed through it: $2Na_2S + 3SO_2 = 2Na_2S_2O_3 + S$. The polysulphides of the alkali metals when left exposed to the air attract oxygen and also form thiosulphates.

Although sulphur, oxidising at a high temperature, only forms a small quantity of sulphuric anhydride, SO_3, and nearly all passes into sulphurous anhydride, still the latter may be converted into the higher oxide, or *sulphuric anhydride*, SO_3, by many methods. Sulphuric anhydride is a solid crystalline substance at the ordinary temperature; it is easily fusible (15°), and volatile (46°), and rapidly attracts moisture. Although it is formed by the combination of sulphurous anhydride with oxygen, it is capable of further combination. Thus it combines with water, hydrochloric acid, ammonia, with many hydrocarbons, and even with sulphuric acid, boric and nitrous anhydrides, &c., and also with bases which burn directly in its vapour, forming sulphates in the presence of traces of moisture (*see* Chapter IX., Note 29). The oxidation of sulphurous anhydride, SO_2, into sulphuric anhydride, SO_3, is effected by passing a mixture of the former and dry oxygen or air over incandescent spongy platinum. An increase of pressure accelerates the reaction (Hanisch). If the product be passed into a cold vessel, crystalline sulphuric anhydride is deposited upon the sides of the vessel, but as it is difficult to avoid all traces of moisture it always contains compounds of its hydrates: $H_2S_2O_7$ and $H_2S_4O_{13}$, whose presence so modifies the properties of the anhydride (Weber) that formerly two modifications of the anhydride were recognised. The same sulphuric anhydride may be obtained from certain anhydrous sulphates, or those which are almost so, which are decomposed by heat, whilst an impure but perfectly anhydrous anhydride is formed by distillation over phosphoric anhydride. For instance, acid sodium sulphate, $NaHSO_4$, and the pyro- or di-sulphate, $Na_2S_2O_7$ (Chapter XII.) formed from it, when ignited evolve sulphuric anhydride. Green vitriol—that is, ferrous sulphate, $FeSO_4$—belongs to the number of those sulphates which easily give off sulphuric anhydride under the action of heat. It contains water of crystallisation and parts with it when it is heated, but the last equivalent of water is driven off with difficulty, just as is the case with magnesium sulphate, $MgSO_4 7H_2O$; however, when thoroughly heated, this evolution of sulphuric anhydride does take place, although not completely, because at a high temperature a portion of it is decomposed by the ferrous oxide (SO_3 + 2FeO), which is converted into ferric oxide, Fe_2O_3, and in consequence part of the sulphuric anhydride is converted into sulphurous anhydride. Thus the products of the decomposition of ferrous sulphate will be: ferric oxide, Fe_2O_3, sulphurous anhydride, SO_2, and sulphuric anhydride, SO_3, according to the equation: $2FeSO_4 = Fe_2O_3 + SO_2 + SO_3$. As water still remains with the ferrous sulphate when it is heated, the result will partially consist of the hydrate H_2SO_4, with anhydride, SO_3, dissolved in it. Sulphuric acid was for a long time prepared in this manner; the process was formerly carried on on a large scale in the neighbourhood of Nordhausen, and hence the sulphuric acid prepared from ferrous sulphate is called *fuming Nordhausen acid*. At the present time the fuming acid is prepared by passing the volatile products of the decomposition of ferrous sulphate through strong sulphuric acid prepared by the ordinary method. The sulphurous anhydride is insoluble in it, but it absorbs the sulphuric anhydride. Sulphuric anhydride may be prepared not only by igniting $FeSO_4$ or sodium pyrosulphate, $Na_2S_2O_7$ (the decomposition proceeds at 600°), but also by heating a mixture of the latter and $MgSO_4$ (Walters); in the former case a stable double salt $MgNa_2(SO_4)_2$ finally remains. It is also obtained by the direct combination of SO_2 and O under the action of spongy platinum or asbestos coated with platinum black (C. Winkler's process). Nordhausen sulphuric acid fumes in air, owing to its containing and easily giving off sulphuric anhydride, and it is therefore also called *fuming sulphuric acid*; these fumes are nothing but the vapour of sulphuric anhydride combining with the moisture in the air and forming non-volatile sulphuric acid (hydrate).

Nordhausen sulphuric acid contains a peculiar compound of SO_3 and H_2SO_4, or *pyrosulphuric acid*; an imperfect anhydride of sulphuric acid, $H_2S_2O_7$, analogous in composition with the salts $Na_2S_2O_7$, $K_2Cr_2O_7$, and bearing the same relation to H_2SO_4 that pyrophosphoric acid does to H_3PO_4. The bond holding the sulphuric acid and anhydride together is unstable. This is obvious from the fact that the anhydride may easily be separated from this compound, by the action of heat. In order to obtain the definite compound, the Nordhausen acid is cooled to 5°, or, better still, a portion of it is distilled until all the anhydride and a certain amount of sulphuric acid have passed over into the distillate, which will then solidify at the ordinary temperature, because the compound H_2SO_4, SO_3 fuses at 35°. Although this substance reacts on water, bases, &c., like a mixture of SO_3 + H_2SO_4, still since a definite compound, $H_2S_2O_7$, exists in a free state and gives salts and a chloranhydride, $S_2O_5Cl_2$, we must admit the existence of a definite pyrosulphuric acid, like pyrophosphoric acid, only that the latter has a far greater stability and is not even converted into a perfect hydrate by water. Further, the salts $M_2S_2O_7$ dissolved in water react in the same manner as the acid salts $MHSO_4$, whilst the imperfect hydrates of phosphoric acid (for example, PHO_3, $H_4P_2O_7$) have independent reactions even in an aqueous solution which distinguish them and their salts from the perfect hydrates.

FIG. 87.—Concentration of sulphuric acid in glass retorts. The neck of each retort is attached to a bent glass tube, whose vertical arm is lowered into a glass or earthenware vessel acting as a receiver for the steam which comes over from the acid, as the former still contains a certain amount of acid.

Sulphuric acid, H_2SO_4, is formed by the combination of its anhydride, SO_3, and water, with the evolution of a large amount of heat; the reaction $SO_3 + H_2O$ develops 21,300 heat units. The method of its preparation on a large scale, and most of the methods employed for its formation, are dependent on the oxidation of sulphurous anhydride, and the formation of sulphuric anhydride, which forms sulphuric acid under the action of water. The technical method of its manufacture has been described in Chapter VI. The acid obtained *from the lead chambers* contains a considerable amount of water, and is also impure owing to the presence of oxides of nitrogen, lead compounds, and certain impurities from the burnt sulphur which have come over in a gaseous and vaporous state (for example, arsenic compounds). For practical purposes, hardly any notice is taken of the majority of these impurities, because they do not interfere with its general qualities. Most frequently endeavours are only made to remove, as far as possible, all the water which can be expelled. That is, the object is to obtain the hydrate, H_2SO_4, from the dilute acid (60 per cent.), and this is effected by evaporation by means of heat. Every given mixture of water and sulphuric acid begins to part with a certain amount of aqueous vapour when heated to a certain definite temperature. At a low temperature either there is no evaporation of water, or there can even be an absorption of moisture from the air. As the removal of the water proceeds, the vapour tension of the residue decreases for the same temperature, and therefore the more dilute the acid the lower the temperature at which it gives up a portion of its water. In consequence of this, the removal of water from dilute solutions of sulphuric acid may be easily carried on (up to 75 p.c. H_2SO_4) in lead vessels, because at low temperatures dilute sulphuric acid does not attack lead. But as the acid becomes more concentrated the temperature at which the water comes over becomes higher and higher, and then the acid begins to act on lead (with the evolution of sulphuretted hydrogen and conversion of the lead into sulphate), and therefore lead vessels cannot be employed for the complete removal of the water. For this purpose the evaporation is generally carried on in glass or platinum retorts, like those depicted in figs. 87 and 88.

FIG. 88.—Concentration of sulphuric acid in platinum retorts.

The concentration of sulphuric acid in glass retorts is not a continuous process, and consists of heating the dilute 75 per cent. acid until it ceases to give off aqueous vapour, and until acid containing 93–98 per cent. H_2SO_4 (66° Baumé) is obtained—and this takes place when the temperature reaches 320° and the density of the residue reaches 1·847 (66° Baumé). The platinum vessels designed for the continuous concentration of sulphuric acid consist of a still *B*, furnished with a still head *E*, a connecting pipe *E F*, and a syphon tube *H R*, which draws off the sulphuric acid concentrated in the boiler. A stream of sulphuric acid previously concentrated in lead retorts to a density of about 60° Baumé—*i.e.* to 75 per cent. or a sp. gr. of 1·7—runs continuously into the retort through a syphon funnel *E'*. The apparatus is fed from above, because the acid freshly supplied is lighter than that which has already lost water, and also because the water is more easily evaporated from the freshly supplied acid at the surface. The platinum retort is heated, and the steam coming off is condensed in a worm *F G*, whilst as fresh dilute acid is supplied to the boiler the acid already concentrated is drawn off through the syphon tube *H B*, which is furnished with a regulating cock by means of which the outflow of the concentrated acid from the bottom of the retort can be so regulated that it will always present one and the same specific gravity, corresponding with the strength required. For this purpose the acid flowing from the syphon is collected in a receiver *R*, in which a hydrometer, indicating its density, floats; if its density be less than 66° Baumé, the regulating cock is closed sufficiently to retard the outflow of sulphuric acid, so as to lengthen the time of its evaporation in the retort.

Strictly speaking, *sulphuric acid is not volatile*, and at its so-called boiling-point it really decomposes into its anhydride and water; its boiling-point (338°) being nothing else but its temperature of decomposition. The products of this decomposition are substances boiling much below the temperature of the decomposition of sulphuric acid. This conclusion with regard to the process of the distillation of sulphuric acid may be deduced from Bineau's observations on the vapour-density of sulphuric acid. This density referred to hydrogen proved to be half that which sulphuric acid should have according to its molecular weight, H_2SO_4, in which case it should be 49, whilst the observed density was equal to 24·5. Besides which, Marignac showed that the first portions of the sulphuric acid distilling over contain less of the elements of water than the portion which remains behind, or which distils over towards the end. This is explained by the fact that on distillation the sulphuric acid is decomposed, but a portion of the water proceeding from its decomposition is retained by the remaining mass of sulphuric acid, and therefore at first a mixture of sulphuric acid and sulphuric anhydride—*i.e.* fuming sulphuric acid—is obtained in the distillate. It is possible by repeating the distillation several times and only collecting the first portions of the distillate, to obtain a distinctly fuming acid. To obtain the definite hydrate H_2SO_4 it is necessary to refrigerate a highly concentrated acid, of as great a purity as possible, to which a small quantity of sulphuric anhydride has been previously added. Sulphuric acid containing a small quantity (a fraction of a per cent. by weight) of water only freezes at a very low temperature, while the pure normal acid, H_2SO_4, solidifies when it is cooled below 0°, and therefore the normal acid first crystallises out from the concentrated sulphuric acid. By repeating the refrigeration several times, and pouring off the unsolidified portion, it is possible to obtain a pure *normal hydrate*, H_2SO_4, which melts at 10°·4. Even at 40° it gives off distinct fumes—that is, it begins to evolve sulphuric anhydride, which volatilises, and therefore even in a dry atmosphere the hydrate H_2SO_4 becomes weaker, until it contains 1½ p.c. of water.

In a concentrated form sulphuric acid is commercially known as *oil of vitriol*, because for a long time it was obtained from green vitriol and because it has an oily appearance and flows from one vessel into another in a thick and somewhat sluggish stream, like the majority of oily substances, and in this clearly differs from such liquids as water, spirit, ether, and the like, which exhibit a far greater mobility. Among its reactions the first to be remarked is its faculty for the formation of many compounds. We already know that it combines with its anhydride, and with the sulphates of the alkali metals; that it is soluble in water, with which it forms more or less stable compounds. Sulphuric acid, when mixed with water, develops a very considerable amount of heat.

Besides the normal hydrate H_2SO_4, *another definite hydrate*, H_2SO_4,H_2O (84·48 per cent. of the normal hydrate, and 15·52 per cent. of water) is known; it crystallises[50 bis] extremely easily in large six-sided prisms, which form above 0°—namely, at about +8°·5; when heated to 210° it loses water. If the hydrates H_2SO_4 and H_2SO_4,H_2O exist at low temperatures as definite crystalline compounds, and if pyrosulphuric acid, $H_2SO_4SO_3$, has the same property, and if they all decompose with more or less ease on a rise of temperature, with the disengagement of either SO_3 or H_2O, and in their ordinary form present all the properties of simple solutions, it follows that between sulphuric anhydride, SO_3, and water, H_2O, there exists a consecutive series of homogeneous liquids or solutions, among which we must distinguish *definite compounds*, and therefore it is quite justifiable to look for other definite compounds between SO_3 and H_2O, beyond the conditions for a change of state. In this respect we may be guided by the variation of properties of any kind, proceeding concurrently with a variation in the composition of a solution.

But only a few properties have been determined with sufficient accuracy. In those properties which have been determined for many solutions of sulphuric acid, it is actually seen that the above-mentioned definite compounds are distinguished by distinctive marks of change. As an example we may cite the variation of the specific gravity with a variation of temperature (namely $K = ds/dt$, if s be the sp. gr. and t the temperature). For the normal hydrate, H_2SO_4, this factor is easily determined from the fact that—

$$s = 18528 - 10·65t + 0·013t^2,$$

where s is the specific gravity at t (degrees Celsius) if the sp. gr. of water at $4° = 10,000$. Therefore $K = 10·65 - 0·026t$. This means that at $0°$ the sp. gr. of the acid H_2SO_4 decreases by $10·65$ for every rise of a degree of temperature, at $10°$ by $10·39$, at $20°$ by $10·13$, at $30°$ by $9·87$. And for solutions containing slightly more anhydride than the acid H_2SO_4 (i.e. for fuming sulphuric acid), as well as for solutions containing more water, K is greater than for the acid H_2SO_4. Thus for the solution $SO_3, 2H_2SO_4$, at $10°$ $K = 11·0$. On diluting the acid H_2SO_4 K again increases until the formation of the solution H_2SO_4, H_2O ($K = 11·1$ at $10°$), and then, on further dilution with water, it again decreases. Consequently both hydrates H_2SO_4 and H_2SO_4, H_2O are here expressed by an alteration of the magnitude of K.

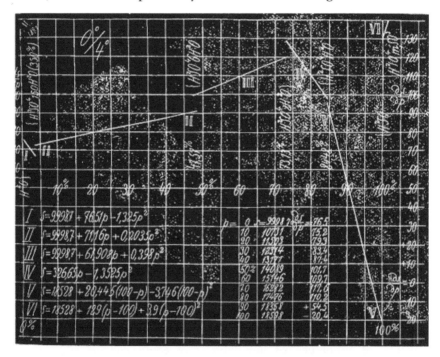

FIG. 89.—Diagram showing the variation of the factor (ds/dp) of the specific gravity of solutions of sulphuric acid. The percentage quantities of the acid, H_2SO_4, are laid out on the axes of abscissæ. The ordinates are the factors or rises in sp. gr. (water at 4 = 10,000) with the increase in the quantity of H_2SO_4.

This shows that in liquid solutions it is possible by studying the variation of their properties (without a change of physical state) to recognise the presence or formation of definite hydrate compounds, and therefore an exact investigation of the properties of solutions, of their specific gravity for instance, should give direct indications of such compounds. The mean result of the most trustworthy determinations of this nature is given in the following tables. The first of these tables gives the specific gravities (in vacuo, taking the sp. gr. of water at 4° = 1), at 0° (column 3), 15° (column 4), and 30° (column 5),[53 bis] for solutions having the composition $H_2SO_4 + nH_2O$ (the value of n is given in the first column), and containing p (column 2) per cent. (by weight in vacuo) of H_2SO_4.[53 tri]

n	p	0°	15°	30°
100	5·16	1·0374	1·0341	1·0292
50	9·82	1·0717	1·0666	1·0603
25	17·88	1·1337	1·1257	1·1173
15	26·63	1·2040	1·1939	1·1837
10	35·25	1·2758	1·2649	1·2540
8	40·50	1·3110	1·2649	1·2998
6	47·57	1·3865	1·3748	1·3622
5	52·13	1·4301	1·4180	1·4062
4	57·65	1·4881	1·4755	1·4631
3	64·47	1·5635	1·5501	1·5370
2	73·13	1·6648	1·6500	1·6359
1	84·48	1·7940	1·7772	1·7608
0·5	91·59	1·8445	1·8284	1·8128
H_2SO_4	100	1·8529	1·8372	1·8221

In the second table the first column gives the percentage amount p (by weight) of H_2SO_4, the second column the weight in grams (S_{15}) of a litre of the solution at 15° (at 4° the weight of a litre of water = 1,000 grams), the third column, the variation (dS/dt) of this weight for a rise of 1°, the fourth column, the variation dS/dp of this weight (at 15°) for a rise of 1 per cent. of H_2SO_4, the fifth column, the difference between the weight of a litre at

0° and 15° ($S_0 - S_{15}$), and the sixth column, the difference between the weight of a litre at 15° and 30° ($S_{15} - S_{30}$).

p	S_{15}	dS_{15}/dt	dS_{15}/dp	S_0-S_{15}	$S_{15}-S_{15}$
0	999·15	0·148	7·0	0·7	3·4
5	1033·0	0·27	6·8	3·1	5·0
10	1067·7	0·28	7·1	5·2	6·4
20	1141·9	0·58	7·7	8·6	8·9
30	1221·3	0·69	8·2	10·4	10·4
40	1306·6	0·75	8·8	11·3	11·2
50	1397·9	0·79	9·9	11·9	11·8
60	1501·2	0·86	10·8	13·0	12·7
70	1613·1	0·93	11·6	14·1	13·8
80	1731·4	1·04	11·0	15·8	15·4
90	1819·9	1·08	5·4	16·4	16·0
95	1837·6	1·03	+1·7	15·8	15·1
100	1837·2	1·03	−1·9	15·7	15·1

The figures in these tables give the means of finding the amount of H_2SO_4 contained in a solution from its specific gravity, and also show that 'special points' in the lines of variation of the specific gravity with the temperature and percentage composition correspond to certain definite compounds of H_2SO_4 with OH_2. This is best seen in the variation of the factors (dS/dt and dS/dp) with the temperature and composition (columns 3, 4, second table). We have already mentioned how the factor of temperature points to the existence of hydrates, H_2SO_4 and H_2SO_4,H_2O. As regards the factor dS/dp (giving the increase of sp. gr. with an increase of 1 per cent. H_2SO_4) the following are the three most salient points: (1) In passing from 98 per cent. to 100 per cent. the factor is negative, and at 100 per cent. about −0·0019 (*i.e.* at 99 per cent. the sp. gr. is about 1·8391, and at 100 per cent. about 1·8372, at 15°, the amount of H_2SO_4 has increased whilst the sp. gr. has decreased), but as soon as a certain amount of SO_3 is added to the definite compound H_2SO_4 (and 'fuming' acid formed) the specific gravity rises (for example, for H_2SO_4 0·136 SO_3 the sp. gr. at 15° = 1·866), that is the factor becomes positive (and, in fact, greater by +0·01), so that the formation of the definite hydrate H_2SO_4 is accompanied by a distinct and considerable break in the continuity of the factor[55 bis]; (2) The factor (dS/dp) in increasing in its passage from dilute to concentrated solutions, attains a maximum value (at 15° about 0·012) about $H_2SO_4 2H_2O$, *i.e.* at about the hydrate corresponding to the form SX_6; proper to the compounds of sulphur, for $S(OH)_6 = H_2SO_4 2H_2O$; the same hydrate corresponds to the composition of gypsum $CaSO_4 2H_2O$, and to it also corresponds the greatest contraction and rise of temperature in mixing H_2SO_4 with H_2O (*see* Chapter I., Note 28); (3) The variation of the factor (dS/dp) under certain variations in the composition proceeds so uniformly and regularly, and is so different from the variation given under other proportions of H_2SO_4 and H_2O, that the sum of the variations of dS/dp is expressed by a series of straight lines, if the values of p be laid along the axis of abscissæ and those of dS/dp along the ordinates. Thus, for instance, for 15°, at 10 per cent. $dS/dp = 0·0071$, at 20 per cent. = 0·0077, at 30 per cent. = 0·0082, at 40 per cent. = 0·0088, that is, for each 10 per cent. the factor increases by about 0·0006 for the whole of the above range, but beyond this it becomes larger, and then, after passing $H_2SO_4 2H_2O$, it begins to fall rapidly. Such changes in the variation of the factor take place apparently about definite hydrates,[56 bis] and especially about $H_2SO_4 4H_2O$, $H_2SO_4 2H_2O$ and $H_2SO_4 H_2O$. All this indicating as it does the special chemical affinity of sulphuric acid for water, although of no small significance for comprehending the nature of solutions (*see* Chapter I. and Chapter VII.), contains many special points which require detailed investigation, the chief difficulty being that it requires great accuracy in a large number of experimental data.

The great affinity of sulphuric acid for water is also seen from the fact that when the strong acid acts on the majority of *organic substances* containing hydrogen and oxygen (especially on heating) it very frequently *takes up these elements in the form of water.* Thus strong sulphuric acid acting on alcohol, C_2H_6O, removes the elements of water from it, and converts it into olefiant gas, C_2H_4. It acts in a similar manner on wood and other vegetable tissues, which it chars. If a piece of wood be immersed in strong sulphuric acid it turns black. This is owing to the fact that the wood contains carbohydrates which give up hydrogen and oxygen as water to the sulphuric acid, leaving charcoal, or a black mass very rich in it. For example, cellulose, $C_6H_{10}O_5$, acts in this manner.

We have already had frequent occasion to notice the very *energetic acid properties* of sulphuric acid, and therefore we will now only consider a few of their aspects. First of all we must remember that, with calcium, strontium, and especially with barium and lead, sulphuric acid forms very slightly soluble salts, whilst with the majority of other metals it gives more easily soluble salts, which in the majority of cases are able, like sulphuric acid itself, to combine with water to form crystallo-hydrates. Normal sulphuric acid, containing two atoms of hydrogen in its molecule, is able for this reason alone to form two classes of salts, *normal* and *acid*, which it does with great facility *with the alkali metals*. The metals of the alkaline earths and the majority of other metals, if they do form acid sulphates, do so under exceptional conditions (with an excess of strong sulphuric acid), and these salts when formed are decomposable by water—that is, although having a certain degree of physical stability they have no chemical stability. Besides the acid salts $RHSO_4$, sulphuric acid also gives other forms of acid salts. An entire series of salts having the composition $RHSO_4, H_2SO_4$, or for bivalent metals $RSO_4, 3H_2SO_4$, has been prepared. Such salts have been obtained for potassium, sodium, nickel, calcium, silver, magnesium, manganese. They are prepared by dissolving the sulphates in an excess of sulphuric acid and heating the solution until the excess of sulphuric acid is driven off; on cooling, the mass solidifies to a crystalline salt. Besides which, Rose obtained a salt having the composition $Na_2SO_4, NaHSO_4$, and if $HNaSO_4$ be heated it easily forms a salt $Na_2S_2O_7 = Na_2SO_4, SO_3$; hence it is clear that sulphuric anhydride combines with various proportions of bases, just as it combines with various proportions of water.

We have already learned that sulphuric acid displaces the acid from the salts of nitric, carbonic, and many other volatile acids. Berthollet's laws (Chapter X.) explain this by the small volatility of sulphuric acid; and, indeed, in an aqueous solution sulphuric acid displaces the much less soluble boric acid from its compounds—for instance, from borax, and it also displaces silica from its compounds with bases; but both boric anhydride and silica, when fused with sulphates, decompose them, displacing sulphuric anhydride, SO_3, because they are less volatile than sulphuric anhydride. It is also well known that with metals, sulphuric acid forms salts giving off hydrogen (Fe, Zn, &c.), or sulphur dioxide (Cu, Hg, &c.).[58 bis]

The reactions of sulphuric acid *with respect to organic substances* are generally determined by its acid character, when the direct extraction of water, or oxidation at the expense of the oxygen of the sulphuric acid, or disintegration does not take place. Thus the majority of the saturated hydrocarbons, C_nH_{2m}, form with sulphuric acid a special class of *sulphonic acids*, $C_nH_{2m-1}(HSO_3)$; for example, benzene, C_6H_6, forms benzenesulphonic acid, $C_6H_5.SO_3H$, water being separated, for the formation of which oxygen is taken up from the sulphuric acid, for the product contains less oxygen than the sulphuric acid. It is evident from the existence of these acids that the hydrogen in organic compounds is replaceable by the group SO_3H, just as it may be replaced by the radicles Cl, NO_2, CO_2H and others. As the radicle of sulphuric acid or *sulphoxyl*, SO_2OH or SHO_3, contains, like carboxyl (Vol. I., p. [395](#)), one hydrogen (hydroxyl) of sulphuric acid, the resultant substances are acids whose basicity is equal to the number of hydrogens replaced by sulphoxyl. Since also sulphoxyl takes the place of hydrogen, and itself contains hydrogen, the sulpho-acids are equal to a hydrocarbon + SO_3, just as every organic (carboxylic) acid is equal to a hydrocarbon + CO_2. Moreover, here this relation corresponds with actual fact, because many sulphonic acids are obtained by the direct combination of sulphuric anhydride: $C_6H_5,(SO_3H) = C_6H_6 + SO_3$. The sulphonic acids give soluble barium salts, and are therefore easily distinguished from sulphuric acid. They are soluble in water, are not volatile, and when distilled give sulphurous anhydride (whilst the hydroxyl previously in combination with the sulphurous anhydride remains in the hydrocarbon group; thus phenol, $C_6H_5.OH$, is obtained from benzenesulphonic acid), and they are very energetic, because the hydrogen acting in them is of the same nature as in sulphuric acid itself.

Sulphuric acid, as containing a large proportion of oxygen, is a substance which frequently acts as an oxidising agent: in which case it is *deoxidised, forming sulphurous anhydride* and water (or even, although more rarely, sulphuretted hydrogen and sulphur). Sulphuric acid acts in this manner on charcoal, copper, mercury, silver, organic and other substances, which are unable to evolve hydrogen from it directly, as we saw in describing sulphurous anhydride.

Although the hydrate of a higher saline form of oxidation (Chapter [XV.](#)), sulphuric anhydride is capable of further oxidation, and forms a kind of peroxide, just as hydrogen gives hydrogen peroxide in addition to water, or as sodium and potassium, besides the oxides Na_2O and K_2O, give their peroxides, compounds which are in a chemical sense unstable, powerfully oxidising, and not directly able to enter into saline combinations. If the oxides of potassium, barium, &c., be compared to water, then their peroxides must in like manner correspond to hydrogen peroxide, not only because the oxygen contained in them is very mobile and easily liberated, and because their reactions are similar, but also because they can be mutually transformed into each other, and are able to form compounds with each other, with bases and with water, and indeed form a kind of peroxide salts. This is also the character of *persulphuric acid*, discovered in 1878 by Berthelot, and its corresponding anhydride or peroxide of sulphur S_2O_7. It is formed from $2SO_3 + O$ with the absorption of heat (-27 thousand heat units), like ozone from $O_2 + O$ (-29 thousand units of heat), or hydrogen peroxide from $H_2O + O$ (-21 thousand heat units).

Peroxide of sulphur is produced by the action of a silent discharge upon a mixture of oxygen and sulphurous anhydride. With water S_2O_7 gives persulphuric acid, $H_2S_2O_8$. The latter is obtained more simply by mixing strong sulphuric acid (not weaker than $H_2SO_4,2H_2O$) directly with hydrogen peroxide, or by the action of a galvanic current on sulphuric acid mixed with a certain amount of water, and cooled, the electrodes being platinum wires, when persulphuric acid naturally appears at the positive pole. When an acid of the strength $H_2SO_4,6H_2O$ is taken, at first the hydrate of the sulphuric peroxide, S_2O_7,H_2O only is formed; but when the concentration about the positive pole reaches $H_2SO_4,3H_2O$, a mixture of hydrogen peroxide and the hydrate of sulphuric peroxide begins to be formed. Dilute solutions of sulphuric peroxide can be kept better than more concentrated solutions, but the latter may be obtained containing as much as 123 grams of the peroxide to a litre. It is a very instructive fact that hydrogen peroxide is always formed when strong solutions of persulphuric acid break up on keeping. So that the bond between the two peroxides is established both by analysis and synthesis: hydrogen peroxide is able to produce $S_2H_2O_8$, and the latter to produce hydrogen peroxide. A mixture of sulphuric peroxide with sulphuric acid or water is immediately decomposed, with the evolution of oxygen, either when heated or under the action of spongy platinum. The same thing takes place with a solution of baryta, although at first no precipitate is formed and the decomposition of the barium salt, BaS_2O_8, with the formation of $BaSO_4$, only proceeds slowly, so that the solution may be filtered (the barium salt of persulphuric acid is soluble in water). Mercury, ferrous oxide, and the stannous salts, are oxidised by $S_2H_2O_8$. These are all distinct signs of true peroxides. The same common properties (capacity for oxidising, property of forming peroxide of hydrogen, &c.) are possessed by the alkali salts of persulphuric acid, which are obtained by the action of an electric current upon certain sulphates, for instance ammonium or potassium sulphate. The ammonium salt of persulphuric acid, $(NH_4)_2S_2O_8$, is especially easily formed by this means, and is now prepared on a large scale and used (like Na_2O_2 and H_2O_2) for bleaching tissues and fibres.

In order to understand the relation of sulphuric peroxide to sulphuric acid we must first remark that hydrogen peroxide is to be considered, in accordance with the law of substitution, as water, H(OH), in which H is replaced by (OH). Now the relation of $H_2S_2O_8$ to H_2SO_4 is exactly similar. The radicle of sulphuric acid, equivalent to hydrogen, is HSO_4;[65 bis] it corresponds with the (OH) of water, and therefore sulphuric acid,

H(SHO$_4$), gives (SHO$_4$)$_2$ or S$_2$H$_2$O$_8$, in exactly the same manner as water gives (HO)$_2$—*i.e.* H$_2$O$_2$.

The largest part *of the sulphuric acid made* is used for reacting on sodium chloride in the manufacture of sodium carbonate; for the manufacture of the volatile acids, like nitric, hydrochloric, &c., from their corresponding salts; for the preparation of ammonium sulphate, alums, vitriols (copper and iron), artificial manures, superphosphate (Chapter XIX., Note 18) and other salts of sulphuric acid; in the treatment of bone ash for the preparation of phosphorus, and for the solution of metals—for example, of silver in its separation from gold—for cleaning metals from rust, &c. A large amount of oil of vitriol is also used in treatment of organic substances; it is used for the extraction of stearin, or stearic acid, from tallow, for refining petroleum and various vegetable oils, in the preparation of nitro-glycerine (Chapter VI., Notes 37 and 37 bis), for dissolving indigo and other colouring matters, for the conversion of paper into vegetable parchment, for the preparation of ether from alcohol, for the preparation of various artificial scents from fusel oil, for the preparation of vegetable acids, such as oxalic, tartaric, citric, for the conversion of non-fermentable starchy substances into fermentable glucose, and in a number of other processes. It would be difficult to find another artificially-prepared substance which is so frequently applied in the arts as sulphuric acid. Where there are not works for its manufacture, the economical production of many other substances of great technical importance is impossible. In those localities which have arrived at a high technical activity the amount of sulphuric acid consumed is proportionally large; sulphuric acid, sodium carbonate, and lime are the most important of the artificially-prepared agents employed in factories.

Besides the normal acids of sulphur, H$_2$SO$_3$, H$_2$SO$_3$S, and H$_2$SO$_4$, corresponding with sulphuretted hydrogen, H$_2$S, in the same way that the oxy-acids of chlorine correspond with hydrochloric acid, HCl, there exists a peculiar series of acids which are termed *thionic acids*. Their general composition is S$_n$H$_2$O$_6$, where *n* varies from 2 to 5. If $n = 2$, the acid is called dithionic acid. The others are distinguished as trithionic, tetrathionic, and pentathionic acids. Their composition, existence, and reactions are very easily understood if they be referred to the class of the sulphonic acids— that is, if their relation to sulphuric acid be expressed in just the same manner as the relation of the organic acids to carbonic acid. The organic acids, as we saw (Chapter IX.), proceed from the hydrocarbons by the substitution of their hydrogen by carboxyl—that is, by the radicle of carbonic acid, CH$_2$O$_3$ - HO = CHO$_2$. The formation of the acids of sulphur by means of sulphoxyl may be represented in the same manner, HSO$_3$ = H$_2$SO$_4$ - HO. Therefore to hydrogen H$_2$, there should correspond the acids H.SHO₃, sulphurous, and SHO₃.SHO₃ = $S_2H_2O_6$, or dithionic; to SH_2 there should correspond the acids $SH(SHO_3)$ = $H_2S_2O_3$ (thiosulphuric), and $S(SHO_3)_2$ = $H_2S_3O_6$ (trithionic); to S_2H_2 the acids $S_2H(SHO_3)$ = $H_2S_3O_2$ (unknown), and $S_2(SHO_3)_2$ = $H_2S_4O_6$ (tetrathionic); to S_3H_2 the acids $S_3H(SHO_3)$ and $S_3(SHO_3)_2$ = $H_2S_5O_6$ (pentathionic). We know that iodine reacts directly with the hydrogen of sulphuretted hydrogen and combines with it, and if thiosulphuric acid contains the radicle of sulphuretted hydrogen (or hydrogen united with sulphur) of the same nature as in sulphuretted hydrogen, it is not surprising that iodine reacts with sodium thiosulphate and forms sodium tetrathionate. Thus, thiosulphuric acid, $HS(SHO_3)$, when deprived of H, gives a radicle which immediately combines with another similar radicle, forming the tetrathionate $S_2(SO_2HO)_2$. On this view of the structure of the thionic acids and salts, it is also clear how all the thionic acids, like thiosulphuric acid, easily give sulphur and sulphides, with the exception only of dithionic acid, $H_2S_2O_6$, which, judging from the above, stands apart from the series of the other thionic acids. Dithionic acid stands in the same relation to sulphuric acid as oxalic acid does to carbonic acid. Oxalic acid is dicarboxyl, $(CHO_2)_2$ = $C_2H_2O_4$, and so also dithionic acid is disulphoxyl, $(SHO_3)_2$ = $S_2H_2O_6$. Oxalic acid when ignited decomposes into carbonic anhydride and carbonic oxide, CO, and dithionic acid when heated decomposes into sulphuric anhydride and sulphurous anhydride, SO_2, and SO_2 stands in the same relation to SO_3 as CO to CO_2. This also explains the peculiarity of the calcium, barium, and lead, &c. salts of the thionic acids being easily soluble (although the corresponding salts of H_2SO_3, H_2SO_4, and H_2S dissolve with difficulty), because the former are similar to the salts of the sulphonic acids, which are also soluble in water. Thus the thionic acids are *disulphonic acids*, just as many dicarboxylic acids are known—for example, $CH_2(CO_2H)_2$, $C_6H_4(CO_2H)_2$.

Sulphur exhibits an acid character, not only in its compounds with hydrogen and oxygen, but also in those with other elements. The compound of sulphur and carbon has been particularly well investigated. It presents a great analogy to carbonic anhydride, both in its elementary composition and chemical character. This substance is the so-called carbon bisulphide, CS_2, and corresponds with CO_2.

The first endeavours to obtain a compound of sulphur with carbon were unsuccessful, for although sulphur does combine directly with carbon, yet the formation of this compound requires distinctly definite conditions. If sulphur be mixed with charcoal and heated, it is simply driven off from the latter, and not the smallest trace of carbon bisulphide is obtained. The formation of this compound requires that the charcoal should be first heated to a red heat, but not above, and then either the vapour of sulphur passed over it or lumps of sulphur thrown on to the red-hot charcoal, but in small quantities, so as not to lower the temperature of the latter. If the charcoal be heated to a white heat, the amount of carbon bisulphide formed is less. This depends, in the first place, on the carbon bisulphide dissociating at a high temperature. In the second place, Favre and Silberman showed that in the combustion of one gram of carbon bisulphide (the products will be $CO_2 + 2SO_2$) 3,400 heat units are evolved—that is, the combustion of a molecular quantity of carbon bisulphide evolves 258,400 heat units (according to Berthelot, 246,000). From a molecule of carbon bisulphide in grams we may obtain 12 grams of carbon, whose combustion evolves 96,000 heat units, and 64 grams of sulphur, evolving by combustion (into SO_2) 140,800 heat units. Hence we see that the component elements separately evolve less heat by their combustion (237,000 heat units) than carbon bisulphide itself—that is, that heat should be evolved (at the ordinary temperature) and not absorbed in its decomposition, and therefore that the formation of carbon bisulphide from charcoal and sulphur is in all probability accompanied by an absorption of heat. It is therefore not surprising that, like other compounds produced with an absorption of heat (ozone, nitrous oxide, hydrogen peroxide, &c.), carbon bisulphide is unstable and easily converted into the original substances from which it is obtained. And indeed if the vapour of carbon bisulphide be passed through a red-hot tube, it is decomposed—that is, it dissociates—into sulphur and carbon. And this takes place at the temperature at which this substance is formed, just as water decomposes into hydrogen and oxygen at the temperature of its formation. In this absorption of heat in the formation of carbon bisulphide is explained the facility with which it suffers reactions of decomposition, which we shall see in the sequel, and its main difference from the closely analogous carbonic anhydride.

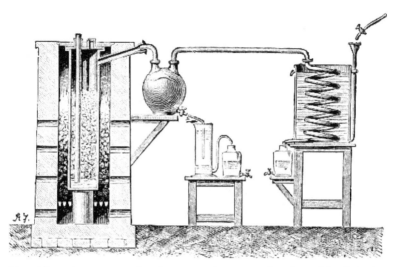

FIG. 90.—Apparatus for the manufacture of carbon bisulphide.

In the laboratory carbon bisulphide is prepared as follows: A porcelain tube is luted into a furnace in an inclined position, the upper extremity of the tube being closed by a cork, and the lower end connected with a condenser. The tube contains charcoal, which is raised to a red heat, and then pieces of sulphur are placed in the upper end. The sulphur melts, and its vapour comes into contact with the red-hot charcoal, when combination takes place; the vapours condense in the condenser, carbon bisulphide being a liquid boiling at 48°. On a large scale the apparatus depicted in fig. 90 is employed. A cast-iron cylinder rests on a stand in a furnace. Wood charcoal is charged into the cylinder through the upper tube closed by a clay stopper, whilst the sulphur is introduced through a tube reaching to the bottom of the cylinder. Pieces of sulphur thrown into this tube fall on to the bottom of the cylinder, and are converted into vapour, which passes through the entire layer of charcoal in the cylinder. The vapour of carbon bisulphide thus formed passes through the exit tube first into a Woulfe's bottle (where the sulphur which has not entered into the reaction is condensed), and then into a strongly-cooled condenser or worm.

Pure carbon bisulphide is a colourless liquid, which refracts light strongly, and has a pure ethereal smell; at 0° its specific gravity is 1·293, and at 15° 1·271. If kept for a long time it seems to undergo a change, especially when it is kept under water, in which it is insoluble. It boils at 48°, and the tension of its vapour is so great that it evaporates very easily, producing cold, and therefore it has to be kept in well-stoppered vessels; it is generally kept under a layer of water, which hinders its evaporation and does not dissolve it.

Carbon bisulphide enters into many combinations, which are frequently closely analogous to the compounds of carbonic anhydride. In this respect it is a *thio-anhydride*—*i.e.* it has the character of the acid anhydrides,[73 bis] like carbonic anhydride, with the difference that the oxygen of the latter is replaced by sulphur. By thio-compounds in general are understood those compounds of sulphur which differ from the compounds of oxygen as carbon bisulphide does from carbonic anhydride—that is, which correspond with the oxygen compounds, but with substitution of sulphur for oxygen. Thus thiosulphuric acid is monothiosulphuric acid—that is, sulphuric acid in which one atom of sulphur replaces one atom of oxygen. With the sulphides of the alkalis and alkaline earths, it forms saline substances corresponding with the carbonates, and these compounds may be termed *thiocarbonates*. For example, the composition of the sodium salt Na_2CS_3 is exactly like that of sodium carbonate. They are formed by the direct solution of carbon bisulphide in aqueous solutions of the sulphides; but they are difficult to obtain in a crystalline form, because they are easily decomposable. When the solutions of these salts are highly concentrated they begin to decompose, with the evolution of sulphuretted hydrogen and the formation of a carbonate, water taking part in the reaction—for example, $K_2CS_3 + 3H_2O = K_2CO_3 + 3H_2S$.

A remarkable example[74 bis] of the thio-compounds is found in *thiocyanic acid*—*i.e.* cyanic acid in which the oxygen is replaced by sulphur, HCNS. We know (Chapter IX.) that with oxygen the cyanides of the alkaline metals RCN give cyanates RCNO; but they also combine with sulphur, and therefore if yellow prussiate of potash be treated as in the preparation of potassium cyanide, and sulphur be added to the mass, potassium thiocyanate, KNCS, is obtained in solution. This salt is much more stable than potassium cyanate; it dissolves without change in water and alcohol, forming colourless solutions from which it easily crystallises on evaporation. It may be kept exposed to air even when in solution; in dissolving in water it absorbs a considerable amount of heat, and forms a starting-point for the preparation of all the thiocyanates, RCNS, and organic compounds in which the metals are replaced by hydrocarbon groups. Such, for example, is volatile mustard oil, C_3H_5CSN (allyl thiocyanate), which gives to mustard its caustic properties. With ferric salts the thiocyanates give an exceedingly brilliant red coloration, which serves for detecting the smallest traces of ferric salts in solution. Thiocyanic acid,

HCNS, may be obtained by a method of double decomposition, by distilling potassium thiocyanate with dilute sulphuric acid. It is a volatile colourless liquid, having a smell recalling that of vinegar, is soluble in water, and may be kept in solution without change.[75 bis]

The sulphur compounds of chlorine Cl_2S and Cl_2S_2 may be regarded on the one hand as products of the metalepsis of the sulphides of hydrogen, H_2S and H_2S_2; and on the other hand of the oxygen compounds of chlorine, because chloride of sulphur, Cl_2S, resembles chlorine oxide, Cl_2O, whilst Cl_2S_2 corresponds with the higher oxide of chlorine; or thirdly, we may see in these compounds the type of the acid chloranhydrides, because they are all decomposed by water, forming hydrochloric acid, and sulphur tetrachloride, SCl_4, is decomposed with the formation of sulphurous anhydride.

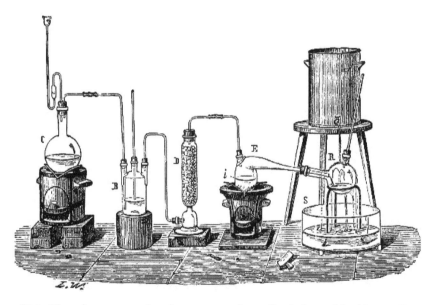

FIG. 91.—Apparatus for the preparation of sulphur chloride, and similar volatile compounds prepared by combustion in a stream of chlorine.

The compounds of sulphur with chlorine are prepared in the apparatus depicted in fig. 91. As sulphur chloride is decomposed by water, the chlorine evolved in the flask C must be dried before coming into contact with the sulphur. It is therefore first passed through a Woulfe's bottle, B, containing sulphuric acid, and then through the cylinder D containing pumice stone moistened with sulphuric acid, and then led into the retort E, in which the sulphur is heated. The compound which is formed distils over into the receiver R. A certain amount of sulphur passes over with the sulphur chloride, but if the resultant distillate be re-saturated with chlorine and distilled no free sulphur remains, the boiling-point rises to 144°, and pure sulphur chloride, S_2Cl_2, is obtained. It has this formula because its vapour density referred to hydrogen is 68. It is also obtained by heating certain metallic chlorides (stannous, mercuric) with sulphur; both the metal and chlorine then combine with the sulphur. Sulphur chloride is a yellowish-brown liquid, which boils at 144°, and has a specific gravity of 1·70 at 0°. It fumes strongly in the air, reacting on the moisture contained therein, and has a heavy chloranhydrous odour. It dissolves sulphur, is miscible with carbon bisulphide, and falls to the bottom of a vessel containing water, by which it is decomposed, forming sulphurous anhydride and hydrochloric acid; but it first forms various lower stages of oxidation of sulphur, because the addition of silver nitrate to the solution gives a black precipitate. With hydrogen sulphide it gives sulphur and hydrochloric acid, and it reacts directly with metals—especially arsenic, antimony, and tin—forming sulphides and chlorides. In the cold, it absorbs chlorine and gives *sulphur dichloride*, SCl_2. The entire conversion into this substance requires the prolonged passage of dry chlorine through sulphur chloride surrounded by a freezing mixture. The distillation of the dichloride must be conducted in a stream of chlorine, as otherwise it partially decomposes into sulphur chloride and chlorine. Pure sulphur dichloride is a reddish-brown liquid, which resembles the lower chloride in many respects; its specific gravity is 1·62; its odour is more suffocating than that of sulphur chloride; it volatilises at 64°.

Thionyl chloride, $SOCl_2$, may be regarded as oxidised sulphur dichloride; it corresponds with sulphur chloride, S_2Cl_2, in which one atom of sulphur is replaced by oxygen. At the same time it is chlorine oxide (hypochlorous anhydride, Cl_2O) combined with sulphur, and also the chloranhydride of sulphurous acid—that is, $SO(HO)_2$, in which the two hydroxyl groups are replaced by two atoms of chlorine, or sulphurous anhydride, SO_2, in which one atom of oxygen is replaced by two atoms of chlorine. All these representations are confirmed by reactions of formation, or decompositions; they all agree with our notions of the other compounds of sulphur, oxygen, and chlorine; hence these definitions are not contradictory to each other. Thus, for instance, thionyl chloride was first obtained by Schiff, by the action of dry sulphurous anhydride on phosphorus pentachloride. On distilling the resultant liquid, thionyl chloride comes over first at 80°, and on continuing the distillation phosphorus oxychloride distils over at above 100°, $PCl_5 + SO_2 = POCl_3 + SOCl_2$. This mode of preparation is direct evidence of the oxychloride character of $SOCl_2$. Würtz obtained the same substance by passing a stream of chlorine oxide through a cold solution of sulphur in sulphur chloride; the chlorine oxide then combined directly with the sulphur, $S + Cl_2O = SOCl_2$, whilst the sulphur chloride remained unchanged (sulphur cannot be combined directly with chlorine oxide, as an explosion takes place). Thionyl chloride is a colourless liquid, with a suffocating acrid smell; it has a specific gravity at 0° of 1·675, and boils at 78°. It sinks in water, by which it is immediately decomposed, like all chloranhydrides—for example, like carbonyl chloride, which corresponds with it: $SOCl_2 + H_2O = SO_2 + 2HCl$.[77 bis]

Normal *sulphuric acid has two corresponding chloranhydrides*; the first, $SO_2(OH)Cl$, is sulphuric acid, $SO_2(HO)_2$, in which one equivalent of HO is replaced by chlorine; the second has the composition SO_2Cl_2—that is, two HO groups are substituted by two of chlorine. The second chloranhydride, or the compound SO_2Cl_2, is called sulphuryl chloride, and the first chloranhydride, SO_2HOCl, may be called chlorosulphonic acid, because it is really an acid; it still retains one hydroxyl of sulphuric acid, and its corresponding salts are known. Thus, potassium chloride absorbs the vapour of sulphuric anhydride, forming a salt, SO_3KCl, corresponding with SO_3HCl as acid. In acting on sodium chloride it forms hydrochloric acid and the salt $NaSO_3Cl$. This first chloranhydride of sulphuric acid, SO_2HOCl, discovered by Williamson, is obtained either by the action of phosphorus pentachloride on sulphuric acid ($PCl_5 + H_2SO_4 = POCl_3 + HCl + HSO_3Cl$), or directly by the action of dry hydrochloric acid on sulphuric anhydride, $SO_3 + HCl = HSO_3Cl$. The most easy and rapid method of its formation is by direct saturation of cold Nordhausen acid with dry hydrochloric acid gas ($SO_3 + HCl = HSO_3Cl$), and distillation of the resultant solution; the distillate then contains HSO_3Cl. It is a colourless fuming liquid, having an acrid odour; it boils at 153° (according to my determination, confirmed by Konovaloff), and its specific gravity at 19° is 1·776. It is immediately decomposed by water, forming hydrochloric and sulphuric acids, as should be the case with a true chloranhydride. In the reactions of this chloranhydride we find the easiest means of introducing the sulphonic group HSO_3 into other compounds, because it is here combined with chlorine. The second chloranhydride of sulphuric acid, or *sulphuryl chloride*, SO_2Cl_2, was obtained by Regnault by the direct action of the sun's ray on a mixture of equal volumes of chlorine and sulphurous oxide. The gases gradually condense into a liquid, combining together as carbonic oxide does with chlorine. It is also obtained when a mixture of the two gases in acetic acid is allowed to stand for some time. The first chloranhydride, SO_3HCl, decomposes when heated at 200° in a closed tube into sulphuric acid and sulphuryl chloride. It boils at 70°, its specific gravity is 1·7, it gives hydrochloric and sulphuric acids with water, fumes in the air, and, judging by its vapour density, does not decompose when distilled.

In the group of the halogens we saw four closely analogous elements—fluorine, chlorine, bromine, and iodine—and we meet with the same number of closely allied analogues in the oxygen group; for besides sulphur this group also includes *selenium* and *tellurium*: O, S, Se, Te. These two groups are very closely allied, both in respect to the magnitudes of their atomic weights and also in the faculty of the elements of both groups for combining with metals. The distinct analogy and definite degree of variance known to us for the halogens, also repeat themselves in the same degree for the elements of the oxygen group. Amongst the halogens fluorine has many peculiarities compared to Cl, Br and I which are more closely analogous, whilst oxygen differs in many respects from S, Se, Te, which possess greater similarities. The analogy in a quantitative respect is perfect in both cases. Thus the halogens combine with H, and the elements of the oxygen group with H_2, forming H_2O, H_2S, H_2Se, H_2Te. The hydrogen compounds of selenium and tellurium are acids like hydrogen sulphide. Selenium, by simple heating in a stream of hydrogen, partially combines with it directly, but seleniuretted hydrogen is more readily decomposable by heat than sulphuretted hydrogen, and this property is still more developed in telluretted hydrogen. Hydrogen selenide and telluride are gases like sulphuretted hydrogen, and, like it, are soluble in water, form saline compounds with alkalis, precipitate metallic salts, are obtained by the action of acids on their compounds with metals, &c. Selenium and tellurium, like sulphur, give two normal grades of combination with oxygen, both of an acid character, of which only the forms corresponding to sulphurous anhydride—namely, selenious anhydride, SeO_2, and tellurous anhydride, TeO_2—are formed directly. These are both solids, obtained by the combustion of the elements themselves and by the action of oxidising agents on them. They form feebly energetic acids, having distinct bibasic properties; however, a characteristic difference from SO_2 is observable both in the physical properties of these compounds and in their stability and capacity for further oxidation, just as in the series of the halogens already known to us, only in an inverse order; in the latter we saw that iodine combines more easily than bromine or chlorine with oxygen, forming more stable oxygen compounds, whereas here, on the contrary, sulphurous anhydride, as we know, is difficultly decomposed, parts with its sulphur with difficulty, and is easily oxidised and especially in its salts, while selenious and tellurous anhydrides are oxidised with difficulty and easily reduced, even by means of sulphurous acid.

Selenium was obtained in 1817 by Berzelius from the sublimate which collects in the first chamber in the preparation of sulphuric acid from Fahlun pyrites. Certain other pyrites also contain small quantities of selenium. Some native selenides, especially those of lead, mercury, and copper, have been found in the Hartz Mountains, but only in small quantities. Pyrites and blendes, in which the sulphur is partially replaced by selenium, still remain the chief source for its extraction. When these pyrites are roasted they evolve selenious anhydride, which condenses in the cooler portions of the apparatus in which the pyrites are roasted, and is partially or wholly reduced by the sulphurous anhydride simultaneously formed. The presence of selenium in ores and sublimates is most simply tested by heating them before the blowpipe, when they evolve the characteristic odour of garlic. Selenium exhibits two modifications, like sulphur: one amorphous and insoluble in carbon bisulphide, the other crystalline and slightly soluble in carbon bisulphide (in 1,000 parts at 45° and 6,000 at 0°), and separating from its solutions in monoclinic prisms. If the red precipitate obtained by the action of sulphurous anhydride on selenious anhydride be dried, it gives a brown powder, having a specific gravity of 4·26, which when heated changes colour and fuses to a metallic mass, which gains lustre as it cools. The selenium acquires different properties according to the rate at which it is cooled from a fused state; if rapidly cooled, it remains amorphous and has the same specific gravity (4·28) as the powder, but if slowly cooled it becomes crystalline and opaque, soluble in carbon bisulphide, and has a specific gravity of 4·80. In this form it fuses at 214° and remains unchanged, whilst the amorphous form, especially above 80°, gradually passes into the crystalline variety. The transition is accompanied by the evolution of heat, as in the case of sulphur; thus the analogy between sulphur and selenium is clearly shown here. In the fused amorphous form selenium presents a brown mass, slightly translucent, with a vitreous fracture, whilst in the crystalline form it has the appearance of a grey metal, with a feeble lustre and a crystalline fracture.[79 bis] Selenium boils at 700°, forming a vapour whose density is only constant at a temperature of about 1,400°, when it is equal to 79·4 (referred to hydrogen)—that is, the molecular formula is then Se_2, like sulphur at an equally high temperature.

Tellurium is met with still more rarely than selenium (it is known in Saxony) in combination with gold, silver, lead, and antimony in the so-called foliated tellurium ore. Bismuth telluride and silver telluride have been found in Hungary and in the Altai. Tellurium is extracted from bismuth telluride by mixing the finely-powdered ore with potassium and charcoal in as intimate a mixture as possible, and then heating in a covered crucible. Potassium telluride, K_2Te, is then formed, because the charcoal reduces potassium tellurite. As potassium telluride is soluble in water, forming a red-brown solution which is decomposed by the oxygen of the atmosphere ($K_2Te + O + H_2O = 2KHO + Te$), the mass formed in the crucible is treated with boiling water and filtered as rapidly as possible, and the resultant solution exposed to the air, by which means the tellurium is precipitated. In a free state tellurium has a perfectly *metallic appearance*; it is of a silver-white colour, crystallises very easily in long brilliant needles; is very brittle, so that it can be easily reduced to powder; but it is a bad conductor of heat and electricity, and in this respect, as in many others, it forms a transition from the metals to the non-metals. Its specific gravity is 6·18, it melts at an incipient red heat, and takes fire when heated in air, like selenium and sulphur, burning with a blue flame, evolving white fumes of tellurous anhydride, TeO_2, and emitting an acrid smell if no selenium be present; but if it be, the odour of the latter preponderates. Alkalis dissolve tellurium when boiled with it, potassium telluride, K_2Te, and potassium tellurite, K_2TeO_3, being formed. The solution is of a red colour, owing to the presence of the telluride, K_2Te; but the colour disappears when the solution is cooled or diluted, the tellurium being all precipitated: $2K_2Te + K_2TeO_3 + 3H_2O = 6KHO + 3Te$.

Footnotes:

The character of sulphur is very clearly defined in the organo-metallic compounds. Not to dwell on this vast subject, which belongs to the province of organic chemistry, I think it will be sufficient for our purpose to compare the physical properties of the ethyl compounds of mercury, zinc, sulphur and oxygen. The composition of all of them is expressed by the general formula $(C_2H_5)_2R$, where R = Hg, Zn, S, or O. They are all volatile: mercury ethyl, $Hg(C_2H_5)_2$, boils at 159°, its sp. gr. is 2·444, molecular volume = 106; zinc ethyl boils at 118°, sp. gr. 1·882, volume 101; ethyl sulphide, $S(C_2H_5)_2$, boils at 90°, sp. gr. 0·825, volume 107; common ether, or ethyl oxide, $O(C_2H_5)_2$, boils at 35°, sp. gr. 0·736, volume 101, in addition to which diethyl itself, $(C_2H_5)_2 = C_4H_{10}$, boils about 0°, sp. gr. about 0·62, volume about 94. Thus the substitution of Hg, S, and O scarcely changes the volume, notwithstanding the difference of the weights; the physical influence, if one may so express oneself, of these elements, which are so very different in their atomic weights, is almost alike.

Therefore in former times sulphur was known as an amphid element. Although the analogy between the compounds of sulphur and oxygen has been recognised from the very birth of modern chemistry (owing, amongst other things, to the fact that the oxides and sulphides are the most widely spread metallic ores in nature), still it has only been clearly expressed by the periodic system, which places both these elements in group VI. Here, moreover, stands out that parallelism which exists between SO_2 and ozone OO_2, between K_2SO_3 and peroxide of potassium K_2O_4 (Volkovitch in 1893 again drew attention to this parallelism).

When in Sicily, I found, near Caltanisetta, a specimen of sulphur with mineral tar. In the same neighbourhood there are naphtha springs and mud volcanoes. It may be that these substances have reduced the sulphur from gypsum.

The chief proof in favour of the origin of sulphur from gypsum is that in treating the deposits for the extraction of the sulphur it is found that the proportion of sulphur to calcium carbonate never exceeds that which it would be had they both been derived from calcium sulphate.

Naturally only those ores of sulphur which contain a considerable amount of sulphur can be treated by this method. With poor ores it is necessary to have recourse to distillation or mechanical treatment in order to separate the sulphur, but its price is so low that this method in most cases is not profitable.

The sulphur obtained by the above-described method still contains some impurities, but it is frequently made use of in this form for many purposes, and especially in considerable quantities for the manufacture of sulphuric acid, and for strewing over grapes. For other purposes, and especially in the preparation of gunpowder, a purer sulphur is required. Sulphur may be purified by distillation. The crude sulphur is called *rough*, and the distilled sulphur *refined*. The arrangement given in fig. 86 is employed for refining sulphur. The rough sulphur is melted in the boiler d, and as it melts it is run through the tube F into an iron retort B heated by the naked flame of the furnace. Here the sulphur is converted into vapour, which passes through a wide tube into the chamber G, surrounded by stone walls and furnished with a safety-valve S.

Flowers of sulphur always contain a certain amount of the oxides of sulphur.

Sulphur may be extracted by various other means. It may be extracted from iron pyrites, FeS_2, which is very widely distributed in nature. From 100 parts of iron pyrites about half the sulphur contained, namely, about 25 parts, may be extracted by heating without the access of air, a lower sulphide of iron, which is more stable under the action of heat, being left behind. Alkali waste (Chapter XII.), containing calcium sulphide and gypsum, $CaSO_4$, may be used for the same purpose, but native sulphur is so cheap that recourse can only be had to these sources when the calcium sulphide appears as a worthless by-product. The most simple process for the extraction of sulphur from alkali waste, in a chemical sense, consists in evolving sulphuretted hydrogen from the calcium sulphide by the action of hydrochloric acid. The sulphuretted hydrogen when burnt gives water and sulphurous anhydride, which reacts on fresh sulphuretted hydrogen with the separation of sulphur. The combustion of the sulphuretted hydrogen may be so conducted that a mixture of $2H_2S$ and SO_2 is straightway formed, and this mixture will deposit sulphur (Chapter XII., Note 14).

Gossage and Chance treat alkali waste with carbonic anhydride, and subject the sulphuretted hydrogen evolved to incomplete combustion (this is best done by passing a mixture of sulphuretted hydrogen and air, taken in the requisite proportions, over red-hot ferric oxide), by which means water and the vapour of sulphur are formed: $H_2S + O = H_2O + S$.

One hundred parts of liquid carbon bisulphide, CS_2, dissolve 16·5 parts of sulphur at -11°, 24 parts at 0°, 37 parts at 15°, 46 parts at 22°, and 181 parts at 55°. The saturated solution boils at 55°, whilst pure carbon bisulphide boils at 47°. The solution of sulphur in carbon bisulphide reduces the temperature, just as in the solution of salts in water. Thus the solution of 20 parts of sulphur in 50 parts of carbon bisulphide at 22° lowers the temperature by 5°; 100 parts of benzene, C_6H_6, dissolves 0·965 part of sulphur at 26°, and 4·377 parts at 71°; chloroform, $CHCl_3$, dissolves 1·2 part of sulphur at 22°, and 16·35 parts at 174°.

If the experiment be made in a vessel with a narrow capillary tube, the sulphur fuses at a lower temperature (occurs, as it were, in a supersaturated state), and solidifying at 90°, appears in a rhombic form (Schützenberger).

If sulphur be cautiously melted in a U tube immersed in a salt bath, and then gradually cooled, it is possible for all the sulphur to remain liquid at 100°. It will now be in a state of superfusion; thus also by careful refrigeration water may be obtained in a liquid state at -10°, and a lump of ice then causes such water to form ice, and the temperature rises to 0°. If a prismatic crystal of sulphur be thrown into one branch of the U tube containing the liquid sulphur at 100°, and an octahedral crystal be thrown into the other branch, then, as Gernez showed, the sulphur in each branch will crystallise in the corresponding form, and both forms are obtained at the same temperature; therefore it is not the influence of temperature only which causes the molecules of sulphur to distribute themselves in one or another form, but also the influence of the crystalline parts already formed. This phenomenon is essentially analogous to the phenomena of supersaturated solutions.

A certain amount of insoluble sulphur remains for a long time in the mass of soft sulphur, changing into the ordinary variety. Freshly-cooled soft sulphur contains about one-third of insoluble sulphur, and after the lapse of two years it still contains about 15 p.c. Flowers of sulphur, obtained by the rapid condensation of sulphur from a state of vapour, also contains a certain amount of insoluble sulphur. *Rapidly distilled and condensed sulphur* also contains some insoluble sulphur. Hence a certain amount of insoluble sulphur is frequently found in roll sulphur. The action of light on a solution of sulphur converts a certain portion into the insoluble modification. Insoluble sulphur is of a lighter colour than the ordinary variety. It is best prepared by vaporising sulphur in a stream of carbonic anhydride, hydrochloric acid, &c., and collecting the vapour in cold water. When condensed in this manner it is nearly all insoluble in carbon bisulphide. It then has the form of hollow spheroids, and is therefore lighter than the common variety: sp. gr. 1·82. An idea of the modifications taking place in sulphur between 110° and 250° may be formed from the fact that at 150° liquid sulphur has a coefficient of expansion of about 0·0005, whilst between 150° and 250° it is less than 0·0003.

Engel (1891), by decomposing a saturated solution of hyposulphite of sodium (Note 29) with HCl in the cold (the sulphur is not precipitated directly in this case), obtained, after shaking up with chloroform and evaporation, crystals of sulphur (sp. gr. 2·135), which, after several hours, passed into the insoluble (in CS_2) state, and in so doing became opaque, and increased in volume. But if a mixture of solution of $Na_2S_2O_3$ and HCl be allowed to stand, it deposits sulphur, which, after sufficient washing, is able to dissolve in water (like the colloid varieties of the metallic sulphides, alumina, boron, and silver), but this colloid *solution of sulphur* soon deposits sulphur insoluble in CS_2.

When a solution of sulphuretted hydrogen in water is decomposed by an electric current the sulphur is deposited on the positive pole, and has therefore an electro-negative character, and this sulphur is soluble in carbon bisulphide. When a solution of sulphurous acid is decomposed in the same manner, the sulphur is deposited on the negative pole, and is therefore electro-positive, and the sulphur so deposited is insoluble in carbon bisulphide. The sulphur which is combined with metals must have the properties of the sulphur contained in sulphuretted hydrogen, whilst the sulphur combined with chlorine is like that which is combined with oxygen in sulphurous anhydride. Hence Berthelot recognises the presence of soluble sulphur in metallic sulphides, and of the insoluble modification of amorphous sulphur in sulphur chloride. Cloez showed that the sulphur precipitated from solutions is either soluble or insoluble, according to whether it separates from an alkaline or acid solution. If sulphur be melted with a small quantity of iodine or bromine, then on pouring out the molten mass it forms amorphous sulphur, which keeps so for a very long time, and is insoluble, or nearly so, in carbon bisulphide. This is taken advantage of in casting certain articles in sulphur, which by this means retain their tenacity for a long time; for example, the discs of electrical machines.

Here, however, it is very important to remark that both benzene and acetylene can exist at the ordinary temperature, whilst the sulphur molecule S_2 only exists at high temperatures; and if this sulphur be allowed to cool, it passes first into S_6 and then into a liquid state. Were it possible to have sulphur at the ordinary temperature in both the above modifications, then in all probability the sulphur in the state S_2 would present totally different properties from those which it has in the form S_6, just as the properties of gaseous acetylene are far from being similar to those of liquid benzene. Sulphur, in the form of S_2, is probably a substance which boils at a much lower temperature than the variety with which we are now dealing. Paterno and Nasini (1888), following the method of depression or fall of the freezing-point in a benzene solution, found that the molecule of sulphur in solution contains S_6.

One must here call attention to the fact that sulphur, with all its analogy to oxygen (which also shows itself in its faculty to give the modification S_2), is also able to give a series of compounds containing more atoms of sulphur than the analogous oxygen compounds do of oxygen. Thus, for instance, compounds of 5 atoms of sulphur with 1 atom of barium, BaS_5, are known, whereas with oxygen only BaO_2 is known. On every side one cannot but see in sulphur a faculty for the union of a greater number of atoms than with oxygen. With oxygen the form of ozone, O_3, is very unstable, the stable form is O_2; whilst with sulphur S_6 is the stable form, and S_2 is exceedingly unstable. Furthermore, it is remarkable that sulphur gives a higher degree of oxidation, H_2SO_4, corresponding, as it were, with its complex composition, if we suppose that in S_6 four atoms of sulphur are replaced by oxygen and one by two atoms of hydrogen. The formulæ of its compounds, K_2SO_4, $K_2S_2O_3$, K_2S_5, BaS_5, and many others, have no analogues among the compounds of oxygen. They all correspond with the form S_6 (one portion of the sulphur being replaced by oxygen and another by metals), which is not attained by oxygen. In this faculty of sulphur to hold many atoms of other substances the same forces appear which cause many atoms of sulphur to form one complex molecule.

In the formation of potassium sulphide, K_2S (that is, in the combination of 32 parts of sulphur with 78 parts of potassium), about 100 thousand heat units are developed. Nearly as much heat is developed in the combination of an equivalent quantity of sodium; about 90 thousand heat units in the formation of calcium or strontium sulphide; about 40 thousand for zinc or cadmium sulphide, and about 20 thousand for iron, cobalt, or nickel sulphide. Less heat is evolved in the combination of sulphur with copper, lead, and silver. According to Thomsen, sulphur develops heat with hydrogen in solutions. The reaction $I_2,Aq,H_2S = 21,830$ calories. But, as the reaction $I_2 + H_2 + Aq$ develops 26,842 calories, it follows that the reaction $H_2 + S$ develops 4,512 calories.

If sulphur be melted in a flask and heated nearly to its boiling point, as Lidoff showed, the addition, drop by drop (from a funnel with a stopcock) of heavy (0·9) naphtha oil (of lubricating oleonaphtha), &c., is followed by a regular evolution of sulphuretted hydrogen. This is analogous to the action of bromine or iodine on paraffin and other oils, because hydrobromic or hydriodic acid is then formed (Chapter XI.) A certain amount of hydrogen sulphide is even formed when sulphur is boiled with water.

However, the matter is really much more complicated. Thus zinc sulphide evolves sulphuretted hydrogen with sulphuric or hydrochloric acids, but does not react with acetic acid and is oxidised by nitric acid. Ferrous sulphide evolves sulphuretted hydrogen with acids, whilst the bisulphide, FeS_2, does not react with acids of ordinary strength. This absence of action depends, among other things, on the form in which the native iron pyrites occurs; it is a crystalline, compact, and very dense substance; and acids in general react with great difficulty on such metallic sulphides. This is seen very clearly in the case of zinc sulphide; if this substance is obtained by double decomposition, it separates as a white precipitate, which evolves sulphuretted hydrogen with great ease when treated with acids. Zinc sulphide is obtained in the same form when zinc is fused with sulphur, but native zinc sulphide—which occurs in compact masses of zinc blende, and has a metallic lustre—is not decomposed or scarcely decomposed by sulphuric acid.

Another source of complication in the behaviour of the metallic sulphides towards acids depends on the action of water, and is shown in the fact that the action varies with different degrees of dilution or proportion of water present. The best known example of this is antimonious sulphide, Sb_2S_3, for strong hydrochloric acid, containing not more water than corresponds with $HCl,6H_2O$, even decomposes native antimony glance, with evolution of sulphuretted hydrogen, whilst dilute acid has no action, and in the presence of an excess of water the reaction $2SbCl_3 + 3H_2S = Sb_2S_3 + 6HCl$ occurs, whilst in the presence of a small amount of water the reaction proceeds in exactly the opposite direction. Here the participation of water in the reaction and its affinity are evident.

The facts that lead sulphide is insoluble in acids, that zinc sulphide is soluble in hydrochloric acid but insoluble in acetic acid, that calcium sulphide is even decomposed by carbonic acid, &c.—all these peculiarities of the sulphides are in correlation with the amount of heat evolved in the reaction of the oxides with hydrogen sulphide and with acids, as is seen from the observations of Favre and Silberman, and from the comparisons made by Berthelot in the Proceedings of the Paris Academy of Sciences, 1870, to which we refer the reader for further details.

Ferrous sulphide is formed by heating a piece of iron to an incipient white heat, and then removing it from the furnace and bringing it into contact with a piece of sulphur. Combination then proceeds, accompanied by the development of heat, and the ferrous sulphide formed fuses. The sulphide of iron thus formed is a black, easily-fusible substance, insoluble in water. When damp it attracts oxygen from the air, and is converted into green vitriol, $FeSO_4$. If all the iron does not combine with the sulphur in the method described above, the action of sulphuric acid will evolve hydrogen as well as hydrogen sulphide.

We will not describe the details of the preparation of sulphuretted hydrogen employed as a reagent in the laboratory, because, in the first place, the methods are essentially the same as in the preparation of hydrogen, and, in the second place, because the apparatus and methods employed are always described in text-books of analytical chemistry. Ferrous sulphide may be advantageously replaced by calcium sulphide or a mixture of calcium and magnesium sulphides. A solution of magnesium hydrosulphide, MgS,H_2S, is very convenient, as at 60° it evolves a stream of pure hydrogen sulphide. A paste, consisting of CuS with crystals of $MgCl_2$ and water, may also be employed, since it only evolves H_2S when heated (Habermann).

[15 bis] Liquid sulphuretted hydrogen is most easily obtained by the decomposition of hydrogen polysulphide, which we shall presently describe, by the action of heat, and in the presence of a small amount of water. If poured into a bent tube, like that described for the liquefaction of ammonia (Chapter VI.), the hydrogen polysulphide is decomposed by heat, in the presence of water, into sulphur and sulphuretted hydrogen, which condenses in the cold end of the tube into a colourless liquid.

Sulphuretted hydrogen is still more soluble in alcohol than in water; one volume at the ordinary temperature dissolves as much as eight volumes of the gas. The solutions in water and alcohol undergo change, especially in open vessels, owing to the fact that the water and alcohol dissolve oxygen from the atmosphere, which, acting on the sulphuretted hydrogen, forms water and sulphur. The solution may be so altered in this manner that every trace of sulphuretted hydrogen disappears. Solutions of sulphuretted hydrogen in glycerine change much more slowly, and may therefore be kept for a long time as reagents. De Forcrand obtained a hydrate, $H_2S,16H_2O$, resembling the hydrates given by many gases.

Some metals evolve hydrogen from sulphuretted hydrogen at the ordinary temperature. For example, the light metals, and copper and silver (especially with the access of air?) among the heavy metals. Hence articles made of silver turn black in the presence of vapours containing sulphuretted hydrogen, because silver sulphide is black. Zinc and cadmium act at a red heat, but not completely.

If sulphuretted hydrogen escapes from a fine orifice into the air, it will burn when lighted, and be transformed into sulphurous anhydride and water. But if it burns in a limited supply of air—for instance, when a cylinder is filled with it and lighted—then only the hydrogen burns, which has, judging from the amount of heat developed in its combustion and from all its properties, a greater affinity for oxygen than sulphur. In this respect the combustion of sulphuretted hydrogen resembles that of hydrocarbons.

Hence bleaching powder and chlorine destroy the disagreeable smell of sulphuretted hydrogen. (For the reaction of hydrogen sulphide and iodine, *see* Chapter XI. p. 504.)

[19 bis] Perfectly dry H_2S (Hughes 1892) has no action upon perfectly dry salts, just as dry HCl does not react with dry NH_3 or metals (Chapter IX., Note 29).

The sulphide P_4S is obtained by cautiously fusing the requisite proportions of common phosphorus and sulphur under water; it is a liquid which solidifies at 0°, and may be distilled without undergoing change, but it fumes in air and easily takes fire. The higher sulphide, P_2S, has similar properties. But little heat is evolved in the formation of these compounds, and it may be supposed that they are formed by the direct conjunction of whole molecules of phosphorus and sulphur; but if the proportion of sulphur be increased, the reaction is accompanied by so considerable a rise of temperature that an explosion takes place, and for the sake of safety red phosphorus must be used, mixed as intimately as possible with powdered sulphur and heated in an atmosphere of carbonic anhydride. The higher compounds are decomposed by water. By increasing the proportion of sulphur, the following compounds have been obtained: P_4S_3 as prisms (fuses at 165°, Rebs), soluble in carbon bisulphide, and unaltered by air and water; *phosphorus trisulphide*, P_2S_3, is the analogue of P_2O_3; it is a light yellow crystalline compound only slightly soluble in carbon bisulphide, fusible and volatile, decomposed into hydrogen sulphide and phosphorous acid by water, and, like the highest compound of sulphur and phosphorus, P_2S_5, it forms thio-salts with potassium sulphide, &c. This *phosphorus pentasulphide* corresponds with phosphoric anhydride; like the trisulphide it gives hydrogen sulphide and phosphoric acid with an excess of water. It reacts in many respects like phosphoric chloride. The sulphide PS_2 is also known; the vapour density of this compound seems to indicate a molecule P_3S_6.

Phosphorus sulphochloride, $PSCl_3$, corresponds with phosphorus oxychloride. It is a colourless, pleasant-smelling liquid, boiling at 124°, and of sp. gr. 1·63; it fumes in air and is decomposed by water: $PSCl_3 + 4H_2O = PH_3O_4 + H_2S + 3HCl$. It is obtained when phosphoric chloride is treated with hydrogen sulphide, hydrochloric acid being also formed; it is also produced by the action of phosphoric chloride on certain sulphides—for example, on antimonious sulphide, also by the (cautious) action of phosphorus on sulphur chloride: $2P + 3S_2Cl_2 = 2PSCl_3 + 4S$, by the action of PCl_5 upon certain sulphides, for example, Sb_2S_3, by the reaction: $3MCl + P_2S_5 = PSCl_3 + M_3PS_4$ (Glatzel, 1893), and in the reaction $3PCl_3 + SOCl_2 = PCl_5 + POCl_3 + PSCl_3$, showing the reducing action of phosphorus trichloride, which is especially clear in the reaction $SO_3 + PCl_3 = SO_2 + POCl_3$. Thorpe and Rodger (1889), by heating $3PbF_2$ or BiF_3 with phosphorus pentasulphide (and also by heating AsF_3 and $PSCl_3$ to 150°), obtained thiophosphoryl fluoride as a colourless, spontaneously inflammable gas (see further on, Note 74 bis, and Chapter XIX., Note 25). The action of $PSCl_3$ upon NaHO gives a salt of monothiophosphoric acid (Würtz, Kubierschky), H_3PSO_3, which gives soluble salts of the alkalis.

Sulphuretted hydrogen does not saturate the alkaline properties of alkali hydroxides, so that a solution of potassium hydroxide will not under any circumstances give a neutral liquid with sulphuretted hydrogen. In this case the sulphuretted hydrogen forms in solution only an acid salt with the potassium: $KHO + H_2S = KHS + H_2O$. It must be supposed that the normal salt is not formed in the solution—that is, that the reaction $2KHO + H_2S = K_2S + 2H_2O$ does not take place. This is seen from the fact that a development of heat, depending on the formation of potassium hydrosulphide, KHS, is remarked when as much hydrogen sulphide is passed into a solution of potassium hydroxide as it will absorb. But if a further quantity of potassium hydroxide be added to the resultant solution, heat is not developed, whilst if alkali be added to potassium acid sulphate or sodium acid carbonate, heat is developed. It must not be concluded from this that H_2S is a monobasic acid, for here there is a question of the decomposing action of water upon K_2S; K_2S and H_2O in reacting on each other should absorb heat if the reaction of KHS upon KHO evolves heat. Furthermore, it must be taken into account that potassium oxide, K_2O, and the anhydrous oxides like it, also do not exist in solutions, for whenever they are formed they immediately react with the water, forming caustic potash, KHO, &c. In the same way, directly potassium sulphide, K_2S, is formed in water it is decomposed into potassium hydroxide and hydrosulphide: $K_2S + H_2O = KHO + KHS$. Potassium sulphide, K_2S, in a solid state corresponds with K_2O, although neither can exist in solution.

During recent years (beginning with Schulze, 1882) it has been found that many metallic sulphides which were considered totally insoluble do, under certain circumstances, form very unstable solutions in water, as already mentioned in Chapter I., Note 57 57. Arsenic sulphide is very easily obtained in the form of a solution (hydrosol). Solutions of copper and cadmium sulphides may also be easily obtained by precipitating their salts CuX_2, or CdX_2, with ammonium sulphide, and washing the precipitate; but they are re-precipitated by the addition of foreign salts.

In reality the preceding reaction should be expressed thus: $FeCl_2$ + $2KHS$ = FeS + $2KCl$ + H_2S (Note 21), because in the presence of water not K_2S but KHS reacts. But as the sulphuretted hydrogen takes no part in the reaction, it is usual to express the formation of such sulphides without taking the hydrogen sulphide proceeding from the potassium or ammonium hydrosulphides into account. It is not usual to employ potassium sulphide but ammonium sulphide—or, to speak more accurately, ammonium hydrosulphide—in order to avoid the formation of a non-volatile salt of potassium and to have, together with the formation of the sulphide, a salt of ammonium which can always be driven off by evaporating the solution and igniting the residue—for instance: $FeCl_2$ + $(NH_4)_2S$ = FeS + $2NH_4Cl$. Thus the metallic sulphides may be divided into three chief classes: (1) *those soluble in water*, (2) *those insoluble in water but reacting with acids*, and (3) *those insoluble both in water and acids*. The third class may be easily subdivided into two groups; to the first group belong those sulphides which correspond with bases or basic oxides, and are therefore unable to play the part of an acid with the sulphides of the alkalis, and are insoluble in NH_4HS, whilst the sulphides of the second group are of an acid character, and give soluble thio-salts with the sulphides of the alkaline metals, in which they play the part of an acid. To this group belong those metals whose corresponding oxides have acid properties. It must be observed, however, that not all metallic acids have corresponding sulphides, partly owing to the fact that certain acids are reducible by sulphuretted hydrogen, especially when their lower degrees of oxidation are of a basic character. Such are, for instance, the acids of chromium, manganese, &c. Sulphuretted hydrogen converts them into lower oxides, having the properties of bases. Those bases which do not combine with feeble acids, such as carbonic acid and hydrogen sulphide, give a precipitate of hydroxide with ammonium sulphide—for example, aluminium salts react in this manner. This difference of the metals in their behaviour towards sulphuretted hydrogen gives a very valuable means of separating them from each other, and *is taken advantage of in analytical chemistry*. If, for instance, the metals of the first and third groups occur together, it is only necessary to convert them into soluble salts, and to act on the solution of the salts with sulphuretted hydrogen; this will precipitate the metals of the third group in the form of sulphides, whilst the metals of the first group will not be in the least acted on. Such a method of separating the metals is considered more fully in analytical chemistry, and we will therefore limit ourselves here to pointing out to which groups the most common metals belong, and the colour which is proper to the sulphide precipitated.

Metals which are precipitated by sulphuretted hydrogen, as sulphides from a solution of their salts, even in the presence of free acid:

The precipitate is soluble in ammonium sulphide:

Platinum (dark brown)

Gold (dark brown)

Tin (yellow and brown)

Antimony (orange)

Arsenic (yellow)

The precipitate is insoluble in ammonium sulphide:

Copper (black)

Silver (black)

Cadmium (yellow)

Mercury (black)

Lead (black)

Metals which are precipitated by ammonium sulphide from neutral solutions, but not precipitated from acid solutions by sulphuretted hydrogen:

The sulphide precipitated is soluble in hydrochloric acid:

Zinc (white)

Manganese (rose colour)

Iron (black)

The sulphide precipitated is not soluble in dilute hydrochloric acid:

Nickel (black)

Cobalt (black)

A hydroxide, and not a sulphide, is precipitated:

Chromium (green)

Aluminium (white)

The metals of the alkalis and of the alkaline earths are not precipitated either by sulphuretted hydrogen or ammonium sulphide. The metals of the alkaline earths when in acid solutions in the form of phosphates and many other salts are precipitated by ammonium sulphide, because the latter neutralises the free acid, with formation of an ammonium salt of the acid and evolution of sulphuretted hydrogen.

Rebs took di-, tri-, tetra-, and pentasulphides of sodium, potassium, and barium, which he prepared by dissolving sulphur in solutions of the normal sulphides; on adding hydrochloric acid he always obtained hydrogen pentasulphide, whence it is evident that $4H_2S_n = (n - 1)H_2S_5 + (5 - n)H_2S$. For example, if H_2S_2 were formed, it would decompose according to the equation $4H_2S_2 = H_2S_5 + 3H_2S$. The hydrogen pentasulphide formed breaks up into hydrogen sulphide and sulphur when brought into contact with water. Previous to Rebs' researches many chemists stated that all polysulphides gave the bisulphide H_2S_2, and Hofmann recognised only hydrogen trisulphide, H_2S_3.

The formation of the polysulphides of hydrogen, H_2S_n is easily understood from the law of substitution, like that of the saturated hydrocarbons, C_nH_{2n+2}, knowing that sulphur gives H_2S, because the molecule of sulphuretted hydrogen may be divided into H and HS. This radicle, HS, is equivalent to H. By substituting this radicle for hydrogen in H_2S we obtain $(HS)HS = H_2S_2$, $(HS)(HS)S = H_2S_3$, &c., in general H_2S_n. The homologues of CH_4, C_nH_{2n+2} are formed in this manner from CH_4, and consequently the polysulphides H_2S_n are the homologues of H_2S. The question arises why in H_2S_n the apparent limit of n is 5—that is, why does the substitution end with the formation of H_2S_5? The answer appears to me to be clearly because in the molecule of sulphur, S_6, there are six atoms of sulphur (Note 11). The forces in one and the other case are the same. In the one case they hold S_6 together, in the other S_5 and H_2; and, judging from H_2S, the two atoms of hydrogen are equal in power and significance to the atom of sulphur. Just as hydrogen peroxide, H_2O_2, expresses the composition of ozone, O_3, in which O is replaced by H_2, so also H_2S_5 corresponds with S_6.

Ammonium sulphide, $(NH_4)_2S$, may be prepared by passing sulphuretted hydrogen into a vessel full of dry ammonia, or by passing both dry gases together into a very cold receiver. In the latter case it is necessary to prevent the access of air, and to have an excess of ammonia. Under these circumstances, two volumes of ammonia combine with one volume of sulphuretted hydrogen, and form a colourless, very volatile, crystalline substance, having a very unpleasant odour, which is very poisonous and exceedingly unstable. When exposed to the air it absorbs oxygen and acquires a yellow colour, and then contains oxygen and polysulphide compounds (because a portion of the hydrogen sulphide gives water and sulphur). It is soluble in water and forms a colourless solution, which, however, in all probability contains free ammonia and the acid salt—that is, ammonium hydrosulphide, NH_4HS, or $(NH_4)_2S,H_2S$. This salt is formed when dry ammonia is mixed with an excess of dry sulphuretted hydrogen. The compound contains equal volumes of the components $NH_3 + H_2S = (NH_4)HS$. It crystallises in an anhydrous state in colourless plates, and may be easily volatilised (dissociating like ammonium chloride), even at the ordinary temperature; it has an alkaline reaction, absorbs oxygen from the air, is soluble in water, and its solution is usually prepared by saturating an aqueous solution of ammonia with sulphuretted hydrogen. According to the ordinary rule, these salts, like other ammonium salts, split up into ammonia and sulphuretted hydrogen when they are distilled.

A solution of ammonium sulphide is able to dissolve sulphur, and it then contains compounds of hydrogen polysulphide and ammonia. Some of these compounds may be obtained in a crystalline form. Thus Fritzsche obtained a compound of ammonia with hydrogen pentasulphide, or ammonium pentasulphide, $(NH_4)_2S_5$, in the following manner: He saturated an aqueous solution of ammonia with sulphuretted hydrogen, added powdered sulphur to it, and passed ammonia gas into the solution, which then absorbed a fresh amount. After this he again passed sulphuretted hydrogen into the solution, and then added sulphur, and then again ammonia. After repeating this several times, orange-yellow crystals of $(NH_4)_2S_5$ separated out from the liquid. These crystals melted at 40° to 50°, and were very unstable.

When a solution of ammonium hydrosulphide, prepared by saturating a solution of ammonia with sulphuretted hydrogen, is exposed to the air, it turns yellow, owing to the presence of an ammonium polysulphide, whose formation is due to the sulphuretted hydrogen being oxidised by the air and converted into water and sulphur, which is dissolved by the ammonium sulphide. In certain analytical reactions it is usual to employ a solution of ammonium sulphide which has been kept for some time and acquired a yellow colour. This yellow sulphide of ammonium deposits sulphur when saturated with acids, whilst a freshly-prepared solution only evolves sulphuretted hydrogen. The yellow solution furthermore contains ammonium thiosulphate, which is derived not only from the oxidation of the ammonium sulphide, but also from the action of the liberated sulphur on the ammonia, just as an alkaline salt of thiosulphuric acid and a sulphide are formed by the action of sulphur on a solution of a caustic alkali.

Potassium sulphide, K_2S, is obtained by heating a mixture of potassium sulphate and charcoal to a bright-red heat. It may be prepared in solution by taking a solution of potassium hydroxide, dividing it into two equal parts, and saturating one portion with sulphuretted hydrogen so long as it is absorbed. This portion will then contain the acid salt KHS (Note 21). The two portions are then mixed together, and potassium sulphide will then be obtained in the solution. This solution has a strongly alkaline reaction, and is colourless when freshly prepared, but it very easily undergoes change when exposed to the air, forming potassium thiosulphate and polysulphides. When the solution is evaporated at low temperatures under the receiver of an air-pump, it yields crystals containing $K_2S,5H_2O$ (heated at 150°, they part with 3 mol. H_2O, and at higher temperatures they lose nearly all their water without evolving sulphuretted hydrogen). When they are ignited in glass vessels they corrode the glass. When a solution of caustic potash, completely saturated with sulphuretted hydrogen, is evaporated under the receiver of an air-pump it forms colourless rhombohedra of *potassium hydrosulphide*, 2(KHS),H$_2$O,K$_2$S,H$_2$S,H$_2$O. These crystals are deliquescent in the air, but do not change in a vacuum when heated up to 170°, and at higher temperatures they lose water but do not evolve sulphuretted hydrogen. The anhydrous compound, KHS, fuses at a dark-red heat into a very mobile yellow liquid, which gradually becomes darker in colour and solidifies to a red mass. It is remarkable that when a solution of the compound KHS is boiled it somewhat easily evolves half its sulphuretted hydrogen, leaving potassium sulphide, K$_2$S, in solution; and a solution of the latter in water is also able to evolve sulphuretted hydrogen on prolonged boiling, but the evolution cannot be rendered complete, and, therefore, at a certain temperature, a solution of potassium sulphide will not be capable of absorbing sulphuretted hydrogen at all. From this we must conclude that potassium hydroxide, water, and sulphuretted hydrogen form a system whose complex equilibrium is subject to the laws of dissociation, depends on the relative mass of each substance, on the temperature, and the dissociation pressure of the component elements. Potassium sulphide is not only soluble in water, but also in alcohol.

Berzelius showed that in addition to potassium sulphide there also exist potassium bisulphide, K$_2$S$_2$; trisulphide, K$_2$S$_3$; tetrasulphide, K$_2$S$_4$; and pentasulphide, K$_2$S$_5$. According to the researches of Schöne, the last three are the most stable. These different compounds of potassium and sulphur may be prepared by fusing potassium hydroxide or carbonate with an excess of sulphur in a porcelain crucible in a stream of carbonic anhydride. At about 600° potassium pentasulphide is formed; this is the highest sulphur compound of potassium. When heated to 800° it loses one-fifth of its sulphur and gives the tetrasulphide, which at this temperature is stable. At a bright-red heat—namely, at about 900°—the trisulphide is formed. This compound may be also formed by igniting potassium carbonate in a stream of carbon bisulphide, in which case a compound, K$_2$CS$_3$, is first formed corresponding to potassium carbonate, and carbonic anhydride is evolved. On further ignition this compound splits up into carbon and potassium trisulphide, K$_2$S$_3$. The tetrasulphide may also be obtained in solution if a solution of potassium sulphide be boiled with the requisite amount of sulphur without access of air. This solution yields red crystals of the composition K$_2$S$_4$,2H$_2$O when it is evaporated in a vacuum. These crystals are very hygroscopic, easily soluble in water, but very sparingly in alcohol; when ignited they give off water, sulphuretted hydrogen, and sulphur. If a solution of potassium sulphide be boiled with an excess of sulphur it forms the pentasulphide, which, however, is decomposed on prolonged boiling into sulphuretted hydrogen and potassium thiosulphate: K$_2$S$_5$ + 3H$_2$O = K$_2$S$_2$O$_3$ + 3H$_2$S. A substance called *liver of sulphur* was formerly frequently used in chemistry and medicine. Under this name is known the substance which is formed by boiling a solution of caustic potash with an excess of flowers of sulphur. This solution contains a mixture of potassium pentasulphide and thiosulphate, $6KHO + 12S = 2K_2S_5 + K_2S_2O_3 + 3H_2O$. The substance obtained by fusing potassium carbonate with an excess of sulphur was also known as liver of sulphur. If this mixture be heated to an incipient dark-red heat it will contain potassium thiosulphate, but at higher temperatures potassium sulphate is formed. In either case a polysulphide of potassium is also present. The sulphides of sodium, for example Na_2S, $NaHS$, &c., in many respects closely resemble the corresponding potassium compounds.

The metals of the alkaline earths, like those of the alkalis, form several compounds with sulphur; thus calcium forms compounds with one and with five atoms of sulphur. There are doubtless also intermediate sulphides. If sulphuretted hydrogen be passed over ignited lime it forms water and *calcium sulphide*, which may also be formed by heating calcium sulphate with charcoal, whilst if sulphur be heated with lime or with calcium carbonate, then naturally oxygen compounds (calcium thiosulphate and sulphate) are formed at the same time as calcium sulphide. The prolonged action of the vapour of carbon bisulphide, especially when mixed with carbonic anhydride, on strongly ignited calcium carbonate entirely converts it into sulphide. Calcium sulphide is generally obtained as an almost colourless, opaque, brittle mass, which is infusible at a white heat, and is soluble in water. The act of solution (as with K_2S, Note 21) is partly accompanied by a double decomposition with the water. When heated, dry calcium sulphide does not absorb oxygen from the air. An excess of water decomposes it, like many other metallic sulphides, precipitating lime (as a product of the decomposition the lime hinders the action of the water upon the CaS; see soda refuse, Chapter XII., Note 12), and forming a hydrosulphide, CaH_2S_2, in solution. This compound is also formed by passing sulphuretted hydrogen through an aqueous solution of calcium sulphide or lime. Its solution, like that of calcium sulphide, has an alkaline reaction. It decomposes when evaporated, and absorbs oxygen from the air. *Calcium pentasulphide*, CaS_5, is not known in a pure state, but may be obtained in admixture with calcium thiosulphate by boiling a solution of lime or calcium sulphide with sulphur: $3CaH_2O_2 + 12S = 2CaS_5 + CaS_2O_3 + 3H_2O$. A similar compound in an impure form is formed by the action of air on alkali waste, and is used for the preparation of thiosulphates.

Many of the sulphides of the metals of the alkaline earths are phosphorescent—that is, they have the faculty of *emitting light*, after having been subjected to the action of sunlight, or of any bright source of light (Canton phosphorus, &c.). The luminosity lasts some time, but it is not permanent, and gradually disappears. This phosphorescent property is inherent, in a greater or less degree, to nearly all substances (Becquerel), but for a very short time, whilst with calcium sulphide it is comparatively durable, lasting for several hours, and Dewar (1894) showed that it is far more intense at very low temperatures (for instance, in bodies cooled in liquid oxygen to -182°). It is due to the excitation of the surfaces of substances by the action of light, and is determined by those rays which exhibit a chemical action. Hence daylight or the light of burning magnesium, &c., acts more powerfully than the light of a lamp, &c. Warnerke has shown that a small quantity of magnesium lighted near the surface of a phosphorescent substance rapidly excites the greatest possible intensity of luminosity; this enabled him to found a method of measuring the intensity of light—*i.e.* to obtain a constant unit of light—and to apply it to photography. The nature of the change which is accomplished on the surface of the luminous substance is at present unknown, but in any case it is a renewable one, because the experiment may be repeated for an infinite number of times and takes place in a vacuum. The intensity and tint of the light emitted depend on the method of preparation of the calcium sulphide, and on the degree of ignition and purity of the calcium carbonate taken. According to the observations of Becquerel, the presence of compounds of manganese, bismuth, &c., sodium sulphide (but not potassium sulphide), &c., although in minute traces, is perfectly indispensable. This gives reason for thinking that the formation (in the dark) and decomposition (in light) of double salts like MnS,Na_2S perhaps form the chemical cause of the phenomena. Compounds of strontium and barium have this property to even a greater extent than calcium sulphide. These compounds may be prepared as in the following example: A mixture of sodium thiosulphate and strontium chloride is prepared; a double decomposition takes place between the salts, and, on the addition of alcohol, strontium thiosulphate, SrS_2O_3, is precipitated, which, when ignited, leaves strontium sulphide behind. The strontium sulphide thus prepared emits (when dry) a greenish yellow light. It contains a certain amount of sulphur, sodium sulphide, and strontium sulphate. By ignition at various temperatures, and by different methods of preparation, it is possible to obtain mixtures which emit different coloured lights.

As examples, we will describe the sulphides of arsenic, antimony, and mercury. Arsenic trisulphide, or *orpiment*, As_2S_3, occurs native, and is obtained pure when a solution of arsenious anhydride in the presence of hydrochloric acid comes into contact with sulphuretted hydrogen (there is no precipitate in the absence of free acid). A beautiful yellow precipitate is then obtained: $As_2O_3 + 3H_2S = 3H_2O + As_2S_3$; it fuses when heated, and volatilises without decomposition. As_2S_3 is easily obtained in a colloid form (Chapter I., Note 57). When fused it forms a semi-transparent, yellow mass, and it is thus that it enters the market. The specific gravity of native orpiment is 3·4, and that of the artificially-fused mass is 2·7. It is used as a yellow pigment, and owing to its insolubility in water and acids it is less injurious than the other compounds corresponding to arsenious acid. According to the type AsX_2, realgar, AsS, is known, but it is probable that the true composition of this compound is As_4S_4—that is, it presents the same relation to orpiment as liquid phosphuretted hydrogen does to gaseous. *Realgar* (*Sandaraca*) occurs native as brilliant red crystals of specific gravity 3·59, and may be prepared artificially by fusing arsenic and sulphur in the proportions indicated by its formulæ. It is prepared in large quantities by distilling a mixture of sulphur and arsenical pyrites. Like orpiment it dissolves in calcium sulphide, and even in caustic potash. It is used for signal lights and fireworks, because it deflagrates and gives a large and very brilliant white flame with nitre.

With antimony, sulphur gives a tri- and a pentasulphide. The former, Sb_2S_3, which corresponds with antimonious oxide, occurs native (Chapter XIX.) in a crystalline form; its sp. gr. is then 4·9, and it presents brilliant rhombic crystals of a grey colour, which fuse when heated. A substance of the same composition is obtained as an amorphous orange powder by passing sulphuretted hydrogen into an acid solution of antimonious oxide. In this respect antimonious oxide again reacts like arsenious acid, and the sulphides of both are soluble in ammonium and potassium sulphides, and, especially in the case of arsenious sulphide, are easily obtained in colloidal solutions. By prolonged boiling with water, antimonious sulphide may be entirely converted into the oxide, hydrogen sulphide being evolved (Elbers). Native antimony sulphide, or the orange precipitated trisulphide when fused with dry, or boiled with dissolved, alkalis, forms a dark-coloured mass (Kermes mineral) formerly much used in medicine, which contains a mixture of antimonious sulphide and oxide. There are also compounds of these substances. A so-called antimony vermilion is much used as a dye; it is prepared by boiling sodium thiosulphate (six parts) with antimony trichloride (five parts) and water (fifty parts). This substance probably contains an oxysulphide of antimony—that is, a portion of the oxygen in the oxide of antimony in it is replaced by sulphur. Red antimony ore, and antimony glass, which is obtained by fusing the trisulphide with antimonious oxide, have a similar composition, Sb_2OS_2. In the arts, the *antimony pentasulphide*, Sb_2S_5, is the most frequently used of the sulphur compounds of antimony. It is formed by the action of acids on the so-called Schlippe's salt, which is a *sodium thiorthantimonate*, $SbS(NaS)_3$, corresponding with (Chapter XIX., Note 41 bis) orthantimonic acid, $SbO(OH)_3$, with the replacement of oxygen by sulphur. It is obtained by boiling finely-powdered native antimony trisulphide with twice its weight of sodium carbonate, and half its weight of sulphur and lime, in the presence of a considerable quantity of water. The processes taking place are as follows:—The sodium carbonate is converted into hydroxide by the lime, and then forms sodium sulphide with the sulphur; the sodium sulphide then dissolves the antimony sulphide, which in this form already combines with the greatest amount of sulphur, so that a compound is formed corresponding with antimony pentasulphide dissolved in sodium sulphide. The solution is filtered and crystallised, care being taken to prevent access of air, which oxidises the sodium sulphide. This salt crystallises in large, yellowish crystals, which are easily soluble in water and have the composition $Na_3SbS_4,9H_2O$. When heated they lose their water of crystallisation and then fuse without alteration; but when in solution, and even in crystalline form, this salt turns brown in air, owing to the oxidation of the sulphur and the breaking up of the compound. As it is used in medicine, especially in the preparation of antimony pentasulphide, it is kept under a layer of alcohol, in which it is insoluble. Acids precipitate antimony pentasulphide from a solution of this salt, as an orange powder, insoluble in acids and very frequently used in medicine (*sulfur auratum antimonii*). This substance when heated evolves vapours of sulphur, and leaves antimony trisulphide behind.

Mercury forms compounds with sulphur of the same types as it does with oxygen. Mercurous sulphide, Hg_2S, easily splits up into mercury and mercuric sulphide. It is obtained by the action of potassium sulphide on mercurous chloride, and also by the action of sulphuretted hydrogen on solutions of salts of the type HgX. Mercuric sulphide, HgS, corresponding with the oxide, is cinnabar; it is obtained as a black precipitate by the action of an excess of sulphuretted hydrogen on solutions of mercuric salts. It is insoluble in acids, and is therefore precipitated in their presence. If a certain amount of water containing sulphuretted hydrogen be added to a solution of mercuric chloride, it first gives a white precipitate of the composition $Hg_3S_2Cl_2$—that is, a compound $HgCl,2HgS$, a sulphochloride of mercury like the oxychloride. But in the presence of an excess of sulphuretted hydrogen, the black precipitate of mercuric sulphide is formed. In this state it is not crystalline (the red variety is formed by the prolonged action of polysulphides of ammonium upon the black HgS), but if it be heated to its temperature of volatilisation it forms a red crystalline sublimate which is identical with native cinnabar. In this form its specific gravity is 8·0, and it forms a red powder, owing to which it is used as a red pigment (vermilion) in oil, pastel, and other paints. It is so little attacked by reagents that even nitric acid has no action on it, and the gastric juices do not dissolve it, so that it is not poisonous. When heated in air, the sulphur burns away and leaves metallic mercury. On a large scale cinnabar is usually prepared in the following manner: 300 parts of mercury and 115 parts of sulphur are mixed together as intimately as possible and poured into a solution of 75 parts of caustic potash in 425 parts of water, and the mixture is heated at 50° for several hours. Red mercury sulphide is thus formed, and separates out from the solution. The reaction which takes place is as follows: A soluble compound, K_2HgS_2, is first formed; this compound is able to separate in colourless silky needles, which are soluble in the caustic potash, but are decomposed by water, and at 50°; this solution (perhaps by attracting oxygen from the air) slowly deposits HgS in a crystalline form.

Spring conducted an interesting research (at Liège, 1894) upon the conversion of the black amorphous sulphide of mercury, HgS, into red crystalline cinnabar. This research formed a sequel to Spring's classical researches on the influence of high pressures upon the properties of solids and their capacity for mutual combination. He showed, among other things, that ordinary solids and even metals (for instance, Pb), after being considerably compressed under a pressure of 20,000 atmospheres, return on removal of the pressure to their original density like gases. But this is only true when the compressed solid is not liable to an allotropic variation, and does not give a denser variety. Thus prismatic sulphur (sp. gr. 1·9) passes under pressure into the octahedral (sp. gr. 2·05) variety. Black HgS (precipitated from solution) has a sp. gr. 7·6, while that of the red variety is 8·2, and therefore it might be expected that the former would pass into the latter under pressure, but experiments both at the ordinary and a higher temperature did not give the looked-for result, because even at a pressure of 20,000 atmospheres the black sulphide was not compressed to the density of cinnabar (a pressure of as much as 35,000 atmospheres was necessary, which could not be attained in the experiment). But Spring prepared a black HgS, which had a sp. gr. of 8·0, and this, under a pressure of 2,500 atmospheres, passed into cinnabar. He obtained this peculiar black variety of HgS (sp. gr. 8·0) by distilling cinnabar in an atmosphere of CO_2, when the greater portion of the HgS is redeposited in the form of cinnabar. Under the action of a solution of polysulphide of ammonium, this variety of HgS passes more slowly into the red variety than the precipitated variety does, while under pressure the conversion is comparatively easy.

It is worthy of remark, that Linder and Picton obtained complex compounds of many of the sulphides of the heavy metals (Ca, Hg, Sb, Zn, Cd, Ag, Au) with H_2S, for example $H_2S,7CuS$ (by the action of H_2S upon the hydrate of oxide of copper), $H_2S,9CuS$ (in the presence of acetic acid and with an excess of H_2S), &c. Probably we have here a sort of 'solid' solution of H_2S in the metallic sulphides.

CH_4 gives CH_4O or $CH_3(OH)$, wood spirit; CH_4O_2 or $CH_2(OH)_2$, which decomposes into water and CH_2O—that is, methylene oxide or formaldehyde; $CH_4O_3 = CH(OH)_3 = H_2O + CHO(OH)$, or formic acid; and $CH_4O_4 = C(OH)_4 = 2H_2O + CO_2$. There are four typical hydrogen compounds, RH, RH_2, RH_3, and RH_4, and each of them has its typical oxide. Beyond H_4 and O_4 combination does not proceed.

Rhombic sulphur, 71,080 heat units; monoclinic sulphur, 71,720 units, according to Thomsen.

[31 bis] However, when sulphur or metallic sulphides burn in an excess of air, there is always formed a certain, although small, amount of SO_3, which gives sulphuric acid with the moisture of the air.

The enormous amount of sulphuric acid now manufactured is chiefly prepared by roasting native pyrites, but a considerable amount of the SO_2 for this purpose is obtained by roasting zinc blende (ZnS) and copper and lead sulphides. A certain amount is also procured from soda refuse (Note 6) and the residues obtained from the purification of coal gas.

[32 bis] Sulphurous anhydride is also obtained by the decomposition of many sulphates, especially of the heavy metals, by the action of heat; but this requires a very powerful heat. This formation of sulphurous anhydride from sulphates is based on the decomposition proper to sulphuric acid itself. When sulphuric acid is strongly heated (for instance, by dropping it upon an incandescent surface) it is decomposed into water, oxygen, and sulphurous anhydride—that is, into those compounds from which it is formed. A similar decomposition proceeds during the ignition of many sulphates. Even so stable a sulphate as gypsum does not resist the action of very high temperatures, but is decomposed in the same manner, lime being left behind. The decomposition of sulphates by heat is accomplished with still greater facility in the presence of sulphur, because in this case the liberated oxygen combines with the sulphur and the metal is able to form a sulphide. Thus when ferrous sulphate (green vitriol) is ignited with sulphur, it gives ferrous sulphide and sulphurous anhydride: $FeSO_4 + 2S = FeS + 2SO_2$, and this reaction may even be used for the preparation of this gas. At 400° sulphuric acid and sulphur give an extremely uniform stream of pure sulphurous anhydride, so that it is best prepared on a manufacturing scale by this method. Iron pyrites, FeS_2, when heated to 150° with sulphuric acid (sp. gr. 1·75) in cast-iron vessels also gives an abundant and uniform supply of sulphurous anhydride.

[32 tri] Mellitic acid is formed at the same time (Verneuille).

The thermochemical data connected with this reaction are as follows: A molecule of hydrogen H_2, in combining with oxygen (O = 16) develops about 69,000 heat units, whilst the molecule of SO_2, in combining with oxygen only develops about 32,000 heat units—that is, about half as much—and therefore those metals which cannot decompose water may still be able to deoxidise sulphuric into sulphurous acid. Those metals which decompose water and sulphuric acid with the evolution of hydrogen, evolve in combining with sixteen parts by weight of oxygen more heat than hydrogen does—for example, K_2, Na_2, Ca develop about or more than

100,000 heat units; Fe, Zn, Mn about 70,000 to 80,000 heat units; whilst those metals which neither decompose water nor evolve hydrogen from sulphuric acid, but are still capable of evolving sulphurous anhydride from it, develop less heat with oxygen than hydrogen, but nearly the same amount, if not more than, sulphurous anhydride develops—for example, Cu and Hg develop about 40,000 and Pb about 50,000 heat units.

That is, it only dissociates and re-forms the original product on cooling.

At a given temperature the pressure of this gas evolved from any salt will be less than that of carbonic anhydride, if we compare the separation of a gas from its salts with the phenomenon of evaporation, as was done in discussing the decomposition of calcium carbonate.

Liquid sulphurous anhydride is used on a large scale (Pictet) for the production of cold.

De la Rive, Pierre, and more especially Roozeboom, have investigated the crystallo-hydrate which is formed by sulphurous anhydride and water at temperatures below 7° under the ordinary pressure, and in closed vessels (at temperatures below 12°). Its composition is $SO_2,7H_2O$, and density 1·2. This hydrate corresponds with the similar hydrate $CO_2,8H_2O$ obtained by Wroblewsky.

[36 bis] Schwicker (1889) by saturating $NaHSO_3$ with potash, or $KHSO_3$ with soda, obtained $NaKSO_3$, in the first instance with H_2O, and in the second instance with $2H_2O$, probably owing to the different media in which the crystals are formed. In general sulphurous acid easily forms double salts.

The normal salts of calcium and magnesium are slightly, and the acid salts easily, soluble in water. These acid sulphites are much used in practice; thus calcium bisulphite is employed in the manufacture of cellulose from sawdust, for mixing with fibrous matter in the manufacture of paper.

This reaction is taken advantage of in removing sulphurous anhydride from a mixture of gases. Lead dioxide, PbO_2, is brown, and when combined with sulphurous anhydride it forms lead sulphate, $PbSO_4$, which is white, so that the reaction is evident both from the change in colour and development of heat. Sulphurous anhydride is slowly decomposed by the action of light, with the separation of sulphur and formation of sulphuric anhydride. This explains the fact that sulphurous anhydride prepared in the dark gives a white precipitate of silver sulphite, Ag_2SO_3, with silver chlorate, $AgClO_4$, but when prepared in the light, even in diffused light, it gives a dark precipitate. This naturally depends on the fact that the sulphur liberated then forms silver sulphide, which is black.

Schönebein observed that the liquid turns yellow, and acquires the faculty of decolorising litmus and indigo. Schützenberger showed that this depends on the formation of a zinc salt of a peculiar and very powerfully-reducing acid, for with cupric salts the yellow solution gives a red precipitate of cuprous hydrate or metallic copper, and it reduces salts of silver and mercury entirely. An exactly similar solution is obtained by the action of zinc on sodium bisulphite without access of air and in the cold. The yellow liquid absorbs oxygen from the air with great avidity, and forms a sulphate. If the solution be mixed with alcohol, it deposits a double sulphite of zinc and sodium, $ZnNa_2(SO_3)_2$, which does not decolorise litmus or indigo. The remaining alcoholic solution deposits colourless crystals in the cold, which absorb oxygen with great energy in the presence of water, but are tolerably stable when dried under the receiver of an air-pump. The solution of these crystals has the above-mentioned decolorising and reducing properties. These crystals contain a sodium salt of a lower acid; their composition was at first supposed to be $HNaSO_2$, but it was afterwards proved that they do not contain hydrogen, and present the composition $Na_2S_2O_4$ (Bernthsen). The same salt is formed by the action of a galvanic current on a solution of sodium bisulphite, owing to the action of the hydrogen at the moment of its liberation. If SO_2 resembles CO_2 in its composition, then hyposulphurous acid $H_2S_2O_4$ resembles oxalic acid $H_2C_2O_4$. Perhaps an analogue of formic acid SH_2O_2 will be discovered.

The instability of this salt is very great, and may be compared to that of the compound of ferrous sulphate with nitric oxide, for when heated under the contact influence of spongy platinum, charcoal, &c., it splits up into potassium sulphate and nitrous oxide. At 130° the dry salt gives off nitric oxide, and re-forms potassium sulphite. The free acid has not yet been obtained. These salts resemble the series of *sulphonitrites* discovered by Frémy in 1845. They are obtained by passing sulphurous anhydride through a concentrated and strongly alkaline aqueous solution of potassium nitrite. They are soluble in water, but are precipitated by an excess of alkali. The first product of the action has the composition $K_3NS_3HO_9$. It is then converted by the further action of sulphurous anhydride, cold water, and other reagents into a series of similar complex salts, many of which give well-formed crystals. One must suppose that the chief cause of the formation of these very complex compounds is that they contain unsaturated compounds, NO, KNO_2, and $KHSO_3$, all of which are subject to oxidation and further combination, and therefore easily combine among each other. The decomposition of these compounds, with the evolution of ammonia, when their solutions are heated is due to the fact that the molecule contains the deoxidant, sulphurous anhydride, which reduces the nitrous acid, $NO(OH)$, to ammonia. In my opinion the composition of the sulphonitrites may be very simply referred to the composition of ammonia, in which the hydrogen is partly replaced by the radicle of the sulphates. If we represent the composition of potassium sulphate as $KO.KSO_3$, the group KSO_3 will be equivalent (according to the law of substitution) to HO and to hydrogen. It combines with hydrogen, forming the potassium acid sulphite, $KHSO_3$. Hence the group KSO_3 may also replace the hydrogen in ammonia. Judging by my analysis (1870) the extreme limit of this substitution, $N(HSO_3)_3$, agrees with that of the sulphonitrite, which is easily formed, simultaneously with alkali, by the action of potassium sulphite on potassium nitrite, according to the equation $3K(KSO_3) + KNO_2 + 2H_2O = N(KSO_3)_3 + 4HKO$. The researches of Berglund, and especially of Raschig (1887), fully verified my conclusions, and showed that we must distinguish the following types of salts, corresponding with ammonia, where X stands for the sulphonic group, HSO_3, in which the hydrogen is replaced by potassium; hence $X = KSO_3$: (1) NH_2X, (2) NHX_2, (3) NH_3, (4) $N(OH)XH$, (5) $N(OH)X_2$, (6) $N(OH)_2X$, just as $NH_2(OH)$ is hydroxylamine, $NH(OH)_2$, is the hydrate of nitrous oxide, and $N(OH)_3$ is orthonitrous acid, as follows from the law of substitution. This class of compounds is in most intimate relation with the series of sulphonitrous compounds, corresponding with 'chamber crystals' and their acids, which we shall consider later.

In the sulphuric acid chambers the lower oxides of nitrogen and sulphur take part in the reaction. They are oxidised by the oxygen of the air, and form nitro-sulphuric acid—for example, $2SO_2 + N_2O_3 + O_2 + H_2O = 2NHSO_5$. This compound dissolves in strong sulphuric acid without changing, and when this solution is diluted (when the sp. gr. falls to 1·5), it splits up into sulphuric acid and nitrous anhydride, and by the action of sulphurous anhydride is converted into nitric oxide, which by itself (in the absence of nitric acid or oxygen) is insoluble in sulphuric acid. These reactions are taken advantage of in retaining the oxides of nitrogen in the Gay-Lussac coke-towers, and for extracting the absorbed oxides of nitrogen from the resultant solution in the Glover tower. Although nitric oxide is not absorbed by sulphuric acid, it reacts (Rose, Brüning) on its anhydride, and forms sulphurous anhydride and a crystalline substance, $N_2S_2O_9 = 2NO + 3SO_3 - SO_2 = N_2O_3 2SO_3$. This may be regarded as the anhydride of nitro-sulphuric acid, because $N_2S_2O_9 = 2NHSO_5 - H_2O$; like nitro-sulphuric acid, it is decomposed by water into nitro-sulphuric acid and nitrous anhydride. Since boric and arsenious anhydrides, alumina and other oxides of the form R_2O_3 are able to combine with sulphuric anhydride to form similar compounds decomposable by water, the above compound does not present any exceptional phenomenon. The substance $NOClSO_3$ obtained by Weber by the action of nitrosyl chloride upon sulphuric anhydride belongs to this class of compounds.

[41 bis] Many double salts of thiosulphuric acid are known, for instance, $PbS_2O_3,3Na_2S_2O_3,12H_2O$; $CaS_2O_3,3K_2S_2O_3,5H_2O$, &c. (Fortman, Schwicker, Fock, and others).

Thus when alkali waste, which contains calcium sulphide, undergoes oxidation in the air it first forms a calcium polysulphide, and then calcium thiosulphate, CaS_2O_3. When iron or zinc acts on a solution of sulphurous acid, besides the hyposulphurous acid first formed, a mixture of sulphite and thiosulphate is obtained (Note 39), $3SO_2 + Zn_2 = ZnSO_3 + ZnS_2O_3$. In this case, as in the formation of hyposulphurous acid, there is no hydrogen liberated. One of the most common methods for preparing thiosulphates consists in the *action of sulphur on the alkalis*. The reaction is accomplished by the formation of sulphides and thiosulphates, just as the reaction of chlorine on alkalis is accompanied by the formation of hypochlorites and chlorides; hence in this respect the thiosulphates hold the same position in the order of the compounds of sulphur as the hypochlorites do among the chlorine compounds. The reaction of caustic soda on an excess of sulphur may be expressed thus: $6NaHO + 12S = 2Na_2S_5 + Na_2S_2O_3 + 3H_2O$. Thus sulphur is soluble in alkalis. On a large scale sodium thiosulphate, $Na_2S_2O_3$, is prepared by first heating sodium sulphate with charcoal, to form sodium sulphide, which is then dissolved in water and treated with sulphurous anhydride. The reaction is complete when the solution has become slightly acid. A certain amount of caustic alkali is added to the slightly acid solution; a portion of the sulphur is thus precipitated, and the solution is then boiled and evaporated when the salt crystallises out. The saturation of the solution of sodium sulphide by sulphurous anhydride is carried on in different ways—for example, by means of coke-towers, by causing the solution of sulphide to trickle over the coke, and the sulphurous anhydride, obtained by burning sulphur, to pass up the coke-tower from below. An excess of sulphurous anhydride must be avoided, as otherwise sodium trithionate is formed. Sodium thiophosphate is also prepared by the double decomposition of the soluble calcium thiosulphate with sodium sulphate or carbonate, in which case calcium sulphate or carbonate is precipitated. The calcium thiosulphate is prepared by the action of sulphurous anhydride on either calcium sulphide or alkali waste. A dilute solution of calcium thiosulphate may be obtained by treating alkali waste which has been exposed to the action of air with water. On evaporation, this solution gives crystals of the salt containing $CaS_2O_3,5H_2O$. A solution of calcium thiosulphate must be evaporated with great care, because otherwise the salt breaks up into sulphur and calcium sulphide. Even the crystallised salt sometimes undergoes this change.

The crystals of sodium thiosulphate are stable, do not effloresce and at $0°$ dissolve in one part of water, and at $20°$ in $0·6$ part. The solution of this salt does not undergo any change when boiled for a short time, but after prolonged boiling it deposits sulphur. The crystals fuse at 56°, and lose all their water at 100°. When the dry salt is ignited it gives sodium sulphide and sulphate. With acids, a solution of the thiosulphate soon becomes cloudy and deposits an exceedingly fine powder of sulphur (Note 10). If the amount of acid added be considerable, it also evolves sulphurous anhydride: $H_2S_2O_3 = H_2O + S + SO_2$. Sodium thiosulphate has many practical uses; it is used in photography for dissolving silver chloride and bromide. Its solvent action on silver chloride may be taken advantage of in extracting this metal as chloride from its ores. In dissolving, it forms a double salt of silver and sodium: $AgCl + Na_2S_2O_3 = NaCl + AgNaS_2O_3$. Sodium thiosulphate is an *antichlor*—that is, a substance which hinders the destructive action of free chlorine owing to its being very easily oxidised by chlorine into sulphuric acid and sodium chloride. The reaction with iodine is different, and is remarkable for the accuracy with which it proceeds. The iodine takes up half the sodium from the salt and converts it into a tetrathionate; $2Na_2S_2O_3 + I_2 = 2NaI + Na_2S_4O_6$, and hence this reaction is employed for the determination of free iodine. As iodine is expelled from potassium iodide by chlorine, it is possible also to determine the amount of chlorine by this method if potassium iodide be added to a solution containing chlorine. And as many of the higher oxides are able to evolve iodine from potassium iodide, or chlorine from hydrochloric acid (for example, the higher oxides of manganese, chromium, &c.), it is also possible to determine the amounts of these higher oxides by means of sodium thiosulphate and liberated iodine. This forms the basis of the iodometric method of volumetric analysis. The details of these methods will be found in works on analytical chemistry.

On adding a solution of a *lead salt* gradually to a solution of sodium thiosulphate a white precipitate of lead thiosulphate, PbS_2O_3, is formed (a soluble double salt is first formed, and if the action be rapid, lead sulphide). When this substance is heated at 200°, it undergoes a change and takes fire. Sodium thiosulphate in solution rapidly reduces cupric salts to cuprous salts by means of the sulphurous acid contained in the thiosulphate, but the resultant cuprous oxide is not precipitated, because it passes into the state of a thiosulphate and forms a double salt. These double cuprous salts are excellent reducing agents. The solution when heated gives a black precipitate of copper sulphide.

The following formulæ sufficiently explain the position held by thiosulphuric acid among the other acids of sulphur:

Sulphurous acid $SO_2H(OH)$

Sulphuric acid	$SO_2OH(OH)$
Thiosulphuric acid	$SO_2SH(OH)$
Hyposulphurous acid	$SO_2H(SO_2H)$
Dithionic acid	$SO_2OH(SO_2OH)$

At one time it was thought that all the salts of thiosulphuric acid only existed in combination with water, and it was then supposed that their composition was $H_4S_2O_4$, or H_2SO_2, but Popp obtained the anhydrous salts.

Nordhausen sulphuric acid may serve as a very simple means for the preparation of sulphuric anhydride. For this purpose the Nordhausen acid is heated in a glass retort, whose neck is firmly fixed in the mouth of a well-cooled flask. The access of moisture is prevented by connecting the receiver with a drying-tube. On heating the retort the vapours of sulphuric anhydride will pass over into the receiver, where they condense; the crystals of anhydride thus prepared will, however, contain traces of sulphuric acid—that is, of the hydrate. By repeatedly distilling over phosphoric anhydride, it is possible to obtain the pure anhydride, SO_3, especially if the process be carried on without access of air in a closed vessel.

The ordinary sulphuric anhydride, which is imperfectly freed from the hydrate, is a snow-white, exceedingly volatile substance, which crystallises (generally by sublimation) in long silky prisms, and only gives the pure anhydride when carefully distilled over P_2O_5. Freshly prepared crystals of almost pure anhydride fuse at 16° into a colourless liquid having a specific gravity at 26° = 1·91, and at 47° = 1·81; it volatilises at 46°. After being kept for some time the anhydride, even containing only small traces of water, undergoes a change of the following nature: A small quantity of sulphuric acid combines by degrees with a large proportion of the anhydride, forming polysulphuric acids, H_2SO_4, nSO_3, which fuse with difficulty (even at 100°, Marignac), but decompose when heated. In the entire absence of water this rise in the fusing point does not occur (Weber), and then the anhydride long remains liquid, and solidifies at about +15°, volatilises at 40°, and has a specific gravity 1·94 at 16°. We may add that Weber (1881), by treating sulphuric anhydride with sulphur, obtained a blue lower oxide of sulphur, S_2O_3. Selenium and tellurium also give similar products with SO_3, $SeSO_3$, and $TeSO_3$. Water does not act upon them.

Pyrosulphuric chloranhydride, or *pyrosulphuryl chloride*, $S_2O_5Cl_2$, corresponds to pyrosulphuric acid, in the same way that sulphuryl chloride, SO_2Cl_2, corresponds to sulphuric acid. The composition $S_2O_5Cl_2 = SO_2Cl_2$

+ SO_3. It is obtained by the action of the vapour of sulphuric anhydride on sulphur chloride: $S_2Cl_2 + 5SO_3 = 5SO_2 + S_2O_5Cl_2$. It is also formed (and not sulphuryl chloride, SO_2Cl_2, Michaelis) by the action of phosphorus pentachloride in excess on sulphuric acid (or its first chloranhydride, SHO_3Cl). It is an oily liquid, boiling at about 150°, and of sp. gr. 1·8. According to Konovaloff (Chapter VII.), its vapour density is normal. It should be noticed that the same substance is obtained by the action of sulphuric anhydride on sulphur tetrachloride, and also on carbon tetrachloride, and this substance is the last product of the metalepsis of CH_4, and therefore the comparison of SCl_2 and S_2Cl_2 with products of metalepsis (*see* later) also finds confirmation in particular reactions. Rose, who obtained pyrosulphuryl chloride, $S_2O_5Cl_2$, regarded it as $SCl_6,5SO_3$, for at that time an endeavour was always made to find two component parts of opposite polarity, and this substance was cited as a proof of the existence of a hexachloride, SCl_6. Pyrosulphuryl chloride is decomposed by cold water, but more slowly than chlorosulphuric acid and the other chloranhydrides.

The relation between pyrosulphuric acid and the normal acid will be obvious if we express the latter by the formula $OH(SO_3H)$, because the sulphonic group (SO_3H) is then evidently equivalent to OH, and consequently to H, and if we replace both the hydrogens in water by this radicle we shall obtain $(SO_3H)_2O$—that is, pyrosulphuric acid.

The removal of the water, or concentration to almost the real acid, H_2SO_4, is effected for two reasons: in the first place to avoid the expense of transit (it is cheaper to remove the water than to pay for its transit), and in the second place because many processes—for instance, the refining of petroleum—require a strong acid free from an excess of water, the weak acid having no action. When in the manufacture of chamber acid, both the Gay-Lussac tower (cold, situated at the end of the chambers) and the Glover tower (hot, situated at the beginning of the plant, between the chambers and ovens for the production of SO_2) are employed, a mixture of nitrose (*i.e.* the product of the Gay-Lussac tower) and chamber acid containing about 60 p.c. H_2SO_4, is poured into the Glover tower, where under the action of the hot furnace gases containing SO_2, and the water held in the chamber acid (1) N_2O_3 is evolved from the nitrose; (2) water is expelled from the chamber acid; (3) a portion of the SO_2 is converted into H_2SO_4; and (4) the furnace gases are cooled. Thus, amongst other things, the Glover tower facilitates the concentration of the chamber acid (removal of H_2O), but the product generally contains many impurities.

The difficulty with which the last portions of water are removed is seen from the fact that the boiling becomes very irregular, totally ceasing at one moment, then suddenly starting again, with the rapid formation of a considerable amount of steam, and at the same time bumping and even overturning the vessel in which it is held. Hence it is not a rare occurrence for the glass retorts to break during the distillation; this causes platinum retorts to be preferred, as the boiling then proceeds quite uniformly.

According to Regnault, the vapour tensions (in millimetres of mercury) of the water given off by the hydrates of sulphuric acid, H_2SO_4, nH_2O, are—

	$t=5°$	$15°$	$30°$
$n=1$	0·1	0·1	0·2
2	0·4	0·7	1·5
3	0·9	1·6	4·1
4	1·3	2·8	7·0
5	2·1	4·2	10·7
7	3·2	6·2	15·6
9	4·1	8·0	19·6
11	4·4	9·0	22·2
17	5·5	10·6	26·1

According to Lunge, the vapour tension of the aqueous vapour given off from solutions of sulphuric acid containing p per cent. H_2SO_4, at $t°$, equals the barometric pressure 720 to 730 mm.

$p =$	10	20	30	40	50	60	70	80	85	90	95
$t =$	102°	105°	108°	114°	124°	141°	170°	207°	233°	262°	295°

The latter figures give the temperature at which water is easily expelled from solutions of sulphuric acid of different strengths. But the evaporation begins sooner, and concentration may be carried on at lower temperatures if a stream of air be passed through the acid. Kessler's process is based upon this (Note 48).

The greatest part of the sulphuric acid is used in the soda manufacture, in the conversion of the common salt into sulphate. For this purpose an acid having a density of 60° Baumé is amply sufficient. Chamber acid has a density up to 1·57 = 50° to 51° Baumé; it contains about 35 per cent. of water. About 15 per cent. of this water can be removed in leaden stills, and nearly all the remainder may be expelled in glass or platinum vessels. Acid of 66° Baumé, = 1·847, contains about 96 per cent. of the hydrate H_2SO_4. The density falls with a greater or less proportion of water, the maximum density corresponding with 97½ per cent. of the hydrate H_2SO_4. The concentration of H_2SO_4 in platinum retorts has the disadvantage that sulphuric acid, upwards of 90 per cent. in strength, does corrode platinum, although but slightly (a few grams per tens of tons of acid). The retorts therefore require repairing, and the cost of the platinum exceeds the price obtained for concentrating the acid from 90 per cent. to 98 per cent. (in factories the acid is not concentrated beyond this by evaporation in the air). This inconvenience has lately (1891, by Mathey) been eliminated by coating the inside of the platinum retorts with a thin (0·1 to 0·02 mm.) layer of gold which is 40 times less corroded by sulphuric acid than platinum. Négrier (1890) carries on the distillation in porcelain dishes, Blond by heating a thin platinum wire immersed in the acid by means of an electric current, but the most promising method is that of Kessler (1891), which consists in passing hot air over sulphuric acid flowing in a thin stream in stone vessels, so that there is no boiling but only evaporation at moderate temperatures: the transference of the heat is direct (and not through the sides of the vessels), which economises the fuel and prevents the distilling vessels being damaged.

When, by evaporation of the water, sulphuric acid attains a density of 66° Baumé (sp. gr. 1·84), it is impossible to concentrate it further, because it then distils over unchanged. *The distillation of sulphuric acid* is not generally carried on on a large scale, but forms a laboratory process, employed when particularly pure acid is required. The distillation is effected either in platinum retorts furnished with corresponding condensers and receivers, or in glass retorts. In the latter case, great caution is necessary, because the boiling of sulphuric acid itself is accompanied by still more violent jerks and greater irregularity than even the evaporation of the last portions of water contained in the acid. If the glass retort which holds the strong sulphuric acid to be distilled be heated directly from below, it frequently jerks and breaks. For greater safety the heating is not effected from below, but at the sides of the retort. The evaporation then does not proceed in the whole mass, but only from the upper portions of the liquid, and therefore goes on much more quietly. The acid may be made to boil quietly also by surrounding the retort with good conductors of heat—for example, iron filings, or by immersing a bunch of platinum wires in the acid, as the bubbles of sulphuric acid vapour then form on the extremities of the wires.

Thus it appears that so common, and apparently so stable, a compound as sulphuric acid decomposes even at a low temperature with separation of the anhydride, but this decomposition is restricted by a limit, corresponding to the presence of about 1½ p.c. of water, or to a composition of nearly $H_2O,12H_2SO_4$.

Now there is no reason for thinking that this substance is a definite compound; it is an equilibrated system which does not decompose under ordinary circumstances below 338°. Dittmar carried on the distillation under pressures varying between 30 and 2,140 millimetres (of mercury), and he found that the composition of the residue hardly varies, and contains from 99·2 to 98·2 per cent. of the normal hydrate, although at 30 mm. the temperature of distillation is about 210° and at 2,140 mm. it is 382°. Furthermore, it is a fact of practical importance that under a pressure of two atmospheres the distillation of sulphuric acid proceeds very quietly.

Sulphuric acid may be *purified* from the majority of its impurities by distillation, if the first and last portions of the distillate be rejected. The first portions will contain the oxides of nitrogen, hydrochloric acid, &c., and the last portions the less volatile impurities. The oxides of nitrogen may be removed by heating the acid with charcoal, which converts them into volatile gases. Sulphuric acid may be freed from arsenic by heating it with manganese dioxide and then distilling. This oxidises all the arsenic into non-volatile arsenic acid. Without a preliminary oxidation it would partially remain as volatile arsenious acid, and might pass over into the distillate. The arsenic may also be driven off by first reducing it to arsenious acid, and then passing hydrochloric acid gas through the heated acid. It is then converted into arsenious chloride, which volatilises.

The amount of heat developed by the mixture of sulphuric acid with water is expressed in the diagram on p. 77, Volume I., by the middle curve, whose abscissæ are the percentage amounts of acid (H_2SO_4) in the resultant solution, and ordinates the number of units of heat corresponding with the formation of 100 cubic centimetres of the solution (at 18°). The calculations on which the curve is designed are based on Thomsen's determinations, which show that 98 grams or a molecular amount of sulphuric acid, in combining with m molecules of water (that is, with $m=18$ grams of water), develop the following number of units of heat, R:—

$m =$	1	2	3	5	9	19	49	100	200
R =	6379	9418	11137	13108	14952	16256	16684	16859	17066
$c =$	0·432	0·470	0·500	0·576	0·701	0·821	0·914	0·954	0·975
T =	127°	149°	146°	121°	82°	45°	19°	9°	5°

c stands for the specific heat of $H_2SO_4 mH_2O$ (according to Marignac and Pfaundler), and T for the rise in temperature which proceeds from the mixture of H_2SO_4 with mH_2O. The diagram shows that contraction and rise of temperature proceed almost parallel with each other.

[50 bis] Pickering (1890) showed (*a*) that dilute solutions of sulphuric acid containing up to $H_2SO_4 + 10H_2O$ deposit ice (at $-0°·12$ when there is $2{,}000H_2O$ per H_2SO_4, at $-0°·23$ when there is $1{,}000H_2O$, at $-1°·04$ when there is $200H_2O$, at $-2°·12$ when there is $100H_2O$, at $-4°·5$ when there is $50H_2O$, at $-15°·7$ when there is $20H_2O$, and at $-61°$ when the composition of the solution is $H_2SO_4 + 10H_2O$); (*b*) that for higher concentrations crystals separate out at a considerable degree of cold, having the composition $H_2SO_4 4H_2O$, which melt at $-24°·5$, and if either water or H_2SO_4 be added to this compound the temperature of crystallisation falls, so that a solution of the composition $12H_2SO_4 + 100H_2O$ gives crystals of the above hydrate at $-70°$, $15H_2SO_4 + 100H_2O$ at $-47°$, $30H_2SO_4 + 100H_2O$ at $-32°$, $40H_2SO_4 + 100H_2O$ at $-52°$; (*c*) that if the amount of H_2SO_4 be still greater, then a hydrate $H_2SO_4 H_2O$ separates out and melts at $+8°·5$, while the addition of water or sulphuric acid to it lowers the temperature of crystallisation so that the crystallisation of $H_2SO_4 H_2O$ from a solution of the composition $H_2SO_4 + 1·73H_2O$ takes place at $-22°$, $H_2SO_4 + 1·5H_2O$ at $-6°·5$, $H_2SO_4 + 1·2H_2O$ at $+3°·7$, $H_2SO_4 + 0·75H_2O$ at $+2°·8$, $H_2SO_4 + 0·5H_2O$ at $-16°$; (*d*) that when there is less than $40H_2O$ per $100H_2SO_4$, refrigeration separates out the normal hydrate H_2SO_4, which melts at $+10°·35$, and that a solution of the composition $H_2SO_4 + 0·35H_2O$ deposits crystals of this hydrate at $-34°$, $H_2SO_4 + 0·1H_2O$ at $-4°·1$, $H_2SO_4 + 0·05H_2O$ at $+4°·9$, while fuming acid of the composition $H_2SO_4 + 0·05SO_3$ deposits H_2SO_4 at about $+7°$. Thus the temperature of the separation of crystals clearly distinguishes the above four regions of solutions, and in the space between $H_2SO_4 + H_2O$ and $+25H_2O$ a particular hydrate $H_2SO_4 4H_2O$ separates out, discovered by Pickering, the isolation of which deserves full attention and further research. I may add here that the existence of a hydrate $H_2SO_4 4H_2O$ was pointed out in my work, *The Investigation of Aqueous Solutions*, p. 120 (1887), upon the basis that it has at all temperatures a smaller value for the coefficient of expansion k in the formula $S_t = S_0/(1 - kt)$ than the adjacent (in composition) solutions of sulphuric acid. And for solutions approximating to $H_2SO_4 10H_2O$ in their composition, k is constant at all temperatures (for more dilute solutions the value of k increases with t and for more concentrated solutions it decreases). This solution (with $10H_2O$) forms the point of transition between more dilute solutions which deposit ice (water) when refrigerated and those which give crystals of $H_2SO_4 4H_2O$. According to R. Pictet (1894) the solution $H_2SO_4 10H_2O$ freezes at $-88°$ (but no reference is made as to what separates out), *i.e.* at a lower temperature than all the other

solutions of sulphuric acid. However, in respect to these last researches of R. Pictet (for 88·88 p.c. H_2SO_4 -55°, for $H_2SO_4H_2O$ +3·5°, for $H_2SO_42H_2O$ -70°, for $H_2SO_44H_2O$ -40°, &c.) it should be remarked that they offer some quite improbable data; for example, for $H_2SO_475H_2O$ they give the freezing point as 0°, for $H_2SO_4300H_2O$ +4°·5, and even for $H_2SO_41000H_2O$ +0°·5, although it is well known that a small amount of sulphuric acid lowers the temperature of the formation of ice. I have found by direct experiment that a frozen solidified solution of H_2SO_4 + $300H_2O$ melted completely at 0°.

With an excess of snow, the hydrate H_2SO_4,H_2O, like the normal hydrate, gives a freezing mixture, owing to the absorption of a large amount of heat (the latent heat of fusion). In melting, the molecule H_2SO_4 absorbs 960 heat units, and the molecule $H_2SO_4H_2O$ 3,680 heat units. If therefore we mix one gram molecule of this hydrate with seventeen gram molecules of snow, there is an absorption of 18,080 heat units, because $17H_2O$ absorbs 17 × 1,430 heat units, and the combination of the monohydrate with water evolves 9,800 heat units. As the specific heat of the resultant compound $H_2SO_4,18H_2O$ = 0·813, the fall of temperature will be -52°·6. And, in fact, a very low temperature may be obtained by means of sulphuric acid.

For example, if it be taken that at 19° the sp. gr. of pure sulphuric acid is 1·8330, then at 20° it is 1·8330 - (20 - 19)10·13 = 1·8320.

Unfortunately, notwithstanding the great number of fragmentary and systematic researches which have been made (by Parks, Ure, Bineau, Kolbe, Lunge, Marignac, Kremers, Thomsen, Perkin, and others) for determining the relation between the sp. gr. and composition of solutions of sulphuric acid, they contain discrepancies which amount to, and even exceed, 0·002 in the sp. gr. For instance, at 15°·4 the solution of composition $H_2SO_43H_2O$ has a sp. gr. 1·5493 according to Perkin (1886), 1·5501 according to Pickering (1890), and 1·5525 according to Lunge (1890). The cause of these discrepancies must be looked for in the methods employed for determining the composition of the solutions—*i.e.* in the inaccuracy with which the percentage amount of H_2SO_4 is determined, for a difference of 1 p.c. corresponds to a difference of from 0·0070 (for very weak solutions) to 0·0118 (for a solution containing about 73 p.c.) in the specific gravity (that is the factor ds/dp) at 15°. As it is possible to determine the specific gravity with an accuracy even exceeding 0·0002, the specific gravities given in the adjoining tables are only averages and most probable data in which the error, especially for the 30–80 p.c. solutions cannot be less than 0·0010 (taking water at 4° as 1).

[53 bis] Judging from the best existing determinations (of Marignac, Kremers, and Pickering) for solutions of sulphuric acid (especially those containing more than 5 p.c. H_2SO_4) within the limits of 0° and 30° (and even to 40°), the variation of the sp. gr. with the temperature t may (within the accuracy of the existing determinations) be perfectly expressed by the equation $S_t = S_0 + At + Bt^2$. It must be added that (1) three specific gravities fully determine the variation of the density with t; (2) $ds/dt = A + 2Bt$—i.e. the factor of the temperature is expressed by a straight line; (3) the value of A (if p be greater than 5 p.c.) is negative, and numerically much greater than B; (4) the value of B for dilute solutions containing less than 25 p.c. is negative; for solutions approximating to $H_2SO_4 3H_2O$ in their composition it is equal to 0, and for solutions of greater concentration B is positive; (5) the factor ds/dp for all temperatures attains a maximum value about $H_2SO_4 H_2O$; (6) on dividing ds/dt by S_0, and so obtaining the coefficient of expansion k (*see* Note 53), a minimum is obtained near H_2SO_4 and $H_2SO_4 4H_2O$, and a maximum at $H_2SO_4 H_2O$ for all temperatures.

[53 tri] These data (as well as those in the following table) have been recalculated by me chiefly upon the basis of Kremer's, Pickering's, Perkin's, and my own determinations; all the requisite corrections have been introduced, and I have reason for thinking that in each of them the probable error (or difference from the true figures, now unknown) of the specific gravity does not exceed ±0·0007 (if water at 4° = 1) for the 25–80 p.c. solutions, and ±0·0002 for the more dilute or concentrated solutions.

The factor dS/dp passes through 0, that is, the specific gravity attains a maximum value at about 98 p.c. This was discovered by Kohlrausch, and confirmed by Chertel, Pickering, and others.

Naturally under the condition that there is no other ingredient besides water, which is sufficiently true. For commercial acid, whose specific gravity is usually expressed in degrees of Baumé's hydrometer, we may add that at 15°

Specific gravity	1	1·1	1·2	1·3	1·4	1·5	1·6	1·7	1·8
Degree Baumé	0	13	24	33·3	41·2	48·1	54·1	59·5	64·2

66° Baumé (the strongest commercial acid or oil of vitriol) corresponds to a sp. gr. 1·84.

By employing the second table (by the method of interpolation) the specific gravity, at a given temperature (from 0° to 30°) can be found for any percentage amount of H_2SO_4, and therefore conversely the percentage of H_2SO_4 can be found from the specific gravity.

[55 bis] Whether similar (even small) breaks in the continuity of the factor dS/dp exist or not, for other hydrates (for instance, for $H_2SO_4H_2O$ and $H_2SO_44H_2O$) cannot as yet be affirmed owing to the want of accurate data (Note 53). In my investigation of this subject (1887) I admit their possibility, but only conditionally; and now, without insisting upon a similar opinion, I only hold to the existence of a distinct break in the factor at H_2SO_4, being guided by C. Winkler's observations ond the specific gravities of fuming sulphuric acid.

In 1887, on considering all the existent observations for a temperature 0°, I gave the accompanying scheme (p. 243) of the variation of the factor ds/dp at 0°.

I did not then (1887) give this scheme an absolute value, and now after the appearance of two series of new determinations (Lunge and Pickering in 1890), which disagree in many points, I think it well to state quite clearly: (1) that Lunge's and Pickering's new determinations have not added to the accuracy of our data respecting the variation of the specific gravity of solutions of sulphuric acid; (2) that the sum total of existing data does not negative (within the limit of experimental accuracy) the possibility of a rectilinear and broken form for the factors ds/dp; (3) that the supposition of 'special points' in ds/dp, indicating definite hydrates, finds confirmation in all the latest determinations; (4) that the supposition respecting the existence of hydrates determining a break of the factor ds/dp is in in way altered if, instead of a series of broken straight lines, there be a continuous series of curves, nearly approaching straight lines; and (5) that this subject deserves (as I mentioned in 1887) new and careful elaboration, because it concerns that foremost problem in our science—solutions—and introduces a special method into it—that is, the study of differential variations in a property which is so easily observed as the specific gravity of a liquid.

[56 bis] These hydrates are: (*a*) $H_2SO_4 = SO_3H_2O$ (melts at + 10°·4); (*b*) $H_2SO_4H_2O = SO_32H_2O$ (crystallo-hydrate, melts at +8°·5); (*c*) $H_2SO_42H_2O$ (is apparently not crystallisable); (*d*) one of the hydrates between $H_2SO_46H_2O$ and $H_2SO_43H_2O$, most probably $H_2SO_44H_2O = SO_35H_2O$, for it crystallises at -24°·5 (Note 50 bis); and (*e*) a certain hydrate with a large proportion of water, about $H_2SO_4150H_2O$. The existence of the last is inferred from the fact that the factor ds/dp first falls, starting from water, and then rises, and this change takes place when p is less than 5 p.c. Certainly a change in the variation of ds/dp or ds/dt does take place in the neighbourhood of these five hydrates (Pickering, 1890, recognised a far greater number of hydrates). I think it well to add that if the composition of the solutions be expressed by the percentage amount of molecules— $nSO_3 + (100 - n)H_2O$ we find that for H_2SO_4, $n = 50$, for $H_2SO_42H_2O$ n

= 25 = 50/2, for $H_2SO_4H_2O$, $n = 33·333 = 50·\frac{2}{3}$, while for $H_2SO_44H_2O$, $n = 16·666 = 50·\frac{1}{3}$—*i.e.* that the chief hydrates are distributed symmetrically between H_2O and H_2SO_4. Besides which I may mention that my researches (1887) upon the abrupt changes in the factor for solutions of sulphuric acid, and upon the correspondence of the breaks of ds/dp with definite hydrates, received an indirect confirmation not only in the solutions of HNO_3, HCl, C_2H_6O, C_3H_8O, &c., which I investigated (in my work cited in Chapter I., Note 19), but also in the careful observations made by Professor Cheltzoff on the solutions of $FeCl_3$ and $ZnCl_2$ (Chapter XVI., Note 4) which showed the existence in these solutions of an almost similar change in ds/dp as is found in sulphuric acid. The detailed researches (1893) made by Tourbaba on the solutions of many organic substances are of a similar nature. Besides which, H. Crompton (1888), in his researches on the electrical conductivity of solutions of sulphuric acid, and Tammann, in his observations on their vapour tension, found a correlation with the hydrates indicated as above by the investigation of their specific gravities. The influence of mixtures of a definite composition upon the chemical relations of solutions is even exhibited in such a complex process as electrolysis. V. Kouriloff (1891) showed that mixtures containing about 3 p.c., 47 p.c. and 73 p.c. of sulphuric acid—*i.e.* whose composition approaches that of the hydrates $H_2SO_4150H_2O$, $H_2SO_46H_2O$ and $H_2SO_42H_2O$—exhibit certain peculiarities in respect to the amount of peroxide of hydrogen formed during electrolysis. Thus a 3 p.c. solution gives a maximum amount of peroxide of hydrogen at the negative pole, as compared with that given by other neighbouring concentrations. Starting from 3 p.c., the formation of peroxide of hydrogen ceases until a concentration of 47 p.c. is reached.

Cellulose, for instance unsized paper or calico, is dissolved by strong sulphuric acid. Acid diluted with about half its volume of water converts it (if the action be of short duration) into vegetable parchment (Chapter I., Note 18). The action of dilute solutions of sulphuric acid converts it into hydro-cellulose, and the fibre loses its coherent quality and becomes brittle. The prolonged action of strong sulphuric acid chars the cellulose while dilute acid converts it into glucose. If sulphuric acid be kept in an open vessel, the organic matter of the dust held in the atmosphere falls into it and blackens the acid. The same thing happens if sulphuric acid be kept in a bottle closed by a cork; the cork becomes charred, and the acid turns black. However, the chemical properties of the acid undergo only a very slight change when it turns black. Sulphuric acid which is considerably diluted with water does not produce the above effects, which clearly shows their dependence on the affinity of the sulphuric acid for water. It is evident from the preceding that strong sulphuric acid will act as a powerful poison; whilst, on the other hand, when very dilute it is employed in certain medicines and as a fertiliser for plants.

Weber (1884) obtained a series of salts $R_2O,8SO_3nH_2O$ for K, Rb, Cs, and Tl.

[58 bis] Ditte (1890) divides all the metals into two groups with respect to sulphuric acid; the first group includes silver, mercury, copper, lead, and bismuth, which are only acted upon by hot concentrated acid. In this case sulphurous anhydride is evolved without any by-reactions. The second group contains manganese, nickel, cobalt, iron, zinc, cadmium, aluminium, tin, thallium, and the alkali metals. They react with sulphuric acid of any concentration at any temperature. At a low temperature hydrogen is disengaged, and at higher temperatures (and with very concentrated acid) hydrogen and sulphurous anhydride are simultaneously evolved.

For example, the action of hot sulphuric acid on nitrogenous compounds, as applied in Kjeldahl's method for the estimation of nitrogen (Volume I. p. 249). It is obvious that when sulphuric acid acts as an oxidising agent it forms sulphurous anhydride.

The action of sulphuric acid on the alcohols is exactly similar to its action on alkalis, because the alcohols, like alkalis, react on acids; a molecule of alcohol with a molecule of sulphuric acid separates water and forms an *acid* ethereal salt—that is there is produced an ethereal compound corresponding with acid salts. Thus, for example, the action of sulphuric acid, H_2SO_4, on ordinary alcohol, C_2H_5OH, gives water and sulphovinic acid, $C_2H_5HSO_4$—that is, sulphuric acid in which one atom of hydrogen is replaced by the radicle C_2H_5 of ethyl alcohol, $SO_2(OH)(OC_2H_5)$, or, what is the same thing, the hydrogen in alcohol is replaced by the radicle (sulphoxyl) of sulphuric acid, $C_2H_5O.SO_2(OH)$.

We will mention the following difference between the sulphonic acids and the ethereal acid sulphates (Note 59): the former re-form sulphuric acid with difficulty and the latter easily. Thus sulphovinic acid when heated with an excess of water is reconverted into alcohol and sulphuric acid. This is explained in the following manner. Both these classes of acids are produced by the substitution of hydrogen by SO_3H, or the univalent radicle of sulphuric acid, but in the formation of ethereal acid sulphates the SO_3H replaces the hydrogen of the hydroxyl in the alcohol, whilst in the formation of the sulphonic acids the SO_3H replaces the hydrogen of a hydrocarbon. This difference is clearly evidenced in the existence of two acids of the composition $SO_4C_2H_6$. The one, mentioned above, is sulphovinic acid or alcohol, $C_2H_5.OH$, in which the hydrogen of the hydroxyl is replaced by sulphoxyl = $C_2H_5.OSO_3H$, whilst the other is alcohol, in which one atom of the hydrogen in ethyl, C_2H_5, is replaced by the sulphonic group—that is = $(C_2H_4)SO_3H \cdot OH$. The latter is called isethionic acid. It is more stable than sulphovinic acid. The details as to these interesting compounds must be looked for in works on organic chemistry, but I think it necessary to note one of the general methods of formation of these acids. The sulphites of the alkalis—for example, K_2SO_3—when heated with the halogen products of metalepsis, give a halogen salt and a salt of a sulphonic acid. Thus methyl iodide, CH_3I, derived from marsh gas, CH_4, when heated to 100° with a solution of potassium sulphite, K_2SO_3, gives potassium iodide, KI, and potassium methylsulphonate, CH_3SO_3K—that is a salt of the sulphonic acid. This shows that the sulphonic acid may be referred to sulphurous acid, and that there is a resemblance between sulphuric and sulphurous acid, which clearly reveals itself here in the formation of one product from them both.

The reaction $BaO + O$ develops 12,000 heat units, whilst the reaction $H_2O + O$ absorbs 21,000 heat units.

Schöne obtained a compound of peroxide of barium with peroxide of hydrogen. If barium peroxide be dissolved in hydrochloric (or acetic) acid, or if a solution of hydrogen peroxide be diluted with a solution of barium hydroxide, a pure hydrate is precipitated having the composition $BaO_2,8H_2O$ (sometimes the composition is taken as $BaO_2,6H_2O$). This fact was already known to Thénard. Schöne showed that if hydrogen peroxide be in excess, a crystalline compound of the two peroxides, $BaO_2H_2O_2$, is precipitated. Schöne also obtained small well-formed crystals of the same composition by adding a solution of ammonia to an acid solution of barium peroxide (containing a barium salt and hydrogen peroxide or a compound of BaO_2 with the acid). Thus barium peroxide combines with both water and hydrogen peroxide. This is a very important fact for the comprehension of the composition of other peroxides. Moreover, if the peroxides are able to give hydrates they can also form corresponding salts, *i.e.* they can combine with bases and acids, as was afterwards found to be the case on further research into this subject.

Anhydrous *sulphuric peroxide*, S_2O_7, is obtained by the prolonged (8 to 10 hours) action of a silent discharge of considerable intensity on a mixture of oxygen and sulphurous anhydride; the vapour of sulphuric peroxide, S_2O_7, condenses as liquid drops, or after being cooled to 0° in the form of long prismatic crystals, resembling those of sulphuric anhydride. The anhydrous compound S_2O_7 (and also the hydrated compound) cannot be preserved long, as it splits up into oxygen and sulphuric anhydride. Direct experiment shows that a mixture of equal volumes of sulphurous anhydride and oxygen leaves a residue of a quarter of the oxygen taken, or half of the whole volume, which indicates the formula S_2O_7. This substance is soluble in water, and it then gives a hydrate, probably having the composition

$S_2O_7,H_2O = 2SHO_4$. This solution oxidises the salts SnX_2, potassium iodide, and others, which renders it possible to prove that the solution actually contains one atom of oxygen capable of effecting oxidation to two molecules of sulphuric anhydride.

In order to fully demonstrate the reality of a peroxide form for acids, it should be mentioned that some years ago Brodie obtained the so-called *acetic peroxide*, $(C_2H_2O)_2O_2$, by the action of barium peroxide on acetic anhydride, $(C_2H_3O)_2O$. Its corresponding hydrate is also known. This shows that true peroxides and their hydrates, with reactions similar to those of hydrogen peroxide, are possible for acids. A similar higher oxide has long been known for chromium, and Berthelot obtained a like compound for nitric acid (Chapter VI., Note 26).

When an acid of the strength $H_2SO_46H_2O$ is taken, at first only the hydrate of the sulphuric peroxide, $S_2O_7H_2O$, is formed, but when the concentration at the positive pole reaches $H_2SO_43H_2O$, a mixture of hydrogen peroxide and the hydrate of sulphuric peroxide begins to be formed. A state of equilibrium is ultimately arrived at when the amounts of these substances correspond to the proportion $S_2O_7 : 2H_2O_2$, which, as it were, answers to a new hydrate, $S_2O_92H_2O$. But its existence cannot be admitted because the sulphuric peroxide can be easily distinguished from the hydrogen peroxide in the solution owing to the fact that the former does not act on an acid solution of potassium permanganate, whilst the hydrogen peroxide disengages both its own oxygen and that of the permanganic acid, converting it into manganous oxide. Their common property of liberating iodine from an acid solution of the potassium iodide enables the sum of the active oxygen in them both to be determined.

If a solution of sulphuric acid which has been first subjected to electrolysis be neutralised with potash or baryta, the salt which is formed begins to decompose rapidly with the evolution of oxygen (Berthelot, 1890). On saturating with caustic baryta, the solution of the salt formed may be separated from the sulphate of barium, and then the composition of the resultant compound, BaS_2O_8, may be determined from the amount of oxygen disengaged. Marshall (1891) studied the formation of this class of compounds more fully; he subjected a saturated solution of bisulphate of potassium to electrolysis with a current of 3–3½ ampères; before electrolysis dilute sulphuric acid is added to the liquid surrounding the negative pole, and during electrolysis the solution at the anode is cooled. The electrolysis is continued without interruption for two days, and a white crystalline deposit separates at the anode. To avoid decomposition, the latter is not filtered through paper, but through a perforated platinum plate, and dried on a porous tile. The mother liquor, with the addition of a fresh solution of bisulphate of potassium, is again subjected to electrolysis and the crystals formed at the anode are again collected, &c. The salt so obtained may be recrystallised by dissolving it in hot water and rapidly cooling the solution after filtration; a small proportion of the salt is decomposed by this treatment. Rapid cooling is followed by the formation of small columnar crystals; slow cooling gives large prismatic crystals. The composition of the salt is determined either by igniting it, when it forms sulphate of potassium, or else by titrating the active oxygen with permanganate: its composition was found to correspond to the salt of persulphuric acid, $K_2S_2O_8$. The solution of the salt has a neutral reaction, and does not give a precipitate with salts of other metals. $K_2S_2O_8$ is the most insoluble of the salts of persulphuric acid. With nitrate of silver it forms persulphate of silver, which gives peroxide of silver under the action of water according to the equation $Ag_2S_2O_8 + 2H_2O = Ag_2O_2 + 2H_2SO_4$. With an alkaline solution of a cupric salt (Fehling's solution) it forms a red precipitate of peroxide of copper. Manganese and cobalt salts give precipitates of MnO_2 and Co_2O_3. Ferrous salts are rapidly oxidised, potassium iodide slowly disengages iodine at the ordinary temperature. All these reactions indicate the powerful oxidising properties of $K_2S_2O_8$. In oxidising in the presence of water it gives a residue of $KHSO_4$. The decomposition of the dry salt begins at 100° but is not complete even at 250°. The freshly prepared salt is inodorous, but after being kept in a closed vessel it evolves a peculiar smell different from that of ozone. The ammonium salt of persulphuric acid, $(NH_4)_2S_2O_8$, is obtained in a similar manner. It is soluble to the extent of 58 parts per 100 parts by weight of water. The decomposition of the ammonium salt by the hydrated oxide of barium gives the barium salt, $BaS_2O_8 4H_2O$, which is soluble to the extent of 52·2 parts in 100 parts of water at 0°. The crystals do not deliquesce in the air and decompose in the course of several days; they decompose most rapidly in perfectly dry air. Solutions of the pure salt decompose slowly at the ordinary temperature; on boiling barium sulphate is gradually precipitated, oxygen being liberated simultaneously. To completely decompose this salt it is necessary to boil its solution for a long time. Alcohol dissolves the solid salt; the anhydrous salt does not separate from the alcoholic solution, but a hydrate containing one molecule of water, $BaS_2O_8 H_2O$, which is soluble in water but insoluble in absolute alcohol. Solid barium persulphate decomposes even when slightly heated. The free acid, which may serve for the preparation of other salts, is obtained by treating the barium salt with sulphuric acid. The lead salt, PbS_2O_8, has been obtained from the free acid; it crystallises with two or three molecules of water. It is soluble in water, deliquesces in the air, and with alkalis gives a precipitate of the hydrated oxide which rapidly oxidises into the binoxide.

Traube, before Marshall's researches, thought that the electrolysis of solutions of sulphuric acid did not give persulphuric acid but a persulphuric oxide having the composition SO_4. On repeating his former researches (1892) Traube obtained a persulphuric oxide by the electrolysis of a 70 per cent. solution of sulphuric acid, and he separated it from the solution by means of barium phosphate. Analysis showed that this substance corresponded to the above composition SO_4, and therefore Traube considers it very likely that the salts obtained by Marshall corresponded to an acid $H_2SO_4 + SO_4$, *i.e.* that the indifferent oxide, SO_4, can combine with sulphuric acid and form peculiar saline compounds.

[65 bis] Or one of those supposed ions which appear at the positive pole in the decomposition of sulphuric acid by the action of a galvanic current.

If this be true one would expect the following peroxide hydrates: for phosphoric acid, $(H_2PO_4)_2 = H_4P_2O_8 = 2H_2O + 2PO_3$; for carbonic acid, $(HCO_3)_2 = H_2C_2O_6 = H_2O + C_2O_5$; and for lead the true peroxide will be also Pb_2O_5, &c. Judging from the example of barium peroxide (Note 62), these peroxide forms will probably combine together. It seems to me that the compounds obtained by Fairley for uranium are very instructive as elucidating the peroxides. In the action of hydrogen peroxide in an acid solution on uranium oxide, UO_3, there is formed a uranium peroxide, $UO_4,4H_2O$ (U = 240), but hydrogen peroxide acts on uranium oxide in the presence of caustic soda; on the addition of alcohol a crystalline compound containing $Na_4UO_8,4H_2O$ is precipitated, which is doubtless a compound of the peroxides of sodium, Na_2O_2, and uranium, UO_4. It is very possible that the first peroxide, $UO_4,4H_2O$, contains the elements of hydrogen peroxide and uranium peroxide, U_2O_7, or even $U(OH)_6,H_2O_2$, just as the peroxide form lately discovered by Spring for tin perhaps contains Sn_2O_3,H_2O_2.

This view was communicated by me in 1870 to the Russian Chemical Society.

Dithionic acid, $H_2S_2O_6$, is distinguished among the thionic acids as containing the least proportion of sulphur. It is also called hyposulphuric acid, because its supposed anhydride, S_2O_5, contains more O than sulphurous oxide, SO_2 or S_2O_4, and less than sulphuric anhydride, SO_3 or S_2O_6. Dithionic acid, discovered by Gay-Lussac and Welter, is known as a hydrate and as salts, but not as anhydride. The method for preparing dithionic acid usually employed is by the action of finely-powdered manganese dioxide on a solution of sulphurous anhydride. On shaking, the smell of the latter disappears, and the manganese salt of the acid in question passes into solution; $MnO_2 + 2SO_2 = MnS_2O_6$. If the temperature be raised, the dithionate splits up into sulphurous anhydride and manganese sulphate, $MnSO_4$. Generally owing to this a mixture of manganese sulphate and dithionate is obtained in the solution. They may be separated by mixing the solution of the manganese salts with a solution of barium hydroxide, when a precipitate of manganese hydroxide and barium sulphate is obtained. In this manner barium dithionate only is obtained in solution. It is purified by crystallisation, and separates as $BaS_2O_6,2H_2O$; this is then dissolved in water, and decomposed with the requisite amount of sulphuric acid. Dithionic acid, $H_2S_2O_6$, then remains in solution. By concentrating the resultant solution under the receiver of an air-pump it is possible to obtain a liquid of sp. gr. 1·347, but it still contains water, and on further evaporation the acid decomposes into sulphuric acid and sulphurous anhydride: $H_2S_2O_6 = H_2SO_4 + SO_2$. The same decomposition takes place if the solution be slightly heated. Like all the thionic acids, dithionic acid is readily attacked by oxidising agents, and passes into sulphurous acid. No dithionate is able to withstand the action of heat, even when very slight, without giving off sulphurous anhydride: $K_2S_2O_6 = K_2SO_4 + SO_2$. The alkali dithionates have a neutral reaction (which indicates the energetic nature of the acid) are soluble in water, and in this respect present a certain resemblance to the salts of nitric acid (their anhydrides are: N_2O_5 and S_2O_5). Klüss (1888) described many of the salts of dithionic acid.

Langlois, about 1840, obtained a peculiar thionic acid by heating a strong solution of acid potassium sulphite with flowers of sulphur to about 60°, until the disappearance of the yellow coloration first produced by the solution of the sulphur. On cooling, a portion of the sulphur was precipitated, and crystals of a salt of *trithionic acid*, $K_2S_3O_6$ (partly mixed with potassium sulphate), separated out. Plessy afterwards showed that the action of sulphurous acid on a thiosulphate also gives sulphur and trithionic acid: $2K_2S_2O_3 + 3SO_2 = 2K_2S_3O_6 + S$. A mixture of potassium acid sulphite and thiosulphate also gives a trithionate. It is very possible that a reaction of the same kind occurs in the formation of trithionic acid by Langloid's method, because potassium sulphite and sulphur yield potassium thiosulphate. The potassium thiosulphate may also be replaced by potassium sulphide, and on passing sulphurous anhydride through the solution thiosulphate is first formed and then trithionate: $4KHSO_3 + K_2S + 4SO_2 = 3K_2S_3O_6 + 2H_2O$. The sodium salt is not formed under the same circumstances as the corresponding potassium salt. The sodium salt does not crystallise and is very unstable: the barium salt is, however, more stable. The barium and potassium salts are anhydrous, they give neutral solutions and decompose when ignited, with the evolution of sulphur and sulphurous anhydride, a sulphate being left behind, $K_2S_3O_6 = K_2SO_4 + SO_2 + S$. If a solution of the potassium salt be decomposed by means of hydrofluosilicic or chloric acid, the insoluble salts of these acids are precipitated and trithionic acid is obtained in solution, which however very easily breaks up on concentration. The addition of salts of copper, mercury, silver, &c., to a solution of a trithionate is followed, either immediately or after a certain time, by the formation of a black precipitate of the sulphides whose formation is due to the decomposition of the trithionic acid with the transference of its sulphur to the metal.

Tetrathionic acid, $H_2S_4O_6$, in contradistinction to the preceding acids, is much more stable in the free state than in the form of salts. In the latter form it is easily converted into trithionate, with liberation of sulphur. Sodium tetrathionate was obtained by Fordos and Gélis, by the action of iodine on a solution of sodium thiosulphate. The reaction essentially consists in the iodine taking up half the sodium of the thiosulphate, inasmuch as the latter contains $Na_2S_2O_3$, whilst the tetrathionate contains NaS_2O_3 or $Na_2S_4O_6$, so that the reaction is as follows: $2Na_2S_2O_3 + I_2 = 2NaI + Na_2S_4O_6$. It is evident that tetrathionic acid stands to thiosulphuric acid in exactly the same relation as dithionic acid does to sulphurous acid; for the same amount of the other elements in dithionate, KSO_3, and tetrathionate, KS_2O_3, there is half as much metal as in sulphite, K_2SO_3, and thiosulphate, $K_2S_2O_3$. If in the above reaction the sodium thiosulphate be replaced by the lead salt PbS_2O_3, the sparingly-soluble lead iodide PbI_2 and the soluble salt PbS_4O_6 are obtained. Moreover the lead salt easily gives tetrathionic acid itself ($PbSO_4$ is precipitated). The solution of tetrathionic acid may be evaporated over a water-bath, and afterwards in a vacuum, when it gives a colourless liquid, which has no smell and a very acid reaction. When dilute it may be heated to its boiling-point, but in a concentrated form it decomposes into sulphuric acid, sulphurous anhydride, and sulphur: $H_2S_4O_6 = H_2SO_4 + SO_2 + S_2$.

Pentathionic acid, $H_2S_5O_6$, also belongs to this series of acids. But little is known concerning it, either as hydrate or in salts. It is formed, together with tetrathionic acid, by the direct action of sulphurous acid on sulphuretted hydrogen in an aqueous solution; a large proportion of sulphur being precipitated at the same time: $5SO_2 + 5H_2S = H_2S_5O_6 + 5S + 4H_2O$.

If, as was shown above, the thionic acids are disulphonic acids, they may be obtained, like other sulphonic acids, by means of potassium sulphite and sulphur chloride. Thus Spring demonstrated the formation of potassium trithionate by the action of sulphur dichloride on a strong solution of potassium sulphite: $2KSO_3K + SCl_2 = S(SO_3K)_2 + 2KCl$. If sulphur chloride be taken, sulphur also is precipitated. The same trithionate is formed by heating a solution of double thiosulphates; for example, of $AgKS_2O_3$. Two molecules of the salts then form silver sulphide and potassium trithionate. If the thiosulphate be the potassium silver salt $SO_3K(AgS)$, then the structure of the trithionate must necessarily be $(SO_3K)_2S$. Previous to Spring's researches, the action of iodine on sodium thiosulphate was an isolated accidentally discovered reaction; he, however, showed its general significance by testing the action of iodine on mixtures of different sulphur compounds. Thus with iodine, I_2, the mixture $Na_2S + Na_2SO_3$ forms $2NaI + Na_2S_2O_3$, whilst the mixture $Na_2S_2O_3 + Na_2SO_3 + I_2$ gives $2NaI + Na_2S_3O_6$—that is, trithionic acid stands in the same relation to thiosulphuric acid as the latter does to sulphuretted hydrogen. We adopt the same mode of representation: by replacing one hydrogen in H_2S by sulphuryl we obtain thiosulphuric acid, $HSO_3.HS$, and by replacing a second hydrogen in the latter again by sulphuryl we obtain trithionic acid, $(HSO_3)_2S$. Furthermore, Spring showed that the action of sodium amalgam on the thionic acids causes reverse reactions to those above indicated for iodine. Thus sodium thiosulphate with Na_2 gives $Na_2S + Na_2SO_3$, and Spring showed that the sodium here is not a simple element taking up sulphur, but itself enters into double decomposition, replacing sulphur; for on taking a potassium salt and acting on it with sodium, $KSO_3(SK) + NaNa = KSO_3Na + (SK)Na$. In a similar way sodium dithionate with sodium gives sodium sulphite: $(NaSO_3)_2 + Na_2 = 2NaSO_3Na$; sodium trithionate forms $NaSO_3Na$ and $NaSO_3.SNa$, and tetrathionate forms sodium thiosulphate, $(NaSO_3)S_2(NaSO_3) + Na_2 = 2(NaSO_3)(NaS)$.

In all the oxidised compounds of sulphur we may note the presence of the elements of sulphurous anhydride, SO_2, the only product of the combustion of sulphur, and in this sense the compounds of sulphur containing one SO_2 are—

$$SO_2{<}^{H}_{HO} \qquad SO_2{<}^{HO}_{HO} \qquad SO_2{<}^{C_6H_5}_{HO} \qquad SO_2{<}^{HS}_{HO}$$

Sulphurous acid · Sulphuric acid · Benzene sulphonic acid · Thiosulphuric acid

while, according to this mode of representation, the thionic acids are—

$$\begin{matrix}HO\\SO_2\\SO_2\\HO\end{matrix} \qquad \begin{matrix}HO\\SO_2\\SO_2\\HO\end{matrix}S \qquad \begin{matrix}HO\\SO_2\\SO_2\\HO\end{matrix}S_2 \qquad \begin{matrix}HO\\SO_2\\SO_2\\HO\end{matrix}S_3$$

Dithionic · Trithionic · Tetrathionic · Pentathionic

Hence it is evident that SO_2 has (whilst CO_2 has not) the faculty for combination, and aims at forming SO_2X_2. These X_2 can $= O$, and the question naturally suggests itself as to whether the O_2 which occurs in SO_2 is not of the same nature as this oxygen which adds itself to SO_2—that is, whether SO_2 does not correspond with the more general type SX_4, and its compounds with the type SX_6? To this we may answer 'Yes' and 'No'—

'Yes' in the general sense which proceeds from the investigation of the majority of compounds, especially metals, where RO corresponds with RCl_2, RX_2; 'No' in the sense that sulphur does not give either SH_4, SH_6, or SCl_6, and therefore the stages SX_4 and SX_6 are only observable in oxygen compounds. With reference to the type SX_6 a hydrate, $S(HO)_6$, might be expected, if not SCl_6. And we must recognise this hydrate from a study of the compounds of sulphuric acid with water. In addition to what has been already said respecting the complex acids formed by sulphur, I think it well to mention that, according to the above view, still more complex oxygen acids and salts of sulphur may be looked for. For instance, the salt $Na_2S_4O_8$ obtained by Villiers (1888) is of this kind. It is formed together with sodium trithionate and sulphur, when SO_2 is passed through a cold solution of $Na_2S_2O_3$, which is then allowed to stand for several days at the ordinary temperature: $2Na_2S_2O_3 + 4SO_2 = Na_2S_4O_8 + Na_2S_3O_6 + S$. It may be assumed here, as in the thionic acids, that there are two sulphoxyls, bound together not only by S, but also by SO_2, or what is almost the same thing, that the sulphoxyl is combined with the residue of trithionic acid, *i.e.* replaces one aqueous residue in trithionic acid.

Even light decomposes carbon bisulphide, but not to the extent of separating carbon; under the action of the sun's rays it is decomposed into sulphur and solid substance which is considered to be carbon monosulphide; it is of a red colour, and its sp. gr. is 1·66. (The formation of a red liquid compound C_3S_2 has also been remarked.) Thorpe (1889) observed a complete decomposition of carbon bisulphide under the action of a liquid alloy of potassium and sodium; it is accompanied by an explosion and the deposition of carbon and sulphur. A similar complete decomposition of carbon bisulphide is also accomplished by the action of mercury fulminate (Chapter XVI., Note [26](#)), and is due to the fact that *at the ordinary temperature* (at which carbon bisulphide is not produced) *the decomposition* of carbon bisulphide takes place with the development of heat—that is, it presents an exothermal reaction, like the decomposition of all explosives. It is very possible that at a higher temperature, when carbon bisulphide is formed, the *combination* of carbon with sulphur is also an exothermal reaction—that is, heat is developed. If this should be the case, carbon bisulphide would present a most instructive example in thermochemistry.

The fact should not be lost sight of that sulphur and charcoal are solids at the ordinary temperature, whilst carbon bisulphide is a very volatile liquid, and consequently, in the act of combination, referred to the ordinary temperature (Note [69](#)), there is, as it were, a passage into a liquid state, and this requires the absorption of heat. And furthermore, the molecule of sulphur contains at least six atoms, and the molecule of carbon in all probability (Chapter VIII.) a very considerable number of atoms; thus the reaction of sulphur on charcoal may be expressed in the following manner: $3C_n + nS_6 = 3nCS_2$—that is, from $n + 3$ molecules there proceed $3n$ molecules, and as n must be very considerable, $3n$ must be greater than $3 + n$, which indicates a decomposition in the formation of carbon bisulphide, although the reaction at first sight appears as one of combination. This decomposition is seen also from the volumes in the solid and liquid states. Carbon bisulphide has a sp. gr. of 1·29; hence its molecular volume is 59. But the volume of carbon, even in the form of charcoal, is not more than 6, and the volume of S_2 is 30; hence 36 volumes after combination give 59 volumes—an expansion takes place, as in decompositions.

Carbon bisulphide, as prepared on a large scale, is generally very impure, and contains not only sulphur, but, more especially, other impurities which give it a very disagreeable odour. The best method of purifying this malodorous carbon bisulphide is to shake it up with a certain amount of mercuric chloride, or even simply with mercury, until the surface of the metal ceases to turn black. After this the carbon bisulphide must be poured off and distilled over a water-bath, after mixing with some oil to retain the impurities.

If carbon bisulphide be evaporated under the receiver of an air-pump, or by means of a current of air, it is possible to obtain a temperature as low as -60°, and the carbon bisulphide does not solidify at this temperature. However, if a series of air-bubbles be passed through it by means of bellows, a crystalline white substance remains which volatilises below 0°: this a hydrate, $H_2O, 2CS_2$; it easily decomposes into water and carbon bisulphide. It is formed in the above experiment by the moisture held in the air passed through the carbon bisulphide, and the fall of temperature.

Strong alcohol is miscible in all proportions with carbon bisulphide, but dilute alcohol only in a definite amount, owing to its diminished solubility from the presence of the water in it. Ether, hydrocarbons, fatty oils, and many other organic substances are soluble with great ease in carbon bisulphide. This is taken advantage of in practice for extracting the fatty oils from vegetable seeds, such as linseed, palm-nuts, or from bones, &c. The preparation of vegetable oils is usually done by pressing the seeds under a press, but the residue always contains a certain amount of oil. These traces of oil can, however, be removed by treatment with carbon bisulphide. In this manner a solution is obtained which when heated easily parts with all the carbon bisulphide, leaving the non-volatile fatty oil behind, so that the same carbon bisulphide may be condensed and used over again for the same purpose. It also dissolves iodine, bromine, indiarubber, sulphur, and tars.

Carbon bisulphide, especially at high temperatures, very often acts by its elements in a manner in which carbon and sulphur alone are not able to react, which will be understood from what has been said above respecting its endothermal origin. If it be passed over red-hot metals—even over copper, for instance, not to mention sodium, &c.—it forms a sulphide of the metal and deposits charcoal, and if the vapour be passed over incandescent metallic oxides it forms metallic sulphides and carbonic anhydride (and sometimes a certain amount of sulphurous anhydride). Lime and similar oxides give under these circumstances a carbonate and a sulphide—for example, $CS_2 + 3CaO = 2CaS + CaCO_3$. The sulphides obtained by this means are often well crystallised, like those found in nature—for example, lead and antimony sulphides.

[73 bis] And just as $COCl_2$ corresponds to CO_2, so also the chloranhydride, $CSCl_2$, or *thiophosgene*, corresponds to CS_2.

If instead of a sulphide we take an alkali hydroxide, a thiocarbonate is also formed, together with a carbonate—thus, $3BaH_2O_2 + 3CS_2 = 2BaCS_3 + BaCO_3 + 3H_2O$. From the instability of the thiocarbonates of the alkaline metals we can clearly see the reason of the difficulty with which the salts of the heavier metals are formed, whose basic properties are incomparably weaker than those of the alkali metals. However, these salts may be obtained by double decomposition. Ammonia in reacting on carbon bisulphide gives, besides products like those formed by other alkalis, a whole series of products of as complex a structure as those substances which are produced by the action of carbonic anhydride on ammonia. In the ninth chapter we examined the formation of the ammonium carbonates, and saw the transition from them into the cyanides. It is not surprising after this that the action of carbon bisulphide on ammonia not only produces the above-mentioned salts, but also amidic compounds corresponding with them, in which the oxygen is wholly or partially replaced by sulphur. Thus ammonium dithiocarbamate is very easily obtained if carbon bisulphide be added to an alcoholic solution of ammonia, and the mixture cooled in a closed vessel. The salt then separates out in minute yellow crystals, $CN_2H_6S_2$.

Carbon bisulphide not only forms compounds with the metallic sulphides, but also with sulphuretted hydrogen—that is, it forms *thiocarbonic acid*, H_2CS_3. This is obtained by carefully mixing solutions of thiocarbonates with dilute hydrochloric acid. It then separates in an oily layer, which easily decomposes in the presence of water into sulphuretted hydrogen and carbon bisulphide, just as the corresponding carbonic acid (hydrate) decomposes into water and carbonic anhydride. Carbon bisulphide combines not only with sodium sulphide, but also with the bisulphide, Na_2S_2, not, however, with the trisulphide, Na_2S_3.

The relation of carbon bisulphide to the other carbon compounds presents many most interesting features which are considered in organic chemistry. We will here only turn our attention to one of the compounds of this class. Ethyl sulphide, $(C_2H_5)_2S$, combines with ethyl iodide, C_2H_5I, forming a new molecule, $S(C_2H_5)_3I$. If we designate the hydrocarbon group, for instance ethyl, C_2H_5, by Et, the reaction would be expressed by the following equation : $Et_2S + EtI = SEt_3I$. This compound is of a saline character, corresponds with salts of the alkalis, and is closely analogous to ammonium chloride. It is soluble in water; when heated it again splits up into its components EtI and Et_2S, and with silver hydroxide gives a hydroxide, $Et_3S \cdot OH$, having the property of a distinct and energetic alkali, resembling caustic ammonia. Thus the compound group SEt_3 combines, like potassium or ammonium, with iodine, hydroxyl, chlorine, &c. The hydroxide $SEt_3 \cdot OH$ is soluble in water, precipitates metallic salts, saturates acids, &c. Hence sulphur here enters into a relation towards other elements similar to that of nitrogen in ammonia and ammonium salts, with only this difference, that nitrogen retains, besides iodine, hydroxyl, and other groups, also H_4 or Et_4 (for example, NH_4Cl, NEt_3HI, NEt_4I), whilst sulphur only retains Et_3. Compounds of the formula SH_3X are however unknown, only the products of substitution SEt_3X, &c. are known. The distinctly alkaline properties of the hydroxide, triethylsulphine hydroxide, SEt_3OH, and also the sharply-defined properties of the corresponding hydroxide, tetraethylammonium hydroxide, NEt_4OH, depend naturally not only on the properties of the nitrogen and sulphur entering into their composition, but also on the large proportion of hydrocarbon groups they contain. Judging from the existence of the ethylsulphine compounds, it might be imagined that sulphur forms a compound, SH_4, with hydrogen; but no such compound is known, just as NH_5 is unknown, although NH_4Cl exists.

[74 bis] Thorpe and Rodger (1889), by heating a mixture of lead fluoride and phosphorus pentasulphide to 250° in an atmosphere of dry nitrogen, obtained gaseous *phosphorus fluosulphide*, or *thiophosphoryl fluoride*, PSF_3, corresponding with $POCl_3$. This colourless gas is converted into a colourless liquid by a pressure of eleven atmospheres; it does not act on dry mercury, and takes fire spontaneously in air or oxygen, forming phosphorus pentafluoride, phosphoric anhydride, and sulphurous anhydride. It is soluble in ether, but is decomposed by water: $PSF_3 + 4H_2O = H_2S + H_3PO_4 + 3HF$ (Note 20).

Although mustard oil may be obtained from the thiocyanates, it is only an isomer of allyl thiocyanate proper, as is explained in Organic Chemistry.

[75 bis] Sulphur can only replace half the oxygen in CO_2, as is seen in *carbon oxysulphide*, or monothiocarbonic anhydride COS. This substance was obtained by Than, and is formed in many reactions. A certain amount is

obtained if a mixture of carbonic oxide and the vapour of sulphur be passed through a red-hot tube. When carbon tetrachloride is heated with sulphurous anhydride, this substance is also formed; but it is best obtained in a pure form by decomposing potassium thiocyanate with a mixture of equal volumes of water and sulphuric acid. A gas is then evolved containing a certain amount of hydrocyanic acid, from which it may be freed by passing it over wool containing moistened mercuric oxide, which retains the hydrocyanic acid. The reaction is expressed by the equation: $2KCNS + 2H_2SO_4 + 2H_2O = K_2SO_4 + (NH_4)_2SO_4 + 2COS$. It is also formed by passing the vapour of carbon bisulphide over alumina or clay heated to redness (Gautier; silicon sulphide is then formed). COS is also formed by passing phosgene over a long layer of asbestos mixed with cadmium sulphide at 270°; $CdS + COCl_3 = CdCl_2 + COS$ (Nuricsán, 1892). The pure gas has an aromatic odour, is soluble in an equal volume of water, which, however, acts on it, so that it must be collected over mercury. When slightly heated, carbon oxysulphide decomposes into sulphur and carbonic oxide. It burns in air with a pale blue flame, explodes with oxygen, and yields potassium sulphide and carbonate with potassium hydroxide: $COS + 4KHO = K_2CO_3 + K_2S + 2H_2O$.

There is no reason for seeing any contradiction or mutual incompatibility in these three views, because every analogy is more or less modified by a change of elements. Thus, for instance, it cannot be expected that the product of the metalepsis of hydrogen sulphide would resemble the corresponding products of water in all respects, because water has not the acid properties of hydrogen sulphide. In the days of dualism and electrical polarity it was supposed that the sulphur varied in its nature: in hydrogen sulphide or potassium sulphide it was considered to be negative, and in sulphurous anhydride or sulphur dichloride positive. It then appeared evident that sulphur dichloride would have no point of analogy with potassium sulphide. But metalepsis, or its expression in the law of substitution, necessitates such opinions being laid aside. If we can compare CO_2, CH_4, CCl_4, $CHCl_3$, $CH_3(OH)$ with each other, we cannot recognise any difference in the sulphur in SH_2, SCl_2, SK_2, or in general SX_2, for otherwise we should have to acknowledge as many different states of sulphur, carbon, or hydrogen as there are compounds of sulphur, carbon, or hydrogen. The essential truth of the matter is that all the elements in a molecule play their part in the reactions into which it enters. Often this appears to be contradicted in the result—for example, hydrogen alone may be replaced; but it is not this hydrogen alone that has determined the reaction; all the elements present have participated in it. This may be made clearer by the following rough illustration. Supposing two regiments of soldiers were fighting against each other, and that several men were lost by one of the regiments; no one could say that it was only these men who took part in the engagement. The other men fired and the bullets flew over the heads of their opponents. It was not only those who fell who fought, although they only were removed from the field of battle; the fighting proceeded among the masses, but only those few were disabled who went forward and were more conspicuous &c.; not that the remainder did not take part in the action; they also fought and were an object of attack, only they remained sound and unhurt. Hydrogen is lighter than other elements and its atoms more mobile; it subjects itself more frequently and easily to reactions; but it is not it alone which reacts, it is even less liable to attack than other elements. It participates in exceedingly diverse reactions, not indeed because the hydrogen itself varies, but because one atom of it puts itself forward, another is hidden, one is united with carbon, another feebly held by sulphur, one stands or moves in the neighbourhood of oxygen, another is joined to a hydrocarbon. All hydrogen atoms are equal, and equally serve as an object of attack for the atoms of molecules encountering them, but those only are removed from the sphere of action which are nearer the surface of a molecule, which are more mobile, or held by a less sum of forces. So also sulphur is one and the same in sulphur dichloride, in sulphurous or sulphuric anhydride, in hydrogen sulphide, in potassium sulphide, but it reacts differently, and those elements which are with it also vary in their reactions because they are with it, and it varies its reactions because it is with them. It is possible to seize on a character common to substances quantitatively and qualitatively analogous to each other. It may be admitted that an element in certain forms is not able to enter into reactions into which in other forms it enters willingly, if only the requisite conditions are encountered; but it must not therefore be concluded that an element changes its essential quality in these different cases. The preceding remarks touch on questions which are subject to much argument among chemists, and I mention them here in order to show the treatment of those most important problems of chemistry which lie at the basis of this treatise.

The observed vapour density of sulphur dichloride referred to hydrogen is 53·3, and that given by the formula is 51·5. The smaller molecular weight explains its boiling point being lower than that of sulphur chloride, S_2Cl_2. The reactions of both these compounds are very similar. Sulphur converts the dichloride, SCl_2, into the monochloride, S_2Cl_2. In one point the dichloride differs distinctly from the monochloride—that is, in its capacity for easily giving up chlorine and decomposing. Even light decomposes it into chlorine and the monochloride. Hence it acts on many substances in the same manner as chlorine, or substances which easily part with the latter, such as phosphoric or antimonic chloride. In distinction to these, however, sulphur dichloride would appear to distil without any considerable decomposition, judging by the vapour density. But this is not a valid conclusion, for if there be a decomposition, then $2SCl_2 = S_2Cl_2 + Cl_2$; now the density of sulphur chloride = 67·5, and of chlorine = 35·5, and consequently a mixture of equal volumes of the two = 51·5, just the same as an equal volume of sulphur dichloride. *Therefore the distillation of sulphur dichloride is probably nothing but its decomposition.* Hence the compound SCl_2, which is stable at the ordinary temperature, decomposes at 64°. In the cold it absorbs a further amount of chlorine, corresponding to SCl_4, but even at -10° a portion of the absorbed chlorine is given off—that is, dissociation takes place. Thus the tetrachloride is even less stable than the dichloride.

[77 bis] Hartog and Sims (1893) obtained thionyl bromide, $SOBr_2$, by treating $SOCl_2$ with sodium bromide; it is a red liquid, sp. gr. 2·62, and decomposes at 150°.

Pyrosulphuryl chloride, $S_2O_5Cl_2$. See Note 44. Thorpe and Kirman, by treating SO_3 with HF, obtained $SO_2(OH)F$, as a liquid boiling at 163°, but which decomposed with greater facility and then gave SO_2F_2.

The acids of sulphur naturally have their corresponding ammonium salts, and the latter their amides and nitriles. It will be readily understood how vast a field for research is presented by the series of compounds of sulphur and nitrogen, if we only remember that to carbonic and formic acids there corresponds, as we saw (Chapter IX.), a vast series of derivatives corresponding with their ammonium salts. To sulphuric acid there correspond two ammonium salts, $SO_2(HO)(NH_4O)$ and $SO_2(NH_4O)_2$; three amides: the acid amide $SO_2(HO)(NH_2)$, or sulphamic acid, the normal saline compound $SO_2(NH_4O)(NH_2)$, or ammonium sulphamate, and the normal amide $SO_2(NH_2)_2$, or sulphamide (the analogue of urea); then the acid nitrile, SON(HO), and two neutral nitriles, $SON(NH_2)$ and SN_2. There are similar compounds corresponding with sulphurous acid, and therefore its nitriles will be, an acid, SN(HO), its salt, and the normal compound, $SN(NH_2)$. Dithionic and the other acids of sulphur should also have their corresponding amides and nitriles. Only a few examples are known, which we will briefly describe. Sulphuric acid forms salts of very great stability with ammonia, and ammonium sulphate is one of the commonest ammoniacal compounds. It is obtained by the direct action of ammonia on sulphuric acid, or by the action of the latter on ammonium carbonate; it separates from its solutions in an anhydrous state, like potassium sulphate, with which it is isomorphous. Hence, the composition of crystals of ammonium sulphate is $(NH_4)_2SO_4$. This salt fuses at 140°, and does not undergo any change when heated up to 180°. At higher temperatures it does not lose water, but parts with half its ammonia, and is converted into the acid salt, HNH_4SO_4; and this acid salt, on further heating, undergoes a further decomposition, and splits up into nitrogen, water, and acid ammonium sulphite, HNH_4SO_3. At the ordinary temperature the normal salt is soluble in twice its weight of water and at the boiling-point of water in an equal weight. In its faculty for combinations this salt exhibits a great resemblance to potassium sulphate, and, like it, easily forms a number of double salts; the most remarkable of which are the ammonia alums, $NH_4AlS_2O_8,12H_2O$, and the double salts formed by the metals of the magnesium group, having, for example, the composition $(NH_4)_2MgS_2O_8,6H_2O$. Ammonium sulphate does not give an amide when heated, perhaps owing to the faculty of sulphuric anhydride to retain the water combined with it with great force. But the amides of sulphuric acid may be very conveniently prepared from sulphuric anhydride. Their formation by this method is very easily understood because an amide is equal to an ammonium salt less water, and if the anhydride be taken it will give an amide directly with ammonia. Thus, if dry ammonia be passed into a vessel surrounded by a freezing mixture and containing sulphuric anhydride, it forms a white powdery mass called sulphatammon, having the composition $SO_3,2H_3N$, and resembling the similar compound of carbonic acid, $CO_2,2NH_3$. This substance is naturally the ammonium salt of sulphamic acid, $SO_2(NH_4O)NH_2$. It is slowly acted on by water, and may therefore be obtained in solution, in which it slowly reacts with barium chloride, which proves that with water it still forms ammonium sulphate. If this substance be carefully dissolved in water and evaporated, it yields well-formed crystals, whose solution no longer gives a precipitate with barium chloride. This is not due to the presence of impurities, but to a change in the nature of the substance, and therefore Rose calls the crystalline modification *parasulphatammon*. Platinum chloride only precipitates half the nitrogen as platinochloride from solutions of sulphat- and parasulphatammon, which shows that they are ammonium salts, $SO_2(NH_4O)(NH_2)$. It may be that the reason of the difference in the two modifications is connected with the fact that two different substances of the composition $N_2H_4SO_2$ are possible: one is the amide $SO_2(NH_2)_2$ corresponding with the normal salt, and the other is the salt of the nitrile acid corresponding with acid ammonium sulphate—that is, $SON(ONH_4)$ corresponds with the acid $SON(OH) = SO_2(NH_4O)OH - 2H_2O$. Hence there may here be a difference of the same nature as between urea and ammonium cyanate. Up to the present, the isomerism indicated above has been but little investigated, and might be the subject of interesting researches.

If in the preceding experiment the ammonia, and not the sulphuric anhydride, be taken in excess, a soluble substance of the composition $2SO_2,3NH_3$ is formed. This compound, obtained by Jacqueline and investigated by Voronin, doubtless also contains a salt of sulphamic acid—that is, of the amide corresponding with the acid ammonium sulphate =

$HNH_4SO_4 - H_2O = (NH_2)SO_2(OH)$. Probably it is a compound of sulphatammon with sulphamic acid. Thus it has an acid reaction, and does not give a precipitate with barium chloride.

With normal sulphate of ammonium, an amide of the composition $N_2H_4SO_2$ should correspond, which should bear the same relation to sulphuric acid as urea bears to carbonic acid. This amide, known as *sulphamide*, is obtained by the action of dry ammonia on the sulphuryl chloride, SO_2Cl_2, just as urea is obtained by the action of ammonia on carbonyl chloride, $SO_2Cl_2 + 4NH_3 = N_2H_4SO_2 + 2NH_4Cl$. The ammonium chloride is separated from the resultant sulphamide with great difficulty. Cold water, acting on the mixture, dissolves them both; the cold solution does not gives precipitate with barium chloride. Alkalis act on it slowly, as they do on urea; but on boiling, especially in the presence of alkalis or acids, it easily recombines with water, and gives an ammonium salt. V. Traube (1892) obtained sulphamide by the reaction of sulphuryl, dissolved in chloroform, upon ammonia. The resultant precipitate dissolves when shaken up with water, and the solution (after boiling with the oxides or lead or silver) is evaporated, when a syrupy liquid remains. With nitrate of silver the latter gives a solid compound, which, when decomposed by hydrochloric acid, gives free sulphamide in large colourless crystals, having the composition $SO_2(NH_2)_2$. This substance fuses at 81°, begins to decompose below 100°, and is entirely decomposed above 250°; it is soluble in water, and the solution has a neutral reaction and bitter taste. When heated with acids, sulphamide gradually decomposes, forming sulphuric acid and ammonia. If the silver compound obtained by the action of sulphamide on nitrate of silver be heated at 170°-180° until ammonia is no longer evolved, and the residue be extracted with water acidulated with nitric acid, a salt separates out from the solution, answering in its composition to sulphamide, SO_2NAg, which = the amide - NH_3 = $SO_2N_2H_4 - NH_3 = SO_2NH$. The action of sulphuryl chloride (and of the other chloranhydrides of sulphur) on ammonium carbonate always, as Mente showed (1888), results in the formation of the salt $NH(SO_3NH_4)_2$.

The nitriles corresponding with sulphuric acid are not as yet known with any certainty. The most simple nitrile corresponding with sulphuric acid should have the composition $N_2H_8SO_4 - 4H_2O = N_2S$. This would be a kind of cyanogen corresponding with sulphuric acid. On comparing sulphurous acid with carbonic acid, we saw that they present a great analogy in many respects, and therefore it might be expected that nitrile compounds having the composition NHS and N_2S_2 would be found. The latter of these compounds is well known, and was obtained by Soubeiron, by the action of dry ammonia on sulphur chloride. This substance corresponds with cyanogen (paracyanogen), and is known as *nitrogen sulphide*, N_2S_2. It is formed according to the equation $3SCl_2 + 8NH_3 = N_2S_2 + S + 6NH_4Cl$. The free sulphur and nitrogen sulphide are dissolved by acting on the product with carbon bisulphide, the nitrogen sulphide being much less soluble than the sulphur. It is a yellow substance, which is excessively irritating to the eyes and nostrils. It explodes when rubbed with a hard substance, being naturally decomposed with the evolution of nitrogen; but when heated it fuses without decomposing, and only decomposes with explosion at 157°. It is insoluble in water, and only slightly so in alcohol, ether, and carbon bisulphide; 100 parts of the latter dissolve 1·5 part of nitrogen sulphide at the boiling point. This solution on cooling deposits it in minute transparent prisms of a golden yellow colour.

Selenious anhydride, SeO_2, is a volatile solid, which crystallises in prisms soluble in water. It is best procured by the action of nitric acid on selenium. The well-known researches of Nilson (1874) showed that the salts of selenious acid easily form acid salts, and are so characteristic in many respects that they may even serve for judging the analogy of types of oxides. Thus the oxides of the composition RO give normal salts of the composition $RSeO_3,2H_2O$, where R = Mn, Co, Ni, Cu, Zn. The salts of magnesium, barium, and calcium contain a different quantity of water, as do also the salts of the oxides R_2O_3. We here turn attention to the fact that beryllium gives a normal salt, $BeSeO_3,2H_2O$, and not a salt analogous to those of aluminium, scandium, $Sc_2(SeO_3)_3,H_2O$, yttrium, $Y_2(SeO_3)_2,12H_2O$, and other oxides of the form R_2O_3, which speaks in favour of the formula BeO.

Tellurous anhydride is also a colourless solid, which crystallises in octahedra; it also, when heated, first fuses and then volatilises. It is insoluble in water, and the decomposition of its salts gives a hydrate, H_2TeO_3, which is insoluble.

It is a very characteristic circumstance that selenious and tellurous anhydrides are very easily *reduced* to selenium and tellurium. This is not only effected by metals like zinc, or by sulphuretted hydrogen, which are powerful deoxidisers, but even by sulphurous anhydride, which is able to precipitate selenium and tellurium from solutions of the selenites and tellurites, and even of the acids themselves, which is taken advantage of in obtaining these elements and separating them from sulphur.

Sulphuric acid, as we know, rarely acts as an oxidising agent. It is otherwise with selenic and telluric acids, H_2SeO_4 and H_2TeO_4, which are powerful oxidising agents—that is, are easily reduced in many circumstances either into the lower oxide or even to selenium and tellurium. A powerful oxidising agent is required in order to convert selenious and tellurous anhydrides into selenic and telluric anhydrides, and, moreover, it must be employed in excess. If chlorine be passed through a solution of potassium selenide, K_2Se, telluride, K_2Te, selenite, K_2SeO_3, or tellurite, K_2TeO_3, it acts as an oxidiser in the presence of the water, forming potassium selenate, K_2SeO_4, or tellurate, K_2TeO_4. The same salts are formed by fusing the lower oxides with nitre. These salts are isomorphous with the corresponding sulphates, and cannot therefore be separated from them by crystallisation. The salts of potassium, sodium, magnesium, copper, cadmium, &c. are soluble like the sulphates, but those of barium and calcium are insoluble, in perfect analogy with the sulphates. When copper selenate, $CuSeO_4$, is treated with sulphuretted hydrogen (CuS is precipitated), *selenic acid* remains in solution. On evaporation and drying in vacuo at 180° it gives a syrupy liquid, which may be concentrated to almost the pure acid, H_2SeO_4, having a specific gravity of 2·6. Cameron and Macallan (1891) showed that pure H_2SeO_4 only remains liquid in a state of superfusion whilst the solidified acid melts at +58°, the solid acid crystallises well, its sp. gr. is then 2·95. The hydrate H_2SeO_4,H_2O melts at +25°. The acid in a superfused state has a sp. gr. 2·36 and the solid 2·63. Like sulphuric acid strong selenic acid attracts moisture from the atmosphere; it is not decomposed by sulphurous acid, but oxidises hydrochloric acid (like nitric, chromic, and manganic acids), evolving chlorine and forming selenious acid, $H_2SeO_4 + 2HCl = H_2SeO_3 + H_2O + Cl_2$. *Telluric acid*, H_2TeO_4, is obtained by fusing tellurous anhydride with potassium hydroxide and chlorate; the solution, containing potassium tellurate, is then precipitated with barium chloride, and the barium tellurate, $BaTeO_4$ obtained in the precipitate is decomposed by sulphuric acid. A solution of telluric acid is thus obtained, which on evaporation yields colourless prisms, soluble in water, and containing $TeH_2O_4,2H_2O$. Two equivalents of water are driven off at 160°; on further heating the last equivalent of water is expelled, and then oxygen is given off. It also gives chlorine with hydrochloric acid, like selenic acid. Its salts also correspond with those of sulphuric acid. It must, however, be remarked that telluric and selenic acids are able to give poly-acid salts with much greater ease than sulphuric acid. Thus, for example, there are known for telluric acid not only $K_2TeO_4,5H_2O$ and $KHTeO_4,3H_2O$, but also $KHTeO_4,H_2TeO_4,H_2O = K_2TeO_4,3H_2TeO_4,2H_2O$. This salt is easily obtained from acid solutions of the preceding salts and is less soluble in water. As selenious anhydride is volatile and gives similar poly-salts, it may be surmised that selenious, tellurous, selenic, and telluric anhydrides are polymeric as compared with sulphurous and sulphuric anhydrides, for which reason it would be desirable to determine the vapour density of selenious anhydride. It would probably correspond with Se_2O_4 or Se_3O_6.

In order to show the very close analogy of selenium to sulphur, I will quote two examples. Potassium cyanide dissolves selenium, as it does sulphur, forming potassium selenocyanate, KCNSe, corresponding with potassium thiocyanate. Acids precipitate selenium from this solution, because selenocyanic acid, H$_2$CNSe, when in a free state is immediately decomposed. A boiling solution of sodium sulphite dissolves selenium, just as it would sulphur, forming a salt analogous to thiosulphate of sodium, namely, sodium selenosulphate, Na$_2$SSeO$_3$. Selenium is separated from a solution of this salt by the action of acid.

[79 bis] Muthmann, in his researches upon the allotropic forms of selenium, pointed out (1889) a peculiar modification, which appears, as it were, as a transition between crystalline and amorphous selenium. It is obtained together with the crystalline variety by slowly evaporating a solution of selenium in bisulphide of carbon, and differs from the crystalline variety in the form of its crystals; it passes into the latter modification when heated. Schultz also obtained selenium (like Ag, see Chapter XXIV.) in a soluble form, but these researches are not so conclusive as those upon soluble silver, and we shall therefore not consider them more fully.

The tellurium thus prepared is impure, and contains a large amount of selenium. The latter may be removed by converting the mixture into the salts of potassium, and treating this with nitric acid and barium nitrate, when barium selenate only is precipitated, whilst the barium tellurate remains in solution. This method does not, however, give a pure product, and it appears to be best to separate the selenium from the tellurium in a metallic form; this is done by boiling the impure potassium tellurate with hydrochloric acid, which converts it into potassium tellurite, from which the tellurium is reduced by sulphurous anhydride. The metal thus obtained is then fused and distilled in a stream of hydrogen; the selenium volatilises first, and then the tellurium, owing to its being much less volatile than the former. Nevertheless, tellurium is also volatile, and may be separated in this manner from less volatile metals, such as antimony. Brauner determined the atomic weight of pure tellurium, and found it to be 125, but showed (1889) that tellurium purified by the usual method, even after distillation, contains a large amount of impurities.

The decomposition proceeds in the above order in the cold, but in a hot solution with an excess of potassium hydroxide it proceeds inversely. A similar phenomenon takes place when tellurium is fused with alkalis, and it is therefore necessary in order to obtain potassium telluride to add charcoal.

Selenium and tellurium form higher compounds with chlorine with comparative ease. For selenium, SeCl$_2$ and SeCl$_4$ are known, and for tellurium TeCl$_2$ and TeCl$_4$. The tetrachlorides of selenium and tellurium are formed by passing chlorine over these elements. Selenium tetrachloride, $SeCl_4$, is a crystalline, volatile mass which gives selenious anhydride and hydrochloric acid with water. Tellurium tetrachloride is much less volatile, fuses easily, and is also decomposed by water. Both elements form similar compounds with bromine. Tellurium tetrabromide is red, fuses to a brown liquid, volatilises, and gives a crystalline salt, $K_2TeBr_6,3H_2O$, with an aqueous solution of potassium bromide.

CHAPTER XXI
CHROMIUM, MOLYBDENUM, TUNGSTEN, URANIUM, AND MANGANESE

Sulphur, selenium, and tellurium belong to the uneven series of the sixth group. In the even series of this group there are known *chromium, molybdenum, tungsten, and uranium*; these give acid oxides of the type RO_3, like SO_3. Their acid properties are less sharply defined than those of sulphur, selenium, and tellurium, as is the case with all elements of the even series as compared with those of the uneven series in the same group. But still the oxides CrO_3, MoO_3, WO_3, and even UO_3, have clearly defined acid properties, and form salts of the composition MO,nRO_3 with bases MO. In the case of the heavy elements, and especially of uranium, the type of oxide, UO_3, is less acid and more basic, because in the even series of oxides the element with the highest atomic weight always acquires a more and more pronounced basic character. Hence UO_3 shows the properties of a base, and gives salts UO_2X_2. The basic properties of chromium, molybdenum, tungsten, and uranium are most clearly expressed in the lower oxides, which they all form. Thus chromic oxide, Cr_2O_3, is as distinct a base as alumina, Al_2O_3.

Of all these elements *chromium* is the most widely distributed and the most frequently used. It gives chromic anhydride, CrO_3, and chromic oxide, Cr_2O_3—two compounds whose relative amounts of oxygen stand in the ratio 2 : 1. Chromium is, although somewhat rarely, met with in nature as a compound of one or the other type. The red chromium ore of the Urals, lead chromate or crocoisite $PbCrO_4$, was the source in which chromium was discovered by Vauquelin, who gave it this name (from the Greek word signifying colour) owing to the brilliant colours of its compounds; the chromates (salts of chromic anhydride) are red and yellow, and the chromic salts (from Cr_2O_3) green and violet. The red lead chromate is, however, a rare chromium ore found only in the Urals and in a few other localities. Chromic oxide, Cr_2O_3, is more frequently met with. In small quantities it forms the colouring matter of many minerals and rocks—for example, of some serpentines. The commonest ore, and the chief source of the chromium compounds, is the *chrome iron ore* or chromite, which occurs in the Urals and Asia Minor, California, Australia, and other localities. This is magnetic iron ore, FeO,Fe_2O_3, in which the ferric oxide is replaced by chromic oxide, its composition being FeO,Cr_2O_3. Chrome iron ore crystallises in octahedra of sp. gr. 4·4; it has a feeble metallic lustre, is of a greyish-black colour, and gives a brown powder. It is very feebly acted on by acids, but when fused with potassium acid sulphate it gives a soluble mass, which contains a chromic salt, besides potassium sulphate and ferrous sulphate. In practice the treatment of chrome iron ore is mainly carried on for the preparation of chromates, and not of chromic salts, and therefore we will trace the history of the element by beginning with chromic acid, and especially with the working up of the chrome iron ore into *potassium dichromate*, $K_2Cr_2O_7$, as the most common salt of this acid. It must be remarked that chromic anhydride, CrO_3, is only obtained in an anhydrous state, and is distinguished for its capacity for easily giving anhydro-salts with the alkalis, containing one, two, and even three equivalents of the anhydride to one equivalent of base. Thus among the potassium salts there is known the normal or yellow chromate, K_2CrO_4, which corresponds to, and is perfectly isomorphous with, potassium sulphate, easily forms isomorphous mixtures with it, and is not therefore suitable for a process in which it is necessary to separate the salt from a mixture containing sulphates. As in the presence of a certain excess of acid, the dichromate, $K_2Cr_2O_7 = 2K_2CrO_4 + 2HX - 2KX - H_2O$, is easily formed from K_2CrO_4, the object of the manufacturer is to produce such a dichromate, the more so as it contains a larger proportion of the elements of chromic acid than the normal salt. Finely-ground chrome iron ore, when heated with an alkali, absorbs oxygen almost as easily (Chapter III., Note 7) as a mixture of the oxides of manganese with an alkali. This absorption is due to the presence of chromic oxide, which is oxidised into the anhydride, and then combines with the alkali $Cr_2O_3 + O_3 = 2CrO_3$. As the oxidation and formation of the chromate proceeds, the mass turns *yellow*. The iron is also oxidised, but does not give ferric acid, because the capacity of the chromium for oxidation is incomparably greater than that of the iron.

A mixture of lime (sometimes with potash) and chrome iron ore is heated in a reverberatory furnace, with free access of air and at a red heat for several hours, until the mass becomes yellow; it then contains normal calcium chromate, $CaCr\_O_4$, which is insoluble in water in the presence of an excess of lime.[1 bis] The resultant mass is ground up, and treated with water and sulphuric acid. The excess of lime forms gypsum, and the soluble calcium dichromate, $CaCr_2O_7$, together with a certain amount of iron, pass into solution. The solution is poured off, and chalk added to it; this precipitates the ferric oxide (the ferrous oxide is converted into ferric oxide in the furnace) and forms a fresh quantity of gypsum, while the chromic acid remains in solution—that is, it does not form the sparingly-soluble normal salt (1 part soluble in 240 parts of water). The solution then contains a fairly pure calcium dichromate, which by double decomposition gives other chromates; for example, with a solution of potassium sulphate it gives a precipitate of calcium sulphate and a solution of potassium dichromate, which crystallises when evaporated.

Potassium dichromate, $K_2Cr_2O_7$, easily crystallises from acid solutions in red, well-formed prismatic crystals, which fuse at a red heat and evolve oxygen at a very high temperature, leaving chromic oxide and the normal salt, which undergoes no further change: $2K_2Cr_2O_7 = 2K_2CrO_4 + Cr_2O_3 + O_3$. At the ordinary temperature 100 parts of water dissolve 10 parts of this salt, and the solubility increases as the temperature rises. It is most important to note that the dichromate does not contain water, it is $K_2CrO_4 + CrO_3$; the acid salt corresponding to potassium acid sulphate, $KHSO_4$, does not exist. It does not even evolve heat when dissolving in water, but on the contrary produces cold, *i.e.* it does not form a very stable compound with water. The solution and the salt itself are poisonous, and act as powerful oxidising agents, which is the character of chromic acid in general. When heated with sulphur or organic substances, with sulphurous anhydride, hydrogen sulphide, &c., this salt is deoxidised, yielding chromic compounds.[2 bis] Potassium dichromate is used in the arts and in chemistry as a source for the preparation of all other chromium compounds. It is converted into yellow pigments by means of double decomposition with salts of lead, barium, and zinc. When solutions of the salts of these metals are mixed with potassium dichromate (in dyeing generally mixed with soda, in order to obtain normal salts), they are precipitated as insoluble normal salts; for example, $2BaCl_2 + K_2Cr_2O_7 + H_2O = 2BaCrO_4 + 2KCl + 2HCl$. It follows from this that these salts are insoluble in dilute acids, but the precipitation is not complete (as it would be with the normal salt). The barium and zinc salts are of a lemon yellow colour; the lead salt has a still more intense colour passing into orange. Yellow cotton prints are dyed with this pigment. The silver salt, Ag_2CrO_4, is of a bright red colour.

When potassium dichromate is mixed with potassium hydroxide or carbonate (carbonic anhydride being disengaged in the latter case) it forms the *normal* salt, K_2CrO_4, known as *yellow chromate of potassium*. Its specific gravity is 2·7, being almost the same as that of the dichromate. It absorbs heat in dissolving; one part of the salt dissolves in 1·75 part of water at the ordinary temperature, forming a yellow solution. When mixed even with such feeble acids as acetic, and more especially with the ordinary acids, it gives the dichromate, and Graham obtained a trichromate, $K_2Cr_3O_{10} = K_2CrO_4,2CrO_3$, by mixing a solution of the latter salt with an excess of nitric acid.

Chromic anhydride is obtained by preparing a saturated solution of potassium dichromate at the ordinary temperature, and pouring it in a thin stream into an equal volume of pure sulphuric acid. On mixing, the temperature naturally rises; when slowly cooled, the solution deposits chromic anhydride in needle-shaped crystals of a red colour sometimes several centimetres long. The crystals are freed from the mother liquor by placing them on a porous tile.[4 bis] It is very important at this point to call attention to the fact that a hydrate of chromic anhydride is never obtained in the decomposition of chromic compounds, but always the *anhydride*, CrO_3. The corresponding hydrate, CrO_4H_2, or any other hydrate, is not even known. Nevertheless, it must be admitted that chromic acid is bibasic, because it forms salts isomorphous or perfectly analogous with the salts formed by sulphuric acid, which is the best example of a bibasic acid. A clear proof of the bibasicity of CrO_3 is seen in the fact that the anhydride and salts give (when heated with sodium chloride and sulphuric acid) a volatile chloranhydride, CrO_2Cl_2, containing two atoms of chlorine as a bibasic acid should. Chromic anhydride is a red crystalline substance, which is converted into a black mass by heat; it fuses at 190°, and disengages oxygen above 250°, leaving a residue of chromium dioxide, CrO_2, and, on still further heating, chromic oxide, Cr_2O_3. Chromic anhydride is exceedingly soluble in water, and even attracts moisture from the air, but, as was mentioned above, it does not form any definite compound with water. The specific gravity of its crystals is 2·7, and when fused it has a specific gravity 2·6. The solution presents perfectly defined acid properties. It liberates carbonic anhydride from carbonates; gives insoluble precipitates of the chromates with salts of barium, lead, silver, and mercury.

The action of hydrogen peroxide on a solution of chromic acid or of potassium dichromate gives a blue solution, which very quickly becomes colourless with the disengagement of oxygen. Barreswill showed that this is due to the formation of a *perchromic anhydride*, Cr_2O_7, corresponding with sulphur peroxide. This peroxide is remarkable from the fact that it very easily dissolves in ether and is much more stable in this solution, so that, by shaking up hydrogen peroxide mixed with a small quantity of chromic acid, with ether, it is possible to transfer all the blue substance formed to the ether.[6 bis]

With oxygen acids, chromic acid evolves oxygen; for example, with sulphuric acid the following reaction takes place: $2CrO_3 + 3H_2SO_4 = Cr_2(SO_4)_3 + O_3 + 3H_2O$. It will be readily understood from this that *a mixture of chromic acid* or *of its salts with sulphuric acid* forms an excellent *oxidising agent*, which is frequently employed in chemical laboratories and even for technical purposes as a means of oxidation. Thus hydrogen sulphide and sulphurous anhydride are converted into sulphuric acid by this means. Chromic acid is able to act as a powerful oxidising agent because it passes into chromic oxide, and in so doing disengages half of the oxygen contained in it: $2CrO_3 = Cr_2O_3 + O_3$. Thus chromic anhydride itself is a powerful oxidising agent, and is therefore employed instead of nitric acid in galvanic batteries (as a depolariser), the hydrogen evolved at the carbon being then oxidised, and the chromic acid converted into a non-volatile product of deoxidation, instead of yielding, as nitric acid does, volatile lower oxides of offensive odour. Organic substances are more or less perfectly oxidised by means of chromic anhydride, although this generally requires the aid of heat, and does not proceed in the presence of alkalis, but generally *in the presence of acids*. In acting on a solution of potassium iodide, chromic acid, like many oxidising agents, liberates iodine; the reaction proceeds in proportion to the amount of CrO_3 present, and may serve for determining the amount of CrO_3, since the amount of iodine liberated can be accurately determined by the iodometric method (Chapter XX., Note 42). If chromic anhydride be ignited in a stream of ammonia, it gives chromic oxide, water, and nitrogen. In all cases when chromic acid acts as an oxidising agent in the presence of acids and under the action of heat, the product of its deoxidation is a chromic salt, CrX_3, which is characterised by the green colour of its solution, so that the *red* or yellow *solution* of a salt of chromic acid is then transformed into a *green solution* of a chromic salt, derived from chromic oxide, Cr_2O_3, which is closely analogous to Al_2O_3, Fe_2O_3, and other bases of the composition R_2O_3. This analogy is seen in the insolubility of the anhydrous oxide, in the gelatinous form of the colloidal hydrate, in the formation of alums, of a volatile chloride of chromium, &c.[7 bis]

Chromic oxide, Cr_2O_3, rarely found, and in small quantities, in chrome ochre, is formed by the oxidation of chromium and its lower oxides, by the reduction of chromates (for example, of ammonium or mercuric chromate) and by the decomposition (splitting up) of the saline compounds of the oxide itself, CrX_3 or Cr_2X_6, like alumina, which it resembles in forming a feeble base easily giving double and basic salts, which are either green or violet.

The reduction of chromic oxide—for instance, in a solution by zinc and sulphuric acid—leads to the formation of chromous oxide, CrO, and its salts, CrX_2, of a blue colour (*see* Notes 7 and 7[bis]). The further reduction of oxide of chromium and its corresponding compounds gives *metallic chromium*. Deville obtained it (probably containing carbon) by reducing chromic oxide with carbon, at a temperature near the melting point of platinum, about 1750°, but the metal itself does not fuse at this temperature. Chromium has a steel-grey colour and is very hard (sp. gr. 5·9), takes a good polish, and dissolves in hydrochloric acid, but cold dilute sulphuric and nitric acids have no action upon it. Bunsen obtained metallic chromium by decomposing a solution of chromic chloride, Cr_2Cl_6, by a galvanic current, as scales of a grey colour (sp. gr. 7·3). Wöhler obtained crystalline chromium by igniting a mixture of the anhydrous chromic chloride Cr_2Cl_6 (*see* Note 7 bis) with finely-divided zinc, and sodium and potassium chlorides, at the boiling-point of zinc. When the resultant mass has cooled the zinc may be dissolved in dilute nitric acid, and grey crystalline chromium (sp. gr. 6·81) is left behind. Frémy also prepared crystalline chromium by the action of the vapour of sodium on anhydrous chromic chloride in a stream of hydrogen, using the apparatus shown in the accompanying drawing, and placing the sodium and the chromic chloride in separate porcelain boats. The tube containing these boats is only heated when it is quite full of dry hydrogen. The crystals of metallic chromium obtained in the tube are grey cubes having a considerable hardness and withstanding the action of powerful acids, and even of aqua regia. The chromium obtained by Wöhler by the action of a galvanic current is, on the contrary, acted on under these conditions. The reason of this difference must be looked for in the presence of impurities, and in the crystalline structure. But in any case, among the properties of metallic chromium, the following may be considered established: it is white in colour, with a specific gravity of about 6·7, is extremely hard in a crystalline form, is not oxidised by air at the ordinary temperature, and with carbon it forms alloys like cast iron and steel.

FIG. 92.—**Apparatus for the preparation of metallic chromium by igniting chromic chloride and sodium in a stream of hydrogen.**

The two analogues of chromium, *molybdenum* and *tungsten* (or wolfram), are of still rarer occurrence in nature, and form acid oxides, RO_3, which are still less energetic than CrO_3. Tungsten occurs in the somewhat rare minerals, *scheelite*, $CaWO_4$, and *wolfram*; the latter being an isomorphous mixture of the normal tungstates of iron and manganese, $(MnFe)WO_4$. Molybdenum is most frequently met with as *molybdenite*, MoS_2, which presents a certain resemblance to graphite in its physical properties and softness. It also occurs, but much more rarely, as a yellow lead ore, $PbMoO_4$. In both these forms molybdenum occurs in the primary rocks, in granites, gneiss, &c., and in iron and copper ores in Saxony, Sweden, and Finland. Tungsten ores are sometimes met with in considerable masses in the primary rocks of Bohemia and Saxony, and also in England, America, and the Urals. The preliminary treatment of the ore is very simple; for example, the sulphide, MoS_2, is roasted, and thus converted into sulphurous anhydride and molybdic anhydride, MoO_3, which is then dissolved in alkalis, generally in ammonia. The ammonium molybdate is then treated with acids, when the sparingly soluble molybdic acid is precipitated. Wolfram is treated in a different manner. Most frequently the finely-ground ore is repeatedly boiled with hydrochloric and nitric acids, and the resultant solutions (of salts of manganese and iron) poured off, until the dark brown mass of ore disappears, whilst the tungstic acid remains, mixed with silica, as an insoluble residue; it is treated also with ammonia, and is thus converted into soluble ammonium tungstate, which passes into solution and yields tungstic acid when treated with acids. This hydrate is then ignited, and leaves tungstic anhydride. The general character of molybdic and tungstic anhydrides is analogous to that of chromic anhydride; they are anhydrides of a feebly acid character, which easily give polyacid salts and colloid solutions.[8 bis]

Hydrogen (which does not directly form compounds with Cr, Mo, and W) reduces molybdic and tungstic anhydride at a red heat; and this forms the means of obtaining metallic molybdenum and tungsten. *Both metals* are infusible, and both under the action of heat form compounds with carbon and iron (the addition of tungsten to steel renders the latter ductile and hard). Molybdenum forms a grey powder, which scarcely aggregates under a most powerful heat, and has a specific gravity of 8·5. It is not acted on by the air at the ordinary temperature, but when ignited it is first converted into a brown, and then into a blue oxide, and lastly into molybdic anhydride. Acids do not act on it—that is, it does not liberate hydrogen from them, not even from hydrochloric acid—but strong sulphuric acid disengages sulphurous anhydride, forming a brown mass, containing a lower oxide of molybdenum. Alkalis in solution do not act on molybdenum, but when fused with it hydrogen is given off, which shows, as does its whole character, the acid properties of the metal. The properties of tungsten are almost identical; it is infusible, has an iron-grey colour, is exceedingly hard, so that it even scratches glass. Its specific gravity is 19·1 (according to Roscoe), so that, like uranium, platinum, &c., it is one of the heaviest metals.[9 bis] Just as sulphur and chromium have their corresponding persulphuric and perchromic acids, $H_2S_2O_8$ and H_2CrO_8, having the properties of peroxides, and corresponding to peroxide of hydrogen, so also molybdenum and tungsten are known to give *permolybdic* and *pertungstic* acids, $H_2Mo_2O_8$ and $H_2W_2O_8$, which have the properties of true peroxides, *i.e.* easily disengage iodine from KI and chlorine from HCl, easily part with their oxygen, and are formed by the action of peroxide of hydrogen, into which they are readily reconverted (hence they may be regarded as compounds of H_2O_2 with $2MoO_3$ and $2WO_3$), &c. Their formation (Boerwald 1884, Kemmerer 1891) is at once seen in the coloration (not destroyed by boiling), which is obtained on mixing a solution of the salts with peroxide of hydrogen, and on treating, for instance, molybdic acid with a solution of peroxide of hydrogen (Péchard 1892). The acid then forms an orange-coloured solution, which after evaporation in vacuo leaves $Mo_2H_2O_8 4H_2O$ as a crystalline powder, and loses $4H_2O$ at 100°, beyond which it decomposes with the evolution of oxygen.[9 tri]

Uranium, U = 240, has the highest atomic weight of all the analogues of chromium, and indeed of all the elements yet known. Its highest salt-forming oxide, UO_3, shows very feeble acid properties. Although it gives sparingly-soluble yellow compounds with alkalis, which fully correspond with the dichromates—for example, $Na_2U_2O_7 = Na_2O,2UO_3$,—yet it more frequently and easily reacts with acids, HX, forming fluorescent yellowish-green salts of the composition UO_2X_2, and in this respect uranic trioxide, UO_3, differs from chromic anhydride, CrO_3, although the latter is able to give the oxychloride, CrO_2Cl_2. In molybdenum and tungsten, however, we see a clear transition from chromium to uranium. Thus, for example, chromyl chloride, CrO_2Cl_2, is a brown liquid which volatilises without change, and is completely decomposed by water; molybdenum oxychloride, MoO_2Cl_2, is a crystalline substance of a yellow colour, which is volatile and soluble in water (Blomstrand), like many salts. Tungsten oxychloride, WO_2Cl_2, stands still nearer to uranyl chloride in its properties; it forms yellow scales on which water and alkalis act, as they do on many salts (zinc chloride, ferric chloride, aluminium chloride, stannic chloride, &c.), and perfectly corresponds with the difficultly volatile salt, UO_2Cl_2 (obtained by Peligot by the action of chlorine on ignited uranium dioxide, UO_2), which is also yellow and gives a yellow solution with water, like all the salts UO_2X_2. The property of uranic oxide, UO_3, of forming salts UO_2X_2 is shown in the fact that the hydrated oxide of uranium, $UO_2(HO)_2$, which is obtained from the nitrate, carbonate, and other salts by the loss of the elements of the acid, is easily soluble in acids, as well as in the fact that the lower grades of oxidation of uranium are able, when treated with nitric acid, to form an easily crystallisable uranyl nitrate, $UO_2(NO_3)_2, 6H_2O$; this is the most commonly occurring uranium salt.

Uranium, which gives an oxide, UO_3, and the corresponding salt UO_2X_2 and dioxide UO_2, to which the salts UX_4 correspond, is rarely met with in nature. Uranite or the double orthophosphate of uranic oxide, $R(UO_2)H_2P_2O_8, 7H_2O$, where R = Cu or Ca, uranium-vitriol $U(SO_4)_2, H_2O$, samarakite, and æschynite, are very rarely found, and then only in small quantities. Of more frequent and abundant occurrence is the non-crystalline, earthy brown uranium ore known as *pitchblende* (sp. gr. 7·2), which is mainly composed of the intermediate oxide, $U_3O_8 = UO_2,2UO_3$. This ore is found at Joachimsthal in Bohemia and in Cornwall. It usually contains a number of different impurities, chiefly sulphides and arsenides of lead and iron, as well as lime and silica compounds. In order to expel the arsenic and sulphur it is roasted, ground, washed with dilute hydrochloric acid, which does not dissolve the uranoso-uranic oxide, U_3O_8, and the residue is dissolved in nitric acid, which transforms the uranium oxide into the nitrate, $UO_2(NO_3)_2$.

It must be observed that the oxide of uranium, first distinguished by Klaproth (1789), was for a long time regarded as able to give metallic uranium under the action of charcoal and other reducing agents (with the aid of heat). But the substance thus obtained was only the *uranium dioxide*, UO_2. The compound nature of this dioxide, or the presence of oxygen in it, was demonstrated by Peligot (1841), by igniting it with charcoal in a stream of chlorine. He thus obtained a volatile *uranium tetrachloride*, UCl_4, which, when heated with sodium, gave *metallic uranium* as a grey metal, having a specific gravity of 18·7, and liberating hydrogen from acids, with the formation of green uranous salts, UX_4, which act as powerful reducing agents.

As the salts of uranic oxide are reduced in the absence of organic matter by the action of light, and as they impart a characteristic coloration to glass, they find a certain application in photography and glass work.

If we compare together the highly acid elements, sulphur, selenium, and tellurium, of the uneven series, with chromium, molybdenum, tungsten, and uranium of the even series, we find that the resemblance of the properties of the higher form RO_3 does not extend to the lower forms, and even entirely disappears in the elements, for there is not the smallest resemblance between sulphur and chromium and their analogues in a free state. In other words, this means that the small periods, like Na, Mg, Al, Si, P, S, Cl, containing seven elements, do not contain any near analogues of chromium, molybdenum, &c., and therefore their true position among the other elements must be looked for only in those large periods which contain two small periods, and whose type is seen in the period containing: K, Ca, Sc, Ti, V, Cr, Mn, Fe, Co, Ni, Cu, Zn, Ga, Ge, As, Se, Br. These large periods contain Ca and Zn, giving RO, Sc, and Ga of the third group, Ti and Ge giving RO_2, V and As forming R_2O_5, Cr and Se of the sixth group, Mn and Br of the seventh group, and the remaining elements, Fe, Co, Ni, form connective members of the intermediate eighth group, to the description of the representatives of which we shall turn in the following chapters. We will now proceed to describe *manganese*, Mn = 55, as an element of the seventh group of the even series, directly following after Cr = 52, which corresponds with Br = 80 to the same degree that Cr does with Se = 79. For chromium, selenium, and bromine very close analogues are known, but for manganese as yet none have been obtained—that is, it is the only representative of the even series in the seventh group. In placing manganese with the halogens in one group, the periodic system of the elements only requires that it should bear an analogy to the halogens in the higher type of oxidation—*i.e.* in the salts and acids—whilst it requires that as great a difference should be expected in the lower types and elements as there exists between chromium or molybdenum and sulphur or selenium. And this is actually the case. The elements of the seventh group form a higher salt-forming oxide, R_2O_7, and its corresponding hydrate, HRO_4, and salts—for example, $KClO_4$. Manganese in the form of potassium permanganate, $KMnO_4$, actually presents a great analogy in many respects to potassium perchlorate, $KClO_4$. The analogy of the crystalline form of both salts was shown by Mitscherlich. The salts of permanganic acid are also nearly all soluble in water, like those of perchloric acid, and if the silver salt of the latter, $AgClO_4$, be sparingly soluble in water, so also is silver permanganate, $AgMnO_4$. The specific volume of potassium perchlorate is equal to 55, because its specific gravity = 2·54; the specific volume of potassium permanganate is equal to 58, because its specific gravity = 2·71. So that the volumes of equivalent quantities are in this instance approximately the same whilst the atomic volumes of chlorine (35·5/1·3 = 27) and manganese (55/7·5) are in the ratio 4 : 1. In a free state the higher acids $HClO_4$ and $HMnO_4$ are both soluble in water and volatile, both are powerful oxidisers—in a word, their analogy is still closer than that of chromic and sulphuric acids, and those points of distinction which they present also appear among the nearest analogues—for example, in sulphuric and telluric acids, in hydrochloric and hydriodic acids, &c. Besides Mn_2O_7 manganese gives a lower grade of oxidation, MnO_3, analogous to sulphuric and chromic trioxides, and with it corresponds potassium manganate, K_2MnO_4, isomorphous with potassium sulphate. In the still lower grades of oxidation, Mn_2O_3 and MnO, there is hardly any similarity to chlorine, whilst every point of resemblance disappears when we come to the elements themselves—*i.e.* to manganese and chlorine—for manganese is a metal, like iron, which combines directly with chlorine to form a saline compound, $MnCl_2$, analogous to magnesium chloride.

Manganese belongs to the number of metals widely distributed in nature, especially in those localities where iron occurs, whose ores frequently contain compounds of manganous oxide, MnO, which presents a resemblance to ferrous oxide, FeO, and to magnesia. In many minerals magnesia and the oxides allied to it are replaced by manganous oxide; calcspars and magnesites—*i.e.* $R''CO_3$ in general—are frequently met with containing manganous carbonate, which also occurs in a separate state, although but rarely. The soil also and the ash of plants generally contain a small quantity of manganese. In the analysis of minerals it is generally found that manganese occurs together with magnesia, because, like it, manganous oxide remains in solution in the presence of ammoniacal salts, not being precipitated by reagents. The property of this manganous oxide, MnO, of passing into the higher grades of oxidation under the influence of heat, alkalis, and air, gives an easy means not only of discovering the presence of manganese in admixture with magnesia, but also of separating these two analogous bases. Magnesia is not able to give higher grades of oxidation, whilst manganese gives them with great facility. Thus, for instance, an *alkaline* solution of sodium hypochlorite produces a precipitate of manganese dioxide in a solution of a manganous salt: $MnCl_2 + NaClO + 2NaHO = MnO_2 + H_2O + 3NaCl$; whilst magnesia is not changed under these circumstances, and remains in the form of $MgCl_2$. If the magnesia be precipitated owing to the presence of alkali, it may be dissolved in acetic acid, in which manganese dioxide is insoluble. The presence of small quantities of manganese may also be recognised by the green coloration which alkalis acquire when heated with manganese compounds in the air. This green coloration depends on the property of manganese of giving a green alkaline manganate: $MnCl_2 + 4KHO + O_2 = K_2MnO_4 + 2KCl + 2H_2O$. Thus *the faculty of oxidising in the presence of alkalis* forms an essential character of manganese. The higher grades of oxidation containing Mn_2O_7 and MnO_3 are quite unknown in nature, and even MnO_2 is not so widely spread in nature as the ores composed of manganous compounds which are met with nearly everywhere. The most important ore of manganese is its dioxide, or so-called *peroxide*, MnO_2, which is known in mineralogy as *pyrolusite*. Manganese also occurs as an oxide corresponding with magnetic iron ore, $MnO,Mn_2O_3 = Mn_3O_4$, forming the mineral known as *hausmannite*. The oxide Mn_2O_3 also occurs in nature as the anhydrous mineral *braunite*, and in a hydrated form, Mn_2O_3,H_2O, called *manganite*. Both of these often occur as an admixture in pyrolusite. Besides which, manganese is met with in nature as a rose-coloured mineral, *rhodonite*, or silicate, $MnSiO_3$. Very fine and rich deposits of manganese ores have been found in the Caucasus, the Urals, and along the Dnieper. Those at the Sharapansky district of the Government of Kutais and at Nicopol on the Dnieper are particularly rich. A large quantity of the ore (as much as 100,000 tons yearly) is exported from these localities.

Thus manganese gives oxides of the following forms: MnO, manganous oxide, and manganous salts, MnX_2, corresponding with the base, which resembles magnesia and ferrous oxide in many respects; Mn_2O_3, a very feeble base, giving salts, MnX_3, analogous to the aluminium and ferric salts, easily reduced to MnX_2; MnO_2, dioxide, generally called peroxide, an almost indifferent oxide, or feebly acid; MnO_3, manganic anhydride, which forms salts resembling potassium sulphate;[18 bis] Mn_2O_7, permanganic anhydride, giving salts analogous to the perchlorates.

All the oxides of manganese when heated with acids give salts, MnX_2, corresponding with the lower grade of oxidation, *manganous oxide*, MnO. Manganic oxide, Mn_2O_3, is a feebly energetic base; it is true that it dissolves in hydrochloric acid and gives a dark solution containing the salt $MnCl_3$, but the latter when heated evolves chlorine and gives a salt corresponding with manganous oxide $MnCl_2$—*i.e.* at first: $Mn_2O_3 + 6HCl = 3H_2O + Mn_2Cl_6$, and then the Mn_2Cl_6 decomposes into $2MnCl_2 + Cl_2$. None of the remaining higher grades of oxidation have a basic character, but *act as oxidising agents in the presence of acids*, disengaging oxygen and passing into salts of the lower grade of oxidation of manganese, MnO. Owing to this circumstance, *the manganous salts* are often obtained; they are, for instance, left in the residue when the dioxide is used for the preparation of oxygen and chlorine.

As the salts of manganous oxide MnX_2 closely resemble (and are isomorphous with) the salts of magnesia MgX_2 in many respects (with the exception of the fact that MnX_2 are rose coloured and are easily oxidised in the presence of alkalis), we will not dwell upon them, but limit ourselves to illustrating the chemical character of manganese by describing the metal and its corresponding acids. The fact alone that the oxides of manganese are not reduced to the metal when ignited in hydrogen (whilst the oxides of iron give metallic iron under these circumstances), but only to manganous oxide, MnO, shows that manganese has a considerable affinity for oxygen—that is, it is difficult to reduce. This may be effected, however, by means of charcoal or sodium at a very high temperature. A mixture of one of the oxides of manganese with charcoal or organic matter gives fused *metallic manganese* under the powerful heat developed by coke with an artificial draught. The metal was obtained for the first time in this manner by Gahn, after Pott, and more especially Scheele, had in the last century shown the difference between the compounds of iron and manganese (they were previously regarded as being the same). Manganese is prepared by mixing one of its oxides in a finely-divided state with oil and soot; the resultant mass is then first ignited in order to decompose the organic matter, and afterwards strongly heated in a charcoal crucible. The manganese thus obtained, however, contains, as a rule, a considerable amount of silicon and other impurities. Its specific gravity varies between 7·2 and 8·0. It has a light grey colour, a feebly metallic lustre, and although it is very hard it can be scratched by a file. It rapidly oxidises in air, being converted into a black oxide; water acts on it with the evolution of hydrogen—this decomposition proceeds very rapidly with boiling water, and if the metal contain carbon.

It has been shown above that if manganese dioxide, or any lower oxide of manganese, be heated with an alkali in the presence of air, the mixture absorbs oxygen, and forms an alkaline manganate of a green colour: $2KHO + MnO_2 + O = K_2MnO_4 + H_2O$. Steam is disengaged during the ignition of the mixture, and if this does not take place there is no absorption of oxygen. The oxidation proceeds much more rapidly if, before igniting in air, potassium chlorate or nitre be added to the mixture, and this is the method of preparing *potassium manganate*, K_2MnO_4. The resultant mass dissolved in a small quantity of water gives a dark green solution, which, when evaporated under the receiver of an air-pump over sulphuric acid, deposits green crystals of exactly the same form as potassium sulphate—namely, six-sided prisms and pyramids. The composition of the product is not changed by being re-dissolved, if perfectly pure water free from air and carbonic acid be taken. But in the presence of even very feeble acids the solution of this salt changes its colour and becomes red, and deposits manganese dioxide. The same decomposition takes place when the salt is heated with water, but when diluted with a large quantity of unboiled water manganese dioxide does not separate, although the solution turns red. This change of colour depends on the fact that potassium manganate, K_2MnO_4, whose solution is green, is transformed into potassium permanganate, $KMnO_4$, whose solution is of a red colour. The reaction proceeding under the influence of acids and a large quantity of water is expressed in the following manner: $3K_2MnO_4 + 2H_2O = 2KMnO_4 + MnO_2 + 4KHO$. If there is a large proportion of acid and the decomposition is aided by heat, the manganese dioxide and potassium permanganate are also decomposed, with formation of manganous salt. Exactly the same decomposition as takes place under the action of acids is also accomplished by magnesium sulphate, which reacts in many cases like an acid. When water holding atmospheric oxygen in solution acts on a solution of potassium manganate, the oxygen combines directly with the manganate and forms potassium permanganate, without precipitating manganese dioxide, $2K_2MnO_4 + O + H_2O = 2KMnO_4 + 2KHO$. Thus a solution of potassium manganate undergoes a very characteristic change in colour and passes from green to red; hence this salt received the name of *chameleon mineral*.

Potassium permanganate, $KMnO_4$, crystallises in well-formed, long red prisms with a bright green metallic lustre. In the arts the potash is frequently replaced by soda, and by other alkaline bases, but no salt of permanganic acid crystallises so well as the potassium salt, and therefore this salt is exclusively used in chemical laboratories. One part of the crystalline salt dissolves in 15 parts of water at the ordinary temperature.

The solution is of a very deep *red colour*, which is so intense that it is still clearly observable after being highly diluted with water. In a solid state it is decomposed by heat, with evolution of oxygen, a residue consisting of the lower oxides of manganese and potassium oxide being left.[22 bis] A mixture of permanganate of potassium, phosphorous and sulphur takes fire when struck or rubbed, a mixture of the permanganate with carbon only takes fire when heated, not when struck. The instability of the salt is also seen in the fact that its solution is decomposed by peroxide of hydrogen, which at the same time it decomposes. A number of substances reduce potassium permanganate to manganese dioxide (in which case the red solution becomes colourless). Many organic substances (although far from all, even when boiled in a solution of permanganate) act in this manner, being oxidised at the expense of a portion of its oxygen. Thus, a solution of sugar decomposes a cold solution of potassium permanganate. In the presence of an excess of alkali, with a small quantity of sugar, the reduction leads to the formation of potassium manganate, because $2KMnO_4 + 2KHO = O + 2K_2MnO_4 + H_2O$. With a considerable amount of sugar and a more prolonged action, the solution turns brown and precipitates manganese dioxide or even oxide. In the oxidation of many organic bodies by an alkaline solution of $KMnO_4$ generally three-eighths of the oxygen in the salt are utilised for oxidation: $2KMnO_4 = K_2O + 2MnO_2 + O_3$. A portion of the alkali liberated is retained by the manganese dioxide, and the other portion generally combines with the substance oxidised, because the latter most frequently gives an acid with an excess of alkali. A solution of potassium iodide acts in a similar manner, being converted into potassium iodate at the expense of the three atoms of oxygen disengaged by two molecules of potassium permanganate.

In the presence of acids, potassium permanganate acts as an oxidising agent with still greater energy than in the presence of alkalis. At any rate, a greater proportion of oxygen is then available for oxidation, namely, not $3/8$, as in the presence of alkalis, but $5/8$, because in the first instance manganese dioxide is formed, and in the second case manganous oxide, or rather the salt, MnX_2, corresponding with it. Thus, for instance, in the presence of an excess of sulphuric acid, the decomposition is accomplished in the following manner: $2KMnO_4 + 3H_2SO_4 = K_2SO_4 + 2MnSO_4 + 3H_2O + 5O$. This decomposition, however, does not proceed directly on mixing a solution of the salt with sulphuric acid, and crystals of the salt even dissolve in oil of vitriol without the evolution of oxygen, and this solution only decomposes by degrees after a certain time. This is due to the fact that sulphuric acid liberates free permanganic acid from the permanganate, which acid is stable in solution. But if, in the presence of acids and a permanganate, there is a substance capable of absorbing oxygen—for instance, capable of passing into a higher grade of oxidation—then the reduction of the permanganic acid into manganous oxides sometimes proceeds directly at the ordinary temperature. This reduction is very clearly seen, because the solutions of potassium permanganate are red whilst the manganous salts are almost colourless. Thus, for instance, nitrous acid and its salts are converted into nitric acid and decolorise the acid solution of the permanganate. Sulphurous anhydride and its salts immediately decolorise potassium permanganate, forming sulphuric acid. Ferrous salts, and in general salts of lower grades of oxidation capable of being oxidised in solution, act in exactly the same manner. Sulphuretted hydrogen is also oxidised to sulphuric acid; even mercury is oxidised at the expense of permanganic acid, and decolorises its solution, being converted into mercuric oxide. Moreover, the end point of these reactions may easily be seen, and therefore, having first determined the amount of active oxygen in one volume of a solution of potassium permanganate, and knowing how many volumes are required to effect a given oxidation, it is easy to determine the amount of an oxidisable substance in a solution from the amount of permanganate expended (Marguerite's method).

The oxidising action of $KMnO_4$, like all other chemical reactions, is not accomplished instantaneously, but only gradually. And, as the course of the reaction is here easily followed by determining the amount of salt unchanged in a sample taken at a given moment, the oxidising reaction of potassium permanganate, in an acid liquid, was employed by Harcourt and Esson (1865) as one of the first cases for the investigation of the laws of the *rate of chemical change* as a subject of great importance in chemical mechanics. In their experiments they took oxalic acid, $C_2H_2O_{-4}$, which in oxidising gives carbonic anhydride, whilst, with an excess of sulphuric acid, the potassium permanganate is converted into manganous sulphate, $MnSO_4$, so that the ultimate oxidation will be expressed by the equation: $5C_2H_2O_4 + 2MnKO_4 + 3H_2SO_4 = 10CO_2 + K_2SO_4 + 2MnSO_4 + 8H_2O$. The influence of the relative amount of sulphuric acid is seen from the annexed table, which gives the measure of reaction p per 100 parts of potassium permanganate, taken four minutes after mixing, using n molecules of sulphuric acid, H_2SO_4, per $2KMnO_4 + 5C_2H_2O_4$:

$n =$ 2 4 6 8 12 16 22

$p =$ 22 36 51 63 77 86 22

showing that in a given time (4 minutes) the oxidation is the more perfect the greater the amount of sulphuric acid taken for given amounts of $KMnO_4$ and $C_2H_2O_4$. It is obvious also that the temperature and relative amount of every one of the acting and resulting substances should show its influence on the relative velocity of reaction; thus, for instance, direct experiment showed the influence of the admixture of manganous sulphate. When a large proportion of oxalic acid (108 molecules) was taken to a large mass of water and to 2 molecules of permanganate 14 molecules of manganous sulphate were added, the quantity x of the potassium permanganate acted on (in percentages of the potassium permanganate taken) in t minutes (at 16°) was as follows:

$t =$ 2   5   8   11   14   44   47   53   61   68

$x =$ 5·2 12·1 18·7 25·1 31·3 68·4 71·7 75·8 79·8 83·0

These figures show that the rate of reaction—that is, the quantity of permanganate changed in one minute—decreases proportionally to the decrease in the amount of unchanged potassium permanganate. At the commencement, about 2·6 per cent. of the salt taken was decomposed in the course of one minute, whilst after an hour the rate was about 0·5 per cent. The same phenomena are observed in every case which has been investigated, and this branch of theoretical or physical chemistry, now studied by many, promises to explain the course of chemical transformations from a fresh point of view, which is closely allied to the doctrine of affinity, because the rate of reaction, without doubt, is connected with the magnitude of the affinities acting between the reacting substances.

Footnotes:

The working of the Ural chrome iron ore into chromium compounds has been firmly established in Russia, thanks to the endeavours of P. K. Ushakoff, who constructed large works for this purpose on the river Kama, near Elabougi, where as much as 2,000 tons of ore are treated yearly, owing to which the importation of chromium preparations into Russia has ceased.

[1 bis] But the calcium chromate is soluble in water in the presence of an excess of chromic acid, as may be seen from the fact that a solution of chromic acid dissolves lime.

There are many variations in the details of the manufacturing processes, and these must be looked for in works on technical chemistry. But we may add that the chromate may also be obtained by slightly roasting briquettes of a mixture of chrome iron and lime, and then leaving the resultant mass to the action of moist air (oxygen is absorbed, and the mass turns yellow).

[2 bis] The oxidising action of potassium dichromate on organic substances at the ordinary temperature is especially marked under the action of light. Thus it acts on gelatin, as Poutven discovered; this is applied to photography in the processes of photogravure, photo-

lithography, pigment printing, &c. Under the action of light this gelatin is oxidised, and the chromic anhydride deoxidised into chromic oxide, which unites with the gelatin and forms a compound insoluble in warm water, whilst where the light has not acted, the gelatin remains soluble, its properties being unaffected by the presence of chromic acid or potassium dichromate.

Ammonium and sodium dichromates are now also prepared on a large scale. The sodium salts may be prepared in exactly the same manner as those of potassium. The normal salt combines with ten equivalents of water, like Glauber's salt, with which it is isomorphous. Its solution above 30° deposits the anhydrous salt. Sodium dichromate crystals contain $Na_2Cr_2O_7,2H_2O$. The *ammonium salts of chromic acid* are obtained by saturating the anhydride itself with ammonia. The dichromate is obtained by saturating one part of the anhydride with ammonia, and then adding a second part of anhydride and evaporating under the receiver of an air-pump. On ignition, the normal and acid salts leave chromic oxide. Potassium ammonium chromate, NH_4KCrO_4, is obtained in yellow needles from a solution of potassium dichromate in aqueous ammonia; it not only loses ammonia and becomes converted into potassium dichromate when ignited, but also by degrees at the ordinary temperature. This shows the feeble energy of chromic acid, and its tendency to form stable dichromates. Magnesium chromate is soluble in water, as also is the strontium salt. The calcium salt is also somewhat soluble, but the barium salt is almost insoluble. The isomorphism with sulphuric acid is shown in the chromates by the fact that the magnesium and ammonium salts form double salts containing six equivalents of water, which are perfectly isomorphous with the corresponding sulphates. The magnesium salt crystallises in large crystals containing seven equivalents of water. The beryllium, cerium, and cobalt salts are insoluble in water. Chromic acid dissolves manganous carbonate, but on evaporation the solution deposits manganese dioxide, formed at the expense of the oxygen of the chromic acid. Chromic acid also oxidises ferrous oxide, and ferric oxide is soluble in chromic acid.

One of the chromates most used by the dyer is the insoluble yellow lead chromate, $PbCrO_4$ (Chapter XVIII., Note 46), which is precipitated on mixing solutions of PbX_2 with soluble chromates. It easily forms a basic salt, having the composition $PbO,PbCrO_4$, as a crystalline powder, obtained by fusing the normal salt with nitre and then rapidly washing in water. The same substance is obtained, although impure and in small quantity, by treating lead chromate with neutral potassium chromate, especially on boiling the mixture; and this gives the possibility of attaining, by means of these materials, various tints of lead chromate, from yellow to red, passing through different orange shades. The decomposition which takes place (incompletely) in this case is as follows: $2PbCrO_4 + K_2CrO_4 = PbCrO_4,PbO + K_2Cr_2O_7$—that is, potassium dichromate is formed in solution.

The sulphuric acid should not contain any lower oxides of nitrogen, because they reduce chromic anhydride into chromic oxide. If a solution of a chromate be heated with an excess of acid—for instance, sulphuric or hydrochloric acid—oxygen or chlorine is evolved, and a solution of a chromic salt is formed. Hence, under these circumstances, chromic acid cannot be obtained from its salts. One of the first methods employed consisted in converting its salts into volatile *chromium hexafluoride*, CrF_6. This compound, obtained by Unverdorben, may be prepared by mixing lead chromate with fluor spar in a dry state, and treating the mixture with fuming sulphuric acid in a platinum vessel: $PbCrO_4 + 3CaF_2 + 4H_2SO_4 = PbSO_4 + 3CaSO_4 + 4H_2O + CrF_6$. Fuming sulphuric acid is taken, and in considerable excess, because the chromium fluoride which is formed is very easily decomposed by water. It is volatile, and forms a very caustic, poisonous vapour, which condenses when cooled in a dry platinum vessel into a red, exceedingly volatile liquid, which fumes powerfully in air. The vapours of this substance when introduced into water are decomposed into hydrofluoric acid and chromic anhydride: $CrF_6 + 3H_2O = CrO_3 + 6HF$. If very little water be taken the hydrofluoric acid volatilises, and chromic anhydride separates directly in crystals. The chloranhydride of chromic acid, CrO_2Cl_2 (Note 5), is also decomposed in the same manner. A solution of chromic acid and a precipitate of barium sulphate are formed by treating the insoluble barium chromate with an equivalent quantity of sulphuric acid. If carefully evaporated, the solution yields crystals of chromic anhydride. Fritzsche gave a very convenient method of preparing chromic anhydride, based on the relation of chromic to sulphuric acid. At the ordinary temperature the strong acid dissolves both chromic anhydride and potassium chromate, but if a certain amount of water is added to the solution the chromic anhydride separates, and if the amount of water be increased the precipitated chromic anhydride is again dissolved. The chromic anhydride is almost all separated from the solution when it contains two equivalents of water to one equivalent of sulphuric acid. Many methods for the preparation of chromic anhydride are based on this fact.

[4 bis] They cannot be filtered through paper or washed, because the chromic anhydride is reduced by the filter-paper, and is dissolved during the process of washing.

Berzelius observed, and Rose carefully investigated, this remarkable reaction, which occurs between chromic acid and sodium chloride in the presence of sulphuric acid. If 10 parts of common salt be mixed with 12 parts of potassium dichromate, fused, cooled, and broken up into lumps, and placed in a retort with 20 parts of fuming sulphuric acid, it gives rise to a violent reaction, accompanied by the formation of brown fumes of *chromic chloranhydride*, or *chromyl chloride*, CrO_2Cl_2, according to the reaction: $CrO_3 + 2NaCl + H_2SO_4 = Na_2SO_4 + H_2O + CrO_2Cl_2$. The addition of an excess of sulphuric acid is necessary in order to retain the water. The same substance is always formed when a metallic chloride is heated with chromic acid, or any of its salts, in the presence of sulphuric acid. The formation of this volatile substance is easily observed from the brown colour which is proper to its vapour. On condensing the vapour in a dry receiver a liquid is obtained having a sp. gr. of 1·9, boiling at 118°, and giving a vapour whose density, compared with hydrogen, is 78, which corresponds with the above formula. Chromyl chloride is decomposed by heat into chromic oxide, oxygen, and chlorine: $2CrO_2Cl_2 = Cr_2O_3 + 2Cl_2 + O$; so that it is able to act simultaneously as a powerful oxidising and chlorinating agent, which is taken advantage of in the investigation of many, and especially of organic, substances. When treated with water, this substance first falls to the bottom, and is then decomposed into hydrochloric and chromic acids, like all chloranhydrides: $CrO_2Cl_2 + H_2O = CrO_3 + 2HCl$. When brought into contact with inflammable substances it sets fire to them; it acts thus, for instance, on phosphorus, sulphur, oil of turpentine, ammonia, hydrogen, and other substances. It attracts moisture from the atmosphere with great energy, and must therefore be kept in closed vessels. It dissolves iodine and chlorine, and even forms a solid compound with the latter, which depends upon the faculty of chromium to form its higher oxide, Cr_2O_7. The close analogy in the physical properties of the chloranhydrides, CrO_2Cl_2 and SO_2Cl_2, is very remarkable, although sulphurous anhydride is a gas, and the corresponding oxide, CrO_2, is a non-volatile solid. It may be imagined, therefore, that chromium dioxide (which will be mentioned in the following note) presents a polymerised modification of the substance having the composition CrO_2; in fact, this is obvious from the method of its formation.

If three parts of potassium dichromate be mixed with four parts of strong hydrochloric acid and a small quantity of water, and gently warmed, it all passes into solution, and no chlorine is evolved; on cooling, the liquid deposits red prismatic crystals, known as *Peligot's salt*, very stable in air. This has the composition KCl,CrO_3, and is formed according to the equation $K_2Cr_2O_7 + 2HCl = 2KCl,CrO_3 + H_2O$. It is evident that this is the first chloranhydride of chromic acid, $HCrO_3Cl$, in which the hydrogen is replaced by potassium. It is decomposed by water, and on evaporation the solution yields potassium dichromate and hydrochloric acid. This is a fresh instance of the reversible reactions so frequently encountered. With sulphuric acid Peligot's salt forms chromyl chloride. The latter circumstance, and the fact that Geuther produced Peligot's salt from potassium chromate and chromyl chloride, give reason for thinking that it is a compound of these two substances: $2KCl,CrO_3 = K_2CrO_4 + CrO_2Cl_2$. It is also sometimes regarded as potassium dichromate in which one atom of oxygen is replaced by chlorine—that is, $K_2Cr_2O_6Cl_2$, corresponding with $K_2Cr_2O_7$. When heated it parts with all its chlorine, and on further heating gives chromic oxide.

This intermediate degree of oxidation, CrO_2, may also be obtained by mixing solutions of chromic salts with solutions of chromates. The brown precipitate formed contains a compound, Cr_2O_3,CrO_3, consisting of equivalent amounts of chromic oxide and anhydride. The brown precipitate of chromium dioxide contains water. The same substance is formed by the imperfect deoxidation of chromic anhydride by various reducing agents. Chromic oxide, when heated, absorbs oxygen, and appears to give the same substance. Chromic nitrate, when ignited, also gives this substance. When this substance is heated it first disengages water and then oxygen, chromic oxide being left. It corresponds with manganese dioxide, $Cr_2O_3,CrO_3 = 3CrO_2$. Krüger treated chromium dioxide with a mixture of sodium chloride and sulphuric acid, and found that chlorine gas was evolved, but that chromyl chloride was not formed. Under the action of light, a solution of chromic acid also deposits the brown dioxide. At the ordinary temperature chromic anhydride leaves a brown stain upon the skin and tissues, which probably proceeds from a decomposition of the same kind. Chromic anhydride is soluble in alcohol containing water, and this solution is decomposed in a similar manner by light. Chromium dioxide forms K_2CrO_4 when treated with H_2O_2 in the presence of KHO.

[6 bis] Now that persulphuric acid $H_2S_2O_8$ is well known it might be supposed that perchromic anhydride, Cr_2O_7, would correspond to perchromic acid, $H_2Cr_2O_8$, but as yet it is not certain whether corresponding salts are formed. Péchard (1891) on adding an excess of H_2O_2 and baryta water to a dilute solution of CrO_2 (8 grm. per litre), observed the formation of a yellow precipitate, but oxygen was disengaged at the same time and the precipitate (which easily exploded when dried) was found to contain, besides an admixture of BaO_2, a compound $BaCrO_5$, and this = $BaO_2 + CrO_3$, and does not correspond to perchromic acid. The fact of its decomposing with an explosion, and the mode of its preparation, proves, however, that this is a similar derivative of peroxide of hydrogen to persulphuric acid (Chapter XX.)

As a mixture of potassium dichromate and sulphuric acid is usually employed for oxidation, the resultant solution generally contains a double sulphate of potassium and chromium—that is, *chrome alum*, isomorphous with ordinary alum—$K_2Cr_2O_7 + 4H_2SO_4 + 20H_2O = O_3 + K_2Cr_2(SO_4)_4,24H_2O$ or $2(KCr(SO_4)_2,12H_2O)$. It is prepared by dissolving potassium dichromate in dilute sulphuric acid; alcohol is then added and the solution slightly heated, or sulphurous anhydride is passed through it. On the addition of alcohol to a cold mixture of potassium dichromate and sulphuric acid, the gradual disengagement of pleasant-smelling volatile products of the oxidation of alcohol, and especially of aldehyde, C_2H_4O, is remarked. If the temperature of decomposition does not exceed 35°, a *violet* solution of chrome alum is obtained, but if the temperature be higher, a solution of the same alum is obtained of a *green* colour. As chrome alum requires for solution 7 parts of water at the ordinary temperature, it follows that if a somewhat strong solution of potassium dichromate be taken (4 parts of water and 1½ of sulphuric acid to 1 part of dichromate), it will give so concentrated a solution of chrome alum that on cooling, the salt will separate without further evaporation. *If the liquid*, prepared as above or in any instance of the deoxidation of chromic acid, *be heated* (the oxidation naturally proceeds more rapidly) somewhat strongly, for instance, to the boiling-point of water, or if the violet solution already formed be raised to the same temperature, it acquires a bright *green colour*, and on evaporation the same mixture, which at lower temperatures so easily gives cubical crystals of chrome alum, *does not give any crystals whatever. If the green solution be kept*, however, *for several weeks* at the ordinary temperature, it deposits *violet crystals* of chrome alum. The green solution, when evaporated, gives a non-crystalline mass, and the violet crystals lose water at 100° and turn green. It must be remarked that the transition of the green modification into the violet is accompanied by a decrease in volume (Lecoq de Boisbaudran, Favre). If the green mass formed at the higher temperature be evaporated to dryness and heated at 30° in a current of air, it does not retain more then 6 equivalents of water. Hence Löwel, and also Schrötter, concluded that the green and violet modifications of the alum depend on different degrees of combination with water, which may be likened to the different compounds of sodium sulphate with water and to the different hydrates of ferric oxide.

However, the question in this case is not so simple, as we shall afterwards see. Not chrome alum alone, but *all the chromic salts*, give two, if not three, *varieties*. At least, there is no doubt about the existence of two—a *green* and a *violet modification*. The green chromic salts are obtained by heating solutions of the violet salts, the violet solutions are produced on keeping solutions of the green salts for a long time. The conversion of the violet salts into green by the action of heat is itself an indication of the possibility of explaining the different modifications by their containing different proportions (or states) of water, and, moreover, by the green salts having a less amount of water than the violet. However, there are other explanations. Chromic oxide is a base like alumina, and is therefore able to give both acid and basic salts. It is supposed that the difference between the green and violet salts is due to this fact. This opinion of Krüger is based on the fact that alcohol separates out a salt from the green solution which contains less sulphuric acid than the normal violet salt. On the other hand, Löwel showed that all the acid cannot be separated from the green chromic salts by suitable reagents, as easily as it can be from the same solution of the violet salts; thus barium salts do not precipitate all the sulphuric acid from solutions of the green salts. According to other researches the cause of the varieties of the chromic salts lies in a difference in the bases they contain—that is, it is connected with a modification of the properties of the oxide of chromium itself. This only refers to the hydroxides, but as hydroxides themselves are only special forms of salts, the differences observed as yet in this direction between the hydroxides only confirm the generality of the difference observed in the chromic compounds (*see* Note 7 bis).

The salts of chromic oxide, like those of alumina, are easily decomposed, give basic and double salts, and have an acid reaction, as chromic oxide is a feeble base. Potassium and sodium hydroxides give a *precipitate* of the hydroxide with chromic salts, CrX_3. The violet and green salts give a *hydroxide soluble in an excess of the reagent*; but the hydroxide is held in solution by very feeble affinities, so that it is partially separated by heat and dilution with water, and completely so on boiling. In an alkaline solution, chromic hydroxide is easily converted into chromic acid by the action of lead dioxide, chlorine, and other oxidising agents. If the chromic oxide occurs together with such oxides as magnesia, or zinc oxide, then on precipitation it separates out from its solution in combination with these oxides, forming, for example, ZnO,Cr_2O_3. Viard obtained compounds of Cr_2O_3 with the oxides of Mg, Zn, Cd, &c.) On precipitating the violet solution of chrome alum with ammonia, a precipitate containing $Cr_2O_3,6H_2O$ is obtained, whilst the precipitate from the boiling solution with caustic potash was a hydrate containing four equivalents of water. When fused with borax chromic salts give a green glass. The same coloration is communicated to ordinary glass by the presence of traces of chromic oxide. A chrome glass containing a large amount of chromic oxide may be ground up and used as a green pigment. Among the hydrates of oxide of chromium *Guignet's green* forms one of the widely-used green pigments which have been substituted for the poisonous arsenical copper pigments, such as Schweinfurt green, which formerly was much used. Guignet's green has an extremely bright green colour, and is distinguished for its great stability, not only under the action of light but also towards reagents; thus it is not altered by alkaline solutions, and even nitric acid does not act on it. This pigment remains unchanged up to a temperature of 250°; it contains $Cr_2O_3,2H_2O$, and generally a small amount of alkali. It is prepared by fusing 3 parts of boric acid with 1 part of potassium dichromate; oxygen is disengaged, and a green glass, containing a mixture of the borates of chromium and potassium, is obtained. When cool this glass is ground up and treated with water, which extracts the boric acid and alkali and leaves the above-named chromic hydroxide behind. This hydroxide only parts with its water at a red heat, leaving the anhydrous oxide.

The chromic hydroxides lose their water by ignition, and in so doing become spontaneously incandescent, like the ordinary ferric hydroxide (Chapter XXII.). It is not known, however, whether all the modifications of chromic oxide show this phenomenon. The anhydrous *chromic oxide*, Cr_2O_3, is exceedingly difficultly soluble in acids, if it has passed through the above recalescence. But if it has parted with its water, or the greater part of it, and not yet undergone this self-induced incandescence (has not lost a portion of its energy), then it is soluble in acids. It is not reduced by hydrogen. It is easily obtained in various crystalline forms by many methods. The chromates of mercury and ammonium give a very convenient method for its preparation, because when ignited they leave chromic oxide behind. In the first instance oxygen and mercury are disengaged, and in the second case nitrogen and water: $2Hg_2CrO_4 = Cr_2O_3 + O_5 + 4Hg$ or $(NH_4)_2Cr_2O_7 = Cr_2O_3 + 4H_2O + N_2$. The second reaction is very energetic, and the mass of salt burns spontaneously if the temperature be sufficiently high. A mixture of potassium sulphate and chromic oxide is formed by heating potassium dichromate with an equal weight of sulphur: $K_2Cr_2O_7 + S = K_2SO_4 + Cr_2O_3$. The sulphate is easily extracted by water, and there remains a bright green residue of the oxide, whose colour is more brilliant the lower the temperature of the decomposition. The oxide thus obtained is used as a green pigment for china and enamel. The anhydrous chromic oxide obtained from chromyl chloride, CrO_2Cl_2, has a specific gravity of 5·21, and forms almost black crystals, which give a green powder. They are hard enough to scratch glass, and have a metallic lustre. The crystalline form of chromic oxide is identical with that of the oxide of iron and alumina, with which it is isomorphous.

[7 bis] The most important of the compounds corresponding with chromic oxide is *chromic chloride*, Cr_2Cl_6, which is known in an anhydrous and in a hydrated form. It resembles ferric and aluminic chlorides in many respects. There is a great difference between the anhydrous and the hydrated chlorides; the former is insoluble in water, the latter easily dissolves, and on evaporation its solution forms a hygroscopic mass which is very unstable and easily evolves hydrochloric acid when heated with water. The anhydrous form is of a violet colour, and Wöhler gives the following method for its preparation: an intimate mixture is prepared of the anhydrous chromic oxide with carbon and organic matter, and charged into a wide infusible glass or porcelain tube which is heated in a combustion

furnace; one extremity of the tube communicates with an apparatus generating chlorine which is passed through several bottles containing sulphuric acid in order to dry it perfectly before it reaches the tube. On heating the portion of the tube in which the mixture is placed and passing chlorine through, a slightly volatile sublimate of chromic chloride, $CrCl_3$ or Cr_2Cl_6, is formed. This substance forms *violet tabular crystals*, which may be distilled in dry chlorine without change, but which, however, require a red heat for their volatilisation. These crystals are greasy to the touch and insoluble in water, but if they be powdered and boiled in water for a long time they pass into *a green solution*. Strong sulphuric acid does not act on the anhydrous salt, or only acts with exceeding slowness, like water. Even aqua regia and other acids do not act on the crystals, and alkalis only show a very feeble action. The specific gravity of the crystals is 2·99. When fused with sodium carbonate and nitre they give sodium chloride and potassium chromate, and when ignited in air they form green chromic oxide and evolve chlorine. On ignition in a stream of ammonia, chromic chloride forms sal-ammoniac and chromium nitride, CrN (analogous to the nitrides BN, AlN). Mosberg and Peligot showed that when chromic chloride is ignited in hydrogen, it parts with one-third of its chlorine, forming chromous chloride, $CrCl_2$—that is, there is formed from a compound corresponding with chromic oxide, Cr_2O_3, a compound answering to the *suboxide*, chromous oxide, CrO—just as hydrogen converts ferric chloride into ferrous chloride with the aid of heat. *Chromous chloride*, $CrCl_2$, forms colourless crystals easily soluble in water, which in dissolving evolve a considerable amount of heat, and form a blue liquid, capable of absorbing oxygen from the air with great facility, being converted thereby into a chromic compound.

The blue solution of chromous chloride may also be obtained by the action of metallic zinc on the green solution of the hydrated chromic chloride; the zinc in this case takes up chlorine just as the hydrogen did. It must be employed in a large excess. Chromic oxide is also formed in the action of zinc on chromic chloride, and if the solution remain for a long time in contact with the zinc the whole of the chromium is converted into chromic oxychloride. Other chromic salts are also reduced by zinc into *chromous salts*, CrX_2, just as the ferric salts FeX_3 are converted into ferrous salts FeX_2 by it. The chromous salts are exceedingly unstable and easily oxidise and pass into chromic salts; hence the reducing power of these salts is very great. From cupric salts they separate cuprous salts, from stannous salts they precipitate metallic tin, they reduce mercuric salts into mercurous and ferric into ferrous salts. Moreover, they absorb oxygen from the air directly. With potassium chromate they give a brown precipitate of chromium dioxide or of chromic oxide, according to the relative amounts of the substances taken: $CrO_3 + CrO = 2CrO_2$ or $CrO_3 + 3CrO = 2Cr_2O_3$.

Aqueous ammonia gives a blue precipitate, and in the presence of ammoniacal salts a blue liquid is obtained which turns red in the air from oxidation. This is accompanied by the formation of compounds analogous to those given by cobalt (Chapter XXII.). A solution of chromous chloride with a hot saturated solution of sodium acetate, $C_2H_3NaO_2$, gives, on cooling, transparent red crystals of chromous acetate, $C_4H_6CrO_4,H_2O$. This salt is also a powerful reducing agent, but may be kept for a long time in a vessel full of carbonic anhydride.

The insoluble anhydrous *chromic chloride* $CrCl_3$ very easily *passes into solution* in the presence of a trace (0·004) of *chromous chloride* $CrCl_2$. This remarkable phenomenon was observed by Peligot and explained by Löwel in the following manner: chromous chloride, as a lower stage of oxidation, is capable of absorbing both oxygen and chlorine, combining with various substances. It is able to decompose many chlorides by taking up chlorine from them; thus it precipitates mercurous chloride from a solution of mercuric chloride, and in so doing passes into chromic chloride: $2CrCl_2 + 2HgCl_2 = Cr_2Cl_6 + 2HgCl$. Let us suppose that the same phenomenon takes place when the anhydrous chromic chloride is mixed with a solution of chromous chloride. The latter will then take up a portion of the chlorine of the former, and pass into a soluble hydrate of chromic chloride (hydrochloride of oxide of chromium), and the original anhydrous chromic chloride will pass into chromous chloride. The chromous chloride re-formed in this manner will then act on a fresh quantity of the chromic chloride, and in this manner transfer it entirely into solution as hydrate. This view is confirmed by the fact that other chlorides, capable of absorbing chlorine like chromous chloride, also induce the solution of the insoluble chromic chloride—for example, ferrous chloride, $FeCl_2$, and cuprous chloride. The presence of zinc also aids the solution of chromic chloride, owing to its converting a portion of it into chromous chloride. The solution of chromic chloride in water obtained by these methods is perfectly identical with that which is formed by dissolving chromic hydroxide in hydrochloric acid. On evaporating the *green solution* obtained in this manner, it gives a green mass, containing water. On further heating it leaves a soluble chromic oxychloride, and when ignited it first forms an insoluble oxychloride and then chromic oxide; but no anhydrous chromic chloride, Cr_2Cl_6, is formed by heating the aqueous solution of chromic chloride, which forms an important fact in support of the view that the green solution of chromic chloride is nothing else but hydrochloride of oxide of chromium. At 100° the composition of the green hydrate is $Cr_2Cl_6,9H_2O$, and on evaporation at the ordinary temperature over H_2SO_4 crystals are obtained with 12 equivalents of water; the red mass obtained at 120° contains $Cr_2O_3,4Cr_2Cl_6,24H_2O$. The greater portion of it is soluble in water, like the mass which is formed at 150°. The latter contains

$Cr_2O_3,2Cr_2Cl_6,9H_2O = 3(Cr_2OCl_4,3H_2O)$—that is, it presents the same composition as chromic chloride in which one atom of oxygen replaces two of chlorine. And if the hydrate of chromic chloride be regarded as $Cr_2O_3,6HCl$, the substance which is obtained should be regarded as $Cr_2O_3,4HCl$ combined with water, H_2O. The addition of alkalis—for example, baryta—to a solution of chromic chloride immediately produces a precipitate, which, however, re-dissolves on shaking, owing to the formation of one of the oxychlorides just mentioned, which may be regarded as *basic salts*. Thus we may represent the product of the change produced on chromic chloride under the influence of water and heat by the following formulæ: first $Cr_2O_3,6HCl$ or $Cr_2Cl_6,3H_2O$ is formed, then $Cr_2O_3,4HCl,H_2O$ or $Cr_2OCl_4,3H_2O$, and lastly $Cr_2O_3,2HCl,2H_2O$ or $Cr_2O_2Cl_2,3H_2O$. In all three cases there are 2 equivalents of chromium to at least 3 equivalents of water. These compounds may be regarded as being intermediate between chromic hydroxide and chloride; chromic chloride is Cr_2Cl_6, the first oxychloride $Cr_2(OH)_2Cl_4$, the second $Cr_2(OH)_4Cl_2$, and the hydrate $Cr_2(OH)_6$—that is, the chlorine is replaced by hydroxyl.

It is very important to remark two circumstances in respect to this: (1) That the whole of the chlorine in the above compounds is not precipitated from their solutions by silver nitrate; thus the normal salt of the composition $Cr_2Cl_6,9H_2O$ only gives up two-thirds of its chlorine; therefore Peligot supposes that the normal salt contains the oxychloride combined with hydrochloric acid: $Cr_2Cl_6 + 2H_2O = Cr_2O_2Cl_2,4HCl$, and that the chlorine held as hydrochloric acid reacts with the silver, whilst that held in the oxychloride does not enter into reaction, just as we observe a very feebly-developed faculty for reaction in the anhydrous chromic chloride; and (2) if the green aqueous solution of $CrCl_3$ be left to stand for some time, it ultimately turns violet; in this form the whole of the chlorine is precipitated by $AgNO_3$, whilst boiling re-converts it into the green variety. Löwel obtained the violet solution of hydrochloride of chromic oxide by decomposing the violet chromic sulphate with barium chloride. Silver nitrate precipitates all the chlorine from this violet modification; but if the violet solution be boiled and so converted into the green modification, silver nitrate then only precipitates a portion of the chlorine.

Recoura (1890–1893) obtained a crystallohydrate of violet chromium sulphate, $Cr_2(SO_4)_3$, with 18 or 15 H_2O. By boiling a solution of this crystallohydrate, he converted it into the green salt, which, when treated with alkalis, gave a precipitate of $Cr_2O_3,2H_2O$, soluble in $2H_2SO_4$ (and not 3), and only forming the basic salt, $Cr_2(OH)_2(SO_4)_2$. He therefore concludes that the green salts are basic salts. The cryoscopic determinations made by A. Speransky (1892) and Marchetti (1892) give a greater 'depression' for the violet than the green salts, that is, indicate a greater molecular weight for the green salts. But as Étard, by heating the violet sulphate to 100°, converted it into a green salt of the same composition, but with a smaller amount of H_2O, it follows that the formation of a basic salt alone is insufficient to explain the difference between the green and violet varieties, and this is also shown by the fact that $BaCl_2$ precipitates the whole of the sulphuric acid of the violet salt, and only a portion of that of the green salt. A. Speransky also showed that the molecular electro-conductivity of the green solutions is less than that of the violet. It is also known that the passage of the former into the latter is accompanied by an increase of volume, and, according to Recoura, by an evolution of heat also.

Piccini's researches (1894) throw an important light upon the peculiarities of the green chromium trichloride (or chromic chloride); he showed (1) that AgF (in contradistinction to the other salts of silver) precipitates all the chlorine from an aqueous solution of the green variety; (2) that solutions of green $CrCl_3,6H_2O$ in ethyl alcohol and acetone precipitate all their chlorine when mixed with a similar solution of $AgNO_3$; (3) that the rise of the boiling-point of the ethyl alcohol and acetone green solutions of $CrCl_3,6H_2O$ (Chapter VII., Note 27 bis) shows that i in this case (as in the aqueous solutions of $MgSO_4$ and $HgCl_2$) is nearly equal to 1, that is, that they are like solutions of non-conductors; (4) that a solution of green $CrCl_3$ in methyl alcohol at first precipitates about $\frac{1}{8}$ of its chlorine (an aqueous solution about $\frac{2}{3}$) when treated with $AgNO_3$, but after a time the whole of the chlorine is precipitated; and (5) that an aqueous solution of the green variety gradually passes into the violet, while a methyl alcoholic solution preserves its green colour, both of itself and also after the whole of the chlorine has been precipitated by $AgNO_3$. If, however, in an aqueous or methyl alcoholic solution only a portion of the chlorine be precipitated, the solution gradually turns violet. In my opinion the general meaning of all these observations requires further elucidation and explanation, which should be in harmony with the theory of solutions. Recoura, moreover, obtained compounds of the green salt, $Cr_2(SO_4)_3$, with 1, 2, and 3 molecules of H_2SO_4, K_2SO_4, and even a compound $Cr_2(SO_4)_3H_2CrO_4$. By neutralising the sulphuric acid of the compounds of $Cr_2(SO_4)_3$ and H_2SO_4 with caustic soda, Recoura obtained an evolution of 33 thousand calories per each 2NaHO, while free H_2SO_4 only gives 30·8 thousand calories. Recoura is of opinion that special *chromo sulphuric acids*, for instance $(CrSO_4)H_2SO_4 = \frac{1}{2}Cr_2(SO_4)_3H_2SO_4$, are formed. With a still larger excess of sulphuric acid, Recoura obtained salts containing a still greater number of sulphuric acid radicles, but even this method does not explain the difference between the green and violet salts.

These facts must naturally be taken into consideration in order to arrive at any complete decision as to the cause of the different modifications of the chromic salts. We may observe that the green modification of chromic chloride does not give double salts with the metallic chlorides, whilst the violet variety forms compounds $Cr_2Cl_6,2RCl$ (where R = an alkali metal), which are obtained by heating the chromates with an excess of hydrochloric acid and evaporating the solution until it acquires a violet colour. As the result of all the existing researches on the green and violet chromic salts, it appears to me most probable that their difference is determined by the feeble basic character of chromic oxide, by its faculty of giving basic salts, and by the colloidal properties of its hydroxide (these three properties are mutually connected), and moreover, it seems to me that the relation between the green and violet salts of chromic oxide best answers to the relation of the purpureo to the luteo cobaltic salts (Chapter XXII., Note 35). This subject cannot yet be considered as exhausted (*see* Note 7).

We may here observe that with tin the chromic salts, CrX_3, give at low temperatures CrX_2 and SnX_2, whilst at high temperatures, on the contrary, CrX_2 reduces the metal from its salts SnX_2. The reaction, therefore, belongs to the class of reversible reactions (Beketoff).

Poulenc obtained anhydrous CrF_3 (sp. gr. 3·78) and CrF_2 (sp. gr. 4·11) by the action of gaseous HF upon $CrCl_2$. A solution of fluoride of chromium is employed as a mordant in dyeing. Recoura (1890) obtained green and violet varieties of $Cr_2Br_6,6H_2O$. The green variety can only be kept in the presence of an excess of HBr in the solution; if alone its solution easily passes into the violet variety with evolution of heat.

The reduction of metallic chromium proceeds with comparative ease in aqueous solutions. Thus the action of sodium amalgams upon a strong solution of Cr_2Cl_6 gives (first $CrCl_2$) an amalgam of chromium from which the mercury may be easily driven off by heating (in hydrogen to avoid oxidation), and there remains a spongy mass of easily oxidizable chromium. Plaset (1891), by passing an electric current through a solution of chrome alum mixed with a small amount of H_2SO_4 and K_2SO_4, obtained hard scales of chromium of a bluish-white colour possessing great hardness and stability (under the action of water, air, and acids). Glatzel (1890) reduced a mixture of $2KCl + Cr_2Cl_6$ by heating it to redness with shavings of magnesium. The metallic chromium thus obtained has the appearance of a fine light-grey powder which is seen to be crystalline under the microscope; its sp. gr. at 16° is 6·7284. It fuses (with anhydrous borax) only at the highest temperatures, and after fusion presents a silver-white fracture. The strongest magnet has no action upon it.

Moissan (1893) obtained chromium by reducing the oxide Cr_2O_3 with carbon in the electrical furnace (Chapter VIII., Note 17) in 9–10 minutes with a current of 350 ampères and 50 volts. The mixture of oxide and carbon gives a bright ingot weighing 100–110 grams. A current of 100 ampères and 50 volts completes the experiment upon a smaller quantity of material in 15 minutes; a current of 30 ampères and 50 volts gave an ingot of 10 grams in 30–40 minutes. The resultant carbon alloy is more or less rich in chromium (from 87·37–91·7 p.c.). To obtain the metal free from carbon, the alloy is broken into large lumps, mixed with oxide of chromium, put into a crucible and covered with a layer of oxide. This mixture is then heated in the electric furnace and the pure metal is obtained. This reduction can also be carried on with chrome iron ore $FeOCr_2O_3$ which occurs in nature. In this case a homogeneous alloy of iron and chromium is obtained. If this alloy be thrown in lumps into molten nitre, it forms insoluble sesquioxide of iron and a soluble alkaline chromate. This alloy of iron and chromium dissolved in molten steel (chrome steel) renders it hard and tough, so that such steel has many valuable applications. The alloy, containing about 3 p.c. Cr and about 1·3 p.c. carbon, is even harder than the ordinary kinds of tempered steel and has a fine granular fracture. The usual mode of preparing the ferrochromes for adding to steel is by fusing powdered chrome iron ore under fluxes in a graphite crucible.

[8 bis] The atomic composition of the tungsten and molybdenum compounds is taken as being identical with that of the compounds of sulphur and chromium, because (1) both these metals give two oxides in which the amounts of oxygen per given amount of metal stand in the ratio 2 : 3; (2) the higher oxide is of the latter kind, and, like chromic and sulphuric anhydrides, it has an acid character; (3) certain of the molybdates are isomorphous with the sulphates; (4) the specific heat of tungsten is 0·0334, consequently the product of the atomic weight and specific heat is 6·15, like that of the other elements—it is the same with molybdenum, $96·0 \times 0·0722 = 6·9$; (5) tungsten forms with chlorine not only compounds WCl_6, WCl_5, and $WOCl_4$, but also WO_2Cl_2, a volatile substance the analogue of chromyl chloride, CrO_2Cl_2, and sulphuryl chloride, SO_2Cl_2. Molybdenum gives the chlorine compounds, $MOCl_2$, $MOCl_3$(?), $MOCl_4$ (fuses at 194°, boils at 268°; according to Debray it contains $MOCl_5$), $MoOCl_4$, MoO_2Cl_2, and $MoO_2(OH)Cl$. The existence of tungsten hexachloride, WCl_6, is an excellent proof of the fact that the type SX_6 appears in the analogues of sulphur as in SO_3; (6) the vapour density accurately determined for the chlorine compounds $MoCl_4$, WCl_6, WCl_5, $WOCl_4$ (Roscoe) leaves no doubt as to the molecular composition of the compounds of tungsten and molybdenum, for the observed and calculated results entirely agree.

Tungsten is sometimes called scheele in honour of Scheele, who discovered it in 1781 and molybdenum in 1778. Tungsten is also known as wolfram; the former name was the name given to it by Scheele, because he extracted it from the mineral then known as tungsten and now called scheelite, CaWO₄. The researches of Roscoe, Blomstrand and others have subsequently thrown considerable light on the whole history of the compounds of molybdenum and tungsten.

The ammonium salts of tungsten and molybdic acids when ignited leave the anhydrides, which resemble each other in many respects. *Tungsten anhydride*, WO₃, is a yellowish substance, which only fuses at a strong heat, and has a sp. gr. of 6·8. It is insoluble both in water and acid, but solutions of the alkalis, and even of the alkali carbonates, dissolve it, especially when heated, forming alkaline salts. *Molybdic anhydride*, MoO₃, is obtained by igniting the acid (hydrate) or the ammonium salt, and forms a white mass which fuses at a red heat, and solidifies to a yellow crystalline mass of sp. gr. 4·4; whilst on further heating in open vessels or in a stream of air this anhydride *sublimes* in pearly scales—this enables it to be obtained in a tolerably pure state. Water dissolves it in small quantities—namely, 1 part requires 600 parts of water for its solution. The hydrates of molybdic anhydride are *soluble also in acids* (a hydrate, H₂MoO₄, is obtained from the nitric acid solution of the ammonium salt), which forms one of their distinctions from the tungstic acids. But after ignition molybdic anhydride is insoluble in acids, like tungstic anhydride; alkalis dissolve this anhydride, easily forming molybdates. Potassium bitartrate dissolves the anhydride with the aid of heat. None of the acids yet considered by us form so many different salts with one and the same base (alkali) as molybdic and tungstic acids. The composition of these salts, and their properties also, vary considerably. The most important discovery in this respect was made by Marguerite and Laurent, who showed that the salts which contain a large proportion of tungstic acid are easily soluble in water, and ascribed this property to the fact that tungstic acid may be obtained *in several states*. The common tungstates, obtained with an excess of alkali, have an alkaline reaction, and on the addition of sulphuric or hydrochloric acid first deposit an acid salt and then a hydrate of tungstic acid, which is insoluble both in water and acids; but if instead of sulphuric or hydrochloric acids, we add acetic or phosphoric acid, or if the tungstate be saturated with a fresh quantity of tungstic acid, which may be done by boiling the solution of the alkali salt with the precipitated tungstic acid, a solution is obtained which, on the addition of sulphuric or a similar acid, does not give a precipitate of tungstic acid at the ordinary or at higher temperatures. The solution then contains peculiar salts of tungstic acid, and if there be an excess of acid it also contains tungstic acid itself; Laurent, Riche, and others called it *metatungstic acid*, and it is still known by this name. Those salts which with acids immediately give the insoluble tungstic acid have the composition $R_2WO_4,RHWO_4$, whilst those which give the soluble metatungstic acid contain a far greater proportion of the acid elements. Scheibler obtained the (soluble) metatungstic acid itself by treating the soluble barium (meta) tetratungstate, $BaO,4WO_3$, with sulphuric acid. Subsequent research showed the existence of a similar phenomenon for molybdic acid. There is no doubt that this is a case of colloidal modifications.

Many chemists have worked on the various salts formed by molybdic and tungstic acids. The tungstates have been investigated by Marguerite, Laurent, Marignac, Riche, Scheibler, Anthon, and others. The molybdates were partially studied by the same chemists, but chiefly by Struvé and Svanberg, Delafontaine, and others. It appears that for a given amount of base the salts contain one to eight equivalents of molybdic or tungstic anhydride; *i.e.* if the base have the composition RO, then the highest proportion of base will be contained by the salts of the composition $ROWO_3$ or $ROMoO_3$—that is, by those salts which correspond with the normal acids H_2WO_4 and H_2MoO_4, of the same nature as sulphuric acid; but there also exist salts of the composition $RO,2WO_3$, $RO,3WO_3$ … $RO,8WO_3$. The water contained in the composition of many of the acid salts is often not taken into account in the above. The properties of the salts holding different proportions of acids vary considerably, but one salt may be converted into another by the addition of acid or base with great facility, and the greater the proportion of the elements of the acid in a salt, the more stable, within a certain limit, is its solution and the salt itself.

The most common ammonium molybdate has the composition $(NH_4HO)_6,H_2O,7MoO_3$ (or, according to Marignac and others, NH_4HMoO_4), and is prepared by evaporating an ammoniacal solution of molybdic acid. It is used in the laboratory for precipitating phosphoric acid, and is purified for this purpose by mixing its solution with a small quantity of magnesium nitrate, in order to precipitate any phosphoric acid present, filtering, and then adding nitric acid and evaporating to dryness. A pure ammonium molybdate free from phosphoric acid may then be extracted from the residue.

Phosphoric acid forms insoluble compounds with the oxides of uranium and iron, tin, bismuth, &c., having feeble basic and even acid properties. This perhaps depends on the fact that the atoms of hydrogen in phosphoric acid are of a very different character, as we saw above. Those atoms of hydrogen which are replaced with difficulty by ammonium, sodium, &c., are probably easily replaced by feebly energetic acid groups— that is, the formation of particular complex substances may be expected to take place at the expense of these atoms of the hydrogen of phosphoric acid and of certain feeble metallic acids; and these substances will still be acids, because the hydrogen of the phosphoric acids and metallic acids, which is easily replaced by metals, is not removed by their mutual combination, but remains in the resultant compound. Such a conclusion is verified in the *phosphomolybdic acids* obtained (1888) by Debray. If a solution of ammonium molybdate be acidified, and a small amount of a solution (it may be acid) containing orthophosphoric acid or its salts be added to it (so that there are at least 40 parts of molybdic acid present to 1 part of phosphoric acid), then after a period of twenty-four hours the whole of the phosphoric acid is separated as a yellow precipitate, containing, however, not more than 3 to 4 p.c. of phosphoric anhydride, about 3 p.c. of ammonia, about 90 p.c. of molybdic anhydride, and about 4 p.c. of water. The formation of this precipitate is so distinct and so complete that this method is employed for the discovery and separation of the smallest quantities of phosphoric acid. Phosphoric acid was found to be present in the majority of rocks by this means. The precipitate is soluble in ammonia and its salts, in alkalis and phosphates, but is perfectly insoluble in nitric, sulphuric, and hydrochloric acids in the presence of ammonium molybdate. The composition of the precipitate appears to vary under the conditions of its precipitation, but its nature became clear when the acid corresponding with it was obtained. If the above-described yellow precipitate be boiled in aqua regia, the ammonia is destroyed, and an acid is obtained in solution, which, when evaporated in the air, crystallises out in yellow oblique prisms of approximately the composition $P_2O_5,20MoO_3,26H_2O$. Such an unusual proportion of component parts is explained by the above-mentioned considerations. We saw above that molybdic acid easily gives salts $R_2OnMoO_3mH_2O$, which we may imagine to correspond to a hydrate $MoO_2(HO)_2nMo_3mH_2O$. And suppose that such a hydrate reacts on orthophosphoric acid, forming water and compounds of the composition $MoO_2(HPO_4)nMoO_3mH_2O$ or $MoO_3(H_2PO_4)_2nMoO_3mH_2O$; this is actually the composition of phosphomolybdic acid. Probably it contains a portion of the hydrogen replaceable by metals of both the acids H_3PO_4 and of H_2MoO_4. The crystalline acid above is probably $H_3MoPO_7,9MoO_3,12H_2O$. This acid is really tribasic, because its aqueous solution precipitates salts of potassium, ammonium, rubidium (but not lithium and sodium) *from acid solutions*, and gives a *yellow* precipitate of the composition $R_3MoPO_7,9MoO_3,8H_2O$, where R = NH_4. Besides these, salts of another composition may be obtained, as would be expected from the preceding. These salts are only stable in acid solutions (which is naturally due to their containing an excess of acid oxides), whilst under the action of alkalis they give *colourless* phosphomolybdates of the composition $R_3MoPO_3,MoO_2,3H_2O$. The corresponding salts of potassium, silver, ammonium, are easily soluble in water and crystalline.

Phosphomolybdic acid is an example of the *complex inorganic acids* first obtained by Marignac and afterwards generalised and studied in detail by Gibbs. We shall afterwards meet with several examples of such acids, and we will now turn attention to the fact that they are usually formed by weak polybasic acids (boric, silicic, molybdic, &c.), and in certain respects resemble the cobaltic and such similar complex compounds, with which we shall become acquainted in the following chapter. As an example we will here mention certain complex compounds containing molybdic and tungstic acids, as they will illustrate the possibility of a considerable complexity in the composition of salts. The action of ammonium molybdate upon a dilute solution of purpureocobaltic salts (*see* Chapter XXII.) acidulated with acetic acid gives a salt which after drying at 100° has the composition $Co_2O_310NH_37MoO_33H_2O$. After ignition this salt leaves a residue having the composition $2CoO_7MoO_3$. An analogous compound is also obtained for tungstic acid, having the composition $Co_2O_310NH_310WO_39H_2O$. In this case after ignition there remains a salt of the composition CoO_5WO_3 (Carnot, 1889). Professor Kournakoff, by treating a solution of potassium and sodium molybdates, containing a certain amount of suboxide of cobalt, with bromine obtained salts having the composition: $3K_2OCo_2O_312MoO_320H_2O$ (light green) and $3K_2OCo_2O_310Mo_310H_2O$ (dark green). Péchard (1893) obtained salts of the four complex phosphotungstic acids by evaporating equivalent mixtures of solutions of phosphoric acid and metatungstic acid (*see* further on): phosphotrimetatungstic acid $P_2O_512WO_348H_2O$, phosphotetrametatungstic acid $P_2O_516WO_369H_2O$, phosphopentametatungstic acid $P_2O_520WO_3H_2O$, and phosphohexametatungstic acid $P_2O_524WO_359H_2O$. Kehrmann and Frankel described still more complex salts, such as: $3Ag_2O_4BaOP_2O_522WO_3H_2O, 5BaO_2K_2OP_2O_322WO_348H_2O$. Analogous double salts with $22WO_3$ were also obtained with KSr, KHg, BaHg, and NH_4Pb. Kehrmann (1892) considers the possibility of obtaining an unlimited number of such salts to be a general characteristic of such compounds. Mahom and Friedheim (1892) obtained compounds of similar complexity for molybdic and arsenic acids.

For tungstic acid there are known: (1) Normal salts—for example, K_2WO_4; (2) the so-called acid salts have a composition like $3K_2O,7WO_3,6H_2O$ or $K_6H_8(WO_4)_7,2H_2O$; (3) the tritungstates like $Na_3O,3WO_3,3H_2O = Na_2H_4(WO_4)_3,H_2O$. All these three classes of salts are soluble in water, but are precipitated by barium chloride, and with acids in solution give an insoluble hydrate of tungstic acid; whilst those salts which are enumerated below do not give a precipitate either with acids or with the salts of the heavy metals, because they form soluble salts even with barium and lead. They are generally called metatungstates. They all contain water and a larger proportion of acid elements than the preceding salts; (4) the tetratungstates, like $Na_2O,4WO_3,10H_2O$ and $BaO,4WO_3,9H_2O$ for example; (5) the octatungstates—for example, $Na_2O,8WO_3,24H_2O$. Since the metatungstates lose so much water at 100° that they leave salts whose composition corresponds with an acid, $3H_2O,4WO_3$—that is, $H_6W_4O_{15}$—whilst in the meta salts only 2 hydrogens are replaced by metals, it is assumed, although without much ground, that these salts contain a particular soluble metatungstic acid of the composition $H_6W_4O_{15}$.

As an example we will give a short description of the sodium salts. The normal salt, Na_2WO_4, is obtained by heating a strong solution of sodium carbonate with tungstic acid to a temperature of 80°; if the solution be filtered hot, it crystallises in rhombic tabular crystals, having the composition $Na_2WO_4,2H_2O$, which remain unchanged in the air and are easily soluble in water. When this salt is fused with a fresh quantity of tungstic acid, it gives a ditungstate, which is soluble in water and separates from its solution in crystals containing water. The same salt is obtained by carefully adding hydrochloric acid to the solution of the normal salt so long as a precipitate does not appear, and the liquid still has an alkaline reaction. This salt was first supposed to have the composition $Na_2W_2O_7,4H_2O$, but it has since been found to contain (at 100°) $Na_6W_7O_{24},16H_2O$—that is, it corresponds with the similar salt of molybdic acid.

(If this salt be heated to a red heat in a stream of hydrogen, it loses a portion of its oxygen, acquires a metallic lustre, and turns a golden yellow colour, and, after being treated with water, alkali, and acid, leaves golden yellow leaflets and cubes which are very like gold. This very remarkable substance, discovered by Wöhler, has, according to Malaguti's analysis, the composition $Na_2W_3O_9$; that is, it, as it were, contains a double tungstate of tungsten oxide, WO_2, and of sodium, Na_2WO_4,WO_2WO_3. The decomposition of the fused sodium salt is best effected by finely-divided tin. This substance has a sp. gr. 6·6; it conducts electricity like metals, and like them has a metallic lustre. When brought into contact with zinc and sulphuric acid it disengages hydrogen, and it becomes covered with a coating of copper in a solution of copper sulphate in the presence of zinc—that is, notwithstanding its complex composition it presents to a certain extent the appearance and reactions of the metals. It is not acted on by aqua regia or alkaline solutions, but it is oxidised when ignited in air.)

The ditungstate mentioned above, deprived of water (having undergone a modification similar to that of metaphosphoric acid), after being treated with water, leaves an anhydrous, sparingly soluble tetratungstate, $Na_2WO_4,3WO_3$, which, when heated at 120° in a closed tube with water, passes into an easily soluble metatungstate. It may therefore be said that the metatungstates are hydrated compounds. On boiling a solution of the above-mentioned salts of sodium with the yellow hydrate of tungstic acid they give a solution of metatungstate, which is the hydrated tetratungstate. Its crystals contain $Na_2W_4O_{13},10H_2O$. After the hydrate of tungstic acid (obtained from the ordinary tungstates by precipitation with an acid) has stood a long time in contact with a solution (hot or cold) of sodium tungstate, it gives a solution which is not precipitated by hydrochloric acid; this must be filtered and evaporated over sulphuric acid in a desiccator (it is decomposed by boiling). It first forms a very dense solution (aluminium floats in it) of sp. gr. 3·0, and octahedral crystals of *sodium metatungstate*, $Na_2W_4O_{13},10H_2O$, sp. gr. 3·85, then separate. It effloresces and loses water, and at 100° only two out of the ten equivalents of water remain, but the properties of the salt remain unaltered. If the salt be deprived of water by further heating, it becomes insoluble. At the ordinary temperature one part of water dissolves ten parts of the metatungstate. The other metatungstates are easily obtained from this salt. Thus a strong and hot solution, mixed with a like solution of barium chloride, gives on cooling crystals of barium metatungstate, $BaW_4O_{13},9H_2O$. These crystals are dissolved without change in water containing hydrochloric acid, and also in hot water, but they are partially decomposed by cold water, with the formation of a solution of metatungstic acid and of the normal barium salt $BaWO_4$.

In order to explain the difference in the properties of the salts of tungstic acid, we may add that a mixture of a solution of tungstic acid with a solution of silicic acid does not coagulate when heated, although the silicic acid alone would do so; this is due to the formation of a silicotungstic acid, discovered by Marignac, which presents a fresh example of a complex acid. A solution of a tungstate dissolves gelatinous silica, just as it does gelatinous tungstic acid, and when evaporated deposits a crystalline salt of silicotungstic acid. This solution is not precipitated either by acids (a clear analogy to the metatungstates) or by sulphuretted hydrogen, and corresponds with a series of salts. These salts contain one equivalent of silica and 8 equivalents of hydrogen or metals, in the same form as in salts, to 12 or 10 equivalents of tungstic anhydride; for example the crystalline potassium salt has the composition $K_8W_{12}SiO_{42},14H_2O =$ $4K_2O,12WO_3,SiO_2,14H_2O$. Acid salts are also known in which half of the metal is replaced by hydrogen. The complexity of the composition of such complex acids (for example, of the phosphomolybdic acid) involuntarily leads to the idea of polymerisation, which we were obliged to recognise for silica, lead oxide, and other compounds. This polymerisation, it seems to me, may be understood thus: a hydrate A (for example, tungstic acid) is capable of combining with a hydrate B (for example, silica or phosphoric acid, with or without the disengagement of water), and by reason of this faculty it is capable of polymerisation—that is, A combines with A—combines with itself—just as aldehyde, C_2H_4O, or the cyanogen compounds are able to combine with hydrogen, oxygen, &c., and are liable to polymerisation. On this view the molecule of tungstic acid is probably much more complex than we represent it; this agrees with the easy volatility of such compounds as the chloranhydrides, CrO_2Cl_2, MoO_2Cl_2, the analogues of the volatile sulphuryl chloride, SO_2Cl_2, and with the non-volatility, or difficult volatility, of chromic and molybdic anhydrides, the analogues of the volatile sulphuric anhydride. Such a view also finds a certain confirmation in the researches made by Graham on the *colloidal* state of tungstic acid, because colloidal properties only appertain to compounds of a very complex composition. The observations made by Graham on the colloidal state of tungstic and molybdic acids introduced much new matter into the history of these substances. When sodium tungstate, mixed in a dilute solution with an equivalent quantity of dilute hydrochloric acid, is placed in a dialyser, hydrochloric acid and sodium chloride pass through the membrane, and a solution of tungstic acid remains in the dialyser. Out of 100 parts of tungstic acid about 80 parts remain in the dialyser. The solution has a bitter, astringent taste, and does not yield gelatinous tungstic acid (hydrogel) either when heated or on the addition of acids or salts. It may also be evaporated to dryness; it then forms a vitreous mass of the *hydrosol* of *tungstic acid*, which adheres strongly to the walls of the vessel in which it has been evaporated, and is perfectly soluble in water. It does not even lose its solubility after having been heated to 200°, and only becomes insoluble when heated to a red heat, when it loses about 2½ p.c. of water. The dry acid, dissolved in a small quantity of water, forms a gluey mass, just like gum arabic, which is one of the representatives of the hydrosols of colloidal substances. The solution, containing 5 p.c of the anhydride, has a sp. gr. of 1·047; with 20 p.c., of 1·217; with 50 p.c., of 1·80; and with 80 p.c., of 3·24. The presence of a polymerised trioxide in the form of hydrate, $H_2OW_3O_9$ or $H_2O_4WO_3$, must then be recognised in the solution: this is confirmed by Sabaneeff's cryoscopic determinations (1889). A similar stable solution of molybdic acid is obtained by the dialysis of a mixture of a strong solution of sodium molybdate with hydrochloric acid (the precipitate which is formed is re-dissolved). If $MoCl_4$ be precipitated by ammonia and washed with water, a point is reached at which perfect solution takes place, and the molybdic acid forms a colloid solution which is precipitated by the addition of ammonia (Muthmann). The addition of alkali to the solutions of the hydrosols of tungstic and molybdic acids immediately results in the re-formation of the ordinary tungstates and molybdates. There appears to be no doubt but that the same transformation is accomplished in the passage of the ordinary tungstates into the metatungstates as takes place in the passage of tungstic acid itself from an insoluble into a soluble state; but this may be even actually proved to be the case, because Scheibler obtained a solution of tungstic acid, before Graham, by decomposing barium metatungstate ($BaO_4WO_3,9H_2O$) with sulphuric acid. By treating this salt with sulphuric acid in the amount required for the precipitation of the baryta, Scheibler obtained a solution of metatungstic acid which, when containing 43·75 p.c. of acid, had a sp. gr. of 1·634, and with 27·61 p.c. a sp. gr. of 1·327—that is, specific gravities corresponding with those found by Graham.

Péchard found that as much heat is evolved by neutralising metatungstic acid as with sulphuric acid.

Questions connected with the metamorphoses or modifications of tungstic and molybdic acids, and the polymerisation and colloidal state of substances, as well as the formation of complex acids, belong to that class of problems the solution of which will do much towards attaining a true comprehension of the mechanism of a number of chemical reactions. I think, moreover, that questions of this kind stand in intimate connection with the theory of the formation of solutions and alloys and other so-called indefinite compounds.

Moissan (1893) studied the compounds of Mo and W formed with carbon in the electrical furnace (they are extremely hard) from a mixture of the anhydrides and carbon. Poleck and Grützner obtained definite compounds FeW_2 and FeW_2C_3 for tungsten. Metallic W and Mo displace Ag from its solutions but not Pb. There is reason for believing that the sp. gr. of pure molybdenum is higher than that (8·5) generally ascribed to it.

[9 bis] We may conclude our description of tungsten and molybdenum by stating that their sulphur compounds have an acid character, like carbon bisulphide or stannic sulphide. If sulphuretted hydrogen be passed through a solution of a molybdate it does not give a precipitate unless sulphuric acid be present, when a dark brown precipitate of *molybdenum trisulphide*, MoS_3, is formed. When this sulphide is ignited without access of air it gives the bisulphide MoS_2; the latter is not able to combine with potassium sulphide like the trisulphide MoS_3, which forms a salt, K_2MoS_4, corresponding with K_2MoO_4. This is soluble in water, and separates out from its solution in red crystals, which have a metallic lustre and reflect a green light. It is easily obtained by heating the native bisulphide, MoS_2, with potash, sulphur, and a small amount of charcoal, which serves for deoxidising the oxygen compounds. Tungsten gives similar compounds, R_2WS_4, where R = NH_4, K, Na. They are decomposed by acids, with the separation of tungsten trisulphide, WS_3, and molybdenum trisulphide, MoS_3. Rideal (1892) obtained W_2N_3 by heating WO_3 in NH_3. This compound exhibited the general properties of metallic nitrides.

[9 tri] When peroxide of hydrogen acts upon a solution of potassium molybdate well-formed yellow crystals belonging to the triclinic system separate out in the cold. When these crystals are heated in vacuo they first lose water and then decompose, leaving a residue composed of the salt originally taken. They are soluble in water but insoluble in alcohol. Their composition is represented by the formula $K_2Mo_2O_8 2H_2O$. An ammonium salt is obtained by evaporating peroxide of hydrogen with ammonium molybdate. The following salts have also been obtained by the action of peroxide of hydrogen upon the corresponding molybdates: $Na_2Mo_2O_6 6H_2O$—in yellow prismatic crystals; $MgMo_2O_8 10H_2O$—stellar needles; $BaMoO_8 2H_2O$—in microscopic yellow octahedra. A corresponding sodium pertungstate has been obtained by Péchard by boiling sodium tungstate with a solution of peroxide of hydrogen for several minutes. The solution rapidly turns yellow, and no longer gives a precipitate of tungstic anhydride when treated with nitric acid. When evaporated in vacuo the solution leaves a thick syrupy liquid from which ray-like crystals separate out; these crystals are more soluble in water than the salt originally taken. When heated they also lose water and oxygen. Their composition answers to the formula $M_2W_2O_8 2H_2O$, where M = Na, NH_4, &c. The permolybdates and pertungstates have similar properties. When treated with oxygen acids they give peroxide of hydrogen, and disengage chlorine and iodine from hydrochloric acid and potassium iodide.

Piccini (1891) showed that peroxide of hydrogen not only combines with the oxygen compounds of Mo and W, but also with their fluo- compounds, among which ammonium fluo-molybdate $MoO_2F_2 2NH_4$ and others have long been known. (A few new salts of similar composition have been obtained by F. Moureu in 1893.) The action of peroxide of hydrogen upon these compounds gives salts containing a larger amount of oxygen; for instance, a solution of $MoO_2F_2 2KFH_2O$ with peroxide of hydrogen gives a yellow solution which after cooling separates out yellow crystalline flakes of $MoO_3F_2 2KFH_2O$, resembling the salt originally taken in their external appearance. By employing a similar method Piccini also obtained: $MoO_3F_2 2RbFH_2O$—yellow monoclinic crystals; $MoO_3F_2, 2CsFH_2O$,—yellow flakes, and the corresponding tungstic compounds. All these salts react like peroxide of hydrogen.

In speaking of these compounds I for my part think it may be well to call attention to the fact that, in the first place, the composition of Piccini's oxy-fluo compounds does not correspond to that of permolybdic and pertungstic acid. If the latter be expressed by formulæ with one equivalent of an element, they will be $HMoO_4$ and HWO_4, and the oxy-fluo form corresponding to them should have the composition MoO_3F and WO_3F

while it contains MO_3F_2 and WO_3F_2, *i.e.* answers as it were to a higher degree of oxidation, MoH_2O_3 and W_3HO_3. But if permolybdic acid be regarded as $2MoO_3 + H_2O_2$, *i.e.* as containing the elements of peroxide of hydrogen, then Piccini's compound will also be found to contain the original salts + H_2O; for example, from $MoO_2F_22KFH_2O$ there is obtained a compound $MoO_2F_22KFH_2O_2$, *i.e.* instead of H_2O they contain H_2O_2. In the second place the capacity of the salts of molybdenum and tungsten to retain a further amount of oxygen or H_2O_2 probably bears some relation to their property of giving complex acids and of polymerising which has been considered in Note 8 bis. There is, however, a great chemical interest in the accumulation of data respecting these high peroxide compounds corresponding to molybdic and tungstic acids. With regard to the peroxide form of uranium, *see* Chapter XX., Note 66.

Uranium trioxide, or uranic oxide, shows its feeble basic and acid properties in a great number of its reactions. (1) Solutions of uranic salts give yellow precipitates with alkalis, but these precipitates do not contain the hydrate of the oxide, but compounds of it with bases; for example, $2UO_2(NO_3)_2 + 6KHO = 4KNO_3 + 3H_2O + K_2U_2O_7$. There are other *urano-alkali compounds* of the same constitution; for example, $(NH_4)_2U_2O_7$ (known commercially as uranic oxide), MgU_2O_7, BaU_2O_7. They are the analogues of the dichromates. Sodium uranate is the most generally used under the name of uranium yellow, $Na_2U_2O_7$. It is used for imparting the characteristic yellow-green tint to glass and porcelain. Neither heat nor water nor acids are able to extract the alkali from sodium uranate, $Na_2U_2O_7$, and therefore it is a true insoluble salt, of a yellow colour, and clearly indicates the acid character (although feeble) of uranic oxide. (2) The carbonates of the alkaline earths (for instance, barium carbonate) precipitate uranic oxide from its salts, as they do all the salts of feeble bases; for example, R_2O_3. (3) The *alkaline carbonates*, when added to solutions of uranic salts, give a *precipitate, which is soluble in an excess of the reagent*, and particularly so if the acid carbonates be taken. This is due to the fact that (4) the uranyl salts *easily form double salts* with the salts of the alkali metals, including the salts of ammonium. Uranium, in the form of these double salts, often gives salts of well-defined crystalline form, although the simple salts are little prone to appear in crystals. Such, for example, are the salts obtained by dissolving potassium uranate, $K_2U_2O_7$, in acids, with the addition of potassium salts of the same acids. Thus, with hydrochloric acid and potassium chloride a well-formed crystalline salt, $K_2(UO_2)Cl_4,2H_2O$, belonging to the monoclinic system, is produced. This salt decomposes in dissolving in pure water. Among these double salts we may mention the double carbonate with the alkalis, $R_4(UO_2)(CO_3)_3$ (equal to $2R_2CO_3 + UO_2CO_3$); the acetates, $R(UO_2)(C_2H_3O_2)_3$—for instance, the sodium salt, $Na(UO_2)(C_2H_3O_2)_3$, and the potassium salt, $K(UO_2)(C_2H_3O_2)_3,H_2O$; the sulphates, $R_2(UO_2)(SO_4)_3,2H_2O$, &c. In the preceding formula R = K, Na, NH_4, or R_2 = Mg, Ba, &c. *This property of giving comparatively stable double salts indicates feebly developed basic properties*, because double salts are mainly formed by salts of distinctly basic metals (these form, as it were, the basic element of a double salt) and salts of feebly energetic bases (these form the acid element of a double salt), just as the former also give acid salts; the acid of the acid salts is replaced in the double salts by the salt of the feebly energetic base, which, like water, belongs to the class of intermediate bases. For this reason barium does not give double salts with alkalis as magnesium does, and this is why double salts are more easily formed by potassium than by lithium in the series of the alkali metals. (5) The most remarkable property, proving the feeble energy of uranic oxide as a base, is seen in the fact that when their composition is compared with that of other salts those of uranic oxide *always appear as basic salts*. It is well known that a normal salt, R_2X_6, corresponds with the oxide R_2O_3, where X = Cl, NO_3, &c., or X_2 = SO_4, CO_3, &c.; but there also exist basic salts of the same type where X = HO or X_2 = O. We saw salts of all kinds among the salts of aluminium, chromium, and others. With uranic oxide no salts are known of the types UX_6 (UCl_6, $U(SO_4)_3$, alums, &c., are not known), nor even salts, $U(HO)_2X_4$ or UOX_4, but it always forms salts of the type $U(HO)_4X_2$, or UO_2X_2. Judging from the fact that nearly all the salts of uranic oxide retain water in crystallising from their solutions, and that this water is difficult to separate from them, it may be thought to be water of hydration. This is seen in part from the fact that the composition of many of the salts of uranic oxide may then be expressed without the presence of water of crystallisation; for instance, $U(HO)_4K_2Cl_4$ (and the salt of NH_4, $U(HO)_4K_2(SO_4)_2$, $U(HO)_4(C_2H_3O_2)_2$. Sodium uranyl acetate however does not contain water.

Uranyl nitrate, or uranium nitrate, $UO_2(NO_3)_2,6H_2O$, crystallises from its solutions in transparent yellowish-green prisms (from an acid solution), or in tabular crystals (from a neutral solution), which effloresce in the air and are easily soluble in water, alcohol, and ether, have a sp. gr. of 2·8, and fuse when heated, losing nitric acid and water in the process. If the salt itself (Berzelius) or its alcoholic solution (Malaguti) be heated up to the temperature at which oxides of nitrogen are evolved, there then remains a mass which, after being evaporated with water, leaves uranyl hydroxide, $UO_2(HO)_2$ (sp. gr. 5·93), whilst if the salt be ignited there remains the dioxide, UO_2, as a brick-red powder, which on further heating loses oxygen and forms the dark olive uranoso-uranic oxide, U_3O_8. The solution of the nitrate obtained from the ore is purified in the following manner: sulphurous anhydride is first passed through it in order to reduce the arsenic acid present into arsenious acid; the solution is then heated to 60°, and sulphuretted hydrogen passed through it; this precipitates the lead, arsenic, and tin, and certain other metals, as sulphides, insoluble in water and dilute nitric acid. This liquid is then filtered and evaporated with nitric acid to crystallisation, and the crystals are dissolved in ether. Or else the solution is first treated with chlorine in order to convert the ferrous chloride (produced by the action of the hydrogen sulphide) into ferric chloride, the oxides are then precipitated by ammonia, and the resultant precipitate, containing the oxides Fe_2O_3, UO_3, and compounds of the latter with potash, lime, ammonia, and other bases present in the solution (the latter being due to the property of uranic oxide of combining with bases), is washed and dissolved in a strong, slightly-heated solution of ammonium carbonate, which dissolves the uranic oxide but not the ferric oxide. The solution is filtered, and on cooling deposits a well-crystallising *uranyl ammonium carbonate*, $UO_2(NH_4)_4(CO_3)_3$, in brilliant monoclinic crystals which on exposure to air slowly give off water, carbonic anhydride, and ammonia; the same decomposition is readily effected at 300°, the residue then consisting of uranic oxide. This salt is not very soluble in water, but is readily so in ammonium carbonate; it is obvious that it may readily be converted into all the other salts of oxides of uranium. Uranium salts are also purified in the form of *acetate*, which is very sparingly soluble, and is therefore directly precipitated from a strong solution of the nitrate by mixing it with acetic acid.

We may also mention the *uranyl phosphate*, $HUPO_6$, which must be regarded as an orthophosphate in which two hydrogens are replaced by the radicle uranyl, UO_2, *i.e.* as $H(UO_2)PO_4$. This salt is formed as a hydrated gelatinous yellow precipitate, on mixing a solution of uranyl nitrate with disodium phosphate. The precipitation occurs in the presence of acetic acid, but not in the presence of hydrochloric acid. If moreover an excess of an ammonium salt be present, the ammonia enters into the composition of the bright yellow gelatinous precipitate formed, in the proportion $UO_2NH_4PO_4$. This precipitate is not soluble in water and acetic acid, and its solution in inorganic acids when boiled entirely expels all the phosphoric acid. This fact is taken advantage of for removing phosphoric acids from solutions—for instance, from those containing salts of calcium and magnesium.

Uranium dioxide, or *uranyl*, UO_2, which is contained in the salts UO_2X_2, has the appearance and many of the properties of a metal. Uranic oxide may be regarded as uranyl oxide, $(UO_2)O$, its salts as salts of this uranyl; its hydroxide, $(UO_2)H_2O_2$, is constituted like CaH_2O_2. The green oxide of uranium, uranoso-uranic oxide (easily formed from uranic salts by the loss of oxygen), $U_3O_8 = UO_2,2UO_3$, when ignited with charcoal or hydrogen (dry) gives a brilliant crystalline substance of sp. gr. about 11·0 (Urlaub), whose appearance resembles that of metals, and decomposes steam at a red heat with the evolution of hydrogen; it does not, however, decompose hydrochloric or sulphuric acid, but is oxidised by nitric acid. The same substance (i.e. uranium dioxide UO_2) is also obtained by igniting the compound $(UO_2)K_2Cl_4$ in a stream of hydrogen, according to the equation $UO_2K_4Cl_4 + H_2 = UO_2 + 2HCl + 2KCl$. It was at first regarded as the metal. In 1841 Peligot found that it contained oxygen, because carbonic oxide and anhydride were evolved when it was ignited with charcoal in a stream of chlorine, and from 272 parts of the substance which was considered to be metal he obtained 382 parts of a volatile product containing 142 parts of chlorine. From this it was concluded that the substance taken contained an equivalent amount of oxygen. As 142 parts of chlorine correspond with 32 parts of oxygen, it followed that 272 - 32 = 240 parts of metal were combined in the substance with 32 parts of oxygen, and also in the chlorine compound obtained with 142 parts of chlorine. These calculations have been made for the now accepted atomic weight of uranium (U = 240, *see* Note 14). Peligot took another atomic weight, but this does not alter the principle of the argument.

Uranium tetrachloride, uranous chloride, UCl_4, corresponds with uranous oxide as a base. It was obtained by Peligot by igniting uranic oxide mixed with charcoal in a stream of *dry* chlorine: $UO_3 + 3C + 2Cl_2 = UCl_4 + 3CO$. This green volatile compound (Note 12) crystallises in regular octahedra, is very hygroscopic, easily soluble in water, with the development of a considerable amount of heat, and no longer separates out from its solution in an anhydrous state, but disengages hydrochloric acid when evaporated. The solution of uranous chloride in water is green. It is also formed by the action of zinc and copper (forming cuprous chloride) on a solution of uranyl chloride, UO_2Cl_2, especially in the presence of hydrochloric acid and sal-ammoniac. Solutions of uranyl salts are converted into uranous salts by the action of various reducing agents, and among others by organic substances or by the action of light, whilst the salts UX_4 are converted into uranyl salts, UO_2X_2, by exposure to air or by oxidising agents. Solutions of the green uranyl salts act as powerful reducing agents, and give a brown precipitate of the uranous hydroxide, UH_4O_4, with potash and other alkalis. This hydroxide is easily soluble in acids but not in alkalis. On ignition it does not form the oxide UO_2, because it decomposes water, but when the higher oxides of uranium are ignited in a stream of hydrogen or with charcoal they yield uranous oxide. Both it and the chloride UCl_4, dissolve in strong sulphuric acid, forming a green salt, $U(SO_4)_2,2H_2O$. The same salt, together with uranyl sulphate, $UO_2(SO_4)$, is formed when the green oxide, U_3O_8, is dissolved in hot sulphuric acid. The salts obtained in the latter instance may be separated by adding alcohol to the solution, which is left exposed to the light; the alcohol reduces the uranyl salt to uranous salt, an excess of acid being required. An excess of water decomposes this salt, forming a basic salt, which is also easily produced under other circumstances, and contains $UO(SO_4),2H_2O$ (which corresponds to the uranic salt).

The atomic weight of uranium was formerly taken as half the present one, U = 120, and the oxides U_2O_3, suboxide UO, and green oxide U_3O_4, were of the same types as the oxides of iron. With a certain resemblance to the elements of the iron group, uranium presents many points of distinction which do not permit its being grouped with them. Thus uranium forms a very stable oxide, U_2O_3(U = 120), but does not give the corresponding chloride U_2Cl_6 (Roscoe, however, in 1874 obtained UCl_5, like $MoCl_5$ and WCl_5), and under those circumstances (the ignition of oxide of uranium mixed with charcoal, in a stream of chlorine), when the formation of this compound might be expected, it gives (U = 120) the chloride UCl_2, which is characterised by its volatility; this is not a property, to such an extent, of any of the bichlorides, RCl_2, of the iron group.

The alteration or doubling of the atomic weight of uranium—*i.e.* the recognition of U = 240—was made for the first time in the first (Russian) edition of this work (1871), and in my memoir of the same year in Liebig's *Annalen*, on the ground that with an atomic weight 120, uranium could not be placed in the periodic system. I think it will not be superfluous to add the following remarks on this subject: (1) In the other groups (K—Rb—Cs, Ca—Sr—Ba, Cl—Br—I) the acid character of the oxides decreases and their basic character increases with the rise of atomic weight, and therefore we should expect to find the same in the group Cr—Mo—W—U, and if CrO_3, MoO_3, WO_3 be the anhydrides of acids then we indeed find a decrease in their acid character, and therefore uranium trioxide, UO_3, should be a very feeble anhydride, but its basic properties should also be very feeble. Uranic oxide does indeed show these properties, as was pointed out above (Note 10). (2) Chromium and its analogues, besides the oxides RO_3, also form lower grades of oxidation RO_2, R_2O_3, and the same is seen in uranium; it forms UO_3, UO_2, U_2O_3 and their compounds. (3) Molybdenum and tungsten, in being reduced from RO_3, easily and frequently give an intermediate oxide of a blue colour, and uranium shows the same property; giving the so-called green oxide which, according to present views, must be regarded as $U_3O_8 = UO_2 2UO_3$, analogous to Mo_3O_8. (4) The higher chlorides, RCl_6, possible for the elements of this group, are either unstable (WCl_6) or do not exist at all (Cr); but there is one single lower volatile compound, which is decomposed by water, and liable to further reduction into a non-volatile chlorine product and the metal. The same is observed in uranium, which forms an easily volatile chloride, UCl_4, decomposed by water. (5) The high sp. gr. of uranium (18·6) is explained by its analogy to tungsten (sp. gr. 19·1). (6) For uranium, as for chromium and tungsten, yellow tints predominate in the form RO_3, whilst the lower forms are green and blue. (7) Zimmermann (1881) determined the vapour densities of uranous bromide, UBr_4, and chloride, UCl_4 (19·4 and 13·2), and they were found to correspond to the formulæ given above—that is, they confirmed the higher atomic weight $U = 240$. Roscoe, a great authority on the metals of this group, was the first to accept the proposed atomic weight of uranium, $U = 240$, which since Zimmermann's work has been generally recognised.

Uranium glass, obtained by the addition of the yellow salt $K_2U_2O_7$ to glass, has a green yellow fluorescence, and is sometimes employed for ornaments; it absorbs the violet rays, like the other salts of uranic oxide—that is, it possesses an absorption spectrum in which the violet rays are absent. The index of refraction of the absorbed rays is altered, and they are given out again as greenish-yellow rays; hence, compounds of uranic acid, when placed in the violet portion of the spectrum, emit a greenish-yellow light, and this forms one of the best examples (another is found in a solution of quinine sulphate) of the phenomenon of fluorescence. The rays of light which pass through uranic compounds do not contain the rays which excite the phenomena of fluorescence and of chemical transformation, as the researches of Stokes prove.

The comparison of potassium permanganate with potassium perchlorate, or of potassium manganate with potassium sulphate, shows directly that many of the physical and chemical properties of substances do not depend on the nature of the elements, but on the atomic types in which they appear, on the kind of movements, or on the positions in which the atoms forming the molecule occur.

If, however, we compare the spectra (Vol. I. p. 565) of chlorine, bromine, and iodine with that of manganese, a certain resemblance or analogy is to be found connecting manganese both to iron and to chlorine, bromine, and iodine.

The name 'peroxide' should only be retained for those *highest* oxides (and MnO_2 stands between MnO and MnO_3) which either by a direct method of double decomposition are able to give hydrogen peroxide or contain a larger proportion of oxygen than the base or the acid, just as hydrogen peroxide contains more oxygen than water. Their type will be H_2O_2, and they are exemplified by barium peroxide, BaO_2, and sulphur peroxide, S_2O_7, &c. Such a dioxide as MnO_2 is, in all probability, a salt—that is, a manganous manganate, MnO_3MnO, and also, as a basic salt of a feeble base, capable of combining with alkalis and acids. Hence the name of manganese peroxide should be abandoned, and replaced by manganese dioxide. PbO_2 is better termed lead dioxide than peroxide. Bisulphide of manganese, MnS_2, corresponding to iron pyrites, FeS_2, sometimes occurs in nature in fine octahedra (and cube combinations), for instance, in Sicily; it is called Hauerite.

[18 bis] On comparing the manganates with the permanganates—for example, K_2MnO_4 with $KMnO_4$—we find that they differ in composition by the abstraction of one equivalent of the metal. Such a relation in composition produced by oxidation is of frequent occurrence—for instance, $K_4Fe(CN)_6$ in oxidising gives $K_3Fe(CN)_6$; H_2SO_4 in oxidising gives persulphuric acid, HSO_4, or $H_2S_7O_8$; H_2O forms HO or H_2O_2, &c.

In the preparation of oxygen from the dioxide by means of H_2SO_4, $MnSO_4$ is formed; in the preparation of chlorine from HCl and MnO_2, $MnCl_2$ is obtained. These two manganous salts may be taken as examples of compounds MnX_2. Manganous sulphate generally contains various impurities, and also a large amount of iron salt (from the native MnO_2), from which it cannot be freed by crystallisation. Their removal may, however, be effected by mixing a portion of the liquid with a solution of sodium carbonate; a precipitate of manganous carbonate is then formed. This precipitate is collected and washed, and then added to the remaining mass of the impure solution of manganous sulphate; on heating the solution with this precipitate, the whole of the iron is precipitated as oxide. This is due to the fact that in the solution of the manganese dioxide in sulphuric acid the whole of the iron is converted into the ferric state (because the dioxide acts as an oxidising agent), which, as an exceedingly feeble base precipitated by calcium carbonate and other kindred salts, is also precipitated by manganous carbonate. After being treated in this manner, the solution of manganous sulphate is further purified by crystallisation. If it be a bright red colour, it is due to the presence of higher grades of oxidation of manganese; they may be destroyed by boiling the solution, when the oxygen from the oxides of manganese is evolved and a very faintly coloured solution of manganous sulphate is obtained. This salt is remarkable for the facility with which it gives various combinations with water. By evaporating the almost colourless solution of *manganous sulphate* at very low temperatures, and by cooling the saturated solution at about 0°, crystals are obtained containing 7 atoms of water of crystallisation, $MnSO_4,7H_2O$, which are isomorphous with cobaltous and ferrous sulphates. These crystals, even at 10°, lose 5 p.c. of water, and completely effloresce at 15°, losing about 20 p.c. of water. By evaporating a solution of the salt at the ordinary temperature, but not above 20°, crystals are obtained containing 5 mol. H_2O, which are isomorphous with copper sulphate; whilst if the crystallisation be carried on between 20° and 30°, large transparent prismatic crystals are formed containing 4 mol. H_2O (see Nickel). A boiling solution also deposits these crystals together with crystals containing 3 mol. H_2O, whilst the first salt, when fused and boiled with alcohol, gives crystals containing 2 mol. H_2O. Graham obtained a monohydrated salt by drying the salt at about 200°. The last atom of water is eliminated with difficulty, as is the case with all salts like $MnSO_4 nH_2O$. The crystals containing a considerable amount of water are rose-coloured, and the anhydrous crystals are colourless. The solubility of $MnSO_4,4H_2O$ (Chapter I., Note 24) per 100 parts of water is: at 10°, 127 parts; at 37°·5, 149 parts; at 75°, 145 parts; and at 101°, 92 parts. Whence it is seen that at the boiling-point this salt is less soluble than at lower temperatures, and therefore a solution saturated at the ordinary temperature becomes turbid when boiled. Manganous sulphate, being analogous to magnesium sulphate, is decomposed, like the latter, when ignited, but it does not then leave manganous oxide, but the intermediate oxide, Mn_3O_4. It gives double salts with the alkali sulphates. With aluminium sulphate it forms fine radiated crystals, whose composition resembles that of the alums—namely, $MnAl_2(SO_4)_4,24H_2O$. This salt is easily soluble in water, and occurs in nature.

Manganous chloride, MCl_2, crystallises with 4 mol. H_2O, like the ferrous salt, and not with 6 mol. H_2O like many kindred salts—for example, those of cobalt, calcium, and magnesium; 100 parts of water dissolve 38 parts of the anhydrous salt at 10° and 55 parts at 62°. Alcohol also dissolves manganous chloride, and the alcoholic solution burns with a red flame. This salt, like magnesium chloride, readily forms double salts. A solution of borax gives a dirty rose-coloured precipitate having the composition $MnH_4(BO_3)_2 H_2O$, which is used as a drier in paint-making. Potassium cyanide produces a yellowish-grey precipitate, MnC_2N_2, with manganous salts, soluble in an excess of the reagent, a double salt, $K_4MnC_6N_6$, corresponding with potassium ferrocyanide, being formed. On evaporation of this solution, a portion of the manganese is oxidised and precipitated, whilst a salt corresponding to Gmelin's red salt, K_3,MnC_6N_6 (*see* Chapter XXII.), remains in solution. Sulphuretted hydrogen does not precipitate salts of manganese, not even the acetate, but ammonium sulphide gives a flesh-coloured precipitate, MnS; at 320° this sulphide of manganese passes into a green variety (Antony). Oxalic acid in strong solutions of manganous salts gives a white precipitate of the oxalate, MnC_2O_4. This precipitate is insoluble in water, and is used for the preparation of manganous oxide itself because it decomposes like oxalic acid when ignited (in a tube without access of air), with the formation of carbonic anhydride, carbonic oxide, and manganous oxide. *Manganous oxide* thus obtained is a green powder, which however oxidises with such facility that it burns in air when brought into contact with an incandescent substance, and passes into the red intermediate oxide Mn_3O_4. In solutions of manganous salts, alkalis produce a precipitate of the hydroxide MnH_2O_2, which rapidly absorbs oxygen in the presence of air and gives the brown intermediate oxide, or, more correctly speaking, its hydrate.

Manganous oxide, besides being obtained by the above-described method from manganous oxalate, may also be obtained by igniting the higher oxides in a stream of hydrogen, and also from manganese carbonate. The manganous oxide ignited in the presence of hydrogen acquires a great density, and is no longer so easily oxidised. It may also be obtained in a crystalline form, if during the ignition of the carbonate or higher oxide a trace of dry hydrochloric acid gas be passed into the current of hydrogen. It is thus obtained in the form of transparent emerald green crystals of the regular system, and in this state is easily soluble in acids.

Manganous oxide in oxidising gives the *red oxide of manganese*, Mn_5O_4. This is the most stable of all the oxides of manganese; it is not only stable at the ordinary but also at a high temperature—that is, it does not absorb or disengage oxygen spontaneously. When ignited, all the higher oxides of manganese pass into it by losing oxygen, and manganous oxide by absorbing oxygen. This oxide does not give any distinct salts, but it dissolves in sulphuric acid, forming a dark red solution, which contains both manganous and manganic (of the *oxide*, Mn_2O_3) sulphates. The latter with potassium sulphate gives a manganese alum, in which the alumina is replaced by its isomorphous oxide of manganese. But this alum, like the solution of the intermediate oxide in sulphuric acid, evolves oxygen and leaves a manganous salt when slightly heated.

Manganese dioxide is still less basic than the oxide, and disengages oxygen or a halogen in the presence of acids, forming manganous salts, like the oxide. However, if it be suspended in ether, and hydrochloric acid gas passed into the mixture, which is kept cool, the ether acquires a green colour, owing to the formation of tetrachloride of manganese, $MnCl_4$, corresponding with the dioxide which passes into solution. It is however very unstable, being exceedingly easily decomposed with the evolution of chlorine. The corresponding fluoride, MnF_4, obtained by Nicklés is much more stable. At all events, manganese dioxide does not exhibit any well-defined basic character, but has rather an acid character, which is particularly shown in the compounds MnF_4 and $MnCl_4$ just mentioned, and in the property of manganese dioxide of combining with alkalis. If the higher grades of oxidation of manganese be deoxidised in the presence of alkalis, they frequently give the dioxide combined with the alkali—for example, in the presence of potash a compound is formed which contains $K_2O,5MnO_2$, which shows the weak acid character of this oxide. When ignited in the presence of sodium compounds manganese dioxide frequently forms $Na_2O,8MnO_2$ and $Na_2O,12MnO_2$, and lime when heated with MnO_2 gives from $CaO,3MnO_2$ to $(CaO)_2,MnO_2$ (Rousseau) according to the temperature. Besides which, perhaps, MnO_2 is a saline compound, containing $MnOMnO_3$ or $(MnO)_3Mn_2O_7$, and there are reactions which support such a view (Spring, Richards, Traube, and others); for instance it is known that manganous chloride and potassium permanganate give the dioxide in the presence of alkalis.

Manganese dioxide may be obtained from manganous salts by the action of oxidising agents. If manganous hydroxide or carbonate be shaken up in water through which chlorine is passed, the hypochlorite of the metal is not formed, as is the case with certain other oxides, but manganese dioxide is precipitated: $2MnO_2H_2 + Cl_2 = MnCl_2 + MnO_2,H_2O + H_2O$. Owing to this fact, hypochlorites in the presence of alkalis and acetic acid when added to a solution of manganous salts give hydrated manganese dioxide, as was mentioned above. Manganous nitrate also leaves manganese dioxide when heated to 200°. It is also obtained from manganous and manganic salts of the alkalis, when they are decomposed in the presence of a small amount of acid; the practical method of converting the salts MnX_2 into the higher grades of oxidation is given in Chapter II., Note 6.

Other chemists have obtained manganese by different methods, and attributed different properties to it. This difference probably depends on the presence of carbon in different proportions. Deville obtained manganese by subjecting the pure dioxide, mixed with pure charcoal (from burnt sugar), to a strong heat in a lime crucible until the resultant metal fused. The metal obtained had a rose tint, like bismuth, and like it was very brittle, although exceedingly hard. It decomposed water at the ordinary temperature. Brunner obtained manganese having a specific gravity of about 7·2, which decomposed water very feebly at the ordinary temperature, did not oxidise in air, and was capable of taking a bright polish, like steel; it had the grey colour of cast iron, was very brittle, and hard enough to scratch steel and glass, like a diamond. Brunner's method was as follows: He decomposed the manganese fluoride (obtained as a soluble compound by the action of hydrofluoric acid on manganese carbonate) with sodium, by mixing these substances together in a crucible and covering the mixture with a layer of salt and fluor spar; after which the crucible was first gradually heated until the reaction began, and then strongly heated in order to fuse the metal separated. Glatzel (1889) obtained 25 grms. of manganese, having a grey colour and sp. gr. 7·39, by heating a mixture of 100 grms. of $MnCl_2$ with 200 grms. KCl and 15 grms. Mg to a bright white heat. Moissan and others, by heating the oxides of manganese with carbon in the electric furnace, obtained carbides of manganese—for example, Mn_3C—and remarked that the metal volatilised in the heat of the voltaic arc. Metallic manganese is, however, not prepared on a large scale, but only its alloys with carbon (they readily and rapidly oxidise) and *ferro-manganese* or a coarsely crystalline alloy of iron, manganese and carbon, which is smelted in blast-furnaces like pig-iron (*see* Chapter XXII.) This ferro-manganese is employed in the manufacture of steel by Bessemer's and other processes (see Chapter XXII.) and for the manufacture of manganese bronze. However, in America, Green and Wahl (1895) obtained almost pure metallic manganese on a large scale. They first treat the ore of MnO_2 with 30 p.c. sulphuric acid (which extracts all the oxides of iron present in the ore), and then heat it in a reducing flame to convert it into MnO, which they mix with a powder of Al, lime and CaF_2 (as a flux), and heat the mixture in a crucible lined with magnesia; a reaction immediately takes place at a certain temperature, and a metal of specific gravity 7·3 is obtained, which only contains a small trace of iron.

Manganese gives two compounds with *nitrogen*, Mn_5N_2 and Mn_3N_2. They were obtained by Prelinger (1894) from the amalgam of manganese Mn_2Hg_5 (obtained on a mercury anode by the action of an electric current upon a solution of $MnCl_2$); the mercury may be removed from this amalgam by heating it in an atmosphere of hydrogen, and then metallic manganese is obtained as a grey porous mass of specific gravity 7·42. If this amalgam be heated in dry nitrogen it gives Mn_5N_2 (grey powder, sp. gr. 6·58), but if heated in an atmosphere of NH_3 it gives (as also does Mn_5N_2) Mn_3N_2, (a dark mass with a metallic lustre, sp. gr. 6·21), which, when heated in nitrogen is converted into Mn_5N_2, and if heated in hydrogen evolves NH_3 and disengages hydrogen from a solution of NH_4Cl. At all events, manganese is a metal which decomposes water more easily than iron, nickel, and cobalt.

Volume I. p. 157, Note 7.

It was known to the alchemists by this name, but the true explanation of the change in colour is due to the researches of Chevillot, Edwards, Mitscherlich, and Forchhammer. The change in colour of potassium manganate is due to its instability and to its splitting up into two other manganese compounds, a higher and a lower: $3MnO_3 = Mn_2O_7 + MnO_2$. Manganese trioxide is really decomposed in this manner by the action of water (see later): $3MnO_3 + H_2O = 2MnHO_4 + MnO_2$ (Franke, Thorpe, and Humbly). The instability of the salt is proved by the fact of its being deoxidised by organic matter, with the formation of manganese dioxide and alkali, so that, for instance, a solution of this salt cannot be filtered through paper. The presence of an excess of alkali increases the stability of the salt; when heated it breaks up in the presence of water, with the evolution of oxygen.

The method of preparing *potassium permanganate* will be understood from the above. There are many recipes for preparing this substance, as it is now used in considerable quantities both for technical and laboratory purposes. But in all cases the essence of the methods is one and the same: a mixture of alkali with any oxide of manganese (even manganous hydroxide, which may be obtained from manganous chloride) is first heated in the presence of air or of an oxidising substance (for the sake of rapidity, with potassium chlorate); the resultant mass is then treated with water and heated, when manganese dioxide is precipitated and potassium permanganate remains in solution. This solution may be boiled, as the liquid will contain free alkali; but the solution cannot be evaporated to dryness, because a strong solution, as well as the solid salt, is decomposed by heat.

By adding a dilute solution of manganous sulphate to a boiling mixture of lead dioxide and dilute nitric acid, the whole of the manganese may be converted into permanganic acid (Crum).

[22 bis] The solution of this salt with an excess of impure commercial alkali generally acquires a green tint.

A solution of potassium permanganate gives a beautiful absorption spectrum (Chapter XIII.) If the light in passing through this solution loses a portion of its rays in it (if one may so account for it), this is partially explained by the increased oxidising power which the solution then acquires. We may here also remark that a dilute solution of permanganate of potassium forms a colourless solution with nickel salts, because the green colour of the solution of nickel salts is complementary to the red. Such a decolorised solution, containing a large proportion of nickel and a small proportion of manganese, decomposes after a time, throws down a precipitate, and re-acquires the green colour proper to the nickel salts. The addition of a solution of a cobalt salt (rose-red) to the nickel salt also destroys the colour of both salts.

If sulphuric acid is allowed to act on potassium permanganate without any special precautions, a large amount of oxygen is evolved (it may even explode and inflame), and a violet spray of the decomposing permanganic acid is given off. But if the pure salt (*i.e.* free from chlorine) be dissolved in pure well-cooled sulphuric acid, without any rise in temperature, a green-coloured liquid settles at the bottom of the vessel. This liquid does not contain any sulphuric acid, and consists of permanganic anhydride, Mn_2O_7

(Aschoff, Terreil). It is impossible to prepare any considerable quantity of the anhydride by this method, as it decomposes with an explosion as it collects, evolving oxygen and leaving red oxide of manganese. *Permanganic anhydride*, Mn_2O_7, in dissolving in sulphuric acid, gives a green solution, which (according to Franke, 1887) contains a compound Mn_2SO_{10} = $(MnO_3)_2SO_4$—that is, sulphuric acid in which both hydrogens are replaced by the group MnO_3, which is combined with OK in permanganate of potassium. This mixture with a small quantity of water gives Mn_2O_7, according to the equation: $(MnO_3)_2SO_4 + H_2O = H_2SO_4 + Mn_2O_7$, and when heated to 30° it gives *manganese trioxide*, $(MnO_3)_2SO_4 + H_2O = 2MnO_2 + H_2SO_4 + O$. Pure manganese trioxide is obtained if the solution of $(MnO_3)_2SO_4$ be poured in drops on to sodium carbonate. Then, together with carbonic anhydride, a spray of manganese trioxide passes over, which may be collected in a well-cooled receiver, and this shows that the reaction proceeds according to the equation: $(MnO_3)_2SO_4 + Na_2CO_3 = Na_2SO_4 + 2MnO_3 + CO_2 + O$ (Thorpe). The trioxide is decomposed by water, forming manganese dioxide and a solution of *permanganic acid*: $3MnO_3 + H_2O = MnO_2 + 2HMnO_4$. The same acid is obtained by dissolving permanganic anhydride in water.

Barium permanganate when treated with sulphuric acid gives the same acid. This barium salt may be prepared by the action of barium chloride on the difficultly soluble silver permanganate, $AgMnO_4$, which is precipitated on mixing a strong solution of the potassium salt with silver nitrate. The solution of permanganic acid forms a bright red liquid which reflects a dark violet tint. A dilute solution has exactly the same colour as that of the potassium salt. It deposits manganese dioxide when exposed to the action of light, and also when heated above 60°, and this proceeds the more rapidly the more dilute the solution. It shows its oxidising properties in many cases, as already mentioned. Even hydrogen gas is absorbed by a solution of permanganic acid; and charcoal and sulphur are also oxidised by it, as they are by potassium permanganate. This may be taken advantage of in analysing gunpowder, because when it is treated with a solution of potassium permanganate, all the sulphur is converted into sulphuric acid and all the charcoal into carbonic anhydride. Finely-divided platinum immediately decomposes permanganic acid. With potassium iodide it liberates iodine (which may afterwards be oxidised into iodic acid) (Mitscherlich, Fromherz, Aschoff, and others). Ammonia does not form a corresponding salt with free permanganic acid, because it is oxidised with evolution of nitrogen. The oxidising action of permanganic acid in a strong solution may be accompanied by flame and the formation of violet fumes of permanganic acid; thus a strong solution of it takes fire when brought into contact with paper, alcohol, alkaline sulphides, fats, &c.

We may add that, according to Franke, 1 part of potassium permanganate with 13 parts of sulphuric acid at 100° gives brown crystals of the salt $Mn_2(SO_4)_3, H_2SO_4, 4H_2O$, which gives a precipitate of hydrated manganese dioxide, $H_2MnO_3 = MnO_2H_2O$, when treated with water.

Spring, by precipitating potassium permanganate with sodium sulphite and washing the precipitate by decantation, obtained a soluble colloidal manganese oxide, whose composition was the mean between Mn_2O_3 and MnO_2—namely, $Mn_2O_3, 4(MnO_2H_2O)$.

For rapid and accurate determinations of this kind, advantage is taken of those methods of chemical analysis which are known as 'titrations' (volumetric analysis), and consist in measuring the volume of solutions of known strength required for the complete conversion of a given substance. Details respecting the theory and practice of titration, in which potassium permanganate is very frequently employed, must be looked for in works on analytical chemistry.

The measurements of velocity and acceleration serve for determining the measure of forces in mechanics, but in that case the velocities are magnitudes of length or paths passed over in a unit of time. The velocity of chemical change embodies a conception of quite another kind. In the first place, the velocities of reactions are magnitudes of the masses which have entered into chemical transformations; in the second place, these velocities can only be relative quantities. Hence the conception of 'velocity' has quite a different meaning in chemistry from what it has in mechanics. Their only common factor is time. If dt be the increment of time and dx the quantity of a substance changed in this space of time, then the fraction (or quotient) dx/dt will express the rate of the reaction. The natural conclusion, come to both by Harcourt and Esson, and previously to them (1850) by Wilhelmj (who investigated the rate of conversion, or inversion, of sugar in its passage into glucose), consists in establishing that this velocity is proportional to the quantity of substances still unchanged—*i.e.* that $dx/dt = C(A - x)$, where C is a constant coefficient of proportionality, and where A is the quantity of a substance taken for reaction at the moment when $t = 0$ and $x = 0$—that is, at the beginning of the experiment, from which the time t and quantity x of substance changed is counted. On integrating the preceding equation we obtain $\log(A/(A - x)) = kt$, where k is a new constant, if we take ordinary (and not natural) logarithms. Hence, knowing A, x, and t, for each reaction, we find k, and it proves to be a constant quantity. Thus from the figures cited in the text for the reaction $2KMnO_4$ + $108C_2H_2O_4$ + $14MnSO_4$, it may be calculated that $k = 0.0114$; for example, $t = 44$, $x = 68.4$ (A = 100), whence $kt = 0.5004$ and $k = 0.0114$, (*see also* Chapter XIV., Note 3, and Chapter XVII., Note 25 bis).

The researches made by Hood, Van't Hoff, Ostwald, Warder, Menschutkin, Konovaloff, and others have a particular significance in this direction. Owing to the comparative novelty of this subject, and the absence of applicable as well as indubitable deductions, I consider it impossible to enter into this province of theoretical chemistry, although I am quite confident that its development should lead to very important results, especially in respect to chemical equilibria, for Van't Hoff has already shown that the limit of reaction in reversible reactions is determined by the attainment of equal velocities for the opposite reactions.

CHAPTER XXII
IRON, COBALT, AND NICKEL

Judging from the atomic weights, and the forms of the higher oxides of the elements already considered, it is easy to form an idea of the seven groups of the periodic system. Such are, for instance, the typical series Li, Be, B, C, N, O, F, or the third series, Na, Mg, Al, Si, P, S, Cl. The seven usual types of oxides from R_2O to R_2O_7 correspond with them (Chapter XV.) The position of the eighth group is quite separate, and is determined by the fact that, as we have already seen, in each group of metals having a greater atomic weight than potassium a distinction ought to be made between the elements of the even and uneven series. The series of even elements, commencing with a strikingly alkaline element (potassium, rubidium, cæsium), together with the uneven series following it, and concluding with a haloid (chlorine, bromine, iodine), forms a large period, the properties of whose members repeat themselves in other similar periods. The elements of the eighth group are situated between the elements of the even series and the elements of the uneven series following them. And for this reason elements of the eighth group are found in the middle of each large period. The properties of the elements belonging to it, in many respects independent and striking, are shown with typical clearness in the case of iron, the well-known representative of this group.

Iron is one of those elements which are not only widely diffused in the crust of the earth, but also throughout the entire universe. Its oxides and their various compounds are found in the most diverse portions of the earth's crust; but here iron is always found combined with some other element. Iron is not found on the earth's surface in a free state, because it easily oxidises under the action of air. It is occasionally found in the native state in meteorites, or aerolites, which fall upon the earth.

Meteoric iron is formed outside the earth. Meteorites are fragments which are carried round the sun in orbits, and fall upon the earth when coming into proximity with it during their motion in space. The meteoric dust, on passing through the upper parts of the atmosphere, and becoming incandescent from friction with the gases, produces that phenomenon which is familiar under the name of falling stars. Such is the doctrine concerning meteorites, and therefore the fact of their containing rocky (siliceous) matter and metallic iron shows that outside the earth the elements and their aggregation are in some degree the same as upon the earth itself.

The most widely diffused terrestrial compound of iron is iron bisulphide, FeS_2, or *iron pyrites*. It occurs in formations of both aqueous and igneous origin, and sometimes in enormous masses. It is a substance having a greyish-yellow colour, with a metallic lustre, and a specific gravity of 5·0; it crystallises in the regular system.[2 bis]

The oxides are the principal ores used for producing metallic iron. The majority of the ores contain ferric oxide, Fe_2O_3, either in a free state or combined with water, or else in combination with ferrous oxide, FeO. The species and varieties of iron ores are numerous and diverse. Ferric oxide in a separate form appears sometimes as crystals of the rhombohedric system, having a metallic lustre and greyish steel colour; they are brittle, and form a red powder, specific gravity about 5·25. Ferric oxide in type of oxidation and properties resembles alumina; it is, however, although with difficulty, soluble in acids even when anhydrous. The crystalline oxide bears the name of *specular iron ore*, but ferric oxide most often occurs in a non-crystalline form, in masses having a red fracture, and is then known as *red hæmatite*. In this form, however, it is rather a rare ore, and is principally found in veins. The hydrates of ferric oxide, ferric hydroxides, are most often found in aqueous or stratified formations, and are known as *brown hæmatites*; they generally have a brown colour, form a yellowish-brown powder, and have no metallic lustre but an earthy appearance. They easily dissolve in acids and diffuse through other formations, especially clays (for instance, ochre); they sometimes occur in reniform and similar masses, evidently of aqueous origin. Such are, for instance, the so-called bog or lake and peat ores found at the bottom of marshes and lakes, and also under and in peat beds. This ore is formed from water containing ferrous carbonate in solution, which, after absorbing oxygen, deposits ferric hydroxide. In rivers and springs, iron is found in solution as ferrous carbonate through the agency of carbonic acid: hence the existence of chalybeate springs containing $FeCO_3$. This ferrous carbonate, or *siderite*, is either found as a non-crystalline product of evidently aqueous origin, or as a crystalline spar called *spathic iron ore*. The reniform deposits of the former are most remarkable; they are called spherosiderites, and sometimes form whole strata in the jurassic and carboniferous formations. *Magnetic iron ore*, $Fe_3O_4 = FeO,Fe_2O_3$, in virtue of its purity and practical uses, is a very important ore; it is a compound of the ferrous and ferric oxides, is naturally magnetic, has a specific gravity of 5·1, crystallises in well-formed crystals of the regular system, is with difficulty soluble in acids, and sometimes forms enormous masses, as, for instance, Mount Blagodat in the Ural. However, in most cases—for instance, at Korsak-Mogila (to the north of Berdiansk and Nogaiska, near the Sea of Azov), or at Krivoi Rog (to the west of Ekaterinoslav)—the magnetic iron ore is mixed with other iron ores. In the Urals, the Caucasus (without mentioning Siberia), and in the districts adjoining the basin of the Don, Russia possesses the richest iron ores in the world. To the south of Moscow, in the Governments of Toula and Nijninovgorod, in the Olonetz district, and in the Government of Orloffsky (near Zinovieff in the district of Kromsky), and in many other places, there are likewise abundant supplies of iron ores amongst the deposited aqueous formations; the siderite of Orloffsky, for instance, is distinguished by its great purity.

Iron is also found in the form of various other compounds—for instance, in certain silicates, and also in some phosphates; but these forms are comparatively rare in nature in a pure state, and have not the industrial importance of those natural compounds of iron previously mentioned. In small quantities iron enters into the composition of every kind of *soil* and all rocky formations. As ferrous oxide, FeO, is isomorphous with magnesia, and ferric oxide, Fe_2O_3, with alumina, isomorphous substitution is possible here, and hence minerals are not unfrequently found in which the quantity of iron varies considerably; such, for instance, are pyroxene, amphibole, certain varieties of mica, &c. Although much iron oxide is deleterious to the growth of vegetation, still plants do not flourish without iron; it enters as an indispensable component into the composition of all higher *organisms*; in the ash of plants we always find more or less of its compounds. It also occurs in blood, and forms one of the colouring matters in it; 100 parts of the blood of the highest organisms contain about 0·05 of iron.

The *reduction* of the ores of iron into metallic iron is in principle very simple, because when the oxides of iron are strongly heated with charcoal, hydrogen, carbonic oxide, and other reducing agents, they easily give metallic iron. But the matter is rendered more difficult by the fact that the iron does not melt at the heat developed by the combustion of the charcoal, and therefore it does not separate from those mechanically mixed impurities which are found in the iron ore. This is obviated by the following very remarkable property of iron: at a high temperature it is capable of combining with a small quantity (from 2 to 5 p.c.) of carbon, and then forms *cast iron*, which easily *melts* in the heat developed by the combustion of charcoal in air. For this reason metallic iron is not obtained directly from the ore, but is only formed after the further treatment of the cast iron; the first product extracted from the ore being cast iron. The fused mass disposes itself in the furnace below the slag—that is, the impurities of the ore fused by the heat of the furnace. If these impurities did not fuse they would block up the furnace in which the ore was being smelted, and the continuous smelting of the cast iron would not be possible; it would be necessary periodically to cool the furnace and heat it up again, which means a wasteful expenditure of fuel, and hence in the production of cast iron, the object in view is to obtain all the earthy impurities of the ore in the shape of a fused mass or slag. Only in rare cases does the ore itself form a mass which fuses at the temperature employed, and these cases are objectionable if much iron oxide is carried away in the slag. The impurities of the ores most often consist of certain mixtures—for instance, a mixture of clay and sand, or a mixture of limestone and clay, or quartz, &c. These impurities do not separate of themselves, or do not fuse. The difficulty of the industry lies in forming an easily-fusible slag, into which the whole of the foreign matter of the ore would pass and flow down to the bottom of the furnace above the heavier cast iron. This is effected by mixing certain *fluxes* with the ore and charcoal. A flux is a substance which, when mixed with the foreign matter of the ore, forms a fusible vitreous mass or slag. The flux used for silica is limestone with clay; for limestone a definite quantity of silica is used, the best procedure having been arrived at by experiment and by long practice in iron smelting and other metallurgical processes.

Thus the following materials have to be introduced into the furnace where the smelting of the iron ore is carried on: (1) the iron ore, composed of oxide of iron and foreign matter; (2) the flux required to form a fusible slag with the foreign matter; (3) the carbon which is necessary (*a*) for reducing, (*b*) for combining with the reduced iron to form cast iron, (*c*) principally for the purpose of combustion and the heat generated thereby, necessary not only for reducing the iron and transforming it into cast iron, but also for melting the slag, as well as the cast iron—and (4) the air necessary for the combustion of the charcoal. The air is introduced after a preparatory heating in order to economise fuel and to obtain the highest temperature. The air is forced in under pressure by means of a special blast arrangement. This permits of an exact regulation of the heat and rate of smelting. All these component parts necessary for the smelting of iron must be contained in a vertical, that is, *shaft furnace*, which at the base must have a receptacle for the accumulation of the slag and cast iron formed, in order that the operation may proceed without interruption. The walls of such a furnace ought to be built of fireproof materials if it be designed to serve for the continuous production of cast iron by charging the ore, fuel, and flux into the mouth of the furnace, forcing a blast of air into the lower part, and running out the molten iron and slag from below. The whole operation is conducted in furnaces known as *blast furnaces*. The annexed illustration, fig. 93 (which is taken by kind permission from Thorpe's Dictionary of Applied Chemistry), represents the vertical section of such a furnace. These furnaces are generally of large dimensions—varying from 50 to 90 feet in height. They are sometimes built against rising ground in order to afford easy access to the top where the ore, flux, and charcoal or coke are charged.

The *cast iron* formed in blast furnaces is not always of the same quality. When slowly cooled it is soft, has a grey colour, and is not completely soluble in acids. When treated with acids a residue of graphite remains; it is known as *grey* or soft cast iron. This is the general form of the ordinary cast iron used for casting various objects, because in this state it is not so brittle as in the shape of *white cast iron*, which does not leave particles of graphite when dissolved, but yields its carbon in the form of hydrocarbons. This white cast iron is characterised by its whitish-grey colour, dull lustre, the crystalline structure of its fracture (more homogeneous than that of grey iron), and such hardness that a file will hardly cut it. When white cast iron is produced (from manganese ore) at high temperatures (and with an excess of lime), and containing little sulphur and silica but a considerable amount of carbon (as much as 5 p.c.), it acquires a coarse crystalline structure which increases in proportion to the amount of manganese, and it is then known under the name of 'spiegeleisen' (and 'ferro-manganese').

Cast iron is a material which is either suitable for direct application for casting in moulds or else for working up into *wrought iron* and *steel*. The latter principally differ from cast iron in their containing less carbon—thus, steel contains from 1 p.c. to 0·5 p.c. of carbon and far less silicon and manganese than cast iron; wrought iron does not generally contain more than 0·25 p.c. of carbon and not more than 0·25 p.c. of the other impurities. Thus the essence of the working up of cast iron into steel and wrought iron consists in the removal of the greater part of the carbon and other elements, S, P, Mn, Si, &c. This is effected by means of oxidation, because the oxygen of the atmosphere, oxidising the iron at a high temperature, forms solid oxides with it; and the latter, coming into contact with the carbon contained in the cast iron, are deoxidised, forming wrought iron and carbonic oxide, which is evolved from the mass in a gaseous form. It is evident that the oxidation must be carried on with a molten mass in a state of agitation, so that the oxygen of the air may be brought into contact with the whole mass of carbon contained in the cast iron, or else the operation is effected by means of the addition of oxygen compounds of iron (oxides, ores, as in Martin's process). Cast iron melts much more easily than wrought iron and steel, and, therefore, as the carbon separates, the mass in the furnace (in puddling) or hearth (in the bloomery process) becomes more and more solid; moreover the degree of hardness forms, to a certain extent, a measure of the amount of carbon separated, and the operation may terminate either in the formation of steel or wrought iron. In any case, the iron used for industrial purposes contains impurities. *Chemically pure iron* may be obtained by precipitating iron from a solution (a mixture of ferrous sulphate with magnesium sulphate or ammonium chloride) by the prolonged action of a feeble galvanic current; the iron may be then obtained as a dense

mass. This method, proposed by Böttcher and applied by Klein, gives, as R. Lenz showed, iron containing occluded hydrogen, which is disengaged on heating. This galvanic deposition of iron is used for making galvanoplastic *clichés*, which are distinguished for their great hardness.

Electro-deposited iron is brittle, but if heated (after the separation of the hydrogen) it becomes soft. If pure ferric hydroxide, which is easily prepared by the precipitation of solutions of ferric salts by means of ammonia, be heated in a stream of hydrogen, it forms, first of all, a dull black powder which ignites spontaneously in air (pyrophoric iron), and then a grey powder of pure iron. The powdery substance first obtained is an iron suboxide; when thrown into the air it ignites, forming the oxide Fe_3O_4. If the heating in hydrogen be continued, more water and pure iron, which does not ignite spontaneously, will be obtained. If a small quantity of iron be fused in the oxyhydrogen flame (with an excess of oxygen) in a piece of lime and mixed with powdered glass, pure molten iron will be formed, because in the oxyhydrogen flame iron melts and burns, but the substances mixed with the iron oxidise first. The oxidised impurities here either disappear (carbonic anhydride) in a gaseous form, or turn into slag (silica, manganese, oxide, and others)—that is, fuse with the glass. Pure iron has a silvery white colour and a specific gravity of 7·84; it melts at a temperature higher than the melting-points of silver, gold, nickel, and steel, *i.e.* about 1400°-1500° and below the melting point of platinum (1750°). But pure iron becomes soft at a temperature considerably below that at which it melts, and may then be easily forged, welded, and rolled or drawn into sheets and wire.[11 bis] Pure iron may be rolled into an exceedingly thin sheet, weighing less than a sheet of ordinary paper of the same size. This ductility is the most important property of iron in all its forms, and is most marked with sheet iron, and least so with cast iron, whose ductility, compared with wrought iron, is small, but it is still very considerable when compared with other substances—such, for instance, as rocks.

The chemical properties of iron have been already repeatedly mentioned in preceding chapters. Iron *rusts* in air at the ordinary temperature—that is to say, it becomes covered with a layer of iron oxides. Here, without doubt, the moisture of the air plays a part, because in dry air iron does not oxidise at all, and also because, more particularly, ammonia is always found in iron rust; the ammonia must arise from the action of the hydrogen of the water, at the moment of its separation, on the nitrogen of the air. Highly-polished steel does not rust nearly so readily, but if moistened with water, it easily becomes coated with rust. As rust depends on the access of moisture, iron may be preserved from rust by coating it with substances which prevent the moisture having access to it. Thus arises the practice of covering iron objects with paraffin, varnish, oil, paints, or enamelling it with a glassy-looking flux possessing the same coefficient of expansion as iron, or with a dense scoria (formed by the heat of superheated steam), or with a compact coating of various metals. Wrought iron (both as sheet iron and in other forms), cast iron, and steel are often coated with tin, copper, lead, nickel, and similar metals, which prevent contact with the air. These metals preserve iron very effectually from rust if they form a completely compact surface, but in those places where the iron becomes exposed, either accidentally or from wear, rust appears much more quickly than on a uniform iron surface, because, towards these metals (and also towards the rust), the iron will then behave as an electro-positive pole in a galvanic couple, and hence will attract oxygen. A coating of zinc does not produce this inconvenience, because iron is electro-negative with reference to zinc, in consequence of which galvanised iron does not easily rust, and even an iron boiler containing some lumps of zinc rusts less than one without zinc. Iron oxidises at a high temperature, forming *iron scale*, Fe_3O_4, composed of ferrous and ferric oxides, and, as has been seen, decomposes water and acids with the evolution of hydrogen. It is also capable of decomposing salts and oxides of other metals, which property is applied in the arts for the extraction of copper, silver, lead, tin, &c. For this reason iron is soluble in the solutions of many salts—for instance, in cupric sulphate, with precipitation of copper and formation of ferrous sulphate. When iron *acts on acids* it always *forms compounds* FeX_2—that is, corresponding to the suboxide FeO—and answering to magnesium compounds—and hence two atoms of hydrogen are replaced by one atom of iron. Strongly oxidising acids like nitric acid may transform the ferrous salt which is forming into the higher degree of oxidation or ferric salt (corresponding with the sesquioxide, Fe_2O_3), but this is a secondary reaction. Iron, although easily soluble in dilute nitric acid, loses this property when plunged into strong fuming nitric acid; after this operation it even loses the property of solubility in other acids until the external coating formed by the action of the strong nitric acid is mechanically removed. This condition of iron is termed the passive state. *The passive condition* of iron depends on the formation, on its surface, of a coating of oxide due to the iron being acted on by the lower oxides of nitrogen contained in the fuming nitric acid. Strong nitric acid which does not contain these lower oxides, does not render iron passive, but it is only necessary to add some alcohol or other reducing agent which forms these lower oxides in the nitric acid, and the iron will assume the passive state.

Iron readily combines with non-metals—for instance, with chlorine, iodine, bromine, sulphur, and even with phosphorus and carbon; but on the other hand the property of combining with metals is but little developed in it—that is to say, it does not easily form alloys. Mercury, which acts on most metals, does not act directly on iron, and the *iron amalgam*, or solution of iron in mercury, which is used for electrical machines, is only obtained in a particular way—namely, with the co-operation of a sodium amalgam, in which the iron dissolves and by means of which it is reduced from solutions of its salts.

When iron acts on acids it forms ferrous salts of the type FeX_2, and in the presence of air and oxidising agents they change by degrees into ferric salts of the type FeX_3. This faculty of passing from the ferrous to the ferric state is still further developed in ferrous hydroxide. If sodium hydroxide be added to a solution of ferrous sulphate or green vitriol, $FeSO_4$, a white precipitate of ferrous hydroxide, FeH_2O_2, is obtained; but on exposure to the air, even under water, it turns green, becomes grey, and finally turns brown, which is due to the oxidation that it undergoes. Ferrous hydroxide is very sparingly soluble in water; the solution has, however, a distinct alkaline reaction, which is due to its being a fairly energetic basic oxide. In any case, ferrous oxide is far more energetic than ferric oxide, so that if ammonia be added to a solution containing a mixture of a ferrous and ferric salt, at first ferric hydroxide only will be precipitated. If barium carbonate, $BaCO_3$, be shaken up in the cold with ferrous salts, it does not precipitate them—that is, does not change them into ferrous carbonate; but it completely separates all the iron from the ferric salts in the cold, according to the equation $Fe_2Cl_6 + 3BaCO_3 + 3H_2O = Fe_2O_3,3H_2O + 3BaCl_2 + 3CO_2$. If ferrous hydroxide be boiled with a solution of potash, the water is decomposed, hydrogen is evolved, and the ferrous hydroxide is oxidised. The ferrous salts are in all respects similar to the salts of magnesium and zinc; they are isomorphous with them, but differ from them in that the ferrous hydroxide is not soluble either in aqueous potash or ammonia. In the presence of an excess of ammonium salts, however, a certain proportion of the iron is not precipitated by alkalis and alkali carbonates, which fact points to the formation of double ammonium salts. The ferrous salts have a dull *greenish* colour, and form solutions also of a pale green colour, whilst the ferric salts have a *brown* or reddish-brown colour. The ferrous salts, being capable of oxidation, form very active reducing agents—for instance, under their action gold chloride, $AuCl_3$, deposits metallic gold, nitric acid is transformed into lower oxides, and the highest oxides of manganese also pass into the lower forms of oxidation. All these reactions take place with especial ease in the presence of an excess of acid. This depends on the fact that the ferrous oxide, FeO (or salt), acting as a reducing agent, turns into ferric oxide, Fe_2O_3 (or salt), and in the ferric state it requires more acid for the formation of a normal salt than in the ferrous condition. Thus in the normal ferrous sulphate, $FeSO_4$, there is one equivalent of iron to one equivalent of sulphur (in the sulphuric radicle), but in the neutral ferric salt, $Fe_2(SO_4)_3$, there is one equivalent of iron to one and a half of sulphur in the form of the elements of sulphuric acid.

The most simple oxidising agent for transforming ferrous into ferric salts is chlorine in the presence of water—for instance, $2FeCl_2 + Cl_2 = Fe_2Cl_6$, or, generally speaking, $2FeO + Cl_2 + H_2O = Fe_2O_3 + 2HCl$. When such a transformation is required it is best to add potassium chlorate and hydrochloric acid to the ferrous solution; chlorine is formed by their mutual reaction and acts as an oxidising agent. Nitric acid produces a similar effect, although more slowly. Ferrous salts may be completely and rapidly oxidised into ferric salts by means of chromic acid or permanganic acid, $HMnO_4$, in the presence of acids—for example, $10FeSO_4 + 2KMnO_4 + 8H_2SO_4 = 5Fe_2(SO_4)_3 + 2MnSO_4 + K_2SO_4 + 8H_2O$. This reaction is easily observed by the change of colour, and its termination is easily seen, because potassium permanganate forms solutions of a bright red colour, and when added to a solution of a ferrous salt the above reaction immediately takes place *in the presence of acid*, and the solution then becomes colourless, because all the substances formed are only faintly coloured in solution. Directly all the ferrous compound has passed into the ferric state, any excess of permanganate which is added communicates a red colour to the liquid (see Chapter XXI.)

Thus when ferrous salts are acted on by oxidising agents, they pass into the ferric form, and under the action of reducing agents the reverse reaction occurs. Sulphuretted hydrogen may, for instance, be used for this complete transformation, for under its influence ferric salts are reduced with separation of sulphur—for example, $Fe_2Cl_6 + H_2S = 2FeCl_2 + 2HCl + S$. Sodium thiosulphate acts in a similar way: $Fe_2Cl_6 + Na_2S_2O_3 + H_2O = 2FeCl_2 + Na_2SO_4 + 2HCl + S$. Metallic iron or zinc, in the presence, of acids, or sodium amalgam, &c., acts like hydrogen, and has also a similar reducing action, and this furnishes the best method for reducing ferric salts to ferrous salts—for instance, $Fe_2Cl_6 + Zn = 2FeCl_2 + ZnCl_2$. Thus *the transition from ferrous salts to ferric salts and vice versâ is always possible.*

Ferric oxide, or *sesquioxide of iron*, Fe_2O_3, is found in nature, and is artificially prepared in the form of a red powder by many methods. Thus after heating green vitriol a red oxide of iron remains, called colcothar, which is used as an oil paint, principally for painting wood. The same substance in the form of a very fine powder (rouge) is used for polishing glass, steel, and other objects. If a mixture of ferrous sulphate with an excess of common salt be strongly heated, crystalline ferric oxide will be formed, having a dark violet colour, and resembling some natural varieties of this substance. When iron pyrites is heated for preparing sulphurous anhydride, ferric oxide also remains behind; it is used as a pigment. On the addition of alkalis to a solution of ferric salts, a brown precipitate of ferric hydroxide is formed, which when heated (even when boiled in water, that is, at about 100°, according to Tomassi) easily parts with the water, and leaves red anhydrous ferric oxide. Pure ferric oxide does not show any magnetic properties, but when heated to a white heat it loses oxygen and is converted into the magnetic oxide. Anhydrous ferric oxide which has been heated to a high temperature is with difficulty soluble in acids (but it is soluble when heated in strong acids, and also when fused with potassium hydrogen sulphate), whilst ferric hydroxide, at all events that which is precipitated from salts by means of alkalis, is very readily soluble in acids. The precipitated *ferric hydroxide* has the composition $2Fe_2O_3 3H_2O$, or $Fe_4H_6O_9$. If this ordinary hydroxide be rendered anhydrous (at 100°), at a certain moment it becomes incandescent—that is, loses a certain quantity of heat. This self-incandescence depends on internal displacement produced by the transition of the easily-soluble (in acids) variety into the difficultly-soluble variety, and does not depend on the loss of water, since the anhydrous oxide undergoes the same change. In addition to this there exists a ferric hydroxide, or hydrated oxide of iron, which, like the strongly-heated anhydrous iron oxide, is difficultly soluble in acids. This hydroxide on losing water, or after the loss of water, does not undergo such self-incandescence, because no such state of internal displacement occurs (loss of energy or heat) with it as that which is peculiar to the ordinary oxide of iron. The ferric hydroxide which is difficultly soluble in acids has the composition Fe_2O_3, H_2O. This hydroxide is obtained by a prolonged ebullition of water in which ferric hydroxide prepared by the oxidation of ferrous oxide is suspended, and also sometimes by similar treatment of the ordinary hydroxide after it has been for a long time in contact with water. The transition of one hydroxide to another is apparent by a change of colour; the easily-soluble hydroxide is redder, and the sparingly-soluble hydroxide more yellow in colour.

The normal salts of the composition Fe_2X_6 or FeX_3 correspond with ferric oxide—for example, the exceedingly volatile *ferric chloride*, Fe_2Cl_6, which is easily prepared in the anhydrous state by the action of chlorine on heated iron. Such also is the *normal ferric nitrate*, $Fe_2(NO_3)_6$; it is obtained by dissolving iron in an excess of nitric acid, taking care as far as possible to prevent any rise of temperature. The normal salt separates from the brown solution when it is concentrated under a bell jar over sulphuric acid. This salt, $Fe_2(NO_3)_6, 9H_2O$, then crystallises in well-formed and perfectly colourless crystals, which deliquesce in air, melt at 35°, and are soluble in and decomposed by water. The decomposition may be seen from the fact that the solution is brown and does not yield the whole of the salt again, but gives partly basic salt. The normal salt (only stable in the presence of an excess of HNO_3) is completely decomposed with great facility by heating with water, even at 130°, and this is made use of for removing iron (and also certain other oxides of the form R_2O_3) from many other bases (of the form RO) whose nitrates are far more stable. The ferric salts, FeX_3, in passing into ferrous salts, act as oxidising agents, as is seen from the fact that they not only liberate S from SH_2, but also iodine from KI like many oxidising agents.[25 bis]

Iron forms one other oxide besides the ferric and ferrous oxides; this contains twice as much oxygen as the former, but is so very unstable that it can neither be obtained in the free state nor as a hydrate. Whenever such conditions of double decomposition occur as should allow of its separation in the free state, it decomposes into oxygen and ferric oxide. It is known in the state of salts, and is only stable in the presence of alkalis, and forms salts with them which have a decidedly alkaline reaction; it is therefore a feebly acid oxide. Thus when small pieces of iron are heated with nitre or potassium chlorate a potassium salt of the composition K_2FeO_4 is formed, and therefore the hydrate corresponding with this salt should have the composition H_2FeO_4. It is called *ferric acid*. Its anhydride ought to contain FeO_3 or Fe_2O_6—twice as much oxygen as ferric oxide. If a solution of potassium ferrate be mixed with acid, the free hydrate ought to be formed, but it immediately decomposes ($2K_2FeO_4 + 5H_2SO_4 = 2K_2SO_4 + Fe_2(SO_4)_3 + 5H_2O + O_3$), oxygen being evolved. If a small quantity of acid be taken, or if a solution of potassium ferrate be heated with solutions of other metallic salts, ferric oxide is separated—for instance:

$$2CuSO_4 + 2K_2FeO_4 = 2K_2SO_4 + O_3 + Fe_2O_3 + 2CuO.$$

Both these oxides are of course deposited in the form of hydrates. This shows that not only the hydrate H_2FeO_4, but also the salts of the heavy metals corresponding with this higher oxide of iron, are not formed by reactions of double decomposition. The solution of potassium ferrate naturally acts as a powerful oxidising agent; for instance, it transforms manganous oxide into the dioxide, sulphurous into sulphuric acid, oxalic acid into carbonic anhydride and water, &c.

Iron thus combines with oxygen in three proportions: RO, R_2O_3, and RO_3. It might have been expected that there would be intermediate stages RO_2 (corresponding to pyrites FeS_2) and R_2O_5, but for iron these are unknown.[26 bis] The lower oxide has a distinctly basic character, the higher is feebly acid. The only one which is stable in the free state is ferric oxide, Fe_2O_3; the suboxide, FeO, absorbs oxygen, and ferric anhydride, FeO_3, evolves it. It is also the same for other elements; the character of each is determined by the relative degree of stability of the known oxides. The salts FeX_2 correspond with the suboxide, the salts FeX_3 or Fe_2X_6 with the sesquioxide, and FeX_6 represents those of ferric acid, as its potassium salt is $FeO_2(OK)_2$, corresponding with K_2SO_4, K_2MnO_4, K_2CrO_4, &c. Iron therefore forms compounds of the types FeX_2, FeX_3, and FeX_6, but this latter, like the type NX_5, does not appear separately, but only when X represents heterogeneous elements or groups; for instance, for nitrogen in the form of $NO_2(OH)$, NH_4Cl, &c., for iron in the form of $FeO_2(OK)_2$. But still the type FeX_6 exists, and therefore FeX_2 and FeX_3 are compounds which, like ammonia, NH_3, are capable of further combinations up to FeX_6; this is also seen in the property of ferrous and ferric salts of forming compounds with water of crystallisation, besides double and basic salts, whose stability is determined by the quality of the elements included in the types FeX_2 and FeX_3.[26 tn] It is therefore to be expected that there should be complex compounds derived from ferrous and ferric oxides. Amongst these the series of cyanogen compounds is particularly interesting; their formation and character is not only determined by the property which iron possesses of forming complex types, but also by the similar faculty of the cyanogen compounds, which, like nitriles (Chapter IX.), have clearly developed properties of polymerisation and in general of forming complex compounds.

In the cyanogen compounds of iron, two degrees might be expected: $Fe(CN)_2$, corresponding with ferrous oxide, and $Fe(CN)_3$, corresponding with ferric oxide. There are actually, however, many other known compounds, intermediate and far more complex. They correspond with the double salts so easily formed by metallic cyanides. The two following double salts are particularly well known, very stable, often used, and easily prepared. *Potassium ferrocyanide* or *yellow prussiate of potash*, a double salt of cyanide of potassium and ferrous cyanide, has the composition $FeC_2N_2,4KCN$; its crystals contain 3 mol. of water: $K_4FeC_6N_6,3H_2O$. The other is *potassium ferricyanide* or *red prussiate of potash*. It is also known as *Gmelin's salt*, and contains cyanide of potassium with ferric cyanide; its composition is $Fe(CN)_3,3KCN$ or $K_3FeC_6N_6$. Its crystals do not contain water. It is obtained from the first by the action of chlorine, which removes one atom of the potassium. A whole series of other ferrocyanic compounds correspond with these ordinary salts.

Before treating of the preparation and properties of these two remarkable and very stable salts, it must be observed that with ordinary reagents neither of them gives the same double decompositions as the other ferrous and ferric salts, and they both present a series of remarkable properties. Thus these salts have a neutral reaction, are unchanged by air, dilute acids, or water, unlike potassium cyanide and even some of its double salts. When solutions of these salts are treated with caustic alkalis, they do not give a precipitate of ferrous or ferric hydroxides, neither are they precipitated by sodium carbonate. This led the earlier investigators to recognise special independent groupings in them. The yellow prussiate was considered to contain the complex radicle FeC_6N_6 combined with potassium, namely with K_4, and K_3 was attributed to the red prussiate. This was confirmed by the fact that whilst in both salts any other metal, even hydrogen, might be substituted for potassium, the iron remained unchangeable, just as nitrogen in cyanogen, ammonium, and nitrates does not enter into double decomposition, being in the state of the complex radicles CN, NH_4, NO_2. Such a representation is, however, completely superfluous for the explanation of the peculiarities in the reactions of such compounds as double salts. If a magnesium salt which can be precipitated by potassium hydroxide does not form a precipitate in the presence of ammonium chloride, it is very clear that it is owing to the formation of a soluble double salt which is not decomposed by alkalis. And there is no necessity to account for the peculiarity of reaction of a double salt by the formation of a new complex radicle. In the same way also, in the presence of an excess of tartaric acid, cupric salts do not form a precipitate with potassium hydroxide, because a double salt is formed. These peculiarities are more easily understood in the case of cyanogen compounds than in all others, because all cyanogen compounds, as unsaturated compounds, show a marked tendency to complexity. This tendency is satisfied in double salts. The appearance of a peculiar character in double cyanides is the more easily understood since in the case of potassium cyanide itself, and also in hydrocyanic acid, a great many peculiarities have been observed which are not encountered in those haloid compounds, potassium chloride and hydrochloric acid, with which it was usual to compare cyanogen compounds. These peculiarities become more comprehensible on comparing cyanogen compounds with ammonium compounds. Thus in the presence of ammonia the reactions of many compounds change considerably. If in addition to this it is remembered that the presence of many carbon (organic) compounds frequently completely disturbs the reaction of salts, the peculiarities of certain double cyanides will appear still less strange, because they contain carbon. The fact that the presence of carbon or another element in the compound produces a change in the reactions, may be compared to the action of oxygen, which, when entering into a combination, also very materially changes the nature of reactions. Chlorine is not detected by silver nitrate when it is in the form of potassium chlorate, $KClO_3$, as it is detected in potassium chloride, KCl. The iron in ferrous and ferric compounds varies in its reactions. In addition to the above-mentioned facts, consideration ought to be given to the circumstance that the easy mutability of nitric acid undergoes modification in its alkali salts, and in general the properties of a salt often differ much from those of the acid. Every double salt ought to be regarded as a peculiar kind of saline compound: potassium cyanide is, as it were, a basic, and ferrous cyanide an acid, element. They may be unstable in the separate state, but form a stable double compound when combined together; the act of combination disengages the energy of the elements, and they, so to speak, saturate each other. Of course, all this is not a definite explanation, but then the supposition of a special complex radicle can even less be regarded as such.

Potassium ferrocyanide, $K_4FeC_6N_6$, is very easily formed by mixing solutions of ferrous sulphate and potassium cyanide. First, a white precipitate of ferrous cyanide, FeC_2N_2, is formed, which becomes blue on exposure to air, but is soluble in an excess of potassium cyanide, forming the ferrocyanide. The same yellow prussiate is obtained on heating animal nitrogenous charcoal or animal matters—such as horn, leather cuttings, &c.—with potassium carbonate in iron vessels,[27 bis] the mass formed being afterwards boiled with water with exposure to air, potassium cyanide first appearing, which gives yellow prussiate. The animal charcoal may be exchanged for wood charcoal, permeated with potassium carbonate and heated in nitrogen or ammonia; the mass thus produced is then boiled in water with ferric oxide. In this manner it is manufactured on the large scale, and is called 'yellow prussiate' ('prussiate de potasse,' Blutlaugensalz).

It is easy to substitute other metals for the potassium in the yellow prussiate. The hydrogen salt or hydroferrocyanic acid, $H_4FeC_6N_6$, is obtained by mixing strong solutions of yellow prussiate and hydrochloric acid. If ether be added and the air excluded, the acid is obtained directly in the form of a white scarcely crystalline precipitate which becomes blue on exposure to air (as ferrous cyanide does from the formation of blue compounds of ferrous and ferric cyanides, and it is on this account used in cotton printing). It is soluble in water and alcohol, but not in ether, has marked acid properties, and decomposes carbonates, which renders it easily possible to prepare ferrocyanides of the metals of the alkalis and alkaline earths; these are readily soluble, have a neutral reaction, and resemble the yellow prussiate. Solutions of these salts form precipitates with the salts of other metals, because the ferrocyanides of the heavy metals are insoluble. Here either the whole of the potassium of the yellow prussiate, or only a part of it, is exchanged for an equivalent quantity of the heavy metal. Thus, when a cupric salt is added to a solution of yellow prussiate, a red precipitate is obtained which still contains half the potassium of the yellow prussiate:

$$K_4FeC_6N_6 + CuSO_4 = K_2CuFeC_6N_6 + K_2SO_4.$$

But if the process be reversed (the salt of copper being then in excess) the whole of the potassium will be exchanged for copper, forming a reddish-brown precipitate, $Cu_2FeC_6N_6, 9H_2O$. This reaction and those similar to it are very sensitive and may be used for detecting metals in solution, more especially as the colour of the precipitate very often shows a marked difference when one metal is exchanged for another. Zinc, cadmium, lead, antimony, tin, silver, cuprous and aurous salts form *white* precipitates;

cupric, uranium, titanium and molybdenum salts *reddish-brown*; those of nickel, cobalt, and chromium, *green* precipitates; *with ferrous salts*, ferrocyanide forms, as has been already mentioned, a *white* precipitate—namely, $Fe_2FeC_6N_6$, or FeC_2N_2—which turns blue on exposure to air, and with ferric salts a *blue precipitate* called *Prussian blue*. Here the potassium is replaced by iron, the reaction being expressed thus: $2Fe_2Cl_6 + 3K_4FeC_6N_6 = 12KCl + Fe_4Fe_3C_{18}N_{18}$, the latter formula expressing the composition of Prussian blue. It is therefore the compound $4Fe(CN)_3 + 3Fe(CN)_2$. The yellow prussiate is prepared in chemical works on a large scale especially for the manufacture of this blue pigment, which is used for dyeing cloth and other fabrics and also as one of the ordinary blue paints. It is insoluble in water, and the stuffs are therefore dyed by first soaking them in a solution of a ferric salt and then in a solution of yellow prussiate. If however an excess of yellow prussiate be present complete substitution between potassium and iron does not occur, and *soluble Prussian blue* is formed; $KFe_2(CN)_6 = KCN,Fe(CN)_2,Fe(CN)_3$. This blue salt is colloidal, is soluble in pure water, but insoluble and precipitated when other salts—for instance, potassium or sodium chloride—are present even in small quantities, and is therefore first obtained as a precipitate.

Potassium ferricyanide, or *red prussiate* of potash, $K_3FeC_6N_6$, is called 'Gmelin's salt,' because this savant obtained it by the action of chlorine on a solution of the yellow prussiate: $K_4FeC_6N_6 + Cl = K_3FeC_6N_6 + KCl$. The reaction is due to the ferrous salt being changed by the action of the chlorine into a ferric salt. It separates from solutions in anhydrous, well-formed prisms of a red colour, but the solution has an olive colour; 100 parts of water, at 10°, dissolve 37 parts of the salt, and at 100°, 78 parts. The red prussiate gives a blue precipitate with ferrous salts, called *Turnbull's blue*, very much like Prussian blue (and the soluble variety), because it also contains ferrous cyanide and ferric cyanide, although in another proportion, being formed according to the equation: $3FeCl_2 + 2K_3FeC_6N_6 = 6KCl + Fe_3Fe_2C_{12}N_{12}$, or $3FeC_2N_2,Fe_2C_6N_6$; in Prussian blue we have Fe_7Cy_{18}, and here Fe_5Cy_{12}. A ferric salt ought to form ferric cyanide $Fe_2C_6N_6$, with red prussiate, but ferric cyanide is soluble, and therefore no precipitate is obtained, and the liquid only becomes brown.

If chlorine and sodium are representatives of independent groups of elements, the same may also be said of iron. Its nearest analogues show, besides a similarity in character, a likeness as regards physical properties and a proximity in atomic weight. Iron occupies a medium position amongst its nearest analogues, both with respect to properties and faculty of forming saline oxides, and also as regards atomic weight. On the one hand, cobalt, 58, and nickel, 59, approach iron, 56; they are metals of a more basic character, they do not form stable acids or higher degrees of oxidation, and are a transition to copper, 63, and zinc, 65. On the other hand, manganese, 55, and chromium, 52, are the nearest to iron; they form both basic and acid oxides, and are a transition to the metals possessing acid properties. In addition to having atomic weights approximately alike, chromium, manganese, iron, cobalt, nickel, and copper have also nearly the same specific gravity, so that the atomic volumes and the molecules of their analogous compounds are also near to one another (see table at the beginning of this volume). Besides this, the likeness between the above-mentioned elements is also seen from the following:

They form suboxides, RO, fairly energetic bases, isomorphous with magnesia—for instance, the salt $RSO_4,7H_2O$, akin to $MgSO_4,7H_2O$, and $FeSO_4,7H_2O$, or to sulphates containing less water; with alkali sulphates all form double salts crystallising with $6H_2O$; all are capable of forming ammonium salts, &c. The lower oxides, in the cases of nickel and cobalt, are tolerably stable, are not easily oxidised (the nickel compound with more difficulty than cobalt, a transition to copper); with manganese, and especially with chromium, they are more easily oxidised than with iron and pass into higher oxides. They also form oxides of the form R_2O_3, and with nickel, cobalt, and manganese this oxide is very unstable, and is more easily reduced than ferric oxide; but, in the case of chromium, it is very stable, and forms the ordinary kind of salts. It is isomorphous with ferric oxide, forms alums, is a feeble base, &c. Chromium, manganese, and iron are oxidised by alkali and oxidising agents, forming salts like Na_2SO_4; but cobalt and nickel are difficult to oxidise; their acids are not known with any certainty, and are, in all probability, still less stable than the ferrates. Cr, Mn and Fe form compounds R_2Cl_6 which are like Fe_2Cl_6 in many respects; in Co this faculty is weaker and in Ni it has almost disappeared. The cyanogen compounds, especially of manganese and cobalt, are very near akin to the corresponding ferrocyanides. The oxides of nickel and cobalt are more easily reduced to metal than those of iron, but those of manganese and chromium are not reduced so easily as iron, and the metals themselves are not easily obtained in a pure state; they are capable of forming varieties resembling cast iron. The metals Cr, Mn, Fe, Co, and Ni have a grey iron colour and are very difficult to melt, but nickel and cobalt can be melted in the reverberatory furnace and are more fusible than iron, whilst chromium is more difficult to melt than platinum (Deville). These metals decompose water, but with greater difficulty as the atomic weight rises, forming a transition to copper, which does not decompose water. All the compounds of these metals have various colours, which are sometimes very bright, especially in the higher stages of oxidation.

These metals of the iron group are often met with together in nature. Manganese nearly everywhere accompanies iron, and iron is always an ingredient in the ores of manganese. Chromium is found principally as chrome ironstone—that is, a peculiar kind of magnetic oxide in which Fe_2O_3 is replaced by Cr_2O_3.

Nickel and cobalt are as inseparable companions as iron and manganese. The similarity between them even extends to such remote properties as magnetic qualities. In this series of metals we find those which are the most magnetic: iron, cobalt, and nickel. There is even a magnetic oxide among the chromium compounds, such being unknown in the other series. Nickel easily becomes passive in strong nitric acid. It absorbs hydrogen in just the same way as iron. In short, in the series Cr, Mn, Fe, Co, and Ni, there are many points in common although there are many differences, as will be seen still more clearly on becoming acquainted with cobalt and nickel.

In nature *cobalt* is principally found in combination with arsenic and sulphur. *Cobalt arsenide*, or *cobalt speiss*, $CoAs_2$, is found in brilliant crystals of the regular system, principally in Saxony. *Cobalt glance*, $CoAs_2CoS_2$, resembles it very much, and also belongs to the regular system; it is found in Sweden, Norway, and the Caucasus. *Kupfernickel* is a nickel ore in combination with arsenic, but of a different composition from cobalt arsenide, having the formula NiAs; it is found in Bohemia and Saxony. It has a copper-red colour and is rarely crystalline; it is so called because the miners of Saxony first mistook it for an ore of copper (*Kupfer*), but were unable to extract copper from it. *Nickel glance*, $NiS_2,NiAs_2$, corresponding with cobalt glance, is also known. Nickel accompanies the ores of cobalt and cobalt those of nickel, so that both metals are found together. The ores of cobalt are worked in the Caucasus in the Government of Elizavetopolsk. Nickel ores containing aqueous hydrated nickel silicate are found in the Ural (Revdansk). Large quantities of a similar ore are exported into Europe from New Caledonia. Both ores contain about 12 per cent. Ni. *Garnierite*, $(RO)_5(SiO_2)_41½H_2O$, where R = Ni and Mg, predominates in the New Caledonian ore. Large deposits of nickel have been discovered in Canada, where the ore (as nickelous pyrites) is free from arsenic. Cobalt is principally worked up into cobalt compounds, but nickel is generally reduced to the metallic state, in which it is now often used for alloys—for instance, for coinage in many European States, and for plating other metals, because it does not oxidise. Cobalt arsenide and cobalt glance are principally used for the preparation of cobalt compounds; they are first sorted by discarding the rocky matter, and then roasted. During this process most of the sulphur and arsenic disappears; the arsenious anhydride volatilises with the sulphurous anhydride and the metal also oxidises. It is a simple matter to obtain nickel and cobalt from their oxides. In order to obtain the latter, solutions of their salts are treated with sodium carbonate and the precipitated carbonates are heated; the suboxides are thus obtained, and these latter are reduced in a stream of hydrogen, or even by heating with ammonium chloride. They easily oxidise when in the state of powder. When the chlorides of nickel and cobalt are heated in a stream of hydrogen, the metal is deposited in brilliant scales. *Nickel is always much more easily and quickly reduced than cobalt*. Nickel melts more easily than cobalt, and this even furnishes a means of testing the heating powers of a reverberatory furnace. Cobalt fuses at a temperature only a little lower than that at which iron does. In general, cobalt is nearer to iron than nickel, nickel being nearer to copper.[32 bis] Both nickel and cobalt have magnetic properties like iron, but Co is less magnetic than Fe, and Ni still less so. The specific gravity of nickel reduced by hydrogen is 9·1 and that of cobalt 8·9. Fused cobalt has a specific gravity of 8·5, the density of ordinary nickel being almost the same. Nickel has a greyish silvery-white colour; it is brilliant and very ductile, so that the finest wire may be easily drawn from it. This wire has a resistance to tension equal to iron wire. The beautiful colour of nickel, and the high polish which it is capable of receiving and retaining, as it does not oxidise, render it a useful metal for many purposes, and in many ways it resembles silver.[32 tri] It is now very common to cover other metals with a layer of nickel (nickel plating). This is done by a process of electro-plating, using a solution of a nickel salt. The colour of cobalt is dark and redder; it is also ductile, and has a greater tensile resistance than iron. Dilute acids act very slowly on nickel and cobalt; nitric acid may be considered as the best solvent for them. The solutions in every case contain salts corresponding with the ferrous salts—that is, the *salts* CoX_2, NiX_2, *correspond with the suboxides* of these metals. These salts in their types are similar to the magnesium salts. The salts of nickel when crystallising with water have a green colour, and form bright green solutions, but in the anhydrous state they most frequently have a yellow colour. The salts of cobalt are generally rose-coloured, and generally blue when in the anhydrous state. Their aqueous solutions are rose-coloured. Cobaltous chloride is easily soluble in alcohol, and forms a solution of an intense blue colour.

If a solution of potassium hydroxide be added to a solution of a cobalt salt, a blue precipitate of the basic salt will be formed. If a solution of a cobalt salt be heated almost to the boiling-point, and the solution be then mixed with a boiling solution of an alkali hydroxide, a *pink precipitate of cobaltous hydroxide*, CoH_2O_2, will be formed. If air be not completely excluded during the precipitation by boiling, the precipitate will also contain brown cobaltic hydroxide formed by the further oxidation of the cobaltous oxide. Under similar circumstances nickel salts form *a green precipitate of nickelous hydroxide*, the formation of which is not hindered by the presence of ammonium salts, but in that case only requires more alkali to completely separate the nickel. The nickelous oxide obtained by heating the hydroxide, or from the carbonate or nitrate, is a grey powder, easily soluble in acids and easily reduced, but the same substance may be obtained in the crystalline form as an ordinary product from the ores; it crystallises in regular octahedra, with a metallic lustre, and is of a grey colour. In this state the nickelous oxide almost resists the action of acids.[34 bis]

It is interesting to note *the relation* of the cobaltous and nickelous hydroxides *to ammonia*; aqueous ammonia dissolves the precipitate of cobaltous and nickelous hydroxide. The blue ammoniacal solution of nickel resembles the same solution of cupric oxide, but has a somewhat reddish tint. It is characterised by the fact that it dissolves silk in the same way as the ammoniacal cupric oxide dissolves cellulose. Ammonia likewise dissolves the precipitate of cobaltous hydroxide, forming a brownish liquid, which becomes darker in air and finally assumes a bright red hue, absorbing oxygen. The admixture of ammonium chloride prevents the precipitation of cobalt salts by ammonia; when the ammonia is added, a brown solution is obtained from which, as in the case of the preceding solution, potassium hydroxide does not separate the cobaltous oxide. Peculiar compounds are produced in this solution; they are comparatively stable, containing ammonia and an excess of oxygen; they bear the name cobaltoamine and cobaltiamine salts. They have been principally investigated by Genth, Frémy, Jörgenson and others. Genth found that when a cobalt salt, mixed with an excess of ammonium chloride, is treated with ammonia and exposed to the air, after a certain lapse of time, on adding hydrochloric acid and boiling, a red powder is precipitated and the remaining solution contains an orange salt. The study of these compounds led to the discovery of a whole series of similar salts, some of which correspond with particular higher degrees of oxidation of cobalt, which are described later.

Nickel does not possess this property of absorbing the oxygen of the air when in an ammoniacal solution. In order to understand this distinction, and in general the relation of nickel, it is important to observe that cobalt more easily forms a higher degree of oxidation—namely, *sesquioxide of cobalt, cobaltic oxide*, Co_2O_3—than nickel, especially in the presence of hypochlorous acid. If a solution of a cobalt salt be mixed with barium carbonate and an excess of hypochlorous acid be added, or chlorine gas be passed through it, then at the ordinary temperature on shaking, the whole of the cobalt will be separated in the form of black cobaltic oxide: $2CoSO_4 + ClHO + 2BaCO_3 = Co_2O_3 + 2BaSO_4 + HCl + 2CO_2$. Under these circumstances nickelous oxide does not immediately form black sesquioxide, but after a considerable space of time it also separates in the form of sesquioxide, Ni_2O_3, but always later than cobalt. This is due to the relative difficulty of further oxidation of the nickelous oxide. It is, however, possible to oxidise it; if, for instance, the hydroxide NiH_2O_2 be shaken in water and chlorine gas be passed through it, then nickel chloride will be formed, which is soluble in water, and insoluble nickelic oxide in the form of a black precipitate: $3NiH_2O_2 + Cl_2 = NiCl_2 + Ni_2O_3,3H_2O$. Nickelic oxide may also be obtained by adding sodium hypochlorite mixed with alkali to a solution of a nickel salt. Nickelic and cobaltic hydrates are black. Nickelic oxide evolves oxygen with all acids, and in consequence of this it is not separated as a precipitate in the presence of acids; thus it evolves chlorine with hydrochloric acid, exactly like manganese dioxide. When nickelic oxide is dissolved in aqueous ammonia it liberates nitrogen, and an ammoniacal solution of nickelous oxide is formed. When heated, nickelic oxide loses oxygen, forming nickelous oxide. Cobaltic oxide, Co_2O_3, exhibits more stability than nickelic oxide, and shows feeble basic properties; thus it is dissolved in acetic acid without the evolution of oxygen.[35 bis] But ordinary acids, especially on heating, evolve oxygen, forming a solution of a cobaltous salt. The presence of a cobaltic salt in a solution of a cobaltous salt may be detected by the brown colour of the solution and the black precipitate formed by the addition of alkali, and also from the fact that such solutions evolve chlorine when heated with hydrochloric acid. Cobaltic oxide may not only be prepared by the above-mentioned methods, but also by heating cobalt nitrate, after which a steel-coloured mass remains which retains traces of nitric acid, but when heated further to incandescence evolves oxygen, leaving a compound of cobaltic and cobaltous oxides, similar to magnetic ironstone. Cobalt (but not nickel) undoubtedly forms besides Co_2O_3 a *dioxide* CoO_2. This is obtained when the cobaltous oxide is oxidised by iodine or peroxide of barium.

Nickel alloys possess qualities which render them valuable for technical purposes, the alloy of nickel with iron being particularly remarkable. This alloy is met with in nature as *meteoric iron*. The Pallasoffsky mass of meteoric iron, preserved in the St. Petersburg Academy, fell in Siberia in the last century; it weighs about 15 cwt. and contains 88 p.c. of iron and about 10 p.c. of nickel, with a small admixture of other metals. In the arts *German silver* is most extensively used; it is an alloy containing nickel, copper, and zinc in various proportions. It generally consists of about 50 parts of copper, 25 parts of zinc, and 25 parts of nickel. This alloy is characterised by its white colour resembling that of silver, and, like this latter metal, it does not rust, and therefore furnishes an excellent substitute for silver in the majority of cases where it is used. Alloys which contain silver in addition to nickel show the properties of silver to a still greater extent. Alloys of nickel are used for currency, and if rich deposits of nickel are discovered a wide field of application lies before it, not only in a pure state (because it is a beautiful metal and does not rust) but also for use in alloys. Steel vessels (pressed or forged out of sheet steel) covered with nickel have such practical merits that their manufacture, which has not long commenced, will most probably be rapidly developed, whilst nickel steel, which exceeds ordinary steel in its tenacity, has already proved its excellent qualities for many purposes (for instance, for armour plate).

Until 1890 no compound of cobalt or nickel was known of sufficient volatility to determine the molecular weights of the compounds of these metals; but in 1890 Mr. L. Mond, in conducting (together with Langer and Quincke) his researches on the action of nickel upon carbonic oxide (Chapter IX., Note 24 bis), observed that nickel gradually volatilises in a stream of carbonic oxide; this only takes place at low temperatures, and is seen by the coloration of the flame of the carbonic oxide. This observation led to the discovery of a remarkable volatile *compound of nickel and carbonic oxide*, having as molecular composition $Ni(CO)_4$, as determined by the vapour density and depression of the freezing point. Cobalt and many other metals do not form volatile compounds under these conditions, but iron gives a similar product (Note 26 bis). $Ni(CO)_4$ is prepared by taking finely divided Ni (obtained by reducing NiO by heating it in a stream of hydrogen, or by igniting the oxalate NiC_2O_4) and passing (at a temperature below 50°, for even at 60° decomposition may take place and an explosion) a stream of CO over it; the latter carries over the vapour of the compound, which condenses (in a well-cooled receiver) into a perfectly colourless extremely mobile liquid, boiling without decomposition at 43°, and crystallising in needles at -25° (Mond and Nasini, 1891). Liquid $Ni(CO)_4$ has a sp. gr. 1·356 at 0°, is insoluble in water, dissolves in alcohol and benzene, and burns with a very smoky flame due to the liberation of Ni. The vapour when passed through a tube heated to 180° and above deposits a brilliant coating of metal, and disengages CO. If the tube be strongly heated the decomposition is accompanied by an explosion. If $Ni(CO)_4$ as vapour be passed through a solution of $CuCl_2$, it reduces the latter to metal; it has the same action upon an ammoniacal solution of AgCl, strong nitric acid oxidises $Ni(CO)_4$, dilute solutions of acids have no action; if the vapour be passed through strong sulphuric acid, CO is liberated, chlorine gives NiCl and $COCl_2$; no simple reactions of double decomposition are yet known for $Ni(CO)_4$, however, so that its connection with other carbon compounds is not clear. Probably the formation of this compound could be applied for extracting nickel from its ores.

Footnotes:

The composition of meteoric iron is variable. It generally contains nickel, phosphorus, carbon, &c. The schreibersite of meteoric stones contains Fe_4Ni_2P.

Comets and the rings of Saturn ought now to be considered as consisting of an accumulation of such meteoric cosmic particles. Perhaps the part played by these minute bodies scattered throughout space is much more important in the formation of the largest celestial bodies than has hitherto been imagined. The investigation of this branch of astronomy, due to Schiaparelli, has a bearing on the whole of natural science.

The question arises as to why the iron in meteorites is in a free state, whilst on earth it is in a state of combination. Does not this tend to show that the condition of our globe is very different from that of the rest? My answer to this question has been already given in Volume I. p. 377, Note 57. It is my opinion that inside the earth there is a mass similar in composition to meteorites—that is, containing rocky matter and metallic iron, partly carburetted. In conclusion, I consider it will not be out of place to add the following explanations. According to the theory of the distribution of pressures (see my treatise, *On Barometrical Levelling*, 1876, pages 48 *et seq.*) in an atmosphere of mixed gases, it follows that two gases, whose densities are d and d_1, and whose relative quantities or partial pressures at a certain distance from the centre of gravity are h and h_1, will, when at a greater distance from the centre of attraction, present a different ratio of their masses $x : x_1$—that is, of their partial pressures—which may be found by the equation $d_1(\log(h) - \log(x)) = d(\log(h_1) - \log(x_1))$. If, for instance, $d : d_1 = 2 : 1$, and $h = h_1$ (that is to say, the masses are equal at the lower height) = 1000, then when $x = 10$ the magnitude of x_1 will not be 10 (*i.e.* the mass of a gas at a higher level whose density = 1 will not be equal to the mass of a gas whose density = 2, as was the case at a lower level), but much greater—namely, $x_1 = 100$—that is, the lighter gas will predominate over a heavier one at a higher level. Therefore, when the whole mass of the earth was in a state of vapour, the substances having a greater vapour density accumulated about the centre and those with a lesser vapour density at the surface. And as the vapour densities depend on the atomic and molecular weights, those substances which have small atomic and molecular weights ought to have accumulated at the surface, and those with high atomic and molecular weights, which are the least volatile and the easiest to condense, at the centre. Thus it becomes apparent why such light elements as hydrogen, carbon, nitrogen, oxygen, sodium, magnesium, aluminium, silicon, phosphorus, sulphur, chlorine, potassium, calcium, and their compounds predominate at the surface and largely form the earth's crust. There is also now much iron in the sun, as spectrum analysis shows, and therefore it must have entered into the composition of the earth and other planets, but would have accumulated at the centre, because the density of its vapour is certainly large and it easily condenses. There was also oxygen near the centre of the earth, but not sufficient to combine with the iron. The former, as a much lighter element, principally accumulated at the surface, where we at the present time find all oxidised compounds and even a remnant of free oxygen. This gives the possibility not only of explaining in accordance with cosmogonic theories the predominance of oxygen compounds on the surface of the earth, with the occurrence of unoxidised iron in the interior of the earth and in meteorites, but also of understanding why the density of the whole earth (over 5) is far greater than that of the rocks (1 to 3) composing its crust. And if all the preceding arguments and theories (for instance the supposition that the sun, earth, and all the planets were formed of an elementary homogeneous mass, formerly composed of vapours and gases) be true, it must be admitted that the interior of the earth and other planets contains metallic (unoxidised) iron, which, however, is only found on the surface as aerolites. And then assuming that aerolites are the fragments of planets which have crumbled to pieces so to say during cooling (this has been held to be the case by astronomers, judging from the paths of aerolites), it is readily understood why they should be composed of metallic iron, and this would explain its occurrence in the depths of the earth, which we assumed as the basis of our theory of the formation of naphtha (Chapter VIII., Notes 57–60).

[2 bis] Immense deposits of iron pyrites are known in various parts of Russia. On the river Msta, near Borovitsi, thousands of tons are yearly collected from the detritus of the neighbouring rocks. In the Governments of Toula, Riazan, and in the Donets district continuous layers of pyrites occur among the coal seams. Very thick beds of pyrites are also known in many parts of the Caucasus. But the deposits of the Urals are particularly vast, and have been worked for a long time. Amongst these I will only indicate the deposits on the Soymensky estate near the Kishteimsky works; the Kaletinsky deposits near the Virhny-Isetsky works (containing 1–2 p.c. Cu); on the banks of the river Koushaivi near Koushvi (3–5 p.c. Cu), and the deposits near the Bogoslovsky works (3–5 p.c. Cu). Iron pyrites (especially that containing copper which is extracted after roasting) is now chiefly employed for roasting, as a source of SO_2, for the manufacture of chamber sulphuric acid (Vol. I. p. 291), but the remaining oxide of iron is perfectly suitable for smelting into pig iron, although it gives a sulphurous pig iron (the sulphur may be easily removed by subsequent treatment, especially with the aid of ferro-manganese in Bessemer's process). The great technical importance of iron pyrites leads to its sometimes being imported from great distances; for instance, into England from Spain. Besides which, when heated in closed retorts FeS_2 gives sulphur, and if allowed to oxidise in damp air, green vitriol, $FeSO_4$.

The hydrated ferric oxide is found in nature in a dual form. It is somewhat rarely met with in the form of a crystalline mineral called *göthite*, whose specific gravity is 4·4 and composition $Fe_2H_3O_4$, or $FeHO_2$—that is, one of oxide of iron to one of water, Fe_2O_3,H_2O; frequently found as brown ironstone, forming a dense mass of fibrous, reniform deposits containing $2Fe_2O_3,3H_2O$—that is, having a composition $Fe_4H_6O_9$. In bog ore and other similar ores we most often find a mixture of this hydrated ferric oxide with clay and other impurities. The specific gravity of such formations is rarely as high as 4·0.

The ores of iron, similarly to all substances extracted from veins and deposits, are worked according to mining practice by means of vertical, horizontal, or inclined shafts which reach and penetrate the veins and strata containing the ore deposits. The mass of ore excavated is raised to the surface, then sorted either by hand or else in special sorting apparatus (generally acting with water to wash the ore), and is subjected to roasting and other treatment. In every case the ore contains foreign matter. In the extraction of iron, which is one of the cheapest metals, the dressing of an ore is in most cases unprofitable, and only ores rich in metal are worked— namely, those containing at least 20 p.c. It is often profitable to transport very rich and pure ores (with as much as 70 p.c. of iron) from long distances. The details concerning the working and extraction of metals will be found in special treatises on metallurgy and mining.

The reduction of iron oxides by hydrogen belongs to the order of reversible reactions (Chapter II.), and is therefore determined by a limit which is here expressed by the attainment of the same pressure as in the case where hydrogen acts on iron oxides, and as in the case where (at the same temperature) water is decomposed by metallic iron. The calculations referring to this matter were made by Henri Sainte-Claire Deville (1870). Spongy iron was placed in a tube having a temperature t, one end of which was connected with a vessel containing water at 0° (vapour tension = 4·6 mm.) and the other end with a mercury pump and pressure gauge which determined the limiting tension attained by the dry hydrogen p (subtracting the tension of the water vapour from the tension observed). A tube was then taken containing an excess of iron oxide. It was filled with hydrogen, and the tension p_1 observed of the residual hydrogen when the water was condensed at 0°.

$t =$ 200° 440° 860° 1040°

$p =$ 95·9 25·8 12·8 9·2 mm.

$p_1 =$ — — 12·8 9·4 mm.

The equality of the pressure (tension) of the hydrogen in the two cases is evident. The hydrogen here behaves like the vapour of iron or of its oxide.

By taking ferric oxide, Fe_2O_3, Moissan observed that at 350° it passed into magnetic oxide, Fe_3O_4, at 500° into ferrous oxide, FeO, and at 600° into metallic iron. Wright and Luff (1878), whilst investigating the reduction of oxides, found that (a) the temperature of reaction depends on the condition of the oxide taken—for instance, precipitated ferric oxide is reduced by hydrogen at 85°, that obtained by oxidising the metal or from its nitrate at 175°; (*b*) when other conditions are the same, the reduction by carbonic oxide commences earlier than that by hydrogen, and the reduction by hydrogen still earlier than that by charcoal; (*c*) the reduction is effected with greater facility when a greater quantity of heat is evolved during the reaction. Ferric oxide obtained by heating ferrous sulphate to a red heat begins to be reduced by carbonic oxide at 202°, by hydrogen at 260°, by charcoal at 430°, whilst for magnetic oxide, Fe_3O_4, the temperatures are 200°, 290°, and 450° respectively.

The primitive methods of iron manufacture were conducted by intermittent processes in hearths resembling smiths' fires. As evidenced by the uninterrupted action of the steam boiler, or the process of lime burning, and the continuous preparation and condensation of sulphuric acid or the uninterrupted smelting of iron, every industrial process becomes increasingly profitable and complete under the condition of the continuous action, as far as possible, of all agencies concerned in the production. This continuous method of production is the first condition for the profitable production on the large scale of nearly all industrial products. This method lessens the cost of labour, simplifies the supervision of the work, renders the product uniform, and frequently introduces a very great economy in the expenditure of fuel and at the same time presents the simplicity and perfection of an equilibrated system. Hence every manufacturing operation should be a continuous one, and the manufacture of pig iron and sulphuric acid, which have long since become so, may be taken as examples in many respects. A study of these two manufactures should form the commencement of an acquaintance with all the contemporary methods of manufacturing both from a technical and economical point of view.

The composition of slag suitable for iron smelting most often approaches the following: 50 to 60 p.c. SiO_2, 5 to 20 Al_2O_3, the rest of the mass consisting of MgO, CaO, MnO, FeO. Thus the most fusible slag (according to the observations of Bodeman) contains the alloy $Al_2O_3,4CaO,7SiO_2$. On altering the quantity of magnesia and lime, and especially of the alkalis (which increases the fusibility) and of silica (which decreases it), the temperature of fusion changes with the relation between the total quantity of oxygen and that in the silica. Slags of the composition RO,SiO_2 are easily fusible, have a vitreous appearance, and are very common. Basic slags approach the composition $2RO,SiO_2$. Hence, knowing the composition and quantity of the foreign matter in the ore, it is at once easy to find the quantity and quality of the flux which must be added to form a suitable slag. The smelting of iron is rendered more complex by the fact that the silica, SiO_2, which enters into the slag and fluxes is capable of forming a slag with the iron oxides. In order that the least quantity of iron may pass into the slag, it is necessary for it to be reduced before the temperature is attained at which the slags are formed (about 1000°), which is effected by reducing the iron, not with charcoal itself, but with carbonic oxide. From this it will be understood how the progress of the whole treatment may be judged by the properties of the slags. Details of this complicated and well-studied subject will be found in works on metallurgy.

The section of a blast furnace is represented by two truncated cones joined at their bases, the upper cone being longer than the lower one; the lower cone is terminated by the hearth, or almost cylindrical cavity in which the cast iron and slag collect, one side being provided with apertures for drawing off the iron and slag. The air is blown into the blast furnace through special pipes, situated over the hearth, as shown in the section. The air previously passes through a series of cast-iron pipes, heated by the combustion of the carbonic oxide obtained from the upper parts of the furnace, where it is formed as in a 'gas-producer.' The blast furnace acts continuously until it is worn out; the iron is tapped off twice a day, and the furnace is allowed to cool a little from time to time so as not to be spoilt by the increasing heat, and to enable it to withstand long usage.

Blast furnaces worked with charcoal fuel are not so high, and in general give a smaller yield than those using coke, because the latter are worked with heavier charges than those in which charcoal is employed. Coke furnaces yield 20,000 tons and over of pig iron a year. In the United States there are blast furnaces 30 metres high, and upwards of 600 cubic metres capacity, yielding as much as 130,000 tons of pig iron, requiring a blast of about 750 cubic metres of air per minute, heated to 600°, and consuming about 0·85 part of coke per 1 part of pig iron produced. At the present time the world produces as much as 30 million tons of pig iron a year, about 9/10 of which is converted into wrought iron and steel. The chief producers are the United States (about 10 million tons a year) and England (about 9 million tons a year); Russia yields about 1⅕ million tons a year. The world's production has doubled during the last 20 years, and in this respect the United States have outrun all other countries. The reason of this increase of production must be looked for in the increased demand for iron and steel for railway purposes, for structures (especially ship-building), and in the fact that: (*a*) the cost of pig iron has fallen, thanks to the erection of large furnaces and a fuller study of the processes taking place in them, and (*b*) that every kind of iron ore (even sulphurous and phosphoritic) can now be converted into a homogeneous steel.

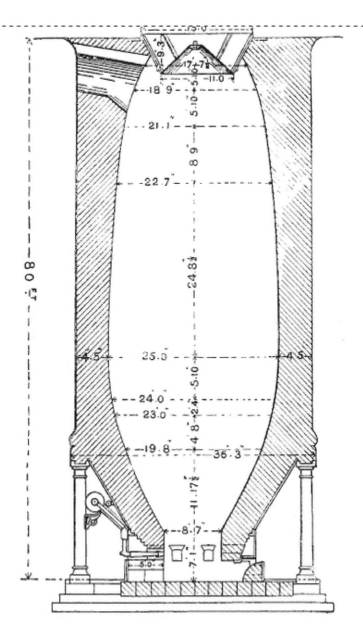

FIG. 93.—Vertical section of a modern Cleveland blast furnace capable of producing 300 to 1,000 tons of pig iron weekly. The outer casing is of riveted iron plates, the furnace being lined with refractory firebrick. It is closed at the top by a 'cap and cone' arrangement, by means of which the charge can be fed into the furnace at suitable intervals by lowering the moveable cone.

In order to more thoroughly grasp the chemical process which takes place in blast furnaces, it is necessary to follow the course of the material charged in at the top and of the air passing through the furnace. From 50 to 200 parts of carbon are expended on 100 parts of iron. The ore, flux, and coke are charged into the top of the furnace, in layers, as the cast iron is formed in the lower parts and flowing down to the bottom causes the whole contents of the furnace to subside, thus forming an empty space at the top, which is again filled up with the afore-mentioned mixture. During its downward course this mixture is subjected to increasing heat. This rise of temperature first drives off the moisture of the ore mixture, and then leads to the formation of the products of the dry distillation of coal or charcoal. Little by little the subsiding mass attains a temperature at which the heated carbon reacts with the carbonic anhydride passing upwards through the furnace and transforms it into carbonic oxide. This is the reason why carbonic anhydride is not evolved from the furnace, but only carbonic oxide. As regards the ore itself, on being heated to about 600° to 800° it is reduced at the expense of the carbonic oxide ascending the furnace, and formed by the contact of the carbonic anhydride with the incandescent charcoal, so that the reduction in the blast furnace is without doubt brought about *by* the formation and decomposition of *carbonic oxide* and not by carbon itself—thus, $Fe_2O_3 + 3CO = Fe_2 + 3CO_2$. The reduced iron, on further subsidence and contact with carbon, forms cast iron, which flows to the bottom of the furnace. In these lower layers, where the temperature is highest (about 1,300°), the foreign matter of the ore finally forms slag, which also is fusible, with the aid of fluxes. The air blown in from below, through the so-called *tuyeres*, encounters carbon in the lower layers of the furnace, and burns it, converting it into carbonic anhydride. It is evident that this develops the highest temperature in these lower layers of the furnace, because here the combustion of the carbon is effected by heated and compressed air. This is very essential, for it is by virtue of this high temperature that the process of forming the slag and of forming and fusing the cast iron are effected simultaneously in these lower portions of the furnace. The carbonic acid formed in these parts rises higher, encounters incandescent carbon, and forms with it carbonic oxide. This heated carbonic oxide acts as a reducing agent on the iron ore, and is reconverted by it into carbonic anhydride; this gas meets with more carbon, and again forms carbonic oxide, which again acts as a reducing agent. The final transformation of the carbonic anhydride into carbonic oxide is effected in those parts of the furnace where the reduction of the oxides of iron does not take place, but where the temperature is still high enough to reduce the carbonic anhydride. The ascending mixture of carbonic oxide and nitrogen, CO_2, &c., is then withdrawn through special lateral apertures formed in the upper cold parts of the furnace walls, and is conducted through pipes to those stoves which are used for heating the air, and also sometimes into other furnaces used for the further processes of iron manufacture. The fuel of blast furnaces consists of wood charcoal (this is the most expensive material, but the pig iron produced is the purest, because charcoal does not contain any sulphur, while coke does), anthracite (for instance, in Pennsylvania, and in Russia at Pastouhoff's works in the Don district), coke, coal, and even wood and peat. It must be borne in mind that the utilisation of naphtha and naphtha refuse would probably give very profitable results in metallurgical processes.

The process just described is accompanied by a series of other processes. Thus, for instance, in the blast furnace a considerable quantity of cyanogen compounds are formed. This takes place because the nitrogen of the air blast comes into contact with incandescent carbon and various alkaline matters contained in the foreign matter of the ores. A considerable quantity of potassium cyanide is formed when wood charcoal is employed for iron smelting, as its ash is rich in potash.

The specific gravity of white cast iron is about 7·5. Grey cast iron has a much lower specific gravity, namely, 7·0. Grey cast iron generally contains less manganese and more silica than white; but both contain from 2 to 3 p.c. of carbon. The difference between the varieties of cast iron depends on the condition of the carbon which enters into the composition of the iron. In white cast iron the carbon is in combination with the iron—in all probability, as the compound CFe_4 (Abel and Osmond and others extracted this compound, which is sometimes called 'carbide,' from tempered steel, which stands to unannealed steel as white cast iron does to grey), but perhaps in the state of an indefinite chemical compound resembling a solution. In any case the compound of the iron and carbon in white cast iron is chemically very unstable, because when slowly cooled it decomposes, with separation of graphite, just as a solution when slowly cooled yields a portion of the substance dissolved. The separation of carbon in the form of graphite on the conversion of white cast iron into grey is never complete, however slowly the separation be carried on; part of the carbon remains in combination with the iron in the same state in which it exists in white cast iron. Hence when grey cast iron is treated with acids, the whole of the carbon does not remain in the form of graphite, but a part of it is separated as hydrocarbons, which proves the existence of chemically-combined carbon in grey cast iron. It is sufficient to re-melt grey cast iron and to cool it quickly to transform it into white cast iron. It is not carbon alone that influences the properties of cast iron; when it contains a considerable amount of sulphur, cast iron remains white even after having been slowly cooled. The same is observed in cast iron very rich in manganese (5 to 7 p.c.), and in this latter case the fracture is very distinctly crystalline and brilliant. When cast iron contains a large amount of manganese, the quantity of carbon may also be increased. Crystalline varieties of cast iron rich in manganese are in practice called ferro-manganese (p. 310), and are prepared for the Bessemer process. Grey cast iron not having an uniform structure is much more liable to various changes than dense and thoroughly uniform white cast iron, and the latter oxidises much more slowly in air than the former. White cast iron is not only used for conversion into wrought iron and steel, but also in those cases where great hardness is required, although it be accompanied by a certain brittleness; for instance, for making rollers, plough-shares, &c.

This direct process of separating the carbon from cast iron is termed *puddling*. It is conducted in reverberatory furnaces. The cast iron is placed on the bed of the furnace and melted; through a special aperture, the puddler stirs up the oxidising mass of cast iron, pressing the oxides into the molten iron. This resembles kneading dough, and the process introduced in England became known as puddling. It is evident that the puddled mass, or bloom, is a heterogeneous substance obtained by mixing, and hence one part of the mass will still be rich in carbon, another will be poor, some parts will contain oxide not reduced, &c. The further treatment of the puddled mass consists in hammering and drawing it out into flat pieces, which on being hammered become more homogeneous, and when several pieces are welded together and again hammered out a still more homogeneous mass is obtained. The quality of the steel and iron thus formed depends principally on their uniformity. The want of uniformity depends on the oxides remaining inside the mass, and on the variable distribution of the carbon throughout the mass. In order to obtain a more homogeneous metal for manufacturing articles out of steel, it is drawn into thin rods, which are tied together in bundles and then again hammered out. As an example of what may be attained in this direction, imitation Damascus steel may be cited; it consists of twisted and plaited wire, which is then hammered into a dense mass. (Real damascened wootz steel may be made by melting a mixture of the best iron with graphite ($1/12$) and iron rust; the article is then corroded with acid, and the carbon remains in the form of a pattern.)

Steel and wrought iron are manufactured from cast iron by puddling. They are, however, obtained not only by this method but also by the *bloomery process*, which is carried out in a fire similar to a blacksmith's forge, fed with charcoal and provided with a blast; a pig of cast iron is gradually pushed into the fire, and portions of it melt and fall to the bottom of the hearth, coming into contact with an air blast, and are thus oxidised. The bloom thus formed is then squeezed and hammered. It is evident that this process is only available when the charcoal used in the fire does not contain any foreign matter which might injure the quality of the iron or steel—for instance, sulphur or phosphorus—and therefore only wood charcoal may be used with impunity, from which it follows that this process can only be carried on where the manufacture of iron can be conducted with this fuel. Coal and coke contain the above-mentioned impurities, and would therefore produce iron of a brittle nature, and thus it would be necessary to have recourse to puddling, where the fuel is burnt on a special hearth, separate from the cast iron, whereby the impurities of the fuel do not come into contact with it. The manufacture of steel from cast iron may also be conducted in fires; but, in addition to this, it is also now prepared by many other methods. One of the long-known processes is called *cementation*, by which steel is prepared from wrought iron but not from cast iron. For this process strips of iron are heated red-hot for a considerable time whilst immersed in powdered charcoal; during this operation the iron at the surface combines with the charcoal, which however does not penetrate; after this the iron strips are re-forged, drawn out again, and cemented anew, repeating this process until a steel of the desired quality is formed—that is, containing the requisite proportion of carbon. The *Bessemer* process occupies the front rank among the newer methods (since 1856); it is so called from the name of its inventor. This process consists in running melted cast iron into converters (holding about 6 tons of cast iron)—that is, egg-shaped receivers, fig. 94, capable of revolving on trunnions (in order to charge in the cast iron and discharge the steel), and forcing a stream of air through small apertures at a considerable pressure. Combustion of the iron and carbon at an elevated temperature then takes place, resulting from the bubbles of oxygen thus penetrating the mass of the cast iron. The carbon, however, burns to a greater extent than the iron, and therefore a mass is obtained which is much poorer in carbon than cast iron. As the combustion proceeds very rapidly in the mass of metal, the temperature rises to such an extent that even the wrought iron which may be formed remains in a molten condition, whilst the steel, being more fusible than the wrought iron, remains very liquid. In half an hour the mass is ready. The purest possible cast iron is used in the Bessemer process, because sulphur and phosphorus do not burn out like carbon, silicon, and manganese.

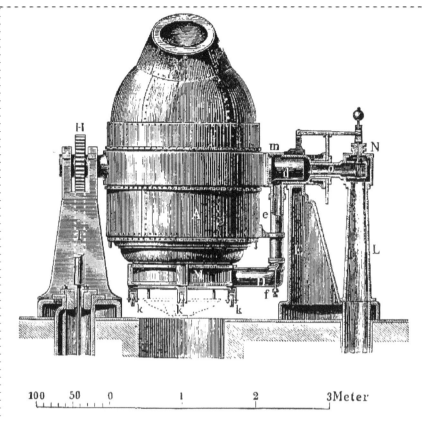

FIG. 94.—Bessemer converter, constructed of iron plate and lined with ganister. The air is carried by the tubes, L, O, D to the bottom, M, from which it passes by a number of holes into the converter. The converter is rotated on the trunnion d by means of the rack and pinion H, when it is required either to receive molten cast iron from the melting furnaces or to pour out the steel.

The presence of manganese enables the sulphur to be removed with the slag, and the presence of lime or magnesia, which are introduced into the lining of the converter, facilitates the removal of the phosphorus. This basic Bessemer process, or *Thomas Gilchrist process*, introduced about 1880, enables ores containing a considerable amount of phosphorus, which had hitherto only been used for cast iron, to be used for making wrought iron and steel. Naturally the greatest uniformity will be obtained by re-melting the metal. Steel is re-melted in small wind furnaces, in masses not exceeding 30 kilos; a liquid metal is formed, which may be cast in moulds. A mixture of wrought and cast iron is often used for making cast steel (the addition of a small amount of metallic Al improves the homogeneity of the castings, by facilitating the passage of the impurities into slag). Large steel castings are made by simultaneous fusion in several furnaces and crucibles; in this way, castings up to 80 tons or more, such as large ordnance, may be made. This molten, and therefore homogeneous, steel is called *cast steel*. Of late years the *Martin's process* for the manufacture of steel has come largely into use; it was invented in France about 1860, and with the use of regenerative furnaces it enables large quantities of cast steel to be made at a time. It is based on the melting of cast iron with iron oxides and iron itself—for instance, pure ores, scrap, &c. There the carbon of the cast iron and the oxygen of the oxide form carbonic oxide, and the carbon therefore burns out, and thus cast steel is obtained from cast iron, providing, naturally, that there is a requisite proportion and corresponding degree of heat. The advantage of this process is that not only do the carbon, silicon, and manganese, but also a great part of the sulphur and phosphorus of the cast iron burn out at the expense of the oxygen of the iron oxides. During the last decade the manufacture of steel and its application for rails, armour plate, guns, boilers, &c., has developed to an enormous extent, thanks to the invention of cheap processes for the manufacture of large masses of homogeneous cast steel. Wrought iron may also be melted, but the heat of a blast furnace is insufficient for this. It easily melts in the oxyhydrogen flame. It may be obtained in a molten state directly from cast iron, if the latter be melted with nitre and sufficiently stirred up. Considerable oxidation then takes place inside the mass of cast iron, and the temperature rises to such an extent that the wrought iron formed remains liquid. A method is also known for obtaining wrought iron directly from rich iron ores by the action of carbonic oxide: the wrought iron is then formed as a spongy mass (which forms an excellent filter for purifying water), and may be worked up into wrought iron or steel either by forging or by dissolving in molten cast iron.

Everybody is more or less familiar with the *difference in the properties of steel and wrought iron*. Iron is remarkable for its softness, pliability, and small elasticity, whilst steel may be characterised by its capability of attaining elasticity and hardness if it be cooled suddenly after having been heated to a definite temperature, or, as it is termed, *tempered*. But if tempered steel be re-heated and slowly cooled, it becomes as soft as wrought iron, and can then be cut with the file and forged, and in general can be made to assume any shape, like wrought iron. In this soft condition it is called *annealed steel*. The transition from tempered to annealed steel thus takes place in a similar way to the transition from white to grey cast iron. Steel, when homogeneous, has considerable lustre, and such a fine granular structure that it takes a very high polish. Its fracture clearly shows the granular nature of its structure. The possibility of tempering steel enables it to be used for making all kinds of cutting instruments, because annealed steel can be forged, turned, drawn (under rollers, for instance, for making rails, bars, &c.), filed, &c., and it may then be tempered, ground and polished. The method and temperature of tempering and annealing steel determine its hardness and other qualities. Steel is generally tempered to the required degree of hardness in the following manner: It is first strongly heated (for instance, up to 600°), and then plunged into water—that is, hardened by rapid cooling (it then becomes as brittle as glass). It is then heated until the surface assumes a definite colour, and finally cooled either quickly or slowly. When steel is heated up to 220°, its surface acquires a yellow colour (surgical instruments); it first of all becomes straw-coloured (razors, &c.), and then gold-coloured; then at a temperature of 250° it becomes brown (scissors), then red, then light blue at 285° (springs), then indigo at 300° (files), and finally sea-green at about 340°. These colours are only the tints of thin films, like the hues of soap bubbles, and appear on the steel because a thin layer of oxides is formed over its surface. Steel rusts more slowly than wrought iron, and is more soluble in acids than cast iron, but less so than wrought iron. Its specific gravity is about 7·6 to 7·9.

As regards the formation of steel, it was a long time before the process of cementation was thoroughly understood, because in this case infusible charcoal permeates unfused wrought iron. Caron showed that this permeation depends on the fact that the charcoal used in the process contains alkalis, which, in the presence of the nitrogen of the air, form metallic cyanides; these being volatile and fusible, permeate the iron, and, giving up their carbon to it, serve as the material for the formation of steel. This explanation is confirmed by the fact that charcoal without alkalis or without nitrogen will not cement iron. The charcoal used for cementation acts badly when used over again, as it has lost alkali. The very volatile ammonium cyanide easily conduces to the formation of steel. Although steel is also formed by the action of cyanogen compounds, nevertheless it does not contain more nitrogen than cast or wrought iron (0·01 p.c.), and these latter contain it because their ores contain titanium, which combines directly with nitrogen. Hence the part played by nitrogen in steel is but an insignificant one. It may be useful here to add some information taken from Caron's treatise concerning the influence of foreign matter on the quality of steel. The principal properties of steel are those of tempering and annealing. The compounds of iron with silicon and boron have not these properties. They are more stable than the carbon compound, and this latter is capable of changing its properties; because the carbon in it either enters into combination or else is disengaged, which determines the condition of hardness or softness of steel, as in white and grey cast iron. When slowly cooled, steel splits up into a mixture of soft and carburetted iron; but, nevertheless, the carbon does not separate from the iron. If such steel be again heated, it forms a uniform compound, and hardens when rapidly cooled. If the same steel as before be taken and heated a long time, then, after being slowly cooled, it becomes much more soluble in acid, and leaves a residue of pure carbon. This shows that the combination between the carbon and iron in steel becomes destroyed when subjected to heat, and the steel becomes iron mixed with carbon. Such *burnt* steel cannot be tempered, but may be corrected by continued forging in a heated condition, which has the effect of redistributing the carbon equally throughout the whole mass. After the forging, if the iron is pure and the carbon has not been burnt out, steel is again formed, which may be tempered. If steel be repeatedly or strongly heated, it becomes burnt through and cannot be tempered or annealed; the carbon separates from the iron, and this is effected more easily if the steel contains other impurities which are capable of forming stable combinations with iron, such as silicon, sulphur, or phosphorus. If there be much silicon, it occupies the place of the carbon, and then continued forging will not induce the carbon once separated to re-enter into combination. Such steel is easily burnt through and cannot be corrected; when burnt through, it is hard and cannot be annealed—this is tough steel, an inferior kind. Iron which contains sulphur and phosphorus cements badly, combines but little with carbon, and steel of this kind is brittle, both hot and cold. Iron in combination with the above-mentioned substances cannot be annealed by slow cooling, showing that these compounds are more stable than those of carbon and iron, and therefore they prevent the formation of the latter. Such metals as tin and zinc combine with iron, but not with carbon, and form a brittle mass which cannot be annealed and is deleterious to steel. Manganese and tungsten, on the contrary, are capable of combining with charcoal; they do not hinder the formation of steel, but even remove the injurious effects of other admixtures (by transforming these admixed substances into new compounds and slags), and are therefore ranked with the substances which act beneficially on steel; but, nevertheless, the best steel, which is capable of renewing most often its primitive qualities after burning or hot forging, is the purest. The addition of Ni, Cr, W, and certain other metals to steel renders it very suitable for certain special purposes, and is therefore frequently made use of.

It is worthy of attention that steel, besides temper, possesses many variable properties, a review of which may be made in the classification of the *sorts of steel* (1878, Cockerell). (1) *Very mild steel* contains from 0·05 to 0·20 p.c. of carbon, breaks with a weight of 40 to 50 kilos per square millimetre, and has an extension of 20 to 30 p.c.; it may be welded, like wrought iron, but cannot be tempered; is used in sheets for boilers, armour plate and bridges, nails, rivets, &c., as a substitute for wrought iron; (2) *mild steel*, from 0·20 to 0·35 p.c. of carbon, resistance to tension 50 to 60 kilos, extension 15 to 20 p.c., not easily welded, and tempers badly, used for axles, rails, and railway tyres, for cannons and guns, and for parts of machines destined to resist bending and torsion; (3) *hard steel*, carbon 0·35 to 0·50 p.c., breaking weight 60 to 70 kilos per square millimetre, extension 10 to 15 p.c., cannot be welded, takes a temper; used for rails, all kinds of springs, swords, parts of machinery in motion subjected to friction, spindles of looms, hammers, spades, hoes, &c.; (4) *very hard steel*, carbon 0·5 to 0·65 p.c., tensile breaking weight 70 to 80 kilos, extension 5 to 10 p.c., does not weld, but tempers easily; used for small springs, saws, files, knives and similar instruments.

The properties of ordinary *wrought iron* are well known. The best iron is the most tenacious—that is to say, that which does not break up when struck with the hammer or bent, and yet at the same time is sufficiently hard. There is, however, a distinction between hard and soft iron. Generally the softest iron is the most tenacious, and can best be welded, drawn into wire, sheets, &c. Hard, especially tough, iron is often characterised by its breaking when bent, and is therefore very difficult to work, and objects made from it are less serviceable in many respects. Soft iron is most adapted for making wire and sheet iron and such small objects as nails. Soft iron is characterised by its attaining a fibrous fracture after forging, whilst tough iron preserves its granular structure after this operation. Certain sorts of iron, although fairly soft at the ordinary temperature, become brittle when heated and are difficult to weld. These sorts are less suitable for being worked up into small objects. The variety of the properties of iron depends on the impurities which it contains. In general, the iron used in the arts still contains carbon and always a certain quantity of silicon, manganese, sulphur, phosphorus, &c. A variety in the proportion of these component parts changes the quality of the iron. In addition to this the change which soft wrought iron, having a fibrous structure, undergoes when subjected to repeated blows and vibrations is considerable; it then becomes granular and brittle. This to a certain degree explains the want of stability of some iron objects—such as truck axles, which must be renewed after a certain term of service, otherwise they become brittle. It is evident that there are innumerable intermediate transitions from wrought iron to steel and cast iron.

At the present day the greater part of the cast iron manufactured is converted into steel, generally cast steel (Bessemer's and Martin's). I may add the Urals, Donetz district, and other parts of Russia offer the greatest advantages for the development of an iron industry, because these localities not only contain vast supplies of excellent iron ore, but also coal, which is necessary for smelting it.

According to information supplied by A. T. Skinder's experiments at the Oboukoff Steel Works, 140 volumes of liquid molten steel give 128 volumes of solid metal. By means of a galvanic current of great intensity and dense charcoal as one electrode, and iron as the other, Bernadoss welded iron and fused holes through sheet iron. Soft wrought iron, like steel and soft malleable cast iron, may be melted in Siemens' regenerative furnaces, and in furnaces heated with naphtha.

[11 bis] Gore (1869), Tait, Barret, Tchernoff, Osmond, and others observed that at a temperature approaching 600°—that is, between dark and bright red heat—all kinds of wrought iron undergo a peculiar change called *recalescence*, *i.e.* a spontaneous rise of temperature. If iron be considerably heated and allowed to cool, it may be observed that at this temperature the cooling stops—that is, latent heat is disengaged, corresponding with a change in condition. The specific heat, electrical conductivity, magnetic, and other properties then also change. In tempering, the temperature of recalescence must not be reached, and so also in annealing, &c. It is evident that a change of the internal condition is here encountered, exactly similar to the transition from a solid to a liquid, although there is no evident physical change. It is probable that attentive study would lead to the discovery of a similar change in other substances.

The particles of steel are linked together or connected more closely than those of the other metals; this is shown by the fact that it only breaks with a tensile strain of 50–80 kilos per sq. mm., whilst wrought iron only withstands about 30 kilos, cast iron 10, copper 35, silver 23, platinum 30, wood 8. The elasticity of iron, steel, and other metals is expressed by the so-called *coefficient of elasticity*. Let a rod be taken whose length is L; if a weight, P, be hung from the extremity of it, it will lengthen to l. The less it lengthens under other equal conditions, the more elastic the material, if it resumes its original length when the weight is removed. It has been shown by experiment that the increase in length l, due to elasticity, is directly proportional to the length L and the weight P, and inversely proportional to the section, but changes with the material. The coefficient of elasticity expresses that weight (in kilos per sq. mm.) under which a rod having a square section taken as 1 (we take 1 sq. mm.) acquires double the length by tension. Naturally in practice materials do not withstand such a lengthening, under a certain weight they attain a limit of elasticity, *i.e.* they stretch permanently (undergo deformation). Neglecting fractions (as the elasticity of metals varies not only with the temperature, but also with forging, purity, &c.), the coefficient of elasticity of steel and iron is 20,000, copper and brass 10,000, silver 7,000, glass 6,000, lead 2,000, and wood 1,200.

Paraffin is one of the best preservatives for iron against oxidation in the air. I found this by experiments about 1860, and immediately published the fact. This method is now very generally applied.

See Chapter XVIII., Note 34 bis. Based on the rapid oxidation of iron and its increase in volume in the presence of water and salts of ammonium, a packing is used for water mains and steam pipes which is tightly hammered into the socket joints. This packing consists of a mixture of iron filings and a small quantity of sal-ammoniac (and sulphur) moistened with water; after a certain lapse of time, especially after the pipes have been used, this mass swells to such an extent that it hermetically seals the joints of the pipes.

Here, however, a ferric salt may also be formed (when all the iron has dissolved and the cupric salt is still in excess), because the cupric salts are reduced by ferrous salts. Cast iron is also dissolved.

Powdery reduced iron is passive with regard to nitric acid of a specific gravity of 1·37, but when heated the acid acts on it. This passiveness disappears in the magnetic field. Saint-Edme attributes the passiveness of iron (and nickel) to the formation of nitride of iron on the surface of the metal, because he observed that when heated in dry hydrogen ammonia is evolved by passive iron.

Remsen observed that if a strip of iron be immersed in acid and placed in the magnetic field, it is principally dissolved at its middle part—that is, the acid acts more feebly at the poles. According to Étard (1891) strong nitric acid dissolves iron in making it passive, although the action is a very slow one.

Iron vitriol or *green vitriol*, sulphate of iron or ferrous sulphate, generally crystallises from solutions, like magnesium sulphate, with seven molecules of water, $FeSO_4,7H_2O$. This salt is not only formed by the action of iron on sulphuric acid, but also by the action of moisture and air on iron pyrites, especially when previously roasted ($FeS_2 + O_2 = FeS + SO_2$), and in this condition it easily absorbs the oxygen of damp air ($FeS + O_4 = FeSO_4$). Green vitriol is obtained in many processes as a by-product. Ferrous sulphate, like all the ferrous salts, has a pale greenish colour hardly perceptible in solution. If it be desired to preserve it without change—that is, so as not to contain ferric compounds—it is necessary to keep it hermetically sealed. This is best done by expelling the air by means of sulphurous anhydride or ether; sulphurous anhydride, SO_2, removes oxygen from ferric compounds, which might be formed, and is itself changed into sulphuric acid, and hence the oxidation of the ferrous compound does not take place in its presence. Unless these precautions are taken, green vitriol turns brown, partly changing into the ferric salt. When turned brown, it is not completely soluble in water, because during its oxidation a certain amount of free insoluble ferric oxide is formed: $6FeSO_4 + O_3 = 2Fe_2(SO_4)_3 + Fe_2O_3$. In order to cleanse such mixed green vitriol from the oxide, it is necessary to add some sulphuric acid and iron and boil the mixture; the ferric salt is then transformed into the ferrous state: $Fe_2(SO_4)_3 + Fe = 3FeSO_4$.

Green vitriol is used for the manufacture of Nordhausen sulphuric acid (Chapter XX.), for preparing ferric oxide, in many dye works (for preparing the indigo vats and reducing blue indigo to white), and in many other processes; it is also a very good disinfectant, and is the cheapest salt from which other compounds of iron may be obtained.

The other ferrous salts (excepting the yellow prussiate, which will be mentioned later) are but little used, and it is therefore unnecessary to dwell upon them. We will only mention *ferrous chloride*, which, in the crystalline state, has the composition $FeCl_2,4H_2O$. It is easily prepared; for instance, by the action of hydrochloric acid on iron, and in the anhydrous state by the action of hydrochloric acid gas on metallic iron at a red heat. The anhydrous ferrous chloride then volatilises in the form of colourless cubic crystals. Ferrous oxalate (or the double potassium salt) acts as a powerful reducing agent, and is frequently employed in photography (as a developer).

Ferrous sulphate, like magnesium sulphate, easily forms double salts—for instance, $(NH_4)_2SO_4,FeSO_4,6H_2O$. This salt does not oxidise in air so readily as green vitriol, and is therefore used for standardising $KMnO_4$.

The transformation of ferrous oxide into ferric oxide is not completely effected in air, as then only a part of the suboxide is converted into ferric oxide. Under these circumstances the so-called magnetic oxide of iron is generally produced, which contains atomic quantities of the suboxide and oxide—namely, $FeO,Fe_2O_3 = Fe_3O_4$. This substance, as already mentioned, is found in nature and in iron scale. It is also formed when most ferrous and ferric salts are heated in air; thus, for instance, when ferrous carbonate, $FeCO_3$ (native or the precipitate given by soda in a solution of FeX_2), is heated it loses the elements of carbonic anhydride, and magnetic oxide remains. This oxide of iron is attracted by the magnet, and is on this account called magnetic oxide, although it does not always show magnetic properties. If magnetic oxide be dissolved in any acid—for instance, hydrochloric—which does not act as an oxidising agent, a ferrous salt is first formed and ferric oxide remains, which is also capable of passing into solution. The best way of preparing the hydrate of the magnetic oxide is by decomposing a mixture of ferrous and ferric salts with ammonia; it is, however, indispensable to pour this mixture into the ammonia, and not *vice versâ*, as in that case the ferrous oxide would at first be precipitated alone, and then the ferric oxide. The compound thus formed has a bright green colour, and when dried forms a black powder. Other combinations of ferrous with ferric oxide are known, as are also compounds of ferric oxide with other bases. Thus, for instance, compounds are known containing 4 molecules of ferrous oxide to 1 of ferric oxide, and also 6 of ferrous to 1 of ferric oxide. These are also magnetic, and are formed by heating iron in air. The magnesium compound MgO,Fe_2O_3 is prepared by passing gaseous hydrochloric acid over a heated mixture of magnesia and ferric oxide. Crystalline magnesium oxide is then formed, and black, shiny, octahedral crystals of the above-mentioned composition. This compound is analogous to the aluminates—for instance, to spinel. Bernheim (1888) and Rousseau (1891) obtained many similar compounds of ferric oxide, and their composition apparently corresponds to the hydrates (Note 22) known for the oxide.

Copper and cuprous salts also reduce ferric oxide to ferrous oxide, and are themselves turned into cupric salts. The essence of the reactions is expressed by the following equations: $Fe_2O_3 + Cu_2O = 2FeO + 2CuO$; $Fe_2O_3 + Cu = 2FeO + CuO$. This fact is made use of in analysing copper compounds, the quantity of copper being ascertained by the amount of ferrous salt obtained. An excess of ferric salt is required to complete the reaction. Here we have an example of reverse reaction; the ferrous oxide or its salt in the presence of alkali transforms the cupric oxide into cuprous oxide and metallic copper, as observed by Lovel, Knopp, and others.

We will here mention the reactions by means of which it may be ascertained whether the ferrous compound has been entirely converted into a ferric compound or *vice versâ*. There are two substances which are best employed for this purpose: potassium ferricyanide, $FeK_3C_6N_6$, and potassium thiocyanate, KCNS. The first salt gives with ferrous salts a blue precipitate of an insoluble salt, having a composition $Fe_5C_{12}N_{12}$; but with ferric salts it does not form any precipitate, and only gives a brown colour, and therefore when transforming a ferrous salt into a ferric salt, the completion of the transformation may be detected by taking a drop of the liquid on paper or on a porcelain plate and adding a drop of the ferricyanide solution. If a blue precipitate be formed, then part of the ferrous salt still remains; if there is none, the transformation is complete. The thiocyanate does not give any marked coloration with ferrous salts; but with ferric salts in the most diluted state it forms a bright red soluble compound, and therefore when transforming a ferric salt into a ferrous salt we must proceed as before, testing a drop of the solution with thiocyanate, when the absence of a red colour will prove the total transformation of the ferric salt into the ferrous state, and if a red colour is apparent it shows that the transformation is not yet complete.

The two ferric hydroxides are not only characterised by the above-mentioned properties, but also by the fact that the first hydroxide forms immediately with potassium ferrocyanide, $K_4FeC_6N_6$, a blue colour depending on the formation of Prussian blue, whilst the second hydroxide does not give any reaction whatever with this salt. The first hydroxide is entirely soluble in nitric, hydrochloric, and all other acids; whilst the second sometimes (not always) forms a brick-coloured liquid, which appears turbid and does not give the reactions peculiar to the ferric salts (Péan de Saint-Gilles, Scheurer-Kestner). In addition to this, when the smallest quantity of an alkaline salt is added to this liquid, ferric oxide is precipitated. Thus a colloidal solution is formed (hydrosol), which is exactly similar to silica hydrosol (Chapter XVII.), according to which example the hydrosol of ferric oxide may be obtained.

If ordinary ferric hydroxide be dissolved in acetic acid, a solution of the colour of red wine is obtained, which has all the reactions characteristic of ferric salts. But if this solution (formed in the cold) be heated to the boiling-point, its colour is very rapidly intensified, a smell of acetic acid becomes apparent, and the solution then contains a new variety of ferric oxide. If the boiling of the solution be continued, acetic acid is evolved, and the modified ferric oxide is precipitated. If the evaporation of the acetic acid be prevented (in a closed or sealed vessel), and the liquid be heated for some time, the whole of the ferric hydroxide then passes into the insoluble form, and if some alkaline salt be added (to the hydrosol formed), the whole of the ferric oxide is then precipitated in its insoluble form. This method may be applied for separating ferric oxide from solutions of its salts.

All phenomena observed respecting ferric oxide (colloidal properties, various forms, formation of double basic salts) demonstrate that this substance, like silica, alumina, lead hydroxide, &c., is polymerised, that the composition is represented by $(Fe_2O_3)_n$.

The ferric compound which is most used in practice (for instance, in medicine, for cauterising, stopping bleeding, &c.—Oleum Martis) is *ferric chloride*, Fe_2Cl_6, easily obtained by dissolving the ordinary hydrated oxide of iron in hydrochloric acid. It is obtained in the anhydrous state by the action of chlorine on heated iron. The experiment is carried on in a porcelain tube, and a solid *volatile substance* is then formed in the shape of brilliant violet scales which very readily absorb moisture from the air, and when heated with water decompose into crystalline ferric oxide and hydrochloric acid: $Fe_2Cl_6 + 3H_2O = 6HCl + Fe_2O_3$. Ferric chloride is so volatile that the density of its vapour may be determined. At 440° it is equal to 164·0 referred to hydrogen; the formula Fe_2Cl_6 corresponds with a density of 162·5. An aqueous solution of this salt has a brown colour. On evaporating and cooling this solution, crystals separate containing 6 or 12 molecules of H_2O. Ferric chloride is not only soluble in water, but also in alcohol (similarly to magnesium chloride, &c.) and in ether. If the latter solutions are exposed to the rays of the sun they become colourless, and deposit ferrous chloride, $FeCl_2$, chlorine being disengaged. After a certain lapse of time, the aqueous solutions of ferric chloride decompose with precipitation of a basic salt, thus demonstrating the instability of ferric chloride, like the other salts of ferric oxide (Note 22). This salt is much more stable in the form of double salts, like all the ferric salts and also the salts of many other feeble bases. Potassium or ammonium chloride forms with it very beautiful red crystals of a double salt, having the composition $Fe_2Cl_6,4KCl,2H_2O$. When a solution of this salt is evaporated it decomposes, with separation of potassium chloride.

B. Roozeboom (1892) studied in detail (as for $CaCl_2$, Chapter XIV., Note 50) the separation of different hydrates from saturated solutions of Fe_2Cl_6 at various concentrations and temperatures; he found that there are 4 crystallohydrates with 12, 7, 5, and 4 molecules of water. An orange yellow only slightly hygroscopic hydrate, $Fe_2Cl_6,12H_2O$, is most easily and usually obtained, which melts at 37°; its solubility at different temperatures is represented by the curve BCD in the accompanying figure, where the point B corresponds to the formation, at -55°, of a cryohydrate containing about $Fe_2Cl_6 + 36H_2O$, the point C corresponds to the melting-point (+37°) of the hydrate $Fe_2Cl_6,12H_2O$, and the curve CD to the fall in the temperature of crystallisation with an increase in the amount of salt, or decrease in the amount of water (in the figure the temperatures are taken along the axis of abscissæ, and the amount of n in the formula $nFe_2Cl_6 + 100H_2O$ along the axis of ordinates). When anhydrous Fe_2Cl_6 is added to the above hydrate ($12H_2O$), or some of the water is evaporated from the latter, very hygroscopic crystals of $Fe_2Cl_6,5H_2O$ (Fritsche) are formed; they melt at 56°, their solubility is expressed by the curve HJ, which also presents a small branch at the end J. This again gives the fall in the temperature of crystallisation with an increase in the amount of Fe_2Cl_6. Besides these curves and the solubility of the anhydrous salt expressed by the line KL (up to 100°, beyond which chlorine is liberated), Roozeboom also gives the two curves, EFG and JK, corresponding to the crystallohydrates, $Fe_2Cl_6,7H_2O$ (melts at $+32°·5$, that is lower than any of the others) and $Fe_2Cl_6,4H_2O$ (melts at $73°·5$), which he discovered by a systematic research on the solutions of ferric chloride. The curve AB represents the separation of ice from dilute solutions of the salt.

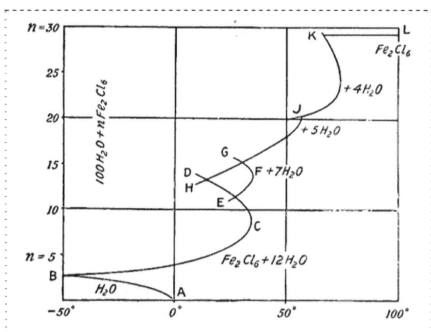

FIG. 95.—Diagram of the solubility of Fe_2Cl_6.

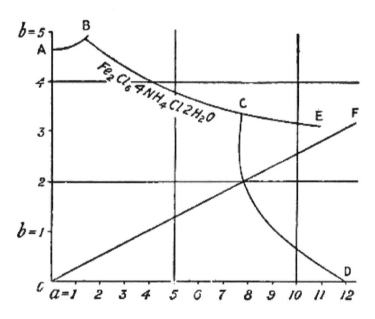

FIG. 96.—Diagram of the formation, at 15°, of the double salt $Fe_2Cl_6 4NH_4Cl 2H_2O$ or $Fe(NH_4)_2Cl_5 H_2O$. (After Roozeboom.)

The researches of the same Dutch chemist upon the conditions of the formation of crystals from the double salt $(NH_4Cl)_4Fe_2Cl_6,2H_2O$ are even more perfect. This salt was obtained in 1839 by Fritsche, and is easily formed from a strong solution of Fe_2Cl_6 by adding sal-ammoniac, when it separates in crimson rhombic crystals, which, after dissolving in water, only deposit again on evaporation, together with the sal-ammoniac.

Roozeboom (1892) found that when the solution contains b molecules of Fe_2Cl_6, and a molecules of NH_4Cl, per 100 molecules H_2O, then at 15° one of the following separations takes place: (1) crystals, $Fe_2Cl_6,12H_2O$, when a varies between 0 and 11, and b between 4·65 and 4·8, or (2) a mixture of these crystals and the double salt, when $a = 1·36$, and $b = 4·47$, or (3) the double salt, $Fe_2Cl_6,4NH_4Cl,2H_2O$, when a varies between 2 and 11·8, and b between 3·1 and 4·56, or (4) a mixture of sal-ammoniac with the iron salt (it crystallises in separate cubes, Retgers, Lehmann), when a varies between 7·7 and 10·9, and b is less than 3·38, or (5) sal-ammoniac, when $a = 11·88$. And as in the double salt, $a : b :: 4 : 1$ it is evident that the double salt only separates out when the ratio $a : b$ is less than 4 : 1 (*i.e.* when Fe_2Cl_6 predominates). The above is seen more clearly in the accompanying figure, where a, or the number of molecules of NH_4Cl per $100H_2O$, is taken along the axis of abscissæ, and b, or the number of molecules of Fe_2Cl_6, along the ordinates. The curves ABCD correspond to saturation and present an iso-therm of 15°. The portion AB corresponds to the separation of chloride of iron (the ascending nature of this curve shows that the solubility of Fe_2Cl_6 is increased by the presence of NH_4Cl, while that of NH_4Cl decreases in the presence of Fe_2Cl_6), the portion BC to the double salt, and the portion CD to a mixture of sal-ammoniac and ferric chloride, while the straight line OF corresponds to the ratio $Fe_2Cl_6,4NH_4Cl$, or $a : b :: 4 : 1$. The portion CE shows that more double salt may be introduced into the solution without decomposition, but then the solution deposits a mixture of sal-ammoniac and ferric chloride (*see* Chapter XXIV. Note 9[bis]). If there were more such well-investigated cases of solutions, our knowledge of double salts, solutions, the influence of water, equilibria, isomorphous mixtures, and such-like provinces of chemical relations might be considerably advanced.

The normal ferric salts are decomposed by heat and even by water, forming basic salts, which may be prepared in various ways. Generally ferric hydroxide is dissolved in solutions of ferric nitrate; if it contains a double quantity of iron the basic salt is formed which contains Fe_2O_3 (in the form of hydroxide) + $2Fe_2(NO_3)_6 = 3Fe_2O(NO_3)_4$, a salt of the type Fe_2OX_4. Probably water enters into its composition. With considerable quantities of ferric oxide, insoluble basic salts are obtained containing various amounts of ferric hydroxide. Thus when a solution of the above-mentioned basic acid is boiled, a precipitate is formed containing $4(Fe_2O_3)_8,2(N_2O_5),3H_2O$, which probably contains $2Fe_2O_2(NO_3)_2 + 2Fe_2O_3,3H_2O$. If a solution of basic nitrate be sealed in a tube and then immersed in boiling water, the colour of the solution changes just in the same way as if a solution of ferric acetate had been employed (Note 22). The solution obtained smells strongly of nitric acid, and on adding a drop of sulphuric or hydrochloric acid the insoluble variety of hydrated ferric oxide is precipitated.

Normal ferric *orthophosphate* is soluble in sulphuric, hydrochloric, and nitric acids, but insoluble in others, such as, for instance, acetic acid. The composition of this salt in the anhydrous state is $FePO_4$, because in orthophosphoric acid there are three atoms of hydrogen, and iron, in the ferric state, replaces the three atoms of hydrogen. This salt is obtained from ferric acetate, which, with disodium phosphate, forms a *white precipitate* of $FePO_4$, containing water. If a solution of ferric chloride (yellowish-red colour) be mixed with a solution of sodium acetate in excess, the liquid assumes an intense brown colour which demonstrates the formation of a certain quantity of ferric acetate; then the disodium phosphate directly forms a white gelatinous precipitate of ferric phosphate. By this means the whole of the iron may be precipitated, and the liquid which was brown then becomes colourless. If this normal salt be dissolved in orthophosphoric acid, the crystalline acid salt $FeH_3(PO_4)_2$ is formed. If there be an excess of ferric oxide in the solution, the precipitate will consist of the basic salt. If ferric phosphate be dissolved in hydrochloric acid, and ammonia be added, a salt is precipitated on heating which, after continued washing in water and heating (to remove the water), has the composition $Fe_4P_2O_{11}$—that is, $2Fe_2O_3,P_2O_5$. In an aqueous condition this salt may be considered as ferric hydroxide, $Fe_2(OH)_6$, in which $(OH)_3$ is replaced by the equivalent group PO_4. Whenever ammonia is added to a solution containing an excess of ferric salt and a certain amount of phosphoric acid, a precipitate is formed containing the whole of the phosphoric acid in the mass of the ferric oxide.

Ferric oxide is characterised as a feeble base, and also by the fact of its forming double salts—for instance, *potassium iron alum*, which has a composition $Fe_2(SO_4)_3,K_2SO_4,24H_2O$ or $FeK(SO_4)_2,12H_2O$. It is obtained in the form of almost colourless or light rose-coloured large octahedra of the regular system by simply mixing solutions of potassium sulphate and the ferric sulphate obtained by dissolving ferric oxide in sulphuric acid.

It would seem that all normal ferric salts are colourless, and that the brown colour which is peculiar to the solutions is really due to basic ferric salts. A remarkable example of the apparent change of colour of salts is represented by the ferrous and ferric oxalates. The former in a dry state has a yellow colour, although as a rule the ferrous salts are green, and the latter is colourless or pale green. When the normal ferric salt is dissolved in water it is, like many salts, probably decomposed by the water into acid and basic salts, and the latter communicates a brown colour to the solution. Iron alum is almost colourless, is easily decomposed by water, and is the best proof of our assertion. The study of the phenomena peculiar to ferric nitrate might, in my opinion, give a very useful addition to our knowledge of the aqueous solutions of salts in general.

[25 bis] The reaction $FeX_3 + KI = FeX_2 + KX + I$ proceeds comparatively slowly in solutions, is not complete (depends upon the mass), and is reversible. In this connection we may cite the following data from Seubert and Rohrer's (1894) comprehensive researches. The investigations were conducted with solutions containing $\frac{1}{10}$ gram— equivalent weights of $Fe_2(SO_4)_3$ (*i.e.* containing 20 grams of salt per litre), and a corresponding solution of KI; the amount of iodine liberated being determined (after the addition of starch) by a solution (also $\frac{1}{10}$ normal) of $Na_2S_2O_3$ (*see* Chapter XX., Note 42). The progress of the reaction was expressed by the amount of liberated iodine in percentages of the theoretical amount. For instance, the following amount of iodide of potassium was decomposed when $Fe_2(SO_4)_3 + 2nKI$ was taken:

$n =$	1	2	3	6	10	20
After 15'	11·4	26·3	40·6	73·5	91·6	96·0
„ 30'	14·0	35·8	47·8	78·5	94·3	97·4
„ 1 hour	19·0	42·7	56·0	84·0	95·7	97·6
„ 10 „	32·6	56·0	75·7	93·2	96·5	97·6
„ 48 „	39·4	67·7	82·6	93·4	96·6	97·6

Similar results were obtained for $FeCl_3$, but then the amount of iodine liberated was somewhat greater. Similar results were also obtained by increasing the mass of FeX_3 per KI, and by replacing it by HI (*see* Chapter XXI., Note 26).

If chlorine be passed through a strong solution of potassium hydroxide in which hydrated ferric oxide is suspended, the turbid liquid acquires a dark pomegranate-red colour and contains potassium ferrate: $10KHO + Fe_2O_3 + 3Cl_2 = 2K_2FeO_4 + 6KCl + 5H_2O$. The chlorine must not be in excess, otherwise the salt is again decomposed, although the mode of decomposition is unknown; however, ferric chloride and potassium chlorate are probably formed. Another way in which the above-described salt is formed is also remarkable; a galvanic current (from 6 Grove elements) is passed through cast-iron and platinum electrodes into a strong solution of potassium hydroxide. The cast-iron electrode is connected with the positive pole, and the platinum electrode is surrounded by a porous earthenware cylinder. Oxygen would be evolved at the cast-iron electrode, but it is used up in oxidation, and a dark solution of potassium ferrate is therefore formed about it. It is remarkable that the cast iron cannot be replaced by wrought iron.

[26 bis] When Mond and his assistants obtained the remarkable volatile compound $Ni(CO)_4$ (described later, Chapter XXII.), it was shown subsequently by Mond and Quincke (1891), and also by Berthelot, that iron, under certain conditions, in a stream of carbonic oxide, also volatilises and forms a compound like that given by nickel. Roscoe and Scudder then showed that when water gas is passed through and kept under pressure (8 atmospheres) in iron vessels a portion of the iron volatilises from the sides of the vessel, and that when the gas is burnt it deposits a certain amount of oxides of iron (the same result is obtained with ordinary coal gas which contains a small amount of CO). To obtain the *volatile compound of iron with carbonic oxide*, Mond prepared a finely divided iron by heating the oxalate in a stream of hydrogen, and after cooling it to 80°-45° he passed CO over the powder. The iron then formed (although very slowly) a volatile compound containing $Fe(CO)_5$ (as though it answered to a very high type, FeX_{10}), which when cooled condenses into a liquid (slightly coloured, probably owing to incipient decomposition), sp. gr. 1·47, which solidifies at -21°, boils at about 103°, and has a vapour density (about 6·5 with respect to air) corresponding to the above formula; it decomposes at 180°. Water and dilute acids do not act upon it, but it decomposes under the action of light and forms a hard, non-volatile crystalline yellow compound $Fe_2(CO)_7$ which decomposes at 80° and again forms $Fe(CO)_5$.

[26 tri] When the molecular Fe_2Cl_6 is produced instead of $FeCl_3$ this complication of the type also occurs.

Some light may be thrown upon the faculty of Fe of forming various compounds with CN, by the fact that Fe not only combines with carbon but also with nitrogen. *Nitride of iron* Fe_2N was obtained by Fowler by heating finely powdered iron in a stream of NH_3 at the temperature of melting lead.

[27 bis] The sulphur of the animal refuse here forms the compound $FeKS_2$, which by the action of potassium cyanide yields potassium sulphide, thiocyanate, and ferrocyanide.

Potassium ferrocyanide may also be obtained from Prussian blue by boiling with a solution of potassium hydroxide, and from the ferricyanide by the action of alkalis and reducing substances (because the red prussiate is a product of oxidation produced by the action of chlorine: a ferric salt is reduced to a ferrous salt), &c. In many works (especially in Germany and France) yellow prussiate is prepared from the mass, containing oxide of iron, and employed for purifying coal gas (Vol. I., p. 361), which generally contains cyanogen compounds. About 2 p.c. of the nitrogen contained in coal is converted into cyanogen, which forms Prussian blue and thiocyanates in the mass used for purifying the gas. On evaporation the solution yields large yellow crystals containing 3 molecules of water, which is easily expelled by heating above 100°. 100 parts of water at the ordinary temperature are capable of dissolving 25 parts of this salt; its sp. gr. is 1·83. When ignited it forms potassium cyanide and iron carbide, FeC_2 (Chapter XIII., Note 12). Oxidising substances change it into potassium ferricyanide. With strong sulphuric acid it gives carbonic oxide, and with dilute sulphuric acid, when heated, prussic acid is evolved according to the equation: $2K_4FeC_6N_6 + 3H_2SO_4 = K_2Fe_2C_6N_6 + 3K_2SO_4 + 6HCN$; hence in the yellow prussiate K_2 replaces Fe.

Skraup obtained this salt both from potassium ferrocyanide with ferric chloride and from ferricyanide with ferrous chloride, which evidently shows that it contains iron in both the ferric and ferrous states. With ferrous chloride it forms Prussian blue, and with ferric chloride Turnbull's blue.

Prussian blue was discovered in the beginning of the last century by a Berlin manufacturer, Diesbach. It was then prepared, as it sometimes is also at present, directly from potassium cyanide obtained by heating animal charcoal with potassium carbonate. The mass thus obtained is dissolved in water, alum is added to the solution in order to saturate the free alkali, and then a solution of green vitriol is added which has previously been sufficiently exposed to the air to contain both ferric and ferrous salts. If the solution of potassium cyanide be mixed with a solution containing both salts, Prussian blue will be formed, because it is a compound of ferrous cyanide, FeC_2N_2, and ferric cyanide, $Fe_2C_6N_6$. A ferric salt with potassium ferrocyanide forms a blue colour, because ferrous cyanide is obtained from the first salt and ferric cyanide from the second. During the preparation of this compound alkali must be avoided, as otherwise the precipitate would contain oxides of iron. Prussian blue has not a crystalline structure; it forms a blue mass with a copper-red metallic lustre. Both acids and alkalis act on it. The action is at first confined to the ferric salt it contains. Thus alkalis form ferric oxide and ferrocyanide in solution: $2Fe_2C_6N_6,3FeC_2N_2 + 12KHO = 2(Fe_2O_3,3H_2O) + 3K_4FeC_6N_6$. Various ferrocyanides may thus be prepared. Prussian blue is soluble in an aqueous solution of oxalic acid, forming blue ink. In air, when exposed to the action of light, it fades; but in the dark again absorbs oxygen and becomes blue, which fact is also sometimes noticed in blue cloth. An excess of potassium ferrocyanide renders Prussian blue soluble in water, although insoluble in various saline solutions—that is, it converts it into the soluble variety. Strong hydrochloric acid also dissolves Prussian blue.

An excess of chlorine must not be employed in preparing this compound, otherwise the reaction goes further. It is easy to find out when the action of the chlorine on potassium ferrocyanide must cease; it is only necessary to take a sample of the liquid and add a solution of a ferric salt to it. If a precipitate of Prussian blue is formed, more chlorine must be added, as there is still some undecomposed ferrocyanide, for the ferricyanide does not give a precipitate with ferric salts. Potassium ferricyanide, like the ferrocyanide, easily exchanges its potassium for hydrogen and various metals by double decomposition. With the salts of tin, silver, and mercury it forms yellow precipitates, and with those of uranium, nickel, cobalt, copper, and bismuth brown precipitates. The lead salt under the action of sulphuretted hydrogen forms lead sulphide and a hydrogen salt or acid, $H_3FeC_6N_6$, corresponding with potassium ferricyanide, which is soluble, crystallises in red needles, and resembles hydroferrocyanic acid, $H_4FeC_6N_6$. Under the action of reducing agents—for instance, sulphuretted hydrogen, copper—potassium ferricyanide is changed into ferrocyanide, especially in the presence of alkalis, and thus forms a rather energetic *oxidising agent*—capable, for instance, of changing manganous oxide into dioxide, bleaching tissues, &c.

It is important to mention a series of readily crystallisable salts formed by the action of nitric acid on potassium and other ferrocyanides and ferricyanides. These salt contain the elements of nitric oxide, and are therefore called *nitro-(nitroso) ferricyanides* (*nitroprussides*). Generally a crystalline sodium salt is obtained, $Na_2FeC_5N_6O,2H_2O$. In its composition this salt differs from the red sodium salt, $Na_3FeC_6N_6$, by the fact that in it one molecule of sodium cyanide, NaCN, is replaced by nitric oxide, NO. In order to prepare it, potassium ferrocyanide in powder is mixed with five-sevenths of its weight of nitric acid diluted with an equal volume of water. The mixture is at first left at the ordinary temperature, and then heated on a water-bath. Here ferricyanide is first of all formed (as shown by the liquid giving a precipitate with ferrous chloride), which then disappears (no precipitate with ferrous chloride), and forms a green precipitate. The liquid, when cooled, deposits crystals of nitre. The liquid is then strained off and mixed with sodium carbonate, boiled, filtered, and evaporated; sodium nitrate and the salt described are deposited in crystals. It separates in prisms of a red colour. Alkalis and salts of the alkaline earths do not give precipitates: they are soluble, but the salts of iron, zinc, copper, and silver form precipitates where sodium is exchanged with these metals. It is remarkable that the sulphides of the alkali metals give with this salt an intense bright purple coloration. This series of compounds was discovered by Gmelin and studied by Playfair and others (1849).

This series to a certain extent resembles the nitro-sulphide series described by Roussin. Here the primary compound consists of black crystals, which are obtained as follows:—Solutions of potassium hydrosulphide and nitrate are mixed, and the mixture is agitated whilst ferric chloride is added, then boiled and filtered; on cooling, *black crystals* are deposited, having the composition $Fe_6S_3(NO)_{10},H_2O$ (Rosenberg), or, according to Demel, $FeNO_2,NH_2S$. They have a slightly metallic lustre, and are soluble in water, alcohol, and ether. They absorb the latter as easily as calcium chloride absorbs water. In the presence of alkalis these crystals remain unchanged, but with acids they evolve nitric oxides. There are several compounds which are capable of interchanging, and correspond with Roussin's salt. Here we enter into the series of the nitrogen compounds which have been as yet but little investigated, and will most probably in time form most instructive material for studying the nature of that element. These series of compounds are as unlike the usual saline compounds of inorganic chemistry as are organic hydrocarbons. There is no necessity to describe these series in detail, because their connection with other compounds is not yet clear, and they have not yet any application.

The residue from the roasting of cobalt ores is called *zafflor*, and is often met with in commerce. From this the purer compounds of cobalt may be prepared. The ores of nickel are also first roasted, and the oxides dissolved in acid, nickelous salts being then obtained.

The further treatment of cobalt and nickel ores is facilitated if the arsenic can be almost entirely removed, which may be effected by roasting the ore a second time with a small addition of nitre and sodium carbonate; the nitre combines with the arsenic, forming an arsenious salt, which may be extracted with water. The remaining mass is dissolved in hydrochloric acid, mixed with a small quantity of nitric acid. Copper, iron, manganese, nickel, cobalt, &c., pass into solution. By passing hydrogen sulphide through the solution, copper, bismuth, lead, and arsenic are deposited as metallic sulphides; but iron, cobalt, nickel, and manganese remain in solution. If an alkaline solution of bleaching powder be then added to the remaining solution, the whole of the manganese will first be deposited in the form of dioxide, then the cobalt as hydrated cobaltic oxide, and finally the nickel also. It is, however, impossible to rely on this method for effecting a complete separation, the more so since the higher oxides of the three above-mentioned metals have all a black colour; but, after a few trials, it will be easy to find how much bleaching powder is required to precipitate the manganese, and the amount which will precipitate all the cobalt. The manganese may also be separated from cobalt by precipitation from a mixture of the solutions of both metals (in the form of the 'ous' salts) with ammonium sulphide, and then treating the precipitate with acetic acid or dilute hydrochloric acid, in which manganese sulphide is easily soluble and cobalt sulphide almost insoluble. Further particulars relating to the separation of cobalt from nickel may be found in treatises on analytical chemistry. In practice it is usual to rely on the rough method of separation founded on the fact that nickel is more easily reduced and more difficult to oxidise than cobalt. The New Caledonian ore is smelted with $CaSO_4$ and $CaCO_3$ on coke, and a metallic regulus is obtained containing all the Ni, Fe, and S. This is roasted with SiO_2, which converts all the iron into slag, whilst the Ni remains combined with the S; this residue on further roasting gives NiO, which is reduced by the carbon to metallic Ni. The Canadian ore (a pyrites containing 11 p.c. Ni) is frequently treated in America (after a preliminary dressing) by smelting it with Na_2SO_4 and charcoal; the resultant fusible Na_2S then dissolves the CuS and FeS_2, while the NiS is obtained in a bottom layer (Bartlett and Thomson's process) from which Ni is obtained in the manner described above.

For manufacturing purposes somewhat impure cobalt compounds are frequently used, which are converted into *smalt*. This is glass containing a certain amount of cobalt oxide; the glass acquires a bright blue colour from this addition, so that when powdered it may be used as a blue pigment; it is also unaltered at high temperatures, so that it used to take the place now occupied by Prussian blue, ultramarine, &c. At present smalt is almost exclusively used for colouring glass and china. To prepare smalt, ordinary impure cobalt ore (zaffre) is fused in a crucible with quartz and potassium carbonate. A fused mass of cobalt glass is thus formed, containing silica, cobalt oxide, and potassium oxide, and a metallic mass remains at the bottom of the crucible, containing almost all the other metals, arsenic, nickel, copper, silver, &c. This metallic mass is called *speiss*, and is used as nickel ore for the extraction of nickel. Smalt usually contains 70 p.c. of silica, 20 p.c. of potash and soda, and about 5 to 6 p.c. of cobaltous oxide; the remainder consisting of other metallic oxides.

[32 bis] All we know respecting the relations of Co and Ni to Fe and Cu confirms the fact that Co is more closely related to Fe and Ni to Cu; and as the atomic weight of Fe = 56 and of Cu = 63, then according to the principles of the periodic system it would be expected that the atomic weight of Co would be about 59–60, whilst that of Ni should be greater than that of Co but less than that of Cu, *i.e.* about 50·5–60·5. However, as

yet the majority of the determinations of the atomic weights of Co and Ni give a different result and show that a lower atomic weight is obtained for Ni than for Co. Thus K. Winkler (1894) obtained (employing metals deposited electrolytically and determining the amount of iodine which combined with them) Ni = 58·72 and Co = 59·37 (if H = 1 and I = 126·53). In my opinion this should not be regarded as proving that the principles of the periodic system cannot be applied in this instance, nor as a reason for altering the position of these elements in the system (*i.e.* by placing Ni after Fe, and Co next to Cu), because in the first place the figures given by different chemists (for instance, Zimmermann, Krüss, and others) are somewhat divergent, and in the second place the majority of the latest modes of determining the atomic weights of Co and Ni aim at finding what weights of these metals react with known weights of other elements without taking into account the faculty they have of absorbing hydrogen; since this faculty is more developed in Ni than in Co the hydrogen (occluded in Ni) should lower the atomic weight of Ni more than that of Co. On the whole, the question of the atomic weights of Co and Ni cannot yet be considered as decided, notwithstanding the numerous researches which have been made; still there can be no doubt that the atomic weights of these two metals are very nearly equal, and greater than that of Fe, but less than that of Cu. This question is of great interest, not only for completing our knowledge of these metals, but also for perfecting our knowledge of the periodic system of the elements.

[32 tri] For instance, the alkalis may be fused in nickel vessels as well as in silver, because they have no action upon either metal. Nickel, like silver, is not acted upon by dilute acids. Only nitric acid dissolves both metals well. Nickel is harder, and fuses at a higher temperature than silver. For castings, a small quantity of magnesium (0·001 part by weight) is added to nickel to render it more homogeneous (just as aluminium is added to steel). Nickel forms many valuable alloys. Steel containing 3 p.c. Ni is particularly valuable, its limit of elasticity is higher and its hardness is greater; it is used for armour plate and other large pieces. The alloys of nickel, especially with copper and zinc (melchior, *see* later), aluminium and silver, although used in certain cases, are now replaced by nickel-plated or nickel-deposited goods (deposited by electricity from a solution of the ammonium salts).

The change of colour is dependent in all probability on the combination with water, or according to others on polymeric transformation. It enables a solution of cobalt chloride to be used as sympathetic ink. If something be written with cobalt chloride on white paper, it will be invisible on account of the feeble colour of the solution, and when dry nothing can be distinguished. If, however, the paper be heated before the fire, the rose-coloured salt will be changed into a less hydrous blue salt, and the writing will become quite visible, but fade away when cool.

The change of colour which takes place in solutions of $CoCl_2$ under the influence not only of solution in water or alcohol, but also of a change of temperature, is a characteristic of all the halogen salts of cobalt. Crystalline iodide of cobalt, CoI_26H_2O, gives a dark red solution between -22° and +20°; above +20° the solution turns brown and passes from olive to green, from +35° to 320° the solution remains green. According to Étard the change of colour is due to the fact that at first the solution contains the hydrate CoI_2H_2O, and that above 35° it contains CoI_24H_2O. These hydrates can be crystallised from the solutions; the former at ordinary temperature and the latter on heating the solution. The intermediate olive colour of the solutions corresponds to the incipient decomposition of the hexahydrated salt and its passage into CoI_24H_2O. A solution of the hexahydrated chloride of cobalt, $CoCl_26H_2O$, is rose-coloured between -22° and +25°; but the colour changes starting from +25°, and passes through all the tints between red and blue right up to 50°; a true blue solution is only obtained at 55° and remains up to 300°. This true blue solution contains another hydrate, $CoCl_22H_2O$.

The dependence between the solubility of the iodide and chloride of cobalt and the temperature is expressed by two almost straight lines corresponding to the hexa- and di-hydrates; the passage of the one into the other hydrate being expressed by a curve. The same character of phenomena is seen also in the variation of the vapour tension of solutions of chloride of cobalt with the temperature. We have repeatedly seen that aqueous solutions (for instance, Chapter XXII., Note 23 for Fe_2Cl_6) deposit different crystallo-hydrates at different temperatures, and that the amount of water in the hydrate decreases as the temperature t rises, so that it is not surprising that $CoCl_22H_2O$ (or according to Potilitzin $CoCl_2H_2O$) should separate out above 55° and $CoCl_26H_2O$ at 25° and below. Nor is it exceptional that the colour of a salt varies according as it contains different amounts of H_2O. But in this instance it is characteristic that the change of colour takes place in solution in the presence of an excess of water. This apparently shows that the actual solution may contain either $CoCl_26H_2O$ or $CoCl_22H_2O$. And as we know that a solution may contain both metaphosphoric PHO_3 and orthophosphoric acid $H_3PO_4 = HPO_3 + H_2O$, as well as certain other anhydrides, the question of the state of substances in solutions becomes still more complicated.

Nickel sulphate crystallises from neutral solutions at a temperature of from 15° to 20° in *rhombic* crystals containing $7H_2O$. Its form approaches very closely to that of the salts of zinc and magnesium. The planes of a vertical prism for magnesium salts are inclined at an angle of 90° 30', for zinc salts at an angle of 91° 7', and for nickel salts at an angle of 91° 10'. Such is also the form of the zinc and magnesium selenates and chromates. Cobalt sulphate containing 7 molecules of water is deposited in crystals of the *monoclinic* system, like the corresponding salts of iron and manganese. The angle of a vertical prism for the iron salt = 82° 20', for cobalt = 82° 22', and the inclination of the horizontal pinacoid to the vertical prism for the iron salt = 99° 2', and for the cobalt salt 99° 36'. All the isomorphous mixtures of the salts of magnesium, iron, cobalt, nickel and manganese have the same form if they contain 7 mol. H_2O and iron or cobalt predominate, whilst if there is a preponderance of magnesium, zinc, or nickel, the crystals have a rhombic form like magnesium sulphate. Hence these sulphates are *dimorphous*, but for some the one form is more stable and for others the other. Brooke, Moss, Mitscherlich, Rammelsberg, and Marignac have explained these relations. Brooke and Mitscherlich also supposed that $NiSO_4,7H_2O$ is not only capable of assuming these forms, but also that of the *tetragonal* system, because it is deposited in this form from acid, and especially from slightly-heated solutions (30° to 40°). But Marignac demonstrated that the tetragonal crystals do not contain 7, but 6, molecules of water, $NiSO_4,6H_2O$. He also observed that a solution evaporated at 50° to 70° deposits monoclinic crystals, but of a different form from ferrous sulphate, $FeSO_4,7H_2O$—namely, the angle of the prism is 71° 52', that of the pinacoid 95° 6'. This salt appears to be the same with 6 molecules of water as the tetragonal. Marignac also obtained magnesium and zinc salts with 6 molecules of water by evaporating their solutions at a higher temperature, and these salts were found to be isomorphous with the monoclinic nickel salt. In addition to this it must be observed that the rhombic crystals of nickel sulphate with $7H_2O$ become turbid under the influence of heat and light, lose water, and change into the tetragonal salt. The monoclinic crystals in time also become turbid, and change their structure, so that the tetragonal form of this salt is the most stable. Let us also add that nickel sulphate in all its shapes forms very beautiful emerald green crystals, which, when heated to 230°, assume a dirty greenish-yellow hue and then contain one molecule of water.

Klobb (1891) and Langlot and Lenoir obtained anhydrous $CoSO_4$ and $NiSO_4$ by igniting the hydrated salt with $(NH_4)_2SO_4$ until the ammonium salt had completely volatilised and decomposed.

We may add that when equivalent aqueous solutions of NiX_2 (green) and CoX_2 (red) are mixed together they give an almost colourless (grey) solution, in which the green and red colour of the component parts disappears owing to the combination of the complementary colours.

A double salt $NiKF_3$ is obtained by heating $NiCl_2$ with $KFHF$ in a platinum crucible; $KCoF_3$ is formed in a similar manner. The nickel salt occurs in fine green plates, easily soluble in water but scarcely soluble in ethyl and methyl alcohol. They decompose into green oxide of nickel and potassium fluoride when heated in a current of air. The analogous salt of cobalt crystallises in crimson flakes.

If instead of potassium fluoride, $CoCl_2$ or $NiCl_2$ be fused with ammonium fluoride, they also form double salts with the latter. This gives the possibility of obtaining anhydrous fluorides NiF_2 and CoF_2. Crystalline fluoride of nickel, obtained by heating the amorphous powder formed by decomposing the double ammonium salt in a stream of hydrofluoric acid, occurs in beautiful green prisms, sp. gr. 4·63, which are insoluble in water, alcohol, and ether; sulphuric, hydrochloric, and nitric acids also have no action upon them, even when heated; NiF_2 is decomposed by steam, with the formation of black oxide, which retains the crystalline structure of the salt. Fluoride of cobalt, obtained as a rose-coloured powder by decomposing the double ammonium salt with the aid of heat in a stream of hydrofluoric acid, fuses into a ruby-coloured mass which bears distinct signs of a crystalline structure; sp. gr. 4·43. The molten salt only volatilises at about 1400°, which forms a clear distinction between CoF_2 and the volatile NiF_2. Hydrochloric, sulphuric, and nitric acids act upon CoF_2 even in the cold, although slowly, while when heated the reaction proceeds rapidly (Poulenc, 1892).

Hydrated suboxide of cobalt (de Schulten, 1889) is obtained in the following manner. A solution of 10 grams of $CoCl_26H_2O$ in 60 c.c. of water is heated in a flask with 250 grams of caustic potash and a stream of coal gas is passed through the solution. When heated the hydrate of the suboxide of cobalt which separates out, dissolves in the caustic potash and forms a dark blue solution. This solution is allowed to stand for 24 hours in an atmosphere of coal gas (in order to prevent oxidation). The crystalline mass which separates out has a composition $Co(OH)_2$, and to the naked eye appears as a violet powder, which is seen to be crystalline under the microscope. The specific gravity of this hydrate is 3·597 at 15°. It does not undergo change in the air; warm acetic acid dissolves it, but it is insoluble in warm and cold solutions of ammonia and sal-ammoniac.

[34 bis] The following reaction may be added to those of the cobaltous and nickelous salts: potassium cyanide forms a precipitate with cobalt salts which is soluble in an excess of the reagent and forms a green solution. On heating this and adding a certain quantity of acid, a double *cobalt cyanide* is formed which corresponds with potassium ferricyanide. Its formation is accompanied with the evolution of hydrogen, and is founded upon the property which cobalt has of oxidising in an alkaline solution, the development of which has been observed in such a considerable measure in the cobaltamine salts. The process which goes on here may be expressed

by the following equation; $CoC_2N_2 + 4KCN$ first forms $CoK_4C_6N_6$, which salt with water, H_2O, forms potassium hydroxide, KHO, hydrogen, H, and the salt, $K_3CoC_6N_6$. Here naturally the presence of the acid is indispensable in consequence of its being required to combine with the alkali. From aqueous solutions this salt crystallises in transparent, hexagonal prisms of a yellow colour, easily soluble in water. The reactions of double decomposition, and even the formation of the corresponding acid, are here completely the same as in the case of the ferricyanide. If a nickelous salt be treated in precisely the same manner as that just described for a salt of cobalt, decomposition will occur.

The cobalt salts may be divided into at least the following classes, which repeat themselves for Cr, Ir, Rh (we shall not stop to consider the latter, particularly as they closely resemble the cobalt salts):—

(a) *Ammonium cobalt salts*, which are simply direct compounds of the cobaltous salts CoX_2 with ammonia, similar to various other compounds of the salts of silver, copper, and even calcium and magnesium, with ammonia. They are easily crystallised from an ammoniacal solution, and have a pink colour. Thus, for instance, when cobaltous chloride in solution is mixed with sufficient ammonia to redissolve the precipitate first formed, octahedral crystals are deposited which have a composition $CoCl_2,H_2O,6NH_3$. These salts are nothing else but combinations with ammonia of crystallisation—if it may be so termed—likening them in this way to combinations with water of crystallisation. This similarity is evident both from their composition and from their capability of giving off ammonia at various temperatures. The most important point to observe is that all these salts contain 6 molecules of ammonia to 1 atom of cobalt, and this ammonia is held in fairly stable connection. Water decomposes these salts. (Nickel behaves similarly without forming other compounds corresponding to the true cobaltic.)

(b) The solutions of the above-mentioned salts are rendered turbid by the action of the air; they absorb oxygen and become covered with a crust of *oxycobaltamine salts*. The latter are sparingly soluble in aqueous ammonia, have a brown colour, and are characterised by the fact that with warm water *they evolve oxygen*, forming salts of the following category: The nitrate may be taken as an example of this kind of salt; its composition is $CoN_2O_7,5NH_3,H_2O$. It differs from cobaltous nitrate, $Co(NO_3)_2$, in containing an extra atom of oxygen—that is, it corresponds with cobalt dioxide, CoO_2, in the same way that the first salts correspond with cobaltous oxide; they contain 5, and not 6, molecules of ammonia, as if NH_3 had been replaced by O, but we shall afterwards meet compounds containing either $5NH_3$ or $6NH_3$ to each atom of cobalt.

(*c*) *The luteocobaltic salts* are thus called because they have a yellow (luteus) colour. They are obtained from the salts of the first kind by submitting them in dilute solution to the action of the air; in this case salts of the second kind are not formed, because they are decomposed by an excess of water, with the evolution of oxygen and the formation of luteocobaltic salts. By the action of ammonia the salts of the fifth kind (roseocobaltic) are also converted into luteocobaltic salts. These last-named salts generally crystallise readily, and have a yellow colour; they are comparatively much more stable than the preceding ones, and even for a certain time resist the action of boiling water. Boiling aqueous potash liberates ammonia and precipitates hydrated cobaltic oxide, $Co_2O_3,3H_2O$, from them. This shows that the luteocobaltic salts correspond with cobaltic oxide, Co_2O_3, and those of the second kind with the dioxide. When a solution of luteocobaltic sulphate, $Co_2(SO_4)_3,12NH_3,4H_2O$, is treated with baryta, barium sulphate is precipitated, and the solution contains luteocobaltic hydroxide, $Co(OH)_3,6NH_3$, which is soluble in water, is powerfully alkaline, absorbs the oxygen of the air, and when heated is decomposed with the evolution of ammonia. This compound therefore corresponds to a solution of cobaltic hydroxide in ammonia. The luteocobaltic salts contain 2 atoms of cobalt and 12 molecules of ammonia—that is, $6NH_3$ to each atom of cobalt, like the salts of the first kind. The CoX_2 salts have a metallic taste, whilst those of luteocobalt and others have a purely saline taste, like the salts of the alkali metals. In the luteo-salts all the X's react (are ionised, as some chemists say) as in ordinary salts—for instance, all the Cl_2 is precipitated by a solution of $AgNO_3$; all the $(SO_4)_3$ gives a precipitate with BaX_2, &c. The double salt formed with $PtCl_4$ is composed in the same manner as the potassium salt, $K_2PtCl_4 = 2KCl + PtCl_4$, that is, contains $(CoCl_3,6NH_3)_2,3PtCl_4$, or the amount of chlorine in the $PtCl_4$ is double that in the alkaline salt. In the rosepentamine (*e*), and rosetetramine (*f*), salts, also all the X's react or are ionised, but in the (*g*) and (*h*) salts only a portion of the X's react, and they are equal to the (*e*) and (*f*) salts minus water; this means that although the water dissolves them it is not combined with them, as PHO_3 differs from PH_3O_3; phenomena of this class correspond exactly to what has been already (Chapter XXI., Note [7]) mentioned respecting the green and violet salts of oxide of chromium.

(*d*) *The fuscocobaltic salts*. An ammoniacal solution of cobalt salts acquires a brown colour in the air, due to the formation of these salts. They are also produced by the decomposition of salts of the second kind; they crystallise badly, and are separated from their solutions by addition of alcohol or an excess of ammonia. When boiled they give up the ammonia and cobaltic oxide which they contain. Hydrochloric and nitric acids give a yellow precipitate with these salts, which turns red when boiled, forming salts of the next category. The following is an example of the composition of two of the fuscocobaltic salts, $Co_2O(SO_4)_2,8NH_3,4H_2O$ and $Co_2OCl_4,8NH_3,3H_2O$. It is evident that the fuscocobaltic salts are ammoniacal compounds of basic cobaltic salts. The normal cobaltic sulphate ought to have the composition $Co_2(SO_4)_3 = Co_2O_3,3SO_3$; the simplest basic salts will be $Co_2O(SO_4)_2 = Co_2O_3)2SO_3$, and $Co_2O_2(SO_4) = Co_2O_3,SO_3$. The fuscocobaltic salts correspond with the first type of basic salts. They are changed (in concentrated solutions) into oxycobaltamine salts by absorption of one atom of oxygen, $Co_2O_2(SO_4)_2$. The whole process of oxidation will be as follows: first of all Co_2X_4, a cobaltous salt, is in the solution (X a univalent haloid, 2 molecules of the salt being taken), then Co_2OX_4, the basic cobaltic salt (4th series), then $Co_2O_2X_4$, the salt of the dioxide (2nd series). The series of basic salts with an acid, 2HX, forms water and a normal salt, Co_2X_6 (in 3, 5, 6 series). These salts are combined with various amounts of water and ammonia. Under many conditions the salts of fuscocobalt are easily transformed into salts of the next series. The salts of the series that has just been described contain 4 molecules of ammonia to 1 atom of cobalt.

(*e*) The *roseocobaltic* (or rosepentamine), $CoX_2H_2O,5NH_3$, *salts*, like the luteocobaltic, correspond with the normal cobaltic salts, but contain less ammonia, and an extra molecule of water. Thus the sulphate is obtained from cobaltous sulphate dissolved in ammonia and left exposed to the air until transformed into a brown solution of the fuscocobaltic salt; when this is treated with sulphuric acid a crystalline powder of the roseocobaltic salt, $Co_2(SO_4)_3,10NH_3,5H_2O$, separates. The formation of this salt is easily understood: cobaltous sulphate in the presence of ammonia absorbs oxygen, and the solution of the fuscocobaltic salt will therefore contain, like cobaltous sulphate, one part of sulphuric acid to every part of cobalt, so that the whole process of formation may be expressed by the equation: $10NH_3 + 2CoSO_4 + H_2SO_4 + 4H_2O + O = Co_2(SO_4)_3,10NH_3,5H_2O$. This salt forms tetragonal crystals of a red colour, slightly soluble in cold, but readily soluble in warm water. When the sulphate is treated with baryta, roseocobaltic hydroxide is formed in the solution, which absorbs the carbonic anhydride of the air. It is obtained from the next series by the action of alkalis.

(*f*) The *rosetetramine cobaltic salts* $CoCl_2,2H_2O,4NH_3$ were obtained by Jörgenson, and belong to the type of the luteo-salts, only with the substitution of $2NH_3$ for H_2O. Like the luteo- and roseo-salts they give double salts with $PtCl_4$, similar to the alkaline double salts, for instance $(Co_2H_2O,4NH_3)2(SO_4)_2Cl_2PtCl_4$. They are darker in colour than the preceding, but also crystallise well. They are formed by dissolving $CoCO_3$ in sulphuric acid (of a given strength), and after NH_3 and carbonate of ammonium have been added, air is passed through the solution (for oxidation) until the latter turns red. It is then evaporated with lumps of carbonate of ammonium, filtered from the precipitate and crystallised. A salt of the composition $Co_2(CO_3)_2(SO_4),(2H_2O,4NH_3)_2$ is thus obtained, from which the other salts may be easily prepared.

(*g*) The *purpureocobaltic salts*, $CoX_3,5NH_3$, are also products of the direct oxidation of ammoniacal solutions of cobalt salts. They are easily obtained by heating the roseocobaltic and luteo-salts with strong acids. They are to all effects the same as the roseocobaltic salts, only anhydrous. Thus, for instance, the purpureocobaltic chloride, $Co_2Cl_6,10NH_3$, or $CoCl_3,5NH_3$, is obtained by boiling the oxycobaltamine salts with ammonia. There is the same distinction between these salts and the preceding ones as between the various compounds of cobaltous chloride with water. In the purpureocobaltic only X_2 out of the X_3 react (are ionised). To the rosetetramine salts (*f*) there correspond the *purpureotetramine* salts, $CoX_3H_2O,4NH_3$. The corresponding chromium purpureopentamine salt, $CrCl_3,5NH_3$ is obtained with particular ease (Christensen, 1893). Dry anhydrous chromium chloride is treated with anhydrous liquid ammonia in a freezing mixture composed of liquid CO_2 and chlorine, and after some time the mixture is taken out of the freezing mixture, so that the excess of NH_3 boils away; the violet crystals then immediately acquire the red colour of the salt, $CrCl_3,5NH_3$, which is formed. The product is washed with water (to extract the luteo-salt, $CrCl_3,6NH_3$), which does not dissolve the salt, and it is then recrystallised from a hot solution of hydrochloric acid.

(*h*) The *praseocobaltic salts*, $CoX_3,4NH_3$, are green, and form, with respect to the rosetetramine salts (*f*), the products of ultimate dehydration (for example, like metaphosphoric acid with respect to orthophosphoric acid, but in dissolving in water they give neither rosetetramine nor tetramine salts. (In my opinion one should expect salts with a still smaller amount of NH_3, of the blue colour proper to the low hydrated compounds of cobalt; the green colour of the prazeo-salts already forms a step towards the blue.) Jörgenson obtained salts for ethylene-diamine, $N_2H_4C_2H_4$ which replaces $2NH_3$. After being kept a long time in aqueous solution they give rosetetramine salts, just as metaphosphoric acid gives orthophosphoric acid, while the rosetetramine salts are converted into prazeo-salts by Ag_2O and NaHO. Here only one X is ionised out of the X_3. There are also basic salts of the same type; but the best known is the chromium salt called the rhodozochromic salt, $Cr_2(OH)_3Cl_3,6NH_3,2H_2O$, which is formed by the prolonged action of water upon the corresponding roseo-salt.

The cobaltamine compounds differ essentially but little from the ammoniacal compounds of other metals. The only difference is that here the cobaltic oxide is obtained from the cobaltous oxide in the presence of ammonia. In any case it is a simpler question than that of the double cyanides. Those forces in virtue of which such a considerable number of ammonia molecules are united with a molecule of a cobalt compound, appertain naturally to the series of those slightly investigated forces which exist even in the highest degrees of combination of the majority of elements. They are the same forces which lead to the formation of compounds containing water of crystallisation, double salts, isomorphous mixtures and complex acids (Chapter XXI., Note 8 bis). The simplest conception, according to my opinion, of cobalt compounds (much more so than by assuming special complex radicles, with Schiff, Weltzien, Claus, and others), may be formed by comparing them with other ammoniacal products. Ammonia, like water, combines in various proportions with a multitude of molecules. Silver chloride and calcium chloride, just like cobalt chloride, absorb ammonia, forming compounds which are sometimes slightly stable, and easily dissociated, sometimes more stable, in exactly the same way as water combines with certain substances, forming fairly stable compounds called hydroxides or hydrates, or less stable compounds which are called compounds with water of crystallisation. Naturally the difference in the properties in both cases depends on the properties of those elements which enter into the composition of the given substance, and on those kinds of affinity towards which chemists have not as yet turned their attention. If boron fluoride, silicon fluoride, &c., combine with hydrofluoric acid, if platinic chloride, and even cadmium chloride, combine with hydrochloric acid, these compounds may be regarded as double salts, because acids are salts of hydrogen. But evidently water and ammonia have the same saline faculty, more especially as they, like haloid acids, contain hydrogen, and are both capable of further combination—for instance, ammonia with hydrochloric acid. Hence it is simpler to compare complex ammoniacal with double salts, hydrates, and similar compounds, but *the ammonio-metallic salts* present a most complete qualitative and quantitative resemblance to *the hydrated salts of metals*. The composition of the latter is $MX_n m H_2O$, where M = metal, X = the haloid, simple or complex, and n and m the quantities of the haloid and so-called water of crystallisation respectively. The composition of the ammoniacal salts of metals is $MX_n m NH_3$. The water of crystallisation is held by the salt with more or less stability, and some salts even do not retain it at all; some part with water easily when exposed to the air, others when heated, and then with difficulty. In the case of some metals all the salts combine with water, whilst with others only a few, and the water so combined may then be easily disengaged. All this applies equally well to the ammoniacal salts, and therefore the combination of ammonia may be termed *the ammonia of crystallisation*. Just as the water which is combined with a salt is held by it with different degrees of force, so it is with ammonia. In combining with $2NH_3, PtCl_2$ evolves 31,000 cals.; while $CaCl_2$ only evolves 14,000 cals.; and the former compound parts with its NH_3 (together with HCl in this case) with more difficulty, only above 200°, while the latter disengages ammonia at 180°. $ZnCl_2,2NH_3$ in forming $ZnCl_2,4NH_3$ evolves only 11,000 cals., and splits up again into its components at 80°. The amount of combined ammonia is as variable as the amount of water of crystallisation—for instance, SnI_48NH_3, $CrCl_28NH_3$, $CrCl_36NH_3$, $CrCl_35NH_3,PtCl_2,4NH_3$, &c. are known. Very often NH_3 is replaceable by OH_2 and conversely. A colourless, anhydrous cupric salt—for instance, cupric sulphate—when combined with water forms blue and green salts, and violet when combined with ammonia. If steam be passed through anhydrous copper sulphate the salt absorbs water and becomes heated; if ammonia be substituted for the water the heating becomes much more intense, and the salt breaks up into a fine violet powder. With water $CuSO_4,5H_2O$ is formed, and with ammonia $CuSO_4,5NH_3$, the number of water and ammonia molecules retained by the salt being the same in each case, and as a proof of this, and that it is not an isolated coincidence, the remarkable fact must be borne in mind that water and ammonia consecutively, molecule for molecule, are capable of supplanting each other, and forming the compounds $CuSO_4,5H_2O$, $CuSO_4,4H_2O,NH_3$; $CuSO_4,3H_2O,2NH_3$; $CuSO_4,2H_2O,3NH_3$; $CuSO_4,H_2O,4NH_3$, and $CuSO_4,5NH_3$. The last of these compounds was obtained by Henry Rose, and my experiments have shown that more ammonia than this cannot be retained. By adding to a strong solution of cupric sulphate sufficient ammonia to dissolve the whole of the oxide precipitated, and then adding alcohol, Berzelius obtained the compound $CuSO_4,H_2O,4NH_3$, &c. The law of substitution also assists in rendering these phenomena clearer, because a compound of ammonia with water forms ammonium hydroxide, NH_4HO, and therefore these molecules combining with one another may also interchange, as being of equal value. In general, those salts form stable ammoniacal compounds which are capable of forming stable compounds with water of crystallisation; and as ammonia is capable of combining with acids, and as some of the salts formed by slightly energetic bases in their properties more closely resemble acids (that is, salts of hydrogen) than those salts containing more energetic bases, we might expect to find more stable and more easily-formed ammonio-metallic salts with metals and their oxides having weaker basic properties than with those which form energetic bases. This explains why the salts of potassium, barium, &c., do not form ammonio-metallic salts, whilst the salts of silver, copper, zinc, &c., easily form them, and the salts RX_3 still more easily and with greater stability. This consideration also accounts for the great stability of the ammoniacal compounds of cupric oxide compared with those of silver oxide, since the former is displaced by the latter. It also enables us to see clearly the distinction which exists in the stability of the cobaltamine salts containing salts corresponding with cobaltous oxide, and those corresponding with higher oxides of cobalt, for the latter are weaker bases than cobaltous oxides. *The nature of the forces and quality of the phenomena occurring during the formation of the most stable substances, and of such compounds as crystallisable compounds, are one and the same, although perhaps exhibited in a different degree.* This, in my opinion, may be best confirmed by examining the compounds of carbon, because for this element the nature of the forces acting during the formation of its compounds is well known. Let us take as an example two unstable compounds of carbon. Acetic acid, $C_2H_4O_2$ (specific gravity 1·06), with water forms the hydrate, $C_2H_4O_2,H_2O$, denser (1·07) than either of the components, but unstable and easily decomposed, generally simply referred to as a solution. Such also is the crystalline compound of oxalic acid, $C_2H_2O_4$, with water, $C_2H_2O_4,2H_2O$. Their formation might be predicted as starting from the hydrocarbon C_2H_6, in which, as in any other, the hydrogen may be exchanged for chlorine, the water residue (hydroxyl), &c. The first substitution product with hydroxyl, $C_2H_5(HO)$, is stable; it can be distilled without alteration, resists a temperature higher than 100°, and then does not give off water. This is ordinary alcohol. The second, $C_2H_4(HO)_2$, can also be distilled without change, but can be decomposed into water and C_2H_4O (ethylene oxide or aldehyde); it boils at about 197°, whilst the first hydrate boils at 78°, a difference of about 100°. The compound $C_2H_3(HO)_3$ will be the third product of such substitution; it ought to boil at about 300°, but does not resist this temperature—it decomposes into H_2O and $C_2H_4O_2$, where only one hydroxyl group remains, and the other atom of oxygen is left in the same condition as in ethylene oxide, C_2H_4O. There is a proof of this. Glycol, $C_2H_4(HO)_2$, boils at 197°, and forms water and ethylene oxide, which boils at 13° (aldehyde, its isomeride, boils at 21°); therefore the product disengaged by the splitting up of the hydrate boils at 184° lower than the hydrate $C_2H_4(HO)_2$. Thus the hydrate $C_2H_3(HO)_3$, which ought to boil at about 300°, splits up in exactly the same way into water and the product $C_2H_4O_2$, which boils at 117°—that is, nearly 183° lower than the hydrate, $C_2H_3(HO)_3$. But this hydrate splits up before distillation. The above-mentioned hydrate of acetic acid is such a decomposable hydrate—that is to say, what is called a solution. Still less stability may be expected from the following hydrates. $C_2H_2(HO)_4$ also splits up into water and a hydrate (it contains two hydroxyl groups) called glycolic acid, $C_2H_2O(HO)_2 = C_2H_4O_3$. The next product of substitution will be $C_2H(HO)_5$; it splits up into water, H_2O, and glyoxylic acid, $C_2H_4O_4$ (three hydroxyl groups). The last hydrate which ought to be obtained from C_2H_6, and ought to contain $C_2(HO)_6$, is the crystalline compound of oxalic acid, $C_2H_2O_4$ (two hydroxyl groups), and water, $2H_2O$, which has been already mentioned. The hydrate $C_2(HO)_6 = C_2H_2O_4,2H_2O$, ought, according to the foregoing reasoning, to boil at about 600° (because the hydrate, $C_2H_4(HO)_2$, boils at about 200°, and the substitution of 4 hydroxyl groups for 4 atoms of hydrogen will raise the boiling-point 400°). It does not resist this temperature, but at a much lower point splits up into water, $2H_2O$, and the hydrate $C_2O_2(HO)_2$, which is also capable of yielding water. Without going into further discussion of this subject, it may be observed that the formation of the hydrates, or compounds with water of crystallisation, of acetic and oxalic acids has thus received an accurate explanation, illustrating the point we desired to prove in affirming that compounds with water of crystallisation are held together by the same forces as those which act in the formation of other complex substances, and that the easy displaceability of the water of crystallisation is only a peculiarity of a local character, and not a radical point of distinction. All the above-mentioned hydrates, C_2X_6, or products of their destruction, are actually obtained by the oxidation of the first hydrate, $C_2H_3(HO)$, or common alcohol, by nitric acid (Sokoloff and others). Hence the forces which induce salts to combine with nH_2O or with NH_3 are undoubtedly of the same order as the forces which govern the formation of ordinary 'atomic' and saline compounds. (A great impediment in the study of the former was caused by the conviction which reigned in the sixties and seventies, that 'atomic' were essentially different from 'molecular' compounds like crystallohydrates, in which it was assumed that there was a combination of entire molecules, as though without the participation of the atomic forces.) If the bond between chlorine and different metals is not equally strong, so also the bond uniting nH_2O and nNH_3 is exceeding variable; there is nothing very surprising in this. And in the fact that the combination of different amounts of NH_3 and H_2O alters the capacity of the haloids X of the salts RX_2 for reaction (for instance, in the luteo-salts all the X_3, while in the purpureo, only 2 out of the 3, and in the prazeo-salts only 1 of the 3 X's reacts), we should see in the first place a phenomenon similar to what we met with in Cr_2Cl_6 (Chapter XXI., Note 7 bis), for in both instances the essence of the difference lies in the removal of water; a molecule $RCl_3,6H_2O$ or $RCl_3,6NH_3$ contains the halogen in a perfectly mobile (ionised) state, while in the molecule $RCl_3,5H_2O$ or $RCl_3,5NH_3$ a portion of the halogen has almost lost its faculty for reacting with $AgNO_3$, just as metalepsical chlorine has lost this faculty which is fully developed in the chloranhydride. Until the reason of this difference be clear, we cannot expect that ordinary points of view and generalisation can give a clear answer. However, we may assume that here the explanation lies in the nature and kind of motion of the atoms in the molecules, although as yet it is not clear how. Nevertheless, I think it well to call attention again (Chapter I.) to the fact that the combination of water, and hence, also, of any other element, leads to most diverse consequences; the water in the gelatinous hydrate of alumina or in the decahydrated Glauber salt is very mobile, and easily reacts like water in a free state; but the same water combined with oxide of calcium, or C_2H_4 (for instance, in C_2H_6O and in $C_4H_{10}O$), or with P_2O_5, has become quite different, and no longer acts like water in a free state. We see the same phenomenon in many other cases—for example, the chlorine in chlorates no longer gives a precipitate of chloride of silver with $AgNO_3$. Thus, although the instance which is found in the difference between the roseo- and purpureo-salts deserves to be fully studied on account of its simplicity, still it is far from being exceptional, and we cannot expect it to be thoroughly explained unless a mass of similar instances, which are exceedingly common among chemical compounds, be conjointly explained. (Among the researches which add to our knowledge respecting the complex ammoniacal compounds, I think it indispensable to call the reader's attention to Prof. Kournakoff's dissertation 'On complex metallic bases,' 1893.)

Kournakoff (1894) showed that the solubility of the luteo-salt, $CoCl_3,6NH_3$, at $0°$ = 4·30 (per 100 of water), at $20°$ = 7·7, that in passing into the roseo-salt, $CoCl_3H_2O_5NH_3$, the solubility rises considerably, and at $0°$ = 16·4, and at $20°$ = about 27, whilst the passage into the purpureo-salt, $CoCl_3,5NH_3$, is accompanied by a great fall in the solubility, namely, at $0°$ = 0·23, and at $20°$ = about 0·5. And as crystallohydrates with a smaller amount of water are usually more soluble than the higher crystallohydrates (Le Chatelier), whilst here we find that the solubility falls (in the purpureo-salt) with a loss of water, that water which is contained in the roseo-salt cannot be compared with the water of crystallisation. Kournakoff, therefore, connects the fall in solubility (in the passage of the roseo- into the purpureo-salts) with the accompanying loss in the reactive capacity of the chlorine.

In conclusion, it may be observed that the elements of the eighth group—that is, the analogues of iron and platinum—according to my opinion, will yield most fruitful results when studied as to combinations with whole molecules, as already shown by the examples of complex ammoniacal, cyanogen, nitro-, and other compounds, which are easily formed in this eighth group, and are remarkable for their stability. This faculty of the elements of the eighth group for forming the complex compounds alluded to, is in all probability connected with the position which the eighth group occupies with regard to the others. Following the seventh, which forms the type RX_7, it might be expected to contain the most complex type, RX_8. This is met with in OsO_4. The other elements of the eighth group, however, only form the lower types RX_2, RX_3, RX_4 ... and these accordingly should be expected to aggregate themselves into the higher types, which is accomplished in the formation of the above-mentioned complex compounds.

[35 bis] Marshall (1891) obtained cobaltic sulphate, $Co_2(SO_4)_3,18H_2O$, by the action of an electric current upon a strong solution of $CoSO_4$.

The action of an alkaline hypochlorite or hypobromite upon a boiling solution of cobaltous salts, according to Schroederer (1889), produces oxides, whose composition varies between Co_3O_5 (Rose's compound) and Co_2O_3, and also between Co_5O_8 and $Co_{12}O_{19}$. If caustic potash and then bromine be added to the liquid, only Co_2O_3 is formed. The action of alkaline hypochlorites or hypo-bromites, or of iodine, upon cobaltic salts, gives a highly-coloured precipitate which has a different colour to the hydrate of the oxide $Co_2(OH)_6$. According to Carnot the precipitate produced by the hypochlorites has a composition $Co_{10}O_{16}$, whilst that given by iodine in the presence of an alkali contains a larger amount of oxygen. Fortmann (1891) re-investigated the composition of the higher oxygen oxide obtained by iodine in the presence of alkali, and found that the greenish precipitate (which disengages oxygen when heated to 100°) corresponds to the formula CoO_2. The reaction must be expressed by the equation: $CoX_2 + I_2 + 4KHO = CoO_2 + 2KX + 2KI + 2H_2O$.

Prior to Fortmann, Rousseau (1889) endeavoured to solve the question as to whether CoO_2 was able to combine with bases. He succeeded in obtaining a barium compound corresponding to this oxide. Fifteen grams of $BaCl_2$ or $BaBr_2$ are triturated with 5–6 grams of oxide of barium, and the mixture heated to redness in a closed platinum crucible; 1 gram of oxide of cobalt is then gradually added to the fused mass. Each addition of oxide is accompanied by a violent disengagement of oxygen. After a short time, however, the mass fuses quietly, and a salt settles at the bottom of the crucible, which, when freed from the residue, appears as black hexagonal, very brilliant crystals. In dissolving in water this substance evolves chlorine; its composition corresponds to the formula $2(CoO_2)BaO$. If the original mass be heated for a long time (40 hours), the amount of dioxide in the resultant mass decreases. The author obtained a neutral salt having the composition CoO_2BaO (this compound $= BaO_2CoO$) by breaking up the mass as it agglomerates together, and bringing the pieces into contact with the more heated surface of the crucible. This salt is formed between the somewhat narrow limits of temperature 1,000°-1,100°; above and below these limits compounds richer or poorer in CoO_2 are formed. The formation of CoO_2 by the action of BaO_2, and the easy decomposition of CoO_2 with the evolution of oxygen, give reason for thinking that it belongs to the class of peroxides (like Cr_2O_7, CaO_2, &c.); it is not yet known whether they give peroxide of hydrogen like the true peroxides. The fact that it is obtained by means of iodine (probably through HIO), and its great resemblance to MnO_2, leads rather to the supposition that CoO_2 is a very feeble saline oxide. The form CoO_2 is repeated in the cobaltic compounds (Note 35), and the existence of CoO_2 should have long ago been recognised upon this basis.

This compound is known as nickel tetra-carbonyl. It appears to me yet premature to judge of the structure of such an extraordinary compound as $Ni(CO)_4$. It has long been known that potassium combines with CO forming $K_n(CO)_n$ (Chapter IX., Note 31), but this substance is apparently saline and non-volatile, and has as little in common with $Ni(CO)_4$ as Na_2H has with SbH_3. However, Berthelot observed that when NiC_4O_4 is kept in air, it oxidises and gives a colourless compound, $Ni_3C_2O_3,10H_2O$, having apparently saline properties. We may add that Schützenberger, on reducing $NiCl_2$ by heating it in a current of hydrogen, observed that a nickel compound partially volatilises with the HCl and gives metallic nickel when heated again. The platinum compound, $PtCl_2(CO)_3$ (Chapter XXIII., Note 11), offers the greatest analogy to $Ni(CO)_4$. This compound was obtained as a volatile substance by Schützenberger by moderately heating (to 235°) metallic platinum in a mixture of chlorine and carbonic oxide. If we designate CO by Y, and an atom of chlorine by X, then taking into account that, according to the periodic system, Ni is an analogue of Pt, a certain degree of correspondence is seen in the composition NiY_4 and PtX_2Y_2. It would be interesting to compare the reactions of the two compounds.

According to its empirical formula oxalate of nickel also contains nickel and carbonic oxide.

The following are the thermo-chemical data (according to Thomsen, and referred to gram weights expressed by the formula, in large calories or thousand units of heat) for the formation of corresponding compounds of Mn, Fe, Co, Ni, and Cu (+ Aq signifies that the reaction proceeds in an excess of water):

	R = Mn	Fe	Co	Ni	Cu
R + Cl_2 + Aq	128	100	95	94	63
R + Br_2 + Aq	106	78	73	72	41
R + I_2 + Aq	76	48	43	41	32
R + O + H_2O	95	68	63	61	38
R + O_2 + SO_2 + nH_2O	193	169	163	163	130
RCl_2 + Aq	+16	18	18	19	11

These examples show that for analogous reactions the amount of heat evolved in passing from Mn to Fe, Co, Ni, and Cu varies in regular sequences as the atomic weight increases. A similar difference is to be found in other groups and series, and proves that thermo-chemical phenomena are subject to the periodic law.

CHAPTER XXIII
THE PLATINUM METALS

The six metals: ruthenium, Ru, rhodium, Rh, palladium, Pd, osmium, Os, iridium, Ir, and platinum, Pt, are met with associated together in nature. Platinum always predominates over the others, and hence they are known as the *platinum metals*. By their chemical character their position in the periodic system is in the eighth group, corresponding with iron, cobalt, and nickel.

The natural transition from titanium and vanadium to copper and zinc by means of the elements of the iron group is demonstrated by all the properties of these elements, and in exactly the same manner a transition from zirconium, niobium, and molybdenum to silver, cadmium, and indium, through ruthenium, rhodium, and palladium, is in perfect accordance with fact and with the magnitude of the atomic weights, as also is the position of osmium, iridium, and platinum between tantalum and tungsten on the one side, and gold and mercury on the other. In all these three cases the elements of smaller atomic weight (chromium, molybdenum, and tungsten) are able, in their higher grades of oxidation, to give acid oxides having the properties of distinct but feebly energetic acids (in the lower oxides they give bases), whilst the elements of greater atomic weight (zinc, cadmium, mercury), even in their higher grades of oxidation, only give bases, although with feebly developed basic properties. The platinum metals present the same intermediate properties such as we have already seen in iron and the elements of the eighth group.

In the platinum metals the intermediate properties *of feebly acid and feebly basic metals* are developed with great clearness, so that there is not one sharply defined acid anhydride among their oxides, although there is a great diversity in the grades of oxidation from the type RO_4 to R_2O. The feebleness of the chemical forces observed in the platinum metals is connected with the ready decomposability of their compounds, with the small atomic volume of the metals themselves, and with their large atomic weight. The oxides of platinum, iridium, and osmium can scarcely be termed either basic or acid; they are capable of combinations of both kinds, each of which is feeble. They are all intermediate oxides.

The atomic weights of platinum, iridium, and osmium are nearly 191 to 196, and of palladium, rhodium, and ruthenium, 104 to 106. Thus, strictly speaking, we have here two series of metals, which are, moreover, perfectly parallel to each other; three members in the first series, and three members in the second—namely, platinum presents an analogy to palladium, iridium to rhodium, and osmium to ruthenium. As a matter of fact, however, the whole *group* of the platinum metals is characterised by *a number of common properties*, both physical and chemical, and, moreover, there are several points of resemblance between the members of this group and those of the *iron* group (Chapter XXII.) The atomic volumes (Table III., column 18) of the elements of this group are *nearly equal* and *very small*. The iron metals have atomic volumes of nearly 7, whilst that of the metals allied to palladium is nearly 9, and of those adjacent to platinum (Pt, Ir, Os) nearly 9·4. This comparatively small atomic volume corresponds with the great infusibility and tenacity proper to all the iron and platinum metals, and to their small chemical energy, which stands out very clearly in the heavy platinum metals. All the platinum metals are very *easily reduced* by ignition and by the action of various reducing agents, in which process oxygen, or a haloid group, is disengaged from their compounds and the metal left behind. This is a property of the platinum metals which determines many of their reactions, and the circumstance of their always being found in nature *in a native state*. In Russia in the Urals (discovered in 1819) and in Brazil (1735) platinum is obtained from alluvial deposits, but in 1892 Professor Inostrantseff discovered a vein deposit of platinum in serpentine near Tagil in the Urals. The facility with which they are reduced is so great that their chlorides are even decomposed by gaseous hydrogen, especially when shaken up and heated under a certain pressure. Hence it will be readily understood that such metals as zinc, iron, &c., separate them from solutions with great ease, which fact is taken advantage of in practice and in the chemical treatment of the platinum metals.[1 bis]

All the platinum metals, like those of the iron group, are grey, with a comparatively feeble metallic lustre, and are very infusible. In this respect they stand in the same order as the metals of the iron series; nickel is more fusible and whiter than cobalt and iron, so also palladium is whiter and more fusible than rhodium and ruthenium, and platinum is comparatively more fusible and whiter than iridium or osmium. The saline compounds of these metals are red or yellow, like those of the majority of the metals of the iron series, and like the latter, the different forms of oxidation present different colours. Moreover, certain complex compounds of the platinum metals, like certain complex compounds of the iron series, either have particular characteristic tints or else are colourless.

The platinum metals are found *in nature associated together in alluvial deposits* in a few localities, from which they are washed, owing to their very considerable density, which enables a stream of water to wash away the sand and clay with which they are mixed. Platinum deposits are chiefly known in the Urals, and also in Brazil and a few other localities. The platinum ore washed from these alluvial deposits presents the appearance of more or less coarse grains, and sometimes, as it were, of semi-fused nuggets.

All the platinum metals give compounds with the halogens, and the highest haloid type of combination for all is RX_4. For the majority of the platinum metals this type is exceedingly unstable; the lower compounds corresponding to the type RX_2, which are formed by the separation of X_2, are more stable. In the type RX_2 the platinum metals form more stable salts, which offer no little resemblance to the kindred compounds of the iron series—for example, to nickelous chloride, $NiCl_2$, cobaltous chloride, $CoCl_2$, &c. This even expresses itself in a similarity of volume (platinous chloride, $PtCl_2$, volume, 46; nickelous chloride, $NiCl_2 = 50$), although in the type RX_2 the true iron metals give very stable compounds, whilst the platinum metals frequently react after the manner of suboxides, decomposing into the metal and higher types, $2RX_2 = R + RX_4$. This probably depends on the facility with which RX_2 decomposes into R and X_2, when X_2 combines with the remaining portion of RX_2.

As in the series iron, cobalt, nickel, nickel gives NiO and Ni_2O_3, whilst cobalt and iron give higher and varied forms of oxidation, so also among the platinum metals, platinum and palladium only give the forms RX_2 and RX_4, whilst rhodium and iridium form another and intermediate type, RX_3, also met with in cobalt, corresponding with the oxide, having the composition R_2O_3, besides which they form an acid oxide, like ferric acid, which is also known in the form of salts, but is in every respect unstable. *Osmium* and *ruthenium*, like manganese, form still higher oxides, and in this respect exhibit the greatest diversity. They not only give RX_2, RX_3, RX_4, and RX_6, but also a still *higher form of oxidation*, RO_4, which is not met with in any other series. This form is exceedingly characteristic, owing to the fact that the oxides, OsO_4 and RuO_4, are volatile and have feebly acid properties. In this respect they most resemble permanganic anhydride, which is also somewhat volatile.

When dissolved in aqua regia ($PtCl_4$ is formed) and liberated from the solution by sal-ammoniac (($NH_4)_2PtCl_6$ is formed) and reduced by ignition (which may be done by Zn and other reducing agents, direct from a solution of $PtCl_4$) platinum[3 bis] forms a powdery mass, known as spongy platinum or platinum black. If this powder of platinum be heated and pressed, or hammered in a cylinder, the grains aggregate or forge together, and form a continuous, though of course not entirely homogeneous, mass. Platinum was formerly, and is even now, worked up in this manner. The platinum money formerly used in Russia was made in this way. Sainte-Claire Deville, in the fifties, for the first time melted platinum in considerable quantities by employing a special furnace made in the form of a small reverberatory furnace, and composed of two pieces of lime, on which the heat of the oxyhydrogen flame has no action. Into this furnace (shown in fig. 34, Vol. I. p. 175)—or, more strictly speaking, into the cavity made in the pieces of lime—the platinum is introduced, and two orifices are made in the lime; through one, the upper, or side orifice, is introduced an oxyhydrogen gas burner, in which either detonating gas or a mixture of oxygen and coal-gas is burnt, whilst the other orifice serves for the escape of the products of combustion and certain impurities which are more volatile than the platinum, and especially the oxidised compounds of osmium, ruthenium, and palladium, which are comparatively easily volatilised by heat. In this manner the platinum is converted into a continuous metallic form by means of fusion, and this method is now used for melting considerable masses of platinum and its alloys with iridium.

To obtain pure platinum, the ore is treated with aqua regia in which only the osmium and iridium are insoluble. The solution contains the platinum metals in the form RCl_4, and in the lower forms of chlorination, RCl_3 and RCl_2, because some of these metals—for instance, palladium and rhodium—form such unstable chlorides of the type RX_4 that they partially decompose even when diluted with water, and pass into the stable lower type of combination; in addition to which the chlorine is very easily disengaged if it comes in contact with substances on which it can act. In this respect platinum resists the action of heat and reducing agents better than any of its companions—that is, it passes with greater difficulty from $PtCl_4$ to the lower compound $PtCl_2$. On this is based the method of preparation of more or less pure platinum. Lime or sodium hydroxide is added to the solution in aqua regia until neutralised, or only containing a very slight excess of alkali. It is best to first evaporate and slightly ignite the solution, in order to remove the excess of acid, and by heating it to partially convert the higher chlorides of the palladium, &c., into the lower. The addition of alkalis completes the reduction, because the chlorine held in the compounds RX_4 acts on the alkali like free chlorine, converting it into a hypochlorite. Thus palladium chloride, $PdCl_4$, for example, is converted into palladious chloride, $PdCl_2$, by this means, according to the equation $PdCl_4 + 2NaHO = PdCl_2 + NaCl + NaClO + H_2O$. In a similar manner iridic chloride, $IrCl_4$, is converted into the trichloride, $IrCl_3$, by this method. When this conversion takes place the platinum still remains in the form of platinic chloride, $PtCl_4$. It is then possible to take advantage of a certain difference in the properties of the higher and lower chlorides of the platinum metals. Thus lime precipitates the lower chlorides of the members of the platinum metals occurring in solution without acting on the platinic chloride, $PtCl_4$, and hence the addition of a large proportion of lime immediately precipitates the associated metals, leaving the platinum itself in solution in the form of a soluble double salt, $PtCl_4,CaCl_2$. A far better and more perfect *separation* is effected *by means of ammonium chloride*, which gives, with platinic chloride, an insoluble yellow precipitate, $PtCl_4,2NH_4Cl$, whilst it forms soluble double salts with the lower chlorides RCl_2 and RCl_3, so that ammonium chloride precipitates the platinum only from the solution obtained by the preceding method. These methods are employed for preparing the platinum which is used for the manufacture of platinum articles, because, having platinum in solution as calcium platinochloride, $PtCaCl_6$, or as the insoluble ammonium platinochloride, $Pt(NH_4)_2Cl_6$, the platinum compound in every case, after drying or ignition, loses all the chlorine from the platinic chloride and leaves finely-divided metallic platinum, which may be converted into homogeneous metal by compression and forging, or by fusion.

Metallic *platinum* in a fused state has a specific gravity of 21; it is grey, softer than iron but harder than copper, exceedingly ductile, and therefore easily drawn into wire and rolled into thin sheets, and may be hammered into crucibles and drawn into thin tubes, &c. In the state in which it is obtained by the ignition of its compounds, it forms a spongy mass, known as spongy platinum, or else as powder (platinum black). In either case it is dull grey, and is characterised, as we already know, by the faculty of absorbing hydrogen and other gases. Platinum is not acted on by hydrochloric, hydriodic, nitric, and sulphuric acids, or a mixture of hydrofluoric and nitric acids. Aqua regia, and any liquid containing chlorine or able to evolve chlorine or bromine, dissolves platinum. Alkalis are decomposed by platinum at a red heat, owing to the faculty of the platinum oxide, PtO_2, formed to combine with alkaline bases, inasmuch as it has a feebly-developed acid character (*see* Note 8). Sulphur, phosphorus (the phosphide, PtP_2, is formed), arsenic and silicon all act more or less rapidly on platinum, under the influence of heat. Many of the metals form alloys with it. Even charcoal combines with platinum when it is ignited with it, and therefore carbonaceous matter cannot be subjected to prolonged and powerful ignition in platinum vessels. Hence a platinum crucible soon becomes dull on the surface in a smoky flame. Platinum also forms alloys with zinc, lead, tin, copper, gold, and silver. Although mercury does not directly dissolve platinum, still it forms a solution or amalgam with spongy platinum in the presence of sodium amalgam; a similar amalgam is also formed by the action of sodium amalgam on a solution of platinum chloride, and is used for physical experiments.

There are *two kinds* of *platinum compounds*, PtX_4 and PtX_2. The former are produced by an excess of halogen in the cold, and the latter by the aid of heat or by the splitting up of the former. The starting-point for the platinum compounds is *platinum tetrachloride*, *platinic chloride*, $PtCl_4$, obtained by dissolving platinum in aqua regia.[7 bis] The solution crystallises in the cold, in a desiccator, in the form of reddish-brown deliquescent crystals which contain hydrochloric acid, $PtCl_4,2HCl,6H_2O$, and behave like a true acid whose salts correspond to the formula R_2PtCl_6—ammonium platinochloride, for example.[7 tri] The hydrochloric acid is liberated from these crystals by gently heating or evaporating the solution to dryness; or, better still, after treatment with silver nitrate a reddish-brown mass remains behind, which dissolves in water, and forms a yellowish-red solution which on cooling deposits crystals of the composition $PtCl_4,8H_2O$. The *tendency* of $PtCl_4$ *to combine* with hydrochloric acid and water—that is, *to form higher crystalline compounds*—is evident in the platinum compounds, and must be taken into account in explaining the properties of platinum and the formation of many other of its complex compounds. Dilute solutions of platinic chloride are yellow, and are completely reduced by hydrogen, sulphurous anhydride, and many reducing agents, which first convert the platinic chloride into the lower compound platinous chloride, $PtCl_2$. That faculty which reveals itself in platinum tetrachloride of combining with water of crystallisation and hydrochloric acid is distinctly marked in its property, with which we are already acquainted, of giving precipitates with the salts of potassium, ammonium, rubidium, &c. In general it *readily forms double salts*, $R_2PtCl_6 = PtCl_4 + 2RCl$, where R is a univalent metal such as potassium or NH_4. Hence the addition of a solution of potassium or ammonium chloride to a solution of platinic chloride is followed by the formation of a yellow precipitate, which is sparingly soluble in water and almost entirely insoluble in alcohol and ether (platinic chloride is soluble in alcohol, potassium iridiochloride, IrK_3Cl_6, *i.e.* a compound of $IrCl_3$, is soluble in water but not in alcohol). It is especially remarkable in this case, that the potassium compounds here, as in a number of other instances, separate in an anhydrous form, whilst the sodium compounds, which are soluble in water and alcohol, form red crystals containing water. The composition $Na_2PtCl_6,6H_2O$ exactly corresponds with the above-mentioned hydrochloric compound. The compounds with barium, $BaPtCl_6,4H_2O$, strontium, $SrPtCl_6,8H_2O$, calcium, magnesium, iron, manganese, and many other metals are all soluble in water.

Platinous chloride, $PtCl_2$, is formed when hydrogen platinochloride, PtH_2Cl_6, is ignited at 300°, or when potassium is heated at 230° in a stream of chlorine. The undecomposed tetrachloride is extracted from the residue by washing it with water, and a greenish-grey or brown insoluble mass of the dichloride (sp. gr. 5·9) is then obtained. It is soluble in hydrochloric acid, giving an acid solution of the composition $PtCl_2,2HCl$, corresponding with the type of double salts PtR_2Cl_4. Although platinous chloride decomposes below 500°, still it is formed to a small extent at higher temperatures. Troost and Hautefeuille, and Seelheim observed that when platinum was strongly ignited in a stream of chlorine, the metal, as it were, slowly volatilised and was deposited in crystals; a volatile chloride, probably platinous chloride, was evidently formed in this case, and decomposed subsequently to its formation, depositing crystals of platinum.

The properties of platinum above-described are repeated more or less distinctly, or sometimes with certain modifications, in the above-mentioned associates and analogues of this metal. Thus although palladium forms $PdCl_4$, this form passes into $PdCl_2$ with extreme ease. Whilst

rhodium and iridium in dissolving in aqua regia also form $RhCl_4$ and $IrCl_4$, but they pass into $RhCl_3$ and $IrCl_3$[9 bis] very easily when heated or when acted upon by substances capable of taking up chlorine (even alkalis, which form bleaching salts). Among the platinum metals, ruthenium and osmium have the most acid character, and although they give $RuCl_4$ and $OsCl_4$ they are easily oxidised to RuO_4, and OsO_4 by the action of chlorine in the presence of water; the latter are volatile and may be distilled with the water and hydrochloric acid, from a solution containing other platinum metals.[9 tri] Thus with respect to the types of combination, all the platinum metals, under certain circumstances, give compounds of the type RX_4—for instance, RO_2, RCl_4, &c. But this is the highest form for only platinum and palladium. The remaining platinum metals further, *like iron, give acids* of the type RO_3 or hydrates, $H_2RO_4 = RO_2(HO)_2$ (the type of sulphuric acid); but they, like ferric and manganic acids, are chiefly known in the form of salts of the composition K_2RO_4 or $K_2R_2O_7$ (like the dichromate). These salts are obtained, like the manganates and ferrates, by fusing the oxides, or even the metals themselves, with nitric, or, better still, with potassium peroxide. They are soluble in water, are easily deoxidised and do not yield the acid anhydrides under the action of acids, but break up, either (like the ferrate) forming oxygen and a basic oxide (iridium and rhodium react in this manner, as they do not give higher forms of oxidation), or passing into a lower and higher form of oxidation—that is, reacting like a manganate (or partly like nitrite or phosphite). Osmium and ruthenium react according to the latter form, as they are capable of giving *higher forms of oxidation*, OsO_4 and RuO_4, and therefore their reactions of decomposition may be essentially represented by the equation: $2OsO_3 = OsO_2 + OsO_4$.

Platinum and its analogues, like iron and its analogues, are able to form complex and comparatively stable cyanogen and ammonia compounds, corresponding with the ferrocyanides and the ammoniacal compounds of cobalt, which we have already considered in the preceding chapter.

If platinous chloride, $PtCl_2$ (insoluble in water), be added by degrees to a solution of potassium cyanide, it is completely dissolved (like silver chloride), and on evaporating the solution deposits rhombic prisms of *potassium platinocyanide*, $PtK_2(CN)_4, 3H_2O$. This salt, like all those corresponding with it, has a remarkable play of colours, due to the phenomena of dichromism, and even polychromism, natural to all the platinocyanides. Thus it is yellow and reflects a bright blue light. It is easily soluble in water, effloresces in air, then turns red, and at 100° orange, when it loses all its water. The loss of water does not destroy its stability—that is, it still remains unchanged, and its stability is further shown by the fact that it is formed when potassium ferrocyanide, $K_4Fe(CN)_6$, is heated with platinum black. This salt, first obtained by Gmelin, shows a neutral reaction with litmus; it is exceedingly stable under the action of air, like potassium ferrocyanide, which it resembles in many respects. Thus the platinum in it cannot be detected by reagents such as sulphuretted hydrogen; the potassium may be replaced by other metals by the action of their salts, so that it corresponds with a whole series of compounds, $R_2Pt(CN)_4$, and it is stable, although the potassium cyanide and platinous salts, of which it is composed, individually easily undergo change. When treated with oxidising agents it passes, like the ferrocyanide, into a higher form of combination of platinum. If salts of silver be added to its solution, it gives a heavy white precipitate of silver platinocyanide, $PtAg_2(CN)_4$, which when suspended in water and treated with sulphuretted hydrogen, enters into double decomposition with the latter and forms insoluble silver sulphide, Ag_2S, and soluble *hydroplatinocyanic acid*, $H_2Pt(CN)_4$. If potassium platinocyanide be mixed with an equivalent quantity of sulphuric acid, the hydroplatinocyanic acid liberated may be extracted by a mixture of alcohol and ether. The ethereal solution, when evaporated in a desiccator, deposits bright red crystals of the composition $PtH_2(CN)_4,5H_2O$. This acid colours litmus paper, liberates carbonic anhydride from sodium carbonate, and saturates alkalis, so that it presents an analogy to hydroferrocyanic acid.

Ammonia, like potassium cyanide, has the faculty of easily reacting with platinum dichloride, forming compounds similar to the platinocyanide and cobaltia compounds, which are comparatively stable. But as ammonia does not contain any hydrogen easily replaceable by metals, and as ammonia itself is able to combine with acids, the PtX_2 plays, as it were, the part of an acid with reference to the ammonia. Owing to the influence of the ammonia, the X_2 in the resultant compound will represent the same character as it has in ammoniacal salts; consequently, the ammoniacal compounds produced from PtX_2 will be salts in which X will be replaceable by various other haloids, just as the metal is replaced in the cyanogen salts; such is the nature of the *platino-ammonium compounds*. PtX_2 forms compounds with $2NH_3$ and with $4NH_3$, and so also PtX_4 gives (not directly from PtX_4 and ammonia, but from the compounds of PtX_2 by the action of chlorine, &c.) similar compounds with $2NH_3$ and with $4NH_3$.

If ammonia acts on a boiling solution of platinous chloride in hydrochloric acid, it produces the green *salt of Magnus* (1829), $PtCl_2,2NH_3$, insoluble in water and hydrochloric acid. But, judging by its reactions, this salt has twice this formula. Thus, Gros (1837), on boiling Magnus's salt with nitric acid, observed that half the chlorine was replaced by the residue of nitric acid and half the platinum was disengaged: $2PtCl_2(NH_3)_2 + 2HNO_3 = PtCl_2(NO_3)_2(NH_3)_4 + 2PtCl_2$. The Gros's salt thus obtained, $PtCl_2(NO_3)_2 4NH_3$ (if Magnus's salt belongs to the type PtX_2, then Gros's salt belongs to the type PtX_4), is soluble in water, and the elements of nitric acid, but not the chlorine, contained in it are capable of easily submitting themselves to double saline decomposition. Thus silver nitrate does not enter into double decomposition with the chlorine of Gros's salt. Most instructive was the circumstance that Gros, by acting on his salt with hydrochloric acid, succeeded in substituting the residue of nitric acid in it by chlorine, and the chlorine thus introduced, easily reacted with silver nitrate. Thus it appeared that Gros's salt contained two varieties of chlorine—one which reacts readily, and the other which reacts with difficulty. The composition of Gros's first salt is $PtCl_2(NH_3)_4(NO_3)_2$; it may be converted into $PtCl_2(NH_3)_4(SO_4)$, and in general into $PtCl_2(NH_3)_4 X_2$.

The salt of Magnus when boiled with a solution of ammonia gives the salt (of Reiset's first base) $PtCl_2(NH_3)_4$, and this, when treated with bromine, forms the salt $PtCl_2Br_2(NH_3)_4$, which has the same composition and reactions as Gros's salt. To Reiset's salts there corresponds a soluble, colourless, crystalline *hydroxide*, $Pt(OH)_2(NH_3)_4$, having the properties of a powerful and very energetic *alkali*; it attracts carbonic anhydride from the atmosphere, precipitates metallic salts like potash, saturates active acids, even sulphuric, forming colourless (with nitric, carbonic, and hydrochloric acids), or yellow (with sulphuric acid), salts of the type $PtX_2(NH_3)_4$. The comparative stability (for instance, as compared with AgCl and NH_3) of such compounds, and the existence of many other compounds analogous to them, endows them with a particular chemical interest. Thus Kournakoff (1889) obtained a series of corresponding compounds containing thiocarbamide, CSN_2H_4, in the place of ammonia, $PtCl_2, 4CSN_2H_4$, and others corresponding with Reiset's salts. Hydroxylamine, and other substances corresponding with ammonia, also give similar compounds. The common properties and composition of such compounds show their entire analogy to the cobaltia compounds (especially for ruthenium and iridium) and correspond to the fact that both the platinum metals and cobalt occur in the same, eighth, group.

Footnotes:

Wells and Penfield (1888) have described a mineral sperryllite found in the Canadian gold-bearing quartz and consisting of platinum diarsenide, $PtAs_2$. It is a noticeable fact that this mineral clearly confirms the position of platinum in the same group as iron, because it corresponds in crystalline form (regular octahedron) and chemical composition with iron pyrites, FeS_2.

[1 bis] Some light is thrown upon the facility with which the platinum compounds decompose by Thomsen's data, showing that in an excess of water (+ Aq) the formation from platinum, of such a double salt as $PtCl_2,2KCl$, is accompanied by a comparatively small evolution of heat (*see* Chapter XXI., Note 40), for instance, $Pt + Cl_2 + 2KCl + Aq$ only evolves about 33,000 calories (hence the reaction, $Pt + Cl_2 + Aq$, will evidently disengage still less, because $PtCl_2 + 2KCl$ evolves a certain amount of heat), whilst on the other hand, $Fe + Cl_2 + Aq$ gives 100,000 calories, and even the reaction with copper (for the formation of the double salt) evolves 63,000 calories.

The largest amount of platinum is extracted in the Urals, about five tons annually. A certain amount of gold is extracted from the washed platinum by means of mercury, which does not dissolve the platinum metals but dissolves the gold accompanying the platinum in its ores. Moreover, the ores of platinum always contain metals of the iron series associated with them. The washed and mechanically sorted ore in the majority of cases contains about 70 to 80 p.c. of platinum, about 5 to 8 p.c. of iridium, and a somewhat smaller quantity of osmium. The other platinum metals—palladium, rhodium, and ruthenium—occur in smaller proportions than the three above named. Sometimes grains of almost pure osmium-iridium, containing only a small quantity of other metals, are found in platinum ores. This *osmium-iridium* may be easily separated from the other platinum metals, owing to its being nearly insoluble in aqua regia, by which the latter are easily dissolved. There are grains of platinum which are magnetic. The grains of osmium-iridium are very hard and malleable, and are therefore used for certain purposes, for instance, for the tips of gold pens.

In characterising the platinum metals according to their relation to the iron metals, it is very important to add two more very remarkable points. The platinum metals are capable of forming a sort of unstable compound with *hydrogen*; they absorb it and only part with it when somewhat strongly heated. This faculty is especially developed in platinum and palladium, and it is very characteristic that nickel, which exactly corresponds with platinum and palladium in the periodic system, should exhibit the same faculty for retaining a considerable quantity of hydrogen (Graham's and Raoult's experiments). Another characteristic property of the platinum metals consists in their easily giving (like cobalt which forms the cobaltic salts) stable and characteristic saline *compounds with ammonia*, and like Fe and Co, double salts with the cyanides of the alkali metals, especially in their lower forms of combination. All the above so clearly brings the elements of the iron series in close relation to the platinum metals, that the eighth group acquires as natural a character as can be required, with a certain originality or individuality for each element.

[3 bis] Platinum was first obtained in the last century from Brazil, where it was called silver (platinus). Watson in 1750 characterised platinum as a separate independent metal. In 1803 Wollaston discovered palladium and rhodium in crude platinum, and at about the same time Tennant distinguished iridium and osmium in it. Professor Claus, of Kazan, in his researches on the platinum metals (about 1840) discovered ruthenium in them, and to him are due many important discoveries with regard to these elements, such as the indication of the remarkable analogy between the series Pd—Rh—Ru and Pt—Ir—Os.

The treatment of platinum ore is chiefly carried on for the extraction of the platinum itself and its alloys with iridium, because these metals offer a greater resistance to the action of chemical reagents and high temperatures than any of the other malleable and ductile metals, and therefore the wire so often used in the laboratory and for technical purposes is made from them, as also are various vessels used for chemical purposes in the laboratory and in works. Thus sulphuric acid is distilled in platinum retorts, and many substances are fused, ignited, and evaporated in the laboratory in platinum crucibles and on platinum foil. Gold and many other substances are dissolved in dishes made of iridium-platinum, because the alloys of platinum and iridium are but slightly attacked when subjected to the action of aqua regia.

The comparatively high density (about 21·5), hardness, ductility, and infusibility (it does not melt at a furnace heat, but only in the oxyhydrogen flame or electric furnace), as well as the fact of its resisting the action of water, air, and other reagents, renders an alloy of 90 parts of platinum and 10 parts of iridium (Deville's platinum-iridium alloy) a most valuable material for making standard weights and measures, such as the metre, kilogram, and pound, and therefore all the newest standards of most countries are made of this alloy.

This process has altered the technical treatment of platinum to a considerable extent. It has in particular facilitated the manufacture of alloys of platinum with iridium and rhodium from the pure platinum ores, since it is sufficient to fuse the ore in order for the greater amount of the osmium to burn off, and for the mass to fuse into a homogeneous, malleable alloy, which can be directly made use of. There is very little ruthenium in the ores of platinum. If during fusion lead be added, it dissolves the platinum (and other platinum metals) owing to its being able to form a very characteristic alloy containing PtPb. If an alloy of the two metals be left exposed to moist air, the excess of lead is converted into carbonate (white lead) in the presence of the water and carbonic acid of the air, whilst the above platinum alloy remains unchanged. The white lead may be extracted by dilute acid, and the alloy PtPb remains unaltered. The other platinum metals also give similar alloys with lead. The fusibility of these alloys enables the platinum metals to be separated from the gangue of the ore, and they may afterwards be separated from the lead by subjecting the alloy to oxidation in furnaces furnished with a bone ash bed, because the lead is then oxidised and absorbed by the bone ash, leaving the platinum metals untouched. This method of treatment was proposed by H. Sainte-Claire Deville in the sixties, and is also used in the analysis of these metals (*see* further on).

For the ultimate purification of platinum from palladium and iridium the metals must be re-dissolved in aqua regia, and the solution evaporated until the residue begins to evolve chlorine. The residue is then re-precipitated with ammonium or potassium chloride. The precipitate may still contain a certain amount of iridium, which passes with greater difficulty from the tetrachloride, $IrCl_4$, into the trichloride, $IrCl_3$, but it will be quite free from palladium, because the latter easily loses its chlorine and passes into palladious chloride, $PdCl_2$, which gives an easily-soluble salt with potassium chloride. The precipitate, containing a small quantity of iridium, is then heated with sodium carbonate in a crucible, when the mass decomposes, giving metallic platinum and iridium oxide. If potassium chloride has been employed, the residue after ignition is washed with water and treated with aqua regia. The iridium oxide remains undissolved, and the platinum easily passes into solution. Only cold and dilute aqua regia must be used. The solution will then contain pure platinic chloride, which forms the starting-point for the preparation of all platinum compounds. Pure platinum for accurate researches (for instance, for the unit of light, according to Violle's method) may be obtained (Mylius and Foerster, 1892) by Finkener's method, by dissolving the impure metal in aqua regia (it should be evaporated to drive off the nitrogen compounds), and adding NaCl so as to form a double sodium salt, which is purified by crystallising with a small amount of caustic soda, washing the crystals with a strong solution of NaCl, and then dissolving them in a hot 1 p.c. solution of soda, repeating the above and ultimately igniting the double salt, previously dried at 120°, in a stream of hydrogen; platinum black and NaCl are then formed. The three following are very sensitive tests (to thousandths of a per cent.) for the presence of Ir, Ru, Rh, Pd (osmium is not usually present in platinum which has once been purified, since it easily volatilises with Cl_2 and CO_2, and in the first treatment of the crude platinum either passes off as OsO_4 or remains undissolved), Fe, Cu, Ag, and Pb: (1) the assay is alloyed with 10 parts of pure lead, the alloy treated with dilute nitric acid (to remove the greater part of the Pb), and dissolved in aqua regia; the residue will consist of Ir and Ru; the Pb is precipitated from the nitric acid solution by sulphuric acid, whilst the remaining platinum metals are reduced from the evaporated solution by formic acid, and the resultant precipitate fused with $KHSO_4$; the Pd and Rh are thus converted into soluble salts, and the former is then precipitated by HgC_2N_2. (2) Iron may be detected by the usual reagents, if the crude platinum be dissolved in aqua regia, and the platinum metals precipitated from the solution by formic acid. (3) If crude platinum (as foil or sponge) be heated in a mixture of chlorine and carbonic oxide it volatilises (with a certain amount of Ir, Pd, Fe, &c.) as $PtCl_2,2CO$ (Note 11), whilst the whole of the Rh, Ag, and Cu it may contain remains behind. Among other characteristic reactions for the platinum metals, we may mention: (1) that rhodium is precipitated from the solution obtained after fusion with $KHSO_4$ (in which Pt does not dissolve) by NH_3, acetic and formic acids; (2) that dilute aqua regia dissolves precipitated Pt, but not Rh; (3) that if the insoluble residue of the platinum metals (Ir, Ru, Os) obtained, after treating with aqua regia, be fused with a mixture of 1 part of KNO_3 and 3 parts of K_2CO_3 (in a gold crucible), and then treated with water, it gives a solution containing the Ru (and a portion of the Ir), but which throws it all down when saturated with chlorine and boiled; (4) that if iridium be fused with a mixture of KHO and KNO_3, it gives a soluble potassium salt, IrK_2O_4 (the solution is blue), which, when saturated with chlorine, gives $IrCl_4$, which is precipitated by NH_4Cl (the precipitate is black), forming a double salt, leaving metallic Ir after ignition; (5) that rhodium mixed with NaCl and ignited in a current of chlorine gives a soluble double salt (from which sal-ammoniac separates Pt and Ir), which gives (according to Jörgensen) a difficultly soluble purpureo-salt (Chapter XXII., Note 35), $Rh_2Cl_3,5NH_3$, when treated with NH_3; in this form the Rh may be easily purified and obtained in a metallic form by igniting in hydrogen; and (6) that palladium, dissolved in aqua regia and dried (NH_4Cl throws down any Pt), gives soluble $PdCl_2$, which forms an easily crystallisable yellow salt, $PdCl_2NH_3$, with ammonia; this salt (Wilm) may be easily purified by crystallisation, and gives metallic Pd when ignited. These reactions illustrate the method of separating the platinum metals from each other.

We have already become acquainted with the effect of finely-divided platinum on many gaseous substances. It is best seen in the so-called *platinum black*, which is a coal-black powder left by the action of sulphuric acid on the alloy of zinc and platinum, or which is precipitated by metallic zinc from a dilute solution of platinum. In any case, finely-divided platinum absorbs gases more powerfully and rapidly the more finely divided and porous it is. Sulphurous anhydride, hydrogen, alcohol, and many organic substances in the presence of such platinum are easily oxidised by the oxygen of the air, although they do not combine with it directly. The absorption of oxygen is as much as several hundred volumes per one volume of platinum, and the oxidising power of such absorbed oxygen is taken advantage of not only in the laboratory but even in manufacturing processes. Asbestos or charcoal, soaked in a solution of platinic chloride and ignited, is very useful for this purpose, because by this means it becomes coated with platinum black. If 50 grams of $PtCl_4$ be dissolved in 60 c.c. of water, and 70 c.c. of a strong (40 p.c.) solution of formic aldehyde added, the mixture cooled, and then a solution of 50 grams of NaHO in 50 grams of water added, the platinum is precipitated. After washing with water the precipitate passes into solution and forms a black liquid containing *soluble colloidal platinum* (Loew, 1890). If the precipitated platinum be allowed to absorb oxygen on the filter, the temperature rises 40°, and a very porous *platinum black* is obtained which vigorously facilitates oxidation.

It is necessary to remark that platinum when alloyed with silver, or as amalgam, is soluble in nitric acid, and in this respect it differs from gold, so that it is possible, by alloying gold with silver, and acting on the alloy with nitric acid, to recognise the presence of platinum in the gold, because nitric acid does not act on gold alloyed with silver.

[7 bis] $PtCl_4$ is also formed by the action of a mixture of HCl vapour and air, and by the action of gaseous chlorine upon platinum.

[7 tri] Pigeon (1891) obtained fine yellow crystals of $PtH_2Cl_6,4H_2O$ by adding strong sulphuric acid to a strong solution of $PtH_2Cl_6,6H_2O$. If crystals of $H_2PtCl_6,6H_2O$ be melted in vacuo (60°) in the presence of anhydrous potash, a red-brown solid hydrate is obtained containing less water and HCl, which parts with the remainder at 200°, leaving anhydrous $PtCl_4$. The latter does not disengage chlorine before 220°, and is perfectly soluble in water.

Nilson (1877), who investigated the platinochlorides of various metals subsequently to Bonsdorff, Topsöe, Clève, Marignac, and others, found that univalent and bivalent metals—such as hydrogen, potassium, ammonium ... beryllium, calcium, barium—give compounds of such a composition that there is always twice as much chlorine in the platinic chloride as in the combined metallic chloride; for example, $K_2Cl_2,PtCl_4$; $BeCl_2,PtCl_4,8H_2O$, &c. Such trivalent metals as aluminium, iron (ferric), chromium, didymium, cerium (cerous) form compounds of the type RCl_3PtCl_4, in which the amounts of chlorine are in the ratio 3:4. Only indium and yttrium give salts of a different composition—namely, $2InCl_3,5PtCl_4,36H_2O$ and $4YCl_3,5PtCl_4,51H_2O$. Such quadrivalent metals as thorium, tin, zirconium give compounds of the type $RCl_4,PtCl_4$, in which the ratio of the chlorine is 1:1. In this manner the valency of a metal may, to a certain extent, be judged from the composition of the double salts formed with platinic chloride.

Platinic bromide, $PtBr_4$, and iodide, PtI_4, are analogous to the tetrachloride, but the iodide is decomposed still more easily than the chloride. If sulphuric acid be added to platinic chloride, and the solution evaporated, it forms a black porous mass like charcoal, which deliquesces in the air, and has the composition $Pt(SO_4)_2$. But this, the only oxygen salt of the type PtX_4, is exceedingly unstable. This is due to the fact that *platinum oxide*, the oxide of the type PtO_2, has a feeble acid character. This is shown in a number of instances. Thus if a strong solution of platinic chloride treated with sodium carbonate be exposed to the action of light or evaporated to dryness and then washed with water, a sodium platinate, $Pt_3Na_2O_7,6H_2O$, remains. The composition of this salt, if we regard it in the same sense as we did the salts of silicic, titanic, molybdic and other acids, will be $PtO(ONa)_2,2PtO_2,6H_2O$—that is, the same type is repeated as we saw in the crystalline compounds of platinum tetrachloride with sodium chloride, or with hydrochloric acid—namely, the type PtX_48Y, where Y is the molecule H_2O,HCl, &c. Similar compounds are also obtained with other alkalis. They will be platinates of the alkalis in which the platinic oxide, PtO_2, plays the part of an acid oxide. Rousseau (1889) obtained different grades of combination $BaOPtO_2$, $3(BaO)2PtO_2$, &c., by igniting a mixture of $PtCl_4$ and caustic baryta. If such an alkaline compound of platinum be treated with acetic acid, the alkali combines with the latter, and a *platinic hydroxide*, $Pt(OH)_4$, remains as a brown mass, which loses water and oxygen when ignited, and in so doing decomposes with a slight explosion. When slightly ignited this hydroxide first loses water and gives the very unstable oxide PtO_2. Platinic sulphide, PtS_2, belongs to the same type; it is precipitated by the action of sulphuretted hydrogen on a solution of platinum tetrachloride. The moist precipitate is capable of attracting oxygen, and is then converted into the sulphate above mentioned, which is soluble in water. This absorption of oxygen and conversion into sulphate is another illustration of the basic nature of PtO_2, so that it clearly exhibits both basic and acid properties. The latter appear, for instance, in the fact that platinic sulphide, PtS_2, gives crystalline compounds with the alkali sulphides.

In comparing the characteristics of the platinum metals, it must be observed that palladium in its form of combination PdX_2 gives saline compounds of considerable stability. Amongst them *palladous chloride* is formed by the direct action of chlorine or aqua regia (not in excess or in dilute solutions) on palladium. It forms a brown solution, which gives a black insoluble precipitate of *palladous iodide*, PdI_2, with solutions of iodides (in this respect, as in many others, palladium resembles mercury in the mercuric compounds HgX_2). With a solution of mercuric cyanide it gives a yellowish white precipitate, palladous cyanide, PdC_2N_2, which is soluble in potassium cyanide, and gives other double salts, $M_2PdC_4N_4$.

That portion of the platinum ore which dissolves in aqua regia and is precipitated by ammonium or potassium chloride does not contain palladium. It remains in solution, because the palladic chloride, $PdCl_4$, is decomposed and the palladous chloride formed is not precipitated by ammonium chloride; the same holds good for all the other lower chlorides of the platinum metals. Zinc (and iron) separates out all the unprecipitated platinum metals (and also copper, &c.) from the solution. The palladium is found in these platinum residues precipitated by zinc. If this mixture of metals be treated with aqua regia, all the palladium will pass into solution as palladous chloride with some platinic chloride. By this treatment the main portion of the iridium, rhodium, &c. remains almost undissolved, the platinum is separated from the mixture of palladous and platinic chlorides by a solution of ammonium chloride, and the solution of palladium is precipitated by potassium iodide or mercuric cyanide. Wilm (1881) showed that palladium may be separated from an impure solution by saturating it with ammonia; all the iron present is thus precipitated, and, after filtering, the addition of hydrochloric acid to the filtrate gives a yellow precipitate of an ammonio-palladium compound, $PdCl_2,2NH_3$, whilst nearly all the other metals remain in solution. *Metallic palladium* is obtained by igniting the ammonio-compound or the cyanide, PdC_2N_2. It occurs native, although rarely, and is a metal of a whiter colour than platinum, sp. gr. 11·4, melts at about 1,500°; it is much more volatile than platinum, partially oxidises on the surface when heated (Wilm obtained spongy palladium by igniting $PdCl_2,2NH_3$, and observed that it gives PdO when ignited in oxygen, and that on further ignition this oxide forms a mixture of Pd_2O and Pd), and loses its absorbed oxygen on a further rise of temperature. It does not blacken or tarnish (does not absorb sulphur) in the air at the ordinary temperature, and is therefore better suited than silver for astronomical and other instruments in which fine divisions have to be engraved on a white metal, in order that the fine lines should be clearly visible. The most remarkable property of palladium, discovered by Graham, consists in its capacity for *absorbing* a large amount of *hydrogen*. Ignited palladium absorbs as much as 940 volumes of hydrogen, or about 0·7 p.c. of its own weight, which closely approaches to the formation of the compound Pd_3H_2, and probably indicates the formation of *palladium hydride*, Pd_2H. This absorption also takes place at the ordinary temperature—for example, when palladium serves as an electrode at which hydrogen is evolved. In absorbing the hydrogen, the palladium does not change in appearance, and retains all its metallic properties, only its volume increases by about 10 p.c.—that is, the hydrogen pushes out and separates the atoms of the palladium from each other, and is itself compressed to $\frac{1}{900}$ of its volume. This compression indicates a great force of chemical attraction, and is accompanied by the evolution of heat (Chapter II., Note 38). The absorption of 1 grm. of hydrogen by metallic palladium (Favre) is accompanied by the evolution of 4·2 thousand calories (for Pt 20, for Na 13, for K 10 thousand units of heat). Troost showed that the dissociation pressure of palladium hydride is inconsiderable at the ordinary temperature, but reaches the atmospheric pressure at about 140°. This subject was subsequently investigated by A. A. Cracow of St. Petersburg (1894), who showed that at first the absorption of hydrogen by the palladium proceeds like solution, according to the law of Dalton and Henry, but that towards the end it proceeds like a dissociation phenomenon in definite compounds; this forms another link between the phenomenon of solution and of the formation of definite atomic compounds. Cracow's observations for a temperature 18°, showed that the electro-conductivity and tension vary until a compound Pd_2H is reached, and namely, that the tension p rises with the volume v of hydrogen absorbed, according to the law of Dalton and Henry—for instance, for

p = 2·1 3·2 5·5 7·7 mm.

v = 14 20 34 47

The maximum tension at 18° is 9 mm. At a temperature of about 140° (in the vapour of xylene) the maximum tension is about 760 mm., and when v = 10–50 vols. the tension (according to Cracow's experiments) stands at 90–450 mm.—that is, increases in proportion to the volume of hydrogen absorbed. But from the point of view of chemical mechanics it is especially important to remark that Moutier clearly showed, through palladium hydride, the similarity of the phenomena which proceed in evaporation and dissociation, which fact Henri Sainte-Claire Deville placed as a fundamental proposition in the theory of dissociation. It is possible upon the basis of the second law of the theory of heat, according to the law of the variation of the tension p of evaporation with the temperature T (counted from -273°), to calculate the latent heat of evaporation L (*see* works on physics) because $424L = T(1/d - 1/D)dp/dt$, where d and D are the weights of cubic measures of the gas (vapour) and liquid. (Thus, for instance, for water, when t = 100°, T = 373, d = 0·605, D = 960, dp/dt = 0·027 m., 13,596 = 367, L = 536, whence 424L = 227,264, and the second portion of the equation 226,144, which is sufficiently near, within the limits of experimental error, *see* Chapter I., Note 11.) The same equation is applicable to the dissociation of Na_2H and K_2H—(Chapter XII., Note 42)—but it has only been verified in this respect for Pd_2H, since Moutier, by calculating the amount of heat L evolved, for t = 20, according to the variation of the tension (dp/dt) obtained 4·1 thousand calories, which is very near the figure obtained experimentally by Favre (*see* Chapter XII., Note 44). The absorbed hydrogen is easily disengaged by ignition or decreased pressure. The resultant compound does not decompose at the ordinary temperature, but when exposed to air the metal sometimes glows spontaneously, owing to the hydrogen burning at the expense of the atmospheric oxygen. The hydrogen absorbed by palladium acts towards many solutions as a reducing agent; in a word, everything here points to the formation of a definite compound and at the same time of a physically-compressed gas, and forms one of the best examples of the bond existing between chemical and physical processes, to which we have many times drawn attention. It must be again remembered that the other metals of the eighth group, even copper, are, like palladium and platinum, able to combine with hydrogen. The permeability of iron and platinum tubes to hydrogen is naturally due to the formation of similar compounds, but palladium is the most permeable.

[9 bis] *Rhodium* is generally separated, together with iridium, from the residues left after the treatment of native platinum, because the palladium is entirely separated from them, and the ruthenium is present in them in very small traces, whilst the osmium at any rate is easily separated, as we shall soon see. The mixture of rhodium and iridium which is left undissolved in dilute aqua regia is dissolved in chlorine water, or by the action of chlorine on a mixture of the metals with sodium chloride. In either case both metals pass into solution. They may be separated by many methods. In either case (if the action be aided by heat) the rhodium is obtained in the form of the chloride $RhCl_3$, and the iridium as iridious chloride, $IrCl_3$. They both form double salts with sodium chloride which are soluble in water, but the iridium salt is also partially soluble in alcohol, whilst the rhodium salt is not. A mixture of the chlorides, when treated with dilute aqua regia, gives iridic chloride, $IrCl_4$, whilst the rhodium chloride, $RhCl_3$, remains unaltered; ammonium chloride then precipitates the iridium as ammonium iridiochloride, $Ir(NH_4)2Cl_6$, and on evaporating the rose-coloured filtrate the rhodium gives a crystalline salt, $Rh(NH_4)_3Cl_6$. Rhodium and its various oxides are dissolved when fused with potassium hydrogen sulphate, and give a soluble double sulphate (whilst iridium remains unacted on); this fact is very characteristic for this metal, which offers in its properties many points of resemblance with the iron metals. When fused with potassium hydroxide and chlorate it is oxidised like iridium, but it is not afterwards soluble in water, in which respect it differs from ruthenium. This is taken advantage of for separating rhodium, ruthenium, and iridium. In any case, rhodium under ordinary conditions always gives salts of the type RX_3, and not of any other type; and not only halogen salts, but also oxygen salts, are known in this type, which is rare among the platinum metals. Rhodium

chloride, $RhCl_3$, is known in an insoluble anhydrous and also in a soluble form (like CrX_3 or salts of chromic oxides), in which it easily gives double salts, compounds with water of crystallisation, and forms rose-coloured solutions. In this form rhodium easily gives double salts of the two types RhM_3Cl_6 and RhM_2Cl_3—for example, $K_5RhCl_6,3H_2O$ and K_2RhCl_5,H_2O. Solutions of the salts (at least, the ammonium salt) of the first kind give salts of the second kind when they are boiled. If a strong solution of potash be added to a red solution of rhodium chloride and boiled, a black precipitate of the hydroxide $Rh(OH)_3$ is formed; but if the solution of potash is added little by little, it gives a yellow precipitate containing more water. This yellow hydrate of rhodium oxide gives a yellow solution when it is dissolved in acids, which only becomes rose-coloured after being boiled. It is obvious a change here takes place, like the transmutations of the salts of chromic oxide. It is also a remarkable fact that the black hydroxide, like many other oxidised compounds of the platinoid metals, does not dissolve in the ordinary oxygen acids, whilst the yellow hydroxide is easily soluble and gives yellow solutions, which deposit imperfectly crystallised salts. Metallic rhodium is easily obtained by igniting its oxygen and other compounds in hydrogen, or by precipitation with zinc. It resembles platinum, and has a sp. gr. of 12·1. At the ordinary temperature it decomposes formic acid into hydrogen and carbonic anhydride, with development of heat (Deville). With the alkali sulphites, the salts of rhodium and iridium of the type RX_3 give sparingly-soluble precipitates of double sulphites of the composition $R(SO_3Na)_3,H_2O$, by means of which these metals may be separated from solution, and also may be separated from each other, for a mixture of these salts when treated with strong sulphuric acid gives a soluble iridium sulphate and leaves a red insoluble double salt of rhodium and sodium. It may be remarked that the oxides Ir_2O_3 and Rh_2O_3 are comparatively stable and are easily formed, and that they also form different double salts (for instance, $IrCl_3,3KCl_3H_2O$, $RhCl_3,2NH_4Cl_4H_2O$, $RhCl_3,3NH_4Cl1½H_2O$) and compounds like the cobaltia compounds (for instance, luteo-salts $RhX_3,6NH_3$, roseo-salts, $RhX_3H_2O_5NH_3$, and purpureo-salts $IrX_3,5NH_3$, &c.) *Iridious oxide*, Ir_2O_3, is obtained by fusing iridious chloride and its compounds with sodium carbonate, and treating the mass with water. The oxide is then left as a black powder, which, when strongly heated, is decomposed into iridium and oxygen; it is easily reduced, and is insoluble in acids, which indicates the feeble basic character of this oxide, in many respects resembling such oxides as cobaltic oxide, ceric or lead dioxide, &c. It does not dissolve when fused with potassium hydrogen sulphate. Rhodium oxide, Rh_2O_3, is a far more energetic base. It dissolves when fused with potassium hydrogen sulphate.

From what has been said respecting the separation of platinum and rhodium it will be understood how the compounds of *iridium*, which is the main associate of platinum, are obtained. In describing the treatment of osmiridium we shall again have an opportunity of learning the method of extraction of the compounds of this metal, which has in recent times found a technical application in the form of its oxide, Ir_2O_3; this is obtained from many of the compounds of iridium by ignition with water, is easily reduced by hydrogen, and is insoluble in acids. It is used in painting on china, for giving a black colour. Iridium itself is more difficultly fusible than platinum, and when fused it does not decompose acids or even aqua regia; it is extremely hard, and is not malleable; its sp. gr. is 22·4. In the form of powder it dissolves in aqua regia, and is even partially oxidised when heated in air, sets fire to hydrogen, and, in a word, closely resembles platinum. Heated in an excess of chlorine it gives iridic chloride, $IrCl_4$, but this loses chlorine at 50°; it is, however, more stable in the form of double salts, which have a characteristic *black* colour—for instance, $Ir(NH_4)_2Cl_6$—but they give iridious chloride, $IrCl_3$, when treated with sulphuric acid.

[9 tri] We have yet to become acquainted with the two remaining associates of platinum—ruthenium and osmium—whose most important property is that they are oxidised even when heated in air, and that they are able to give *volatile* oxides of the form RuO_4 and OsO_4; these have a powerful odour (like iodine and nitrous anhydride). Both these higher oxides are solids; they volatilise with great ease at 100°; the former is yellow and the latter white. They are known as *ruthenic* and *osmic anhydrides*, although their aqueous solutions (they both slowly dissolve in water) do not show an acid reaction, and although they do not even expel carbonic anhydride from potassium carbonate, do not give crystalline salts with bases, and their alkaline solutions partially deposit them again when boiled (an excess of water decomposes the salts). The formulæ OsO_4 and RuO_4 correspond with the vapour density of these oxides. Thus Deville found the vapour density of osmic anhydride to be 128 (by the formula 127·5) referred to hydrogen. Tennant and Vauquelin discovered this compound, and Berzelius, Wöhler, Fritzsche, Struvé, Deville, Claus, Joly, and others helped in its investigation; nevertheless there are still many questions concerning it which remain unsolved. It should be observed that RO_4 is the highest known form for an oxygen compound, and RH_4 is the highest known form for a compound of hydrogen; whilst the highest forms of acid hydrates contain SiH_4O_4, PH_3O_4, SH_2O_4, $ClHO_4$—all with four atoms of oxygen, and therefore in this number there is apparently the limit for the simple forms of combination of hydrogen and oxygen. In combination with *several* atoms of an element, or several elements, there may be more than O_4 or H_4, but a molecule never contains more than four atoms of either O or H to one atom of another element. Thus the simplest forms of

combination of hydrogen and oxygen are exhausted by the list RH_4, RH_3, RH_2, RH, RO, RO_2, RO_3, RO_4. The extreme members are RH_4 and RO_4, and are only met with for such elements as carbon, silicon, osmium, ruthenium, which also give RCl_4 with chlorine. In these extreme forms, RH_4 and RO_4, the compounds are the least stable (compare SiH_4, PH_3, SH_2, ClH, or RuO_4, MoO_3, ZrO_2, SrO), and easily give up part, or even all, their oxygen or hydrogen.

The primary source from which the compounds of ruthenium and osmium are obtained is either *osmiridium* (the osmium predominates, from $IrOs$ to $IrOs_4$, sp. gr. from 16 to 21), which occurs in platinum ores (it is distinguished from the grains of platinum by its crystalline structure, hardness, and insolubility in aqua regia), or else those insoluble residues which are obtained, as we saw above, after treating platinum with aqua regia. Osmium predominates in these materials, which sometimes contain from 30 p.c. to 40 p.c. of it, and rarely more than 4 p.c. to 5 p.c. of ruthenium. The process for their treatment is as follows: they are first fused with 6 parts of zinc, and the zinc is then extracted with dilute hydrochloric acid. The osmiridium thus treated is, according to Fritzsche and Struvé's method, then added to a fused mixture of potassium hydroxide and chlorate in an iron crucible; the mass as it begins to evolve oxygen acts on the metal, and the reaction afterwards proceeds spontaneously. The dark product is treated with water, and gives a solution of osmium and ruthenium in the form of soluble salts, R_2OsO_4 and R_2RuO_4, whilst the insoluble residue contains a mixture of oxides of iridium (and some osmium, rhodium, and ruthenium), and grains of metallic iridium still unacted on. According to Frémy's method the lumps of osmiridium are straightway heated to whiteness in a porcelain tube in a stream of air or oxygen, when the very volatile osmic anhydride is obtained directly, and is collected in a well-cooled receiver, whilst the ruthenium gives a crystalline sublimate of the dioxide, RuO_2, which is, however, very difficultly volatile (it volatilises together with osmic anhydride), and therefore remains in the cooler portions of the tube; this method does not give volatile ruthenic anhydride, and the iridium and other metals are not oxidised or give non-volatile products. This method is simple, and at once gives dry, pure osmic anhydride in the receiver, and ruthenium dioxide in the sublimate. The air which passes through the tube should be previously passed through sulphuric acid, not only in order to dry it, but also to remove the organic and reducing dust. The vapour of osmic anhydride must be powerfully cooled, and ultimately passed over caustic potash. A third mode of treatment, which is most frequently employed, was proposed by Wöhler, and consists in slightly heating (in order that the sodium chloride should not melt) an intimate mixture of osmiridium and common salt in a stream of moist chlorine. The metals then form compounds with chlorine and sodium chloride, whilst the osmium forms the chloride, $OsCl_4$, which reacts with the moisture, and gives osmic anhydride, which is condensed. The ruthenium in this, as in the other processes, does not directly give ruthenic anhydride, but is always extracted as the soluble ruthenium salt, K_2RuO_4, obtained by fusion with potassium hydroxide and chlorate or nitrate. When the orange-coloured ruthenate, K_2RuO_4, is mixed with acids, the liberated ruthenic acid immediately decomposes into the volatile ruthenic anhydride and the insoluble ruthenic oxide: $2K_2RuO_4 + 4HNO_3 = RuO_4 + RuO_2,2H_2O + 4KNO_3$. When once one of the above compounds of ruthenium or osmium is procured it is easy to obtain all the remaining compounds, and by reduction (by metals, hydrogen, formic acid, &c.) the metals themselves.

Osmic anhydride, OsO_4, is very easily deoxidised by many methods. It blackens organic substances, owing to reduction, and is therefore used in investigating vegetable and animal, and especially nerve, preparations under the microscope. Although osmic anhydride may be distilled in hydrogen, still complete reduction is accomplished when a mixture of hydrogen and osmic anhydride is slightly ignited (just before it inflames). If osmium be placed in the flame it is oxidised, and gives vapours of osmic anhydride, which become reduced, and the flame gives a brilliant light. Osmic anhydride deflagrates like nitre on red-hot charcoal; zinc, and even mercury and silver, reduce osmic anhydride from its aqueous solutions into the lower oxides or metal; such reducing agents as hydrogen sulphide, ferrous sulphate, or sulphurous anhydride, alcohol, &c., act in the same manner with great ease.

The lower oxides of osmium, ruthenium, and of the other elements of the platinum series are not volatile, and it is noteworthy that the other elements behave differently. On comparing SO_2, SO_3; As_2O_3, As_2O_5; P_2O_3, P_2O_5; CO, CO_2, &c., we observe a converse phenomenon; the higher oxides are less volatile than the lower. In the case of osmium all the oxides, with the exception of the highest, are non-volatile, and it may therefore be thought that this higher form is more simply constituted than the lower. It is possible that osmic oxide, OsO_2, stands in the same relation to the anhydride as C_2H_4 to CH_4—*i.e.* the lower oxide is perhaps Os_2O_4, or is still more polymerised, which would explain why the lower oxides, having a greater molecular weight, are less volatile than the higher oxides, just as we saw in the case of the nitrogen oxides, N_2O and NO.

Ruthenium and osmium, obtained by the ignition or reduction of their compounds in the form of powder, have a density considerably less than in the fused form, and differ in this condition in their capacity for reaction; they are much more difficultly fused than platinum and iridium, although ruthenium is more fusible than osmium. Ruthenium in powder has a specific gravity of 8·5, the fused metal of 12·2; osmium in powder has a specific gravity of 20·0, and when semi-fused—or, more strictly speaking, agglomerated—in the oxyhydrogen flame, of 21·4, and fused 22·5. The powder of slightly-heated osmium oxidises very easily in the air, and when ignited burns like tinder, directly forming the odoriferous osmic anhydride (hence its name, from the Greek word signifying odour); ruthenium also oxidises when heated in air, but with more difficulty, forming the oxide RuO_2. The oxides of the types RO, R_2O_3, and RO_2 (and their hydrates) obtained by reduction from the higher oxides, and also from the chlorides, are analogous to those given by the other platinum metals, in which respect osmium and ruthenium closely resemble them. We may also remark that ruthenium has been found in the platinum deposits of Borneo in the form of *laurite*, Ru_2S_3, in grey octahedra of sp. gr. 7·0.

For osmium, Moraht and Wischin (1893) obtained free osmic acid, H_2OsO_4, by decomposing K_2OsO_4 with water, and precipitating with alcohol in a current of hydrogen (because in air volatile OsO_4 is formed); with H_2S, osmic acid gives $OsO_3(HS)_2$ at the ordinary temperature.

Debray and Joly showed that ruthenic anhydride, RuO_4, fuses at 25°, boils at 100°, and evolves oxygen when dissolved in potash, forming the salt $KRuO_4$ (not isomorphous with potassium permanganate).

Joly (1891), who studied the ruthenium compounds in greater detail, showed that the easily-formed $KRuO_4$ gives $RuKO_4RuO_3$ when ignited, but it resembles $KMnO_4$ in many respects. In general, Ru has much in common with Mn. Joly (1889) also showed that if KNO_3 be added to a solution of $RuCl_3$ containing HCl, the solution becomes hot, and a salt, $RuCl_3NO_2KCl$, is formed, which enters into double decomposition and is very stable. Moreover, if $RuCl_3$ be treated with an excess of nitric acid, it forms a salt, $RuCl_3NOH_2O$, after being heated (to boiling) and the addition of HCl. The vapour density of RuO_4, determined by Debray and Joly, corresponds to that formula.

Although palladium gives the same types of combination (with chlorine) as platinum, its reduction to RX_2 is incomparably easier than that of platinic chloride, and in the case of iridium it is also very easy. Iridic chloride, $IrCl_4$, acts as an oxidising agent, readily parts with a fourth of its chlorine to a number of substances, readily evolves chlorine when heated, and it is only at low temperatures that chlorine and aqua regia convert iridium into iridic chloride. In disengaging chlorine iridium more often and easily gives the very stable iridious chloride, $IrCl_3$ (perhaps this substance is $Ir_2Cl_6 = IrCl_2,IrCl_4$, insoluble in water, but soluble in potassium chloride, because it forms the double salt K_3IrCl_6), than the dichloride, $IrCl\_2$. This compound, corresponding to IrX_2, is very stable, and corresponds with the *basic oxide*, Ir_2O_3, resembling the oxides Fe_2O_3, Co_2O_3. To this form there correspond ammoniacal compounds similar to those given by cobaltic oxide. Although iridium also gives an acid in the form of the salt $K_2Ir_2O_7$, it does not, like iron (and chromium), form the corresponding chloride, $IrCl_6$. In general, in this as in the other elements, it is impossible to predict the chlorine compounds from those of oxygen. Just as there is no chloride SCl_6, but only SCl_2, so also, although IrO_3 exists, $IrCl_6$ is wanting, the only chloride being $IrCl_4$, and this is unstable, like SCl_2, and easily parts with its chlorine. In this respect rhodium is very much like iridium (as platinum is like palladium). For $RhCl_4$ decomposes with extreme ease, whilst rhodium chloride, $RhCl_3$, is very stable, like many of the salts of the type RhX_3, although like the platinum elements these salts are easily reduced to metal by the action of heat and powerful reagents. There is as close a resemblance between osmium and ruthenium. Osmium when submitted to the action of dry chlorine gives osmic chloride, $OsCl_4$, but the latter is converted by water (as is osmium by moist chlorine) into osmic anhydride, although the greater portion is then decomposed into $Os(HO)_4$ and $4HCl$, like a chloranhydride of an acid. In general this acid character is more developed in osmium than in platinum and iridium. Having parted with chlorine, osmic chloride, $OsCl_4$, gives the unstable trichloride, $OsCl_3$, and the stable soluble dichloride, $OsCl_2$, which corresponds with platinous chloride in its properties and reactions. The relation of ruthenium to the halogens is of the same nature.

This acid character is explained by the influence of the platinum on the hydrogen, and by the attachment of the cyanogen groups. Thus cyanuric acid, $H_3(CN)_3O_3$, is an energetic acid compared with cyanic acid, $HCNO$. And the formation of a compound with five molecules of water of crystallisation, $(PtH_2(CN)_4,5H_2O)$, confirms the opinion that platinum is able to form compounds of still higher types than that expressed in its saline compounds, and, moreover, the combination of hydroplatinocyanic acid with water does not reach the limit of the compounds which appears in $PtCl_4,2HCl,6H_2O$.

A whole series of *platinocyanides* of the common type $PtR_2(CN)_4nH_2O$ is obtained by means of double decomposition with the potassium or hydrogen or silver salts. For example, the salts of sodium and lithium contain, like the potassium salt, three molecules of water. The sodium salt is soluble in water and alcohol. The ammonium salt has the composition $Pt(NH_4)_2(CN)_4,2H_2O$ and gives crystals which reflect blue and rose-coloured light. This ammonium salt decomposes at 300°, with evolution of water and ammonium cyanide, leaving a greenish *platinum dicyanide*, $Pt(CN)_2$, which is insoluble in water and acid but dissolves in potassium cyanide, hydrocyanic acid, and other cyanides. The same platinous cyanide is obtained by the action of sulphuric acid on the potassium salts in the form of a reddish-brown amorphous precipitate. The most characteristic of the platinocyanides are those of the alkaline earths. The magnesium salt $PtMg(CN)_4,7H_2O$ crystallises in regular prisms, whose side faces are of a metallic green colour and terminal planes dark blue. It shows a carmine-red colour along the main axis, and dark red along the lateral axes; it easily loses water, ($2H_2O$), at 40°, and then turns blue (it then contains $5H_2O$, which is frequently the case with the platinocyanides). Its aqueous solution is colourless, and an alcoholic solution deposits yellow crystals. The remainder of the water is given off at 230°. It is obtained by saturating platinocyanic acid with magnesia, or else by double decomposition between the barium salt and magnesium sulphate. The strontium salt, $SrPt(CN)_4,4H_2O$ crystallises in milk-white plates having a violet and green iridescence. When it effloresces in a desiccator, its surfaces have a violet and metallic green iridescence. A colourless solution of the barium salt $PtBa(CN)_4,4H_2O$ is obtained by saturating a solution of hydroplatinocyanic acid with baryta, or by boiling the insoluble copper platinocyanide in baryta water. It crystallises in monoclinic prisms of a yellow colour, with blue and green iridescence; it loses half its water at 100°, and the whole at 150°. The ethyl salt, $Pt(C_2H_5)_2(CN)_4,2H_2O$, is also very characteristic; its crystals are isomorphous with those of the potassium salt, and are obtained by passing hydrochloric acid into an alcoholic solution of hydroplatinocyanic acid. The facility with which they crystallise, the regularity of their forms, and their remarkable play of colours, renders the preparation of the platinocyanides one of the most attractive lessons of the laboratory.

By the action of chlorine or dilute nitric acid, the platinocyanides are converted into salts of the composition $PtM_2(CN)_5$, which corresponds with $Pt(CN)_3,2KCN$—that is, they express the type of a non-existent form of oxidation of platinum, PtX_3 (*i.e.* oxide Pt_2O_3), just as potassium ferricyanide ($FeCy_3,3KCy$) corresponds with ferric oxide, and the ferrocyanide corresponds with the ferrous oxide. The potassium salt of this series contains $PtK_2(CN)_5,3H_2O$, and forms brown regular prisms with a metallic lustre, and is soluble in water but insoluble in alcohol. Alkalis re-convert this compound into the ordinary platinocyanide $K_2Pt(CN)_4$, taking up the excess of cyanogen. It is remarkable that the salts of the type PtM_2Cy_5 contain the same amount of water of crystallisation as those of the type PtM_2Cy_4. Thus the salts of potassium and lithium contain three, and the salt of magnesium seven, molecules of water, like the corresponding salts of the type of platinous oxide. Moreover, neither platinum nor any of its associates gives any cyanogen compound corresponding with the oxide, *i.e.* having the composition PtK_2Cy_6, just as there are no compounds higher than those which correspond to RCy_3nMCy_3 for cobalt or iron. This would appear to indicate the absence of any such cyanides, and indeed, for no element are there yet known any poly-cyanides containing more than three equivalents of cyanogen for one equivalent of the element. The phenomenon is perhaps connected with the faculty of cyanogen of giving tricyanogen polymerides, such as cyanuric acid, solid cyanogen chloride, &c. Under the action of an excess of chlorine, a solution of $PtK_2(CN)_4$ gives (besides PtK_2Cy_5) a product $PtK_2Cy_4Cl_2$, which evidently contains the form PtX_4, but at first the action of the chlorine (or the electrolysis of, or addition of dilute peroxide of hydrogen to, a solution of PtK_2Cy_4, acidulated with hydrochloric acid) produces an easily soluble intermediate salt which crystallises in thin copper-red needles (Wilm, Hadow, 1889). It only contains a small amount of chlorine, and apparently corresponds to a compound $5PtK_2Cy_4 + PtK_2Cy_4Cl_2 + 24H_2O$. Under the action of an excess of ammonia both these chlorine products are converted either completely or in part (according to Wilm ammonia does not act upon PtK_2Cy_4) into $PtCy_2,2NH_3$, *i.e.* a platino-ammonia compound (*see* further on). It is also necessary to pay attention to the fact that ruthenium and osmium—which, as we know, give higher forms of oxidation than platinum—are also able to combine with a larger proportion of potassium cyanide (but not of cyanogen) than platinum. Thus ruthenium forms a crystalline *hydroruthenocyanic acid*, $RuH_4(CN)_6$, which is soluble in water and alcohol, and corresponds with the salts $M_4Ru(CN)_6$. There are exactly similar osmic compounds—for example, $K_4Os(CN)_6,3H_2O$. The latter is obtained in the form of colourless, sparingly-soluble regular tablets on evaporating the solution obtained from a fused mixture of potassium osmiochloride, K_2OsCl_6, and potassium cyanide. These osmic and ruthenic compounds fully correspond with potassium ferrocyanide, $K_4Fe(CN)_6,3H_2O$, not only in their composition but also in their crystalline form and reactions, which again demonstrates the close analogy between iron, ruthenium, and osmium, which we have shown by giving these three elements a similar position (in the eighth group) in the periodic system. For rhodium and iridium only salts of the same type as the ferricyanides, M_3RCy_6, are known, and for palladium only of the type M_2PdCy_4, which are analogous to the platinum salts. In all these examples a *constancy of the types* of the double cyanides is apparent. In the eighth group we have iron, cobalt, nickel, copper, and their analogues ruthenium, rhodium, palladium, silver, and also osmium, iridium, platinum, gold. The double cyanides of iron, ruthenium, osmium have the type $K_4R(CN)_6$; of cobalt, rhodium, iridium, the type $K_3R(CN)_6$; of nickel, palladium, platinum the type $K_2R(CN)_4$ and $K_2R(CN)_5$; and for copper, silver, gold there are known $KR(CN)_2$, so that the presence of 4, 3, 2, and 1 atoms of potassium corresponds with the order of the elements in the periodic system. Those types which we have seen in the ferrocyanides and ferricyanides of iron repeat themselves in all the platinoid metals, and this naturally leads to the conclusion that the formation of similar so-called double salts is of exactly the same nature as that of the ordinary salts. If, in expressing the union of the elements in the oxygen salts, the existence of an *aqueous residue* (hydroxyl group) be admitted, in which the hydrogen is replaced by a metal, we have then only to apply this mode of expression to the double salts and the analogy will be obvious, if only we remember that Cl_2, $(CN)_2$, SO_4, &c., are equivalent to O, as we see in RO, RCl_2, RSO_4, &c. They all $= X_2$, and, therefore, in point of fact, wherever X ($=$ Cl or OH, &c.) can be placed, there (Cl_2H), (SO_4H), &c., can also stand. And as $Cl_2H =$ Cl + HCl and $SO_4H =$ OH + SO_3, &c., it follows that molecules HCl or SO_3, or, in general, whole molecules—for instance, NH_3, H_2O, salts, &c., can annex themselves to a compound containing X. (This is an indirect consequence of the law of substitution which explains the origin of double salts, ammonia compounds, compounds with water of crystallisation, &c., by one general method.) Thus the double salt $MgSO_4,K_2SO_4$, according to this reasoning, *may be* considered as a substance of the same type as $MgCl_2$, namely, as $=$ $Mg(SO_4K)_2$, and the alums as derived from $Al(OH)(SO_4)$, namely, as $Al(SO_4K)(SO_4)$. Without stopping to pursue this digression further, we will apply these considerations to the type of the ferrocyanides and ferricyanides and their platinum analogues. Such a salt as K_2PtCy_4 may accordingly be regarded as $Pt(Cy_2K)_2$, like $Pt(OH)_2$; and such a salt as PtK_2Cy_5 as $PtCy(Cy_2K)_2$, the analogue of $PtX(OH)_2$, or $AlX(OH)_2$, and other compounds of the type RX_3. Potassium ferricyanide and the analogous compounds of cobalt, iridium, and rhodium, belong to the same type, with the same difference as there is between $RX(OH)_2$ and $R(OH)_3$, since $FeK_3Cy_6 = Fe(Cy_2K)_3$. Limiting myself to these considerations, which may partially elucidate the nature of double salts, I will now pass again to the complex saline compounds known for platinum.

(*A*) On mixing a solution of potassium thiocyanate with a solution of potassium platinosochloride, K_2PtCl_4, they form a double thiocyanate, $PtK_2(CNS)_4$, which is easily soluble in water and alcohol, crystallises in red prisms, and gives an orange-coloured solution, which precipitates salts of the heavy metals. The action of sulphuric acid on the lead salt of the same type gives the acid itself, $PtH_2(SCN)_4$, which corresponds with these salts. The type of these compounds is evidently the same as that of the cyanides.

(*B*) *Platinous chloride*, $PtCl_2$, which is insoluble in water, forms *double salts with the metallic chlorides*. These double chlorides are soluble in water, and capable of crystallising. Hence when a hydrochloric acid solution of platinous chloride is mixed with solutions of metallic salts and evaporated it forms crystalline salts of a red or yellow colour. Thus, for example, the potassium salt, PtK_2Cl_4, is red, and easily soluble in water; the sodium salt is also soluble in alcohol; the barium salt, $PtBaCl_4,3H_2O$, is soluble in water, but the silver salt, $PtAg_2Cl_4$, is insoluble in water, and may be used for obtaining the remaining salts by means of double decomposition with their chlorides.

(C) A remarkable example of the complex compounds of platinum was observed by Schützenberger (1868). He showed that finely-divided platinum in the presence of chlorine and carbonic oxide at 250°-300° gives phosgene and a volatile compound containing platinum. The same substance is formed by the action of carbonic oxide on platinous chloride. It decomposes with an explosion in contact with water. Carbon tetrachloride dissolves a portion of this substance, and on evaporation gives crystals of $2PtCl_2,3CO$, whilst the compound $PtCl_2,2CO$ remains undissolved. When fused and sublimed it gives yellow needles of $PtCl_2,CO$, and in the presence of an excess of carbonic oxide $PtCl_2,2CO$ is formed. These compounds are fusible (the first at 250°, the second at 142°, and the third at 195°). In this case (as in the double cyanides) combination takes place, because both carbonic oxide and platinous chloride are unsaturated compounds capable of further combination. The carbon tetrachloride solution absorbs NH_3 and gives $PtCl_2,CO,2NH_3$, and $PtCl_2,2CO,2NH_3$, and these substances are analogous (Foerster, Zeisel, Jörgensen) to similar compounds containing complex amines (for instance, pyridine, C_5H_5N), instead of NH_3, and ethylene, &c., instead of CO, so that here we have a whole series of complex platino-compounds. The compound $PtCl_2CO$ dissolves in hydrochloric acid without change, and the solution disengages all the carbonic oxide when KCN is added to it, which shows that those forces which bind 2 molecules of KCN to $PtCl_2$ can also bind the molecule CO, or 2 molecules of CO. When the hydrochloric acid solution of $PtCl_2CO$ is mixed with a solution of sodium acetate or acetic acid, it gives a precipitate of PtOCO, *i.e.* the Cl_2 is replaced by oxygen (probably because the acetate is decomposed by water). This oxide, PtOCO, splits up into Pt + CO_2 at 350°. PtSCO is obtained by the action of sulphuretted hydrogen upon $PtCl_2CO$. All this leads to the conclusion that the group PtCO is able to assimilate $X_2 = Cl_2$, S, O, &c. (Mylius, Foerster, 1891). Pullinger (1891), by igniting spongy platinum at 250°, first in a stream of chlorine, and then in a stream of carbonic oxide, obtained (besides volatile products) a non-volatile yellow substance which remained unchanged in air and disengaged chlorine and phosgene gas when ignited; its composition was $PtCl_6(CO)_2$, which apparently proves it to be a compound of $PtCl_2$ and $2COCl_2$, as $PtCl_2$ is able to combine with oxychlorides, and forms somewhat stable compounds.

(D) The faculty of platinous chloride for forming stable compounds with divers substances shows itself in the formation of the compound

$PtCl_2,PCl_3$ by the action of phosphorus pentachloride at 250° on platinum powder (Pd reacts in a similar manner, according to Fink, 1892). The product contains both phosphorus pentachloride and platinum, whilst the presence of $PtCl_2$ is shown in the fact that the action of water produces *chlorplatino-phosphorous acid*, $PtCl_2P(OH)_3$.

(E) After the cyanides, the *double salts* of platinum *formed by sulphurous acid* are most distinguished for their stability and characteristic properties. This is all the more instructive, as sulphurous acid is only feebly energetic, and, moreover, in these, as in all its compounds, it exhibits a dual reaction. The salts of sulphurous acid, R_2SO_3, either react as salts of a feeble bibasic acid, where the group SO_3 presents itself as bivalent, and consequently equal to X_2, or else they react after the manner of salts of a monobasic acid containing the same residue, RSO_3, as occurs in the salts of sulphuric acid. In sulphurous acid this residue is combined with hydrogen, $H(SO_3H)$, whilst in sulphuric acid it is united with the aqueous residue (hydroxyl), $OH(SO_3H)$. These two forms of action of the sulphites appear in their reactions with the platinum salts—that is to say, salts of both kinds are formed, and they both correspond with the type PtH_2X_4. The one series of salts contain $PtH_2(SO_3)_2$, and their reactions are due to the bivalent residue of sulphurous acid, which replaces X_2. The others, which have the composition $PtR_2(SO_3H)_4$, contain sulphoxyl. The latter salts will evidently react like acids; they are formed simultaneously with the salts of the first kind, and pass into them. These salts are obtained either by directly dissolving platinous oxide in water containing sulphurous acid, or by passing sulphurous anhydride into a solution of platinous chloride in hydrochloric acid. If a solution of platinous chloride or platinous oxide in sulphurous acid be saturated with sodium carbonate, it forms a white, sparingly soluble precipitate containing $PtNa_2(SO_3Na)_4,7H_2O$. If this precipitate be dissolved in a small quantity of hydrochloric acid and left to evaporate at the ordinary temperature, it deposits a salt of the other type, $PtNa_2(SO_3)_2,H_2O$, in the form of a yellow powder, which is sparingly soluble in water. The potassium salt analogous to the first salt, $PtK_2(SO_3K)_4,2H_2O$, is precipitated by passing sulphurous anhydride into a solution of potassium sulphite in which platinous oxide is suspended. A similar salt is known for ammonium, and with hydrochloric acid it gives a salt of the second kind, $Pt(NH_4)_2(SO_3)_2,H_2O$. If ammonio-chloride of platinum be added to an aqueous solution of sulphurous anhydride, it is first deoxidised, and chlorine is evolved, forming a salt of the type PtX_2; a double decomposition then takes place with the ammonium sulphite, and a salt of the composition $Pt(NH_4)_2Cl_3(SO_3H)$ is formed (in a desiccator). The acid character of this substance is explained by the fact that it contains the elements SO_3H—sulphoxyl, with the hydrogen not yet displaced by a metal. On saturating a solution of this acid with potassium carbonate it gives orange-coloured crystals of a potassium salt of the composition $Pt(NH_4)_2Cl_3(SO_3K)$. Here it is evident that an equivalent of chlorine in $Pt(NH_4)_2Cl_4$ is replaced by the univalent residue of sulphurous acid. Among these salts, that of the composition $Pt(NH_4)Cl_2(SO_3H)_2,H_2O$ is very readily formed, and crystallises in well-formed colourless crystals; it is obtained by dissolving ammonium platinosochloride, $Pt(NH_4)_2Cl_4$, in an aqueous solution of sulphurous acid. The difficulty with which sulphurous anhydride and platinum are separated from these salts indicates the same basic character in these compounds as is seen in the double cyanides of platinum. In their passage into a complex salt, the metal platinum and the group SO_2 modify their relations (compared with those of PtX_2 or SO_2X_2), just as the chlorine in the salts $KClO$, $KClO_3$, and $KClO_4$ is modified in its relations as compared with hydrochloric acid or potassium chloride.

(F) No less characteristic are the *platinonitrites* formed by platinous oxide. They correspond with nitrous acid, whose salts, RNO_2, contain the univalent radicle, NO_2, which is capable of replacing chlorine, and therefore the salts of this kind should form a common type $PtR_2(NO_2)_4$, and such a salt of potassium has actually been obtained by mixing a solution of potassium platinosochloride with a solution of potassium nitrite, when the liquid becomes colourless, especially if it be heated, which indicates the change in the chemical distribution of the elements. As the liquid decolorises it gradually deposits sparingly soluble, colourless prisms of the potassium salt $K_2Pt(NO_2)_4$, which does not contain any water. With silver nitrate a solution of this salt gives a precipitate of silver platinonitrite, $PtAg_2(NO_2)_4$. The silver of this salt may be replaced by other metals by means of double decomposition with metallic chlorides. The sparingly soluble barium salt, when treated with an equivalent quantity of sulphuric acid, gives a soluble acid, which separates, under the receiver of an air-pump, in red crystals; this acid has the composition $PtH_2(NO_2)_4$. To the potassium salt, $K_2Pt(NO_2)_4$, there correspond (Vèzes, 1892) $K_2Pt(NO_2)_4Br_2$ and $K_2Pt(NO_2)_4Cl_2$ and other compounds of the same type K_2PtX_6, where X is partly replaced by Cl or Br and partly by (NO_2), showing a transition towards the type of the double salts like the platino-ammoniacal salts. (The corresponding double sodium nitrite salt of cobalt is soluble in water, while the K, NH_4 and many other salts are insoluble in water, as I was informed by Prof. K. Winkler in 1894).

In all the preceding complex compounds of Pt we see a common type $PtX_2,2MX$ (*i.e.* of double salts corresponding to PtO) or $PtM_2X_4 = Pt(MX_2)_2$, corresponding to $Pt(HO)_2$ with the replacement of O by its equivalent X_2. Two other facts must also be noted. In the first place these X's generally correspond to elements (like chlorine) or groups (like CN, NO_2, SO_3, &c.), which are capable of further combination. In the second place all the compounds of the type PtM$_2$X$_4$ are capable of combining with chlorine or similar elements, and thus passing into compounds of the types PtX$_3$ or PtX$_4$.

The platinum salt and ammonia, when once combined together, are no longer subject to their ordinary reactions but form compounds which are comparatively very stable. The question at once suggests itself to all who are acquainted with these phenomena, as to what is the relation of the elements contained in these compounds. The first explanation is that these compounds are salts of ammonium in which the hydrogen is partially replaced by platinum. This is the view, with certain shades of difference, held by many respecting the platino-ammonium compounds. They were regarded in this light by Gerhardt, Schiff, Kolbe, Weltzien, and many others. If we suppose the hydrogen in 2NH$_4$X to be replaced by bivalent platinum (as in the salts PtX$_2$), we shall obtain NH$_3$ NH$_3$ Pt X X —that is, the compound PtX$_2$,2NH$_3$. The compound with 4NH$_3$ will then be represented by a further substitution of the hydrogen in ammonia by ammonium itself—*i.e.* as NH$_2$(NH$_4$X)$_2$Pt or PtX$_2$,4NH$_3$. A modification of this view is found in that representation of compounds of this kind which is based on atomicity. As platinum in PtX$_2$ is bivalent, has two affinities, and ammonia, NH$_3$, is also bivalent, because nitrogen is quinquivalent and is here only combined with H$_3$, it is evident what bonds should be represented in PtX$_2$,2NH$_3$ and in PtX$_2$,4NH$_3$. In the former, Pt(NH$_3$Cl)$_2$, the nitrogen of each atom of ammonia is united by three affinities with H$_3$, by one with platinum, and by the fifth with chlorine. The other compound is Pt(NH$_3$.NH$_3$Cl)$_2$—that is, the N is united by one affinity with the other N, whilst the remaining bonds are the same as in the first salt. It is evident that this union or chain of ammonias has no obvious limit, and the most essential fault of such a mode of representation is that it does not indicate at all what number of ammonias are capable of being retained by platinum. Moreover, it is hardly possible to admit the bond between nitrogen and platinum in such stable compounds, for these kinds of affinities are, at all events, feeble, and cannot lead to stability, but would rather indicate explosive and easily-decomposed compounds. Moreover, it is not clear why this platinum, which is capable of giving PtX$_4$, does not act with its remaining affinities when the addition of ammonia to PtX$_2$ takes place. These, and certain other considerations which indicate the imperfection of this representation of the structure of the platino-ammonium salts, cause many chemists to incline more to the representations of Berzelius, Claus, Gibbs, and others, who suppose that NH$_3$ is able to combine with substances, to adjoin itself or pair itself with them (this kind of combination is called 'Paarung') without altering the fundamental capacity of a substance for further combinations. Thus, in PtX$_2$,2NH$_3$, the ammonia is the associate of PtX$_2$, as is expressed by the formula N$_2$H$_6$PtX$_2$. Without enlarging on the exposition of the details of this doctrine, we will only mention that it, like the first, does not render it possible to foresee a limit to the compounds with ammonia; it isolates compounds of this kind into a special and artificial class; does not show the connection between compounds of this and of other kinds, and therefore it essentially only expresses the fact of the combination with ammonia and the modification in its ordinary reactions. For these reasons we do not hold to either of these proposed representations of the ammonio-platinum compounds, but regard them from the point of view cited above with reference to double salts and water of crystallisation—that is, we embrace all these compounds under the representation of compounds of complex types. The type of the compound $PtX_2,2NH_3$ is far more probably the same as that of $PtX_2,2Z$—*i.e.* as PtX_4, or, still more accurately and truly, it is a compound of the same type as $PtX_2,2KX$ or $PtX_2,2H_2O$, &c. Although the platinum first entered into PtK_2X_4 as the type PtX_2, yet its character has changed in the same manner as the character of sulphur changes when from SO_2 the compound $SO_2(OH)_2$ is obtained, or when $KClO_4$, the higher form, is obtained from KCl. For us as yet there is no question as to *what* affinities hold X_2 and what hold $2NH_3$, because this is a question which arises from the supposition of the existence of different affinities in the atoms, which there is no reason for taking as a common phenomenon. It seems to me that it is most important *as a commencement* to render clear the analogy in the formation of various complex compounds, and it is this analogy of the ammonia compounds with those of water of crystallisation and double salts that forms the main object of the primary generalisation. We recognise in platinum, at all events, not only the four affinities expressed in the compound $PtCl_4$, but a much larger number of them, if only the *summation of affinities* is actually possible. Thus, in sulphur we recognise not two but a much greater number of affinities; it is clear that at least six affinities can act. So also among the analogues of platinum: osmic anhydride, OsO_4, $Ni(CO)_4$, PtH_2Cl_6, &c. indicate the existence of at least eight affinities; whilst, in chlorine, judging from the compound $KClO_4 = ClO_3(OK) = ClX_7$, we must recognise at least seven affinities, instead of the one which is accepted. The latter mode of calculating affinities is a tribute to that period of the development of science when only the simplest hydrogen compounds were considered, and when all complex compounds were entirely neglected (they were placed under the class of molecular compounds). This is insufficient for the present state of knowledge, because we find that, in complex compounds as in the most simple, the same constant types or modes of equilibrium are repeated, and the character of certain elements is greatly modified in the passage from the most simple into very complex compounds.

Judging from the most complex platino-ammonium compounds $PtCl_4,4NH_3$, we should admit the possibility of the formation of compounds of the type PtX_4Y_4, where $Y_4 = 4X_2 = 4NH_3$, and this shows that those forces which form such a characteristic series of double platinocyanides $PtK_2(CN)_4,3H_2O$, probably also determine the formation of the higher ammonia derivatives, as is seen on comparing—

$PtCl_2$ NH_3 $2Cl_2$ $3NH_3$

$Pt(CN)_2$ KCN   KCN $3H_2O$.

Moreover, it is obviously much more natural to ascribe the faculty for combination with nY to the whole of the acting elements—that is, to PtX_2 or PtX_4, and not to platinum alone. Naturally such compounds are not produced with any Y. With certain X's there only combine certain Y's. The best known and most frequently-formed compounds of this kind are those with water—that is, compounds with water of crystallisation. Compounds with salts are double salts; also we know that similar compounds are also frequently formed by means of ammonia. Salts of zinc, ZnX_2, copper, CuX_2, silver, AgX, and many others give similar compounds, but these and many other *ammonio-metallic* saline compounds are unstable, and readily part with their combined ammonia, and it is only in the elements of the platinum group and in the group of the analogues of iron, that we observe the faculty to form stable ammonio-metallic compounds. It must be remembered that the metals of the platinum and iron groups are able to form several high grades of oxidation which have an acid character, and consequently in the lower degrees of combination there yet remain affinities capable of retaining other elements, and they probably retain ammonia, and hold it the more stably, because all the properties of the platinum compounds are rather acid than basic—that is, PtX_n recalls rather HX or SnX_n or CX_n than KX, CaX_2, BaX_2, &c., and ammonia naturally will rather combine with an acid than with a basic substance. Further, a dependence, or certain connection of the forms of oxidation with the ammonia compounds, is seen on comparing the following compounds:

$PdCl_2,2NH_3,H_2O$ $PdCl_2,4NH_3,H_2O$

$PtCl_2,2NH_3$ $PtCl_4,4NH_3$

$RhCl_3,5NH_3$ $RuCl_2,4NH_3,3H_2O$

$IrCl_3,5NH_3$ $OsCl_2,4NH_3,2H_2O$

We know that platinum and palladium give compounds of lower types than iridium and rhodium, whilst ruthenium and osmium give the highest forms of oxidation; this shows itself in this case also. We have purposely cited the same compounds with $4NH_3$ for osmium and ruthenium as we have for platinum and palladium, and it is then seen that Ru and Os are capable of retaining $2H_2O$ and $3H_2O$, besides Cl_2 and NH_3, which the compounds of platinum and palladium are unable to do. The same ideas which were developed in Note 35, Chapter XXII. respecting the cobaltia compounds are perfectly applicable to the present case, *i.e.* to the *platinia* compounds or ammonia compounds of the platinum metals, among which Rh and Ir give compounds which are perfectly analogous to the cobaltia compounds.

Iridium and rhodium, which easily give compounds of the type RX_3, give compounds (Claus) of the type $IrX_3,5NH_3$, of a rose colour, and $RhX_3,5NH_3$, of a yellow colour. Jörgensen, in his researches on these compounds, showed their entire analogy with the cobalt compounds, as was to be expected from the periodic system.

Subsequently, a whole series of such compounds was obtained with various elements in the place of the (non-reacting) chlorine, and nevertheless they, like the chlorine, reacted with difficulty, whilst the second portion of the X's introduced into such salts easily underwent reaction. This formed the most important reason for the interest which the study of the composition and structure of the platino-ammonium salts subsequently presented to many chemists, such as Reiset, Blomstrand, Peyrone, Raeffski, Gerhardt, Buckton, Clève, Thomsen, Jörgensen, Kournakoff, Verner, and others. The salts $PtX_4,2NH_3$, discovered by Gerhardt, also exhibited several different properties in the two pairs of X's. In the remaining platino-ammonium salts all the X's appear to react alike.

The quality of the X's, retainable in the platino-ammonium salts, may be considerably modified, and they may frequently be wholly or partially replaced by hydroxyl. For example, the action of ammonia on the nitrate of Gerhardt's base, $Pt(NO_3)_4,2NH_3$, in a boiling solution, gradually produces a yellow crystalline precipitate which is nothing else than a *basic hydrate* or *alkali*, $Pt(OH)_4,2NH_3$. It is sparingly soluble in water, but gives directly soluble salts $PtX_4,2NH_3$ with acids. The stability of this hydroxide is such that potash does not expel ammonia from it, even on boiling, and it does not change below 130°. Similar properties are shown by the hydroxide $Pt(OH)_2,2NH_3$ and the oxide $PtO,2NH_3$ of Reiset's second base. But the hydroxides of the compounds containing $4NH_3$ are particularly remarkable. The presence of ammonia renders them soluble and energetic. The brevity of this work does not permit us, however, to mention many interesting particulars in connection with this subject.

Hydroxides are known corresponding with Gros's salts, which contain one hydroxyl group in the place of that chlorine or haloid which in Gros's salts reacts with difficulty, and these hydroxides do not at once show the properties of alkalis, just as the chlorine which stands in the same place does not react distinctly; but still, after the prolonged action of acids, this hydroxyl group is also replaced by acids. Thus, for example, the action of nitric acid on $Pt(NO_3)_2Cl_2,4NH_3$ causes the non-active chlorine to react, but in the product all the chlorine is not replaced by NO_3, but only half, and the other half is replaced by the hydroxyl group: $Pt(NO_3)_2Cl_2,4NH_3 + HNO_3 + H_2O = Pt(NO_3)_3(OH),4NH_3 + 2HCl$; and this is particularly characteristic, because here the hydroxyl group has not reacted with the acid—an evident sign of the non-alkaline character of this residue. I think it may be well to call attention to the fact that the composition of the ammonio-metallosalts very often exhibits a correspondence between the amount of X's and the amount of NH_3, of such a nature that we find they contain either XNH_3 or the grouping X_2NH_3; for example, $Pt(XNH_3)_2$ and $Pt(X_2NH_3)_2$, $Co(X_2NH_3)_3$, $Pt(XNH_3)_4$, &c. Judging from this, the view of the constitution of the double cyanides of platinum given in Note 11 finds some confirmation here, but, in my opinion, all questions respecting the composition (and structure) of the ammoniacal, double, complex, and crystallisation compounds stand connected with the solution of questions respecting the formation of compounds of various degrees of stability, among which a theory of solutions must be included, and therefore I think that the time has not yet come for a complete generalisation of the data which exist for these compounds; and here I again refer the reader to Prof. Kournakoff's work cited in Chapter XXII., Note 35. However, we may add a few individual remarks concerning the platinia compounds.

To the common properties of the platino-ammonium salts, we must add not only their *stability* (feeble acids and alkalis do not decompose them, the ammonia is not evolved by heating, &c.), but also the fact that the ordinary reactions of platinum are concealed in them to as great an extent as those of iron in the ferricyanides. Thus neither alkalis nor hydrogen sulphide will separate the platinum from them. For example, sulphuretted hydrogen in acting on Gros's salts gives sulphur, removes half the chlorine by means of its hydrogen, and forms salts of Reiset's first base. This may be understood or explained by considering the platinum in the molecule as covered, walled up by the ammonia, or situated in the centre of the molecule, and therefore inaccessible to reagents. On this assumption, however, we should expect to find clearly-expressed ammoniacal properties, and this is not the case. Thus ammonia is easily decomposed by chlorine, whilst in acting on the platino-ammonium salts containing PtX_2 and $2NH_3$ or $4NH_3$, chlorine combines and does not destroy the ammonia; it converts Reiset's salts into those of Gros and Gerhardt. Thus from $PtX_2,2NH_3$ there is formed $PtX_2Cl_2,2NH_3$, and from $PtX_2,4NH_3$ the salt of Gros's base $PtX_2Cl_2,4NH_3$. This shows that the amount of chlorine which combines is not dependent on the amount of ammonia present, but is due to the basic properties of platinum. Owing to this some chemists suppose the ammonia to be inactive or passive in certain compounds. It appears to me that these relations, these modifications, in the usual properties of ammonia and platinum are explained directly by their mutual combination. Sulphur, in sulphurous anhydride, SO_2, and hydrogen sulphide, SH_2, is naturally one and the same, but if we only knew of it in the form of hydrogen sulphide, then, having obtained it in the form of sulphurous anhydride, we should consider its properties as hidden. The oxygen in magnesia, MgO, and in nitric peroxide, NO_2, is so different that there is no resemblance. Arsenic no longer reacts in its compounds with hydrogen as it reacts in its compounds with chlorine, and in their compounds with nitrogen all metals modify both their reactions and their physical properties. We are accustomed to judge the metals by their saline compounds with haloid groups, and ammonia by its compounds with acid substances, and here, in the platino-compounds, if we assume the platinum to be bound to the entire mass of the ammonia—to its hydrogen and nitrogen—we shall understand that both the platinum and ammonia modify their characters. Far more complicated is the question why a portion of the chlorine (and other haloid simple and complex groups) in Gros's salts acts in a different manner from the other portion, and why only half of it acts in the usual way. But this also is not an exclusive case. The chlorine in potassium chlorate or in carbon tetrachloride does not react with the same ease with metals as the chlorine in the salts corresponding with hydrochloric acid. In this case it is united to oxygen and carbon, whilst in the platino-ammonium compounds it is united partly to platinum and partly to the platino-ammonium group. Many chemists, moreover, suppose that a part of the chlorine is united directly to the platinum and the other part to the nitrogen of the ammonia, and thus explain the difference of the reactions; but chlorine united to platinum reacts as well with a silver salt as the chlorine of ammonium chloride, NH_4Cl, or nitrosyl chloride, NOCl, although there is no doubt that in this case there is a union between the chlorine and nitrogen. Hence it is necessary to explain the absence of a facile reactive capacity in a portion of the chlorine by the conjoint influence of the platinum and ammonia on it, whilst the other portion may be admitted as being under the influence of the platinum only, and therefore as reacting as in other salts. By admitting a certain kind of stable union in the platino-ammonium grouping, it is possible to imagine that the chlorine does not react with its customary facility, because access to a portion of the atoms of chlorine in this complex grouping is difficult, and the chlorine union is not the same as we usually meet in the saline compounds of chlorine. These are the grounds on which we, in refuting the now accepted explanations of the reactions and formation of the platino-compounds, pronounce the following opinion as to their structure.

In characterising the platino-ammonium compounds, it is necessary to bear in mind that compounds which already contain PtX_4 do not combine directly with NH_3, and that such compounds as $PtX_4,4NH_3$ only proceed from PtX_2, and therefore it is natural to conclude that those affinities and forces which cause PtX_2 to combine with X_2 also cause it to combine with $2NH_3$. And having the compound $PtX_2,2NH_3$, and supposing that in subsequently combining with Cl_2 it reacts with those affinities which produce the compounds of platinic chloride, $PtCl_4$, with water, potassium chloride, potassium cyanide, hydrochloric acid, and the like, we explain not only the fact of combination, but also many of the reactions occurring in the transition of one kind of platino-ammonium salts into another. Thus by this means we explain the fact that (1) $PtX_2,2NH_3$ combines with $2NH_3$, forming salts of Reiset's first base; (2) and the fact that this compound (represented as follows for distinctness), $PtX_2,2NH_3,2NH_3$, when heated, or even when boiled in solution, again passes into $PtX_2,2NH_3$ (which resembles the easy disengagement of water of crystallisation, &c.); (3) the fact that $PtX_2,2NH_3$ is capable of absorbing, under the action of the same forces, a molecule of chlorine, $PtX_2,2NH_3,Cl_2$, which it then retains with energy, because it is attracted, not only by the platinum, but also by the hydrogen of the ammonia; (4) the fact that this chlorine held in this compound (of Gerhardt) will have a position unusual in salts, which will explain a certain (although very feebly-marked) difficulty of reaction; (5) the fact that this does not exhaust the faculty of platinum for further combination (we need only recall the compound $PtCl_4,2HCl,16H_2O$), and that therefore both $PtX_2,2NH_3,Cl_2$ and $PtX_2,2NH_3,2NH_3$ are still capable of combination, whence the latter, with chlorine, gives $PtX_2,2NH_3,2NH_3,Cl_2$, after the type of PtX_4Y_4 (and perhaps higher); (6) the fact that Gros's compounds thus formed are readily reconverted into the salts of Reiset's first base when acted on by reducing agents; (7) the fact that in Gros's salts, $PtX_2,2NH_3(NH_3X)_2$, the newly-attached chlorine or haloid will react with difficulty with salts of silver, &c., because it is attached both to the platinum and to the ammonia, for both of which it has an attraction; (8) the fact that the faculty for further combination is not even yet exhausted in the type of Gros's salts, and that we actually have a compound of Gros's chlorine salt with platinous chloride and with platinic chloride; the salt $PtSO_4,2NH_3,2NH_3,SO_4$ combines further also with H_2O; (9) the fact that such a faculty of combination with new molecules is naturally more developed in the lower forms of combination than in the higher. Hence the salts of Reiset's first base—for example, $PtCl_2,2NH_3,2NH_3$—both combine with water and give precipitates (soluble in water but not in hydrochloric acid) of double salts with many salts of the heavy metals—for example, with lead chloride, cupric chloride, and also with platinic and platinous chlorides (Buckton's salts). The latter compounds will have the composition $PtCl_2,2NH_3,2NH_3,PtCl_2$—that is, the same composition as the salts of Reiset's second base, but it cannot be identical with it. Such an interesting case does actually exist. The first salt, $PtCl_2,4NH_3,PtCl_2$, is green, insoluble in water and in hydrochloric acid, and is known as *Magnus's salt*, and the second, $PtCl_2,2NH_3$, is Reiset's yellow, sparingly soluble (in water). They are polymeric, namely, the first contains twice the number of elements held in the second, and at the same time they easily pass into each other. If ammonia be added to a hot hydrochloric acid solution of platinous chloride, it forms the salt $PtCl_2,4NH_3$, but in the presence of an excess of platinous chloride it gives Magnus's salt. On boiling the latter in ammonia it gives a colourless soluble salt of Reiset's first base, $PtCl_2,4NH_3$, and if this be boiled with water, ammonia is disengaged, and a salt of Reiset's second base, $PtCl_2,2NH_3$, is obtained.

A class of platino-ammonium isomerides (obtained by Millon and Thomsen) are also known. Buckton's salts—for example, the copper salt—were obtained by them from the salts of Reiset's first base, $PtCl_2,4NH_3$, by treatment with a solution of cupric chloride, &c., and therefore, according to our method of expression, Buckton's copper salt will be $PtCl_2,4NH_3,CuCl_2$. This salt is soluble in water, but not in hydrochloric acid. In it the ammonia must be considered as united to the platinum. But if cupric chloride be dissolved in ammonia, and a solution of platinous chloride in ammonium chloride is added to it, a violet precipitate is obtained of the same composition as Buckton's salt, which, however, is insoluble in water, but soluble in hydrochloric acid. In this a portion, if not all, of the ammonia must be regarded as united to the copper, and it must therefore be represented as $CuCl_2,4NH_3,PtCl_2$. This form is identical in composition but different in properties (is isomeric) with the preceding salt (Buckton's). The salt of Magnus is intermediate between them, $PtCl_2,4NH_3,PtCl_2$; it is insoluble in water and hydrochloric acid. These and certain other instances of isomeric compounds in the series of the platino-ammonium salts throw a light on the nature of the compounds in question, just as the study of the isomerides of the carbon compounds has served and still serves as the chief cause of the rapid progress of organic chemistry. In conclusion, we may add that (according to the law of substitution) we must necessarily expect all kinds of intermediate compounds between the platino and analogous ammonia derivatives on the one hand, and the complex compounds of nitrous acid on the other. Perhaps the instance of the reaction of ammonia upon osmic anhydride, OsO_4, observed by Fritsche, Frémy, and others, and more fully studied by Joly (1891), belongs to this class. The latter showed that when ammonia acts upon an alkaline solution of OsO_4 the reaction proceeds according to the equation: $OsO_4 + KHO + NH_3 = OsNKO_3 + 2H_2O$. It might be imagined that in this case the ammonia is oxidised, probably forming the residue of nitrous acid (NO), while the type OsO_4 is deoxidised into OsO_2, and a salt, OsO(NO)(KO), of the type OsX_4 is formed. This salt crystallises well in light yellow octahedra. It corresponds to *osmiamic acid*, OsO(ON)(HO), whose anhydride $[OsO(NO)]_2$, has the composition $Os_2N_2O_5$, which equals $2Os + N_2O_5$ to the same extent as the above-mentioned compound $PtCO_2$ equals $Pt + CO_2$ (*see* Note 11).

CHAPTER XXIV
COPPER, SILVER, AND GOLD

That degree of analogy and difference which exists between iron, cobalt, and nickel repeats itself in the corresponding triad ruthenium, rhodium, and palladium, and also in the heavy platinum metals, osmium, iridium, and platinum. These nine metals form Group VIII. of the elements in the periodic system, being the intermediate group between the even elements of the large periods and the uneven, among which we know zinc, cadmium, and mercury in Group II. Copper, silver, and gold complete this transition, because their properties place them in proximity to nickel, palladium, and platinum on the one hand, and to zinc, cadmium, and mercury on the other. Just as Zn, Cd, and Hg; Fe, Ru, and Os; Co, Rh, and Ir; Ni, Pd, and Pt, resemble each other in many respects, so also do Cu, Ag, and Au. Thus, for example, the atomic weight of copper $Cu = 63$, and in all its properties it stands between $Ni = 59$ and $Zn = 65$. But as the transition from Group VIII. to Group II., where zinc is situated, cannot be otherwise than through Group I., so in copper there are certain properties of the elements of Group I. Thus it gives a suboxide, Cu_2O, and salts, CuX, like the elements of Group I., although at the same time it forms an oxide, CuO, and salts CuX_2, like nickel and zinc. In the state of the oxide, CuO, and the salts, CuX_2, copper is analogous to zinc, judging from the insolubility of the carbonates, phosphates, and similar salts, and by the isomorphism, and other characters. In the cuprous salts there is undoubtedly a great resemblance to the silver salts—thus, for example, silver chloride, AgCl, is characterised by its insolubility and capacity of combining with ammonia, and in this respect cuprous chloride closely resembles it, for it is also insoluble in water, and combines with ammonia and dissolves in it, &c. Its composition is also RCl, the same as AgCl, NaCl, KCl, &c., and silver in many compounds resembles, and is even isomorphous with, sodium, so that this again justifies their being brought together. Silver chloride, cuprous chloride, and sodium chloride crystallise in the regular system. Besides which, the specific heats of copper and silver require that they should have the atomic weights ascribed to them. To the oxides Cu_2O and Ag_2O there are corresponding sulphides Ag_2S and Cu_2S. They both occur in nature in crystals of the rhombic system, and, what is most important, copper glance contains an isomorphous mixture of them both, and retains the form of copper glance with various proportions of copper and silver, and therefore has the composition R_2S where $R = Cu, Ag$.

Notwithstanding the resemblance in the atomic composition of the cuprous compounds, CuX, and silver compounds, AgX, with the compounds of the alkali metals KX, NaX, there is a considerable degree of difference between these two series of elements. This difference is clearly seen in the fact that the alkali metals belong to those elements which combine with extreme facility with oxygen, decompose water, and form the most alkaline bases; whilst silver and copper are oxidised with difficulty, form less energetic oxides, and do not decompose water, even at a rather high temperature. Moreover, they only displace hydrogen from very few acids. The difference between them is also seen in the dissimilarity of the properties of many of the corresponding compounds. Thus cuprous oxide, Cu_2O, and silver oxide, Ag_2O, are insoluble in water: the cuprous and silver carbonates, chlorides, and sulphates are also sparingly soluble in water. The oxides of silver and copper are also easily reduced to metal. This difference in properties is in intimate relation with that difference in the density of the metals which exists in this case. The alkali metals belong to the lightest, and copper and silver to the heaviest, and therefore the distance between the molecules in these metals is very dissimilar—it is greater for the former than the latter (tables in Chapter XV.). From the point of view of the periodic law, this difference between copper and silver and such elements of Group I. as potassium and rubidium, is clearly seen from the fact that copper and silver stand in the middle of those large periods (for example, K, Ca, Sc, Ti, V, Cr, Mn, Fe, Co, Ni, Cu, Zn, Ga, Ge, As, Se, Br) which start with the true metals of the alkalis—that is to say, the analogy and difference between potassium and copper are of the same nature as that between chromium and selenium, or vanadium and arsenic.

Copper is one of the few metals which have long been known in a metallic form. The Greeks and Romans imported copper chiefly from the island of Cyprus—whence its Latin name, *cuprum*. It was known to the ancients before iron, and was used, especially when alloyed with other metals, for arms and domestic utensils. That copper was known to the ancients will be understood from the fact that it occurs, although rarely, in a *native state*, and is easily extracted from its other natural compounds. Among the latter are the oxygen compounds of copper. When ignited with charcoal, they easily give up their oxygen to it, and yield metallic copper; hydrogen also easily takes up the oxygen from copper oxide when heated. Copper occurs in a native state, sometimes in association with other ores, in many parts of the Urals and in Sweden, and in considerable masses in America, especially in the neighbourhood of the great American lakes; and also in Chili, Japan, and China. The oxygen compounds of copper are also of somewhat common occurrence in certain localities; in this respect certain deposits of the Urals are especially famous. The geological period of the Urals (Permian) is characterised by a considerable distribution of copper ores. Copper is met with in the form of *cuprous oxide*, or *suboxide of copper*, Cu_2O, and is then known as *red copper ore*, because it forms red masses which not unfrequently are crystallised in the regular system. It is found much more rarely in the state of *cupric oxide*, CuO, and is then called *black copper ore*. The most common of the oxygenised compounds of copper are the *basic carbonates* corresponding with the oxides. That these compounds are undoubtedly of aqueous origin is apparent, not only from the fact that specimens are frequently found of a gradual transition from the metallic, sulphuretted, and oxidised copper into its various carbonates, but also from the presence of water in their composition, and from the laminar, reniform structure which many of them present. In this respect *malachite* is particularly well known; it is used as a green paint and also for ornaments, owing to the diversity of the shades of colour presented by the different layers of deposited malachite. The composition of malachite corresponds with the basic carbonate containing one molecule of cupric carbonate to one of hydroxide: $CuCO_3,CuH_2O_2$. In this form the copper frequently occurs in admixture with various sedimentary rocks, forming large strata, which confirms the aqueous origin of these compounds. There are many such localities in the Perm and other Governments bounding the Urals. *Blue carbonate of copper*, or *azurite*, is also often met with in the same localities; it contains the same ingredients as malachite, but in a different proportion, its composition being $CuH_2O_2,2CuCO_3$. Both these substances may be obtained artificially by the action of the alkali carbonates on solutions of cupric salts at various temperatures. These native carbonates are often used for the extraction of copper, all the more as they very readily give metallic copper, evolving water and carbonic anhydride when ignited, and leaving the easily-reducible cupric oxide. Copper is, however, still more often met with in the form of the sulphides. The sulphides of copper generally occur in chemical combination with the sulphides of iron. These copper-sulphur compounds (copper pyrites $CuFeS_2$, variegated copper ore Cu_3FeS_3, &c.) generally occur in veins in a rock gangue.

The extraction of copper from its oxide ores does not present any difficulty, because the copper, when ignited with charcoal and melted, is reduced from the impurities which accompany it. This mode of smelting copper ores is carried on in cupola or cylindrical furnaces, fluxes forming a slag being added to the mixture of ore and charcoal. The smelted copper still contains sulphur, iron, and other metallic impurities, from which it is freed by fusion in reverberatory furnaces, with access of air to the surface of the molten metal, as the iron and sulphur are more easily oxidised than the copper. The iron then separates as oxides, which collect in the slag.

Copper is characterised by its red colour, which distinguishes it from all other metals. Pure copper is soft, and may be beaten out by a hammer at the ordinary temperature, and when hot may be rolled into very thin sheets. Extremely thin leaves of copper transmit a green light. The tenacity of copper is also considerable, and next to iron it is one of the most durable metals in this respect. Copper wire of 1 sq. millimetre in section only breaks under a weight of 45 kilograms. The specific gravity of copper is 8·8, unless it contains cavities due to the fact that molten copper absorbs oxygen from the air, which is disengaged on cooling, and therefore gives a porous mass whose density is much less. Rolled copper, and also that which is deposited by the electric current, has a comparatively high density. Copper melts at a bright red heat, about 1050°, although below the temperature at which many kinds of cast iron melt. At a high temperature it is converted into vapour, which communicates a green colour to the flame. Both native copper and that cooled from a molten state crystallise in regular octahedra. Copper is not oxidised in dry air at the ordinary temperature, but when calcined it becomes coated with a layer of oxide, and it does not burn even at the highest temperature. Copper, when calcined in air, forms either the red cuprous oxide or the black cupric oxide, according to the temperature and quantity of air supplied. In air at the ordinary temperature, copper—as everyone knows—becomes coated with a brown layer of oxides or a green coating of basic salts, due to the action of the damp air containing carbonic acid. If this action continue for a prolonged time, the copper is covered with a thick coating of basic carbonate, or the so-called verdigris (the *ærugo nobilis* of ancient statues). This is due to the fact that copper, although scarcely capable of oxidising by itself, *in the presence of water and acids*—even very feeble acids, like carbonic acid—*absorbs oxygen from the air and forms salts*, which is a very characteristic property of it (and of lead). *Copper does not decompose water*, and therefore does not disengage hydrogen from it either at the ordinary or at high temperatures. Nor does copper liberate hydrogen from the oxygen acids; these act on it in two ways: they either give up a portion of their oxygen, forming lower grades of oxidation, or else only react in the presence of air. Thus, when nitric acid acts on copper it evolves nitric oxide, the copper being oxidised at the expense of the nitric acid. In the same way copper converts sulphuric acid into the lower grade of oxidation—into sulphurous anhydride, SO_2. In these cases the copper is oxidised to copper oxide, which combines with the excess of acid taken, and therefore forms a cupric salt, CuX_2. Dilute nitric acid does not act on copper at the ordinary temperature, but when heated it reacts with great ease; dilute sulphuric acid does not act on copper except in presence of air.

Both the oxides of copper, Cu_2O and CuO, are unacted on by air, and, as already mentioned, they both occur in nature.[6 bis] However, in the majority of cases copper is obtained in the form of cupric oxide and its salts—and the copper compounds used industrially generally belong to this type. This is due to the fact that the *cuprous compounds absorb oxygen* from the air and pass into cupric compounds. The cupric compounds may serve as the source for the preparation of cuprous oxide, because many reducing agents are capable of deoxidising the oxide into the suboxide. Organic substances are most generally employed for this purpose, and especially saccharine substances, which are able, in the presence of alkalis, to undergo oxidation at the expense of the oxygen of the cupric oxide, and to give acids which combine with the alkali: $2CuO - O = Cu_2O$. In this case the deoxidation of the copper may be carried further and metallic copper obtained, if only the reaction be aided by heat. Thus, for example, a fine powder of metallic copper may be obtained by heating an ammoniacal solution of cupric oxide with caustic potash and grape sugar. But if the reducing action of the saccharine substance proceed in the presence of a sufficient quantity of alkali in solution, and at not too high a temperature, cuprous oxide is obtained. To see this reaction clearly, it is not sufficient to take any cupric salt, because the alkali necessary for the reaction might precipitate cupric oxide—it is necessary to add previously some substance which will prevent this precipitation. Among such substances, tartaric acid, $C_4H_6O_6$, is one of the best. In the presence of a sufficient quantity of tartaric acid, any amount of alkali may be added to a solution of cupric salt without producing a precipitate, because a soluble double salt of cupric oxide and alkali is then formed. If glucose (for instance, honey or molasses) be added to such an alkaline tartaric solution, and the temperature be slightly raised, it first gives a yellow precipitate (this is cuprous hydroxide, CuHO), and then, on boiling, a red precipitate of (anhydrous) cuprous oxide. If such a mixture be left for a long time at the ordinary temperature, it deposits well-formed crystals of anhydrous cuprous oxide belonging to the regular system.

Cupric chloride, $CuCl_2$, when ignited, gives *cuprous chloride*, $CuCl$—*i.e.* the salt corresponding with suboxide of copper—and therefore cuprous chloride is always formed when copper enters into reaction with chlorine at a high temperature. Thus, for example, when copper is calcined with mercuric chloride, it forms cuprous chloride and vapours of mercury. The same substance is obtained on heating metallic copper in hydrochloric acid, hydrogen being disengaged; but this reaction only proceeds with finely-divided copper, as hydrochloric acid acts very feebly on compact masses of copper, and, in the presence of air, gives cupric chloride. The green solution of cupric chloride is decolorised by metallic copper, cuprous chloride being formed; but this reaction is only accomplished with ease when the solution is very concentrated and in the presence of an excess of hydrochloric acid to dissolve the cuprous chloride. The addition of water to the solution precipitates the cuprous chloride, because it is less soluble in dilute than in strong hydrochloric acid. Many reducing agents which are able to take up half the oxygen from cupric oxide are able, in the presence of hydrochloric acid, to form cuprous chloride. Stannous salts, sulphurous anhydride, alkali sulphites, phosphorous and hypophosphorous acids, and many similar reducing agents, act in this manner. The usual method of preparing cuprous chloride consists in passing sulphurous anhydride into a very strong solution of cupric chloride: $2CuCl_2 + SO_2 + 2H_2O = 2CuCl + 2HCl + H_2SO_4$. Cuprous chloride forms colourless cubic crystals which are insoluble in water. It is easily fusible, and even volatile. Under the action of oxidising agents, it passes into the cupric salt, and it absorbs oxygen from moist air, forming cupric oxychloride, Cu_2Cl_2O. *Aqueous ammonia* easily *dissolves* cuprous chloride as well as cuprous oxide; the solution also turns blue on exposure to the air. Thus an ammoniacal solution of cuprous chloride serves as an excellent absorbent for oxygen; but this solution absorbs not only oxygen, but also certain other gases—for example, carbonic oxide and acetylene.

When copper is oxidised with a considerable quantity of oxygen at a high temperature, or at the ordinary temperature in the presence of acids, and also when it decomposes acids, converting them into lower grades of oxidation (for example, when submitted to the action of nitric and sulphuric acids), it forms *cupric oxide*, CuO, or, in the presence of acids, cupric salts. Copper rust, or that black mass which forms on the surface of copper when it is calcined, consists of cupric oxide. The coating of the oxidised copper is very easily separated from the metallic copper, because it is brittle and very easily peels off, when it is struck or immersed in water. Many copper salts (for instance, the nitrite and carbonate) leave oxide of copper[8 bis] in the form of friable black powder, after being ignited. If the ignition be carried further, Cu_2O may be formed from the CuO.[8 tri] Anhydrous cupric oxide is very easily dissolved in acids, forming cupric salts, CuX_2. They are analogous to the salts MgX_2, ZnX_2, NiX_2, FeX_2, in many respects. On adding potassium or ammonium hydroxide to a solution of a cupric salt, it forms a gelatinous blue precipitate of the hydrated oxide of copper, CuH_2O_2, insoluble in water. The resultant precipitate *is re-dissolved by an excess of ammonia*, and gives a very beautiful azure blue solution, of so intense a colour that the presence of small traces of cupric salts may be discovered by this means. An excess of potassium or sodium hydroxide does not dissolve cupric hydroxide. A hot solution gives a black precipitate of the anhydrous oxide instead of the blue precipitate, and the precipitate of the hydroxide of copper becomes granular, and turns black when the solution is heated. This is due to the fact that the blue hydroxide is exceedingly unstable, and when slightly heated it loses the elements of water and gives the black anhydrous cupric oxide: $CuH_2O_2 = CuO + H_2O$.

Cupric oxide fuses at a strong heat, and on cooling forms a heavy crystalline mass, which is black, opaque, and somewhat tenacious. It is a feebly energetic base, so that not only do the oxides of the metals of the alkalis and alkaline earths displace it from its compounds, but even such oxides as those of lead and silver precipitate it from solutions, which is partially due to these oxides being soluble, although but slightly so, in water. However, cupric oxide, and especially the hydroxide, easily combines with even the least energetic acids, and does not give any compounds with bases; but, on the other hand, *it easily forms basic salts*,[9 bis] and in this respect outstrips magnesium and recalls the oxides of lead or mercury. Hence the hydroxide of copper dissolves in solutions of neutral cupric salts. The cupric salts are generally blue or green, because cupric hydroxide itself is coloured. But some of the salts in the anhydrous state are colourless.

The commonest normal salt is *blue vitriol*—i.e. the normal cupric sulphate. It generally contains five molecules of water of crystallisation, $CuSO_4, 5H_2O$. It forms the product of the action of strong sulphuric acid on copper, sulphurous anhydride being evolved. The same salt is obtained in practice by carefully roasting sulphuretted ores of copper, and also by the action of water holding oxygen in solution on them: $CuS + O_4 = CuSO_4$. This salt forms a by-product, obtained in gold refineries, when the silver is precipitated from the sulphuric acid solution by means of copper. It is also obtained by pouring dilute sulphuric acid over sheet copper in the presence of air, or by heating cupric oxide or carbonate in sulphuric acid. The crystals of this salt belong to the triclinic system, have a specific gravity of 2·19, are of a beautiful blue colour, and give a solution of the same colour. 100 parts of water at 0° dissolve 15, at 25° 23, and at 100° about 45 parts of cupric sulphate, $CuSO_4$.[10 bis] At 100° this salt loses a portion of its water of crystallisation, which it only parts with entirely at a high temperature (220°) and then gives a white powder of the anhydrous sulphate; and the latter, on further calcination, loses the elements of sulphuric anhydride, leaving cupric oxide, like all the cupric salts. The anhydrous (colourless) cupric sulphate is sometimes used for absorbing water; it turns blue in the process. It offers the advantage that it retains both hydrochloric acid and water, but not carbonic anhydride. Cupric sulphate is used for steeping seed corn; this is said to prevent the growth of certain parasites on the plants. In the arts a considerable quantity of cupric sulphate is also used in the preparation of other copper salts—for instance, of certain pigments[11 bis]—and a particularly large quantity is used *in the galvanoplastic process*, which consists in the deposition of copper from a solution of cupric sulphate by the action of a galvanic current, when the metallic copper is deposited on the negative pole and takes the shape of the latter. The description of the processes of galvanoplastic art introduced by Jacobi in St. Petersburg forms a part of applied physics, and will not be touched on here, and we will only mention that, although first introduced for small articles, it is now used for such articles as type moulds (*clichés*), for maps, prints, &c., and also for large statues, and for the deposition of iron, zinc, nickel, gold, silver, &c. on other metals and materials. The beginning of the application of the galvanic current to the practical extraction of metals from solutions has also been established, especially since the dynamo-electric machines of Gramme, Siemens, and others have rendered it possible to cheaply convert the mechanical motion of the steam engine into an electric current. It is to be expected that the application of the electric current, which has long since given such important results in chemistry, will, in the near future, play an important part in technical processes, the example being shown by electric lighting.

The alloys of copper with certain metals, and especially with zinc and tin, are easily formed by directly melting the metals together. They are easily cast into moulds, forged, and worked like copper, whilst they are much more durable in the air, and are therefore frequently used in the arts. Even the ancients used exclusively alloys of copper, and not pure copper, but its alloys with tin or different kinds of bronze (Chapter XVIII., Note 35). The alloys of copper with zinc are called *brass* or 'yellow metal.' Brass contains about 32 p.c. of zinc; generally, however, it does not contain more than 65 p.c. of copper. The remainder is composed of lead and tin, which usually occur, although in small quantities, in brass. Yellow metal contains about 40 p.c. of zinc. The addition of zinc to copper changes the colour of the latter to a considerable degree; with a certain amount of zinc the colour of the copper becomes yellow, and with a still larger proportion of zinc an alloy is formed which has a greenish tint. In those alloys of zinc and copper which contain a larger amount of zinc than of copper, the yellow colour disappears and is replaced by a greyish colour. But when the amount of zinc is diminished to about 20 p.c., the alloy is red and hard, and is called 'tombac.' A contraction takes place in alloying copper with zinc, so that the volume of the alloy is less than that of either metal individually. The zinc volatilises on prolonged heating at a high temperature and the excess of metallic copper remains behind. When heated in the air, the zinc oxidises before the copper, so that all the zinc alloyed with copper may be removed from the copper by this means. An important property of brass containing about 30 p.c. of zinc is that it is soft and malleable in the cold, but becomes somewhat brittle when heated. We may also mention that ordinary copper coins contain, in order to render them hard, tin, zinc, and iron (Cu = 95 p.c.); that it is now customary to add a small amount of phosphorus to copper and bronze, for the same purpose; and also that copper is added to silver and gold in coining, &c. to render it hard; moreover, in Germany, Switzerland, and Belgium, and other countries, a silver-white alloy (melchior, German silver, &c.), for base coinage and other purposes, is prepared from brass and nickel (from 10 to 20 p.c. of nickel; 20 to 30 p.c. zinc: 50 to 70 p.c. copper), or directly from copper and nickel, or, more rarely, from an alloy containing silver, nickel, and copper.[12 bis]

Copper, in its cuprous compounds, is so analogous to *silver*, that were there no cupric compounds, or if silver gave stable compounds of the higher oxide, AgO, the resemblance would be as close as that between chlorine and bromine or zinc and cadmium; but silver compounds corresponding to AgO are quite unknown. Although silver peroxide—which was regarded as AgO, but which Berthelot (1880) recognised as the sesquioxide Ag_2O_3—is known, still it does not form any true salts, and consequently cannot be placed along with cupric oxide. In distinction to copper, silver as a metal does not oxidise under the influence of heat; and its oxides, Ag_2O and Ag_2O_3, easily lose oxygen (*see* Note 8 tri). Silver does *not oxidise* in air at the ordinary pressure, and is therefore classed among the so-called *noble metals*. It has a white colour, which is much purer than that of any other known metal, especially when the metal is chemically pure. In the arts silver is always used alloyed, because chemically-pure silver is so soft that it wears exceedingly easily, whilst when fused with a small amount of copper, it becomes very hard, without losing its colour.

Silver occurs in *nature*, both in a native state and in certain compounds. Native silver, however, is of rather rare occurrence. A far greater quantity of silver occurs in combination with sulphur, and especially in the form of *silver sulphide*, Ag_2S, with lead sulphide or copper sulphide, or the ores of various other metals. The largest amount of silver is extracted from the lead in which it occurs. If this lead be calcined in the presence of air, it oxidises, and the resultant lead oxide, PbO ('litharge' or 'silberglätte,' as it is called), melts into a mobile liquid, which is easily removed. The silver remains in an unoxidised metallic state. This process is called *cupellation*.

Commercial silver generally contains copper, and, more rarely, other metallic impurities also. Chemically *pure silver* is obtained either by cupellation or by subjecting ordinary silver to the following treatment. The silver is first dissolved in nitric acid, which converts it and the copper into nitrates, $Cu(NO_3)_2$ and $AgNO_3$; hydrochloric acid is then added to the resultant solution (green, owing to the presence of the cupric salt), which is considerably diluted with water in order to retain the lead chloride in solution if the silver contained lead. The copper and many other metals remain in solution, whilst the silver is precipitated as silver chloride. The precipitate is allowed to settle, and the liquid is decanted off; the precipitate is then washed and fused with sodium carbonate. A double decomposition then takes place, sodium chloride and silver carbonate being formed; but the latter decomposes into metallic silver, because the silver oxide is decomposed by heat: $Ag_2CO_3 = Ag_2 + O + CO_2$. The silver chloride may also be mixed with metallic zinc, sulphuric acid, and water, and left for some time, when the zinc removes the chlorine from the silver chloride and precipitates the silver as a powder. This finely-divided silver is called 'molecular silver.'

Chemically-pure silver has an exceeding pure white colour, and a specific gravity of 10·5. Solid silver is lighter than the molten metal, and therefore a piece of silver floats on the latter. The fusing-point of silver is about 950° C., and at the high temperature attained by the combustion of detonating gas it volatilises. By employing silver reduced from silver chloride by milk sugar and caustic potash, and distilling it, Stas obtained silver purer than that obtained by any other means; in fact, this was perfectly pure silver. The vapour of silver has a very beautiful green colour, which is seen when a silver wire is placed in an oxyhydrogen flame.

It has long been known (Wöhler) that when nitrate of silver, $AgNO_3$, reacts as an oxidising agent upon citrates and tartrates, it is able under certain conditions to give either a salt of suboxide of silver (see Note [19]) or a red solution, or to give a precipitate of metallic silver reduced at the expense of the organic substances. In 1889 Carey Lea, in his researches on this class of reactions, showed that *soluble silver* is here formed, which he called *allotropic silver*. It may be obtained by taking 200 c.c. of a 10 per cent. solution of $AgNO_3$ and quickly adding a mixture (neutralised with NaHO) of 200 c.c. of a 30 per cent. solution of $FeSO_4$ and 200 c.c. of a 40 per cent. solution of sodium citrate. A lilac precipitate is obtained, which is collected on a filter (the precipitate becomes blue) and washed with a solution of NH_4NO_3. It then becomes soluble in pure water, forming a red perfectly transparent solution from which the dissolved silver is precipitated on the addition of many soluble foreign bodies. Some of the latter—for instance, NH_4NO_3, alkaline sulphates, nitrates, and citrates—give a precipitate which re-dissolves in pure water, whilst others—for instance, $MgSO_4$, $FeSO_4$, $K_2Cr_2O_7$, $AgNO_3$, $Ba(NO_3)_2$ and many others—convert the precipitated silver into a new variety, which, although no longer soluble in water, regains its solubility in a solution of borax and is soluble in ammonia. Both the soluble and insoluble silver are rapidly converted into the ordinary grey-metallic variety by sulphuric acid, although nothing is given off in the reaction; the same change takes place on ignition, but in this case CO_2 is disengaged; the latter is formed from the organic substances which remain (to the amount of 3 per cent.) in the modified silver (they are not removed by soaking in alcohol or water). If the precipitated silver be slightly washed and laid in a smooth thin layer on paper or glass, it is seen that the soluble variety is red when moist and a fine blue colour when dry, whilst the insoluble variety has a blue reflex. Besides these, under special conditions a golden yellow variety may be obtained, which gives a brilliant golden yellow coating on glass; but it is easily converted into the ordinary grey-metallic state by friction or trituration. There is no doubt[18 bis] that there is the same relation between ordinary silver which is perfectly insoluble in water and the varieties of silver obtained by Carey Lea[18 tri] as there is between quartz and soluble silica or between CuS and As_2S_2 in their ordinary insoluble forms and in the state of the colloid solution of their hydrosols (*see* Chapter I., Note 57, and Chapter XVII., Note 25 bis). Here, however, an important step in advance has been made in this respect, that we are dealing with the solution of a simple body, and moreover of a metal—*i.e.* of a particularly characteristic state of matter. And as boron, gold, and certain other simple bodies have already been obtained in a soluble (colloid) form, and as numerous organic compounds (albuminous substances, gum, cellulose, starch, &c.) and inorganic substances are also known in this form, it might be said that the colloid state (of hydrogels and hydrosols) can be acquired, if not by every substance, at all events by substances of most varied chemical character under particular conditions of formation from solutions. And this being the case, we may hope that a further study of soluble colloid compounds, which apparently present various transitions towards emulsions, may throw a new light upon the complex question of solutions, which forms one of the problems of the present epoch of chemical science. Moreover, we may remark that Spring (1890) clearly proved the colloid state of soluble silver by means of dialysis as it did not pass through the membrane.

As regards the capacity of silver for chemical reactions, it is remarkable for its small capacity for combination with oxygen and for its considerable energy of combination with sulphur, iodine, and certain kindred non-metals. *Silver does not oxidise* at any temperature, and its oxide, Ag_2O, is decomposed by heat. It is also a very important fact that silver is not oxidised by oxygen either in the presence of alkalis, even at exceedingly high temperatures, or in the presence of acids—at least, of dilute acids—which properties render it a very important metal in chemical industry for the fusion of alkalis, and also for many purposes in everyday life; for instance, for making spoons, salt-cellars, &c. Ozone, however, oxidises it. Of all acids nitric acid has the greatest action on silver. The reaction is accompanied by the formation of oxides of nitrogen and silver nitrate, $AgNO_3$, which dissolves in water and does not, therefore, hinder the further action of the acid on the metal. The halogen acids, especially hydriodic acid, act on silver, hydrogen being evolved; but this action soon stops, owing to the halogen compounds of silver being insoluble in water and only very slightly soluble in acids; they therefore preserve the remaining mass of metal from the further action of the acid; in consequence of this the action of the halogen acids is only distinctly seen with finely-divided silver. Sulphuric acid acts on silver in the same manner that it does on copper, only it must be concentrated and at a higher temperature. Sulphurous anhydride, and not hydrogen, is then evolved, but there is no action at the ordinary temperature, even in the presence of air. Among the various salts, sodium chloride (in the presence of moisture, air, and carbonic acid) and potassium cyanide (in the presence of air) act on silver more decidedly than any others, converting it respectively into silver chloride and a double cyanide.

Although silver does not directly combine with oxygen, still three different grades of combination with oxygen may be obtained indirectly from the salts of silver. They are all, however, unstable, and decompose into oxygen and metallic silver when ignited. These three oxides of silver have the following composition: *silver suboxide*, Ag_4O, corresponding with the (little investigated) suboxides of the alkali metals; *silver oxide*, Ag_2O, corresponding with the oxides of the alkali metals and the ordinary salts of silver, AgX; and *silver peroxide*, AgO,[19 bis] or, judging from Berthelot's researches, Ag_2O_3. *Silver oxide* is obtained as a brown precipitate (which when dried does not contain water) by adding potassium hydroxide to a solution of a silver salt—for example, of silver nitrate. The precipitate formed seems, however, to be an hydroxide, AgHO, *i.e.* $AgNO_3$ + KHO = KNO_3 + AgHO, and the formation of the anhydrous oxide, 2AgHO = Ag_2O + H_2O, may be compared with the formation of the anhydrous cupric oxide by the action of potassium hydroxide on hot cupric solutions. Silver hydroxide decomposes into water and silver oxide, even at low temperatures; at least, the hydroxide no longer exists at 60°, but forms the anhydrous oxide, Ag_2O.[19 tri] Silver oxide is almost insoluble in water; but, nevertheless, it is undoubtedly a rather powerful basic oxide, because it displaces the oxides of many metals from their soluble salts, and saturates such acids as nitric acid, forming with them neutral salts, which do not act on litmus paper. Undoubtedly water dissolves a small quantity of silver oxide, which explains the possibility of its action on solutions of salts—for example, on solutions of cupric salts. Water in which silver oxide is shaken up has a distinctly alkaline reaction. The oxide is distinguished by its great instability when heated, so that it loses all its oxygen when slightly heated. Hydrogen reduces it at about 80°.[20 bis] The feebleness of the affinity of silver for oxygen is shown by the fact that silver oxide decomposes under the action of light, so that it must be kept in opaque vessels. The silver *salts* are colourless and decompose when heated, leaving metallic silver if the elements of the acid are volatile.[20 tri] They have a peculiar metallic taste, and are exceedingly poisonous; the majority of them are acted on by light, especially in the presence of organic substances, which are then oxidised. The alkaline carbonates give a white precipitate of silver carbonate, Ag_2CO_3, which is insoluble in water, but soluble in ammonia and ammonium carbonate. Aqueous ammonia, added to a solution of a normal silver salt, first acts like potassium hydroxide, but the precipitate dissolves in an excess of the reagent, like the precipitate of cupric hydroxide. Silver oxalate and the halogen compounds of silver are insoluble in water; hydrochloric acid and soluble chlorides give, as already repeatedly observed, a white precipitate of silver chloride in solutions of silver salts. Potassium iodide gives a yellowish precipitate of silver iodide. Zinc separates all the silver in a metallic form from solutions of silver salts. Many other metals and reducing agents—for example, organic substances—also reduce silver from the solutions of its salts.

Silver nitrate, $AgNO_3$, is known by the name of lunar caustic (or *lapis infernalis*); it is obtained by dissolving metallic silver in nitric acid. If the silver be impure, the resultant solution will contain a mixture of the nitrates of copper and silver. If this mixture be evaporated to dryness and the residue carefully fused at an incipient red heat, all the cupric nitrate is decomposed, whilst the greater part of the silver nitrate remains unchanged. On treating the fused mass with water the latter is dissolved, whilst the cupric oxide remains insoluble. If a certain amount of silver oxide be added to the solution containing the nitrates of silver and copper, it displaces all the cupric oxide. In this case it is of course not necessary to take pure silver oxide, but only to pour off some of the solution and to add potassium hydroxide to one portion, and to mix the resultant precipitate of the hydroxides, $Cu(OH)_2$ and $AgOH$, with the remaining portion. By these methods all the copper can be easily removed and pure silver nitrate obtained (its solution is colourless, while the presence of Cu renders it blue), which may be ultimately purified by crystallisation. It crystallises in colourless transparent prismatic plates, which are not acted on by air. They are anhydrous. Its sp. gr. is 4·34; it dissolves in half its weight of water at the ordinary temperature.[22 bis] The pure salt is not acted on by light, but it easily acts in an oxidising manner on the majority of organic substances, which it generally blackens. This is due to the fact that the organic substance is oxidised by the silver nitrate, which is reduced to metallic silver; the latter is thus obtained in a finely-divided state, which causes the black stain. This peculiarity is taken advantage of for marking linen. Silver nitrate is for the same reason used for *cauterising wounds* and various growths on the body. Here again it acts by virtue of its oxidising capacity in destroying the organic matter, which it oxidises, as is seen from the separation of a coating of black metallic powdery silver from the part cauterised.[22 tri] From the description of the preparation of silver nitrate it will have been seen that this salt, which fuses at 218°, does not decompose at an incipient red heat; when cast into sticks it is usually employed for cauterising. On further heating, the fused salt undergoes decomposition, first forming silver nitrite and then metallic silver. With ammonia, silver nitrate forms, on evaporation of the solution, colourless crystals containing $AgNO_3,2HN_3$ (Marignac). In general the salts of silver, like cuprous, cupric, zinc, &c. salts, are able to give several compounds with ammonia; for example, silver nitrate in a dry state absorbs three molecules (Rose). The ammonia is generally easily expelled from these compounds by the action of heat.

Nitrate of silver easily forms double salts like $AgNO_3 2NaNO_3$ and $AgNO_3 KNO_3$. Silver nitrate under the action of water and a halogen gives nitric acid (*see* Vol. I. p. 280, formation of N_2O_5), a halogen salt of silver, and a silver salt of an oxygen acid of the halogen. Thus, for example, a solution of chlorine in water, when mixed with a solution of silver nitrate, gives silver chloride and chlorate. It is here evident that the reaction of the silver nitrate is identical with the reaction of the caustic alkalis, as the nitric acid is all set free and the silver oxide only reacts in exactly the same way in which aqueous potash acts on free chlorine. Hence the reaction may be expressed in the following manner: $6AgNO_3 + 3Cl_2 + 3H_2O = 5AgCl + AgClO_3 + 6NHO_3$.

Silver nitrate, like the nitrates of the alkalis, does not contain any water of crystallisation. Moreover the other salts of silver almost always separate out without any water of crystallisation. The silver salts are further characterised by the fact that they *give neither basic nor acid salts*, owing to which the formation of silver salts generally forms the means of determining the true composition of acids—thus, to any acid H_nX there corresponds a salt Ag_nX—for instance, Ag_3PO_4 (Chapter XIX., Note 15).

Silver gives insoluble and exceedingly stable *compounds with the halogens*. They are obtained by double decomposition with great facility whenever a silver salt comes in contact with halogen salts. Solutions of nitrate, sulphate, and all other kindred salts of silver give a precipitate of silver chloride or iodide in solutions of chlorides and iodides and of the halogen acids, because the halogen salts of silver are insoluble both in water and in other acids. *Silver chloride*, AgCl, is then obtained as a white flocculent precipitate, silver bromide forms a yellowish precipitate, and silver iodide has a very distinct yellow colour. These halogen compounds sometimes occur in nature; they are formed by a dry method—by the action of halogen compounds on silver compounds, especially under the influence of heat. Silver chloride easily fuses at 451° on cooling from a molten state; it forms a somewhat soft horn-like mass which can be cut with a knife and is known as *horn silver*. It volatilises at a higher temperature. Its ammoniacal solution, on the evaporation of the ammonia, deposits crystalline chloride of silver, in octahedra. Bromide and iodide of silver also appear in forms of the regular system, so that in this respect the halogen salts of silver resemble the halogen salts of the alkali metals.

Silver chloride may be decomposed, with the separation of silver oxide, by heating it with a solution of an alkali, and if an organic substance be added to the alkali the chloride can easily be reduced o metallic silver, the silver oxide being reduced in the oxidation of the organic substance. Iron, zinc, and many other metals reduce silver chloride in the presence of water. Cuprous and mercurous chlorides and many organic substances are also able to reduce the silver from chloride of silver. This shows the rather easy decomposability of the halogen compounds of silver. Silver iodide is much more stable in this respect than the chloride. The same is also observed with respect to the *action of light* upon moist AgCl. White silver chloride soon acquires a violet colour when exposed to the action of light, and especially under the direct action of the sun's rays. After being acted upon by light it is no longer entirely soluble in ammonia, but leaves metallic silver undissolved, from which it might be assumed that the action of light consisted in the decomposition of the silver chloride into chlorine and metallic silver and in fact the silver chloride becomes in time darker and darker. Silver bromide and iodide are much more slowly acted on by light, and, according to certain observations, when pure they are even quite unacted on; at least they do not change in weight,[24 bis] so that if they are acted on by light, the change they undergo must be one of a change in the structure of their parts and not of decomposition, as it is in silver chloride. The silver chloride under the action of light changes in weight, which indicates the formation of a volatile product, and the deposition of metallic silver on dissolving in ammonia shows the loss of chlorine. The change does actually occur under the action of light, but the decomposition does not go as far as into chlorine and silver, but only to the formation of a subchloride of silver, Ag_2Cl, which is of a brown colour and is easily decomposed into metallic silver and silver chloride, $Ag_2Cl = AgCl + Ag$. This change of the chemical composition and structure of the halogen salts of silver under the action of light forms the basis of *photography*, because the halogen compounds of silver, after having been exposed to light, give a precipitate of finely-divided silver, of a black colour, when treated with reducing agents.

The insolubility of the halogen compounds of silver forms the basis of many methods used in practical chemistry. Thus by means of this reaction it is possible to obtain salts of other acids from a halogen salt of a given metal, for instance, $RCl_2 + 2AgNO_3 = R(NO_3)_2 + 2AgCl$. The formation of the halogen compounds of silver is very frequently used in the investigation of organic substances; for example, if any product of metalepsis containing iodine or chlorine be heated with a silver salt or silver oxide, the silver combines with the halogen and gives a halogen salt, whilst the elements previously combined with the silver replace the halogen. For instance, ethylene dibromide, $C_2H_4Br_2$, is transformed into ethylene diacetate, $C_2H_4(C_2H_3O_2)_2$, and silver bromide by heating it with silver acetate, $2C_2H_3O_2Ag$. The insolubility of the halogen compounds of silver is still more frequently taken advantage of in determining the amount of silver and halogen in a given solution. If it is required, for instance, to determine the quantity of chlorine present in the form of a metallic chloride in a given solution, a solution of silver nitrate is added to it so long as it gives a precipitate. On *shaking or stirring* the liquid, the silver chloride easily settles in the form of heavy flakes. It is possible in this way to precipitate the whole of the chlorine from a solution, without adding an excess of silver nitrate, since it can be easily seen whether the addition of a fresh quantity of silver nitrate produces a precipitate in the clear liquid. In this manner it is possible to add to a solution containing chlorine, as much silver as is required for its entire precipitation, and to calculate the amount of chlorine previously in solution from the amount of the solution of silver nitrate consumed, if the quantity of silver nitrate in this solution has been previously determined.[25 bis] The atomic proportions and preliminary experiments with a pure salt—for example, with sodium chloride—will give the amount of chlorine from the quantity of silver nitrate. Details of these methods will be found in works on analytical chemistry.[25 tri]

Accurate experiments, and more especially the *researches of Stas* at Brussels, show the proportion in which silver reacts with metallic chlorides. These researches have led to the determination of the *combining weights* of silver, sodium, potassium, chlorine, bromine, iodine, and other elements, and are distinguished for their model exactitude, and we will therefore describe them in some detail. As sodium chloride is the chloride most generally used for the precipitation of silver, since it can most easily be obtained in a pure state, we will here cite the quantitative observations made by Stas for showing the co-relation between the quantities of chloride of sodium and silver which react together. In order to obtain perfectly pure sodium chloride, he took pure rock salt, containing only a small quantity of magnesium and calcium compounds and a small amount of potassium salts. This salt was dissolved in water, and the saturated solution evaporated by boiling. The sodium chloride separated out during the boiling, and the mother liquor containing the impurities was poured off. Alcohol of 65 p.c. strength and platinic chloride were added to the resultant salt, in order to precipitate all the potassium and a certain part of the sodium salts. The resultant alcoholic solution, containing the sodium and platinum chlorides, was then mixed with a solution of pure ammonium chloride in order to remove the platinic chloride. After this precipitation, the solution was evaporated in a platinum retort, and then separate portions of this purified sodium chloride were collected as they crystallised. The same salt was prepared from sodium sulphate, tartrate, nitrate, and from the platinochloride, in order to have sodium chloride prepared by different methods and from different sources, and in this manner ten samples of sodium chloride thus prepared were purified and investigated in their relation to silver. After being dried, weighed quantities of all ten samples of sodium chloride were dissolved in water and mixed with a solution in nitric acid of a weighed quantity of perfectly pure silver. A slightly greater quantity of silver was taken than would be required for the decomposition of the sodium chloride, and when, after pouring in all the silver solution, the silver chloride had settled, the amount of silver remaining in excess was determined by means of a solution of sodium chloride of known strength. This solution of sodium chloride was added so long as it formed a precipitate. In this manner Stas determined how many parts of sodium chloride correspond to 100 parts by weight of silver. The result of ten determinations was that for the entire precipitation of 100 parts of silver, from 54·2060 to 54·2093 parts of sodium chloride were required. The difference is so inconsiderable that it has no perceptible influence on the subsequent calculations. The mean of ten experiments was that 100 parts of silver react with 54·2078 parts of sodium chloride. In order to learn from this the relation between the chlorine and silver, it was necessary to determine the quantity of chlorine contained in 54·2078 parts of sodium chloride, or, what is the same thing, the quantity of chlorine which combines with 100 parts of silver. For this purpose Stas made a series of observations on the quantity of silver chloride obtained from 100 parts of silver. Four syntheses were made by him for this purpose. The first synthesis consisted in the formation of silver chloride by the action of chlorine on silver at a red heat. This experiment showed that 100 parts of silver give 132·841, 132·843 and 132·843 of silver chloride. The second method consisted in dissolving a given quantity of silver in nitric acid and precipitating it by means of gaseous hydrochloric acid passed over the surface of the liquid; the resultant mass was evaporated in the dark to drive off the nitric acid and excess of hydrochloric acid, and the remaining silver chloride was fused first in an atmosphere of hydrochloric acid gas and then in air. In this process the silver chloride was not washed, and therefore there could be no loss from solution. Two experiments made by this method showed that 100 parts of silver give 132·849 and 132·846 parts of silver chloride. A third series of determinations was also made by precipitating a solution of silver nitrate with a certain excess of gaseous hydrochloric acid. The amount of silver chloride obtained was altogether 132·848. Lastly, a fourth determination was made by precipitating dissolved silver with a solution of ammonium chloride, when it was found that a considerable amount of silver (0·3175) had passed into solution in the washing; for 100 parts of silver there was obtained altogether 132·8417 of silver chloride. Thus from the mean of seven determinations it appears that 100 parts of silver give 132·8445 parts of silver chloride—that is, that

32·8445 parts of chlorine are able to combine with 100 parts of silver and with that quantity of sodium which is contained in 54·2078 parts of sodium chloride. These observations show that 32·8445 parts of chlorine combine with 100 parts of silver and with 21·3633 parts of sodium. From these figures expressing the relation between the combining weights of chlorine, silver, and sodium, it would be possible to determine their atomic weights—that is, the combining quantity of these elements with respect to one part by weight of hydrogen or 16 parts of oxygen, if there existed a series of similarly accurate determinations for the reactions between hydrogen or oxygen and one of these elements—chlorine, sodium, or silver. If we determine the quantity of silver chloride which is obtained from silver chlorate, $AgClO_3$, we shall know the relation between the combining weights of silver chloride and oxygen, so that, taking the quantity of oxygen as a constant magnitude, we can learn from this reaction the combining weight of silver chloride, and from the preceding numbers the combining weights of chlorine and silver. For this purpose it was first necessary to obtain pure silver chlorate. This Stas did by acting on silver oxide or carbonate, suspended in water, with gaseous chlorine.

The decomposition of the silver chlorate thus obtained was accomplished by the action of a solution of sulphurous anhydride on it. The salt was first fused by carefully heating it at 243°. The solution of sulphurous anhydride used was one saturated at 0°. Sulphurous anhydride in dilute solutions is oxidised at the expense of silver chlorate, even at low temperatures, with great ease if the liquid be continually shaken, sulphuric acid and silver chloride being formed: $AgClO_3 + 3SO_2 + 3H_2O = AgCl + 3H_2SO_4$. After decomposition, the resultant liquid was evaporated, and the residue of silver chloride weighed. Thus the process consisted in taking a known weight of silver chlorate, converting it into silver chloride, and determining the weight of the latter. The analysis conducted in this manner gave the following results, which, like the preceding, designate the weight in a vacuum calculated from the weights obtained in air: In the first experiment it appeared that 138·7890 grams of silver chlorate gave 103·9795 parts of silver chloride, and in the second experiment that 259·5287 grains of chlorate gave 194·44515 grams of silver chloride, and after fusion 194·4435 grams. The mean result of both experiments, converted into percentages, shows that 100 parts of silver chlorate contain 74·9205 of silver chloride and 25·0795 parts of oxygen. From this it is possible to calculate the combining weight of silver chloride, because in the decomposition of silver chlorate there are obtained three atoms of oxygen and one molecule of silver chloride: $AgClO_3 = AgCl + 3O$. Taking the weight of an atom of oxygen to be 16, we find from the mean result that the equivalent weight of silver chloride is equal to 143·395. Thus if $O = 16$, $AgCl = 143·395$, and as the preceding experiments show that silver chloride contains 32·8445 parts of chlorine per 100 parts of silver, the weight of the atom of silver[26 bis] must be 107·94 and that of chlorine 35·45. The weight of the atom of sodium is determined from the fact that 21·3633 parts of sodium chloride combine with 32·8445 parts of chlorine; consequently Na = 23·05. This conclusion, arrived at by the analysis of silver chlorate, was verified by means of the analysis of potassium chlorate by decomposing it by heat and determining the weight of the potassium chloride formed, and also by effecting the same decomposition by igniting the chlorate in a stream of hydrochloric acid. The combining weight of potassium chloride was thus determined, and another series of determinations confirmed the relation between chlorine, potassium, and silver, in the same manner as the relation between sodium, chlorine, and silver was determined above. Consequently, the combining weights of sodium, chlorine, and potassium could be deduced by combining these data with the analysis of silver chlorate and the synthesis of silver chloride. The agreement between the results showed that the determinations made by the last method were perfectly correct, and did not depend in any considerable degree on the methods which were employed in the preceding determinations, as the combining weights of chlorine and silver obtained were the same as before. There was naturally a difference, but so small a one that it undoubtedly depended on the errors incidental to every process of weighing and experiment. The atomic weight of silver was also determined by Stas by means of the synthesis of silver sulphide and the analysis of silver sulphate. The combining weight obtained by this method was 107·920. The synthesis of silver iodide and the analysis of silver iodate gave the figure 107·928. The synthesis of silver bromide with the analysis of silver bromate gave the figure 107·921. The synthesis of silver chloride and the analysis of silver chlorate gave a mean result of 107·937. Hence there is no doubt that the combining weight of silver is at least as much as 107·9—greater than 107·90 and less than 107·95, and probably equal to the mean = 107·92. Stas determined the combining weights of many other elements in this manner, such as lithium, potassium, sodium, bromine, chlorine, iodine, and also nitrogen, for the determination of the amount of silver nitrate obtained from a given amount of silver gives directly the combining weight of nitrogen. Taking that of oxygen as 16, he obtained the following combining weights for these elements: nitrogen 14·04, silver 107·93, chlorine 35·46, bromine 79·95, iodine 126·85, lithium 7·02, sodium 23·04, potassium 39·15. These figures differ slightly from those which are usually employed in chemical investigations. They must be regarded as the result of the best observations, whilst the figures usually used in practical chemistry are only approximate—are, so to speak, round numbers for the atomic weights which differ so little from the exact figures (for instance, for Ag 108 instead of 107·92, for Na 23 instead of 23·04) that in ordinary determinations and calculations the difference falls within the limits of experimental error inseparable from such determinations.

The exhaustive investigations conducted by Stas on the atomic weights of the above-named elements have great significance in the solution of the problem as to whether the atomic weights of the elements can be expressed in whole numbers if the unit taken be the atomic weight of hydrogen. Prout, at the beginning of this century, stated that this was the case, and held that the atomic weights of the elements are multiples of the atomic weight of hydrogen. The subsequent determinations of Berzelius, Penny, Marchand, Marignac, Dumas, and more especially of Stas, proved this conclusion to be untenable; since a whole series of elements proved to have fractional atomic weights—for example, chlorine, about 35·5. On account of this, Marignac and Dumas stated that the atomic weights of the elements are expressed in relation to hydrogen, either by whole numbers or by numbers with simple fractions of the magnitudes ½ and ¼. But Stas's researches refute this supposition also. Even between the combining weight of hydrogen and oxygen, there is not, so far as is yet known, that simple relation which is required by *Prout's hypothesis*, *i.e.*, taking O = 16, the atomic weight of hydrogen is equal not to 1 but to a greater number somewhere between 1·002 and 1·008 or mean 1·005. Such a conclusion arrived at by direct experiment cannot but be regarded as having greater weight than Prout's supposition (hypothesis) that the atomic weights of the elements are in multiple proportion to each other, which would give reason for surmising (but not asserting) a complexity of nature in the elements, and their common origin from a single primary material, and for expecting their mutual conversion into each other. All such ideas and hopes must now, thanks more especially to Stas, be placed in a region void of any experimental support whatever, and therefore not subject to the discipline of the positive data of science.

Among the platinum metals ruthenium, rhodium, and palladium, by their atomic weights and properties, approach silver, just as iron and its analogues (cobalt and nickel) approach copper in all respects. *Gold* stands in exactly the same position in relation to the heavy platinum metals, osmium, iridium, and platinum, as copper and silver do to the two preceding series. The atomic weight of gold is nearly equal to their atomic weights; it is dense like these metals. It also gives various grades of oxidation, which are feeble, both in a basic and an acid sense. Whilst near to osmium, iridium, and platinum, gold at the same time is able, like copper and silver, to form compounds which answer to the type RX—that is, oxides of the composition R_2O. Cuprous chloride, CuCl, silver chloride, AgCl, and aurous chloride, AuCl, are substances which are very much alike in their physical and chemical properties.[28 bis] They are insoluble in water, but dissolve in hydrochloric acid and ammonia, in potassium cyanide, sodium thiosulphate, &c. Just as copper forms a link between the iron metals and zinc, and as silver unites the light platinum metals with cadmium, so also gold presents a transition from the heavy platinum metals to mercury. Copper gives saline compounds of the types CuX and CuX_2, silver of the type AgX, whilst gold, besides compounds of the type AuX, very easily and most frequently forms those of the type $AuCl_3$. The compounds of this type frequently pass into those of the lower type, just as PtX_4 passes into PtX_2, and the same is observable in the elements which, in their atomic weights, follow gold. Mercury gives HgX_2 and HgX, thallium gives TlX_3 and TlX, lead gives PbX_4 and PbX_2. On the other hand, gold in a qualitative respect differs from silver and copper in the *extreme ease* with which all its compounds are *reduced to metal* by many means. This is not only accomplished by many reducing agents, but also by the action of heat. Thus its chlorides and oxides lose their chlorine and oxygen when heated, and, if the temperature be sufficiently high, these elements are entirely expelled and metallic gold alone remains. Its compounds, therefore, act as oxidising agents.

In nature gold occurs in the primary and chiefly in quartzose rocks, and especially in quartz veins, as in the Urals (at Berezoffsk), in Australia, and in California. The native gold is extracted from these rocks by subjecting them to a mechanical treatment consisting of crushing and washing.[29 bis] Nature has already accomplished a similar disintegration of the hard rocky matter containing gold. These disintegrated rocks, washed by rain and other water, have formed gold-bearing deposits, which are known as *alluvial gold deposits*. Gold-bearing soil is sometimes met with on the surface and sometimes under the upper soil, but more frequently along the banks of dried-up water-courses and running streams. The sand of many rivers contains, however, a very small amount of gold, which it is not profitable to work; for example, that of the Alpine rivers contains 5 parts of gold in 10,000,000 parts of sand. The richest gold deposits are those of Siberia, especially in the southern parts of the Government of Yeniseisk, the South Urals, Mexico, California, South Africa, and Australia, and then the comparatively poorer alluvial deposits of many countries (Hungary, the Alps, and Spain in Europe). The extraction of the gold from alluvial deposits is based on the principle of levigation; the earth is washed, while constantly agitated, by a stream of water, which carries away the lighter portion of the earth, and leaves the coarser particles of the rock and heavier particles of the gold, together with certain substances which accompany it, in the washing apparatus. The extraction of this *washed* gold only necessitates mechanical appliances, and it is not therefore surprising that gold was known to savages and in the most remote period of history. It sometimes occurs in crystals belonging to the regular system, but in the majority of cases in nuggets or grains of greater or less magnitude. It always contains silver (from very small quantities up to 30 p.c., when it is called 'electrum') and certain other metals, among which lead and rhodium are sometimes found.

The separation of the silver from gold is generally carried on with great precision, as the presence of the silver in the gold does not increase its value for exchange, and it can be substituted by other less valuable metals, so that the extraction of the silver, as a precious metal, from its alloy with gold, is a profitable operation. This separation is conducted by different methods. Sometimes the argentiferous gold is melted in crucibles, together with a mixture of common salt and powdered bricks. The greater portion of the silver is thus converted into the chloride, which fuses and is absorbed by the slags, from which it may be extracted by the usual methods. The silver is also extracted from gold by treating it with boiling sulphuric acid, which does not act on the gold but dissolves the silver. But if the alloy does not contain a large proportion of silver it cannot be extracted by this method or at all events the separation will be imperfect, and therefore a fresh amount of silver is added (by fusion) to the gold, in such quantity that the alloy contains twice as much silver as gold. The silver which is added is preferably such as contains gold, which is very frequently the case. The alloy thus formed is poured in a thin stream into water, by which means it is obtained in a granulated form; it is then boiled with strong sulphuric acid, three parts of acid being used to one part of alloy. The sulphuric acid extracts all the silver without acting on the gold. It is best, however, to pour off the first portion of the acid, which has dissolved the silver, and then treat the residue of still imperfectly pure gold with a fresh quantity of sulphuric acid. The gold is thus obtained in the form of powder, which is washed with water until it is quite free from silver. The silver is precipitated from the solution by means of copper, so that cupric sulphate and metallic silver are obtained. This process is carried out in many countries, as in Russia, at the Government mints.

Gold is generally used alloyed with copper; since pure gold, like pure silver, is very soft, and therefore soon worn away. In assaying or determining the amount of pure gold in such an alloy it is usual to add silver to the gold in order to make up an alloy containing three parts of silver to one of gold (this is known as quartation because the alloy contains ¼ of gold), and the resultant alloy is treated with nitric acid. If the silver be not in excess over the gold, it is not all dissolved by the nitric acid, and this is the reason for the quartation. The amount of pure gold (assay) is determined by weighing the gold which remains after this treatment. English gold (= 22 carats) coinage is composed of an alloy containing 91·66 p.c. of gold, but for many articles gold is frequently used containing a larger amount of foreign metals.

Pure gold may be obtained from gold alloys by dissolving in aqua regia, and then adding ferrous sulphate to the solution or heating it with a solution of oxalic acid. These deoxidising agents reduce the gold, but not the other metals. The chlorine combined with the gold then acts like free chlorine. The gold, thus reduced, is precipitated as an exceedingly fine brown powder.[31 bis] It is then washed with water, and fused with nitre or borax. Pure gold reflects a yellow light, and in the form of very thin sheets (gold leaf), into which it can be hammered and rolled,[31 tri] it transmits a bluish-green light. The specific gravity of gold is about 19·5, the sp. gr. of gold coin is about 17·1. It fuses at 1090°—at a higher temperature than silver—and can be drawn into exceedingly fine wires or hammered into thin sheets. With its softness and ductility, gold is distinguished for its tenacity, and a gold wire two millimetres thick breaks only under a load of 68 kilograms. Gold vaporises even at a furnace heat, and imparts a greenish colour to a flame passing over it in a furnace. Gold alloys with copper almost without changing its volume. In its chemical aspect, gold presents, as is already seen from its general characteristics given above, an example of the so-called noble metals—*i.e.* it is incapable of being oxidised at any temperature, and its oxide is decomposed when calcined. Only chlorine and bromine combine directly with it at the ordinary temperature, but many other metals and non-metals combine with it at a red heat—for example, sulphur, phosphorus, and arsenic. Mercury dissolves it with great ease. It dissolves in potassium cyanide in the presence of air; a mixture of sulphuric acid with nitric acid dissolves it with the aid of heat, although in small quantity. It is also soluble in aqua regia and in selenic acid. Sulphuric, hydrochloric, nitric, and hydrofluoric acids and the caustic alkalis do not act on gold, but a mixture of hydrochloric acid with such oxidising agents as evolve chlorine naturally dissolves it like aqua regia.[32 bis]

As regards the compounds of gold, they belong, as was said above, to the types AuX_3 and AuX. *Auric chloride* or *gold trichloride*, $AuCl_3$, which is formed when gold is dissolved in aqua regia, belongs to the former and higher of these types. The solution of this substance in water has a yellow colour, and it may be obtained pure by evaporating the solution in aqua regia to dryness, but not to the point of decomposition. If the evaporation proceed to the point of crystallisation, a compound of gold chloride and hydrochloric acid, $AuHCl_4$, is obtained, like the allied compounds of platinum; but it easily parts with the acid and leaves auric chloride, which fuses into a red-brown liquid, and then solidifies to a crystalline mass. If dry chlorine be passed over gold in powder it forms a mixture of aurous and auric chlorides, but the aurous chloride is also decomposed by water into gold and auric chloride. Auric chloride crystallises from its solutions as $AuCl_3,2H_2O$, which easily loses water, and the dry chloride loses two-thirds of its chlorine at 185°, forming aurous chloride, whilst above 300° the latter chloride also loses its chlorine and leaves metallic gold. Auric chloride is the usual form in which gold occurs in solutions, and in which its salts are used in the arts and for chemical purposes. It is soluble in water, alcohol, and ether. Light has a reducing action on these solutions, and after a time metallic gold is deposited upon the sides of vessels containing the solution. Hydrogen when nascent, and even in a gaseous form, reduces gold from this solution to a metallic state. The reduction is more conveniently and usually effected by ferrous sulphate, and in general by the action of ferrous salts.

If a solution of potassium hydroxide be added to a solution of auric chloride, a precipitate is first formed, which re-dissolves in an excess of the alkali. On being evaporated under the receiver of an air-pump, this solution yields yellow crystals, which present the same composition as the double salts $AuMCl_4$, with the substitution of the chlorine by oxygen—that is to say, *potassium aurate*, $AuKO_2$, is formed in crystals containing $3H_2O$. The solution has a distinctly alkaline reaction. *Auric oxide*, Au_2O_3, separates when this alkaline solution is boiled with an excess of sulphuric acid. But it then still retains some alkali; however, it may be obtained in a pure state as a brown powder by dissolving in nitric acid and diluting with water. The brown powder decomposes below 250° into gold and oxygen. It is insoluble in water and in many acids, but it dissolves in alkalis, which shows the acid character of this oxide. An hydroxide, $Au(OH)_3$ may be obtained as a brown powder by adding magnesium oxide to a solution of auric chloride and treating the resultant precipitate of magnesium aurate with nitric acid. This hydroxide loses water at 100°, and gives auric oxide.

The starting-point of the compounds of the type AuX is *gold monochloride* or *aurous chloride*, AuCl, which is formed, as mentioned above, by heating auric chloride at 185°. Aurous chloride forms a yellowish-white powder; this, when heated with water, is decomposed into metallic gold and auric chloride, which passes into solution: $3AuCl = AuCl_3 + 2Au$. This decomposition is accelerated by the action of light. Hence it is obvious that the compounds corresponding with aurous oxide are comparatively unstable. But this only refers to the simple compounds AuX; some of the complex compounds, on the contrary, form the most stable compounds of gold. Such, for example, is the cyanide of gold and potassium, $AuK(CN)_2$. It is formed, for instance, when finely-divided gold dissolves in the presence of air in a solution of potassium cyanide: $4KCN + 2Au + H_2O + O = 2KAu(CN)_2 + 2KHO$ (this reaction also proceeds with solid pieces of gold, although very slowly). The same compound is formed in solution when many compounds of gold are mixed with potassium cyanide, because if a higher compound of gold be taken, it is reduced by the potassium cyanide into aurous oxide, which dissolves in potassium cyanide and forms $KAu(CN)_2$. This substance is soluble in water, and gives a colourless solution, which can be kept for a long time, and is employed in electro-gilding—that is, for coating other metallic objects with a layer of gold, which is deposited if the object be connected with the negative pole of a battery and the positive pole consist of a gold plate. When an electric current is passed between them, the gold from the latter will dissolve, whilst a coating of gold from the solution will be deposited on the object.

Footnotes:

The perfectly unique position held by copper, silver, and gold in the periodic system of the elements, and the degree of affinity which is found between them, is all the more remarkable, as nature and practice have long isolated these metals from all others by having employed them—for example, for coinage—and determined their relative importance and value in conformity with the order (silver between copper and gold) of their atomic weights, &c.

Cupric sulphate contains 5 molecules of water, $CuSO_4,5H_2O$, and the isomorphous mixtures with $ZnSO_4,7H_2O$ contain either 5 or 7 equivalents, according to whether copper or zinc predominates (Vol. II. p. 6). If there be a large proportion of copper, and if the mixture contain $5H_2O$, the form of the isomorphous mixture (triclinic) will be isomorphous with cupric sulphate, $CuSO_4,5H_2O$, but if a large amount of zinc (or magnesium, iron, nickel, or cobalt) be present the form (rhombic or monoclinic) will be nearly the same as that of zinc sulphate, $ZnSO_4,7H_2O$. Supersaturated solutions of each of these salts crystallise in that form and with that amount of water which is contained in a crystal of one or other of the salts brought in contact with the solution (Chapter XIV., Note 27).

Iron pyrites, FeS_2, very often contain a small quantity of copper sulphide (see Chapter XXII., Note 2 bis), and on burning the iron pyrites for sulphurous anhydride the copper oxide remains in the residue, from which the copper is often extracted with profit. For this purpose the whole of the sulphur is not burnt off from the iron pyrites, but a portion is left behind in the ore, which is then slowly ignited (roasted) with access of air. Cupric sulphate is then formed, and is extracted by water; or what is better and more frequently done, the residue from the roasting of the pyrites is roasted with common salt, and the solution of cupric chloride obtained by lixiviating is precipitated with iron. A far greater amount of copper is obtained from other sulphuretted ores. Among these *copper glance*, Cu_2S, is more rarely met with. It has a metallic lustre, is grey, generally crystalline, and is obtained in admixture with organic matter; so that there is no doubt that its origin is due to the reducing action of the latter on solutions of cupric sulphate. *Variegated copper ore*, which crystallises in octahedra, not infrequently forms an admixture in copper glance; it has a metallic lustre, and is reddish-brown; it has a superficial play of colours, due to oxidation proceeding on its surface. Its composition is Cu_3FeS_3. But the most common and widely-distributed copper ore is *copper pyrites*, which crystallises in regular octahedra; it has a metallic lustre, a sp. gr. of 4·0, and yellow colour. Its composition is $CuFeS_2$. It must be remarked that the sulphurous ores of copper are oxidised in the presence of water containing oxygen in solution, and form cupric sulphate, blue vitriol, which is easily soluble in water. If this water contains calcium carbonate, gypsum and cupric carbonate are formed by double decomposition: $CuSO_4 + CaCO_3 = CuCO_3 + CaSO_4$. Hence copper sulphide in the form of different ores must be considered as the primary product, and the many other copper ores as secondary products, formed by water. This is confirmed by the fact that at the present time the water extracted from many copper mines contains cupric sulphate in solution. From this liquid it is easy to extract cupric oxide by the action of organic matter and various impurities of water. Hence metallic copper is sometimes found in natural products of the modification of copper sulphide and is probably deposited by the action of organic matter present in the water.

Copper ores rich in oxygen are very rare; the sulphur ores are of more common occurrence, but the extraction of the copper from them is much more difficult. The problem here not only consists in the removal of the sulphur, but also in the removal of the iron combined with the sulphur and copper. This is attained by a whole series of operations, after which there still sometimes remains the extraction of the metallic silver which generally accompanies the copper, although in but small quantity. These processes commence with the roasting—*i.e.* calcination—of the ore with access of air, by which means the sulphur is converted into sulphurous anhydride. It should here be remarked that iron sulphide is more easily oxidised than copper sulphide, and therefore the greater part of the iron in the residue from roasting is no longer in the form of sulphide but of oxide of iron. The roasted ore is mixed with charcoal, and siliceous fluxes, and smelted in a cupola furnace. The iron then passes into the slag, because its oxide gives an easily-fusible mass with the silica, whilst the copper, in the form of sulphide, fuses and collects under the slag. The greater part of the iron is removed from the mass by this smelting. The resultant *coarse metal* is again roasted in order to remove the greater part of the sulphur from the copper sulphide, and to convert the metal into oxide, after which the mass is again smelted. These processes are repeated several times, according to the richness of the ore. During these smeltings a portion of the copper is already obtained in a metallic form, because copper sulphide gives metallic copper with the oxide ($CuS + 2CuO = 3Cu + SO_2$). We will not here describe the furnaces used or the details of this process, but the above remarks include the explanation of those chemical processes which are accomplished in the various technical operations which are made use of in the process (for details *see* works on metallurgy).

Besides the smelting of copper there also exist methods for its extraction from solutions in the wet way, as it is called. Recourse is generally had to these methods for poor copper ores. The copper is brought into solution, from which it is separated by means of metallic iron or by other methods (by the action of an electric current). The sulphides are roasted in such a manner that the greater part of the copper is oxidised into cupric sulphate, whilst at the same time the corresponding iron salts are as far as possible decomposed. This process is based on the fact that the copper sulphides absorb oxygen when they are calcined in the presence of air, forming cupric sulphate. The roasted ore is treated with water, to which acid is sometimes added, and after lixiviation the resultant solution containing copper is treated either with metallic iron or with milk of lime, which precipitates cupric hydroxide from the solution. Copper oxide ores poor in metal may be treated with dilute acids in order to obtain the copper oxides in solution, from which the copper is then easily precipitated either by iron or as hydroxide by lime. According to Hunt and Douglas's method, the copper in the ore is converted by calcination into the cupric oxide, which is brought into solution by the action of a mixture of solutions of ferrous sulphate and sodium chloride; the oxide converts the ferrous chloride into ferric oxide, forming copper chlorides, according to the equation $3CuO + 2FeCl_2 = CuCl_2 + 2CuCl + Fe_2O_3$. The cupric chloride is soluble in water, whilst the cuprous chloride is dissolved in the solution of sodium chloride, and therefore all the copper passes into solution, from which it is precipitated by iron.

The same American metallurgists give the following wet method for extracting the Ag and Au occurring in many copper ores, especially in sulphurous ores: (1) The Cu_2S is first converted into oxide by roasting in a calciner; (2) the CuO is extracted by the dilute sulphuric acid obtained in the fourth process, the Cu then passes into solution, while the Ag, Au and oxides of iron remain behind in the residue (from which the noble metals may be extracted); (3) a portion of the copper in solution is converted into $CuCl_2$ (and $CaSO_4$ precipitated) by means of the $CaCl_2$ obtained in the fifth process; (4) the mixture of solutions of $CuSO_4$ and $CuCl_2$ is converted into the insoluble CuCl (salt of the suboxide) by the action of the SO_2 obtained by roasting the ore (in the first operation), sulphuric acid is then formed in the solution, according to the equation: $CuSO_4 + CuCl_2 + SO_2 + 2H_2O = 2H_2SO_4 + 2CuCl$; (5) the precipitated CuCl is treated with lime and water, and gives $CuCl_2$ in solution and CuO in the residue; and lastly (6) the Cu_2O

is reduced to metallic Cu by carbon in a furnace. According to Crooke's method the impure copper regulus obtained by roasting and smelting the ore is broken up and immersed repeatedly in molten lead, which extracts the Ag and Au occurring in the regulus. The regulus is then heated in a reverberatory furnace to run off the lead, and is then smelted for Cu.

The copper brought into the market often contains small quantities of various impurities. Among these there are generally present iron, lead, silver, arsenic, and sometimes small quantities of oxides of copper. As copper, when mixed with a small amount of foreign substances, loses its tenacity to a certain degree, the manufacture of very thin sheet copper requires the use of Chili copper, which is distinguished for its great softness, and therefore when it is desired to have pure copper, it is best to take thin sheet copper, like that which is used in the manufacture of cartridges. But the purest copper is electrolytic copper—that is, that which is deposited from a solution by the action of an electric current.

If the copper contains silver, as is often the case, it is used in gold refineries for the precipitation of silver from its solutions in sulphuric acid. Iron and zinc reduce copper salts, but copper reduces mercury and silver salts. The precipitate contains not only the silver which was previously in solution, but also all that which was in the copper. The silver solutions in sulphuric acid are obtained in the separation of silver from gold by treating their alloys with sulphuric acid, which only dissolves the silver.

Schützenberger showed that when the basic carbonate of copper is decomposed by an electric current it gives, besides the ordinary copper, an allotropic form which grows on the negative platinum electrode, if its surface be smaller than that of the positive copper electrode, in the form of brittle crystalline growths of sp. gr. 8·1. It differs from ordinary copper by giving not nitric oxide but nitrous oxide when treated with nitric acid, and in being very easily oxidised in air, and coated with red shades of colour. It is possible that this is copper hydride, or copper which has occluded hydrogen. Spring (1892) observed that copper reduced from the oxide by hydrogen at the lowest possible temperature was pulverulent, while that reduced from $CuCl_2$ at a somewhat high temperature appeared in bright crystals. The same difference occurs with many other metals, and is probably partly due to the volatility of the metallic chlorides.

This is taken advantage of in practice; for instance, by pouring dilute acids over copper turnings on revolving tables in the preparation of copper salts, such as verdigris, or the basic acetate $2C_4H_6CuO_4,CuH_2O_2,5H_2O$, which is so much used as an oil paint (*i.e.* with boiled oil). The capacity of copper for absorbing oxygen in the presence of acids is so great that it is possible by this means (by taking, for example, thin copper shavings moistened with sulphuric acid) to take up all the oxygen from a given volume of air, and this is even employed for the analysis of air.

The combination of copper with oxygen is not only aided by acids but also by alkalis, although cupric oxide does not appear to have an acid character. Alkalis do not act on copper except in the presence of air, when they produce cupric oxide, which does not appear to combine with such alkalis as caustic potash or soda. But the *action of ammonia* is particularly distinct (Chapter V., Note 2). In the action of a solution of ammonia not only is oxygen absorbed by the copper, but it also acts on the ammonia, and a definite quantity of ammonia is always acted on simultaneously with the passage of the copper into solution. The ammonia is then converted into nitrous acid, according to the reaction: $NH_3 + O_3 = NHO_2 + H_2O$, and the nitrous acid thus formed passes into the state of ammonium nitrite, NH_4NO_2. In this manner three equivalents of oxygen are expended on the oxidation of the ammonia, and six equivalents of oxygen pass over to the copper, forming six atoms of cupric oxide. The latter does not remain in the state of oxide, but combines with the ammonia.

A strong solution of common salt does not act on copper, but a dilute solution of the salt corrodes copper, converting it into oxychloride—that is, in the presence of air. This action of salt water is evident in those cases where the bottoms of ships are coated with sheet copper. From what has been said above it will be evident that copper vessels should not be employed in the preparation of food, because this contains salts and acids which act on copper in the presence of air, and give copper salts, which are poisonous, and therefore the food prepared in untinned copper vessels may be poisonous. Hence tinned vessels are employed for this purpose—that is, copper vessels coated with a thin layer of tin, on which acid and saline solutions do not act.

[6 bis] Copper, besides the cuprous oxide, Cu_2O, and cupric oxide, CuO, gives two known higher forms of oxidation, but they have scarcely been investigated, and even their composition is not well known. *Copper dioxide* (CuO_2, or CuO_2,H_2O, perhaps $CuOH_2O_2$) is obtained by the action of hydrogen peroxide on cupric hydroxide, when the green colour of the latter is changed to yellow. It is very unstable, and is decomposed even by boiling water, with the evolution of oxygen, whilst the action of acids gives cupric salts, oxygen being also disengaged. A still higher *copper peroxide* is formed by heating a mixture of caustic potash, nitre, and metallic copper to a red heat, and by dissolving cupric hydroxide in solutions of the hypochlorites of the alkali metals. A slight heating of the soluble salt formed is enough for it to be decomposed into oxygen and copper dioxide, which is precipitated. Judging from Frémy's researches, the composition of the

copper-potassic compound should be K_2CuO_4. Perhaps this is a compound of the peroxides of potassium, K_2O_2, and of copper, CuO_2.

Colourless solutions of cuprous salts may also be obtained by the action of sulphurous or phosphorous acid and similar lower grades of oxidation on the blue solutions of the cupric salts. This is very clearly and easily effected by means of sodium thiosulphate, $Na_2S_2O_3$, which is oxidised in the process. Cuprous oxide can not only be obtained by the deoxidation of cupric oxide, but also directly from metallic copper itself, because the latter, in oxidising at a red heat in air, first gives cuprous oxide. It is prepared in this manner on a large scale by heating sheet copper rolled into spirals in reverberatory furnaces. Care must be taken that the air is not in great excess, and that the coating of red cuprous oxide formed does not begin to pass into the black cupric oxide. If the oxidised spiral sheet is then unbent, the brittle cuprous oxide falls away from the soft metal. The suboxide obtained in this manner fuses with ease. It is necessary to prevent the access of air during the fusion, and if the mass contains cupric oxide it must be mixed with charcoal, which reduces the latter. Cuprous chloride, CuCl, corresponding with cuprous oxide (as sodium chloride corresponds with sodium oxide), when calcined with sodium carbonate, gives sodium chloride and cuprous oxide, carbonic anhydride being evolved, because it does not combine with the cuprous oxide under these conditions. The reaction can be expressed by the following equation: $2CuCl + Na_2CO_3 = Cu_2O + 2NaCl + CO_2$. The cupric oxide itself, when calcined with finely-divided copper, this copper powder may be obtained by many methods—for instance, by immersing zinc in a solution of a copper salt, or by igniting cupric oxide in hydrogen), gives the fusible cuprous oxide: $Cu + CuO = Cu_2O$. Both the native and artificial cuprous oxide have a sp. gr. of 5·6. It is insoluble in water, and is not acted on by (dry) air. When heated with acids the suboxide forms a solution of a cupric salt and metallic copper—for example, $Cu_2O + H_2SO_4 = Cu + CuSO_4 + H_2O$. However, strong hydrochloric acid does not separate metallic copper on dissolving cuprous oxide, which is due to the fact that the cuprous chloride formed is soluble in strong hydrochloric acid. Cuprous oxide also dissolves in a solution of ammonia, and in the absence of air gives a colourless solution, which turns blue in the air, absorbing oxygen, owing to the conversion of the cuprous oxide into cupric oxide. The blue solution thus formed may be again reconverted into a colourless cuprous solution by immersing a copper strip in it, because the metallic copper then deoxidises the cupric oxide in the solution into cuprous oxide. Cuprous oxide is characterised by the fact that it gives red glasses when fused with glass or with salts forming vitreous alloys. Glass tinted with cuprous oxide is used for ornaments. The access of air must be avoided during its preparation, because the colour then becomes green, owing to the formation of cupric oxide, which colours glass blue. This may even be taken advantage of in testing for copper under the blow-pipe by heating the copper compound with borax in the flame of a blow-pipe; a red glass is obtained in the reducing flame, and a blue glass in the oxidising flame, owing to the conversion of the cuprous into cupric oxide.

Étard (1882), by passing sulphurous anhydride into a solution of cupric acetate, obtained a white precipitate of cuprous sulphite, Cu_2SO_3,H_2O, whilst he obtained the same salt, of a red colour, from the double salt of sodium and copper; but there are not any convincing proofs of isomerism in this case.

The solubility of cuprous chloride in ammonia is due to the formation of compounds between the ammonia and the chloride. In a warm solution the compound $NH_3,2CuCl$ is formed, and at the ordinary temperature $CuCl,NH_3$. This salt is soluble in hydrochloric acid, and then forms a corresponding double salt of cuprous chloride and ammonium chloride. By the action of a certain excess of ammonia on a hydrochloric acid solution of cuprous chloride, very well formed crystals, having the composition $CuCl,NH_3,H_2O$, are obtained. Cuprous chloride is not only soluble in ammonia and hydrochloric acid, but it also dissolves in solutions of certain other salts—for example, in sodium chloride, potassium chloride, sodium thiosulphate, and certain others. All the solutions of cuprous chloride act in many cases as very powerful deoxidising substances; for example, it is easy, by means of these solutions, to precipitate gold from its solutions in a metallic form, according to the equation $AuCl_3 + 3CuCl = Au + 3CuCl_2$.

Among the other compounds corresponding with cuprous oxide, *cuprous iodide*, CuI, is worthy of remark. It is a colourless substance which is insoluble in water and sparingly soluble in ammonia (like silver iodide), but capable of absorbing it, and in this respect it resembles cuprous chloride. It is remarkable from the fact that it is exceedingly easily formed from the corresponding cupric compound CuI_2. A solution of cupric iodide easily decomposes into iodine and cuprous iodide, even at the ordinary temperature, whilst cupric chloride only suffers a similar change on ignition. If a solution of a cupric salt be mixed with a solution of potassium iodide the cupric iodide formed immediately decomposes into free iodine and cuprous iodide, which separates out as a precipitate. In this case the cupric salt acts in an oxidising manner, like, for example, nitrous acid, ozone, and other substances which liberate iodine from iodides, but with this difference, that it only liberates half, whilst they set free the whole of the iodine from potassium iodide: $2KI + CuCl_2 = 2KCl + CuI + I$.

It must also be remarked that cuprous oxide, when treated with hydrofluoric acid, gives an insoluble cuprous fluoride, CuF. Cuprous cyanide is also insoluble in water, and is obtained by the addition of hydrocyanic acid to a solution of cupric chloride saturated with sulphurous anhydride. This cuprous cyanide, like silver cyanide, gives a double soluble salt with potassium cyanide. The double cyanide of copper and potassium is tolerably stable in the air, and enters into double decompositions with various other salts, like those double cyanides of iron with which we are already acquainted.

Copper hydride, CuH, also belongs to the number of the cuprous compounds. It was obtained by Würtz by mixing a hot (70°) solution of cupric sulphate with a solution of hypophosphorous acid, H_3PO_2. The addition of the reducing hypophosphorous acid must be stopped when a brown precipitate makes its appearance, and when gas begins to be evolved. The brown precipitate is the hydrated cuprous hydride. When gently heated it disengages hydrogen; it gives cuprous oxide when exposed to the air, burns in a stream of chlorine, and liberates hydrogen with hydrochloric acid: $CuH + HCl = CuCl + H_2$. Zinc, silver, mercury, lead, and many other heavy metals do not form such a compound with hydrogen, neither under these circumstances nor under the action of hydrogen at the moment of the decomposition of salts by a galvanic current. The greatest resemblance is seen between cuprous hydride and the hydrogen compounds of potassium, sodium, Pd, Ca, and Ba.

[8 bis] The oxide of copper obtained by igniting the nitrate is frequently used for organic analyses. It is hygroscopic and retains nitrogen (1·5 c.c. per gram) when the nitrate is heated in vacuo (Richards and Rogers, 1893).

[8 tri] Oxide of copper is also capable of dissociating when heated. Debray and Joannis showed that it then disengages oxygen, whose maximum tension is constant for a given temperature, providing that fusion does not take place (the CuO then dissolves in the molten Cu_2O); that this loss of oxygen is followed by the formation of suboxide, and that on cooling, the oxygen is again absorbed, forming CuO.

Cupric oxide and many of its salts are able to give definite, although unstable, *compounds with ammonia*. This faculty already shows itself in the fact that cupric oxide, as well as the salts of copper, dissolves in aqueous ammonia, and also in the fact that salts of copper absorb ammonia gas. If ammonia be added to a solution of any cupric salt, it first forms a precipitate of cupric hydroxide, which then dissolves in an excess of ammonia. The solution thus formed, when evaporated or on the addition of alcohol, frequently deposits crystals of salts containing both the elements of the salt of copper taken and of ammonia. Several such compounds are generally formed. Thus cupric chloride, $CuCl_2$, according to Deherain, forms four compounds with ammonia—namely, with one, two, four, and six molecules of ammonia. Thus, for example, if ammonia gas be passed into a boiling saturated solution of cupric chloride, on cooling, small octahedral crystals of a blue colour separate out, containing $CuCl_2,2NH_3,H_2O$. At 150° this substance loses half the ammonia and all the water contained in it, leaving the compound $CuCl_2,NH_3$. Nitrate of copper forms the compound $Cu(NO_3)_2,2NH_3$. This compound remains unchanged on evaporation. Dry cupric sulphate absorbs ammonia gas, and gives a compound containing five molecules of ammonia to one of sulphate (Vol. I., p. 257, and Chapter XXII., Note 35). If this compound is dissolved in aqueous ammonia, on evaporation it deposits a crystalline substance containing $CuSO_4,4NH_3,H_2O$. At 150° this substance loses the molecule of water and one-fourth of its ammonia. On ignition all these compounds part with the remaining ammonia in the form of an ammoniacal salt, so that the residue consists of cupric oxide. Both the hydrated and anhydrous cupric oxide are soluble in aqueous ammonia.

The solution obtained by the action of aqueous ammonia and air on copper turnings (Note 6) is remarkable for its faculty of *dissolving cellulose*, which is insoluble in water, dilute acids, and alkalis. Paper soaked in such a solution acquires the property of not rotting, of being difficultly combustible, and waterproof, &c. It has therefore been applied, especially in England, to many practical purposes—for example, to the construction of temporary buildings, for covering roofs, &c. The composition of the substance held in solution is $Cu(HO)_2,4NH_3$.

If dry ammonia gas be passed over cupric oxide heated to 265°, a portion of the oxide of copper remains unaltered, whilst the other portion gives *copper nitride*, the oxygen of the copper oxide combining with the hydrogen and forming water. The oxide of copper which remains unchanged is easily removed by washing the resultant product with aqueous ammonia. Copper nitride is very stable, and is insoluble; it has the composition Cu_3N (*i.e.* the copper is monatomic here as in Cu_2O), and is an amorphous green powder, which is decomposed when strongly ignited, and gives cuprous chloride and ammonium chloride when treated with hydrochloric acid. Like the other nitrides, copper nitride, Cu_3N, has scarcely been investigated. Granger (1892), by heating copper in the vapour of phosphorus, obtained hexagonal prisms of Cu_5P, which passed into Cu_6P (previously obtained by Abel) when heated in nitrogen. Arsenic is easily absorbed by copper, and its presence (like P), even in small quantities, has a great influence upon the properties of copper—for instance, pure copper wire 1 sq. mm. in section breaks under a load of 35 kilos, while the presence of O·22 p.c. of arsenic raises the breaking load to 42 kilos.

[9 bis] As a comparatively feeble base, oxide of copper easily forms both basic and double salts. As an instance we may mention the double salts composed of the dichloride $CuCl_2,2H_2O$ and potassium chloride. The double salt $CuK_2Cl_4,2H_2O$ crystallises from solutions in *blue* plates, but when heated alone or with substances taking up water easily gives *brown* needles $CuKCl_3$ and at the same time KCl, and this reaction is reversible at 92° as Meyerhoffer (1889) showed (*i.e.* above 92° the simpler double salt is formed and below 92° the more complex salt). With an excess of the copper salt, KCl gives another double salt, $Cu_2KCl_5,4H_2O$, the transition temperature of which is 55°. The instances of equilibria which are encountered in such complex relations (*see* Chapter XIV., Note 25, astrakhanite, and Chapter XXII., Note 23) are embraced by the *law of phases* given by Gibbs (Transactions of the Connecticut Academy of Sciences, 1875–1878, in J. Willard Gibbs' memoir 'On the equilibrium of heterogeneous substances:' and in a clearer and more accessible form in H. W. Bakhuis Roozeboom's papers, Rec. trav. chim., Vol. VI., and in W. Meyerhoffer's memoir *Die Phasenregel und ihre Anwendungen*, 1893, to which sources we refer those desiring fuller information respecting this law). Gibbs calls '*bodies*' substances (simple or compound) capable of forming homogeneous complexes (for instance, solutions or intercombinations) of a varied composition; a *phase*—a mechanically separable portion of such bodies or of their homogeneous complexes (for instance, a vapour, liquid or precipitated solid), *perfect equilibrium*—such a state of bodies and of their complexes as is characterised by a constant pressure at a constant temperature even under a change in the amount of one of the component parts (for instance, of a salt in a saturated solution), while an *imperfect equilibrium* is such a one for which such a change corresponds with a change of pressure (for instance, an unsaturated solution). The law of phases consists in the fact that: *n bodies only give a perfect equilibrium when n + 1 phases participate in that equilibrium*—for example, in the equilibrium of a salt in its saturated solution in water there are two bodies (the salt and water) and three phases, namely, the salt, solution, and vapour, which can be mechanically separated from each other, and to this equilibrium there corresponds a definite tension. At the same time, *n bodies may occur in n + 2 phases, but only at one definite temperature and one pressure*; a change of one of these may bring about another state (perfect or not—equilibrium stable or unstable). Thus water when liquid at the ordinary temperature offers two phases (liquid and vapour) and is in perfect equilibrium (as also is ice below 0°), but water, ice, and vapour (three phases and only one body) can only be in equilibrium at 0°, and at the ordinary pressure; with a change of *t* there will remain either only ice and vapour or only liquid water and vapour; whilst with a rise of pressure not only will the vapour pass into the liquid (there again only remain two phases) but also the temperature of the

formation of ice will fall (by about 7° per 1000 atmospheres). The same laws of phases are applicable to the consideration of the formation of simple or double salts from saturated solutions and to a number of other purely chemical relations. Thus, for example, in the above-mentioned instance, when the bodies are KCl, CuCl$_2$, and H$_2$O, perfect equilibrium (which here has reference to the solubility) consisting of four phases, corresponds to the following seven cases, considering only the phases (above 0°) A = CuCl$_2$,2KCl,2H$_2$O; B = CuCl$_2$KCl; C = CuCl$_2$,2H$_2$O,KCl, solution and vapour: (1) A + B + solution + vapour; (2) A + C + solution + vapour; (3) A + KCl + solution + vapour; (4) A + B + C + vapour (it follows that B + KCl + solution gives A); (5) A + C + KCl + vapour; (6) B + C + solution + vapour; and (7) B + KCl + solution + vapour. Thus above 92° A gives B + KCl. The law of phases by bringing complex instances of chemical reaction under simple physical schemes, facilitates their study in detail and gives the means of seeking the simplest chemical relations dealing with solutions, dissociation, double decompositions and similar cases, and therefore deserves consideration, but a detailed exposition of this subject must be looked for in works on physical chemistry.

The normal *cupric nitrate*, CuN$_2$O$_6$,3H$_2$O, is obtained as a deliquescent salt of a blue colour (soluble in water and in alcohol) by dissolving copper or cupric oxide in nitric acid. It is so easily decomposed by the action of heat that it is impossible to drive off the water of crystallisation from it before it begins to decompose. During the ignition of the normal salt the cupric oxide formed enters into combination with the remaining undecomposed normal salt, and gives a basic salt, CuN$_2$O$_6$,2CuH$_2$O$_2$. The same basic salt is obtained if a certain quantity of alkali or cupric hydroxide or carbonate be added to the solution of the normal salt, which is even decomposed when boiled with metallic copper, and forms the basic salt as a green powder, which easily decomposes under the action of heat and leaves a residue of cupric oxide. The basic salt, having the composition CuN$_2$O$_6$,3CuH$_2$O$_2$, is nearly insoluble in water.

The normal *carbonate of copper*, CuCO$_3$, occurs in nature, although extremely rarely. If solutions of cupric salts be mixed with solutions of alkali carbonates, then, as in the case of magnesium, carbonic anhydride is evolved and basic salts are formed, which vary in composition according to the temperature and conditions of the reaction. By mixing cold solutions, a voluminous blue precipitate is formed, containing an equivalent proportion of cupric hydroxide and carbonate (after standing or heating, its composition is the same as malachite, sp. gr. 3·51: 2CuSO$_4$ + 2Na$_2$CO$_3$ + H$_2$O = CuCO$_3$,CuH$_2$O$_2$ + 2Na$_2$SO$_4$ + CO$_2$. If the resultant blue precipitate be heated in the liquid, it loses water and is transformed into a granular green mass of the composition Cu_2CO_4—i.e. into a compound of the normal salt with anhydrous cupric oxide. This salt of the oxide corresponds with orthocarbonic acid, $C(OH)_4 = CH_4O_4$ where 4H is replaced by 2Cu. On further boiling this salt loses a portion of the carbonic acid, forming black cupric oxide, so unstable is the compound of copper with carbonic anhydride. Another basic salt which occurs in nature, $2CuCO_3,CuH_2O_2$, is known as azurite, or blue carbonate of copper; it also loses carbonic acid when boiled with water. On mixing a solution of cupric sulphate with sodium sesquicarbonate no precipitate is at first obtained, but after boiling a precipitate is formed having the composition of malachite. Debray obtained artificial azurite by heating cupric nitrate with chalk.

[10 bis] Although sulphate of copper usually crystallises with $5H_2O$, that is, differently to the sulphates of Mg, Fe, and Mn, it is nevertheless perfectly isomorphous with them, as is seen not only in the fact that it gives isomorphous mixtures with them, containing a similar amount of water of crystallisation, but also in the ease with which it forms, like all bases analogous to MgO, double salts, $R_2Cu(SO_4)_2,6H_2O$, where R = K, Rb, Cs, of the monoclinic system.

Salts of this kind, like $CuCl_2,2KCl,2H_2O, PtK_2Cy_4$, &c., present a composition CuX_2 if the representation of double salts given in Chapter XXIII., Note 11, be admitted, because they, like $Cu(HO)_2$, contain $Cu(X_2K)_2$, where $X_2 = SO_4$, i.e. the residue of sulphuric acid, which combines with H_2, and is therefore able to replace the H_2 by X_2 or O. A detailed study of the crystalline forms of these salts, made by Tutton (1893) (*see* Chapter XIII., Note 1), showed: (1) that 22 investigated salts of the composition $R_2M(SO_4),6H_2O$, where R = K, Rb, Cs, and M = Mg, Zn, Cd, Mn, Fe, Co, Ni, Cu, present a complete crystallographic resemblance; (2) that in all respects the Rb salts present a transition between the K and Cs salts; (3) that the Cs salts form crystals most easily, and the K salts the most difficultly, and that for the K salts of Cd and Mn it was even impossible to obtain well-formed crystals; (4) that notwithstanding the closeness of their angles, the general appearance (habit) of the potassium compound differs very clearly from the Cs salts, while the Rb salts present a distinct transition in this respect; (5) that the angle of the inclination of one of the axes to the plane of the two other axes showed that in the K salts (angle from 75° to 75° 38') the inclination is least, in the Cs salts (from 72° 52' to 73° 50') greatest, and in the Rb salts (from 73° 57' to 74° 42') intermediate between the two; the replacement of Mg ... Cu produces but a very small change in this angle; (6) that the other angles and the ratio of the axes of the crystals exhibit a similar variation; and (7) that thus the variation of the form is chiefly determined by the atomic weight of the alkaline metal. As an example we cite the magnitude of the inclination of the axes of $R_2M(SO_4)_2,6H_2O$.

R =	K	Rb	Cs
M = Mg	75° 12'	74° 1'	72° 54'
Zn	75° 12'	74° 7'	72° 59'
Cd	—	74° 7'	72° 49'
Mn	—	73° 3'	72° 53'
Fe	75° 28'	74° 16'	73° 8'
Co	75° 5'	73° 59'	72° 52'
Ni	75° 0'	73° 57'	72° 58'
Cu	75° 32'	74° 42'	73° 50'

This shows clearly (within the limits of possible error, which may be as much as 30') the almost perfect identity of the independent crystalline forms notwithstanding the difference of the atomic weights of the diatomic elements, M = Mg, Cu.

In addition to what has been said (Chapter I., Note 65, and Chapter XXII., Note 35) respecting the combination of $CuSO_4$ with water and ammonia, we may add that Lachinoff (1893) showed that $CuSO_4,5H_2O$ loses $4¾H_2O$ at 180°, that $CuSO_4,5NH_3$ also loses $4¾NH_3$ at 320°, and that only $¼H_2O$ and $¼NH_3$ remain in combination with the $CuSO_4$. The last $¼H_2O$ can only be driven off by heating to 200°, and the last $¼NH_3$ by heating to 360°. Ammonia displaces water from $CuSO_4,5H_2O$, but water cannot displace the ammonia from $CuSO_4,5NH_3$. If hydrochloric acid gas be passed over $CuSO_4,5H_2O$ at the ordinary temperature, it first forms $CuSO_4,5H_2O,3HCl$, and then $CuSO_4,2H_2O,2HCl$. When air is passed over the latter compound it passes into $CuSO_4H_2O$ with a small amount of HCl (about $⅛HCl$). At 100° $CuSO_4,5H_2O$ in a stream of hydrochloric acid gas gives $CuSO_4,¼H_2O,2HCl$, and then $CuSO_4,¼H_2O,HCl$, whilst after prolonged heating $CuSO_4$ remains, which rapidly passes into $CuSO_4,5H_2O$ when placed under a bell jar over water. Over sulphuric acid, however, $CuSO_4,5H_2O$ only parts with $3H_2O$, and if $CuSO_4,2H_2O$ be placed over water it again forms $CuSO_4,5H_2O$, and so on.

[11 bis] Commercial blue vitriol generally contains ferrous sulphate. The salt is purified by converting the ferrous salt into a ferric salt by heating the

solution with chlorine or nitric acid. The solution is then evaporated to dryness, and the unchanged cupric sulphate extracted from the residue, which will contain the larger portion of the ferric oxide. The remainder will be separated if cupric hydroxide is added to the solution and boiled; the cupric oxide, CuO, then precipitates the ferric oxide, Fe_2O_3, just as it is itself precipitated by silver oxide. But the solution will contain a small proportion of a basic salt of copper, and therefore sulphuric acid must be added to the filtered solution, and the salt allowed to crystallise. Acid salts are not formed, and cupric sulphate itself has an acid reaction on litmus paper.

Among the alloys of copper resembling brass, *delta metal*, invented by A. Dick (London) is largely used (since 1883). It contains 55 p.c. Cu, and 41 p.c. Zn, the remaining 4 p.c. being composed of iron (as much as 3½ p.c., which is first alloyed with zinc), or of cobalt, and manganese, and certain other metals. The sp. gr. of delta metal is 8·4. It melts at 950°, and then becomes so fluid that it fills up all the cavities in a mould and forms excellent castings. It has a tensile strength of 70 kilos per sq. mm. (gun metal about 20, phosphor bronze about 30). It is very soft, especially when heated to 600°, but after forging and rolling it becomes very hard; it is more difficultly acted upon by air and water than other kinds of brass, and preserves its golden yellow colour for any length of time, especially if well polished. It is used for making bearings, screw propellers, valves, and many other articles. In general the alloys of Cu and Zn containing about ⅔ p.c. by weight of copper were for a long time almost exclusively made in Sweden and England (Bristol, Birmingham). These alloys for the most part are cheaper, harder, and more fusible than copper alone, and form good castings. The alloys containing 45–80 p.c. Cu crystallise in cubes if slowly cooled (Bi also gives crystals). By washing the surface of brass with dilute sulphuric acid, Zn is removed and the article acquires the colour of copper. The alloys approaching Zn_2Cu_3 in their composition exhibit the greatest resistance (under other equal conditions; of purity, forging, rolling, &c.) The addition of 3 p.c. Al, or 5 p.c. Sn, improves the quality of brass. Respecting aluminium bronze *see* Chapter XVII. p. [88](#).

[12 bis] Ball (also Kamensky), 1888, by investigating the electrical conductivity of the alloys of antimony and copper with lead, came to the conclusion that only two definite compounds of antimony and copper exist, whilst the other alloys are either alloys of these two together or with antimony or with copper. These compounds are Cu_2Sb and Cu_4Sb—one corresponds with the maximum, and the other with the minimum, electrical resistance. In general, the resistance offered to an electrical current forms one of the methods by which the composition of definite alloys (for example, Pb_2Zn_7) is often established, whilst the electromotive

force of alloys affords (Laurie, 1888) a still more accurate method—for instance, several definite compounds were discovered by this method among the alloys of copper with zinc and tin; but we will not enter into any details of this subject, because we avoid all references to electricity, although the reader is recommended to make himself acquainted with this branch of science, which has many points in common with chemistry. The study of alloys regarded as solid solutions should, in my opinion, throw much light upon the question of solutions, which is still obscure in many aspects and in many branches of chemistry.

FIG. 97.—Cupel for silver assaying.

FIG. 98.—Clay muffle.

There are not many soft metals; lead, tin, copper, silver, iron, and gold are somewhat soft, and potassium and sodium very soft. The metals of the alkaline earths are sonorous and hard, and many other metals are even brittle, especially bismuth and antimony. But the very slight significance which these properties have in determining the fundamental chemical properties of substances (although, however, of immense importance in the practical applications of metals) is seen from the example shown by zinc, which is hard at the ordinary temperature, soft at 100°, and brittle at 200°.

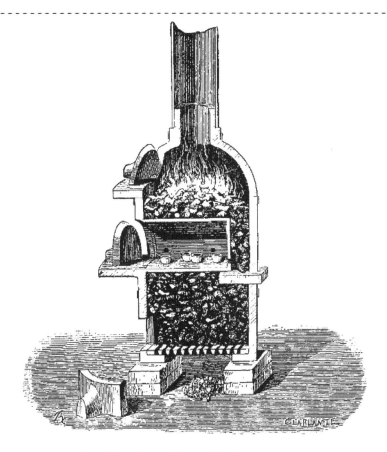

FIG. 99.—Portable muffle furnace.

As the value of silver depends exclusively on its purity, and as there is no possibility of telling the amount of impurities alloyed with it from its external appearance, it is customary in most countries to mark an article with the amount of pure silver it contains after an accurately-made analysis known as the assay of the silver. In France the assay of silver shows the amount of pure silver in 1,000 parts by weight; in Russia the amount of pure silver in 96 parts—that is, the assay shows the number of zolotniks (4·26 grams) of pure silver in one pound (410 grams) of alloyed silver. Russian silver is generally 84 assay—that is, contains 84 parts by weight of pure silver and 12 parts of copper and other metals. French money contains 90 p.c. (in the Russian system this will be 86·4 assay) by weight of silver [English coins and jewellery contain 92·5 p.c. of silver]; the silver rouble is of 83⅓ assay—that is, it contains 86·8 p.c. of silver—and the smaller Russian silver coinage is of 48 assay, and therefore contains 50 p.c. of silver. Silver ornaments and articles are usually made in Russia of 84 and

72 assay. As the alloys of silver and copper, especially after being subjected to the action of heat, are not so white as pure silver, they generally undergo a process known as 'blanching' (or 'pickling') after being worked up. This consists in removing the copper from the surface of the article by subjecting it to a dark-red heat and then immersing it in dilute acid. During the calcination the copper on the surface is oxidised, whilst the silver remains unchanged; the dilute acid then dissolves the copper oxides formed, and pure silver is left on the surface. The surface is dull after this treatment, owing to the removal of a portion of the metal by the acid. After being polished the article acquires the desired lustre and colour, so as to be indistinguishable from a pure silver object. In order to test a silver article, a portion of its mass must be taken, not from the surface, but to a certain depth. The methods of assay used in practice are very varied. The commonest and most often used is that known as *cupellation*. It is based on the difference in the oxidisability of copper, lead, and silver. The cupel is a porous cup with thick sides, made by compressing bone ash. The porous mass of bone ash absorbs the fused oxides, especially the lead oxide, which is easily fusible, but it does not absorb the unoxidised metal. The latter collects into a globule under the action of a strong heat in the cupel, and on cooling solidifies into a button, which may then be weighed. Several cupels are placed in a muffle. A muffle is a semi-cylindrical clay vessel, shown in the accompanying drawing. The sides of the muffle are pierced with several orifices, which allow the access of air into it. The muffle is placed in a furnace, where it is strongly heated. Under the action of the air entering the muffle the copper of the silver alloy is oxidised, but as the oxide of copper is infusible, or, more strictly speaking, difficultly fusible, a certain quantity of lead is added to the alloy; the lead is also oxidised by the air at the high temperature of the muffle, and gives the very fusible lead oxide. The copper oxide then fuses with the lead oxide, and is absorbed by the cupel, whilst the silver remains as a bright white globule. If the weight of the alloy taken and of the silver left on the cupel be determined, it is possible to calculate the composition of the alloy. Thus the essence of cupellation consists in the separation of the oxidisable metals from silver, which does not oxidise under the action of heat. A more accurate method, based on the precipitation of silver from its solutions in the form of silver chloride, is described in detail in works on analytical chemistry.

In America, whence the largest amount of silver is now obtained, ores are worked containing not more than ⅕ p.c. of silver, whilst at ½ p.c. its extraction is very profitable. Moreover, the extraction of silver from ores containing not more than $0·01$ p.c. of this metal is sometimes profitable. The majority of the lead smelted from galena contains silver, which is extracted from it. Thus near Arras, in France, an ore is worked which contains about 65 parts of lead and 0·088 part of silver in 100 parts of ore, which corresponds with 136 parts of silver in 100,000 parts of lead. At Freiberg, in Saxony, the ore used (enriched by mechanical dressing) contains about 0·9 of silver, 160 of lead, and 2 of copper in 10,000 parts. In every case the lead is first extracted in the manner described in Chapter XVIII., and this lead will contain all the silver. Not unfrequently other ores of silver are mixed with lead ores, in order to obtain an argentiferous lead as the product. The extraction of small quantities of silver from lead is facilitated by the fact (Pattinson's process) that molten argentiferous lead in cooling first deposits crystals of pure lead, which fall to the bottom of the cooling vessel, whilst the proportion of silver in the unsolidified mass increases owing to the removal of the crystals of lead. The lead is enriched in this manner until it contains $\frac{1}{400}$ part of silver, and is then subjected to cupellation on a larger scale. According to Park's process, zinc is added to the molten argentiferous lead, and the alloy of Pb and Zn, which first separates out on cooling, is collected. This alloy is found to contain all the silver previously contained in the lead. The addition of 0·5 p.c. of aluminium to the zinc (Rossler and Edelman) facilitates the extraction of the Ag from the resultant alloy besides preventing oxidation; for, after re-melting, nearly all the lead easily runs off (remains fluid), and leaves an alloy containing about 30 p.c. Ag and about 70 p.c. Zn. This alloy may be used as an anode in a solution of $ZnCl_2$, when the Zn is deposited on the cathode, leaving the silver with a small amount of Pb, &c. behind. The silver can be easily obtained pure by treating it with dilute acids and cupelling.

The ores of silver which contain a larger amount of it are: silver glance, Ag_2S (sp. gr. 7·2); argentiferous-copper glance, $CuAgS$; horn silver or chloride of silver, $AgCl$; argentiferous grey copper ore; polybasite, M_9RS_6 (where M = Ag, Cu, and R = Sb, As), and argentiferous gold. The latter is the usual form in which gold is found in alluvial deposits and ores. The crystals of gold from the Berezoffsky mines in the Urals contain 90 to 95 of gold and 5 to 9 of silver, and the Altai gold contains 50 to 65 of gold and 36 to 38 of silver. The proportion of silver in native gold varies between these limits in other localities. Silver ores, which generally occur in veins, usually contain native silver and various sulphur compounds. The most famous mines in Europe are in Saxony (Freiberg), which has a yearly output of as much as 26 tons of silver, Hungary, and Bohemia (41 tons). In Russia, silver is extracted in the Altai and at Nerchinsk (17 tons). The richest silver mines known are in America, especially in Chili (as much as 70 tons), Mexico (200 tons), and more particularly in the Western States of North America. The richness of these mines may be judged from the fact that one mine in the State of Nevada (Comstock, near Washoe and the cities of Gold Hill and Virginia), which was discovered in 1859, gave an output of 400 tons in 1866. In place of cupellation, chlorination may also be employed for extracting silver from its ores. The method of chlorination consists in converting the silver in an ore into silver chloride. This is either done by a wet or by a dry method, roasting the ore with NaCl. When the silver chloride is formed, the extraction of the metal is also done by two methods. The first consists in the silver chloride being reduced to metal by means of iron in rotating barrels, with the subsequent addition of mercury which dissolves the silver, but does not act on the other metals. The mercury holding the silver in solution is distilled, when the silver remains behind. This method is called *amalgamation*. The other method is less frequently used, and consists in dissolving the silver chloride in sodium chloride or in sodium thiosulphate, and then precipitating the silver from the solution. The amalgamation is then carried on in rotating barrels containing the roasted ore mixed with water, iron, and mercury. The iron reduces the silver chloride by taking up the chlorine from it. The technical details of these processes are described in works on metallurgy. The extraction of AgCl by the wet method is carried on (Patera's process) by means of a solution of hyposulphite of sodium which dissolves AgCl (*see* Note 23), or by lixiviating with a 2 p.c. solution of a double hyposulphite of Na and Cu (obtained by adding $CuSO_4$ to $Na_2S_2O_3$). The resultant solution of AgCl is first treated with soda to precipitate $PbCO_3$, and then with Na_2S, which precipitates the Ag and Au. The process should be carried on rapidly to prevent the precipitation of Cu_2S from the solution of $CuSO_4$ and $Na_2S_2O_3$.

There is another practical method which is also suitable for separating the silver from the solutions obtained in photography, and consists in precipitating the silver by oxalic acid. In this case the amount of silver in the solution must be known, and 23 grams of oxalic acid dissolved in 400 grams of water must be added for every 60 grams of silver in solution in a litre of water. A precipitate of silver oxalate, $Ag_2C_2O_4$, is then obtained, which is insoluble in water but soluble in acids. Hence, if the liquid contain any free acid it must be previously freed from it by the addition of sodium carbonate. The resultant precipitate of silver oxalate is dried, mixed with an equal weight of dry sodium carbonate, and thrown into a gently-heated crucible. The separation of the silver then proceeds without an explosion, whilst the silver oxalate if heated alone decomposes with explosion.

According to Stas, the best method for obtaining silver from its solutions is by the reduction of silver chloride dissolved in ammonia by means of an ammoniacal solution of cuprous thiosulphate; the silver is then precipitated in a crystalline form. A solution of ammonium sulphite may be used instead of the cuprous salt.

Silver is very malleable and ductile; it may be beaten into leaves 0·002 mm. in thickness. Silver wire may be made so fine that 1 gram is drawn into a wire 2½ kilometres long. In this respect silver is second only to gold. A wire of 2 mm. diameter breaks under a strain of 20 kilograms.

In melting, silver absorbs a considerable amount of oxygen, which is disengaged on solidifying. One volume of molten silver absorbs as much as 22 volumes of oxygen. In solidifying, the silver forms cavities like the craters of a volcano, and throws off metal, owing to the evolution of the gas; all these phenomena recall a volcano on a miniature scale (Dumas). Silver which contains a small quantity of copper or gold, &c., does not show this property of dissolving oxygen.

The absorption of oxygen by molten silver is, however, an oxidation, but it is at the same time a phenomenon of solution. One cubic centimetre of molten silver can dissolve twenty-two cubic centimetres of oxygen, which, even at 0°, only weighs 0·03 gram, whilst 1 cubic centimetre of silver weighs at least 10 grams, and therefore it is impossible to suppose that the absorption of the oxygen is attended by the formation of any definite compound (rich in oxygen) of silver and oxygen (about 45 atoms of silver to 1 of oxygen) in any other but a dissociated form, and this is the state in which substances in solution must be regarded (Chapter I.)

Le Chatelier showed that at 300° and 15 atmospheres pressure silver absorbs so much oxygen that it may be regarded as having formed the compound Ag_4O, or a mixture of Ag_2 and Ag_2O. Moreover, silver oxide, Ag_2O, only decomposes at 300° under low pressures, whilst at pressures above 10 atmospheres there is no decomposition at 300° but only at 400°.

Stas showed that silver is oxidised by air in the presence of acids. V. d. Pfordten confirmed this, and showed that an acidified solution of potassium permanganate rapidly dissolves silver in the presence of air.

When solutions of $AgNO_3$, $FeSO_4$, sodium citrate, and NaHO are mixed together in the manner described above, they throw down a precipitate of a beautiful lilac colour; when transferred to a filter paper the precipitate soon changes colour, and becomes dark blue. To obtain the substance as pure as possible it is washed with a 5–10 p.c. solution of ammonium nitrate; the liquid is decanted, and 150 c.c. of water poured over the precipitate. It then dissolves entirely in the water. A small quantity of a saturated solution of ammonium nitrate is added to the solution, and the silver in solution again separates out as a precipitate. These alternate solutions and precipitations are repeated seven or eight times, after which the precipitate is transferred to a filter and washed with 95 p.c. alcohol until the filtrate gives no residue on evaporation. An analysis of the substance so obtained showed that it contained from 97·18 p.c. to 97·31 p.c. of metallic silver. It remained to discover what the remaining 2–3 p.c. were composed of. Are they merely impurities, or is the substance some compound of silver with oxygen or hydrogen, or does it contain citric acid in combination which might account for its solubility? The first supposition is set aside by the fact that no gases are disengaged by the precipitate of silver, either under the action of gases or when heated. The second supposition is shown to be impossible by the fact that there is no definite relation between the silver and citric acid. A determination of the amount of silver in solution showed that the amount of citric acid varies greatly for one and the same amount of silver, and there is no simple ratio between them. Among other methods of preparing soluble silver given by Carey Lea, we may mention the method published by him in 1891. $AgNO_3$ is added to a solution of dextrine in caustic soda or potash; at first a precipitate of brown oxide of silver is thrown down, but the brown colour then changes into a reddish chocolate, owing to the reduction of the silver by the dextrine, and the solution turns a deep red. A few drops of this solution turn water bright red, and give a perfectly transparent liquid. The dextrine solution is prepared by dissolving 40 grams of caustic soda and the same amount of ordinary brown dextrine in two litres of water. To this solution is gradually added 28 grams of $AgNO_3$ dissolved in a small quantity of water.

The insoluble allotropic silver is obtained, as was mentioned above, from a solution of silver prepared in the manner described, by the addition of sulphate of copper, iron, barium, magnesium, &c. In one experiment Lea succeeded in obtaining the insoluble allotropic Ag in a crystalline form. The red solution, described above, after standing several weeks, deposits crystals spontaneously in the form of short black needles and thin prisms, the liquid becoming colourless. This insoluble variety, when rubbed upon paper, has the appearance of bright shining green flakes, which polarise light.

The gold variety is obtained in a different manner to the two other varieties. A solution is prepared containing 200 c.c. of a 10 p.c. solution of nitrate of silver, 200 c.c. of a 20 p.c. solution of Rochelle salt, and 800 c.c. of water. Just as in the previous case the reaction consisted in the reduction of the citrate of silver, so in this case it consists in the reduction of the tartrate, which here first forms a red, and then a black precipitate of allotropic Ag, which, when transferred to the filter, appears of a beautiful bronze colour. After washing and drying, this precipitate acquires the lustre and colour peculiar to polished gold, and this is especially remarked where the precipitate comes into contact with glass or china. An analysis of the golden variety gave a percentage composition of 98·750 to 98·749 Ag. Both the insoluble varieties (the blue and gold) have a different specific gravity from ordinary silver. Whilst that of fused silver is 10·50, and of finely-divided silver 10·62, the specific gravity of the blue insoluble variety is 9·58, and of the gold variety 8·51. The gold variety passes into ordinary Ag with great ease. This transition may even be remarked on the filter in those places which have accidentally not been moistened with water. A simple shock, and therefore friction of one particle upon another, is enough to convert the gold variety into normal white silver. Carey Lea sent samples of the gold variety for a long distance by rail packed in three tubes, in which the silver occupied about the quarter of their volume; in one tube only he filled up this space with cotton-wool. It was afterwards found that the shaking of the particles of Ag had completely converted it into ordinary white silver, and that only the tube containing the cotton-wool had preserved the golden variety intact.

The soluble variety of Ag also passes into the ordinary state with great ease, the heat of conversion being, as Prange showed in 1890, about +60 calories.

[18 bis] The opinion of the nature of soluble silver given below was first enunciated in the *Journal of the Russian Chemical Society*, February 1, 1890, Vol. XXII., Note 73. This view is, at the present time, generally accepted, and this silver is frequently known as the 'colloid' variety. I may add that Carey Lea observed the solution of ordinary molecular silver in ammonia without the access of air.

[18 tri] It is, however, noteworthy that ordinary metallic lead has long been considered soluble in water, that boron has been repeatedly obtained in a brown solution, and that observations upon the development of certain bacteria have shown that the latter die in water which has been for some time in contact with metals. This seems to indicate the passage of small quantities of metals into water (however, the formation of peroxide of hydrogen may be supposed to have some influence in these cases).

Silver suboxide (Ag_4O) or argentous oxide is obtained from argentic citrate by heating it to 100° in a stream of hydrogen. Water and argentous citrate are then formed, and the latter, although but slightly soluble in water, gives a reddish-brown solution of colloid silver (Note 18), and when boiled this solution becomes colourless and deposits metallic silver, the argentic salt being again formed. Wöhler, who discovered this oxide, obtained it as a black precipitate by adding potassium hydroxide to the above solution of argentous citrate. With hydrochloric acid the suboxide gives a brown compound, Ag_2Cl. Since the discovery of soluble silver the above data cannot be regarded as perfectly trustworthy; it is probable that a mixture of Ag_2 and Ag_2O was being dealt with, so that the actual existence of Ag_4O is now doubtful, but there can be no doubt as to the formation of a subchloride, Ag_2Cl (see Note 25), corresponding to the suboxide. The same compound is obtained by the action of light on the higher chloride. Other acids do not combine with silver suboxide, but convert it into an argentic salt and metallic silver. In this respect cuprous oxide presents a certain resemblance to these suboxides. But copper forms a suboxide of the composition Cu_4O, which is obtained by the action of an alkaline solution of stannous oxide on cupric hydroxide, and is decomposed by acids into cupric salts and metallic copper. The problems offered by the suboxides, as well as by the peroxides, cannot be considered as fully solved.

[19 bis] *Silver peroxide*, AgO or Ag_2O_3, is obtained by the decomposition of a dilute (10 p.c.) solution of silver nitrate by the action of a galvanic current (Ritter). On the positive pole, where oxygen is usually evolved in the decomposition of salts, brittle grey needles with a metallic lustre, which occasionally attain a somewhat considerable size, are then formed. They are insoluble in water, and decompose with the evolution of oxygen when they are dried and heated, especially up to 150°, and, like lead dioxide, barium peroxide, &c., their action is strongly oxidising. When treated with acids, oxygen is evolved and a salt of the oxide formed. Silver peroxide absorbs sulphurous anhydride and forms silver sulphate. Hydrochloric acid evolves chlorine; ammonia reduces the silver, and is itself oxidised, forming water and gaseous nitrogen. Analyses of the above-mentioned crystals show that they contain silver nitrate, peroxide, and water. According to Fisher, they have the composition $4AgO, AgNO_3, H_2O$, and, according to Berthelot, $4Ag_2O_5, 2AgNO_3, H_2O$.

[19 tri] According to Carey Lea, however, oxide of silver still retains water even at 100°, and only parts with it together with the oxygen. Oxide of silver is used for colouring glass yellow.

The reaction of $Pb(OH)_2$ upon $AgHO$ in the presence of $NaHO$ leads to the formation of a compound of both oxides, $PbOnAg_2O$, from which the oxide of lead cannot be removed by alkalies (Wöhler, Leton). Wöhler, Welch, and others obtained crystalline double salts, R_2AgX_3, by the action of strong solutions of RX of the halogen salts of the alkaline metals upon AgX, where $R = Cs, Rb, K$.

[20 bis] According to Müller, ferric oxide is reduced by hydrogen (see Chapter XXII., Note 5) at 295° (into what ?), cupric oxide at 140°, Ni_2O_3 at 150°; nickelous oxide, NiO, is reduced to the suboxide, Ni_2O, at 195°, and to nickel at 270°; zinc oxide requires so high a temperature for its reduction that the glass tube in which Müller conducted the experiment did not stand the heat; antimony oxide requires a temperature of 215° for its reduction; yellow mercuric oxide is reduced at 130° and the red oxide at

230°; silver oxide at 85°, and platinum oxide even at the ordinary temperature.

[20 tri] A silica compound, Ag_2OSiO_2 is obtained by fusing $AgNO_3$ with silica; this salt is able to decompose with the evolution of oxygen, leaving $Ag + SiO_2$.

If a solution of a silver salt be precipitated by sodium hydroxide, and aqueous ammonia is added drop by drop until the precipitate is completely dissolved, the liquid when evaporated deposits a violet mass of crystalline silver oxide. If moist silver oxide be left in a strong solution of ammonia it gives a black mass, which easily decomposes with a loud explosion, especially when struck. This black substance is called fulminating silver. Probably this is a compound like the other compounds of oxides with ammonia, and in exploding the oxygen of the silver oxide forms water with the hydrogen of the ammonia, which is naturally accompanied by the evolution of heat and formation of gaseous nitrogen, or, as Raschig states, fulminating silver contains NAg_3 or one of the amides (for instance, $NHAg_2 = NH_3 + Ag_2O - H_2O$). Fulminating silver is also formed when potassium hydroxide is added to a solution of silver nitrate in ammonia. The dangerous explosions which are produced by this compound render it needful that great care be taken when salts of silver come into contact with ammonia and alkalis (*see* Chapter XVI., Note 26).

So that we here encounter the following phenomena: copper displaces silver from the solutions of its salts, and silver oxide displaces copper oxide from cupric salts. Guided by the conceptions enunciated in Chapter XV., we can account for this in the following manner: The atomic volume of silver = 10·3, and of copper = 7·2, of silver oxide = 32, and of copper oxide = 13. A greater contraction has taken place in the formation of cupric oxide, CuO, than in the formation of silver oxide, Ag_2O, since in the former (13 - 7 = 6) the volume after combination with the oxygen has increased by very little, whilst the volume of silver oxide is considerably greater than that of the metal it contains [32 - (2 × 10·3) = 11·4]. Hence silver oxide is less compact than cupric oxide, and is therefore less stable; but, on the other hand, there are greater intervals between the atoms in silver oxide than in cupric oxide, and therefore silver oxide is able to give more stable compounds than those of copper oxide. This is verified by the figures and data of their reactions. It is impossible to calculate for cupric nitrate, because this salt has not yet been obtained in an anhydrous state; but the sulphates of both oxides are known. The specific gravity of copper sulphate in an anhydrous state is 3·53, and of silver sulphate 5·36; the molecular volume of the former is 45, and of the latter 58. The group SO_3 in the copper occupies, as it were, a volume 45 - 13 = 32, and in the silver salt a volume 58 - 32 = 26; hence a smaller contraction has taken place in the formation of the copper salt from the oxide than in the formation of the silver salt, and consequently the latter should be more stable than the former. Hence silver oxide is able to decompose the salt of copper oxide, whilst with respect to the metals both salts have been formed with an almost identical contraction, since 58 volumes of the silver salt contain 21 volumes of metal (difference = 37), and 45 volumes of the copper salt contain 7 volumes of copper (difference = 38). Besides which, it must be observed that copper oxide displaces iron oxide, just as silver oxide displaces copper oxide. Silver, copper, and iron, in the form of oxides, displace each other in the above order, but in the form of metals in a reverse order (iron, copper, silver). The cause of this order of the displacement of the oxides lies, amongst other things, in their composition. They have the composition Ag_2O, Cu_2O_2, Fe_2O_3; the oxide containing a less proportion of oxygen displaces that containing a larger proportion, because the basic character diminishes with the increase of contained oxygen.

Copper also displaces mercury from its salts. It may here be remarked that Spring (1888), on leaving a mixture of dry mercurous chloride and copper for two hours, observed a distinct reduction, which belongs to the category of those phenomena which demonstrate the existence of a mobility of parts (*i.e.* atoms and molecules) in solid substances.

[22 bis] The reaction of 1 part by weight of $AgNO_3$ requires (according to Kremers) the following amounts of water: at $0°$, $0·82$ part, at $19°·5$, $0·41$ part, at $54°$, $0·20$ part, at $110°$, $0·09$ part, and, according to Tilden, at $125°$, $0·0617$ part, and at $133°$, $0·0515$ part.

[22 tri] It may be remarked that the black stain produced by the reduction of metallic silver disappears under the action of a solution of mercuric chloride or of potassium cyanide, because these salts act on finely-divided silver.

Silver chloride is almost perfectly insoluble in water, but is somewhat soluble in water containing sodium chloride or hydrochloric acid, or other chlorides, and many salts, in solution. Thus at $100°$, 100 parts of water saturated with sodium chloride dissolve $0·4$ part of silver chloride. Bromide and iodide of silver are less soluble in this respect, as also in regard to other solvents. It should be remarked that *silver chloride dissolves in solutions of ammonia, potassium cyanide, and of sodium thiosulphate*, $Na_2S_2O_3$. Silver bromide is almost perfectly analogous to the chloride, but silver iodide is nearly insoluble in a solution of ammonia. Silver chloride even absorbs dry ammonia gas, forming very unstable ammoniacal compounds. When heated, these compounds (Vol. I. p. 250, Note 8) evolve the ammonia, as they also do under the action of all acids. Silver chloride enters into double decomposition with potassium cyanide, forming a soluble double cyanide, which we shall presently describe; it also forms a soluble double salt, $NaAgS_2O_3$, with sodium thiosulphate.

Silver chloride offers different modifications in the structure of its molecule, as is seen in the variations in the consistency of the precipitate, and in the differences in the action of light which partially decomposes AgCl (*see* Note 25). Stas and Carey Lea investigated this subject, which has a particular importance in photography, because silver bromide also gives *photo-salts*. There is still much to be discovered in this respect, since Abney showed that perfectly dry AgCl placed in a vacuum in the dark is not in the least acted upon when subsequently exposed to light.

Silver bromide and *iodide* (which occur as the minerals bromite and iodite) resemble the chloride in many respects, but the degree of affinity of silver for iodine is greater than that for chlorine and bromine, although less heat is evolved (*see* Note 28 bis). Deville deduced this fact from a number of experiments. Thus silver chloride, when treated with hydriodic acid, evolves hydrochloric acid, and forms silver iodide. Finely-divided silver easily liberates hydrogen when treated with hydriodic acid; it produces the same decomposition with hydrochloric acid, but in a considerably less degree and only on the surface. The difference between silver chloride and iodide is especially remarkable, since the formation of the former is attended with a greater contraction than that of the latter. The volume of AgCl = 26; of chlorine 27, of silver 10, the sum = 37, hence a contraction has ensued; and in the formation of silver iodide an expansion takes place, for the volume of Ag is 10, of I 26, and of AgI 39 instead of 36 (density, AgCl, 5·59; AgI, 5·67). The atoms of chlorine have united with the atoms of silver without moving asunder, whilst the atoms of iodine must have moved apart in combining with the silver. It is otherwise with respect to the metal; the distance between its atoms in the metal = 2·2, in silver chloride = 3·0, and in silver iodide = 3·5; hence its atoms have moved asunder considerably in both cases. It is also very remarkable, as Fizeau observed, that the density of silver iodide increases with a rise of temperature—that is, a contraction takes place when it is heated and an expansion when it is cooled.

In order to explain the fact that in silver compounds the iodide is more stable than the chloride and oxide, Professor N. N. Beketoff, in his 'Researches on the Phenomena of Substitutions' (Kharkoff, 1865), proposed the following original hypothesis, which we will give in almost the words of the author:—In the case of aluminium, the oxide, Al_2O_3, is more stable than the chloride, Al_2Cl_6, and the iodide, Al_2I_6. In the oxide the amount of the metal is to the amount of the element combined with it as 54·8 (Al = 27·3) is to 48, or in the ratio 112 : 100; for the chloride the ratio is = 25 : 100; for the iodide it = 7 : 100. In the case of silver the oxide (ratio = 1350 : 100) is less stable than the chloride (ratio = 304 : 100), and the iodide (ratio of the weight of metal to the weight of the halogen = 85 : 100) is the most stable. From these and similar examples it follows that the most stable compounds are those in which the weights of the combined substances are equal. This may be partly explained by the attraction of similar molecules even after their having passed into combination with others. This attraction is proportional to the product of the acting masses. In silver oxide the attraction of Ag_2 for Ag_2 = 216 × 216 = 46,656, and the attraction of Ag_2 for O = 216 × 16 = 3,456. The attraction of like molecules thus counteracts the attraction of the unlike molecules. The former naturally does not overcome the latter, otherwise there would be a disruption, but it nevertheless diminishes the stability. In the case of an equality or proximity of the magnitude of the combining masses, the attraction of the like parts will counteract the stability of the compound to the least extent—in other words, with an inequality of the combined masses, the molecules have an inclination to return to an elementary state, to decompose, which does not exist to such an extent where the combined masses are equal. There is, therefore, a tendency for large masses to combine with large, and for small masses to combine with small. Hence $Ag_2O + 2KI$ gives $K_2O + 2AgI$. The influence of an equality of masses on the stability is seen particularly clearly in the effect of a rise of temperature. Argentic, mercuric, auric and other oxides composed of unequal masses, are somewhat readily decomposed by heat, whilst the oxides of the lighter metals (like water) are not so easily decomposed by heat. Silver chloride and iodide approach the condition of equality, and are not decomposed by heat. The most stable oxides under the action of heat are those of magnesium, calcium, silicon, and aluminium, since they also approach the condition of equality. For the same reason hydriodic acid decomposes with greater facility than hydrochloric acid. Chlorine does not act on magnesia or alumina, but it acts on lime and silver oxide, &c. This is partially explained by the fact that by considering heat as a mode of motion, and knowing that the atomic heats of the free elements are equal, it must be supposed that the amount of the motion of atoms (their *vis viva*) is equal, and as it is equal to the product of the mass (atomic weight) into the square of the velocity, it follows that the greater the combining weight the smaller will be the square of the velocity, and if the combining weights be nearly equal, then the velocities also will be nearly equal. Hence the greater the difference between the weights of the combined atoms the greater will be the difference between their velocities. The difference between the velocities will increase with the temperature, and therefore the temperature of decomposition will be the sooner attained the greater be the original difference—that is, the greater the difference of the weights of the combined substances. The nearer these weights are to each other, the more analogous the motion of the unlike atoms, and consequently, the more stable the resultant compound.

The instability of cupric chloride and nitric oxide, the absence of compounds of fluorine with oxygen, whilst there are compounds of oxygen with chlorine, the greater stability of the oxygen compounds of iodine than those of chlorine, the stability of boron nitride, and the instability of cyanogen, and a number of similar instances, where, judging from the above argument, one would expect (owing to the closeness of the atomic weights) a stability, show that Beketoff's addition to the mechanical theory of chemical phenomena is still far from sufficient for explaining the true relations of affinities. Nevertheless, in his mode of explaining the relative stabilities of compounds, we find an exceedingly interesting treatment of questions of primary importance. Without such efforts it would be impossible to generalise the complex data of experimental knowledge.

Fluoride of silver, AgF, is obtained by dissolving Ag_2O or Ag_2CO_3 in hydrofluoric acid. It differs from the other halogen salts of silver in being soluble in water (1 part of salt in 0·55 of water). It crystallises from its solution in prisms, $AgFH_2O$ (Marignac), or AgF_2H_2O (Pfaundler), which lose their water in vacuo. Güntz (1891), by electrolising a saturated solution of Ag_2F, obtained *polyfluoride of silver*, Ag_2F, which is decomposed by water into AgF + Ag. It is also formed by the action of a strong solution of AgF upon finely-divided (precipitated) silver.

[24 bis] The changes brought about by the action of light necessitate distinguishing the photo-salts of silver.

In photography these are called 'developers.' The most common developers are: solutions of ferrous sulphate, pyrogallol, ferrous oxalate, hydroxylamine, potassium sulphite, hydroquinone (the last acts particularly well and is very convenient to use), &c. The chemical processes of photography are of great practical and theoretical interest; but it would be impossible in this work to enter into this special branch of chemistry, which has as yet been very little worked out from a theoretical point of view. Nevertheless, we will pause to consider certain aspects of this subject which are of a purely chemical interest, and especially the facts concerning *subchloride of silver*, Ag_2Cl (*see* Note 19), and the photo-salts (Note 23). There is no doubt that under the action of light, AgCl becomes darker in colour, decreases in weight, and probably forms a mixture of AgCl, Ag_2Cl, and Ag. But the isolation of the subchloride has only been recently accomplished by Güntz by means of the Ag_2F, discovered by him (*see* Note 24). Many chemists (and among them Hodgkinson) assumed that an oxychloride of silver was formed by the decomposition of AgCl under the action of light.

Carey Lea's (1889) and A. Richardson's (1891) experiments showed that the product formed does not, however, contain any oxygen at all, and the change in colour produced by the action of light upon AgCl is most probably due to the formation of Ag_2Cl. This substance was isolated by Güntz (1891) by passing HCl over crystals of Ag_2F. He also obtained Ag_2I in a similar manner by passing HI, and Ag_2S by passing H_2S over Ag_2F. Ag_2Cl is best prepared by the action of phosphorus trichloride upon Ag_2F. At the temperature of its formation Ag_2Cl has an easily changeable tint, with shades of violet red to violet black. Under the action of light a similar (isomeric) substance is obtained, which splits up into AgCl + Ag when heated. With potassium cyanide Ag_2Cl gives Ag + AgCN + KCl, whence it is possible to calculate the heat of formation of Ag_2Cl; it = 29·7, whilst the heat of formation of AgCl = 29·2—*i.e.* the reaction 2AgCl = Ag_2Cl + Cl corresponds to an absorption of 28·7 major calories. If we admit the formation of such a compound by the action of light, it is evident that the energy of the light is consumed in the above reaction. Carey Lea (1892) subjected AgCl, AgBr, and AgI to a pressure (of course in the dark) of 3,000 atmospheres, and to trituration with water in a mortar, and observed a change of colour indicating incipient decomposition, which is facilitated under the action of light by the molecular currents set up (Lermontoff, Egoroff). The change of colour of the halogen salts of silver under the action of light, and their faculty of subsequently giving a visible photographic image under the action of 'developers,' must now be regarded as connected with the decomposition of AgX, leading to the formation of Ag_2X, and the different tinted photo-salts must be considered as systems containing such Ag_2X's. Carey Lea obtained photo-salts of this kind not only by the action of light but also in many other ways, which we will enumerate to prove that they contain the products of an incomplete combination of Ag with the halogens, (for the salts Ag_2X must be regarded as such). The photo-salts have been obtained (1) by the imperfect chlorination of silver; (2) by the incomplete decomposition of Ag_2O or Ag_2CO_3 by alternately heating and treating with a halogen acid; (3) by the action of nitric acid or $Na_2S_2O_3$ upon Ag_2Cl; (4) by mixing a solution of $AgNO_3$ with the hydrates of FeO, MnO and CrO, and precipitating by HCl; (5) by the action of HCl upon the product obtained by the reduction of citrate of silver in hydrogen (Note 19), and (6) by the action of milk sugar upon $AgNO_3$ together with soda and afterwards acidulating with HCl. All these reactions should lead to the formation of products of imperfect combination with the halogens and give photo-salts of a similar diversity of colour to those produced by the action of developers upon the halogen salts of silver after exposure to light.

[25 bis] In order to determine when the reaction is at an end, a few drops of a solution of K_2CrO_4 are added to the solution of the chloride.

Before all the chlorine is precipitated as AgCl, the precipitate (after shaking) is white (since Ag_2CrO_4 with 2RCl gives 2AgCl); but when all the chlorine is thrown down Ag_2CrO_4 is formed, which colours the precipitate reddish-brown. In order to obtain accurate results the liquid should be neutral to litmus.

[25 tri] *Silver cyanide*, AgCN, is closely analogous to the haloid salts of silver. It is obtained, in similar manner to silver chloride, by the addition of potassium cyanide to silver nitrate. A white precipitate is then formed, which is almost insoluble in boiling water. It is also, like silver chloride, insoluble in dilute acids. However, it is dissolved when heated with nitric acid, and both hydriodic and hydrochloric acids act on it, converting it into silver chloride and iodide. Alkalis, however, do not act on silver cyanide, although they act on the other haloid salts of silver. Ammonia and solutions of the cyanides of the alkali metals dissolve silver cyanide, as they do the chloride. In the latter case double cyanides are formed—for example, $KAgC_2N_2$. This salt is obtained in a crystalline state on evaporating a solution of silver cyanide in potassium cyanide. It is much more stable than silver cyanide itself. It has a neutral reaction, does not change in the air, and does not smell of hydrocyanic acid. Many acids, in acting on a solution of this double salt, precipitate the insoluble silver cyanide. Metallic silver dissolves in a solution of potassium cyanide in the presence of air, with formation of the same double salt and potassium hydroxide, and when silver chloride dissolves in potassium cyanide it forms potassium chloride, besides the salt $KAgC_2N_2$. This double salt of silver is used in silver plating. For this purpose potassium cyanide is added to its solution, as otherwise silver cyanide, and not metallic silver, is deposited by the electric current. If two electrodes—one positive (silver) and the other negative (copper)—be immersed in such a solution, silver will be deposited upon the latter, and the silver of the positive electrode will be dissolved by the liquid, which will thus preserve the same amount of metal in solution as it originally contained. If instead of the negative electrode a copper object be taken, well cleaned from all dirt, the silver will be deposited in an even coating; this, indeed, forms the mode of *silver plating by the wet method*, which is most often used in practice. A solution of one part of silver nitrate in 30 to 50 parts of water, and mixed with a sufficient quantity of a solution of potassium cyanide to redissolve the precipitate of silver cyanide formed, gives a dull coating of silver, but if twice as much water be used the same mixture gives a bright coating.

Silver plating in the wet way has now replaced to a considerable extent the old process of *dry silvering*, because this process, which consists in dissolving silver in mercury and applying the amalgam to the surface of the objects, and then vaporising the mercury, offers the great disadvantage of the poisonous mercury fumes. Besides these, there is another method of silver plating, based on the direct displacement of silver from its salts by other metals—for example, by copper. The copper reduces the silver from its compounds, and the silver separated is deposited upon the copper. Thus a solution of silver chloride in sodium thiosulphate deposits a coating of silver upon a strip of copper immersed in it. It is best for this purpose to take pure *silver sulphite*. This is prepared by mixing a solution of silver nitrate with an excess of ammonia, and adding a saturated solution of sodium sulphite and then alcohol, which precipitates silver sulphite from the solution. The latter and its solutions are very easily decomposed by copper. Metallic iron produces the same decomposition, and iron and steel articles may be very readily silver-plated by means of the thiosulphate solution of silver chloride. Indeed, copper and similar metals may even be silver-plated by means of silver chloride; if the chloride of silver, with a small amount of acid, be rubbed upon the surface of the copper, the latter becomes covered with a coating of silver, which it has reduced.

Silver plating is not only applicable to metallic objects, but also to glass, china, &c. Glass is silvered for various purposes—for example, glass globes silvered internally are used for ornamentation, and have a mirrored surface. Common looking-glass silvered upon one side forms a mirror which is better than the ordinary mercury mirrors, owing to the truer colours of the image due to the whiteness of the silver. For optical instruments—for example, telescopes—concave mirrors are now made of silvered glass, which has first been ground and polished into the required form. The *silvering of glass* is based on the fact that silver which is reduced from certain solutions deposits itself uniformly in a perfectly homogeneous and continuous but very thin layer, forming a bright reflecting surface. Certain organic substances have the property of reducing silver in this form. The best known among these are certain aldehydes—for instance, ordinary acetaldehyde, C_2H_4O, which easily oxidises in the air and forms acetic acid, $C_2H_4O_2$. This oxidation also easily takes place at the expense of silver oxide, when a certain amount of ammonia is added to the mixture. The oxide of silver gives up its oxygen to the aldehyde, and the silver reduced from it is deposited in a metallic state in a uniform bright coating. The same action is produced by certain saccharine substances and certain organic acids, such as tartaric acid, &c.

The phenomenon which then takes place is described by Stas as follows, in a manner which is perfect in its clearness and accuracy: if silver oxide or carbonate be suspended in water, and an excess of water saturated with chlorine be added, all the silver is converted into chloride, just as is the case with oxide or carbonate of mercury, and the water then contains, besides the excess of chlorine, only pure hypochlorous acid without the least trace of chloric or chlorous acid. If a stream of chlorine be passed into water containing *an excess of silver oxide* or silver carbonate while the liquid is continually agitated, the reaction is the same as the preceding; silver chloride and hypochlorous acid are formed. But this acid does not long remain in a free state: it gradually acts on the silver oxide and gives silver hypochlorite, *i.e.* AgClO. If, after some time, the current of chlorine be stopped but the shaking continued, the liquid loses its characteristic odour of hypochlorous acid, while preserving its energetic decolorising property, because the silver hypochlorite which is formed is easily soluble in water. In the presence of an excess of silver oxide this salt can be kept for several days without decomposition, but it is exceedingly unstable when no excess of silver oxide or carbonate is present. So long as the solution of silver hypochlorite is shaken up with the silver oxide, it preserves its transparency and bleaching property, but directly it is allowed to stand, and the silver oxide settles, it becomes rapidly cloudy and deposits large flakes of silver chloride, so that the black silver oxide which had settled becomes covered with the white precipitate. The liquid then loses its bleaching properties and contains silver chlorate, *i.e.* $AgClO_3$, in solution, which has a slightly alkaline reaction, owing to the presence of a small amount of dissolved oxide. In this manner the reactions which are consecutively accomplished may be expressed by the equations:

$$6Cl_2 + 3Ag_2O + 3H_2O = 6AgCl + 6HClO;$$
$$6HClO + 3Ag_2O = 3H_2O + 6AgClO;$$
$$6AgClO = 4AgCl + 2AgClO_3.$$

Hence, Stas gives the following method for the preparation of silver chlorate: A slow current of chlorine is caused to act on oxide of silver, suspended in water which is kept in a state of continual agitation. The shaking is continued after the supply of chlorine has been stopped, in order that the free hypochlorous acid should pass into silver hypochlorite, and the resultant solution of the hypochlorite is drawn off from the sediment of the excess of silver oxide. This solution decomposes spontaneously into silver chloride and chlorate. The pure silver chlorate, $AgClO_3$, does not change under the action of light. The salt is prepared for further use by drying it in dry air at 150°. It is necessary during drying to prevent the access of any organic matter; this is done by filtering the air through cotton wool, and passing it over a layer of red-hot copper oxide.

[26 bis] The results given by Stas' determinations have recently been recalculated and certain corrections have been introduced. We give in the context the average results of van der Plaats and Thomsen's calculations, as well as in Table III. neglecting the doubtful thousandths.

This hypothesis, for the establishment or refutation of which so many researches have been made, is exceedingly important, and fully deserves the attention which has been given to it. Indeed, if it appeared that the atomic weights of all the elements could be expressed in whole numbers with reference to hydrogen, or if they at least proved to be commensurable with one another, then it could be affirmed with confidence that the elements, with all their diversity, were formed of one material condensed or grouped in various manners into the stable, and, under known conditions, undecomposable groups which we call the atoms of the elements. At first it was supposed that all the elements were nothing else but condensed hydrogen, but when it appeared that the atomic weights of the elements could not be expressed in whole numbers in relation to hydrogen, it was still possible to imagine the existence of a certain material from which both hydrogen and all the other elements were formed. If it should transpire that four atoms of this material form an atom of hydrogen, then the atom of chlorine would present itself as consisting of 142 atoms of this substance, the weight of whose atom would be equal to 0·25. But in this case the atoms of all the elements should be expressed in whole numbers with respect to the weight of the atom of this original material. Let us suppose that the atomic weight of this material is equal to unity, then all the atomic weights should be expressible in whole numbers relatively to this unit. Thus the atom of one element, let us suppose, would weigh m, and of another n, but, as both m and n must be whole numbers, it follows that the atomic weights of all the elements would be commensurable. But it is sufficient to glance over the results obtained by Stas, and to be assured of their accuracy, especially for silver, in order to entirely destroy, or at least strongly undermine, this attractive hypothesis. We must therefore refuse our assent to the doctrine of the building up from a single substance of the elements known to us. This hypothesis is not supported either by any known transformation (for one element has never been converted into another element), or by the commensurability of the atomic weights of the elements. Although the hypothesis of the formation of all the elements from a single substance (for which Crookes has suggested the name protyle) is most attractive in its comprehensiveness, it can neither be denied nor accepted for want of sufficient data. Marignac endeavoured, however, to overcome Stas's conclusions as to the incommensurability of the atomic weights by supposing that in his, as in the determinations of all other observers, there were unperceived errors which were quite independent of the mode of observation—for example, silver nitrate might be supposed to be an unstable substance which changes, under the heatings, evaporations, and other processes to which it is subjected in the reactions for the determination of the combining weight of silver. It might be supposed, for instance, that silver nitrate contains some impurity which cannot be removed by any means; it might also be supposed that a portion of the elements of the nitric acid are disengaged in the evaporation of the solution of silver nitrate (owing to the decomposing action of water), and in its fusion, and that we have not to deal with normal silver nitrate, but with a slightly basic salt, or perhaps an excess of nitric acid which cannot be removed from the salt. In this case the observed combining weight will not refer to an actually definite chemical compound, but to some mixture for which there does not exist any perfectly exact combining relations. Marignac upholds this proposition by the fact that the conclusions of Stas and other observers respecting the combining weights determined with the greatest exactitude very nearly agree with the proposition of the commensurability of the atomic weights—for example, the combining weight of silver was shown to be equal to 107·93, so that it only differs by 0·08 from the whole number 108, which is generally accepted for silver. The combining weight of iodine proved to be equal to 126·85—that is, it differs from 127 by 0·15. The combining weights of sodium, nitrogen, bromine, chlorine, and lithium are still nearer to the whole or round numbers which are generally accepted. But Marignac's proposition will hardly bear criticism. Indeed if we express the combining weights of the elements determined by Stas in relation to hydrogen, the approximation of these weights to whole numbers disappears, because one part of hydrogen in reality does not combine with 16 parts of oxygen, but with 15·92 parts, and therefore we shall obtain, taking $H = 1$, not the above-cited figures, but for silver 107·38, for bromine 79·55, magnitudes which are still further removed from whole numbers. Besides which, if Marignac's proposition were true the combining weight of silver determined by one method—*e.g.* by the analysis of silver chlorate combined with the synthesis of silver chloride—would not agree well with the combining weight determined by another method—*e.g.* by means of the analysis of silver iodate and the synthesis of silver iodide. If in one case a basic salt could be obtained, in the other case an acid salt might be obtained. Then the analysis of the acid salt would give different results from that of the basic salt. Thus Marignac's arguments cannot serve as a support for the vindication of Prout's hypothesis.

In conclusion, I think it will not be out of place to cite the following passage from a paper I read before the Chemical Society of London in 1889 (Appendix II.), referring to the hypothesis of the complexity of the elements recognised in chemistry, owing to the fact that many have endeavoured to apply the periodic law to the justification of this idea 'dating from a remote antiquity, when it was found convenient to admit the existence of many gods but only one matter.'

'When we try to explain the origin of the idea of a unique primary matter, we easily trace that, in the absence of deductions from experiment, it derives its origin from the scientifically philosophical attempt at discovering some kind of unity in the immense diversity of individualities which we see around. In classical times such a tendency could only be satisfied by conceptions about the immaterial world. As to the material world, our ancestors were compelled to resort to some hypothesis, and they adopted the idea of unity in the formative material, because they were not able to evolve the conception of any other possible unity in order to connect the multifarious relations of matter. Responding to the same legitimate scientific tendency, natural science has discovered throughout the universe a unity of plan, a unity of forces, and a unity of matter; and the convincing conclusions of modern science compel every one to admit these kinds of unity. But while we admit unity in many things, we none the less must also explain the individuality and the apparent diversity which we cannot fail to trace everywhere. It was said of old "Give us a fulcrum and it will become easy to displace the earth." So also we must say, "Give us something that is individualised, and the apparent diversity will be easily understood." Otherwise, how could unity result in a multitude.

'After a long and painstaking research, natural science has discovered the individualities of the chemical elements, and therefore it is now capable, not only of analysing, but also of synthesising; it can understand and grasp generality and unity, as well as the individualised and multifarious. The general and universal, like time and space, like force and motion, vary uniformly. The uniform admit of interpolations, revealing every intermediate phase; but the multitudinous, the individualised—such as ourselves, or the chemical elements, or the members of a peculiar periodic function of the elements, or Dalton's multiple proportions—is characterised in another way. We see in it—side by side with a general connecting principle—leaps, breaks of continuity, points which escape from the analysis of the infinitely small—an absence of complete intermediate links. Chemistry has found an answer to the question as to the causes of multitudes, and while retaining the conception of many elements, all submitted to the discipline of a general law, it offers an escape from the Indian Nirvana—the absorption in the universal—replacing it by the individualised. However, the place for individuality is so limited by the all-grasping, all-powerful universal, that it is merely a point of support for the understanding of multitude in unity.'

It might be expected from the periodic law and analogies with the series iron, cobalt, nickel, copper, zinc, that the atomic weights of the elements of the series osmium, iridium, platinum, gold, mercury, would rise in this order, and at the time of the establishment of the periodic law (1869), the determinations of Berzelius, Rose, and others gave the following values for the atomic weights: Os = 200, Ir = 197, Pt = 198, Au = 196, Hg = 200. The fulfilment of the expectations of the periodic law was given in the first place by the fresh determinations (Seubert, Dittmar, and Arthur) of the atomic weight of platinum, which proved to be nearly 196, if O = 16 (as Marignac, Brauner, and others propose); in the second place, by the fact that Seubert proved that the atomic weight of osmium is really less than that of platinum, and approximately Os = 191; and, in the third place, by the fact that after the researches of Krüss, Thorpe, and Laurie there was no doubt that the atomic weight of gold is greater than that of platinum—namely, nearly 197.

[28 bis] In Chapter XXII., Note 40, we gave the thermal data for certain of the compounds of copper of the type CuX_2; we will now cite certain data for the cuprous compounds of the type CuX, which present an analogy to the corresponding compounds AgX and AuX, some of which were investigated by Thomsen in his classical work, 'Thermochemische Untersuchungen' (Vol. iii., 1883). The data are given in the same manner as in the above-mentioned note:

R =	Cu	Ag	Au
R + Cl	+33	+29	+6
R + Br	+25	+23	0
R + I	+16	+14	-6
R + O	+41	+6	-?

Thus we see in the first place that gold, which possesses a much smaller affinity than Ag, evolves far less heat than an equivalent amount of copper, giving the same compound, and in the second place that the combination of copper with one atom of oxygen disengages more heat than its combination with one atom of a halogen, whilst with silver the reverse is the case. This is connected with the fact that Cu_2O is more stable under the action of heat than Ag_2O.

Heavy atoms and molecules, although they may present many points of analogy, are more easily isolated; thus $C_{16}H_{32}$, although, like C_2H_4, it combines with Br_2, and has a similar composition, yet reacts with much greater difficulty than C_2H_4, and in this it resembles gold; the heavy atoms and molecules are, so to say, inert, and already saturated by themselves. Gold in its higher grade of oxidation, Au_2O_3, presents feeble basic properties and weakly-developed acid properties, so that this oxide of gold,

Au_2O_3, may be referred to the class of feeble acid oxides, like platinic oxide. This is not the case in the highest known oxides of copper and silver. But in the lower grade of oxidation, aurous oxide, Au_2O, gold, like silver and copper, presents basic properties, although they are not very pronounced. In this respect it stands very close in its properties, although not in its types of combination (AuX and AuX_3), to platinum (PtX_2 and PtX_4) and its analogues.

As yet the general chemical characteristics of gold and its compounds have not been fully investigated. This is partly due to the fact that very few researches have been undertaken on the compounds of this metal, owing to its inaccessibility for working in large quantities. As the atomic weight of gold is high (Au = 197), the preparation of its compounds requires that it should be taken in large quantities, which forms an obstacle to its being fully studied. Hence the facts concerning the history of this metal are rarely distinguished by that exactitude with which many facts have been established concerning other elements more accessible, and long known in use.

[29 bis] Sonstadt (1872) showed that sea water, besides silver, always contains gold. Munster (1892) showed that the water of the Norwegian fiords contains about 5 milligrams of gold per ton (or 5 milliardths)—*i.e.* a quantity deserving practical attention, and I think it may be already said that, considering the immeasurable amount of sea water, in time means will be discovered for profitably extracting gold from sea water by bringing it into contact with substances capable of depositing gold upon their surface. The first efforts might be made upon the extraction of salt from sea water, and as the total amount of sea water may be taken as about 2,000,000,000,000,000,000 tons, it follows that it contains about 10,000 million tons of gold. The yearly production of gold is about 200 tons for the whole world, of which about one quarter is extracted in Russia. It is supposed that gold is dissolved in sea water owing to the presence of iodides, which, under the action of animal organisms, yield free iodine. It is thought (as Professor Konovaloff mentions in his work upon 'The Industries of the United States,' 1894) that iodine facilitates the solution of the gold, and the organic matter its precipitation. These facts and considerations to a certain extent explain the distribution of gold in veins or rock fissures, chiefly filled with quartz, because there is sufficient reason for supposing that these rocks once formed the ocean bottom. R. Dentrie, and subsequently Wilkinson, showed that organic matter—for instance, cork—and pyrites are able to precipitate gold from its solutions in that metallic form and state in which it occurs in quartz veins, where (especially in the deeper parts of vein deposits) gold is frequently found on the surface of pyrites, chiefly arsenical pyrites. Kazantseff (in Ekaterinburg, 1891) even

supposes, from the distribution of the gold in these pyrites, that it occurred in solution as a compound of sulphide of gold and sulphide of arsenic when it penetrated into the veins. It is from such considerations that the origin of vein and pyritic gold is, at the present time, attributed to the reaction of solutions of this metal, the remains of which are seen in the gold still present in sea water.

However, in recent times, especially since about 1870, when chlorine (either as a solution of the gas or as bleaching powder) and bromine began to be applied to the extraction of finely-divided gold from poor ores (previously roasted in order to drive off arsenic and sulphur, and oxidise the iron), the extraction of gold from quartz and pyrites, by the wet method, increases from year to year, and begins to equal the amount extracted from alluvial deposits. Since the nineties the *cyanide process* (Chapter XIII., Note 13 bis) has taken an important place among the wet methods for extracting gold from its ores. It consists in pouring a dilute solution of cyanide of potassium (about 500 parts of water and 1 to 4 parts of cyanide of potassium per 1,000 parts of ore, the amount of cyanide depending upon the richness of the ore) and a mixture of it with NaCN, (*see* Chapter XIII., Note 12) over the crushed ore (which need not be roasted, whilst roasting is indispensable in the chlorination process, as otherwise the chlorine is used up in oxidising the sulphur, arsenic, &c.) The gold is dissolved very rapidly even from pyrites, where it generally occurs on the surface in such fine and adherent particles that it either cannot be mechanically washed away, or, more frequently is carried away by the stream of water, and cannot be caught by mechanical means or by the mercury used for catching the gold in the sluices. Chlorination had already given the possibility of extracting the finest particles of gold; but the cyanide process enables such pyrites to be treated as could be scarcely worked by other means. The treatment of the crushed ore by the KCN is carried on in simple wooden vats (coated with paraffin or tar) with the greatest possible rapidity (in order that the KCN solution should not have time to change) by a method of systematic lixiviation, and is completed in 10 to 12 hours. The resultant solution of gold, containing $AuK(CN)_2$, is decomposed either with freshly-made zinc filings (but when the gold settles on the Zn, the cyanide solution reacts upon the Zn with the evolution of H_2 and formation of ZnH_2O_2) or by sodium amalgam prepared at the moment of reaction by the action of an electric current upon a solution of NaHO poured into a vessel partially immersed in mercury (the NaCN is renewed continually by this means). The silver in the ore passes into solution, together with the gold, as in amalgamation.

But the particles of gold are sometimes so small that a large amount is lost during the washing. It is then profitable to have recourse to the extraction by chlorine and KCN (Note 30).

In speaking of the extraction of gold the following remarks may not be out of place:

In California advantage is taken of water supplied from high altitudes in order to have a powerful head of water, with which the rocks are directly washed away, thus avoiding the greater portion of the mechanical labour required for the exploitation of these deposits.

The last residues of gold are sometimes extracted from sand by washing them with mercury, which dissolves the gold. The sand mixed with water is caused to come into contact with mercury during the washing. The mercury is then distilled.

Many sulphurous ores, even pyrites, contain a small amount of gold. Compounds of gold with bismuth, $BiAu_2$, tellurium, $AuTe_2$ (calverite), &c., have been found, although rarely.

Among the minerals which accompany gold, and from which the presence of gold may be expected, we may mention white quartz, titanic and magnetic iron ores, and also the following, which are of rarer occurrence: zircon, topaz, garnet, and such like. The concentrated gold washings first undergo a mechanical treatment, and the impure gold obtained is treated for pure gold by various methods. If the gold contain a considerable amount of foreign metals, especially lead and copper, it is sometimes cupelled, like silver, so that the oxidisable metals may be absorbed by the cupel in the form of oxides, but in every case the gold is obtained together with silver, because the latter metal also is not oxidised. Sometimes the gold is extracted by means of mercury, that is, by amalgamation (and the mercury subsequently driven off by distillation), or by smelting it with lead (which is afterwards removed by oxidation) and processes like those employed for the extraction of silver, because gold, like silver, does not oxidise, is dissolved by lead and mercury, and is non-volatile. If copper or any other metal contain gold and it be employed as an anode, pure copper will be deposited upon the cathode, while all the gold will remain at the anode as a slime. This method often amply repays the whole cost of the process, since it gives, besides the gold, a pure electrolytic copper.

[31 bis] Schottländer (1893) obtained gold in a soluble colloid form (the solution is violet) by the action of a mixture of solutions of cerium acetate and NaHO upon a solution of $AuCl_3$. The gold separates out from such a solution in exactly the same manner as Ag does from the solution of

colloid silver mentioned above. There always remains a certain amount of a higher oxide of cerium, CeO_2, in the solution—*i.e.* the gold is reduced by converting the cerium into a higher grade of oxidation. Besides which Krüss and Hofmann showed that sulphide of gold precipitated by the action of H_2S upon a solution of $AuKCy_2$ mixed with HCl easily passes into a colloid solution after being properly washed (like As_2S_3, CuS, &c., Chapter I., Note 57).

[31 tri] Gold-leaf is used for gilding wood (leather, cardboard, and suchlike, upon which it is glued by means of varnish, &c.), and is about 0·003 millimetre thick. It is obtained from thin sheets (weighing at first about ¼ grm. to a square inch), rolled between gold rollers, by gradually hammering them (in packets of a number at once) between sheets of moist (but not wet) parchment, and then, after cutting them into four pieces, between a specially prepared membrane, which, when at the right degree of moisture, does not tear or stick together under the blows of the hammer.

The formation of the alloys Cu + Zn, Cu + Sn, Cu + Bi, Cu + Sb, Pb + Sb, Ag + Pb, Ag + Sn, Au + Zn, Au + Sn, &c., is accompanied by a contraction (and evolution of heat). The formation of the alloys Fe + Sb, Fe + Pb, Cu + Pb, Pb + Sn, Pb + Sb, Zn + Sb, Ag + Cu, Au + Cu, Au + Pb, takes place with a certain increase in volume. With regard to the alloys of gold, it may be mentioned that gold is only slightly dissolved by mercury (about 0·06 p.c., Dudley, 1890); the remaining portion forms a granular alloy, whose composition has not been definitely determined. Aluminium (and silicon) also have the capacity of forming alloys with gold. The presence of a small amount of aluminium lowers the melting point of gold considerably (Roberts-Austen, 1892); thus the addition of 4 p.c. of aluminium lowers it by 14°·28, the addition of 10 p.c. Al by 41°·7. The latter alloy is white. The alloy $AuAl_2$ has a characteristic purple colour, and its melting point is 32°·5 above that of gold, which shows it to be a definite compound of the two metals. The melting points of alloys richer in Al gradually fall to 660°—that is, below that of aluminium (665°).

Heycock and Neville (1892), in studying the triple alloys of Au, Cd, and Sn, observed a tendency in the gold to give compounds with Cd, and by sealing a mixture of Au and Cd in a tube, from which the air had been exhausted, and heating it, they obtained a grey crystalline brittle definite alloy AuCd.

[32 bis] Calderon (1892), at the request of some jewellers, investigated the cause of a peculiar alteration sometimes found on the surface of dead-gold articles, there appearing brownish and blackish spots, which widen and alter their form in course of time. He came to the conclusion that these spots are due to the appearance and development of peculiar micro-

organisms (Aspergillus niger and Micrococcus cimbareus) on the gold, spores of which were found in abundance on the cotton-wool in which the gold articles had been kept.

Stannous chloride as a reducing agent also acts on auric chloride, and gives a red precipitate known as *purple of Cassius*. This substance, which probably contains a mixture or compound of aurous oxide and tin oxide, is used as a red pigment for china and glass. Oxalic acid, on heating, reduces metallic gold from its salts, and this property may be taken advantage of for separating it from its solutions. The oxidation which then takes place in the presence of water may be expressed by the following equation: $2AuCl_3 + 3C_2H_2O_4 = 2Au + 6HCl + 6CO_2$. Nearly all organic substances have a reducing action on gold, and solutions of gold leave a violet stain on the skin.

Auric chloride, like platinic chloride, is distinguished for its clearly-developed property of forming double salts. These double salts, as a rule, belong to the type $AuMCl_4$. The compound of auric chloride with hydrochloric acid mentioned above evidently belongs to the same type. The compounds $2KAuCl_4,5H_2O$, $NaAuCl_4,2H_2O$, $AuNH_4Cl_4,H_2O$, $Mg(AuCl_4)_2,2H_2O$, and the like are easily crystallised in well-formed crystals. Wells, Wheeler, and Penfield (1892) obtained $RbAuCl_4$ (reddish yellow) and $CsAuCl_4$ (golden yellow), and corresponding bromides (dark coloured). $AuBr_3$ is extremely like the chloride. Auric cyanide is obtained easily in the form of a double salt of potassium, $KAu(CN)_4$ by mixing saturated and hot solutions of potassium cyanide with auric chloride and then cooling.

If ammonia be added to a solution of auric chloride, it forms a yellow precipitate of the so-called fulminating gold, which contains gold, chlorine, hydrogen, nitrogen, and oxygen, but its formula is not known with certainty. It is probably a sort of ammonio-metallic compound, $Au_2O_3,4NH_3$, or amide (like the mercury compound). This precipitate explodes at 140°, but when left in the presence of solutions containing ammonia it loses all its chlorine and becomes non-explosive. In this form the composition $Au_2O_3,2NH_3,H_2O$ is ascribed to it, but this is uncertain. Auric sulphide, Au_2S_3, is obtained by the action of hydrogen sulphide on a solution of auric chloride, and also directly by fusing sulphur with gold. It has an acid character, and therefore dissolves in sodium and ammonium sulphides.

Many double salts of suboxide of gold belong to the type AuX—for instance, the cyanide corresponding to the type $AuKX_2$, like PtK_2X_4, with which we became acquainted in the last chapter. We will enumerate several of the representatives of this class of compounds. If auric chloride, $AuCl_3$, be mixed with a solution of sodium thiosulphate, the gold passes into a colourless solution, which deposits colourless crystals, containing a double thiosulphate of gold and sodium, which are easily soluble in water but are precipitated by alcohol. The composition of this salt is $Na_3Au(S_2O_3)_2,2H_2O$. If the sodium thiosulphate be represented as NaS_2O_3Na, the double salt in question will be $AuNa(S_2O_3Na)_2,2H_2O$, according to the type $AuNaX_2$. The solution of this colourless and easily crystallisable salt has a sweet taste, and the gold is not separated from it either by ferrous sulphate or oxalic acid. This salt, which is known as *Fordos and Gelis's salt*, is used in medicine and photography. In general, aurous oxide exhibits a distinct inclination to the formation of similar double salts, as we saw also with PtX_2—for example, it forms similar salts with sulphurous acid. Thus if a solution of sodium sulphite be gradually added to a solution of oxide of gold in sodium hydroxide, the precipitate at first formed re-dissolves to a colourless solution, which contains the double salt $Na_3Au(SO_3)_2 = AuNa(SO_3Na)_2$. The solution of this salt, when mixed with barium chloride, first forms a precipitate of barium sulphite, and then a red barium double salt which corresponds with the above sodium salt.

The oxygen compound of the type AuX, *aurous oxide*, Au_2O, is obtained as a greenish violet powder on mixing aurous chloride with potassium chloride in the cold. With hydrochloric acid this oxide gives gold and auric chloride, and when heated it easily splits up into oxygen and metallic gold.

APPENDIX I

AN ATTEMPT TO APPLY TO CHEMISTRY ONE OF THE PRINCIPLES OF NEWTON'S NATURAL PHILOSOPHY

By PROFESSOR MENDELÉEFF

A LECTURE DELIVERED AT THE ROYAL INSTITUTION OF GREAT BRITAIN ON FRIDAY, MAY 31, 1889

Nature, inert to the eyes of the ancients, has been revealed to us as full of life and activity. The conviction that motion pervaded all things, which was first realised with respect to the stellar universe, has now extended to the unseen world of atoms. No sooner had the human understanding denied to the earth a fixed position and launched it along its path in space, than it was sought to fix immovably the sun and the stars. But astronomy has demonstrated that the sun moves with unswerving regularity through the star-set universe at the rate of about 50 kilometres per second. Among the so-called fixed stars are now discerned manifold changes and various orders of movement. Light, heat, electricity—like sound—have been proved to be modes of motion; to the realisation of this fact modern science is indebted for powers which have been used with such brilliant success, and which have been expounded so clearly at this lecture table by Faraday and by his successors. As, in the imagination of Dante, the invisible air became peopled with spiritual beings, so before the eyes of earnest investigators, and especially before those of Clerk Maxwell, the invisible mass of gases became peopled with particles: their rapid movements, their collisions, and impacts became so manifest that it seemed almost possible to count the impacts and determine many of the peculiarities or laws of their collisions. The fact of the existence of these invisible motions may at once be made apparent by demonstrating the difference in the rate of diffusion through porous bodies of the light and rapidly moving atoms of hydrogen and the heavier and more sluggish particles of air. Within the masses of liquid and of solid bodies we have been forced to acknowledge the existence of persistent though limited motion of their ultimate particles, for otherwise it would be impossible to explain, for example, the celebrated experiments of Graham on diffusion through liquid and colloidal substances. If there were, in our times, no belief in the molecular motion in solid bodies, could the famous Spring have hoped to attain any result by mixing carefully-dried powders of potash, saltpetre and sodium acetate, in order to produce, by pressure, a chemical reaction between these substances through the interchange of their metals, and have derived, for the conviction of the incredulous, a mixture of two hygroscopic though solid salts—sodium nitrate and potassium acetate?

In these invisible and apparently chaotic movements, reaching from the stars to the minutest atoms, there reigns, however, a harmonious order which is commonly mistaken for complete rest, but which is really a consequence of the conservation of that dynamic equilibrium which was first discerned by the genius of Newton, and which has been traced by his successors in the detailed analysis of the particular consequences of the great generalisation, namely, relative immovability in the midst of universal and active movement.

But the unseen world of chemical changes is closely analogous to the visible world of the heavenly bodies, since our atoms form distinct portions of an invisible world, as planets, satellites, and comets form distinct portions of the astronomer's universe; our atoms may therefore be compared to the solar systems, or to the systems of double or of single stars: for example, ammonia (NH_3) may be represented in the simplest manner by supposing the sun, nitrogen, surrounded by its planets of hydrogen; and common salt ($NaCl$) may be looked on as a double star formed of sodium and chlorine. Besides, now that the indestructibility of the elements has been acknowledged, chemical changes cannot otherwise be explained than as changes of motion, and the production by chemical reactions of galvanic currents, of light, of heat, of pressure, or of steam power, demonstrates visibly that the processes of chemical reaction are inevitably connected with enormous though unseen displacements, originating in the movements of atoms in molecules. Astronomers and natural philosophers, in studying the visible motions of the heavenly bodies and of matter on the earth, have understood and have estimated the value of this store of energy. But the chemist has had to pursue a contrary course. Observing in the physical and mechanical phenomena which accompany chemical reactions the quantity of energy manifested by the atoms and molecules, he is constrained to acknowledge that within the molecules there exist atoms in motion, endowed with an energy which, like matter itself, is neither being created nor capable of being destroyed. Therefore, in chemistry, we must seek dynamic equilibrium not only between the molecules, but also in their midst among their component atoms. Many conditions of such equilibrium have been determined, but much remains to be done, and it is not uncommon, even in these days, to find that some chemists forget that there is the possibility of motion in the interior of molecules, and therefore represent them as being in a condition of death-like inactivity.

Chemical combinations take place with so much ease and rapidity, possess so many special characteristics, and are so numerous, that their simplicity and order were for a long time hidden from investigators. Sympathy, relationship, all the caprices or all the fancifulness of human intercourse, seemed to have found complete analogies in chemical combinations, but with this difference, that the characteristics of the material substances—such as silver, for example, or of any other body—remain unchanged in every subdivision from the largest masses to the smallest particles, and consequently these characteristics must be properties of the particles. But the world of heavenly luminaries appeared equally fanciful at man's first acquaintance with it, so much so, that the astrologers imagined a connection between the individualities of men and the conjunctions of planets. Thanks to the genius of Lavoisier and of Dalton, man has been able, in the unseen world of chemical combinations, to recognise laws of the same simple order as those which Copernicus and Kepler proved to exist in the planetary universe. Man discovered, and continues every hour to discover, *what* remains unchanged in chemical evolution, and *how* changes take place in combinations of the unchangeable. He has learned to predict, not only what possible combinations may take place, but also the very existence of atoms of unknown elementary substances, and has besides succeeded in making innumerable practical applications of his knowledge to the great advantage of his race, and has accomplished this notwithstanding that notions of sympathy and affinity still preserve a strong vitality in science. At present we cannot apply Newton's principles to chemistry, because the soil is only being now prepared. The invisible world of chemical atoms is still waiting for the creator of chemical mechanics. For him our age is collecting a mass of materials, the inductions of well-digested facts, and many-sided inferences similar to those which existed for Astronomy and Mechanics in the days of Newton. It is well also to remember that Newton devoted much time to chemical experiments, and while considering questions of celestial mechanics, persistently kept in view the mutual action of those infinitely small worlds which are concerned in chemical evolutions. For this reason, and also to maintain the unity of laws, it seems to me that we must, in the first instance, seek to harmonise the various phases of contemporary chemical theories with the immortal principles of the Newtonian natural philosophy, and so hasten the advent of true chemical mechanics. Let the above considerations serve as my justification for the attempt which I propose to make to act as a champion of the universality of the Newtonian principles, which I believe are competent to embrace every phenomenon in the universe, from the rotation of the fixed stars to the interchanges of chemical atoms.

In the first place I consider it indispensable to bear in mind that, up to quite recent times, only a one-sided affinity has been recognised in chemical reactions. Thus, for example, from the circumstance that red-hot iron decomposes water with the evolution of hydrogen, it was concluded that oxygen had a greater affinity for iron than for hydrogen. But hydrogen, in presence of red-hot iron scale, appropriates its oxygen and forms water, whence an exactly opposite conclusion may be formed.

During the last ten years a gradual, scarcely perceptible, but most important change has taken place in the views, and consequently in the researches, of chemists. They have sought everywhere, and have always found, systems of conservation or dynamic equilibrium substantially similar to those which natural philosophers have long since discovered in the visible world, and in virtue of which the position of the heavenly bodies in the universe is determined. There where one-sided affinities only were at first detected, not only secondary or lateral ones have been found, but even those which are diametrically opposite; yet among these, dynamical equilibrium establishes itself not by excluding one or other of the forces, but regulating them all. So the chemist finds in the flame of the blast furnace, in the formation of every salt, and, with especial clearness, in double salts and in the crystallisation of solutions, not a fight ending in the victory of one side, as used to be supposed, but the conjunction of forces; the peace of dynamic equilibrium resulting from the action of many forces and affinities. Carbonaceous matters, for example, burn at the expense of the oxygen of the air, yielding a quantity of heat, and forming products of combustion, in which it was thought that the affinities of the oxygen with the combustible elements were satisfied. But it appeared that the heat of combustion was competent to decompose these products, to dissociate the oxygen from the combustible elements, and therefore to explain combustion fully it is necessary to take into account the equilibrium between opposite reactions, between those which evolve and those which absorb heat.

In the same way, in the case of the solution of common salt in water, it is necessary to take into account, on the one hand, the formation of compound particles generated by the combination of salt with water, and, on the other, the disintegration or scattering of the new particles formed, as well as of these originally contained. At present we find two currents of thought, apparently antagonistic to each other, dominating the study of solutions: according to the one, solution seems a mere act of building up or association; according to the other, it is only dissociation or disintegration. The truth lies, evidently, between these views; it lies, as I have endeavoured to prove by my investigations into aqueous solutions, in the dynamic equilibrium of particles tending to combine and also to fall asunder. The large majority of chemical reactions which appeared to act victoriously along one line have been proved capable of acting as victoriously even along an exactly opposite line. Elements which utterly decline to combine directly may often be formed into comparatively stable compounds by indirect means, as, for example, in the case of chlorine and carbon; and consequently the sympathies and antipathies which it was thought to transfer from human relations to those of atoms should be laid aside until the mechanism of chemical relations is explained. Let us remember, however, that chlorine, which does not form with carbon the chloride of carbon, is strongly absorbed, or, as it were, dissolved, by carbon, which leads us to suspect incipient chemical action even in an external and purely surface contact, and involuntarily gives rise to conceptions of that unity of the forces of nature which has been so energetically insisted on by Sir William Grove and formulated in his famous paradox. Grove noticed that platinum, when fused in the oxyhydrogen flame, during which operation water is formed, when allowed to drop into water decomposes the latter and produces the explosive oxyhydrogen mixture. The explanation of this paradox, as of many others which arose during the period of chemical renaissance, has led, in our time, to the promulgation by Henri Sainte-Claire Deville of the conception of dissociation and of equilibrium, and has recalled the teaching of Berthollet, which, notwithstanding its brilliant confirmation by Heinrich Rose and Dr. Gladstone, had not, up to that period, been included in received chemical views.

Chemical equilibrium in general, and dissociation in particular, are now being so fully worked out in detail, and supplied in such various ways, that I do not allude to them to develop, but only use them as examples by which to indicate the correctness of a tendency to regard chemical combinations from points of view differing from those expressed by the term hitherto appropriated to define chemical forces, namely, 'affinity.' Chemical equilibria, dissociation, the speed of chemical reactions, thermochemistry, spectroscopy, and, more than all, the determination of the influence of masses and the search for a connection between the properties and weights of atoms and molecules—in one word, the vast mass of the most important chemical researches of the present day—clearly indicate the near approach of the time when chemical doctrines will submit fully and completely to the doctrine which was first announced in the *Principia* of Newton.

In order that the application of these principles may bear fruit it is evidently insufficient to assume that statical equilibrium reigns alone in chemical systems or chemical molecules: it is necessary to grasp the conditions of possible states of dynamical equilibria, and to apply to them kinetic principles. Numerous considerations compel us to renounce the idea of statical equilibrium in molecules, and the recent yet strongly-supported appeals to dynamic principles constitute, in my opinion, the foundation of the modern teaching relating to atomicity, or the valency of the elements, which usually forms the basis of investigations into organic or carbon compounds.

This teaching has led to brilliant explanations of very many chemical relations and to cases of isomerism, or the difference in the properties of substances having the same composition. It has been so fruitful in its many applications and in the foreshadowing of remote consequences, especially respecting carbon compounds, that it is impossible to deny its claims to be ranked as a great achievement of chemical science. Its practical application to the synthesis of many substances of the most complicated composition entering into the structure of organised bodies, and to the creation of an unlimited number of carbon compounds, among which the colours derived from coal tar stand prominently forward, surpass the synthetical powers of Nature itself. Yet this teaching, as applied to the structure of carbon compounds, is not on the face of it directly applicable to the investigation of other elements, because in examining the first it is possible to assume that the atoms of carbon have always a definite and equal number of affinities, whilst in the combinations of other elements this is evidently inadmissible. Thus, for example, an atom of carbon yields only one compound with four atoms of hydrogen and one with four atoms of chlorine in the molecule, whilst the atoms of chlorine and hydrogen unite only in the proportions of one to one. Simplicity is here evident, and forms a point of departure from which it is easy to move forward with firm and secure tread. Other elements are of a different nature. Phosphorus unites with three and with five atoms of chlorine, and consequently the simplicity and sharpness of the application of structural conceptions are lost. Sulphur unites only with two atoms of hydrogen, but with oxygen it enters into higher orders of combination. The periodic relationship which exists among all the properties of the elements—such, for example, as their ability to enter into various combinations—and their atomic weights, indicate that this variation in atomicity is subject to one perfectly exact and general law, and it is only carbon and its near analogues which constitute cases of permanently preserved atomicity. It is impossible to recognise as constant and fundamental properties of atoms, powers which, in substance, have proved to be variable. But by abandoning the idea of permanence, and of the constant saturation of affinities—that is to say, by acknowledging the possibility of free affinities—many retain a comprehension of the atomicity of the elements 'under given conditions;' and on this frail foundation they build up structures composed of chemical molecules, evidently only because the conception of manifold affinities gives, at once, a simple statical method of estimating the composition of the most complicated molecules.

I shall enter neither into details, nor into the various consequences following from these views, nor into the disputes which have sprung up respecting them (and relating especially to the number of isomerides possible on the assumption of free affinities), because the foundation or origin of theories of this nature suffers from the radical defect of being in opposition to dynamics. The molecule, as even Laurent expressed himself, is represented as an architectural structure, the style of which is determined by the fundamental arrangement of a few atoms, whilst the decorative details, which are capable of being varied by the same forces, are formed by the elements entering into the combination. It is on this account that the term 'structural' is so appropriate to the contemporary views of the above order, and that the 'structuralists' seek to justify the tetrahedric, plane, or prismatic disposition of the atoms of carbon in benzene. It is evident that the consideration relates to the statical position of atoms and molecules and not to their kinetic relations. The atoms of the structural type are like the lifeless pieces on a chess board: they are endowed but with the voices of living beings, and are not those living beings themselves; acting, indeed, according to laws, yet each possessed of a store of energy which, in the present state of our knowledge, must be taken into account.

In the days of Haüy, crystals were considered in the same statical and structural light, but modern crystallographers, having become more thoroughly acquainted with their physical properties and their actual formation, have abandoned the earlier views, and have made their doctrines dependent on dynamics.

The immediate object of this lecture is to show that, starting with Newton's third law of motion, it is possible to preserve to chemistry all the advantages arising from structural teaching, without being obliged to build up molecules in solid and motionless figures, or to ascribe to atoms definite limited valencies, directions of cohesion, or affinities. The wide extent of the subject obliges me to treat only a small portion of it, namely of *substitutions*, without specially considering combinations and decompositions, and even then limiting myself to the simplest examples, which, however, will throw open prospects embracing all the natural complexity of chemical relations. For this reason, if it should prove possible to form groups similar, for example, to H_4 or CH_6 as the remnants of molecules CH_4 or C_2H_7 we shall not pause to consider them, because, as far as we know, they fall asunder into two parts, $H_2 + H_2$ or $CH_4 + H_2$, as soon as they are even temporarily formed, and are incapable of separate existence, and therefore can take no part in the elementary act of substitution. With respect to the simplest molecules which we shall select—that is to say, those of which the parts have no separate existence, and therefore cannot appear in substitutions—we shall consider them according to the periodic law, arranging them in direct dependence on the atomic weight of the elements.

Thus, for example, the molecules of the simplest hydrogen compounds—

HF	H₂O	H₃N	H₄C
hydrofluoric acid	water	ammonia	Hmethane

correspond with elements the atomic weights of which decrease consecutively

$$F = 19, O = 16, N = 14, C = 12.$$

Neither the arithmetical order (1, 2, 3, 4 atoms of hydrogen) nor the total information we possess respecting the elements will permit us to interpolate into this typical series one more additional element; and therefore we have here, for hydrogen compounds, a natural base on which are built up those simple chemical combinations which we take as typical. But even they are competent to unite with each other, as we see, for instance, in the property which hydrofluoric acid has of forming a hydrate—that is, of combining with water; and a similar attribute of ammonia, resulting in the formation of a caustic alkali, NH_3,H_2O, or NH_4OH.

Having made these indispensable preliminary observations, I may now attack the problem itself and attempt to explain the so-called structure or rather construction, of molecules—that is to say, their constitution and transformations—without having recourse to the teaching of 'structuralists,' but on Newton's dynamical principles.

Of Newton's three laws of motion, only the third can be applied directly to chemical molecules when regarded as systems of atoms among which it must be supposed that there exist common influences or forces, and resulting compounded relative motions. Chemical reactions of every kind are undoubtedly accomplished by changes in these internal movements, respecting the nature of which nothing is known at present, but the existence of which the mass of evidence collected in modern times forces us to acknowledge as forming part of the common motion of the universe, and as a fact further established by the circumstance that chemical reactions are always characterised by changes of volume or the relations between the atoms or the molecules. Newton's third law, which is applicable to every system, declares that, 'action is also associated with reaction, and is equal to it.' The brevity of conciseness of this axiom was, however, qualified by Newton in a more expanded statement, 'the action of bodies one upon another are always equal, and in opposite directions.' This simple fact constitutes the point of departure for explaining dynamic equilibrium—that is to say, systems of conservancy. It is capable of satisfying even the dualists, and of explaining, without additional assumptions, the preservation of those chemical types which Dumas, Laurent, and Gerhardt created unit types, and those views of atomic combinations which the structuralists express by atomicity or the valency of the elements, and, in connection with them, the various numbers of affinities. In reality, if a system of atoms or a molecule be given, then in it, according to the third law of Newton, each portion of atoms acts on the remaining portion in the same manner, and with the same force as the second set of atoms acts on the first. We infer directly from this consideration that both sets of atoms, forming a molecule, are not only equivalent with regard to themselves, as they must be according to Dalton's law, but also that they may, if united, replace each other. Let there be a molecule containing atoms A B C, it is clear that, according to Newton's law, the action of A on B C must be equal to the action of B C on A, and if the first action is directed on B C, then the second must be directed on A, and consequently then, where A can exist in dynamic equilibrium, B C may take its place and act in a like manner. In the same way the action of C is equal to the action of A B. In one word every two sets of atoms forming a molecule are equivalent to each other, and may take each other's place in other molecules, or, having the power of balancing each other, the atoms or their complements are endowed with the power of replacing each other. Let us call this consequence of an evident axiom 'the principle of substitution,' and let us apply it to those typical forms of hydrogen compounds which we have already discussed, and which, on account of their simplicity, and regularity, have served as starting-points of chemical argument long before the appearance of the doctrine of structure.

In the type of hydrofluoric acid, HF, or in systems of double stars, are included a multitude of the simplest molecules. It will be sufficient for our purpose to recall a few: for example, the molecules of chlorine, Cl_2, and of hydrogen, H_2, and hydrochloric acid, HCl, which is familiar to all in aqueous solution as spirits of salt, and which has many points of resemblance with HF, HBr, HI. In these cases division into two parts can only be made in one way, and therefore the principle of substitution renders it probable that exchanges between the chlorine and the hydrogen can take place, if they are competent to unite with each other. There was a time when no chemist would even admit the idea of any such action; it was then thought that the power of combination indicated a polar difference of the molecules in combination, and this thought set aside all idea of the substitution of one component element by another.

Thanks to the observations and experiments of Dumas and Laurent fifty years ago, such fallacies were dispelled, and in this manner the principle of substitution was exhibited. Chlorine and bromine acting on many hydrogen compounds, occupy immediately the place of their hydrogen, and the displaced hydrogen, with another atom of chlorine or bromine, forms hydrochloric acid or bromide of hydrogen. This takes place in all typical hydrogen compounds. Thus chlorine acts on this principle on gaseous hydrogen—reaction, under the influence of light, resulting in the formation of hydrochloric acid. Chlorine acting on the alkalis, constituted similarly to water, and even on water itself—only, however, under the influence of light and only partially because of the instability of HClO—forms by this principle bleaching salts, which are the same as the alkalis, but with their hydrogen replaced by chlorine. In ammonia and in methane, chlorine can also replace the hydrogen. From ammonia is formed in this manner the so-called chloride of nitrogen, NCl_3, which decomposes very readily with violent explosion on account of the evolved gases, and falls asunder as chlorine and nitrogen. Out of marsh gas, or methane, CH_4, may be obtained consecutively, by this method, every possible substitution, of which chloroform, $CHCl_3$, is the best known, and carbon tetrachloride, CCl_4, the most instructive. But by virtue of the fact that chlorine and bromine act, in the manner shown, on the simplest typical hydrogen compounds, their action on the more complicated ones may be assumed to be the same. This can be easily demonstrated. The hydrogen of benzene, C_6H_6, reacts feebly under the influence of light on liquid bromine, but Gustavson has shown that the addition of the smallest quantity of metallic aluminium causes energetic action and the evolution of large volumes of hydrogen bromide.

If we pass on to the second typical hydrogen compound—that is to say, water—its molecule, HOH, may be split up in two ways: either into an atom of hydrogen and a semi-molecule of hydrogen peroxide, HO, or into oxygen, O, and two atoms of hydrogen, H; and therefore, according to the principle of substitution, it is evident that one atom of hydrogen can exchange with hydrogen oxide, HO, and two atoms of hydrogen, H, with one atom of oxygen, O.

Both these forms of substitution will constitute methods of oxidation—that is to say, of the entrance of oxygen into the compound—a reaction which is so common in nature as well as in the arts, taking place at the expense of the oxygen of the air or by the aid of various oxidising substances or bodies which part easily with their oxygen. There is no occasion to reckon up the unlimited number of cases of such oxidising reactions. It is sufficient to state that in the first of these oxygen is directly transferred, and the position, the chemical function, which hydrogen originally occupied, is, after the substitution, occupied by the hydroxyl. Thus ammonia, NH_3, yields hydroxylamine, $NH_2(OH)$, a substance which retains many of the properties of ammonia.

Methane and a number of other hydrocarbons yield, by substitution of the hydrogen by its oxide, methyl alcohol, $CH_3(OH)$, and other alcohols.

The substitution of one atom of oxygen for two atoms of hydrogen is equally common with hydrogen compounds. By this means alcoholic liquids containing ethyl alcohol, or spirits of wine, $C_2H_5(OH)$, are oxidised until they become vinegar, or acetic acid, $C_2H_3O(OH)$. In the same way caustic ammonia, or the combination of ammonia with water, NH_3,H_2O, or $NH_4(OH)$, which contains a great deal of hydrogen, by oxidation exchanges four atoms of hydrogen for two atoms of oxygen, and becomes converted into nitric acid, $NO_2(OH)$. This process of conversion of ammonium salts into saltpetre goes on in the fields every summer, and with especial rapidity in tropical countries. The method by which this is accomplished, though complex, though involving the agency of all-permeating micro-organisms, is, in substance, the same as that by which alcohol is converted into acetic acid, or glycol, $C_2H_4(OH)_2$, into oxalic acid, if we view the process of oxidation in the light of the Newtonian principles.

But while speaking of the application of the principle of substitution to water, we need not multiply instances, but must turn our attention to two special circumstances which are closely connected with the very mechanism of substitutions.

In the first place, the replacement of two atoms of hydrogen by one atom of oxygen may take place in two ways, because the hydrogen molecule is composed of two atoms, and therefore, under the influence of oxygen, the molecule forming water may separate before the oxygen has time to take its place. It is for this reason that we find, during the conversion of alcohol into acetic acid, that there is an interval during which is formed aldehyde, C_2H_4O, which, as its very name implies, is 'alcohol dehydrogenatum,' or alcohol deprived of hydrogen. Hence aldehyde combined with hydrogen yields alcohol; and united to oxygen, acetic acid.

For the same reason there should be, and there actually are, intermediate products between ammonia and nitric acid, $NO_2(HO)$, containing either less hydrogen than ammonia, less oxygen than nitric acid, or less water than caustic ammonia. Accordingly we find, among the products of the deoxidation of nitric acid and the oxidation of ammonia, not only hydroxylamine, but also nitrous oxide, nitrous and nitric anhydrides. Thus, the production of nitrous acid results from the removal of two atoms of hydrogen from caustic ammonia and the substitution of the oxygen for the hydrogen, $NO(OH)$; or by the substitution, in ammonia, of three atoms of hydrogen by hydroxyl, $N(OH)_3$, and by the removal of water: $N(OH)_3 - H_2O = NO(OH)$. The peculiarities and properties of nitrous acid—as, for instance, its action on ammonia and its conversion, by oxidation, into nitric acid—are thus clearly revealed.

On the other hand, in speaking of the principle of substitution as applied to water, it is necessary to observe that hydrogen and hydroxyl, H and OH, are not only competent to unite, but also to form combinations with themselves, and thus become H_2 and H_2O_2; and such are hydrogen and the peroxide thereof. In general, if a molecule A B exists, then molecules A A and B B can exist also. A direct reaction of this kind does not, however, take place in water, therefore undoubtedly, at the moment of formation, hydrogen reacts on hydrogen peroxide, as we can show at once by experiment; and further because hydrogen peroxide, H_2O_2, exhibits a structure containing a molecule of hydrogen, H_2, and one of oxygen, O_2, either of which is capable of separate existence. The fact, however, may now be taken as thoroughly established, that, at the moment of combustion of hydrogen or of the hydrogen compounds, hydrogen peroxide is always formed, and not only so, but in all probability its formation invariably precedes the formation of water. This was to be expected as a consequence of the law of Avogadro and Gerhardt, which leads us to expect this sequence in the case of equal interactions of volumes of vapours and gases; and in hydrogen peroxide we actually have such equal volumes of the elementary gases.

The instability of hydrogen peroxide—that is to say, the ease with which it decomposes into water and oxygen, even at the mere contact of porous substances—accounts for the circumstance that it does not form a permanent product of combustion, and is not produced during the decomposition of water. I may mention this additional consideration that, with respect to hydrogen peroxide, we may look for its effecting still further substitutions of hydrogen by means of which we may expect to obtain still more highly oxidised water compounds, such as H_2O_3 and H_2O_4. These Schönbein and Bunsen have long been seeking, and Berthelot is investigating them at present. It is probable, however, that the reaction will stop at the last compound, because we find that, in a number of cases, the addition of four atoms of oxygen seems to form a limit. Thus, OsO_4, $KClO_4$, $KMnO_4$, K_2SO_4, Na_3PO_4, and such like, represent the highest grades of oxidation.

As for the last forty years, from the times of Berzelius, Dumas, Liebig, Gerhardt, Williamson, Frankland, Kolbe, Kekulé, and Butleroff, most theoretical generalisations have centred round organic or carbon compounds, we will, for the sake of brevity, leave out the discussion of ammonia derivatives, notwithstanding their simplicity with respect to the doctrine of substitutions; we will dwell more especially on its application to carbon compounds, starting from methane, CH_4, as the simplest of the hydrocarbons, containing in its molecule one atom of carbon. According to the principles enumerated we may derive from CH_4 every combination of the form CH_3X, CH_2X_2, CHX_3, and CX_4, in which X is an element, or radicle, equivalent to hydrogen—that is to say, competent to take its place or to combine with it. Such are the chlorine substitutes already mentioned, such is wood-spirit, $CH_3(OH)$, in which X is represented by the residue of water, and such are numerous other carbon derivatives. If we continue, with the aid of hydroxyl, further substitutions of the hydrogen of methane we shall obtain successively $CH_2(OH)_2$, $CH(OH)_3$, and $C(OH)_4$. But if, in proceeding thus, we bear in mind that $CH_2(OH)_2$ contains two hydroxyls in the same form as hydrogen peroxide, H_2O_2 or $(OH)_2$, contains them—and moreover not only in one molecule, but together, attached to one and the same atom of carbon—so here we must look for the same decomposition as that which we find in hydrogen peroxide, and accompanied also by the formation of water as an independently existing molecule; therefore $CH_2(OH)_2$ should yield, as it actually does, immediately water and the oxide of methylene, CH_2O, which is methane with oxygen substituted for two atoms of hydrogen. Exactly in the same manner out of $CH(OH)_3$ are formed water and formic acid, $CHO(OH)$, and out of $C(OH)_4$ is produced water and carbonic acid, or directly carbonic anhydride, CO_2, which will therefore be nothing else than methane with the double replacement of pairs of hydrogen by oxygen. As nothing leads to the supposition that the four atoms of hydrogen in methane differ one from the other, so it does not matter by what means we obtain any one of the combinations indicated—they will be identical; that is to say, there will be no case of actual isomerism, although there may easily be such cases of isomerism as have been distinguished by the term metamerism.

Formic acid, for example, has two atoms of hydrogen, one attached to the carbon left from the methane, and the other attached to the oxygen which has entered in the form of hydroxyl, and if one of them be replaced by some substance X it is evident that we shall obtain substances of the same composition, but of different construction, or of different orders of movement among the molecules, and therefore endowed with other properties and reactions. If X be methyl, CH_4—that is to say, a group capable of replacing hydrogen because it is actually contained with hydrogen in methane itself—then by substituting this group for the original hydrogen we obtain acetic acid, $CCH_3O(OH)$, out of formic, and by substitution of the hydrogen in its oxide or hydroxyl we obtain methyl formate, $CHO(OCH_3)$. These substances differ so much from each other physically and chemically that at first sight it is hardly possible to admit that they contain the same atoms in identically the same proportions. Acetic acid, for example, boils at a higher temperature than water, and has a higher specific gravity than it, whilst its metameride, methyl formate, is lighter than water, and boils at 30°—that is to say, it evaporates very easily.

Let us now turn to carbon compounds containing two atoms of carbon to the molecule, as in acetic acid, and proceed to evolve them from methane by the principle of substitution. This principle declares at once that methane can only be split up in the four following ways:—

1. Into a group CH_3 equivalent with H. Let us call changes of this nature methylation.

2. Into a group CH_2 and H_2. We will call this order of substitutions methylenation.

3. Into CH and H_3, which commutations we will call acetylenation.

4. Into C and H_4, which may be called carbonation.

It is evident that hydrocarbon compounds containing two atoms of carbon can only proceed from methane, CH_4, which contains four atoms of hydrogen by the first three methods of substitution; carbonation would yield free carbon if it could take place directly, and if the molecule of free carbon—which is in reality very complex, that is to say strongly polyatomic, as I have long since been proving by various means—could contain only C_2 like the molecules O_2, H_2, N_2, and so on.

By methylation we should evidently obtain from marsh gas, ethane, $CH_3CH_3 = C_2H_6$.

By methylenation—that is, by substituting group CH_2 for H_2—methane forms ethylene, $CH_2CH_2 = C_2H_4$.

By acetylenation—that is, by substituting three atoms of hydrogen, H_3, in methane—by the remnant CH, we get acetylene, $CHCH = C_2H_2$.

If we have applied the principles of Newton correctly, there should not be any other hydrocarbons containing two atoms of carbon in the molecule. All these combinations have long been known, and in each of them we can not only produce those substitutions of which an example has been given in the case of methane, but also all the phases of other substitutions, as we shall find from a few more instances, by the aid of which I trust that I shall be able to show the great complexity of those derivatives which, on the principle of substitution, can be obtained from each hydrocarbon. Let us content ourselves with the case of ethane, CH_3CH_3, and the substitution of the hydrogen by hydroxyl. The following are the possible changes:—

1. $CH_3CH_2(OH)$: this is nothing more than spirit of wine, or ethyl alcohol, $C_2H_5(OH)$ or C_2H_6O.

2. $CH_2(OH)CH_2(OH)$: this is the glycol of Würtz, which has shed so much light on the history of alcohol. Its isomeride may be $CH_3CH(OH)_2$, but as we have seen in the case of $CH(OH)_2$, it decomposes, giving off water, and forming aldehyde, CH_3CHO, a substance capable of yielding alcohol by uniting with hydrogen, and of yielding acetic acid by uniting with oxygen.

If glycol, $CH_2(OH)CH_2(OH)$, loses its water, it may be seen at once that it will not now yield aldehyde, CH_3CHO, but its isomeride, CH_2CH_2/O, the oxide of ethylene. I have here indicated in a special manner the oxygen which has taken the place of two atoms of the hydrogen of ethane taken from different atoms of the carbon.

3. $CH_3C(OH)_3$ decomposed as $CH(OH)_3$, forming water and acetic acid, $CH_3CO(OH)$. It is evident that this acid is nothing else than formic acid, $CHO(OH)$, with its hydrogen replaced by methyl. Without examining further the vast number of possible derivatives, I will direct your attention to the circumstance that in dissolving acetic acid in water we obtain the maximum contraction and the greatest viscosity when to the molecule $CH_3CO(OH)$ is added a molecule of water, which is the proportion which would form the hydrate $CH_3C(OH)_3$. It is probable that the doubling of the molecule of acetic acid at temperatures approaching its boiling-point has some connection with this power of uniting with one molecule of water.

4. $CH_2(OH)C(OH)_3$ is evidently an alcoholic acid, and indeed this compound, after losing water, answers to glycolic acid, $CH_2(OH)CO(OH)$. Without investigating all the possible isomerides, we will note only that the hydrate $CH(OH)_2CH(OH)_2$ has the same composition as $CH_2(OH)C(OH)_3$, and although corresponding to glycol, and being a symmetrical substance, it becomes, on parting with its water, the aldehyde of oxalic acid, or the glyoxal of Debus, $CHOCHO$.

5. $CH(OH)_2C(OH_3)$, from the tendency of all the preceding, corresponds with glyoxylic acid, an aldehyde acid, $CHOCO(OH)$, because the group $CO(OH)$, or carboxyl, enters into the compositions of organic acids, and the group CHO defines the aldehyde function.

6. $C(OH)_3C(OH)_3$ through the loss of $2H_2O$ yields the bibasic oxalic acid $CO(OH)CO(OH)$, which generally crystallises with $2H_2O$, following thus the normal type of hydration characteristic of ethane.

Thus, by applying the principle of substitution, we can, in the simplest manner, derive not only every kind of hydrocarbon compound, such as the alcohols, the aldehyde-alcohols, aldehydes, alcohol-acids, and the acids, but also combinations analogous to hydrated crystals which usually are disregarded.

But even those unsaturated substances, of which ethylene, CH_2CH_2, and acetylene, CHCH, are types, may be evolved with equal simplicity. With respect to the phenomena of isomerism, there are many possibilities among the hydrocarbon compounds containing two atoms of carbon, and without going into details it will be sufficient to indicate that the following formulæ, though not identical, will be isomeric substantially among themselves:— CH_3CHX_2 and CH_2XCH_2X, although both contain $C_2H_4X_2$; or CH_2CX_2 and CHXCHX, although both contain $C_2H_2X_2$, if by X we indicate chlorine or generally an element capable of replacing one atom of hydrogen, or capable of uniting with it. To isomerism of this kind belongs the case of aldehyde and the oxide of ethylene, to which we have already referred, because both have the composition C_2H_4O.

What I have said appears to me sufficient to show that the principle of substitution adequately explains the composition, the isomerism, and all the diversity of combination of the hydrocarbons, and I shall limit the further development of these views to preparing a complete list of every possible hydrocarbon compound containing three atoms of carbon in the molecule. There are eight in all, of which only five are known at present.

Among those possible for C_3H_6 there should be two isomerides, propylene and trimethylene, and they are both already known. For C_3H_4 there should be three isomerides: allylene and allene are known, but the third has not yet been discovered; and for C_3H_2 there should be two isomerides, though neither of them is known as yet. Their composition and structure are easily deduced from ethane, ethylene, and acetylene, by methylation, by methylenation, by acetylenation and by carbonation.

1. $C_3H_8 = CH_3CH_2CH_3$ out of CH_3CH_3 by methylation. This hydrocarbon is named propane.

2. $C_3H_6 = CH_3CHCH_2$ out of CH_3CH_3 by methylenation. This substance is propylene.

3. $C_3H_6 = CH_2CH_2CH_2$ out of CH_3CH_3 by methylenation. This substance is trimethylene.

4. $C_3H_4 = CH_3CCH$ out of CH_3CH_3 by acetylenation or from CHCH by methylation. This hydrocarbon is named allylene.

5. $C_3H_4 = $ CHCH CH_2 out of CH_3CH_3 by acetylenation, or from CH_2CH_2 by methylenation, because CH_2CH CH = CHCH CH_2 . This body is as yet unknown.

6. $C_3H_4 = CH_2CCH_2$ out of CH_2CH_2 by methylenation. This hydrocarbon is named allene, or iso-allylene.

7. C_3H_2 = CHCH C out of CH_3CH_3 by symmetrical carbonation, or out of CH_2CH_2 by acetylenation. This compound is unknown.

8. C_3H_2 = CC CH_2 out of CH_3CH_3 by carbonation, or out of CHCH by methylenation. This compound is unknown.

If we bear in mind that for each hydrocarbon serving as a type in the above tables there are a number of corresponding derivatives, and that every compound obtained may, by further methylation, methylenation, acetylenation, and carbonation, produce new hydrocarbons, and these may be followed by a numerous suite of derivatives and an immense number of isomeric substances, it is possible to understand the limitless number of carbon compounds, although they all have the one substance, methane, for their origin. The number of substances is so enormous that it is no longer a question of enlarging the possibilities of discovery, but rather of finding some means of testing them analogous to the well-known two which for a long time have served as gauges for all carbon compounds.

I refer to the law of even numbers and to that of limits, the first enunciated by Gerhardt some forty years ago, with respect to hydrocarbons, namely, that their molecules always contain an even number of atoms of hydrogen. But by the method which I have used of deriving all the hydrocarbons from methane, CH_4, this law may be deduced as a direct consequence of the principle of substitutions. Accordingly, in methylation, CH_3 takes the place of H, and therefore CH_2 is added. In methylenation the number of atoms of hydrogen remains unchanged, and at each acetylenation it is reduced by two, and in carbonation by four, atoms—that is to say, an even number of atoms of hydrogen is always added or removed. And because the fundamental hydrocarbon, methane, CH_4, contains an even number of atoms of hydrogen, all its derivative hydrocarbons will also contain even numbers of hydrogen, and this constitutes the law of even numbers.

The principle of substitutions explains with equal simplicity the conception of the limiting compositions of hydrocarbons C_nH_{2n+2}, which I derived, in 1861, in an empirical manner from accumulated materials available at that time, and on the basis of the limits to combinations worked out by Dr. Frankland for other elements.

Of all the various substitutions the highest proportion of hydrogen is yielded by methylation, because in that operation alone does the quantity of hydrogen increase; hence, taking methane as a point of departure, if we imagine methylation effected (n - 1) times we obtain hydrocarbon compounds containing the highest quantities of hydrogen. It is evident that they will contain CH_4 + (n - 1)CH_2, or C_nH_{2n+2}, because methylation leads to the addition of CH_2 to the compound.

It will thus be seen that by the principle of substitution—that is to say, by the third law of Newton—we are able to deduce, in the simplest manner, not only the individual composition, the isomerism, and relations of substances, but also the general laws which govern their most complex combinations without having recourse either to statical constructions, to the definition of atomicities, to the exclusion of free affinities, or to the recognition of those single, double or treble bonds which are so indispensable to structuralists in the explanation of the composition and construction of hydrocarbon compounds. And yet, by the application of the dynamical principles of Newton, we can attain to that chief and fundamental object, the comprehension of isomerism in hydrocarbon compounds, and the forecasting of the existence of combinations as yet unknown, by which the edifice raised by structural teaching is strengthened and supported. Besides—and I count this for a circumstance of special importance—the process which I advocate will make no difference in those special cases which have been already so well worked out, such as, for example, the isomerism of the hydrocarbons and alcohols, even to the extent of not interfering with the nomenclature which has been adopted, and the structural system will retain all the glory of having worked up, in a thoroughly scientific manner, the store of information which Gerhardt had accumulated about the middle of the fifties, and the still higher glory of establishing the rational synthesis of organic substances. Nothing will be lost to the structural doctrine except its statical origin; and as soon as it will embrace the dynamic principles of Newton, and suffer itself to be guided by them, I believe that we shall attain for chemistry that unity of principle which is now wanting. Many an adept will be attracted to that brilliant and fascinating enterprise, the penetration into the unseen world of the kinetic relations of atoms, to the study of which the last twenty-five years have contributed so much labour and such high inventive faculties.

D'Alembert found in mechanics that if inertia be taken to represent force, dynamic equations may be applied to statical questions, which are thereby rendered more simple and more easily understood.

The structural doctrine in chemistry has unconsciously followed the same course, and therefore its terms are easily adopted; they may retain their present forms provided that a truly dynamical—that is to say, Newtonian—meaning be ascribed to them.

Before finishing my task and demonstrating the possibility of adapting structural doctrines to the dynamics of Newton, I consider it indispensable to touch on one question which naturally arises, and which I have heard discussed more than once. If bromine, the atom of which is eighty times heavier than that of hydrogen, takes the place of hydrogen, it would seem that the whole system of dynamic equilibrium must be destroyed.

Without entering into the minute analysis of this question, I think it will be sufficient to examine it by the light of two well-known phenomena, one of which will be found in the department of chemistry and the other in that of celestial mechanics, and both will serve to demonstrate the existence of that unity in the plan of creation which is a consequence of the Newtonian doctrines. Experiments demonstrate that when a heavy element is substituted for a light one in a chemical compound—for example, for magnesium, in the oxide of that metal, an atom of mercury, which is $8\frac{1}{3}$ times heavier—the chief chemical characteristics or properties are generally, though not always, preserved.

The substitution of silver for hydrogen, than which it is 108 times heavier, does not affect all the properties of the substance, though it does some. Therefore chemical substitutions of this kind—the substitution of light for heavy atoms—need not necessarily entail changes in the original equilibrium; and this point is still further elucidated by the consideration that the periodic law indicates the degree of influence of an increment of weight in the atom as affecting the possible equilibria, and also what degree of increase in the weight of the atoms reproduces some, though not all, of the properties of the substance.

This tendency to repetition—these periods—may be likened to those annual or diurnal periods with which we are so familiar on the earth. Days and years follow each other, but, as they do so, many things change; and in like manner chemical evolutions, changes in the masses of the elements, permit of much remaining undisturbed, though many properties undergo alteration. The system is maintained according to the laws of conservation in nature, but the motions are altered in consequence of the change of parts.

Next, let us take an astronomical case—such, for example, as the earth and the moon—and let us imagine that the mass of the latter is constantly increasing. The question is, what will then occur? The path of the moon in space is a wave-line similar to that which geometricians have named epicycloidal, or the locus of a point in a circle rolling round another circle. But in consequence of the influence of the moon it is evident that the path of the earth itself cannot be a geometric ellipse, even supposing the sun to be immovably fixed; it must be an epicycloidal curve, though not very far removed from the true ellipse—that is to say, it will be impressed with but faint undulations. It is only the common centre of gravity of the earth and the moon which describes a true ellipse round the sun. If the moon were to increase, the relative undulations of the earth's path would increase in amplitude, those of the moon would also change, and when the mass of the moon had increased to an equality with that of the earth, the path would consist of epicycloidal curves crossing each other, and having opposite phases. But a similar relation exists between the sun and the earth, because the former is also moving in space. We may apply these views to the world of atoms, and suppose that in their movements, when heavy ones take the place of those that are lighter, similar changes take place, provided that the system or the molecule is preserved throughout the change.

It seems probable that in the heavenly systems, during incalculable astronomical periods, changes have taken place and are still going on similar to those which pass rapidly before our eyes during the chemical reaction of molecules, and the progress of molecular mechanics may—we hope will—in course of time permit us to explain those changes in the stellar world which have more than once been noticed by astronomers, and which are now so carefully studied. A coming Newton will discover the laws of these changes. Those laws, when applied to chemistry, may exhibit peculiarities, but these will certainly be mere variations on the grand harmonious theme which reigns in nature. The discovery of the laws which produce this harmony in chemical evolution will only be possible, it seems to me, under the banner of Newtonian dynamics, which has so long waved over the domains of mechanics, astronomy, and physics. In calling chemists to take their stand under its peaceful and catholic shadow I imagine that I am aiding in establishing that scientific union which the managers of the Royal Institution wish to effect, who have shown their desire to do so by the flattering invitation which has given me—a Russian—the opportunity of laying before the countrymen of Newton an attempt to apply to chemistry one of his immortal principles.

Footnotes:

Because more than four atoms of hydrogen never unite with one atom of the elements, and because the hydrogen compounds (*e.g.* HCl, H_2S, H_3P, H_4Si) always form their highest oxides with four atoms of oxygen, and as the highest forms of oxides (OsO_4, RuO_4) also contain four of oxygen, and eight groups of the periodic system, corresponding to the highest basic oxides R_2O, RO, R_2O_3, RO_2, R_2O_5, RO_3, R_2O_7, and RO_4, imply the above relationship, and because of the nearest analogues among the elements—such as Mg, Zn, Cd, and Hg; or Cr, Mo, W, and U; or Si, Ge, Sn, and Pt; or F, Cl, Br, and I, and so forth—not more than four are known, it seems to me that in these relationships there lies a deep interest and meaning with regard to chemical mechanics. But because, to my imagination, the idea of unity of design in Nature, either acting in complex celestial systems or among chemical molecules, is very attractive, especially because the atomic teaching at once acquires its true meaning, I will recall the following facts relating to the solar system. There are eight major planets, of which the four inner ones are not only separated from the four outer by asteroids, but differ from them in many respects, as, for example, in the smallness of their diameters and their greater density. Saturn with his ring has eight satellites, Jupiter and Uranus have each four. It is evident that in the solar systems also we meet with these higher numbers four and eight which appear in the combination of chemical molecules.

One more isomeride, $CH_2CH(OH)$, is possible—that is, secondary vinyl alcohol, which is related to ethylene, CH_2CH_2, but derived by the principle of substitution from CH_4. Other isomerides, of the composition C_2H_4O, such, for example, as $CCH_3(OH)$, are impossible, because it would correspond with the hydrocarbon $CHCH_3 = C_2H_4$, which is isomeric with ethylene, and it cannot be derived from methane. If such an isomeride existed it would be derived from CH_2, but such products are, up to the present, unknown. In such cases the insufficiency of the points of departure of the statical structural teaching is shown. It first admits constant atomicity and then rejects it, the facts serving to establish either one or the other view; and therefore it seems to me that we must come to the conclusion that the structural method of reasoning, having done a service to science, has outlived the age, and must be regenerated, as in their time was the teaching of the electro-chemists, the radicalists, and the adherents of the doctrine of types. As we cannot now lean on the views above stated, it is time to abandon the structural theory. They will all be united in chemical mechanics, and the principle of substitution must be looked on only as a preparation for the coming epoch in chemistry, where such cases as the isomerism of fumaric and maleic acids, when explained dynamically, as proposed by Le Bel and Van't Hoff, may yield points of departure.

Conceding variable atomicity, the structuralists must expect an incomparably larger number of isomerides, and they cannot now decline to acknowledge the change of atomicity, were it only for the examples $HgCl$ and $HgCl_2$, CO and CO_2, PCl_3 and PCl_5.

'Essai d'une théorie sur les limites des combinaisons organiques,' par D. Mendeléeff, 2/11 août 1861, *Bulletin de l'Académie i. d. Sc. de St. Pétersbourg*, t. v

APPENDIX II

THE PERIODIC LAW OF THE CHEMICAL ELEMENTS

By PROFESSOR MENDELÉEFF

FARADAY LECTURE DELIVERED BEFORE THE FELLOWS OF THE CHEMICAL SOCIETY IN THE THEATRE OF THE ROYAL INSTITUTION, ON TUESDAY, JUNE 4, 1889

The high honour bestowed by the Chemical Society in inviting me to pay a tribute to the world-famed name of Faraday by delivering this lecture has induced me to take for its subject the Periodic Law of the Elements—this being a generalisation in chemistry which has of late attracted much attention.

While science is pursuing a steady onward movement, it is convenient from time to time to cast a glance back on the route already traversed, and especially to consider the new conceptions which aim at discovering the general meaning of the stock of facts accumulated from day to day in our laboratories. Owing to the possession of laboratories, modern science now bears a new character, quite unknown, not only to antiquity, but even to the preceding century. Bacon's and Descartes' idea of submitting the mechanism of science simultaneously to experiment and reasoning has been fully realised in the case of chemistry, it having become not only possible but always customary to experiment. Under the all-penetrating control of experiment, a new theory, even if crude, is quickly strengthened, provided it be founded on a sufficient basis; the asperities are removed, it is amended by degrees, and soon loses the phantom light of a shadowy form or of one founded on mere prejudice; it is able to lead to logical conclusions, and to submit to experimental proof. Willingly or not, in science we all must submit not to what seems to us attractive from one point of view or from another, but to what represents an agreement between theory and experiment; in other words, to demonstrated generalisation and to the approved experiment. Is it long since many refused to accept the generalisations involved in the law of Avogadro and Ampère, so widely extended by Gerhardt? We still may hear the voices of its opponents; they enjoy perfect freedom, but vainly will their voices rise so long as they do not use the language of demonstrated facts The striking observations with the spectroscope which have permitted us to analyse the chemical constitution of distant worlds, seemed, at first, applicable to the task of determining the nature of the atoms themselves; but the working out of the idea in the laboratory soon demonstrated that the characters of spectra are determined, not directly by the atoms, but by the molecules into which the atoms are packed; and so it became evident that more verified facts must be collected before it will be possible to formulate new generalisations capable of taking their place beside those ordinary ones based upon the conception of simple substances and atoms. But as the shade of the leaves and roots of living plants, together with the relics of a decayed vegetation, favour the growth of the seedling and serve to promote its luxurious development, in like manner sound generalisations—together with the relics of those which have proved to be untenable—promote scientific productivity, and ensure the luxurious growth of science under the influence of rays emanating from the centres of scientific energy. Such centres are scientific associations and societies. Before one of the oldest and most powerful of these I am about to take the liberty of passing in review the twenty years' life of a generalisation which is known under the name of the Periodic Law. It was in March 1869 that I ventured to lay before the then youthful Russian Chemical Society the ideas upon the same subject which I had expressed in my just written 'Principles of Chemistry.'

Without entering into details, I will give the conclusions I then arrived at in the very words I used:—

'1. The elements, if arranged according to their atomic weights, exhibit an evident *periodicity* of properties.

'2. Elements which are similar as regards their chemical properties have atomic weights which are either of nearly the same value (*e.g.* platinum, iridium, osmium) or which increase regularly (*e.g.* potassium, rubidium, cæsium).

'3. The arrangement of the elements, or of groups of elements, in the order of their atomic weights, corresponds to their so-called *valencies* as well as, to some extent, to their distinctive chemical properties—as is apparent, among other series, in that of lithium, beryllium, barium, carbon, nitrogen, oxygen, and iron.

'4. The elements which are the most widely diffused have *small* atomic weights.

'5. The *magnitude* of the atomic weight determines the character of the element, just as the magnitude of the molecule determines the character of a compound.

'6. We must expect the discovery of many yet *unknown* elements—for example, elements analogous to aluminium and silicon, whose atomic weight would be between 65 and 75.

'7. The atomic weight of an element may sometimes be amended by a knowledge of those of the contiguous elements. Thus, the atomic weight of tellurium must lie between 123 and 126, and cannot be 128.

'8. Certain characteristic properties of the elements can be foretold from their atomic weights.

'The aim of this communication will be fully attained if I succeed in drawing the attention of investigators to those relations which exist between the atomic weights of dissimilar elements, which, so far as I know, have hitherto been almost completely neglected. I believe that the solution of some of the most important problems of our science lies in researches of this kind.'

To-day, twenty years after the above conclusions were formulated, they may still be considered as expressing the essence of the now well-known periodic law.

Reverting to the epoch terminating with the sixties, it is proper to indicate three series of data without the knowledge of which the periodic law could not have been discovered, and which rendered its appearance natural and intelligible.

In the first place, it was at that time that the numerical value of atomic weights became definitely known. Ten years earlier such knowledge did not exist, as may be gathered from the fact that in 1860 chemists from all parts of the world met at Karlsruhe in order to come to some agreement, if not with respect to views relating to atoms, at any rate as regards their definite representation. Many of those present probably remember how vain were the hopes of coming to an understanding, and how much ground was gained at that Congress by the followers of the unitary theory so brilliantly represented by Cannizzaro. I vividly remember the impression produced by his speeches, which admitted of no compromise, and seemed to advocate truth itself, based on the conceptions of Avogadro, Gerhardt, and Regnault, which at that time were far from being generally recognised. And though no understanding could be arrived at, yet the objects of the meeting were attained, for the ideas of Cannizzaro proved, after a few years, to be the only ones which could stand criticism, and which represented an atom as— 'the smallest portion of an element which enters into a molecule of its compound.' Only such real atomic weights—not conventional ones—could afford a basis for generalisation. It is sufficient, by way of example, to indicate the following cases in which the relation is seen at once and is perfectly clear:—

$K = 39$ $Rb = 85$ $Cs = 133$

$Ca = 40$ $Sr = 87$ $Ba = 137$

whereas with the equivalents then in use—

K = 39   Rb = 85   Cs = 133

Ca = 20 Sr = 43·5 Ba = 68·5

the consecutiveness of change in atomic weight, which with the true values is so evident, completely disappears.

Secondly, it had become evident during the period 1860–70, and even during the preceding decade, that the relations between the atomic weights of analogous elements were governed by some general and simple laws. Cooke, Cremers, Gladstone, Gmelin, Lenssen, Pettenkofer, and especially Dumas, had already established many facts bearing on that view. Thus Dumas compared the following groups of analogous elements with organic radicles:—

	Diff.		Diff.		Diff.		Diff.
		Mg = 12		P = 81		O = 8	
			$\vert$ 8		$\vert$ 44		$\vert$ 8
Li = 7		Ca = 20		As = 75		S = 16	
	$\vert$ 16		$\vert$ 3 × 8		$\vert$ 44		$\vert$ 3 × 8
Na = 23		Sr = 44		Sb = 119		Se = 40	
	$\vert$ 16		$\vert$ 3 × 8		$\vert$ 2 × 44		$\vert$ 3 × 8
K = 39		Ba = 68		Bi = 207		Te = 64	

and pointed out some really striking relationships, such as the following:—

F = 19.

Cl = 35·5 = 19 + 16·5.

Br = 80 = 19 + 2 × 16·5 + 28.

I = 127 = 2 x 19 + 2 × 16·5 + 2 × 28.

A. Strecker, in his work 'Theorien und Experimente zur Bestimmung der Atomgewichte der Elemente' (Braunschweig, 1859), after summarising the data relating to the subject, and pointing out the remarkable series of equivalents—

Cr = 26·2 Mn = 27·6 Fe = 28 Ni = 29 Co = 30 Cu = 31·7 Zn = 32·5

remarks that: 'It is hardly probable that all the above-mentioned relations between the atomic weights (or equivalents) of chemically analogous elements are merely accidental. We must, however, leave to the future the discovery of the *law* of the relations which appears in these figures.'

In such attempts at arrangement and in such views are to be recognised the real forerunners of the periodic law; the ground was prepared for it between 1860 and 1870, and that it was not expressed in a determinate form before the end of the decade may, I suppose, be ascribed to the fact that only analogous elements had been compared. The idea of seeking for a relation between the atomic weights of all the elements was foreign to the ideas then current, so that neither the *vis tellurique* of De Chancourtois, nor the *law of octaves* of Newlands, could secure anybody's attention. And yet both De Chancourtois and Newlands like Dumas and Strecker, more than Lenssen and Pettenkofer, had made an approach to the periodic law and had discovered its germs. The solution of the problem advanced but slowly, because the facts, but not the law, stood foremost in all attempts; and the law could not awaken a general interest so long as elements, having no apparent connection with each other, were included in the same octave, as for example:—

	H	F	Cl	Co & Ni	Br	Pd	Ir	Pt & Ir
1st octave of Newlands	H	F	Cl	Co & Ni	Br	Pd	Ir	Pt & Ir
7th Ditto	O	S	Fe	Se	Rh & Ru	Te	Au	Os or Th

Analogies of the above order seemed quite accidental, and the more so as the octave contained occasionally ten elements instead of eight, and when two such elements as Ba and V, Co and Ni, or Rh and Ru, occupied one place in the octave. Nevertheless, the fruit was ripening, and I now see clearly that Strecker, De Chancourtois, and Newlands stood foremost in the way towards the discovery of the periodic law, and that they merely wanted the boldness necessary to place the whole question at such a height that its reflection on the facts could be clearly seen.

A third circumstance which revealed the periodicity of chemical elements was the accumulation, by the end of the sixties, of new information respecting the rare elements, disclosing their many-sided relations to the other elements and to each other. The researches of Marignac on niobium, and those of Roscoe on vanadium, were of special moment. The striking analogies between vanadium and phosphorus on the one hand, and between vanadium and chromium on the other, which became so apparent in the investigations connected with that element, naturally induced the comparison of V = 51 with Cr = 52, Nb = 94 with

Mo = 96, and Ta = 192 with W = 194; while, on the other hand, P = 31 could be compared with S = 32, As = 75 with Se = 79, and Sb = 120 with Te = 125. From such approximations there remained but one step to the discovery of the law of periodicity.

The law of periodicity was thus a direct outcome of the stock of generalisations and established facts which had accumulated by the end of the decade 1860–1870; it is an embodiment of those data in a more or less systematic expression. Where, then, lies the secret of the special importance which has since been attached to the periodic law, and has raised it to the position of a generalisation which has already given to chemistry unexpected aid, and which promises to be far more fruitful in the future and to impress upon several branches of chemical research a peculiar and original stamp? The remaining part of my communication will be an attempt to answer this question.

In the first place we have the circumstance that, as soon as the law made its appearance, it demanded a revision of many facts which were considered by chemists as fully established by existing experience. I shall return, later on, briefly to this subject, but I wish now to remind you that the periodic law, by insisting on the necessity for a revision of supposed facts, exposed itself at once to destruction in its very origin. Its first requirements, however, have been almost entirely satisfied during the last 20 years; the supposed facts have yielded to the law, thus proving that the law itself was a legitimate induction from the verified facts. But our inductions from data have often to do with such details of a science so rich in facts, that only generalisations which cover a wide range of important phenomena can attract general attention. What were the regions touched on by the periodic law? This is what we shall now consider.

The most important point to notice is, that periodic functions, used for the purpose of expressing changes which are dependent on variations of time and space, have been long known. They are familiar to the mind when we have to deal with motion in closed cycles, or with any kind of deviation from a stable position, such as occurs in pendulum-oscillations. A like periodic function became evident in the case of the elements, depending on the mass of the atom. The primary conception of the masses of bodies, or of the masses of atoms, belongs to a category which the present state of science forbids us to discuss, because as yet we have no means of dissecting or analysing the conception. All that was known of functions dependent on masses derived its origin from Galileo and Newton, and indicated that such functions either decrease or increase with the increase of mass, like the attraction of celestial bodies. The numerical expression of the phenomena was always found to be proportional to the mass, and in no case was an increase of mass followed by a recurrence of properties such as is disclosed by the periodic law of the elements. This constituted such a novelty in the study of the phenomena of nature that, although it did not lift the veil which conceals the true conception of mass, it nevertheless indicated that the explanation of that conception must be searched for in the masses of the atoms; the more so, as all masses are nothing but aggregations, or additions, of chemical atoms which would be best described as chemical individuals. Let me remark, by the way, that though the Latin word 'individual' is merely a translation of the Greek word 'atom,' nevertheless history and custom have drawn a sharp distinction between the two words, and the present chemical conception of atoms is nearer to that defined by the Latin word than by the Greek, although this latter also has acquired a special meaning which was unknown to the classics. The periodic law has shown that our chemical individuals display a harmonic periodicity of properties dependent on their masses. Now natural science has long been accustomed to deal with periodicities observed in nature, to seize them with the vice of mathematical analysis, to submit them to the rasp of experiment. And these instruments of scientific thought would surely, long since, have mastered the problem connected with the chemical elements, were it not for a new feature which was brought to light by the periodic law, and which gave a peculiar and original character to the periodic function.

If we mark on an axis of abscissæ a series of lengths proportional to angles, and trace ordinates which are proportional to sines or other trigonometrical functions, we get periodic curves of a harmonic character. So it might seem, at first sight, that with the increase of atomic weights the function of the properties of the elements should also vary in the same harmonious way. But in this case there is no such continuous change as in the curves just referred to, because the periods do not contain the infinite number of points constituting a curve, but a *finite* number only of such points. An example will better illustrate this view. The atomic weights—

$$Ag = 108 \quad Cd = 112 \quad In = 113 \quad Sn = 118 \quad Sb = 120 \quad Te = 125 \quad I = 127$$

steadily increase, and their increase is accompanied by a modification of many properties which constitutes the essence of the periodic law. Thus, for example, the densities of the above elements decrease steadily, being respectively—

$$10·5 \quad 8·6 \quad 7·4 \quad 7·2 \quad 6·7 \quad 6·4 \quad 4·9$$

while their oxides contain an increasing quantity of oxygen—

$$Ag_2O \quad Cd_2O_2 \quad In_2O_3 \quad Sn_2O_4 \quad Sb_2O_5 \quad Te_2O_6 \quad I_2O_7$$

But to connect by a curve the summits of the ordinates expressing any of these properties would involve the rejection of Dalton's law of multiple proportions. Not only are there no intermediate elements between silver, which gives AgCl, and cadmium, which gives $CdCl_2$, but, according to the very essence of the periodic law, there can be none; in fact a uniform curve would be inapplicable in such a case, as it would lead us to expect elements possessed of special properties at any point of the curve. The periods of the elements have thus a character very different from those which are so simply represented by geometers. They correspond to points, to numbers, to sudden changes of the masses, and not to a continuous evolution. In these sudden changes destitute of intermediate steps or positions, in the absence of elements intermediate between, say, silver and cadmium, or aluminium and silicon, we must recognise a problem to which no direct application of the analysis of the infinitely small can be made. Therefore, neither the trigonometrical functions proposed by Ridberg and Flavitzky, nor the pendulum-oscillations suggested by Crookes, nor the cubical curves of the Rev. Mr. Haughton, which have been proposed for expressing the periodic law, from the nature of the case, can represent the periods of the chemical elements. If geometrical analysis is to be applied to this subject, it will require to be modified in a special manner. It must find the means of representing in a special way, not only such long periods as that comprising

K Ca Sc Ti V Cr Mn Fe Co Ni Cu Zn Ga Ge As
                            Se Br,

but short periods like the following:—

Na Mg Al Si P S Cl.

In the theory of numbers only do we find problems analogous to ours, and two attempts at expressing the atomic weights of the elements by algebraic formulæ seem to be deserving of attention, although neither of them can be considered as a complete theory, nor as promising finally to solve the problem of the periodic law. The attempt of E. J. Mills (1886) does not even aspire to attain this end. He considers that all atomic weights can be expressed by a logarithmic function,

$$15(n - 0.9375^t),$$

in which the variables n and t are *whole numbers*. Thus, for oxygen, $n = 2$, and $t = 1$, whence its atomic weight is $= 15.94$; in the case of chlorine, bromine, and iodine, n has respective values of 3, 6, and 9, whilst $t = 7, 6$, and 9; in the case of potassium, rubidium, and cæsium, $n = 4, 6$, and 9, and $t = 14, 18$, and 20.

Another attempt was made in 1888 by B. N. Tchitchérin. Its author places the problem of the periodic law in the first rank, but as yet he has investigated the alkali metals only. Tchitchérin first noticed the simple relations existing between the atomic volumes of all alkali metals; they can be expressed, according to his views, by the formula

$$A(2 - 0{\cdot}00535An),$$

where A is the atomic weight, and n is equal to 8 for lithium and sodium, to 4 for potassium, to 3 for rubidium, and to 2 for cæsium. If n remained equal to 8 during the increase of A, the volume would become zero at $A = 46\frac{2}{3}$, and it would reach its maximum at $A = 23\frac{1}{3}$. The close approximation of the number $46\frac{2}{3}$ to the differences between the atomic weights of analogous elements (such as Cs - Rb, I - Br, and so on); the close correspondence of the number $23\frac{1}{3}$ to the atomic weight of sodium; the fact of n being necessarily a whole number, and several other aspects of the question, induce Tchitchérin to believe that they afford a clue to the understanding of the nature of the elements; we must, however, await the full development of his theory before pronouncing judgment on it. What we can at present only be certain of is this: that attempts like the two above named must be repeated and multiplied, because the periodic law has clearly shown that the masses of the atoms increase abruptly, by steps, which are clearly connected in some way with Dalton's law of multiple proportions; and because the periodicity of the elements finds expression in the transition from RX to RX_2, RX_3, RX_4, and so on till RX_8, at which point, the energy of the combining forces being exhausted, the series begins anew from RX to RX_2, and so on.

While connecting by new bonds the theory of the chemical elements with Dalton's theory of multiple proportions, or atomic structure of bodies, the periodic law opened for natural philosophy a new and wide field for speculation. Kant said that there are in the world 'two things which never cease to call for the admiration and reverence of man: the moral law within ourselves, and the stellar sky above us.' But when we turn our thoughts towards the nature of the elements and the periodic law, we must add a third subject, namely, 'the nature of the elementary individuals which we discover everywhere around us.' Without them the stellar sky itself is inconceivable; and in the atoms we see at once their peculiar individualities, the infinite multiplicity of the individuals, and the submission of their seeming freedom to the general harmony of Nature.

Having thus indicated a new mystery of Nature, which does not yet yield to rational conception, the periodic law, together with the revelations of spectrum analysis, have contributed to again revive an old but remarkably long-lived hope—that of discovering, if not by experiment, at least by a mental effort, the *primary matter*—which had its genesis in the minds of the Grecian philosophers, and has been transmitted, together with many other ideas of the classic period, to the heirs of their civilisation. Having grown, during the times of the alchemists up to the period when experimental proof was required, the idea has rendered good service; it induced those careful observations and experiments which later on called into being the works of Scheele, Lavoisier, Priestley, and Cavendish. It then slumbered awhile, but was soon awakened by the attempts either to confirm or to refute the ideas of Prout as to the multiple proportion relationship of the atomic weights of all the elements. And once again the inductive or experimental method of studying Nature gained a direct advantage from the old Pythagorean idea: because atomic weights were determined with an accuracy formerly unknown. But again the idea could not stand the ordeal of experimental test, yet the prejudice remains and has not been uprooted, even by Stas; nay, it has gained a new vigour, for we see that all which is imperfectly worked out, new and unexplained, from the still scarcely studied rare metals to the hardly perceptible nebulæ, have been used to justify it. As soon as spectrum analysis appears as a new and powerful weapon of chemistry, the idea of a primary matter is immediately attached to it. From all sides we see attempts to constitute the imaginary substance *helium* the so much longed for primary matter. No attention is paid to the circumstance that the helium line is only seen in the spectrum of the solar protuberances, so that its universality in Nature remains as problematic as the primary matter itself; nor to the fact that the helium line is wanting amongst the Fraunhofer lines of the solar spectrum, and thus does not answer to the brilliant fundamental conception which gives its real force to spectrum analysis.

And finally, no notice is even taken of the indubitable fact that the brilliancies of the spectral lines of the simple substances vary under different temperatures and pressures; so that all probabilities are in favour of the helium line simply belonging to some long since known element placed under such conditions of temperature, pressure, and gravity as have not yet been realised in our experiments. Again, the idea that the excellent investigations of Lockyer of the spectrum of iron can be interpreted in favour of the compound nature of that element, evidently must have arisen from some misunderstanding. The spectrum of a compound certainly does not appear as a sum of the spectra of its components; and therefore the observations of Lockyer can be considered precisely as a proof that iron undergoes no other changes at the temperature of the sun than those which it experiences in the voltaic arc—provided the spectrum of iron is preserved. As to the shifting of some of the lines of the spectrum of iron while the other lines maintain their positions, it can be explained, as shown by M. Kleiber ('Journal of the Russian Chemical and Physical Society,' 1885, 147), by the relative motion of the various strata of the sun's atmosphere, and by Zöllner's laws of the relative brilliancies of different lines of the spectrum. Moreover, it ought not to be forgotten that if iron were really proved to consist of two or more unknown elements, we should simply have an increase in the number of our elements—not a reduction, and still less a reduction of all of them to one single primary matter.

Feeling that spectrum analysis will not yield a support to the Pythagorean conception, its modern promoters are so bent upon its being confirmed by the periodic law, that the illustrious Berthelot, in his work 'Les origines de l'Alchimie,' 1885, 313, has simply mixed up the fundamental idea of the law of periodicity with the ideas of Prout, the alchemists, and Democritus about primary matter. But the periodic law, based as it is on the solid and wholesome ground of experimental research, has been evolved independently of any conception as to the nature of the elements; it does not in the least originate in the idea of a unique matter; and it has no historical connection with that relic of the torments of classical thought, and therefore it affords no more indication of the unity of matter or of the compound character of our elements, than the law of Avogadro, or the law of specific heats, or even the conclusions of spectrum analysis. None of the advocates of a unique matter have ever tried to explain the law from the standpoint of ideas taken from a remote antiquity when it was found convenient to admit the existence of many gods—and of a unique matter.

When we try to explain the origin of the idea of a unique primary matter, we easily trace that in the absence of inductions from experiment it derives its origin from the scientifically philosophical attempt at discovering some kind of unity in the immense diversity of individualities which we see around. In classical times such a tendency could only be satisfied by conceptions about the immaterial world. As to the material world, our ancestors were compelled to resort to some hypothesis, and they adopted the idea of unity in the formative material, because they were not able to evolve the conception of any other possible unity in order to connect the multifarious relations of matter. Responding to the same legitimate scientific tendency, natural science has discovered throughout the universe a unity of plan, a unity of forces, and a unity of matter, and the convincing conclusions of modern science compel every one to admit these kinds of unity. But while we admit unity in many things, we none the less must also explain the individuality and the apparent diversity which we cannot fail to trace everywhere. It has been said of old, 'Give us a fulcrum, and it will become easy to displace the earth.' So also we must say, 'Give us something that is individualised, and the apparent diversity will be easily understood.' Otherwise, how could unity result in a multitude?

After a long and painstaking research, natural science has discovered the individualities of the chemical elements, and therefore it is now capable not only of analysing, but also of synthesising; it can understand and grasp generality and unity, as well as the individualised and the multifarious. The general and universal, like time and space, like force and motion, vary uniformly; the uniform admit of interpolations, revealing every intermediate phase. But the multitudinous, the individualised—such as ourselves, or the chemical elements, or the members of a peculiar periodic function of the elements, or Dalton's multiple proportions—is characterised in another way: we see in it, side by side with a connecting general principle, leaps, breaks of continuity, points which escape from the analysis of the infinitely small—an absence of complete intermediate links. Chemistry has found an answer to the question as to the causes of multitudes; and while retaining the conception of many elements, all submitted to the discipline of a general law, it offers an escape from the Indian Nirvana—the absorption in the universal, replacing it by the individualised. However, the place for individuality is so limited by the all-grasping, all-powerful universal, that it is merely a point of support for the understanding of multitude in unity.

Having touched upon the metaphysical bases of the conception of a unique matter which is supposed to enter into the composition of all bodies, I think it necessary to dwell upon another theory, akin to the above conception—the theory of the compound character of the elements now admitted by some—and especially upon one particular circumstance which, being related to the periodic law, is considered to be an argument in favour of that hypothesis.

Dr. Pelopidas, in 1883, made a communication to the Russian Chemical and Physical Society on the periodicity of the hydrocarbon radicles, pointing out the remarkable parallelism which was to be noticed in the change of properties of hydrocarbon radicles and elements when classed in groups. Professor Carnelley, in 1886, developed a similar parallelism. The idea of M. Pelopidas will be easily understood if we consider the series of hydrocarbon radicles which contain, say, 6 atoms of carbon:—

I.	II.	III.	IV.	V.	VI.	VII.	VIII.
C_6H_{13}	C_6H_{12}	C_6H_{11}	C_6H_{10}	C_6H_9	C_6H_8	C_6H_7	C_6H_6

The first of these radicles, like the elements of the 1st group, combines with Cl, OH, and so on, and gives the derivatives of hexyl alcohol, $C_6H_{13}(OH)$; but, in proportion as the number of hydrogen atoms decreases, the capacity of the radicles of combining with, say, the halogens increases. C_6H_{12} already combines with 2 atoms of chlorine; C_6H_{11}, with 3 atoms, and so on. The last members of the series comprise the radicles of acids: thus C_6H_8, which belongs to the 6th group, gives, like sulphur, a bibasic acid, $C_6H_8O_2(OH)_2$, which is homologous with oxalic acid. The parallelism can be traced still further, because C_6H_5 appears as a monovalent radicle of benzene, and with it begins a new series of aromatic derivatives, so analogous to the derivatives of the aliphatic series. Let me also mention another example from among those which have been given by M. Pelopidas. Starting from the alkaline radicle of monomethylammonium, $N(CH_3)H_3$, or NCH_6, which presents many analogies with the alkaline metals of the 1st group, he arrives, by successively diminishing the number of the atoms of hydrogen, at a 7th group which contains cyanogen, CN, which has long since been compared to the halogens of the 7th group.

The most important consequence which, in my opinion, can be drawn from the above comparison is that the periodic law, so apparent in the elements, has a wider application than might appear at first sight; it opens up a new vista of chemical evolutions. But, while admitting the fullest parallelism between the periodicity of the elements and that of the compound radicles, we must not forget that in the periods of the hydrocarbon radicles we have a *decrease* of mass as we pass from the representatives of the first group to the next, while in the periods of the elements the mass *increases* during the progression. It thus becomes evident that we cannot speak of an identity of periodicity in both cases, unless we put aside the ideas of mass and attraction, which are the real corner-stones of the whole of natural science, and even enter into those very conceptions of simple substances which came to light a full hundred years later than the immortal principles of Newton.

From the foregoing, as well as from the failures of so many attempts at finding in experiment and speculation a proof of the compound character of the elements and of the existence of primordial matter, it is evident, in my opinion, that this theory must be classed among mere utopias. But utopias can only be combated by freedom of opinion, by experiment, and by new utopias. In the republic of scientific theories freedom of opinions is guaranteed. It is precisely that freedom which permits me to criticise openly the widely-diffused idea as to the unity of matter in the elements. Experiments and attempts at confirming that idea have been so numerous that it really would be instructive to have them all collected together, if only to serve as a warning against the repetition of old failures. And now as to new utopias which may be helpful in the struggle against the old ones, I do not think it quite useless to mention a *phantasy* of one of my students who imagined that the weight of bodies does not depend upon their mass, but upon the character of the motion of their atoms. The atoms, according to this new utopian, may all be homogeneous or heterogeneous, we know not which; we know them in motion only, and that motion they maintain with the same persistence as the stellar bodies maintain theirs. The weights of atoms differ only in consequence of their various modes and quantity of motion; the heaviest atoms may be much simpler than the lighter ones: thus an atom of mercury may be simpler than an atom of hydrogen—the manner in which it moves causes it to be heavier. My interlocutor even suggested that the view which attributes the greater complexity to the lighter elements finds confirmation in the fact that the hydrocarbon radicles mentioned by Pelopidas, while becoming lighter as they lose hydrogen, change their properties periodically in the same manner as the elements change theirs, according as the atoms grow heavier.

The French proverb, *La critique est facile, mais l'art est difficile*, however, may well be reversed in the case of all such ideal views, as it is much easier to formulate than to criticise them. Arising from the virgin soil of newly-established facts, the knowledge relating to the elements, to their masses, and to the periodic changes of their properties has given a motive for the formation of utopian hypotheses, probably because they could not be foreseen by the aid of any of the various metaphysical systems, and exist, like the idea of gravitation, as an independent outcome of natural science, requiring the acknowledgment of general laws, when these have been established with the same degree of persistency as is indispensable for the acceptance of a thoroughly established fact. Two centuries have elapsed since the theory of gravitation was enunciated, and although we do not understand its cause, we still must regard gravitation as a fundamental conception of natural philosophy, a conception which has enabled us to perceive much more than the metaphysicians did or could with their seeming omniscience. A hundred years later the conception of the elements arose; it made chemistry what it now is; and yet we have advanced as little in our comprehension of simple substances since the times of Lavoisier and Dalton as we have in our understanding of gravitation. The periodic law of the elements is only twenty years old; it is not surprising, therefore, that, knowing nothing about the causes of gravitation and mass, or about the nature of the elements, we do not comprehend the *rationale* of the periodic law. It is only by collecting established laws—that is, by working at the acquirement of truth—that we can hope gradually to lift the veil which conceals from us the causes of the mysteries of Nature and to discover their mutual dependency. Like the telescope and the microscope, laws founded on the basis of experiment are the instruments and means of enlarging our mental horizon.

In the remaining part of my communication I shall endeavour to show, and as briefly as possible, in how far the periodic law contributes to enlarge our range of vision. Before the promulgation of this law the chemical elements were mere fragmentary, incidental facts in Nature; there was no special reason to expect the discovery of new elements, and the new ones which were discovered from time to time appeared to be possessed of quite novel properties. The law of periodicity first enabled us to perceive undiscovered elements at a distance which formerly was inaccessible to chemical vision; and long ere they were discovered new elements appeared before our eyes possessed of a number of well-defined properties. We now know three cases of elements whose existence and properties were foreseen by the instrumentality of the periodic law. I need but mention the brilliant discovery of *gallium*, which proved to correspond to eka-aluminium of the periodic law, by Lecoq de Boisbaudran; of *scandium*, corresponding to ekaboron, by Nilson; and of *germanium*, which proved to correspond in all respects to ekasilicon, by Winkler. When, in 1871, I described to the Russian Chemical Society the properties, clearly defined by the periodic law, which such elements ought to possess, I never hoped that I should live to mention their discovery to the Chemical Society of Great Britain as a confirmation of the exactitude and the generality of the periodic law. Now that I have had the happiness of doing so, I unhesitatingly say that, although greatly enlarging our vision, even now the periodic law needs further improvements in order that it may become a trustworthy instrument in further discoveries.

I will venture to allude to some other matters which chemistry has discerned by means of its new instrument, and which it could not have made out without a knowledge of the law of periodicity, and I will confine myself to simple substances and to oxides.

Before the periodic law was formulated the atomic weights of the elements were purely empirical numbers, so that the magnitude of the equivalent, and the atomicity, or the value in substitution possessed by an atom, could only be tested by critically examining the methods of determination, but never directly by considering the numerical values themselves; in short, we were compelled to move in the dark, to submit to the facts, instead of being masters of them. I need not recount the methods which permitted the periodic law at last to master the facts relating to atomic weights, and I would merely call to mind that it compelled us to modify the valencies of *indium* and *cerium*, and to assign to their compounds a different molecular composition. Determinations of the specific heats of these two metals fully confirmed the change. The trivalency of *yttrium*, which makes us now represent its oxide as Y_2O_3 instead of as YO, was also foreseen (in 1870) by the periodic law, and it has now become so probable that Clève, and all other subsequent investigators of the rare metals, have not only adopted it, but have also applied it without any new demonstration to substances so imperfectly known as those of the cerite and gadolinite group, especially since Hillebrand determined the specific heats of lanthanum and didymium and confirmed the expectations suggested by the periodic law. But here, especially in the case of didymium, we meet with a series of difficulties long since foreseen through the periodic law, but only now becoming evident, and chiefly arising from the relative rarity and insufficient knowledge of the elements which usually accompany didymium.

Passing to the results obtained in the case of the rare elements *beryllium*, *scandium*, and *thorium*, it is found that these have many points of contact with the periodic law. Although Avdéeff long since proposed the magnesia formula to represent beryllium oxide, yet there was so much to be said in favour of the alumina formula, on account of the specific heat of the metals and the isomorphism of the two oxides, that it became generally adopted and seemed to be well established. The periodic law, however, as Brauner repeatedly insisted ('Berichte,' 1878, 872; 1881, 53), was against the formula Be_2O_3; it required the magnesia formula BeO—that is, an atomic weight of 9—because there was no place in the system for an element like beryllium having an atomic weight of 13·5. This divergence of opinion lasted for years, and I often heard that the question as to the atomic weight of beryllium threatened to disturb the generality of the periodic law, or, at any rate, to require some important modifications of it. Many forces were operating in the controversy regarding beryllium, evidently because a much more important question was at issue than merely that involved in the discussion of the atomic weight of a relatively rare element: and during the controversy the periodic law became better understood, and the mutual relations of the elements became more apparent than ever before. It is most remarkable that the victory of the periodic law was won by the researches of the very observers who previously had discovered a number of facts in support of the trivalency of beryllium. Applying the higher law of Avogadro, Nilson and Petterson have finally shown that the density of the vapour of the beryllium chloride, $BeCl_2$, obliges us to regard beryllium as bivalent in conformity with the periodic law. I consider the confirmation of Avdéeff's and Brauner's view as important in the history of the periodic law as the discovery of scandium, which, in Nilson's hands, confirmed the existence of ekaboron.

The circumstance that *thorium* proved to be quadrivalent, and Th = 232, in accordance with the views of Chydenius and the requirements of the periodic law, passed almost unnoticed, and was accepted without opposition, and yet both thorium and uranium are of great importance in the periodic system, as they are its last members, and have the highest atomic weights of all the elements.

The alteration of the atomic weight of *uranium* from U = 120 into U = 240 attracted more attention, the change having been made on account of the periodic law, and for no other reason. Now that Roscoe, Rammelsberg, Zimmermann, and several others have admitted the various claims of the periodic law in the case of uranium, its high atomic weight is received without objection, and it endows that element with a special interest.

While thus demonstrating the necessity for modifying the atomic weights of several insufficiently known elements, the periodic law enabled us also to detect errors in the determination of the atomic weights of several elements whose valencies and true position among other elements were already well known. Three such cases are especially noteworthy: those of tellurium, titanium and platinum. Berzelius had determined the atomic weight of *tellurium* to be 128, while the periodic law claimed for it an atomic weight below that of iodine, which had been fixed by Stas at 126·5, and which was certainly not higher than 127. Brauner then undertook the investigation, and he has shown that the true atomic weight of tellurium is lower than that of iodine, being near to 125. For *titanium* the extensive researches of Thorpe have confirmed the atomic weight of Ti = 48, indicated by the law, and already foreseen by Rose, but contradicted by the analyses of Pierre and several other chemists. An equally brilliant confirmation of the expectations based on the periodic law has been given in the case of the series osmium, iridium, platinum, and gold. At the time of the promulgation of the periodic law, the determinations of Berzelius, Rose, and many others gave the following figures:—

$$Os = 200 \quad Ir = 197; \quad Pt = 198; \quad Au = 196.$$

The expectations of the periodic law have been confirmed, first, by new determinations of the atomic weight of *platinum* (by Seubert, Dittmar, and M'Arthur, which proved to be near to 196 (taking O = 16, as proposed by Marignac, Brauner, and others); secondly, by Seubert having proved that the atomic weight of *osmium* is really lower than that of platinum, being near to 191; and thirdly, by the investigations of Krüss, Thorpe and Laurie, proving that the atomic weight of *gold* exceeds that of platinum, and approximates to 197. The atomic weights which were thus found to require correction were precisely those which the periodic law had indicated as affected with errors; and it has been proved, therefore, that the periodic law affords a means of testing experimental results. If we succeed in discovering the exact character of the periodic relationships between the increments in atomic weights of allied elements discussed by Ridberg in 1885, and again by Bazaroff in 1887, we may expect that our instrument will give us the means of still more closely controlling the experimental data relating to atomic weights.

Let me next call to mind that, while disclosing the variation of chemical properties, the periodic law, has also enabled us to systematically discuss many of the physical properties of elementary bodies, and to show that these properties are also subject to the law of periodicity. At the Moscow Congress of Russian Naturalists in August, 1869, I dwelt upon the relations which existed between density and the atomic weight of the elements. The following year Professor Lothar Meyer, in his well-known paper, studied the same subject in more detail, and thus contributed to spread information about the periodic law. Later on, Carnelley, Laurie, L. Meyer, Roberts-Austen, and several others applied the periodic system to represent the order in the changes of the magnetic properties of the elements, their melting points, the heats of formation of their haloid compounds, and even of such mechanical properties as the coefficient of elasticity, the breaking stress, &c., &c. These deductions, which have received further support in the discovery of new elements endowed not only with chemical but even with physical properties, which were foreseen by the law of periodicity, are well known; so I need not dwell upon the subject, and may pass to the consideration of oxides.

In indicating that the gradual increase of the power of elements of combining with oxygen is accompanied by a corresponding decrease in their power of combining with hydrogen, the periodic law has shown that there is a limit of oxidation, just as there is a well-known limit to the capacity of elements for combining with hydrogen. A single atom of an element combines with at most four atoms of either hydrogen or oxygen; and while CH_4 and SiH_4 represent the highest hydrides, so RuO_4 and OsO_4 are the highest oxides. We are thus led to recognise types of oxides, just as we have had to recognise types of hydrides.

The periodic law has demonstrated that the maximum extent to which different non-metals enter into combination with oxygen is determined by the extent to which they combine with hydrogen, and that the sum of the number of equivalents of both must be equal to 8. Thus chlorine, which combines with 1 atom or 1 equivalent of hydrogen, cannot fix more than 7 equivalents of oxygen, giving Cl_2O_7; while sulphur, which fixes 2 equivalents of hydrogen, cannot combine with more than 6 equivalents or 3 atoms of oxygen. It thus becomes evident that we cannot recognise as a fundamental property of the elements the atomic valencies deduced from their hydrides; and that we must modify, to a certain extent, the theory of atomicity if we desire to raise it to the dignity of a general principle capable of affording an insight into the constitution of all compound molecules. In other words, it is only to carbon, which is quadrivalent with regard both to oxygen and hydrogen, that we can apply the theory of constant valency and of bond, by means of which so many still endeavour to explain the structure of compound molecules. But I should go too far if I ventured to explain in detail the conclusions which can be drawn from the above considerations. Still, I think it necessary to dwell upon one particular fact which must be explained from the point of view of the periodic law in order to clear the way to its extension in that particular direction.

The higher oxides yielding salts the formation of which was foreseen by the periodic system—for instance, in the short series beginning with sodium—

$$Na_2O, \quad MgO, \quad Al_2O_3, \quad SiO_2, \quad P_2O_5, \quad SO_3, \quad Cl_2O_7,$$

must be clearly distinguished from the higher degrees of oxidation which correspond to hydrogen peroxide and bear the true character of peroxides. Peroxides such as Na_2O_2, BaO_2, and the like have long been known. Similar peroxides have also recently become known in the case of chromium, sulphur, titanium, and many other elements, and I have sometimes heard it said that discoveries of this kind weaken the conclusions of the periodic law in so far as it concerns the oxides. I do not think so in the least, and I may remark, in the first place, that all these peroxides are endowed with certain properties obviously common to all of them, which distinguish them from the actual, higher, salt-forming oxides, especially their easy decomposition by means of simple contact agencies; their incapability of forming salts of the common type; and their capability of combining with other peroxides (like the faculty which hydrogen peroxide possesses of combining with barium peroxide, discovered by Schoene). Again, we remark that some groups are especially characterised by their capacity of generating peroxides. Such is, for instance, the case in the sixth group, where we find the well-known peroxides of sulphur, chromium, and uranium; so that further investigation of peroxides will probably establish a new periodic function, foreshadowing that molybdenum and tungsten will assume peroxide forms with comparative readiness. To appreciate the constitution of such peroxides, it is enough to notice that the peroxide form of sulphur (so-called persulphuric acid) stands in the same relation to sulphuric acid as hydrogen peroxide stands to water:—

$$H(OH), \text{ or } H_2O, \text{ responds to } (OH)(OH), \text{ or } H_2O_2,$$

and so also—

$$H(HSO_4), \text{ or } H_2SO_4, \text{ responds to } (HSO_4)(HSO_4), \text{ or } H_2S_2O_8.$$

Similar relations are seen everywhere, and they correspond to the principle of substitutions which I long since endeavoured to represent as one of the chemical generalisations called into life by the periodic law. So also sulphuric acid, if considered with reference to hydroxyl, and represented as follows:—

$$HO(SO_2OH),$$

has its corresponding compound in dithionic acid—

$$(SO_2OH)(SO_2OH), \text{ or } H_2S_2O_6.$$

Therefore, also, phosphoric acid, $HO(POH_2O_2)$, has, in the same sense, its corresponding compound in the subphosphoric acid of Saltzer:—

$$(POH_2O_2)(POH_2O_2), \text{ or } H_4P_2O_6;$$

and we must suppose that the peroxide compound corresponding to phosphoric acid, if it be discovered, will have the following structure:—

$$(H_2PO_4)_2 \text{ or } H_4P_2O_8 = 2H_2O + 2PO_3.$$

So far as is known at present, the highest form of peroxides is met with in the peroxide of uranium, UO_4, prepared by Fairley; while OsO_4 is the highest oxide giving salts. The line of argument which is inspired by the periodic law, so far from being weakened by the discovery of peroxides, is thus actually strengthened, and we must hope that a further exploration of the region under consideration will confirm the applicability to chemistry generally of the principles deduced from the periodic law.

Permit me now to conclude my rapid sketch of the oxygen compounds by the observation that the periodic law is especially brought into evidence in the case of the oxides which constitute the immense majority of bodies at our disposal on the surface of the earth.

The oxides are evidently subject to the law, both as regards their chemical and their physical properties, especially if we take into account the cases of polymerism which are so obvious when comparing CO_2, with Si_nO_{2n}. In order to prove this I give the densities s and the specific volumes v of the higher oxides of two short periods. To render comparison easier, the oxides are all represented as of the form R_2O_n. In the column headed Δ the differences are given between the volume of the oxygen compound and that of the parent element, divided by n—that is, by the number of atoms of oxygen in the compound:—

	s.	v.	Δ		s.	v.	Δ
Na_2O	2·6	24	-22	K_2O	2·7	35	-55
Mg_2O_2	3·6	22	-3	Ca_2O	3·15	36	-7
Al_2O_3	4·0	26	+1·3	Sc_2O_3	3·86	35	0
Si_2O_4	2·65	45	5·2	Li_2O_4	4·2	38	+5

P_2O_5	2·39	59	6·2	V_2O_5	3·49	52	6·7
S_2O_6	1·96	82	8·7	Cr_2O_6	2·74	73	9·5

I have nothing to add to these figures, except that like relations appear in other periods as well. The above relations were precisely those which made it possible for me to be certain that the relative density of ekasilicon oxide would be about 4·7; germanium oxide, actually obtained by Winkler, proved, in fact, to have the relative density 4·703.

The foregoing account is far from being an exhaustive one of all that has already been discovered by means of the periodic law telescope in the boundless realms of chemical evolution. Still less is it an exhaustive account of all that may yet be seen, but I trust that the little which I have said will account for the philosophical interest attached in chemistry to this law. Although but a recent scientific generalisation, it has already stood the test of laboratory verification, and appears as an instrument of thought which has not yet been compelled to undergo modification; but it needs not only new applications, but also improvements, further development, and plenty of fresh energy. All this will surely come, seeing that such an assembly of men of science as the Chemical Society of Great Britain has expressed the desire to have the history of the periodic law described in a lecture dedicated to the glorious name of Faraday.

Footnotes:

'Es ist wohl kaum anzunehmen, dass alle im Vorhergehenden hervorgehobenen Beziehungen zwischen den Atomgewichten (oder Aequivalenten) in chemischen Verhältnissen einander ähnliche Elemente bloss zufällig sind. Die Auffindung der in diesen Zahlen *gesetzlichen* Beziehungen müssen wir jedoch der Zukunft überlassen.'

To judge from J. A. R. Newlands's work, *On the Discovery of the Periodic Law*, London, 1884, p. 149; 'On the Law of Octaves' (from the *Chemical News*, 12, 83, August 18, 1865).

That is, a substance having a wave-length equal to 0·0005875 millimetre.

He maintains (on p. 309) that the periodic law requires two new analogous elements, having atomic weights of 48 and 64, occupying positions between sulphur and selenium, although nothing of the kind results from any of the different readings of the law.

It is noteworthy that the year in which Lavoisier was born (1743)—the author of the idea of elements and of the indestructibility of matter—is later by exactly one century than the year in which the author of the theory of gravitation and mass was born (1643 N.S.). The affiliation of the ideas of Lavoisier and those of Newton is beyond doubt.

I foresee some more new elements, but not with the same certitude as before. I shall give one example, and yet I do not see it quite distinctly. In the series which contains Hg = 204, Pb = 206, and Bi = 208, we can imagine the existence (at the place VI-11) of an element analogous to tellurium, which we can describe as dvi-tellurium, Dt, having an atomic weight of 212, and the property of forming the oxide DtO_3. If this element really exists, it ought in the free state to be an easily fusible, crystalline, non-volatile metal of a grey colour, having a density of about 9·3, capable of giving a dioxide, DtO_2, equally endowed with feeble acid and basic properties. This dioxide must give on active oxidation an unstable higher oxide, DtO_3, which should resemble in its properties PbO_2 and Bi_2O_5. Dvi-tellurium hydride, if it be found to exist, will be a less stable compound than even H_2Te. The compounds of dvi-tellurium will be easily reduced, and it will form characteristic definite alloys with other metals.

Let me mention another proof of the bivalency of beryllium which may have passed unnoticed, as it was only published in the Russian chemical literature. Having remarked (in 1884) that the density of such solutions of chlorides of metals, MCl_n, as contain 200 mols. of water (or a large and constant amount of water) regularly increases as the molecular weight of the dissolved salt increases, I proposed to one of our young chemists, M. Burdakoff, that he should investigate beryllium chloride. If its molecule be $BeCl_2$ its weight must be = 80; and in such a case it must be heavier than the molecule of KCl = 74·5, and lighter than that of $MgCl_2$, = 93. On the contrary, if beryllium chloride is a trichloride, $BeCl_3$ = 120, its molecule must be heavier than that of $CaCl_2$ = 111, and lighter than that of $MnCl_2$ = 126. Experiment has shown the correctness of the former formula, the solution $BeCl_2 + 200H_2O$ having (at 15°/4°) a density of 1·0138, this being a higher density than that of the solution $KCl + 200H_2O$ (= 1·0121), and lower than that of $MgCl_2 + 200H_2O$ (= 1·0203). The bivalency of beryllium was thus confirmed in the case both of the dissolved and the vaporised chloride.

I pointed them out in the *Liebig's Annalen*, Supplement Band., viii. 1871, p. 211.

Thus, in the typical small period of

Li, Be, B, C, N, O, F,

we see at once the progression from the alkali metals to the acid non-metals, such as are the halogens.

Liebig's Annalen, Supplement Band., vii. 1870.

A distinct periodicity can also be discovered in the spectra of the elements. Thus the researches of Hartley, Ciamician, and others have disclosed, first, the homology of the spectra of analogous elements: secondly, that the alkali metals have simpler spectra than the metals of the following groups; and thirdly, that there is a certain likeness between the complicated spectra of manganese and iron on the one hand, and the no less complicated spectra of chlorine and bromine on the other hand, and their likeness corresponds to the degree of analogy between those elements which is indicated by the periodic law.

Formerly it was supposed that, being a bivalent element, oxygen can enter into any grouping of the atoms, and there was no limit foreseen as to the extent to which it could further enter into combination. We could not explain why bivalent sulphur, which forms compounds such as

$$S\!\!<\!\!\begin{array}{c}O\\O\end{array}\!\!> \quad \text{and} \quad S\!\!<\!\!\begin{array}{c}O\\O\end{array}\!\!>\!\!O,$$

could not also form oxides such as—

$$S\!\!<\!\!\begin{array}{c}O\!\!-\!\!O\\O\!\!-\!\!O\end{array}\!\!> \quad \text{or} \quad S\!\!<\!\!\begin{array}{c}O\!\!-\!\!O\\O\!\!-\!\!O\end{array}\!\!>\!\!O,$$

while other elements, as, for instance, chlorine, form compounds such as—

$$\text{Cl—O—O—O—O—K}$$

In this sense, oxalic acid, $(COOH)_2$, also corresponds to carbonic acid, $OH(COOH)$, in the same way that dithionic acid corresponds to sulphuric acid, and subphosphoric acid to phosphoric; hence, if a peroxide corresponding to carbonic acid be obtained, it will have the structure of $(HCO_3)_2$, or $H_2C_2O_6 = H_2O + C_2O_5$. So also lead must have a real peroxide, Pb_2O_5.

The compounds of uranium prepared by Fairley seem to me especially instructive in understanding the peroxides. By the action of hydrogen peroxide on uranium oxide, UO_3, a peroxide of uranium, $UO_4,4H_2O$, is obtained ($U = 240$) if the solution be acid; but if hydrogen peroxide act on uranium oxide in the presence of caustic soda, a crystalline deposit is obtained which has the composition $Na_4UO_8,4H_2O$, and evidently is a combination of sodium peroxide, Na_2O_2, with uranium peroxide, UO_4. It is possible that the former peroxide, $UO_4,4H_2O$, contains the elements of hydrogen peroxide and uranium peroxide, U_2O_7, or even $U(OH)_6,H_2O_2$, like the peroxide of tin recently discovered by Spring, which has the constitution Sn_2O_5,H_2O_2.

Δ thus represents the average increase of volume for each atom of oxygen contained in the higher salt-forming oxide. The acid oxides give, as a rule, a higher value of Δ, while in the case of the strongly alkaline oxides its value is usually negative.

APPENDIX III

ARGON, A NEW CONSTITUENT OF THE ATMOSPHERE

WRITTEN BY PROFESSOR MENDELÉEFF IN FEBRUARY 1895

The remarks made in Chapter V., Note 16 bis respecting the newly discovered constituent of the atmosphere are here supplemented by data (taken from the publications of the Royal Society of London) given by the discoverers Lord Rayleigh and Professor Ramsay in January 1895, together with observations made by Crookes and Olszewsky upon the same subject.

This gas, which was discovered by Rayleigh and Ramsay in atmospheric nitrogen, was named *argon* by them, and upon the supposition of its being an element, they gave it the symbol A. But its true chemical nature is not yet fully known, for not only has no compound of it been yet obtained, but it has not even been brought into any reaction. From all that is known about it at the present time, we may conclude with the discoverers that argon belongs to those gases which are permanent constituents of the atmosphere, and that it is a new element. The latter statement, however, requires confirmation. We shall presently see, however, that the negative chemical character of argon (its incapacity to react with any substance), and the small amount of it present in the atmosphere (about 1¼ per cent. by volume in the nitrogen of air, and consequently about 1 per cent. by volume in air), as well as the recent date of its discovery (1894) and the difficulty of its preparation, are quite sufficient reasons for the incompleteness of the existing knowledge respecting this element. But since, so far as is yet known, we are dealing with a normal constituent of the atmosphere[1 bis], the existing data, notwithstanding their insufficiently definite nature, should find a place even in such an elementary work as the present, all the more as the names of Rayleigh, Ramsay, Crookes and Olszewsky, who have worked upon argon, are among the highest in our science, and their researches among the most difficult. These researches, moreover, were directed straight to the goal, which was only partly reached owing to the unusual properties of argon itself.

When it became known (Chapter V., Note 4 bis) that the nitrogen obtained from air (by removing the oxygen, moisture and CO_2, by various reagents) has a greater density than that obtained from the various (oxygen, hydrogen and metallic) compounds of nitrogen, it was a plausible explanation that the latter contained an admixture of hydrogen, or of some other light gas lowering the density of the mixture. But such an assumption is refuted not only by the fact that the nitrogen obtained from its various compounds (after purification) has always the same density (although the supposed impurities mixed with it should vary), but also by Rayleigh and

Ramsay's experiment of artificially adding hydrogen to nitrogen, and then passing the mixture over red-hot oxide of copper, when it was found that the nitrogen regained its original density, *i.e.* that the whole of the hydrogen was removed by this treatment. Therefore the difference in the density of the two varieties of nitrogen had to be explained by the presence of a heavier gas in admixture with the nitrogen obtained from the atmosphere. This hypothesis was confirmed by the fact that Rayleigh and Ramsay having obtained purified nitrogen (by removing the O_2, CO_2, and H_2O), both from ordinary air and from air which had been previously subjected to atmolysis, that is which had been passed through porous tubes (of burnt clay, *e.g.* pipe-stem), surrounded by a rarefied space, and so deprived of its lighter constituents (chiefly nitrogen), found that the nitrogen from the air which had been subjected to atmolysis was heavier than that obtained from air which had not been so treated. This experiment showed that the nitrogen of air contains an admixture of a gas which, being heavier than nitrogen itself, diffuses more slowly than nitrogen through the porous material. It remained, therefore, to separate this impurity from the nitrogen. To do this Rayleigh and Ramsay adopted two methods, converting the nitrogen into solid and liquid substances, either by absorbing the nitrogen by heated magnesium (Chapter V., Note 6, and Chapter XIV., Note 14), with the formation of nitride of magnesium, or else by converting it into nitric acid by the action of electric sparks or the presence of an excess of air and alkali, as in Cavendish's method.[3 bis] In both cases the nitrogen entered into reaction, while the heavier gas mixed with it remained inert, and was thus able to be isolated. That is, the argon could be separated by these means from the excess of atmospheric nitrogen accompanying it. As an illustration we will describe how argon was obtained from the atmospheric nitrogen by means of magnesium. To begin with, it was discovered that when atmospheric nitrogen was passed through a tube containing metallic magnesium heated to redness, its specific gravity rose to 14·88. As this showed that part of the gas was absorbed by the magnesium, a mercury gasometer filled with atmospheric nitrogen was taken, and the gas drawn over soda-lime, P_2O_5, heated magnesium and then through tubes containing red-hot copper oxide, soda-lime and phosphoric anhydride to a second mercury gasometer. Every time the gas was repassed through the tubes, it decreased in volume and increased in density. After repeating this for ten days 1,500 c.c. of gas were reduced to 200 cc., and the density increased to 16·1 (if that of $H_2 = 1$ and $N_2 = 14$). Further treatment of the remainder brought the density up to 19·09. After adding a small quantity of oxygen and repassing the gas through the apparatus, the density rose to 20·0. To obtain argon by this process Ramsay and Rayleigh (employing a mercury air pump and mercury gasometers) once treated about 150 litres of atmospheric nitrogen. On another occasion they treated 7,925 c.c. of air by the oxidation method and obtained 65 c.c. of argon, which corresponds to 0·82 per cent. The density of the argon obtained by this means was nearly 19·7, while that obtained by the magnesium method varied between 19·09 and 20·38.

Thus the first positive and very important fact respecting argon is that its specific gravity is nearly 20—that is, that it is 20 times heavier than hydrogen, while nitrogen is only 14 times and oxygen 16 times heavier than hydrogen. This explains the difference observed by Rayleigh between the densities of nitrogen obtained from its compounds and from the atmosphere (Chapter V., Note 4 bis). At 0° and 760 mm. a litre of the former gas weighs 1·2505 grm., while a litre of the latter weighs 1·2572, or taking H = 1, the density of the first = 13·916, and of the latter = 13·991. If the density of argon be taken as 20, it is contained in atmospheric nitrogen to the extent of about 1·23 per cent. by volume, whilst air contains about 0·97 per cent. by volume.

When argon had been isolated the question naturally arose, was it a new homogeneous substance having definite properties or was it a mixture of gases? The former may now be positively asserted, namely, that argon is a peculiar gas previously unknown to chemistry. Such a conviction is in the first place established by the fact that argon has a greater number of negative properties, a smaller capacity for reaction, than any other simple or compound body known. The most inert gas known is nitrogen, but argon far exceeds it in this respect. Thus nitrogen is absorbed at a red heat by many metals, with the formation of nitrides, while argon, as is seen in the mode of its preparation and by direct experiment, does not possess this property. Nitrogen, under the action of electric sparks, combines with hydrogen in the presence of acids and with oxygen in the presence of alkalis, while argon is unable to do so, as is seen from the method of separation from nitrogen. Rayleigh and Ramsay also proved that argon is unable to react with chlorine (dry or moist) either directly or under the action of an electric discharge, or with phosphorus or sulphur, at a red heat. Sodium, potassium, and tellurium may be distilled in an atmosphere of argon without change. Fused caustic soda, incandescent soda-lime, molten nitre, red-hot peroxide of sodium, and the polysulphides of calcium and sodium also do not react with argon. Platinum black does not absorb it, and spongy platinum is unable to excite its reaction with oxygen or chlorine. Aqua regia, bromine water, and a mixture of hydrochloric acid and $KMnO_4$ were also without action upon argon. Besides which it is evident from the method of its preparation that it is not acted upon by red-hot oxide of copper. All these facts exclude any possibility of argon containing any already known body, and prove it to be the most inert of all the gases known. But besides these negative points, the independency of argon is confirmed by four observed positive properties possessed by it, which are:—

1. The spectrum of argon observed by Crookes under a low pressure (in Geissler-Plücker tubes) distinguishes it from other gases. It was proved by this means that the argon obtained by means of magnesium is identical with that which remains after the conversion of the atmospheric nitrogen into nitric acid. Like nitrogen, argon presents two spectra produced at different potentials of the induced current, one being orange-red, the other steel-blue; the latter is obtained under a higher degree of rarefaction and with a battery of Leyden jars. Both the spectra of argon (in contradistinction to those of nitrogen) are distinguished by clearly defined lines. The red (ordinary) spectrum of argon has two particularly brilliant and characteristic red lines (not far from the bright red line of lithium, on the opposite side to the orange band) having wave-lengths 705·64 and 696·56 (*see* Vol. I., p. 565). Between these bright lines there are in addition lines with wave lengths 603·8, 565·1, 561·0, 555·7, 518·58, 516·5, 450·95, 420·10, 415·95 and 394·85. Altogether 80 lines have been observed in this spectrum and 119 in the blue spectrum, of which 26 are common to both spectra.

2. According to Rayleigh and Ramsay the solubility of argon in water is approximately 4 volumes in 100 volumes of water at 13°. Thus argon is nearly 2½ times more soluble than nitrogen, and its solubility approaches that of oxygen. Direct experiment proves that nitrogen obtained from air from boiled water is heavier than that obtained straight from the atmosphere. This again is an indirect proof of the presence of argon in air.

3. The ratio k of the two specific heats (at a constant pressure and at a constant volume) of argon was determined by Rayleigh and Ramsay by the method of the velocity of sound (*see* Chapter XIV., Note 7 and Chapter VII., Note 26) and was found to be nearly 1·66, that is greater than for those gases whose molecules contain two atoms (for instance, CO, H_2, N_2, air, &c., for which k is nearly 1·4) or those whose molecules contain three atoms (for instance, CO_2, N_2O, &c., for which k is about 1·3), but closely approximate to the ratio of the specific heats of mercury vapour (Kundt and Warburg, $k = 1·67$). And as the molecule of mercury vapour contains one atom, so it may be said that argon is a simple gaseous body whose molecule contains one atom. A compound body should give a smaller ratio. The experiments upon the liquefaction of argon, which we shall presently describe, speak against the supposition that argon is a mixture of two gases. The importance of the results in question makes one wish that the determinations of the ratio of the specific heats (and other physical properties) might be confirmed with all possible accuracy. If we admit, as we are obliged to do for the present, that argon is a new element, its density shows that its atomic weight must be nearly 40, that is, near to that of K =

39 and Ca = 40, which does not correspond to the existing data respecting the periodicity of the properties of the elements in dependence upon their atomic weights, for there is no reason on the basis of existing data for admitting any intermediate elements between Cl = 35·5 and K = 39, and all the positions above potassium in the periodic system are occupied. This renders it very desirable that the velocity of sound in argon should be re-determined.

4. Argon was liquefied by Professor Olszewsky, who is well known for his classical researches upon liquefied gases. These researches have an especial interest since they show that argon exhibits a perfect constancy in its properties in the liquid and critical states, which almost disposes of the supposition that it contains a mixture of two or more unknown gases. As the first experiments showed, argon remains a gas under a pressure of 100 atmospheres and at a temperature of -90°; this indicated that its critical temperature was probably below this temperature, as was indeed found to be the case when the temperature was lowered to -128°·6 by means of liquid ethylene. At this temperature argon easily liquefies to a colourless liquid under 38 atmospheres. The meniscus begins to disappear at between -119°·8 and -121°·6, mean -121° at a pressure of 50·6 atmospheres. The vapour tension of liquid argon at -128°·6, is 38·0 atmospheres, at -187° it is one atmosphere, and at -189°·6 it solidifies to a colourless substance like ice. The specific gravity of liquid argon at about -187° is nearly 1·5, which is far above that of other liquefied gases of very low absolute boiling point.

The discovery of argon is one of the most remarkable chemical acquisitions of recent times, and we trust that Lord Rayleigh and Professor Ramsay, who made this wonderful discovery, will further elucidate the true nature of argon, as this should widen the fundamental principles of chemistry, to which the chemists of Great Britain have from early times made such valuable contributions. It would be premature now to give any definite opinions upon so new a subject. Only one thing can be said; argon is so inert that its rôle in nature cannot be considerable, notwithstanding its presence in the atmosphere. But as the atmosphere itself plays such a vast part in the life of the surface of the earth, every addition to our knowledge of its composition must directly or indirectly react upon the sum total of our knowledge of nature.

Footnotes:

From the Greek Ἀργὸν—inert.

[1 bis] In Note 16 bis, Chapter V., I mentioned that, judging from the specific gravity of argon, it might possibly be polymerised nitrogen, N_3, bearing the same relationship to nitrogen, N_2, that ozone, O_3, bears to ordinary oxygen. If this idea were confirmed, still one would not imagine

that argon was formed from the atmospheric nitrogen by those reactions by which it was obtained by Rayleigh and Ramsay, but rather that it arises from the nitrogen of the atmosphere under natural conditions. Although this proposition is not quite destroyed by the more recent results, still it is contradicted by the fact that the ratio of the specific heats of argon was found to be 1·66, which, as far as is now known, could not be the case for a gas containing 3 atoms in its molecule, since such gases (*see* Chapter XIV., Note 7) give the ratio approximately 1·3 (for example, CO_2). In abstaining from further conclusions, for they must inevitably be purely conjectural, I consider it advisable to suggest that in conducting further researches upon argon it might be well to subject it to as high a temperature as possible. And the possibility of nitrogen polymerising is all the more admissible from the fact that the aggregation of its atoms in the molecule is not at all unlikely, and that polymerised nitrogen, judging from many examples, might be inert if the polymerisation were accompanied by the evolution of heat. In the following footnotes I frequently return to this hypothesis, not only because I have not yet met any facts definitely contradictory to it, but also because the chief properties of argon agree with it to a certain extent.

The chief difficulty in investigating argon lies in the fact that its preparation requires the employment of a large quantity of air, which has to be treated with a number of different reagents, whose perfect purity (especially that of magnesium) will always be doubtful, and argon has not yet been transferred to a substance in which it could be easily purified. Perhaps the considerable solubility of argon in water (or in other suitable liquids, which have not apparently yet been tried) may give the means of doing so, and it may be possible, by collecting the air expelled from boiling water, to obtain a richer source of argon than ordinary air.

It might also be supposed that this heavy gas is separated by the copper when the latter absorbs the oxygen of the air; but such a supposition is not only improbable in itself, but does not agree with the fact that nitrogen may be obtained from air by absorbing the oxygen by various other substances in solution (for instance, by the lower oxides of the metals, like FeO) besides red-hot copper, and that the nitrogen obtained is always just as heavy. Besides which, nitrogen is also set free from its oxides by copper, and the nitrogen thus obtained is lighter. Therefore it is not the copper which produces the heavy gas—*i.e.* argon.

[3 bis] It is worthy of note that Cavendish obtained a small residue of gas in converting nitrogen into nitric acid; but he paid no attention to it, although probably he had in his hands the very argon recently discovered.

When in these experiments, instead of atmospheric nitrogen the gas obtained from its compound was taken, an inert residue of a heavy gas, having the properties of argon, was also remarked, but its amount was very small. Rayleigh and Ramsay ascribe the formation of this residue to the fact that the gas in these experiments was collected over water, and a portion of the dissolved argon in it might have passed into the nitrogen. As the authors of this supposition did not prove it by any special experiments, it forms a weak point in their classical research. If it be admitted that argon is N_3, the fact of its being obtained from the nitrogen of compounds might be explained by the polymerisation of a portion of the nitrogen in the act of reaction, although it is impossible to refute Rayleigh and Ramsay's hypothesis of its being evolved from the water employed in the manipulation of the gases. Three thousand volumes of nitrogen extracted from its compounds gave about three volumes of argon, while thirty volumes were yielded by the same amount of atmospheric nitrogen.

The preparation of argon by the conversion of nitrogen into nitric acid is complicated by the necessity of adding a large proportion of oxygen and alkali, of passing an electric discharge through the mixture for a long period, and then removing the remaining oxygen. All this was repeatedly done by the authors, but this method is far more complex, both in practice and theory, than the preparation of argon by means of magnesium. From 100 volumes of air subjected to conversion into HNO_3, 0·76 volume of argon were obtained after absorbing the excess of oxygen.

In these and the following experiments the magnesium was placed in an ordinary hard glass tube, and heated in a gas furnace to a temperature almost sufficient to soften the glass. The current of gas must be very slow (a tube containing a small quantity of sulphuric acid served as a meter), as otherwise the heat evolved in the formation of the Mg_3N_2 (Chapter XIV., Note 14) will melt the tube.

The greatest brilliancy of the spectrum of argon is obtained at a tension of 3 mm., while for nitrogen it is about 75 mm. (Crookes). In Chapter V., Note 16 bis, it is said that the same blue line observed in the spectrum of argon is also observed in the spectrum of nitrogen. This is a mistake, since there is no coincidence between the blue lines of the argon and nitrogen spectra. However, we may add that for nitrogen the following moderately bright lines are known of wave-lengths 585, 574, 544, 516, 457, 442, 436, and 426, which are repeated in the spectra (red and blue) of argon, judging by Crookes' researches (1895); but it is naturally impossible to assert that there is perfect identity until some special comparative work has been done in this subject, which is very desirable, and more especially for the bluish-violet portion of the spectrum, more particularly between the lines 442–436, as these lines are distinguished by their brilliancy in both the argon and nitrogen spectra. The above-mentioned supposition of argon being polymerised nitrogen (N_3), formed from nitrogen (N_2), with the evolution of heat, might find some support should it be found after careful comparison that even a limited number of spectral lines coincided.

At first the spectrum of argon exhibits the nitrogen lines, but after a certain time these lines disappear (under the influence of the platinum, and also of Al and Mg, but with the latter the spectrum of hydrogen appears) and leave a pure argon spectrum. It does not appear clear to me whether a polymerisation here takes place or a simple absorption. Perhaps the elucidation of this question would prove important in the history of argon. It would be desirable to know, for instance, whether the volume of argon changes when it is first subjected to the action of the electric discharge.

Crookes supposes that argon contains a mixture of two gases, but as he gives no reasons for this, beyond certain peculiarities of a spectroscopic character, we will not consider this hypothesis further.

This portion of Rayleigh and Ramsay's researches deserves particular attention as, so far, no gaseous substance is known whose molecule contains but one atom. Were it not for the above determinations, it might be thought that argon, having a density 20, has a complex molecule, and may be a compound or polymerised body, for instance, N_3 or NX_n, or in general X_n; but as the matter stands, it can only be said that either (1) argon is a new, peculiar, and quite unusual elementary substance, since there is no reason for assuming it to contain two simple gases, or (2) the magnitude, k (the ratio of the specific heats) does not only depend upon the number of atoms contained in the molecules, but also upon the store of internal energy (internal motion of the atoms in the molecule). Should the latter be admitted, it would follow that the molecules of very active gaseous elements would correspond to a smaller k than those of other gases having an equal number of atoms in their molecule. Such a gas is chlorine, for which $k = 1·33$ (Chapter XIV., Note). For gases having a small chemical energy, on the contrary, a larger magnitude would be expected for k. I think these questions might be partially settled by determining k for ozone (O_3) and sulphur (S_6) (at about 500°). In other words, I would suggest, though only provisionally, that the magnitude, $k = 1·6$, obtained for argon might prove to agree with the hypothesis that argon is N_3, formed from N_2 with the evolution of heat or loss of energy. Here argon gives rise to questions of primary importance, and it is to be hoped that further research will throw some light upon them. In making these remarks, I only wish to clear the road for further progress in the study of argon, and of the questions depending on it. I may also remark that if argon is N_3 formed with the evolution of heat, its conversion into nitrogen, N_2, and into nitride compounds (for instance, boron nitride or nitride of titanium) might only take place at a very high temperature.

Without having the slightest reason for doubting the accuracy of Rayleigh and Ramsay's determinations, I think it necessary to say that as yet (February 1895) I am only acquainted with the short memoir of the above chemists in the 'Proceedings of the Royal Society,' which does not give any description of the methods employed and results obtained, while at the end (in the general conclusions) the authors themselves express some doubt as to the simple nature of argon. Moreover, it seems to me that (Note 10) there must be a dependence of k upon the chemical energy. Besides which, it is not clear what density of the gas Rayleigh and Ramsay took in determining k. (If argon be N_3, its density would be near to 21.) Hence I permit myself to express some doubt as to whether the molecule of argon contains but one atom.

If it should be found that k for argon is less than 1·4, or that k is dependent upon the chemical energy, it would be possible to admit that the molecule of argon contains not one, but several atoms—for instance, either N_3 (then the density would be 21, which is near to the observed density) or X_6, if X stand for an element with an atomic weight near to 6·7. No elements are known between H = 1 and Li = 7, but perhaps they may exist. The hypothesis A = 40 does not admit argon into the periodic system. If the molecule of argon be taken as A_2—*i.e.* the atomic weight as A = 20— argon apparently finds a place in Group VIII., between F = 19 and Na = 23; but such a position could only be justified by the consideration that elements of small atomic weight belong to the category of typical elements which offer many peculiarities in their properties, as is seen on comparing N with the other elements of Group V., or O with those of Group VI. Apart from this there appears to me to be little probability, in the light of the periodic law, in the position of an inert substance like argon in Group VIII., between such active elements as fluorine and sodium, as the representatives of this group by their atomic weights and also by their properties show distinct transitions from the elements of the last groups of the uneven series to the elements of the first groups of the even series—for instance,

Group VI. VII.   VIII.   I.   II.

     Cr Mn Fe, Co, Ni Cu Zn

While if we place argon in a similar manner,

  VI.   VII.   VIII.   I.   II.

O = 16   F = 19   A = 20   Na = 23   Mg = 24

although from a numerical point of view there is a similar sequence to the above, still from a chemical and physical point of view the result is quite different, as there is no such resemblance between the properties of O, F and Na, Mg, as between Cr, Mn, and Cu, Zn. I repeat that only the typical character of the elements with small atomic weights can justify the atomic weight A = 20, and the placing of argon in Group VIII. amongst the typical elements; then N, O, F, A are a series of gases.

It appears to me simpler to assume that argon contains N_3, especially as argon is present in nitrogen and accompanies it, and, as a matter of fact, none of the observed properties of argon are contradictory to this hypothesis.

These observations were written by me in the beginning of February 1895, and on the 29th of that month I received a letter, dated February 25, from Professor Ramsay informing me that 'the periodic classification entirely corresponds to its (argon's) atomic weight, and that it even gives a fresh proof of the periodic law,' judging from the researches of my English friends. But in what these researches consisted, and how the above agreement between the atomic weight of argon and the periodic system was arrived at, is not referred to in the letter, and we remain in expectation of a first publication of the work of Lord Rayleigh and Professor Ramsey. [For more complete information see papers read before the Royal Society, January 31, 1895, February 13, March 10, and May 21, 1896, and a paper published in the Chemical Society's Transactions, 1895, p. 684. For abstracts of these and other papers on argon and helium, and correspondence, see 'Nature,' 1895 and 1896.

There only remains the very remote possibility that argon consists of a mixture of two gases having very nearly the same properties.

The following data, given by Olszewsky, supplement the data given in Chapter II., Note 29, upon liquefied gases.

	(t_c)	(p_c)	t	t_1	s
N_2	-146°	35	-194°·4	-214°	0·885
CO	-139°·5	35·5	-190°	-207°	?

	t_c	p_c	t	t_1	s
A	-121°	50·6	-187°	-189°·6	1·5
O_2	-118°·8	50·8	-182°·7	?	1·124
NO	-93°·5	71·2	-153°·6	-167°	?
CH_4	-81°·8	54·9	-164°	-158°·8	0·415

where t_c is the absolute (critical) boiling point, p_c the pressure (critical) in atmospheres corresponding to it, t the boiling point (under a pressure of 760 mm.), t_1 the melting point, and s the specific gravity in a liquid state at t.

The above shows that argon in its properties in a liquid state stands near to oxygen (as it also does in its solubility), but that all the temperatures relating to it (t_c, t, and t_1) are higher than for nitrogen. This fully answers, not only to the higher density of argon, but also to the hypothesis that it contains N_3. And as the boiling point of argon differs from that of nitrogen and oxygen by less than 10°, and its amount is small, it is easy to understand how Dewar (1894), who tried to separate it from liquid air and nitrogen by fractional distillation, was unable to do so. The first and last portions were identical, and nitrogen from air showed no difference in its liquefaction from that obtained from its compounds, or from that which had been passed through a tube containing incandescent magnesium. Still, it is not quite clear why both kinds of nitrogen, after being passed over the magnesium in Dewar's experiments, exhibited an almost similar alteration in their properties, independent of the appearance of a small quantity of hydrogen in them.

Concluding Remarks (March 31, 1895).—The 'Comptes rendus' of the Paris Academy of Sciences of March 18, 1895, contains a memoir by Berthelot upon the reaction of argon with the vapour of benzene under the action of a silent discharge. In his experiments, Berthelot succeeded in treating 83 per cent. of the argon taken for the purpose, and supplied to him by Ramsay (37 c.c. in all). The composition of the product could not be determined owing to the small amount obtained, but in its outward appearance it quite resembled the product formed under similar conditions by nitrogen. This observation of the famous French chemist to some extent supports the supposition that argon is a polymerised variety of nitrogen whose molecule contains N_3, while ordinary nitrogen contains N_2. Should this supposition be eventually verified, the interest in argon will not only not lessen, but become greater. For this, however, we must wait for further observations and detailed experimental data from Rayleigh and Ramsay.

The latest information obtained by me from London is that Professor Ramsay, by treating cleveite (containing PbO, UO_3, Y_2O_3, &c.) with sulphuric acid, obtained argon, and, judging by the spectrum, helium also. The accumulation of similar data may, after detailed and diversified research, considerably increase the stock of chemical knowledge which, constantly widening, cannot be exhaustively treated in these 'Principles of Chemistry,' although very probably furnishing fresh proof of the 'periodicity of the elements.'